SYMMETRIES IN SCIENCE II

SYMMETRIES IN SCIENCE II

Edited by
Bruno Gruber
and Romuald Lenczewski
Southern Illinois University
Carbondale, Illinois

SPRINGER SCIENCE+BUSINESS MEDIA, LLC

Library of Congress Cataloging in Publication Data

Symmetries in science II.

"Proceedings of the symposium Symmetries in Science II, held March 24–26, 1986, at Southern Illinois University, Carbondale, Illinois"—T.p. verso.

Bibliography: p.

Includes index.

1. Symmetry—Congresses. 2. Symmetry (Physics)—Congresses. I. Gruber, Bruno, 1936– . II. Lenczewski, Romuald. III. Title: Symmetries in science 2. IV. Title: Symmetries in science two.

Q172.5.S95S923 1986 530.1 86-22673

ISBN 978-1-4757-1474-6 ISBN 978-1-4757-1472-2 (eBook)

DOI 10.1007/978-1-4757-1472-2

Proceedings of the symposium Symmetries in Science II, held March 24–26, 1986, at Southern Illinois University, Carbondale, Illinois

A portion of the expenses for this symposium was provided through gifts from Mr. James R. Brigham, the Norge Corporation, and the Southern Illinoisan.

Originally published by Plenum Press, New York in 1986

FOREWORD

The Symposium "Symmetries in Science II" was held at Southern Illinois University, Carbondale, during the period March 24-26, 1986, following the Einstein Centennial Symposium "Symmetries in Science" after a lapse of seven years. As it was the case for the original Symposium, the 1986 Symposium was truly interdisciplinary and truly international. I wish to thank all participants who made the effort to come to Carbondale, Illinois, from all over the world.

At this point I also wish to express my sincere thanks to Dr. Albert Somit, President of Southern Illinois University at Carbondale, and Dr. John C. Guyon, Vice President for Academic Affairs and Research at Southern Illinois University at Carbondale. Their generous support and encouragement was instrumental in getting the Symposium organized.

In addition I wish to thank Associate Vice President Charles B. Klasek, Dr. Russell R. Dutcher, Dean of the College of Science, John H. Yopp, Associate Dean, College of Science, Dr. Subir K. Bose, Chairman of the Physics Department, Dr. James Tyrrell, Chairman of the Chemistry Department, Dr. Jared H. Dorn, Director of International Programs and Services, Dr. Rhonda Jo Vinson, Director of International and Economic Development, Dr. Tommy T. Dunagan, Vice President of Sigma Xi at Southern Illinois University, Dr. George Garoian, Professor of Zoology, Dr. Ann Phillippi, Assistant Professor of Zoology and Dr. Linda R. Gannon, Coordinator of Women's Studies, for their support and assistance.

Finally, I wish to thank those who worked with me on a day to day basis, namely Dr. Anne Carman, Acting President of the Southern Illinois University Foundation, Dr. Susan S. Rehwaldt, Assistant to the President and above all Dr. Michael R. Dingerson, Acting Associate Vice President for Academic Affairs and Research and Dean of the Graduate School and Mr. Charles J. Lerner, Vice President of the Bank of Carbondale, the two co-chairmen of the organizational committee, for the untiring cooperation. It was truly their common effort which made the Symposium a great success.

Bruno Gruber

CONTENTS

THE MYSTERIOUS WORLD OF SYMMETRY IN PHYSICS

F. Iachello

A. W. Wright Nuclear Structure Laboratory
Yale University
New Haven, Connecticut 06511

ABSTRACT

I describe the various ways in which the concept of symmetry is used in physics today. Starting from the notion of symmetry in art, I discuss briefly its applications to physics.

1. THE NOTION OF SYMMETRY

The notion of symmetry as applied to the arts is a very old one. The word symmetry, from the greek σύμμετρον, had originally the meaning of well-proportioned, well-ordered. It applied originally to spatial objects, such as paintings, sculptures and architectural designs. The Sumerians seem to have developed (about 3,000 B.C.) some of the first examples of spatial symmetries. I will refer to these as geometric symmetries. However, it was only in the Greek world that the notion of symmetry became the central notion in the arts. One of the oldest examples of a geometric symmetry in an artifact from Greece is shown in Fig. 1. Here, a floor pattern found at the Megaron in Tiryns and dated from around 1,200 B.C., is shown as an example of reflection symmetry about an axis going through the middle of the figure from top to bottom (Weyl, 1952). As time went on, the notion of symmetry grew more complex. Fig. 2 shows a portion of a decoration from Greece where symmetry is now of the translation plus reflection type (Weyl, 1952). In addition to reflection symmetries about axes going through the middle of the palmettes from top to bottom, there is now a symmetry of translation to the right or left by an amount corresponding to the distance between two palmettes.

These early ideas were codified during the Golden Age of Greek civilization and spread from the visual arts to music, philosophy and even the natural behavior. Thus the sculptor Polykleitos wrote a book on proportions, the Pythagoreans considered the circle in a plane and the sphere in space as the most perfect geometric figures and Aristotle based most of his writings on the notion of symmetry. Perhaps, the most elaborate and complete example of this codified point of view is provided by the Greek temple. This structure was built according to a set of well specified rules most of which based on the notion of symmetry, although no longer as simple as those described above. Fig. 3 shows the plan of a Greek temple, the Temple D at Selinus, Sicily, ca. 500 B.C. The temple is is called esastylos since the number of columns on the short side is six.

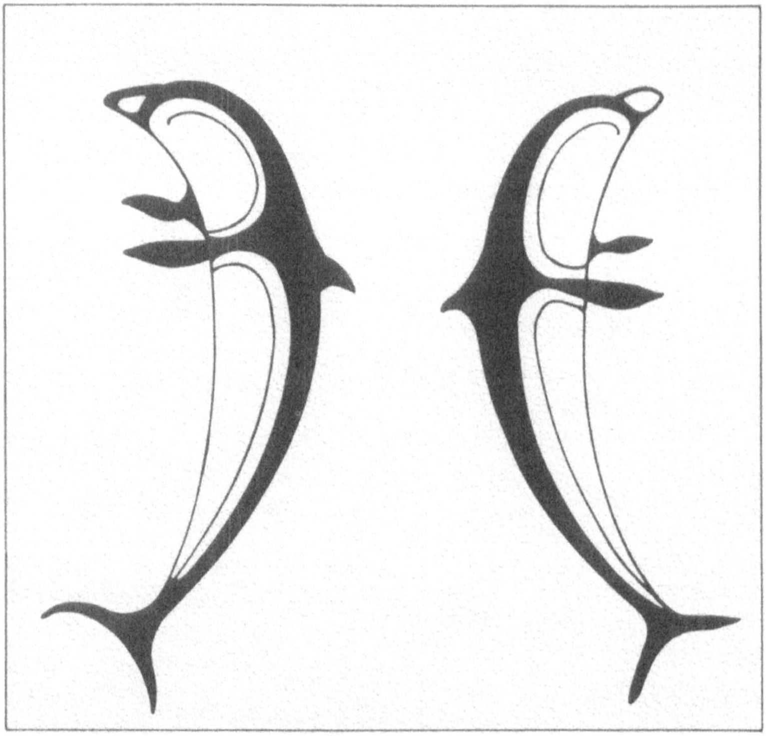

Fig. 1. Design found on a floor at the Megaron in Tiryns, Greece, ca. 1,200 B.C. (after Weyl).

Fig. 2. Late helladic design found in Greece, ca.1,000 B.C. (after Weyl).

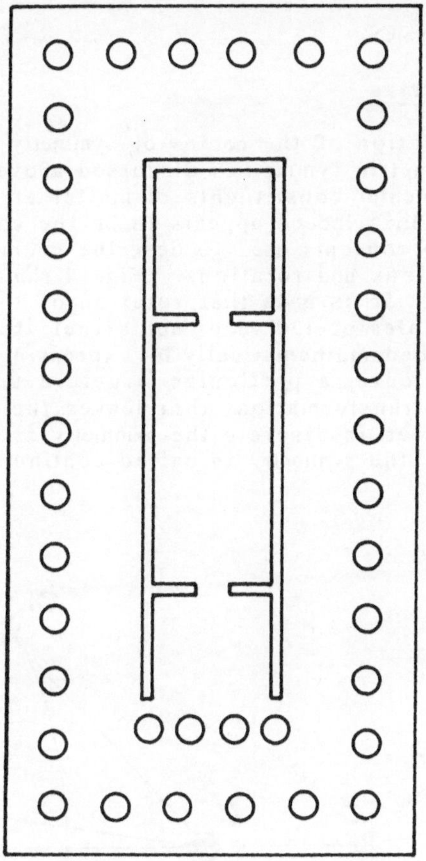

Fig. 3. Plan of Temple D at Selinus, Sicily, ca. 500 B.C..

The number of columnns on the long side is 13 and given by the canonical rule 2n+1, where n in the number of columns in the short side. In addition to reflection symmetry each individual part of the temple, such as the columns, has additional symmetries. A very recurrent symmetry is rotational symmetry, which means that a particular pattern or architectural design does not change upon rotation of a certain angle ϕ around a given axis. The notion of symmetry was transmitted by the Greeks to the Romans. Vitruvius wrote a text on architecture in which he employed very many of the ideas developed by the Greeks. Throughout the middle ages the notion of symmetry in art somewhat disappeared and in many cases it was replaced by the opposite notion, that of chaos. Chaos, from the Greek χάος, means disorganized. In many instances chaos was used as a guideline for architectural design. For example, while in the Greek temple, all columns were identical, thus displaying permutational symmetry with respect to the exchange of two columns, in several medieval cloisters, the columns were on purpose built one different from the other.

With the birth of Renaissance in Italy, the notion of symmetry reappeared in art. A typical example of the way of thinking of that period is the Book of Andrea Palladio "I Quattro Libri dell'Architettura", Venice, 1570. This notion has remained with us ever since.

2. SYMMETRIES IN PHYSICS

2.1. Geometric Symmetries

The first application of the **notion** of symmetry to physics was along the **lines of** the geometric symmetries discussed above. It was natural to think that the fundamental constituents of matter arrange themselves **into** geometric **patterns. This** indeed appears to be the case **for** crystals and molecules. The three concepts used to describe geometric symmetries are: reflections, translations and rotations. Fig. 4 shows the twisted $H_3C - C\,Cl_3$ molecule. It is **seen** that rotation of the molecule through angles **that** are multiples **of** 120" does not affect **its** structure. Symmetries are described mathematically by expressing the set of transformations that leaves a particular structure unchanged. In the case of Fig. 4 the **set of** transformations that leaves the **figure** unchanged **is** called C_3. **When** the **set is** discrete the symmetry **is** called also discrete, when **it is** continuous the symmetry **is** called continuous.

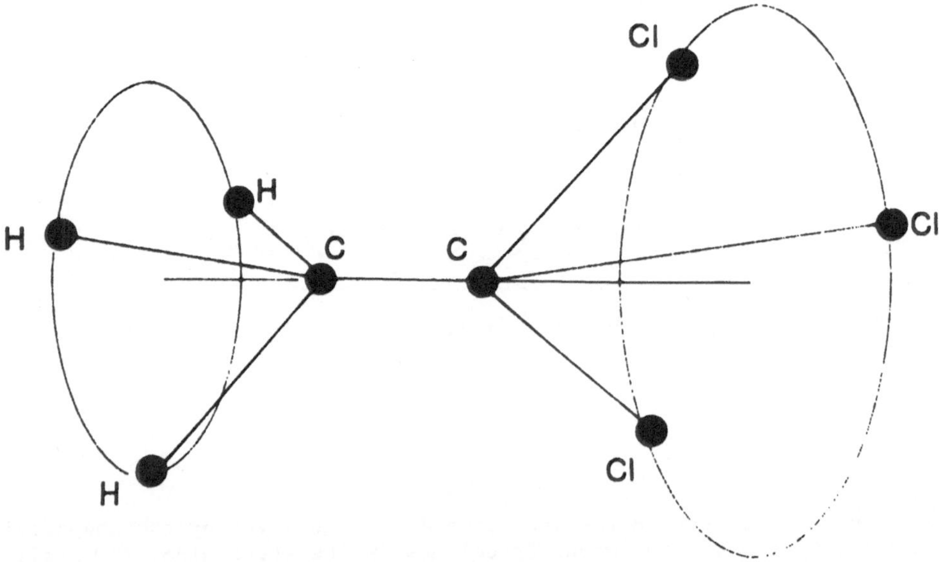

Fig. 4. The twisted $H_3C - C\,Cl_3$ molecule.

2.2. Space-Time (or Fundamental) Symmetries

For a **long** time, discrete geometric symmetries, were the only ones to be extensively employed in physics. Although the Greeks had employed continuous symmetries and indeed **Aristotle** thought that the world was built on spheres (**the** most **symmetric** body in a three-dimensional space), the extensive use of continuous symmetries in physics came only at the **beginning** of the 20th Century and brought in one of the **major** revolution in physics. The revolution was that of **considering** space and time **as** being the **same** and one entity and thus enlarging the three-dimensional world of space coordinates, **x,y,z,** with a fourth **dimension,** the time t. One could thus consider transformations which mix space and time. These transformations, called Lorentz transformations, relate the transformed corrdinates, for example **x'** and t', to the original coordinates, **x** and t. For a body moving in the **x direction,** with velocity **v,** the Lorentz transformation is

$$\begin{cases} x' = \dfrac{x-vt}{\sqrt{1-v^2/c^2}} , \\ \\ t' = \dfrac{t-(v/c^2)x}{\sqrt{1-v^2/c^2}} , \end{cases} \tag{2.1}$$

where c is the the speed of light. The introduction of symmetries relating space and time, led to major developments, in particular the theory of special relativity. Because of their importance, space-time symmetries are often called fundamental symmetries.

While geometric, i.e. space, symmetries can be displayed graphically and thus can be easily recognized, space-time symmetries (and even more so the symmetries to be discussed in the following paragraphs) cannot be easily shown. One can only investigate the consequences that the symmetries have on physical phenomena. One of the consequences that Lorentz symmetry has is the fact that the length of a body moving with velocity v appears to be contracted in the reference frame moving with the body. One of the few artists who has attempted to capture graphically some of the symmetry concepts employed in modern physics is the Dutch artist, M. C. Escher. I will therefore employ some of his etchings to illustrate these concepts. Fig. 5 illustrates the concept of space contraction. A body moving outward from the center with ever increasing speed appears contracted. The velocity v has an upper limit c, the speed of light. When v reaches c the dimension of the body shrinks to zero.

Fig. 5. Circle limit III (1959) (after Escher).
(©M.C. Escher Heirs, c/o Cordon Art, Baarn, Holland)

2.3. Permutational Symmetries

Another important type of **symmetry** used by physicists **is** permutational symmetry. **This** symmetry **is** relevant to the description of a **set of identical objects.** For example, the physical properties of the **set** of two **objects** 1 and 2 may **remain** the **same if** one interchanges the two **objects**

$$1 \leftrightarrow 2 \qquad (2.2)$$

Permutational symmetry became particularly important in physics with the development of quantum mechanics. Properties of bodies, such as atoms, nuclei, etc. are described in quantum mechanics by a wave function ψ. For two bodies, the wave function ψ will depend on the coordinates 1 and 2, **i.e.** 1   2 **It** has been found that **under** interchange of bodies 1 and 2, there are two possible types of symmetry,

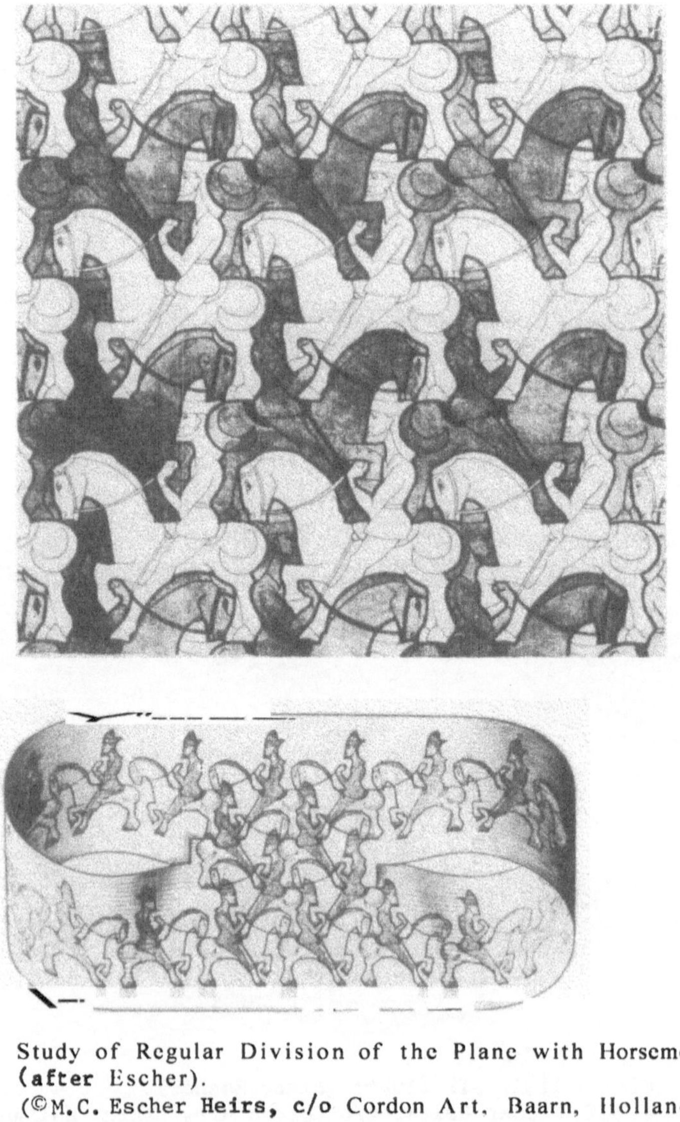

Fig. 6. Study of Regular Division of the Plane with Horsemen (1946) (**after** Escher).
(©M.C. Escher **Heirs, c/o** Cordon Art, Baarn, Holland)

$$\psi(1,2) = \psi(2,1) \quad \text{(bosons)} \quad ,$$
$$\psi(1,2) = -\,\psi(2,1) \quad \text{(fermions)} \quad . \tag{2.3}$$

Objects such that their wave function remains the same under interchange are called bosons, while objects such that the wave function change sign are called fermions. The set of Permutations is denoted usually by S_n. The notion of permutational symmetry is illustrated in Fig. 6. This figure is invariant under Permutation of the horsemen. There are in fact in this figure two permutational symmetries since their are two distinct types of horsemen, the black and the white. Permutational symmetries are of particular importance in the description of physical systems composed of several identical particles, such as complex atoms (with many electrons) and complex nuclei (composed of several protons and neutrons).

2.4. Dynamic Symmetries

As time went on, and with the increasing sophistication of both theoretical and experimental techniques, other types of symmetry were introduced in physics. An important new type is that some times called dynamic symmetry. The notion of this type of symmetry is rather complex and it is illustrated schematically in Fig. 7. Consider the perfect figure in

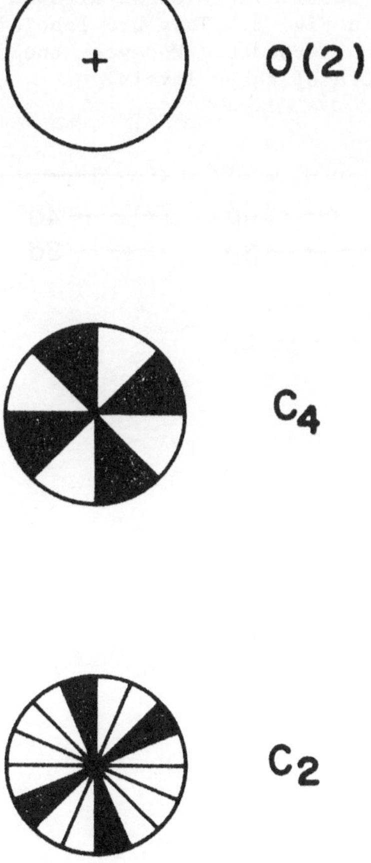

Fig. 7. Illustration of dynamic symmetries by regular breaking of rotational invariance.

the plane, i.e. the circle. This figure is invariant under the set of rotations around an axis perpendicular to the page and going through the center of the circle. This set is called O(2) in the figure. Now, it may happen that, because of some reason, usually of dynamic origin, the symmetry is broken, but in a very ordered way which leaves the figure invariant under a smaller set of transformations. For example, in Fig. 7 the figure in the middle, a wheel with parts of one metal (white) and parts of another metal (black) is invariant only under the set of rotations by angles multiples of 90°, called C_4. The C_4 symmetry may or may not be further broken to a symmetry by additional (dynamical) constraints as indicated in the lower part of Fig. 7. The ordered breaking of a symmetry is usually called a dynamic symmetry. Dynamic symmetries are very useful because they provide a detailed understanding of complex physical systems. Although the first example of a dynamic symmetry in physics is rather old (Pauli, 1926; Fock, 1935), and a major application of this concept had been made by Elliott (1958), this type of symmetry became very popular only after Gell-Mann (1961) and Ne'eman (1961) used it to describe the properties of the particles called hadrons.

While geometric symmetries act on the familiar three-dimensional space, dynamic symmetries usually act on some abstract space and hence they appear rather mysterious to non-physicists. Again, one can display visually only the effects of these symmetries. For systems described by quantum mechanics, the most convenient way to display the effects of dynamic symmetries is by showing the allowed quantum mechanical states. This is usually called a level diagram. The lowest allowed energy levels of the hydrogen atom are shown in Fig. 8. They are labeled by certain letters and numbers for classification purposes. However, the important point to note here is the regular pattern of these levels.

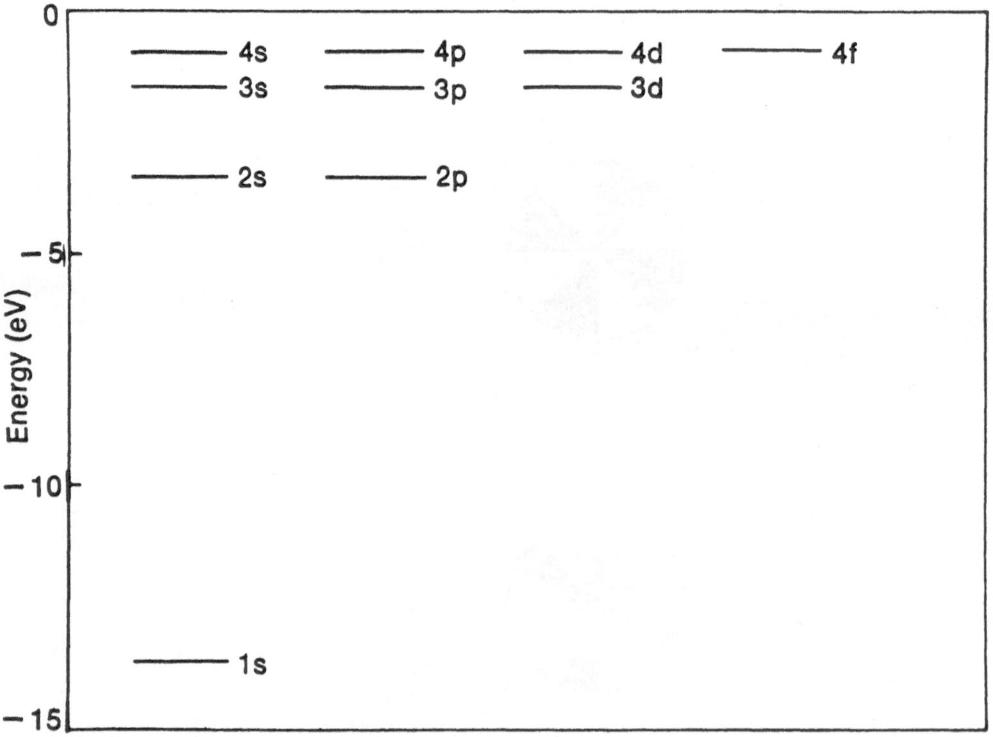

Fig. 8. Level diagram of the hydrogen atom.

The level diagram of the hydrogen atom was the first example of a dynamic symmetry in physics. Since then, several others have been found. A notable example is provided by the level diagram of the so-called elementary particles. Fig. 9 shows the level diagram of some particles called baryons. Again here the point to note is the regular pattern. Dynamic symmetries usually lead to a deeper understanding of a physical system. For example, the dynamic symmetries of baryons led to the developement of the quark model, in which the baryons themselves are built out of more fundamental objects called quarks.

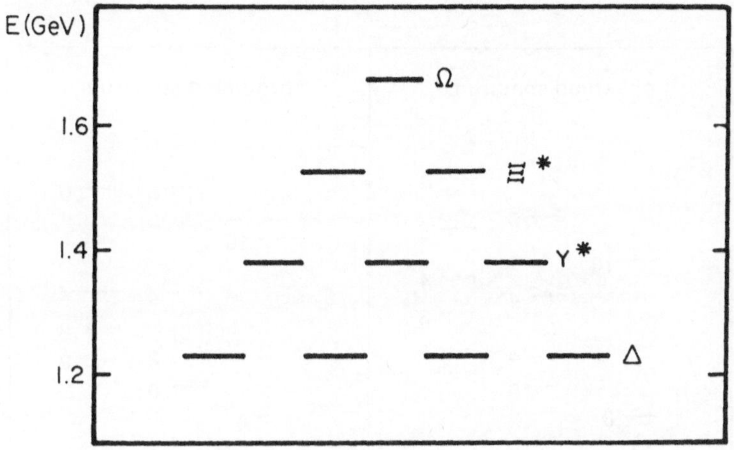

Fig. 9. Level diagram of the baryon decuplet.

Dynamic symmetries may be rather elaborate and related to combinations of objects rather than to single objects. An example is provided by the properties of atomic nuclei. It appears that the constituents of these nuclei, protons and neutrons, bind together in pairs. There are two types of pairs, called s and d respectively. Nuclei can then be treated as an assembly of pairs and unpaired particles, Fig. 10. The dynamic symmetries of the pairs provide a classification scheme for the energy levels of

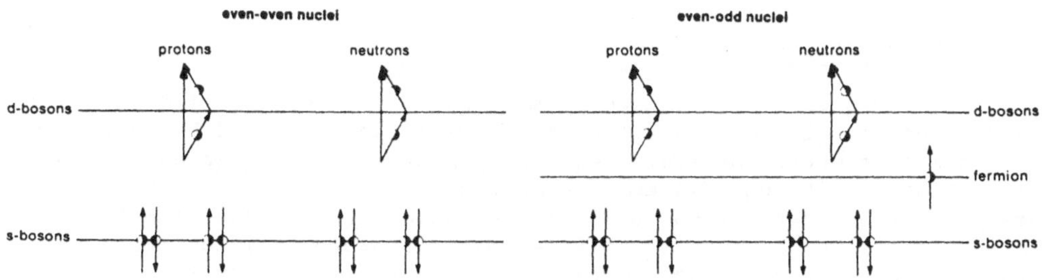

Fig. 10. Illustration of the binding of particles into pairs in atomic nuclei.

nuclei. An example is shown in Fig. 11. This figure also illustrates another concept. Most symmetries encountered in physics (and in art) are not exact but broken. For example, in Fig. 11, one can observe a small difference between the observed level diagram and that predicted on the basis of the dynamic symmetry. Similarly, if one returns to Fig. 1, one observes a small difference between the right and left hand side of the figure. The fins on the right fish are slightly different from those of the left fish. Despite the fact that most dynamic symmetries are broken, nontheless they provide a major tool for understanding complex structures. For example, the dynamic symmetries of the pairs have led to the development of a model, called the interacting boson model, which accounts for a good fraction of the properties of atomic nulei (Arima and Iachello, 1975).

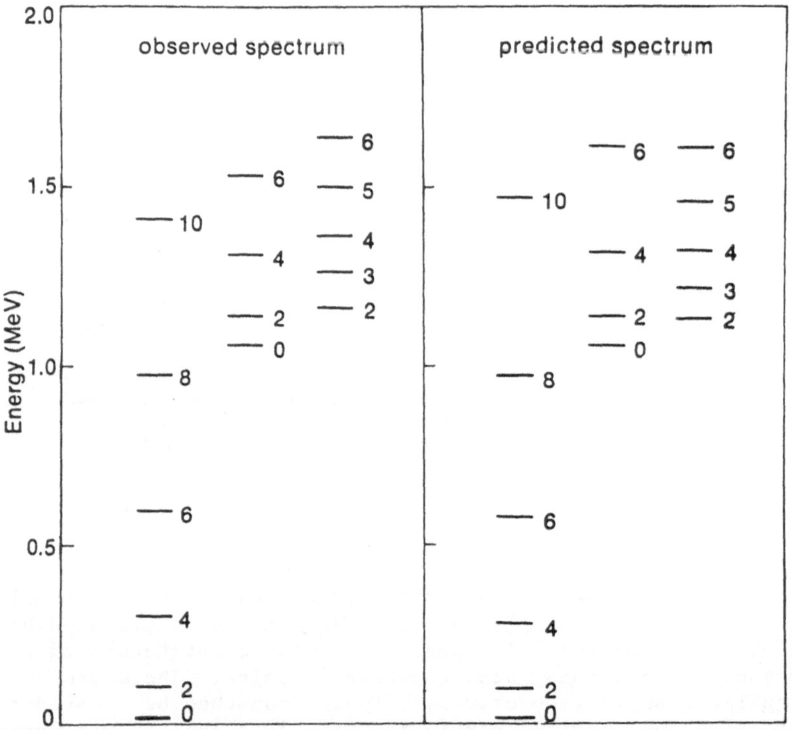

Fig. 11. Level diagram of a medium mass nucleus. On the left the observed spectrum, on the right that predicted on the basis of dynamic symmetry.

2.5. Gauge Symmetries

A final type of symmetry used by physicists is the so-called gauge symmetry. Again, although this type of symmetry is quite old being present in the laws of electrodynamics developed by Maxwell in the second part of the 19th Century, its use did not become popular until recently when it became clear that most (if not all) of the basic interactions of nature possess this type of symmetry. Gauge symmetries are best illustrated by returning to Maxwell eleçtrodynamics. This is described in terms of a scalar, ϕ, and a vector, $\vec{A}$, potential. The laws of electrogynamics are invariant under the set of transformations

$$\vec{A} \rightarrow \vec{A} + \vec{\nabla}\Lambda \quad ,$$
$$\phi \rightarrow \phi - \frac{1}{c}\frac{\partial\Lambda}{\partial t} \quad . \tag{2.4}$$

This invarience is called gauge symmetry. When combined with quantum mechanics, electrodynamics becomes quantum electrodynamics (QED). It has been found that also the weak and the strong interactions are governed by gauge symmetries. The presently accepted theory of strong interactions, called quantum chromodynamics (QCD) is one of these examples.

3. RECENT DEVELOPMENTS

3.1. Supersymmetries

In 1971 a new notion appeared in physics (Ramond, 1971). This is an

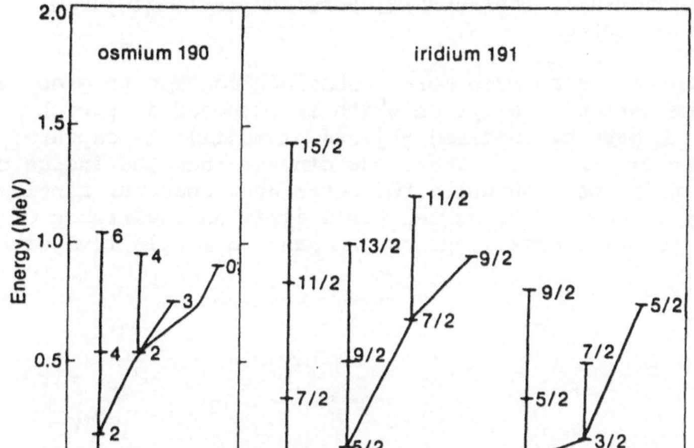

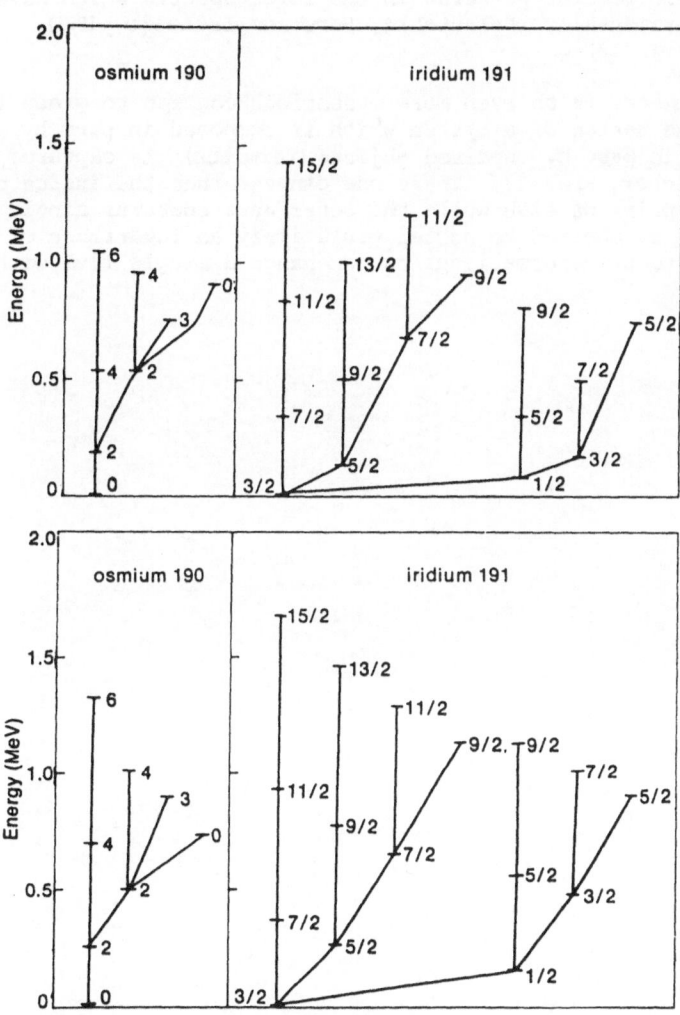

Fig. 12. Level diagram of the nuclei osmium 190 and iridium 191. The observed (above) and theoretically predicted (below) level sequence is shown as an example of a dynamic supersymmetry in nuclear physics.

attempt to further unify the laws of physics by using even more elaborate forms of symmetry. One of the results of permutational symmetry is that, as discussed bove, there exist in nature two types of fundamental objects, bosons and fermions. The symmetries discussed in the previous Sect. 2 only deal either with bosons or with fermions. One may inquire what happens if bosons are exchanged with fermions and viceversa. This new type of symmetry has been called supersymmetry. Up to now, there have been two main applications of supersymmetries in physics. The first application is to space-time supersymmetries, similar to those discussed in Sect. 2.2. These were introduced, among others, by Wess and Zumino (1974). The second application is to dynamic supersymmetries along the lines of those discussed in Sect. 2.4. An example of these dynamic supersymmetries is provided again by the level diagrams of nuclei. We have said before that some particles in nuclei join in pairs, while others remain single. It turns out that the pairs behave as bosons while the single particles behave as fermions. The invariance of the theory under the set of transformations of bosons into bosons, fermions into fermions and bosons into fermions produce certain regular patterns in the level spectra which have been observed experimentally (Balantekin, Bars and Iachello, 1981). An example is shown in Fig. 12.

Supersymmetry is an even more mysterious concept to grasp than normal symmetry. The notion of a system which is composed in part by pairs (bosons) and in part by unpaired objects (fermions) is captured again in a drawing by Escher, Fig. 13. There one can see that the inside of the figure is formed by pairs of fish while the outer part contains single fish. Supersymmetry as applied to nuclei would imply an invariance of the figure with respect to transformations that exchange a single fish with a pair.

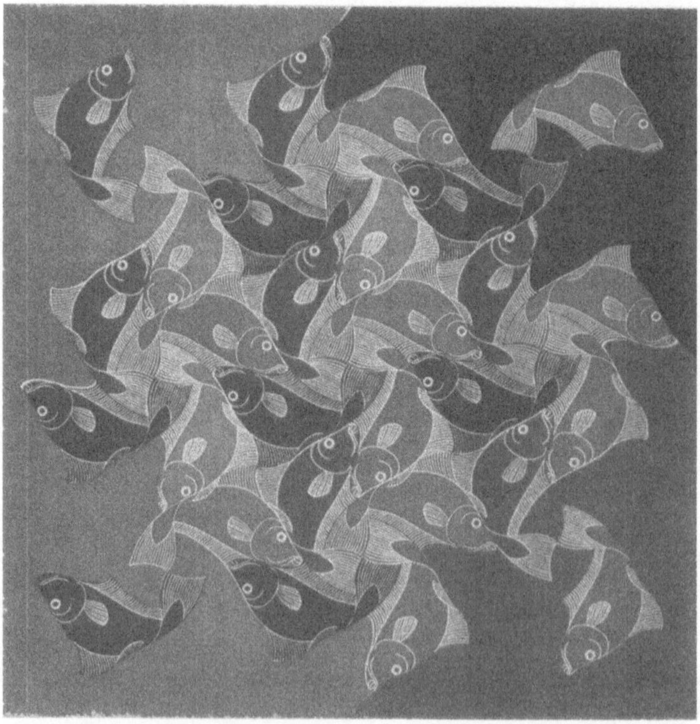

Fig. 13. Fish. Woodcut on cloth, ca. 1942 (after Escher). (©M.C. Escher Heirs, c/o Cordon Art, Baarn, Holland)

3.2. Quasi-symmetries

Another notion which **is beginning** to appear in physics **is** that of quasi-symmetries. These are cases in which, although strictly speaking there **is** no symmetry, nontheless some regularity remains. **The major** example of **application** of **this notion** to physics **is** to geometric quasi-symmetries (**Levine** and Steinhardt, 1984). **Being** of the geometric type, **it is** relatively easy to display visually quasi-symmetries. An example **is shown** in Fig. 14, called Penrose **tiling.** It **is** seen here that translational symmetry **is** not strictly present, even on a long range scale. Nonetheless, a set of rules exists by means of which one can construct the figure. Quasi-symmetries have led to the concept of quasi-crystals.

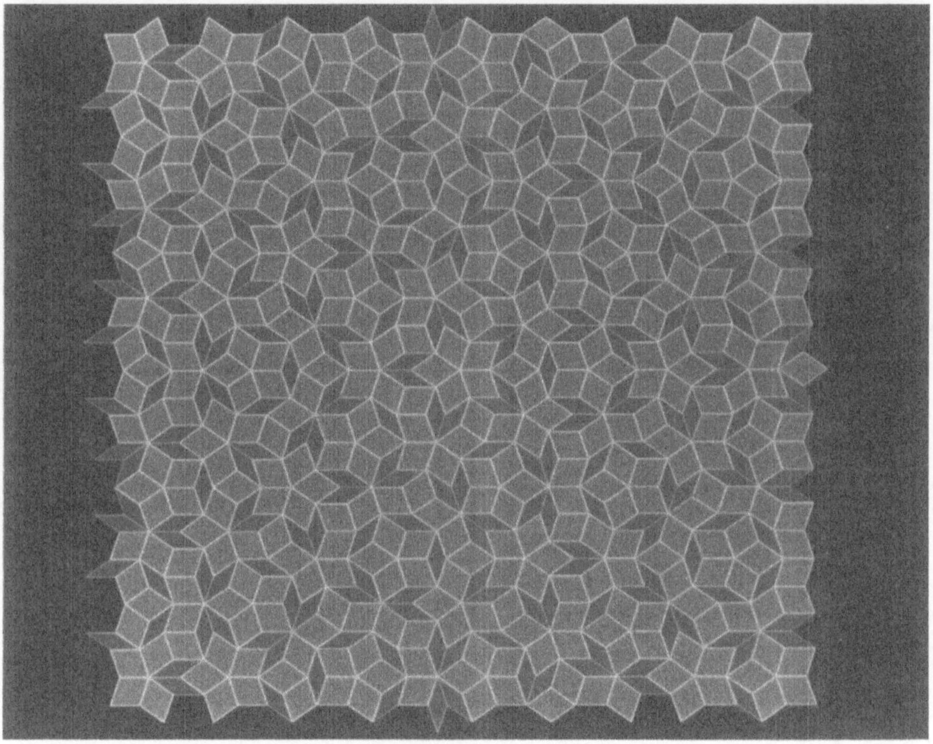

Fig. 14. Penrose tiling I (after Steinhardt).

4. CONCLUSIONS

The notion of symmetry **is** used in physics in many ways which range from the very simple to the very elaborate. Roth in physics and in art the simplest form of symmetry **is** perhaps reflection **symmetry (or** bilateral symmetry). Weyl attributes the importance of this symmetry to the fact that the human body also possesses bilateral symmetry **(right-left symmetry).** Starting from this simple symmetry one can construct very elaborate ones again both in physics and in art. I have discussed in the **previous sections** some applications to physics. The best examples of elaborate **symmetries** in art are probably provided by the moorish architecture found in the south of Spain. Fig. 15 shows an example from Cordoba.

The world of symmetries in physics may appear to the non-physicists as a mysterious world, especially that of symmetries which do not act on the familiar three-dimensional space. The abstractness of this world is captured by the lithograph shown in Fig. 16, and it is made evident by the interweaving of the columns. Nonetheless, the basic motivation for the use of the notion of symmetry in physics is the ancient motivation, particularly present in the Greek world, that beauty is bound up with symmetry. This equation, exemplified by the beauty of the Greek temple, Fig. 17, can be (and has been) further extended following a sentence attributed to P.A.M. Dirac that "if a theory of nature is beautiful, it must be true".

Fig. 15. The door of Al-Hakan at La Mezquita in Cordoba (Moorish art).

Fig. 16. Belvedere, 1958. Lithograph (after Escher).
(©M.C. Escher Heirs, c/o Cordon Art, Baarn, Holland)

Fig. 17. The Parthenon, Athens, Greece.

ACKNOWLEDGEMENTS

Figs. 5,6,13 and 16 were taken from the book "The Infinite World of M. C. Escher", Abradale Press/Harry N. Abrams, Inc., New York (1984). I wish to thank the publishers for letting me reproduce the figures. Fig. 14 was kindly given to me by Paul Steinhardt. I wish to thank him for his courtesy.

This work was perfomed in part under Department of Energy Contract No. DE-AC-02-76 ER 03074.

References

Arima, A., and F. Iachello. 1975. Collective nuclear states as representations of a SU(6) group, Phys. Rev. Lett. 35 : 1069.

Balantekin, A. B., I. Bars and F. Iachello. 1981. U(6/4) dynamical supersymmetry in nuclei. Phys. Rev. Lett. 47 : 19.

Elliott, J. P. 1958. Collective motion in the nuclear shell model. I. Classification schemes for states of mixed configurations. Proc. Roy. Soc. A245 : 128.

Fock, V. 1935. Zur Theorie des Wasserstoffatoms. Z. Phyzik 98 : 145.

Gell-Mann, M. 1961. The eightfold way: a theory of strong interaction physics. California Institute of Technology, Report CTSL -20.

Levine, D., and P. J. Steinhardt. 1984. Quasicrystals: a new class of ordered structures. Phys. Rev. Lett. 53: 2477.

Ne'eman, Y. 1961. Derivation of strong interactions from a gauge invariance. Nucl. Phys. 26 : 222.

Pauli, W. 1926. Über des Wassenstoffspektrum von Standpunkt der neuen Quantenmechanik. Z. Physik 36 : 336.

Ramond, P. 1971. Dual theory for free fermions. Phys. Rev. D3: 2415.

Weyl, H. 1952. Symmetry. Princeton University Press, Princeton, N. J.

Wess, J. and B. Zumino, 1974. Supergauge transformations in four dimensions. Nucl. Phys. B70 : 39.

DE SITTER STRUCTURED CONNECTION AND GAUGE TRANSLATIONS

R. R. Aldinger

Department of Physics
Eastern Illinois University
Charleston, Illinois 61920

INTRODUCTION

A local gauge field description of space-time is discussed using fiber bundle techniques as a theoretical framework. The basic idea is to endow ordinary Minkowski space, M_4, with a somewhat richer structure than that implied by relativity by attaching to each position $x \in M_4$ a copy of a four-dimensional micro-space of constant curvature characterized by an elementary subatomic length parameter R of the order of a Fermi, thus allowing for additional internal degrees-of-freedom. Therefore, two sets of variables are introduced: (i) the usual space-time variables x which determine an element of M_4 and (ii) a second set ξ which are elements of an internal space F (which is identified with the fiber of a bundle constructed over M_4).

Consequently, we consider a fiber bundle[1,2] E(B,F,G,P) constructed over a four-dimensional base manifold B (which is taken to be the usual Minkowski space, M_4), possessing a four-dimensional fiber F, and associated with the principal bundle P=P(B,G) (on which the connection is defined). The structural (gauge) group G of the bundle plays the role of an internal symmetry group and therefore determines the possible motions of the internal degrees-of-freedom belonging to the fibers.

As fiber we choose a four-dimensional pseudo-Riemannian space of constant (negative) curvature $R: F=SO(4,1)/SO(3,1)$ (i.e., a de Sitter space) which contains, at each point of contact with base manifold M_4, a tangent space which is isomorphic to Minkowski space. The structural group of the bundle is a de Sitter SO(4,1) which contains a Lorentz subgroup and a four parameter family of transformations (the de Sitter "boosts"), which in the limit $R \to \infty$ correspond to translations.

SO(4,1) CONNECTION AND CURVATURE

The four-dimensional pseudo-Riemannian space of constant curvature (de Sitter space), SO(4,1)/SO(3,1), on which an SO(4,1) acts as the symmetry group of motion may be embedded into a five-dimensional pseudo-Euclidean space, $E^5_{4,1}$, and is specified by a four-dimensional hypersurface which is noncompact in time and compact in the space directions. In terms of the coordinates of $E^5_{4,1}$, $\xi^A = (\xi^0, \xi^1, \xi^2, \xi^3, \xi^4)$, the hypersurface may be expressed as:

$$\xi^A \xi_A = \xi^A \xi^B \eta_{AB} = -R^2 \qquad A,B = i,j,4 \quad i,j = 0,1,2,3 \tag{1}$$

where the de Sitter metric $\eta_{AB} = \text{diag}\,(1,-1,-1,-1,-1)$ and R is a fixed length which characterizes the radius of the space SO(4,1)/SO(3,1).

The de Sitter SO(4,1) is the group of hyperbolic rotations in $E^5_{4,1}$ which leaves the form of Eq.(1) invariant. A basis for the Lie algebra $\mathcal{G}$ of G = SO(4,1) is given by:

$$\lambda_{AB} = -\lambda_{BA} \tag{2}$$

which obey:

$$[\lambda_{AB}\,,\,\lambda_{CD}] = -i(\eta_{AC}\lambda_{BD} + \eta_{BD}\lambda_{AC} - \eta_{AD}\lambda_{BC} - \eta_{BC}\lambda_{AD}). \tag{3}$$

The ten generators λ_{AB} are completely specified by a set of differential operators acting in $E^5_{4,1}$ where:

$$\lambda_{AB} = -i(\xi_A \partial_B - \xi_B \partial_A). \tag{4}$$

In order to display the subgroup reduction of $SO(4,1)_{\lambda_{AB}}$ we now introduce the de Sitter "boosts":

$$\pi_i \equiv \frac{1}{R}\lambda_{4i} \qquad i = 0,1,2,3 \tag{5}$$

where λ_{4i} generate de Sitter rotations in the 4-i plane and π_i are the corresponding generators of translation along the de Sitter fibers. Therefore, the commutation relation of Eq.(3) decomposes in the following way:

$$[\lambda_{ij}, \lambda_{k\ell}] = -i(\eta_{ik}\lambda_{j\ell} + \eta_{j\ell}\lambda_{ik} - \eta_{i\ell}\lambda_{jk} - \eta_{jk}\lambda_{i\ell}), \tag{6a}$$

$$[\pi_i, \lambda_{jk}] = -i(\eta_{ik}\pi_j - \eta_{ij}\pi_k), \tag{6b}$$

$$[\pi_i, \pi_j] = i\,\frac{1}{R^2}\lambda_{ij}, \tag{6c}$$

where the $SO(3,1)_{\lambda_{ij}}$ metric $\eta_{ij} = \text{diag}\,(1,-1,-1,-1)$. These relations specify the $SO(3,1)_{\lambda_{ij}}$ "stability" subalgebra of $\mathcal{G}$ spanned by the λ_{ij} (which generate Lorentz rotations about the auxiliary space axis ξ^4 while leaving the point of contact between base space and fiber, $\overset{\sigma}{\xi}{}^A = (0,0,0,0,-R)$ fixed) along with the four-dimensional vector subspace spanned by the π_i

(which generate the tangent subspace of SO(4,1)/SO(3,1) at the contact point, $\overset{\circ}{\xi}{}^A$. Therewith, the Lie algebra has decomposed according to:

$$\mathcal{G} = \mathcal{L} \oplus \mathcal{T} \tag{7}$$

where the elements of $\mathcal{L}$ generate the "stability" subgroup $SO(3,1)_{\lambda_{ij}}$ and those of $\mathcal{T}$ generate the tangent subspace at $\overset{\circ}{\xi}{}^A$.

The de Sitter structured connection in the principal fiber bundle $L^R(M_4) \simeq P(M_4, SO(4,1))$ is a matrix-valued four-vector field in Minkowski space with matrices defining a representation of $\mathcal{G}$ and has the form:

$$\Gamma^{g}_{\mu}(x) = \tfrac{1}{2}\Gamma_{\mu}^{AB}(x)\lambda_{AB}, \tag{8}$$

where g denotes a $\mathcal{G}$-valued element. The ten generators λ_{AB} carry the dependence on the (internal space) group coordinates and satisfy Eq.(3) while the gauge potentials $\Gamma^{AB}_{\mu}(x)$ depend only on the (external space) Minkowski coordinates and represent the 40 de Sitter rotation coefficients of an SO(4,1) connection which determine the nature of the local frame when making an infinitesimal transformation in M_4. The gauge potentials, when related to a specific cross-section (gauge), describe the observable change in the frame when one follows that cross-section relative to the horizontal direction.

The SO(4,1) horizontal lift of a tangent vector ∂_μ at $x \in M_4$ to an arbitrary point $u \in P(M_4, SO(4,1))$ is expressed as:

$$D^{g}_{\mu} = \partial_\mu + i\Gamma^{g}_{\mu}(x). \tag{9}$$

The horizontal vector defines the gauge curvature field of the de Sitter connection where:

$$[D^{g}_{\mu}, D^{g}_{\nu}] = i\mathcal{R}^{g}_{\mu\nu}(x), \tag{10}$$

$$\mathcal{R}^{g}_{\mu\nu}(x) \equiv \partial_\mu\Gamma^{g}_{\nu}(x) - \partial_\nu\Gamma^{g}_{\mu}(x) + i[\Gamma^{g}_{\mu}(x), \Gamma^{g}_{\nu}(x)]. \tag{11}$$

The requirement of local de Sitter gauge invariance yields the following transformation law for the g-valued gauge potentials typical for a connection:

$$\overline{\Gamma}^{g}_{\mu}(x) = A^{-1}(x)\Gamma^{g}_{\mu}(x)A(x) - iA^{-1}(x)\partial_\mu A(x), \tag{12}$$

where: $A(x) \in SO(4,1)$.

Using the Lie algebra decomposition given by Eq.(7), the SO(4,1) connection may be expressed as:

$$\Gamma^{g}_{\mu}(x) = \Gamma^{\ell}_{\mu}(x) + \Gamma^{t}_{\mu}(x) \tag{13}$$

$$= \tfrac{1}{2}\Gamma^{ij}_{\mu}(x)\lambda_{ij} + \Gamma^{4i}_{\mu}(x)\lambda_{4i} \tag{14}$$

where the de Sitter (g-valued) connection has decomposed naturally into $SO(3,1)_{\lambda_{ij}}$ (ℓ-valued) and de Sitter "boost" (t-valued) components. The $\Gamma^{ij}_{\mu}(x)$ are associated with $SO(3,1)_{\lambda_{ij}}$ rotations about the auxiliary

ξ^4 axis leaving the contact point, $\overset{o}{\xi}{}^A$, fixed whereas the four-vector gauge potentials, $\Gamma_\mu^{4i}(x)$, are associated with translations in the four-dimensional tangent space of SO(4,1)/SO(3,1) at $\overset{o}{\xi}{}^A$, both transformations evolving in an abstract four-dimensional space SO(4,1)/SO(3,1) of constant curvature R.

CONTRACTION OF SO(4,1)

In the Inönü-Wigner[3] contraction limit when one takes the radius of curvature of the de Sitter space $R\to\infty$, the de Sitter "boosts" π_i, go over into the generators of translation, $P_i = -i\partial_i$, of a flat (abstract) Minkowski space where:

$$[P_i, P_j] = 0, \tag{15a}$$

$$[P_i, \lambda_{jk}] = -i(\eta_{ik}P_j - \eta_{ij}P_k), \tag{15b}$$

$$[\lambda_{ij},\lambda_{k\ell}] = -i(\eta_{ik}\lambda_{j\ell} + \eta_{j\ell}\lambda_{ik} - \eta_{i\ell}\lambda_{jk} - \eta_{jk}\lambda_{i\ell}), \tag{15c}$$

which are the well-known relations defining a representation of the Poincaré group, $ISO(3,1)_{\lambda_{ij}P_i}$. Therefore, in contracting $SO(4,1)_{\lambda_{ij}\pi_i}$ with respect to its stability subgroup $SO(3,1)_{\lambda_{ij}}$, the de Sitter bundle:

$$T_R(M_4) \simeq E(M_4,\ SO(4,1)/SO(3,1),\ SO(4,1),\ L^R(M_4)) \tag{16}$$

goes into, in the $R\to\infty$ limit, the affine tangent bundle:

$$T_A(M_4) \simeq E(M_4,\ ISO(3,1)/SO(3,1),\ ISO(3,1),\ A(M_4)) \tag{17}$$

and the associated principal bundle:

$$L_R(M_4) \simeq P(M_4, SO(4,1)) \tag{18}$$

contracts into the bundle of affine frames:

$$A(M_4) \simeq P(M_4,\ ISO(3,1)), \tag{19}$$

over space-time. $A(M_4)$ is a principal fiber bundle possessing an ISO(3,1) as its structural group which is the semi-direct product of the homogeneous Lorentz group SO(3,1) and a four-dimensional Minkowski space which is spanned by the generators of translation P_i.

In the limit $R\to\infty$ the Lie algebra of G: $\mathcal{G}=\mathcal{L}\oplus\mathcal{T}$ contracts into:

$$\mathcal{G} \xrightarrow{R\to\infty} \tilde{\mathcal{G}} = \mathcal{L} \oplus \tilde{\mathcal{T}} \tag{20}$$

where $\tilde{\mathcal{G}}$ is the Lie algebra of ISO(3,1) and is the direct sum of the Lie algebra of the stability subgroup SO(3,1) and the lie algebra of the four-dimensional real vector space which itself is isomorphic to the Abelian group of translations. Therefore, the g-valued connection form $\Gamma^g(x) = \Gamma^\ell(x) + \Gamma^t(x)$ contracts into:

$$\Gamma^g(x) \xrightarrow{R\to\infty} \tilde{\Gamma}^g(x) = \Gamma^\ell(x) + \tilde{\Gamma}^t(x) \tag{21}$$

or:

$$\Gamma^g_\mu(x) \xrightarrow{R\to\infty} \tilde{\Gamma}^g_\mu(x) = \Gamma^\ell_\mu(x) + \tilde{\Gamma}^t_\mu(x) \tag{22}$$

$$= \tfrac{1}{2}\Gamma^{ij}_\mu(x)\lambda_{ij} + \gamma^i_\mu(x)P_i \tag{23}$$

where the $\gamma^i_\mu(x)$ determine a field of gauge translations in $T_A(M_4)$.

SOLDERING AND GAUGE TRANSLATIONS

The affine tangent bundle, $T_A(M_4)$, (obtained from the de Sitter bundle, $T^R(M_4)$, in the $R\to\infty$ contraction limit) is the eight-dimensional differentiable manifold which is locally the direct product of the "physical" space-time manifold M_4 with some "abstract" Minkowski space, ISO(3,1)/SO(3,1). These two Minkowski spaces are, in general, completely unrelated where M_4 is the usual physical space where distance is measured in cm whereas the abstract space ISO(3,1)/SO(3,1) (fiber space) has no particular physical relevance. However, a precise physical meaning may be attached to ISO(3,1)/SO(3,1) if one considers $T_A(M_4)$ to be "soldered" to M_4, thereby identifying the tangent space of ISO(3,1)/SO(3,1) at the origin of ISO(3,1)/SO(3,1), (which is left invariant by the action of the stability subgroup SO(3,1)) with the local tangent space of M_4. The necessity of "soldering" arises from physical considerations alone and must be introduced when one considers space-time associated structural groups.

Consider the theory of gravitation which may be described by the bundle of linear frames over M_4 which possesses as structural group the general lineral group $G\ell(4,R)$ (which contains the Lorentz SO(3,1) as a metric-preserving subgroup).[5,6] The usual linear connection is a connection in the bundle of linear frames (and also, therefore, in the Lorentz frame bundle, $L(M_4)$). However, the bundle of linear frames (and also $L(M_4)$) has a special property not possessed by other principal fiber bundles that are unrelated to space-time (i.e., the usual Yang-Mills type) which requires the additional concept of "soldering" of the local fibers F_x to base space at $x \in M_4$. Namely the principal fiber bundle $P(R^4, G\ell(4,R))$, constructed over a four-dimensional differentiable manifold possessing as structural group $G\ell(4,R)$, is isomorphic to the bundle of linear bases of R^4 if and only if there exists an R^4-valued 1-form on P such that[7]:

$$\Theta(X_p a) = a^{-1}\Theta(X_p)a \tag{24}$$

where X_p is a tangent vector at $p \in P$ and $a \in G\ell(4,R)$. On P, Θ is called the canonical form of P and the principal fiber bundle is said to be "soldered" to its base space thereby identifying points in the fiber bundle with the local frames of the base manifold.

Conversely, if one starts with an abstractly defined (space-time) fiber bundle P on which an R^4-valued 1-form satisfying Eq.(24) is defined, one may consider it as the canonical form thereby completely specifying an embedding of P into the bundle of linear frames of space-time. Such a form of soldering is found to be naturally present in a theory with translations included in the structural group where the translational part of the connection form can serve in the capacity of the canonical form.

Consider the contracted form of the de Sitter connection (i.e., the Poincaré connection) given by Eq.(21). Here the $\tilde{\Gamma}^t(x)$ will generate four fields which are related to the group of translations along the (Poincaré) fibers, but as long as these "vertical" translations of $A(M_4)$ are not somehow related to the physical translations as carried out in the base manifold, their particular form is not specified any further. On the other hand, if the transformation properties of $\tilde{\Gamma}^t(x)$ satisfy Eq.(24), it may naturally be defined as the canonical form, thereby transferring a physical meaning to the measurement of distance directly to the fibers. Therewith, a dynamical significance has been attached to the fiber space coordinates ξ since making a change in the external (base manifold) position forces, via the "soldering mechanism", a corresponding transformation of ξ along the locally attached fiber F_x over $x \in M_4$.

On the contracted bundle, $A(M_4)$, the identification of the R^4-valued 1-form $\tilde{\Gamma}^t(x)$ as the canonical form implies that the corresponding gauge potentials $\gamma^i_\mu(x)$ should characterize a local change of the observable Lorentz coordinates when an infinitesimal shift of the origin of the frame is made, and therefore, should be interpreted as being proportional to the components of the canonical connection (i.e., the space-time tetrad fields, $h^i_\mu(x)$, which determine a relation between the Lorentz orthonormal frames and the natural frames of the general coordinate system where $\partial_\mu = h^i_\mu(x)X_i$).[8]

On the de Sitter bundle, $L^R(M_4)$, the translational gauge degrees-of-freedom are contained in the vector subspace spanned by the $\pi_i \equiv \frac{1}{R}\lambda_{4i}$. Furthermore, there is a subbundle of $L^R(M_4)$ (in which the fibers are determined by the Lorentz subgroup of $G = SO(4,1)$) on which $\Gamma^\ell(x)$ is taken as the (ℓ-valued) connection form while $\Gamma^t(x)$ is an R^4-valued 1-form which satisfies the transformation properties of Eq.(24).

Therefore, on $L^R(M_4)$ one may identify $\Gamma^t(x)$ as the canonical form of soldering, thus providing an embedding of the subbundle into the bundle of linear frames and the SO(4,1) gauge potentials are then interpreted as being proportional to the components of the canonical form:

$$\Gamma^{4i}_\mu(x) = g h^i_\mu(x) \tag{25}$$

where g is a proportionality constant with the dimension of inverse length. Due to the introduction of the de Sitter "boosts" we have that:

$$\Gamma^{4i}_{\mu}(x)\lambda_{4i} = gh^{i}_{\mu}(x)R\pi_{i} = h^{i}_{\mu}(x)\pi_{i} \tag{26}$$

where we have interpreted the constant g as the inverse radius of the de Sitter space.

With the identification of the canonical form on $L^R(M_4)$ we have that the de Sitter structured connection may be written as:

$$\Gamma^{\mathfrak{g}}_{\mu}(x) = \Gamma^{\ell}_{\mu}(x) + \Gamma^{t}_{\mu}(x) \tag{27}$$

$$= \tfrac{1}{2}\Gamma^{ij}_{\mu}(x)\lambda_{ij} + h^{i}_{\mu}(x)\pi_{i}, \tag{28}$$

while the SO(4,1) horizontal lift becomes:

$$D^{\mathfrak{g}}_{\mu} = \partial_{\mu} + i\Gamma^{\ell}_{\mu} + i\Gamma^{t}_{\mu} \tag{29}$$

(where the explicit x dependence has been suppressed).

According to the natural decomposition of the $\mathfrak{g}$ -valued horizontal lift into rotational and translational parts, the curvature of the SO(4,1) connection, Eq.(10) becomes:

$$[D^{\mathfrak{g}}_{\mu}, D^{\mathfrak{g}}_{\nu}] = i\mathcal{R}^{\mathfrak{g}}_{\mu\nu} = iQ^{\ell}_{\mu\nu} + iS^{t}_{\mu\nu}, \tag{30}$$

where:

$$\mathcal{R}^{\mathfrak{g}}_{\mu\nu} = \tfrac{1}{2}\mathcal{R}^{AB}_{\mu\nu}\lambda_{AB}, \tag{31a}$$

$$Q^{\ell}_{\mu\nu} = \tfrac{1}{2}Q^{ij}\lambda_{ij}, \tag{31b}$$

$$S^{t}_{\mu\nu} = S^{4i}_{\mu\nu}\lambda_{4i}, \tag{31c}$$

and:

$$Q^{\ell}_{\mu\nu} \equiv R^{\ell}_{\mu\nu} + i[\Gamma^{t}_{\mu}, \Gamma^{t}_{\nu}] \tag{32a}$$

$$R^{\ell}_{\mu\nu} \equiv \partial_{\mu}\Gamma^{\ell}_{\nu} - \partial_{\nu}\Gamma^{\ell}_{\mu} + i[\Gamma^{\ell}_{\mu}, \Gamma^{\ell}_{\nu}] \tag{32b}$$

$$S^{t}_{\mu\nu} \equiv \partial_{\mu}\Gamma^{t}_{\nu} - \partial_{\nu}\Gamma^{t}_{\mu} + i[\Gamma^{\ell}_{\mu}, \Gamma^{t}_{\nu}] + i[\Gamma^{t}_{\mu}, \Gamma^{\ell}_{\nu}]. \tag{32c}$$

Therefore, the de Sitter gauge curvature field has decomposed into rotational and translational components, $Q^{\ell}_{\mu\nu}$ and $S^{t}_{\mu\nu}$, respectively. The $R^{\ell}_{\mu\nu}$ part of $Q^{\ell}_{\mu\nu}$ is the usual Lorentz-valued curvature of the linear connection while the presence of the second term in $Q^{\ell}_{\mu\nu}$ is a direct consequence of the non-Abelian character of the translational part of the de Sitter connection. Furthermore, the interpretation of the Γ^{4i}_{μ} as the space-time tetrad fields together with the commutation relations of Eq.(6b) leads to the well-known fact that the "translational curvature", $S^{t}_{\mu\nu}$, can be interpreted as the torsion of the connection.

GAUGE CONSTRAINTS

The choice of a specific de Sitter gauge (cross-section on $L^R(M_4)$)

leads to certain gauge-fixing relations (constraints on the guage potentials). From these constraints it follows that certain terms may be absent from the set of expressions making up the de Sitter curvature field.

For example, if $R^{\ell}_{\mu\nu}(x) = 0$ over a local neighborhood in M_4, then there exists a natural horizontal cross-section where $\Gamma^{ij}_{\mu}(x) = 0$ and, as a consequence, directions have a global meaning in that region. In the case of translations, one may interpret the $\Gamma^{4i}_{\mu}(x)$ as characterizing an observable shift of the origin of $L^R(M_4)$ in the local tangent space at $x \in M_4$. If, in some particular gauge, it can be found that $\Gamma^{4i}_{\mu}(x) = 0$, we then have the uninteresting case where no change of position is observed when one shifts in the ∂_μ direction. For this gauge choice, $\Gamma^{t}_{\mu}(x) = 0$ and the entire manifold reduces to a single point in M_4 where $S^{t}_{\mu\nu} = 0$.

On the other hand, for a "flat" space-time we assume that both $R^{\ell}_{\mu\nu}$ and $S^{t}_{\mu\nu}$ are zero implying that the preferred cross-section is the one where $\Gamma^{ij}_{\mu}(x) = 0$ and $\Gamma^{4i}_{\mu}(x) = 0$ (i.e., the one that follows the horizontal direction). However, what is detected is the (nontrivial) case where a change of position is observed, i.e., $\Gamma^{4i}_{\mu}(x) \neq 0$. Therefore, it should be possible to obtain a specific class of Lorentz-connected cross-sections such that $\Gamma^{ij}_{\mu}(x) = 0$ while $\Gamma^{4i}_{\mu}(x) \neq 0$ and fulfilling:

$$R^{\ell}_{\mu\nu}(x) = 0, \tag{33a}$$

$$S^{t}_{\mu\nu} = \partial_\mu \Gamma^{t}_{\nu} - \partial_\nu \Gamma^{t}_{\mu} = 0, \tag{33b}$$

$$Q^{\ell}_{\mu\nu} = i[\Gamma^{t}_{\mu}, \Gamma^{t}_{\nu}] \tag{33c}$$

thereby reducing Eq.(30) to:

$$[D^{g}_{\mu}, D^{g}_{\nu}] = i[\Gamma^{t}_{\mu}, \Gamma^{t}_{\nu}]. \tag{34}$$

In a previous work[9] a quantized version of the de Sitter fiber bundle structure as presented in the present paper was investigated and, for the specific Lorentz-valued gauge choice discussed above, the quantum mechanical analogue of the (reduced) curvature relations given by Eq.(33) was obtained resulting in the mathematically consistent and experimentally verifiable hadron model of the quantum relativistic rotator.[10]

REFERENCES

1. S. Kobayashi and K. Nomizu, "Foundations of Differential Geometry. Vol. I," Wiley, New York (1963).
2. W. Drechsler and M. E. Mayer, "Fiber Bundle Techniques in Gauge Theories," Vol. 67, Lecture Notes in Physics, Springer-Verlag, Heidelberg (1977).

3. E. Inönü and E. P. Wigner, On the contraction of groups and their representations, *Proc. N.A.S.*,. 39:510 (1953).

4. W. Drechsler, Group contraction in a fiber bundle with a Cartan connection, *J. Math. Phys.*, 18:1358 (1977).

5. R. Utiyama, Invariant theoretical interpretation of interaction, *Phys. Rev.*, 101:1597 (1955).

6. T. W. B. Kibble, Lorentz invariance and the gravitational field, *J. Math. Phys.* 2:212 (1961).

7. A. Trautman, Fiber bundles associated with space-time, *Rep. Math. Phys.* 1:29 (1970).

8. P. K. Smrz, A new unified field theory based on de Sitter gauge invariance, *Acta Phys. Pol.* 10:1025 (1979).

9. R. R. Aldinger, Quantum de Sitter fiber bundle interpretation of hadron extension, *Int. J. Theor. Phys.* 25:527 (1986).

10. R. R. Aldinger, A. Bohm, P. Kielanowski, M. Loewe, P. Magnollay, N. Mukunda, W. Drechsler and S. R. Komy, Relativistic rotator I: quantum observables and constrained hamiltonian mechanics, *Phys. Rev. D* 28:3020 (1983).

THE ALGEBRAIC APPROACH TO SCATTERING

AND THE "EUCLIDEAN CONNECTION"

Y. Alhassid

A. W. Wright Nuclear Structure Laboratory, Yale University

New Haven, Connecticut 06511

ABSTRACT

Starting from a dynamical group describing a scattering problem I explain how the S-matrix can be calculated algebraically through the technique of Euclidean connection. The S-matrix corresponding to the dynamic group SO(3,2) is constructed explicitly and provides a realistic model for the analysis of heavy-ion collisions.

1. Introduction

Dynamcial symmetries have proved to be useful in the analysis of bound state problems in a variety of fields from elementary particle physics to nuclear and molecular physics[1]. In scattering problems however, the only known example of a dynamical symmetry has been non-relativistic Coulomb scattering[2] where the symmetry group is SO(3,1), the analytic continuation of the bound state symmetry group SO(4)[3]. We have recently started extending the methods of spectrum generating algebras to the continuum[4]. The novel feature in the algebraic description of scattering states is the use of non-compact groups[5]. The latter are known to possess unitary representations that are continuous and infinite-dimensional and are thus natural candidates for describing continuous spectra.

The description of scattering states in terms of continuous representations of non-compact groups G is not sufficient for solving the scattering problem. The quantity of interest here is the scattering matrix and the major problem is how to calculate it starting from a dynamical group G. While in the conventional approach the S-matrix is defined through the asymptotic behaviour of the scattering wave-function, it was not clear how to incorporate the notion of an asymptotic limit into the algebraic approach, in which a state is described simply as an abstract vector in some group representation space.

Let us make a small digression and inspect the partial wave Coulomb S-matrix at a given momentum k and angular momentum ℓ

$$S_\ell(k) = \frac{\Gamma(\ell+1+i\beta/k)}{\Gamma(\ell+1-i\beta/k)} \quad . \qquad (1.1)$$

Here $\beta = \alpha Z_1 Z_2 \mu c/\hbar$ where α is the fine structure constant ; Z_1, Z_2 the colliding charges and μ their reduced mass. The simple recursion relation that the Γ function satisfies

$$\Gamma(z+1) = z\Gamma(z) \quad , \qquad (1.2)$$

suggests that (1.1) has an "algebraic" structure and the question arising is whether it has any relation to the symmetry group SO(3,1). This was a subject of many interesting investigations[6] but the methods employed were specific to the Coulomb problem.

We have suggested the construction of an algebraic framework to characterize asymptotic behavior[4,7]. An asymptotic group F describes the symmetries in the absence of interactions. As explained below, the S-matrix is then found by connecting the representations of G with those of F. Since for non-relativistic problems F is usually related to Euclidean groups, we call the above method of finding the S-matrix the Euclidean connection.

The algebraic approach to scattering and the implications of dynamic symmetries are discussed in a paper by F. Iachello[8]. Here I shall focus my discussion on the technique of Euclidean connection, which is crucial to the S-matrix calculation. The technique is demonstrated in the context of a dynamic group SO(3,2) which corresponds to a Coulomb force modified by a short range interaction, and is therefore well suited for the analysis of elastic heavy-ion collisions[9]. Here a dynamic symmetry occurs for the chain

$$SO(3,2) \supset SO(3) \otimes SO(2) \quad . \qquad (1.3)$$

The SO(3) describes the spatial rotational symmetry while the SO(2) subalgebra is used to label the interaction strength v. A scattering state is then labelled by $|\omega,\ell,m,v\rangle$, where ℓ and m are the angular momentum and its projection and $\omega(\omega+3)$ is the Casimir eigenvalue of SO(3,2). Here ω describes a continuous principal series representation

$$\omega = -\frac{3}{2} + if(k) \quad , \qquad (1.4)$$

where f(k) is a real function of the momentum k and is determined from the relation between the Hamiltonian and the Casimir invariant.

2. Asymptotic Symmetries

A particle scattered (in n-dimensional space) behaves asymptotically as a free particle so that the asymptotic spatial symmetries include translations in addition to rotations. Thus the relevant symmetry group is the Euclidean group in n dimensions E(n), which contains the linear momentum $\vec{p}$ along with the angular momentum (the generators of the rotational group SO(n)). We assume that there is an additional internal E(2) symmetry containing the SO(2) generator v_3 (used to label the interaction strength v) and two internal translations v_1, v_2. For a three-dimensional scattering problem the asymptotic dynamical group is therefore $F = E(3) \otimes E(2)$, where

$$E(3) = \begin{Bmatrix} SO(3) = \{\ell_1, \ell_2, \ell_3\} \\ p_1, p_2, p_3 \end{Bmatrix} ; \quad E(2) = \begin{Bmatrix} SO(2) = \{v_3\} \\ v_1, v_2 \end{Bmatrix} . \tag{2.1}$$

Incoming and outgoing free waves of energy k^2 belong then to the "-k" and "+k" representations[10] of E(3), respectively, and are labelled by the angular momentum ℓ and its projection m, and the interaction strength v. An SO(3,2) scattering state $|\omega,\ell,m,v\rangle$ can then be expanded (asymptotically) into $E(3) \otimes E(2)$ states

$$|\omega,\ell,m,v\rangle = A_{\ell v}(k)\,|-k,\ell,m,v\rangle + B_{\ell v}(k)\,|+k,\ell,m,v\rangle \tag{2.2}$$

The partial wave S-matrix is given by $S_\ell(k) = (-)^{\ell+1}\, B_{\ell v}(k)/A_{\ell v}(k)$.

3. The Euclidean Connection

The coefficients $A_{\ell v}$ and $B_{\ell v}$ in the expansion (2.2) are determined by the connection between the SO(3,2) representation and the $E(3) \otimes E(2)$ representations. It is then necessary to express the generators of SO(3,2) as function of the $E(3) \otimes E(2)$ generators within a given representation +k or -k of the latter. That such a connection must exist is made plausible by the following physical argument. Any SO(3,2) generator commutes with the SO(3,2) Casimir invariant and therefore leaves the energy invariant. Asymptotically this means that it commutes the $E(3) \otimes E(2)$ Casimir invariants and thus is a function of the generators of $E(3) \otimes E(2)$.

To construct the SO(3,2) generators from those of $E(3) \otimes E(2)$, we assume that the $SO(3) \otimes SO(2)$ subgroup is the same for both groups. We then have to reconstruct only the non-compact generators $K_{i\alpha}$ $(i=1,2,3\ ;\ \alpha=4,5)$ of SO(3,2) which mix the spatial and internal degrees of freedom. It is obvious that the $K_{i\alpha}$ have to form an $SO(3) \otimes SO(2)$ tensor. We define two SO(3) vectors $\vec{D}$ and $\vec{F}$ by $D_i = K_{i4}$ and $F_i = K_{i5}$ (i=1,2,3). Considering combinations of SO(3) vectors such as $\vec{p}$ and $\vec{\ell} \times \vec{p}$, and taking into account the SO(2) transformation properties, we find for the +k representation

$$\vec{D} = [(v_3 v_2 - i\omega\, v_1)\vec{p} - v_1\, \vec{\ell} \times \vec{p}]/k$$

$$\vec{F} = [(-v_3 v_1 - i\omega\, v_2)\vec{p} - v_2\, \vec{\ell} \times \vec{p}]/k \quad . \tag{3.1}$$

The coefficients of the various terms were determined by requiring that $\vec{D}$ and $\vec{F}$ satisfy the SO(3,2) commutation rules and that the proper Casimir invariant $\omega(\omega+3)$ is obtained. Eqs. (3.1) are called the "Euclidean connection".

4. The S-matrix

Recursion relations in ℓ and v for the expansion coefficients $A_{\ell v}$ and $B_{\ell v}$ can now be obtained algebraically by using the Euclidean connection (3.1). To this end we operate with $D_\pm = D_1 \pm iD_2$ and $F_\pm = F_1 \pm iF_2$ on the l.h.s. of (2.2) and with its Euclidean connection (3.1) on the r.h.s. using the standard expressions for the action of the Euclidean generators on a basis,

$$P_\pm |k,\ell,m,v\rangle = \pm\sqrt{\frac{(\ell\mp m)(\ell\mp m-1)}{4\ell^2-1}}\,(ik)\,|k,\ell-1,m\pm 1,v\rangle \pm\sqrt{\frac{(\ell\pm m+1)(\ell\pm m+2)}{4(\ell+1)^2-1}}\,(ik)\,|k,\ell+1,m\pm 1,v\rangle \quad . \tag{4.1}$$

In such a procedure, ℓ and v can change by 0,±1. The recursion relations can be obtained more economically by constructing shift operators which shift ℓ and v by given amounts (+1 or −1). Such operators are quadratic in the group generators. For instance

$$X_{++} = (D_- + iF_-)\,L_+ + (D_3 + iF_3)\,(L_3 + \ell + 1) \tag{4.2}$$

shifts ℓ and v by +1

$$X_{++}\,|\omega,\ell,m,v\rangle \propto |\omega,\ell+1,m,v+1\rangle \quad . \tag{4.3}$$

Operating with X_{++} on the l.h.s. of (2.2) and using (4.2), (3.1) and (4.1) on the r.h.s. we get a recursion formula for $R_{\ell v}(k) = B_{\ell v}(k)/A_{\ell v}(k)$:

$$R_{\ell+1,v+1}(k) = -\,\frac{\ell+3/2+v+if(k)}{\ell+3/2+v-if(k)}\,R_{\ell,v}(k) \quad . \tag{4.4}$$

Similarly, using operators of the type X_{-+}, X_{+-} and X_{--}, we find

$$R_{\ell-1,v+1}(k) = -\,\frac{\ell-1/2-v-if(k)}{\ell-1/2-v+if(k)}\,R_{\ell,v}(k)$$

$$R_{\ell+1,v-1}(k) = -\frac{\ell+3/2-v+if(k)}{\ell+3/2-v-if(k)} R_{\ell,v}(k)$$

$$R_{\ell-1,v-1}(k) = -\frac{\ell-1/2+v-if(k)}{\ell-1/2+v+if(k)} R_{\ell,v}(k) \tag{4.5}$$

The solution of (4.4) and (4.5) is given by

$$S_\ell(k) = \frac{\Gamma\left(\frac{\ell+v+3/2+if(k)}{2}\right)\ \Gamma\left(\frac{\ell-v+3/2+if(k)}{2}\right)}{\Gamma\left(\frac{\ell+v+3/2-if(k)}{2}\right)\ \Gamma\left(\frac{\ell-v+3/2-if(k)}{2}\right)} e^{i\phi(k)} \tag{4.6}$$

where $\phi(k)$ is a real phase. For $v=1/2$ (4.6) reduces to the Coulomb phase assuming $\phi(k)=(2\ell n2)f(k)$. We have thus derived Eq.(2.19) of Ref. 8. For an appropriate choice of v this S-matrix corresponds to a Coulomb potential modified by a short range interaction, a form well suited for analysis of heavy-ion reactions.

5. Contractions and Expansions

We now intend to clarify in more detail the relation between the algebras G and F. We shall show[11] that they are closely related by the group-theoretical procedures known as contraction and expansion[12].

To contract the SO(3,2) algebra we define

$$P_{i\alpha}^{\varepsilon} = \varepsilon K_{i\alpha} \qquad (i=1,2,3\ ,\ \alpha=4,5) \tag{5.1}$$

for small ε, while the SO(3) ⊗ SO(2) subalgebra is left unchanged. In the limit $\varepsilon \to 0$ the $P_{i\alpha}$ (we omit ε) become an abelian invariant subalgebra and a new algebra is formed

$$ISO(3,2) = \{\vec{\ell}, v_3\ ,\ P_{i\alpha}\} \quad , \tag{5.2}$$

known as the inhomogenuous special orthogonal group in 3+2 dimensions. This algebra, obtained by a contraction of the SO(3,2) algebra is related to the asymptotic dynamical group E(3) ⊗ E(2) by the following identification

$$P_{i4} = p_i v_1 \quad ,$$

$$P_{i5} = p_i v_2 \quad . \tag{5.3}$$

Thus, the contracted (5.2) algebra is in the enveloping algebra of E(3) ⊗ E(2).

In the contraction process, the Casimir invariant of SO(3,2) $C_2 = -\sum_{i,\alpha} K_{i\alpha}^2 + \vec{\ell}^2 + v_3^2$ becomes

$$\varepsilon^2 C^2 + \sum_{i,\alpha} P_{i\alpha}^2 = \vec{p}^2 \tag{5.4}$$

namely, the Casimir invariant $\vec{p}^2$ of E(3) ⊗ E(2) (note that we choose $v_1^2+v_2^2 = 1$) whose eigenvalue is the energy k^2.

The inverse process of recovering the original algebra from its contraction is known as <u>expansion</u>. In an expansion the generators of the algebra are expressed as non-linear functions of the contracted ones and therefore provide the Euclidean conection. In fact, the connection formula (3.1) can be rewritten as

$$\begin{aligned} \vec{D} &= \{\tfrac{1}{2i}\,[C_2(SO(3)+C_2(SO(2))\ ,\ v_1\vec{p}] + f(k)v_1\vec{p}\}/k \\ \vec{F} &= \{\tfrac{1}{2i}\,[C_2(SO(3)+C_2(SO(2))\ ,\ v_2\vec{p}] + f(k)v_2\vec{p}\}/k \end{aligned} \tag{5.5}$$

Here the Casimir invariants of the maximal compact subgroup SO(3) ⊗ SO(2) of SO(3,2), which is left unchanged in the contraction, are used to reconstruct the SO(3,2) algebra, by commutation with the tensor $\{v_1\vec{p}, v_2\vec{p}\}$.

Equation (5.5) provides a technique to construct Euclidean connections that can be generalized to scattering problems in n dimensions with a dynamical symmetry

$$SO(n,m) \supset SO(n) \otimes SO(m) \tag{5.6}$$

Here SO(m) describes the internal symmetries of the interaction. Denoting by L_{ij} the generators of the spatial SO(n), by $M_{\alpha\beta}$ the generators of SO(m) and by $K_{i\alpha}$ the non-compact rotations of SO(n,m) we have an expansion formula[13]

$$\begin{aligned} K_{i\alpha} &= \{\tfrac{1}{2i}\,[C_2(SO(n))+C_2(SO(m))\ ,\ P_{i\alpha}] + f(k)\ P_{i\alpha}\}/k \\ &=(L_{ij}\ P_{i\alpha} - M_{\alpha\beta}\ P_{i\beta} - i\omega P_{i\alpha})/k \quad , \end{aligned} \tag{5.7}$$

where $\omega = -\frac{n+m-2}{2} + if(k)$. In (5.7) the $P_{i\alpha}$ are ISO(n,m) generators obtained by the contraction of SO(n,m).

6. Conclusions

We have demonstrated how, starting from a dynamical group G of the scattering problem, the S-matrix can be calculated algebraically by using the method of Euclidean connection.

Several extensions of the Euclidean connection techniques will be useful. One such extension is the study of other types of dynamical symmetries and their associated S-matrices. Modifications are necessary when other degrees of freedom such as spin or more complicated interactions are introduced. Other interesting extensions occurs when the asymptotic symmetry is not necessarily that of a free particle, but corresponds to a distorted wave in a solvable potential. The technique of expansion may then be used to obtain the additional phase shift when the potential is modified.

Finally, we have described a method to obtain a closed expression for the S-matrix when a dynamical symmetry occurs. When this dynamical symmetry is broken, a numerical algebraic algorithm for calculating the S-matrix has yet to be developed.

Acknowledgements

I would like to thank my collaborators in this work F. Iachello, F. Gürsey and J. Wu. I also thank A. Frank for his collaboration on Sec. 5 and J. Engel for his collaboration on the Euclidean connection in its earlier stages. This work was supported in part by the Department of Energy Contract No. DE-AC-02-76 ER 03074. The author acknowledges receipt of an Alfred P. Sloan fellowship.

References

1. For a review see A. Böhm and Y. Ne'eman, "Twenty Years of Dynamical Groups and Spectra Generating Algebras", in Proceedings of the International Conference on Group Theoretical Methods in Physics, Yurmula, U.S.S.R. (1985), V. Manko ed., VNU Publishers, Netherlands (1986). See also Refs. 1-6 in Ref. 8.

2. A. Böhm, Nuovo Cimento 43, 665 (1966); M. Bander and C. Itzykson, Res. Mod. Phys. 38, 346 (1966).

3. W. Pauli, Z. Phyzik 36, 336 (1926).

4. Y. Alhassid, F. Gürsey and F. Iachello, Phys. Rev. Lett. 50, 873 (1983); Ann. Phys. (N.Y.) 148, 346 (1983); 167, 181 (1986), J. Wu, F. Iachello and Y. Alhassid, Ann. Phys., to be published.

5. For a discussion of non-compact groups see B. G. Wybourne "Classical Groups for Physicists", Wiley, New York, 1974 and references therein.

6. L. C. Biedenharn and P. J. Brussaard in "Coulomb Excitation", p. 107, Clarendon Press, Oxford, 1965; D. Zwanziger, J. Math Phys. 8, 1858 (1967); A. M. Perelomov and V. S. Popov, Sov. Phys. JETP 27, 967 (1968); A. O. Barut and W. Rasmussen, J. Phys. B 6, 1713 (1973).

7. Y. Alhassid, J. Engel and J. Wu, Phys. Rev. Lett. 53, 17 (1984).

8. F. Iachello, these proceedings.

9. Y. Alhassid, F. Iachello and J. Wu, Phys. Rev. Lett. 56, 271 (1986).

10. W. Miller, Jr. "On Lie Algebras and Some Special Functions of Mathematical Physics", American Mathematical Society, Providence, R. I. (1964).

11. A. Frank, Y. Alhassid and F. Iachello, Yale preprint YNT-86-06.

12. For a review of contractions and expansions, see R. Gilmore, "Lie Groups, Lie Algebras and Some of Their Applications" (Wiley, N. Y. 1974) and references therein.

13. K. B. Wolf and C. P. Boyer, J. Math. Phys. 15, 2096 (1974).

SYMMETRY AND DYNAMICS: TWO DISTINCT METHODOLOGIES FROM KEPLER TO SUPERSYMMETRY

A. O. Barut

Department of Physics
University of Colorado
Boulder, Colorado 80309

Symmetry and Dynamics are different ways of formulating the laws of physics, not necessarily one derivable from the other; sometimes conflicting, sometimes complementary to each other, often answering to different typs of questions, together necessary for a more complete understanding of nature. The scope and limitations of both methods are investigated.

1. KEPLER AND THE SYMMETRY IN THE ORDER OF PLANETS

The lasting achievement of Johannes Kepler (1571-1633) is the establishment of his three laws of planetary motion, the last one he found after many years of continuous and painful effort. Yet, these were not the goal that he originally set out to find. He was trying to find in the motion of the planets the same proportions of perfection as found in the harmonic sounds (Pythagoras), or in regular polyhedra of Plato. "Geometria est archetypus pulchritudinis mundi" - geometry is the archetype (Urbild) of the beauty of the world. He was trying to discover that the five regular solids could be inserted between the orbits of the six planets:

> "The intense pleasure I received from this discovery can never be told in words. I regretted no more the time wasted; I tired of no labour; I shunned no toil of reckoning, days and nights spent in calculation, until I could see whether my hypothesis would agree with the orbits of Copernicus or whether my joy was to vanish into air".

Kepler could not prove completely his hypothesis. It is believed to be irrelevelant today, after the establishment of the universal law of gravitation (inverse square law) by Newton for which the three laws of Kepler were the main pillars. It is also believed to be wrong after the discovery of three more planets and the satellites of the planets. (There are no more than five regular solids, but at least three more half-regular solids.) Kepler has even been ridiculed later for his hypothesis:

> "... an idea so crazy by modern standards that it does not even make sense" (P. T. Mathews. Inaugural Lecture, Imperial College 1962).

However, Kepler was after a global description of the order in the planetary motion, not a local dynamical scheme, a cause and effect relationship, or what we call today, a time evolution equation of motion. [Johannes Kepler actually also talks about invisible "magnetic effects" pulling between the planets, anticipating universal gravitation and even beyond]. And the question posed by Kepler is still unsolved today. We may think that we understand completely the laws of the planets. But we do not understand why there are nine planets with their satellites, their distances from the sun, their masses, angular momentum, etc. These questions are usually put aside as due to complicated initial conditions on the formation of planets. However, it is not possible, to start with dynamical equations, integrate them from the beginning of the solar system and derive the simple laws on the the number and distances of the planets. But the final result may be amenable to a simple satisfactory description in terms of symmetry. In the words of Hermann Weyl, "the objective world is (does not become)".

The empirical Titus - Bode formula[1]

$$R = 0.4 + (0.3).2^n \tag{1}$$

gives the length of the major semiaxis of the planets in astronomical units (1 A.U. = the major semiaxis of the earth's orbit) with $n = -\infty$ for M (Mercury), 0 for V (Venus), 1 and 2 for E (Earth) and M (Mars), n = 3 for the Asteroids, 4, 5, 6, 7 for J (Jupiter), S (Saturn), U, (Uranus), P (Pluto). The formula does not account for the orbit of Neptune!

Bode's law has unfortunately no theoretical basis. Nor are the n values regular.

One can try to fit the plenatary radii by a law which is of the type of hydrogen atom.[2] In the equation of motion

$$mv^2/R = GmM/R^2 \quad , \quad \text{or } v_n^2 R_n = GM = \Gamma \tag{2}$$

If we assume a "quantization"

$$v_n R_n = n\sigma \tag{3}$$

which is not an angular momentum quantization but angular momentum per unit mass. Then

$$v_n = \frac{\Gamma}{\sigma} \cdot \frac{1}{n} \rightarrow R_n = \frac{\sigma^2}{\Gamma} \cdot n^2 \tag{4}$$

also

$$T_n = 2\pi \frac{\sigma^3}{\Gamma^2} n^3 \quad , \qquad E_n^{kin} = \frac{m}{2} \frac{\Gamma^2}{\sigma^2} \frac{1}{n^2} \quad , \qquad U = -\frac{\Gamma^2 m}{\sigma^2 n^2} \quad ,$$

$$E_n = -\frac{1}{2} m_n \frac{\Gamma^2}{\sigma^2} \cdot \frac{1}{n^2} \tag{5}$$

A very good fit can be obtained if

$$\sigma = 9.22 \times 10^{14} \ m^2/s$$

But the n values are somewhat arbitrary:

	M	V	E	M	...	J	S	U	N	P
n	3	4	5	6	...	11	15	21	26	31

One can argue about the missing n values, or try to identify them with the satellites of the planets. But the scheme is still not satisfactory. Nor again, is there a reason for the above quantization.

However if one just plots the logarithm of the velocities and the distances, there is no doubt that an approximate linear law holds. (Fig. 1). It could have been after all quite irregular. The variations can be attributed perhaps to elliptic orbits, etc. It seems more like that the velocity is "quantized" according to an exponential law:

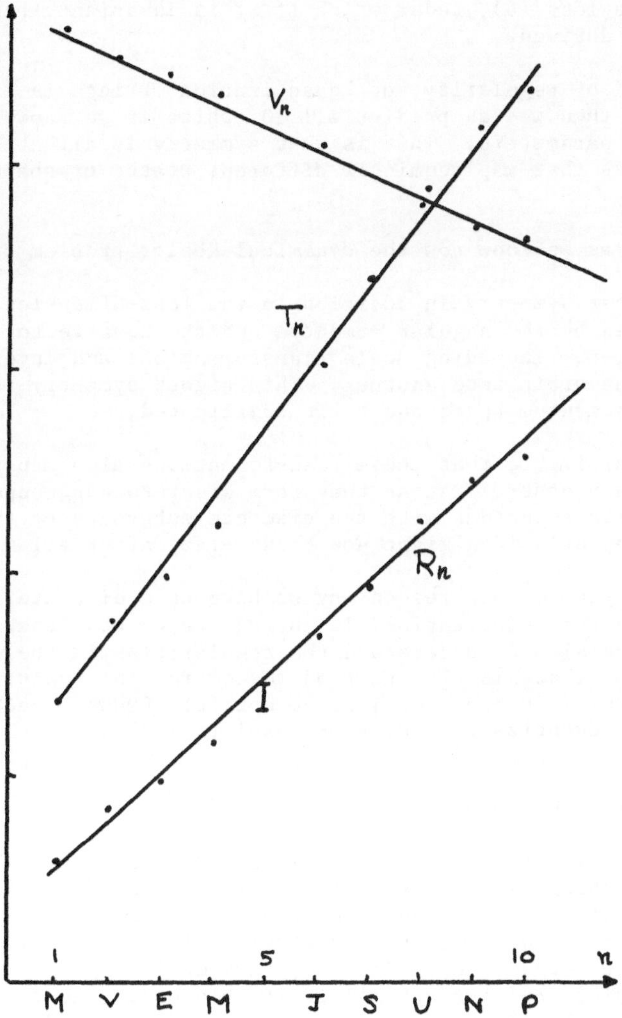

Fig. 1. The logarithms of planetary velocities, periods and distances. (n=5 is the position for Asteroids).

$$v_n = v_o e^{-\lambda n} \qquad (6)$$

so that

$$\log v_n = \log v_o - \lambda n ; \quad \log R_n = \log R_o + 2\lambda n; \quad \log T_n = \log T_o + 3\lambda n$$

where

$$R_o = \frac{\Gamma}{v_o^2} , \qquad T_o = 2\pi \frac{\Gamma}{v_o^3}$$

A time and space dilation

$$t \rightarrow e^{3\lambda n} t \quad , \quad x \rightarrow e^{2\lambda n} x \tag{7}$$

gives the equations (6), under which t^2/x^3 is invariant, hence Kepler's second law is derived.

This kind of regularity, or "quantization" brings in one new constant λ. But then we can predict all 10 orbits if we know one of them, instead of 10 parameters. This is what symmetry is all about: to find transformations that map seemingly different states or shapes into each other.

Actually as we know now the dynamical Kepler problem of $\frac{1}{r^2}$ - force law has a higher symmetry in addition to the four-dimensional rotation group generated by the angular momentum and the Lenz vector, namely the conformal symmetry including scale transformations and inversions. Thus one can map one orbit into another. This higher symmetry is most simply expressed in momentum space and a bit complicated.

It is conceivable that these transformations also map the regular solids into each other. Rather they form discrete subgroups which for low lying orbits coincides with the discrete subgroups of the regular polyhedra. Kepler's idea might not be so crazy after all.[3]

There is yet no good reason why we have this discrete scale transformations (7) or "quantization" in the occurence of planets. (Of course, we have also to understand the regularities in the masses - hence energies). But there is also no good reason for the quantization of angular momentum in H-atom, we just accept it! (Quant. Mechanics describes this quantization, does not explain it?)

2. A DUALITY OR DYCHOTOMY

This case history shows two distinct and different methodologies, or approaches in Theoretical physics to understand and describe the nature: dynamical equations, cause and effect, objects and forces and interactions on the one hand, and, the description of the symmetry of a final situation on the other hand. In fact, this duality goes on in every domain of physics, and through all of its history. The two approaches are sometime complementary, sometimes opposing, but seldom combinable to a single comprehensive framework.

In fact the thesis of this paper is that there are inherent basic limitations in both approaches which often prevents us to go from one kind of description to the other. We must for all foreseeable future work with both method in a complementary fashion. The Symmetry appears as the closure of a dynamical theory which like an asymptotic series can be approached but never reached, but may be perceived or conceived immediately for independent reasoning or assumptions.

I see a dichotomy, at least a duality, in the interplay of Symmetry and Dynamics is physics. In one direction we have the derivability of symmetry from an underlying dynamics, and in the other direction we have the inference of dynamics from globally recognized symmetries.

The understanding of the physical world does not seem to be a single smooth process: Find the basic laws or axioms, the most basic and most universally valid (maximum range), then simply deduce everything from these axioms. We may be even in conflict with Gödel's theorem, because in every axiom system with a finite number of laws there are undecidable statements. Provability (or deducibility) is much narrower than truth! I think it is worth to try to formulate these problems even in our present day research. In fact they become more critical in the domain of fundamental particles as we shall see where symmetries may lead to temporary or false dynamical models.

Let me consider another case history.

3. CRYSTAL SYMMETRY

We might think that the basic Schrödinger dynamics of electrons and nuclei describes all the physical and chemical properties of matter. It is said that we completely know the laws of atomic and molecular physics, as it was said (Laplace) than we can predict everything in a mechanical world from the laws of mechanics.

Yet it has not been possible to derive from these first principles the simple and beautiful symmetries of crystals. Nor the simple observables like colour, let alone, smell. One might argue that this is a complication of many-body dynamics, and that we know in principle that it must happen. I doubt that even with the fastest computers we will ever be able to arrive at the crystal symmetry. For one thing it is not clear at all, as Schrödinger observed[4] how to define the observables like length of a crystal side or the angle in terms of the n-body wave function

$$\psi(\vec{x}_1, \vec{x}_2, \vec{x}_3, \ldots.) \quad .$$

What is the observable "colour" in terms of the atomic coordinates and the coordinates of the electromagnetic field?

As an evolution process we must show that during the growth of a crystal each atom must exactly know where to go and which next atom to which next site. Or the atoms in the gas or liquid phase must make many trials and the correct one will stick to correct sites as to produce the final crystal structure which will minimize energy or other quantities. Of course, we know actually that the dynamics is discontinuous. The original dynamical equations loose their predictive power; they cannot be continued in time, due to the three body collisions, for example, in classical mechanics, or phase transitions occur.

Many observables are singular functions, not analytic. In classical mechanics, we can analytically continue beyond two-body collisions (for 1/r potentials), but not beyond three-body collisions. This is an essential limitation of predictibility - a blow to the world view a la Laplace!

The three-body problem is not solvable, in the sense that the constants of motion are not analytic functions of canonical coordinates. However, on the other side of the dychotomy crystal symmetry has a precise description from first principles: the three dimensionality of space, foliation of space, 32 point groups, 230 space groups, 1450 magnetic groups, as subgroups of the rotation group and the Euclidian group, respectively. The latters include the translations and time reversal of spin, respectively.

What the Symmetry theory cannot do however is to predict which substance will crystallize in what form and what the lattice constants will be. And not all of these groups are realized in Nature.

4. FUNDAMENTAL PARAMETERS AND UNPREDICTIBILITY RELATIONS

Here is a general feature: The symmetry theories contain new parameters. Phenomenological parameters one might say. In a broader sense actually all socalled phenomenological theories are of the type of symmetry theories in the sense that one perceives some global or general or macroscopic features of the phenomenon and begins to describes these features quantitatively. Now the more fundamental dynamical laws have fewer parameters (these are called or elevated to the status of the constants of Nature!) but the passage to phenomena may involve inherent impossibilities.

There seem to be some new kind of uncertainty relations, or unpredictibility relations if we start from dynamics.

(i) One is the already mentioned unpredictibility beyond the three body collisions and unsolvability.

(ii) The other happens in deterministic nonlinear dynamical equations. There are certain solutions which are so sensitive to the initial conditions that an extremely small change in the initial conditions can cause tremendous changes in later stages. The later behaviour cannot be distinguished from a completely stochastic behaviour. I should add here another practical uncertainty; as almost all studies of this kind of nonlinear behaviour is based on numerical results, a small change in the initial conditions cannot be separated from unavoidable computer cutoff!

Coming back to the initial conditions, we must realize that only half of the physics is the establishment of fundamental dynamical laws, the other half is the "physics of initial conditions" and solving the dynamical equations. By initial conditions we also mean the boundary conditions. For example, stationary solutions (in a certain time interval) do not have initial conditions in time but the boundary conditions at infinity for example. Now the initial and boundary conditions cannot be known with absolute accuracy both theoretically and experimentally.

The mathematicians are fond of the Cauchy Problem: Given the values of the field and its gradient on a space like surface, we can predict its values for all time in the future and everywhere. But it is impossible to give these values on the whole of the space-like surface, let alone with absolute accuracy. We can only know its values and its gradient inside the light cone in some past.

Then there are problems when the initial conditions go back all the way to the initial big bang (if there is one!), as in the case of black-body background radiation or neutrinos.

The generalization of this duality between crystal symmetry and underlying particle (or field dynamicsl) is the question: CAN WE REDUCE CHEMISTRY AND BIOLOGY TO PHYSICS? A similar question within biology itself is: can we reduce the classification of species (symmetries) to evolution?

It seems that this problem of reducibility is not very well understood. Apart from the limitations that we talked about, there is the fact that there may be undecidable statements. Or the number of steps of proving a statement from the axioms becomes so large that it will be impossible to do it, or completely uninteresting.

5. SYMMETRY-DYNAMICS DUALITY IN THE DOMAIN OF ATOMIC, NUCLEAR AND PARTICLE PHYSICS

How do we fare in microscopic world, in small scale localized phenomena, in elementary particle processes, between symmetry and dynamics? Can we derive one from other? What is the role of each one in our understanding of the presently most basic entities of Physics?

S. Hawkings proclaimed some years ago in his Cambridge Inaugural address "The end of Theoretical Physics". By this he meant the so called unification of electromagnetic, strong, weak and gravitational interactions in a formal field theory Lagrangian. This is more like putting all these theories side by side rather than "Unification" because the strengths of these interactions are not derived. The simplest of such theories has some 98 basic fields or particles, and 20 - 25 parameters. Most of these particles are not observed in nature, some are made unobservable (confined). Most of the observed particles are unstable. We should avoid the temptation of presenting a frame as the "theory of everything" with only "the details missing". The detail derivations are paramount in physics.

Let me go first a little back and see what we can and cannot do with basic quantum field theories (QFT).

Quantumelectrodynamics

The best theory physicists have ever invented, it is said, is Quantumelectrodynamics (QED), the theory of interactions of electrons (or other charged particles) with photons. A QFT is by definition based on perturbation theory. And in perturbation theory alone it is not possible to grasp the bound state of two charged particles, say a H-atom. You need to take infinitely many terms (or graphs) and the convergences of such a series is unknown (or is it undecidable?)

Some kind of a phase transition to bound states takes place. At this stage we inject some symmetry information contained in the bound state theory of Schrödinger equation, and then continue with perturbation theory from there.

The general situation is something like this:

Elementary particle interactions .....→..... ....→.... ....→.... ...→...

↑ inject bound states	↑ inject molecules	↑ inject crystal symmetry

There is another limitation: We are at the stage now in QED, where in order to improve or check the agreement between theory and experiment one decimal out of some 12 (for example in g-2 of the electron), we need to calculate many thousands of graphs and integrals and renormalize them. This is just the statement that a proof (or deducibility) may take such a large number of steps that we may approach theoretical or practical undecidability. Note these are numerical calculations.

In addition, in its present form of QFT electrodynamics is basically mathematically unsound (there are infinities to be subtracted).

So the situation is not very satisfactory.

However I believe that at least this part of electrodynamics, namely the grasping of bound states may be satisfactorily understood within electrodynamics, but with other nonperturbative methods then QFT. I shall come back to this point at the end.

Periodic Table of Elements

Let us proceed to the understanding of atoms (elements) in the Schrödinger dynamics. The final result should be the symmetry of the Periodic Table of Elements of Mendeleyef and Lothar Meyer. This system has very remarkable group theoretical properties. The derivation of this symmetry by the naive idea of filling the alredy existing levels of one-body Coulomb problem does not work (Madelung rule!). The complicated numerical Hartree Fock calculations may indicate, but do not achieve readily the cleanness and simplicity of the final result.

The final result is a single irreducible representation of the dynamical group SO(4,2) same as the one used in classifying the levels of the H-atom but with different physical interpretation of the group generators. The analog of the quantum numbers n, ℓ, m, s of the H-atom, now called ν, λ, μ, σ label different elements. The atomic number Z can be expressed as a function of these quantum numbers by separate formulas for neutral and highly ionized atoms.[5] The group structure reveals a double-shell structure of the periodic Table, not easily derivable from dynamics. The SO(4,2) representation contains two irreducible representations of the subgroup SO(3,2) corresponding to the oscillar levels. Many atomic properties, for example, ionization potential, are erratic when plotted in the periodic Table, but they are smooth in each sub-shell (Fig. 2) so that phenomenological mass and energy formulas can be written down similar to the mass formulas in nuclear and particle physics, thus showing again the occurence of global phenomenological parameters in the symmetry theories.

Superconductivity

The striking exact results of superconducting behavior, like flux quantization, the frequency of the AC Josephson effect, the nature of vortices are simple, beautiful, and precise results, in fact so precise that they are used to determine the values of fundamental constants. They can be derived on the basis of topology, or, in a more technical language, of the spontaneous breaking of gauge symmetry U(1) to the

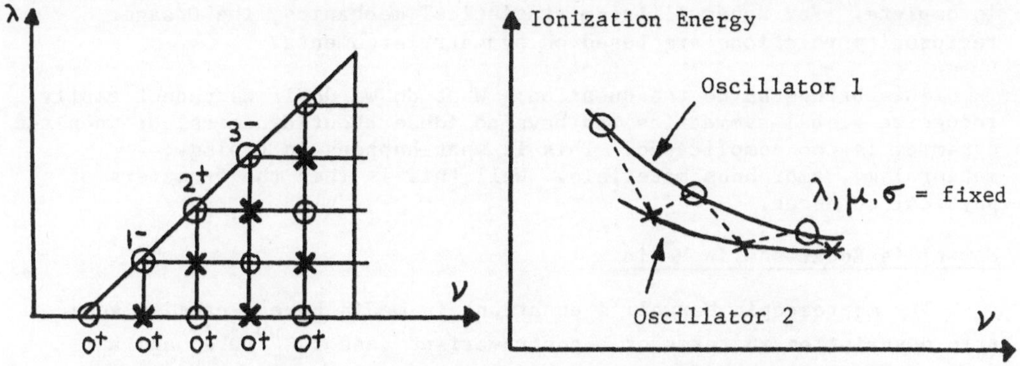

Fig. 2. Two double shells of the Periodic Table of Elements according to the reduction of SO(4,2) representations with respect to SO(3,2), and the smooth mass formulas in each shell.

finite sub-group Z_2. These symmetry behavior is described (phenomenologically) by a new order parameter field ψ. But the microscopic dynamical theory operates with the concepts appropriate to constituent electrons and lattice and it is very difficult to obtain the final symmetry from the microscopic many-body BCS-theory.

Nuclear Physics

In the case of Nuclear physics symmetry considerations of the Nuclear Chart have suggested the shell model of nuclei. Many nuclear levels are understood from the point of view of the symmetries of the nuclear shape (Bohr-Mottelson model: rotations and oscillations of the liquid drop). But the microscopic or dynamical derivation of nuclear physics is still missing. One can not even calculate yet the binding energy of the simplest nucleus, the deuteron from a fundamental dynamical theory. First one has of course to decide what the real fundamental constituents are, the nucleons or the quarks, or something else.

More recently many new partial dynamical groups for nuclear levels have been recognized and studied.[6] The mass formulas involve phenomenological parameters. Typically they are a linear combination of the Casimir operators of a subgroup chain with constant coefficients to be fitted. The question is what we can infere from these dynamical groups about the internal dynamics of nuclei.

6. OTHER PHYSICAL THEORIES

Thermodynamics and Statistical Mechanics

Thermodynamics is an exact beautiful macroscopic theory, which can now be formulated in a precise geometric way in terms of differential forms. For general thermodynamic relations we do not need any underlying dynamical theory. However, it must and does contain phenomenological parameters (specific heats, compressibilities, etc.) and we go to dynamics to understand these parameters.

Next the all powerful universally valid law of canonical distribution of equilibrium statistical mechanics is derived from general symmetry-type arguments, for example, from the principle of maximum entropy. Its derivation from particle dynamics is very difficult and incomplete. For nonequilibrium statistical mechanics, the Onsager reciprocity relations are based on symmetry arguments.

This brings us to the question: What do we do if we cannot easily recognize global symmetries and have no ideas about dynamics, or when the dynamics is too complicated. This is what happens in biology, meteorology, amorphous materials. Well this is then the frontiers of physical sciences.

Maxwell's Equations in Media

The macroscopic Maxwell's equations in media have a precise geometric description in terms of a contravariant tensor $G^{\mu\nu}(\vec{D},\vec{H})$ and a covariant tensor $F_{\mu\nu}(\vec{E},\vec{B})$. They are even covariant in general relativity and can be derived from an action principle. But we need a phenomenological set of constitutive equations connecting $(\vec{D},\vec{H})$ to $(\vec{E},\vec{B})$ and involving properties of the medium and of the surrounding metric $g_{\mu\nu}$. The derivation of these equations from the microscopic dynamical equations of constituents is a very complicated process, as one can expect.

The macroscopic laws of optics, for example, the reflection from a mirror, are very simple, but their derivation from microscopic dynamics via the scattering of individual photons or electrons from the atoms or electron of the mirror has, as far as I know, not been done. This is particularly significant when we come to fundamental quantummechanical experiments with single photons. What is the behavior of a single isolated quantum particle in a mirror?

Gravitation and Symmetries of Dynamical Systems

Is gravitation a symmetry theory or a dynamical theory as we distinguished it? Gravitation is observed sofar in macroscopic bodies (or one macroscopic body, the earth, and one neutron, for example). But to the extend that it treats these bodies as mass points it is a dynamical theory. It has been known since Heinrich Hertz (1894) that a dynamical motion of a particle in a potential is equivalent to a geodesic of a Riemannian space in one-higher dimension. This idea is used in general relativity. But the geometry, the Riemanmian space, is determined by the matter itself. The geometrical setup here is an expression of dynamics. We can endow the space-time manifold with different types of geometries depending upon the convenience of the dynamics.

Up to now we used the concept of "symmetry" in the sense of general regularities of objects and phenomena. The notion of geometry or symmetry of dynamical equations is a bit different. Every set of differential equations has its own symmetries in the sense of the transformation theory of Sophus Lie. These are the symmetries in the space of solutions of the differential equations: Transformations that allow to obtain new solutions from the old ones.

7. PARTICLE PHYSICS

In the understanding of the many new meson and baryon states of the 1950's and 60's symmetry ideas preceeded dynamical ideas. An octet symmetry relative to the quantum numbers of low lying mesons and baryons was recognized[7] which can be interpreted as stemming from three basic exchange relations.[8] It was then recognized that the internal symmetry group SU(3) gave exactly the quantum numbers of these hadrons and the octet could be identified with the weight diagram of this group.[9]

Here we come to a new twist, a new instance how the symmetry has influenced the subsequent inference of the dynamics. From the SU(3)-group structure of the baryon multiplets one has extracted the two three-dimensional fundamental representations of this group (quarks and antiquarks), and then took them to be the real physical constituents of the underlying dynamics of hadrons, in spite of their fractional charges unusual spin-statistics properties and nonobservability. This had never occured before. We do not take in the case of atomic, nuclear physics or periodic table the fundamental representations of the symmetry groups as constituents. In fact leptons can be also grouped into octets and we do not represent them as built from the fundamental representations. But if we had followed the same approach in atomic and nuclear physics, and we could, namely going from the symmetry group of atomic and nuclear multiplets to their fundamental representations and taking these objects as the constituents we would end up with an entirely different dynamics. For example, from the SO(4)-symmetry group of H-levels we can mathematically build all the levels of the atom from 4 "quarks". But the dynamics will differ as soon as we try to put forces between these 4 (unobservable) "H-atom quarks".[10]

This development is an example of the problems of infering the dynamics from symmetry. It is a general inverse problem where we do not know even the basic equations. In the original inverse problem of "determining the shape of a bell by means of the sound that it emits" (Sir Arthur Schuster, 1882) the equations were known. Given the mass specturm of a quantum system and quantum numbers we cannot uniquely determine internal dynamics, much more information is needed (form factors, magnetic moments, etc.). Conversely, assuming the quark model it has not been possible to calculate the masses of the low-lying hadrons in order to derive the symmetry.

Whether this socalled flavour group SU(3) (generalized later to SU(4), SU(5), SU(6)) is really a Lie symmetry group of some underlying internal "space" is debatable. We physicists use the notion of a flavor group in a rather inprecise way when it comes to describing the particle multiplets. First of all we use algebras rather than the groups themselves. Secondly we use mostly some low dimensional representations of these groups. A Lie group has infinitely many representations and only low lying few representations are realized in nature. Now for these low dimensional representations the representations of finite subgroups coincide with the representation of the algebra. In fact it is possible to understand the flavor multiplet structures and the symmetries of the S-matrix completely in terms of the finite groups, more precisely, the group of permutations of the constituents.[11]

The flavour groups are no longer considered to be fundamental symmetry groups.

Many hadron properties (mass formulas, form factors, structure functions, polarizabilities) can be accounted on the basis of dynamical groups, infinite-component wave equations, or general Hamiltonians describing the hadron as a single relativistic composite system with internal degrees of freedom.[12] But these approaches again contain necessarily phenomenological parameters which in principle an underlying dynamics should give us.

It is generally stated in particle physics in recent times that all fundamental theories of nature are or should be gauge theories, generalizing the U(1)-gauge principle of electromagnetism to other nonabelian groups in which there are multiplets of matter (particle) fields and another multiplets of gauge fields mediating the forces between the particles. It is implied that gauge theories indicate the fundamental nature of the assumed particles and fields; the world of particles and fields can be very complex, but the gauge symmetry idea is very simple.

It is worth to emphasize however that gauge theories can also be formulated with composite objects, no fundamental role being implied. I cite two examples. The system pH, pp, HH where p is the proton and H the H-atom can interact via photon exchange, electron exchange and electron-positron exchange (Fig.3)

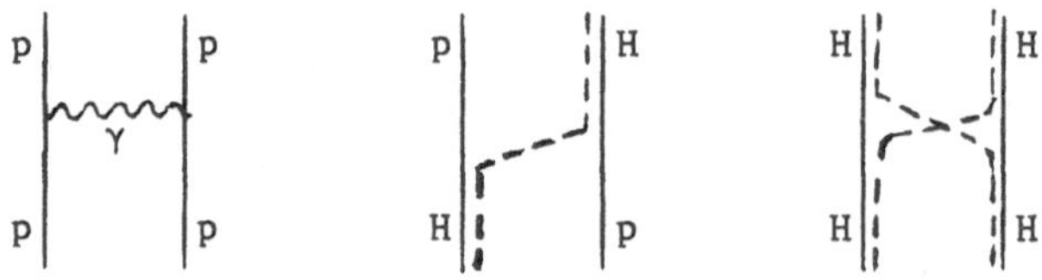

Fig. 3. The doublet of proton (p) and hydrogen (H) and the exchange of 3 agents: γ, e, (e^+e^-)

Thus we could assign p and H to an SU(2)-multiplet and γ, e and (e^+,e^-) to a multiplet of 3 exchanged particles. The gauge symmetry is broken and e, and (e^+e^-) acquire mass, γ does not.

The same scheme as Fig. 3 applies to the system p and n if we view n as n = (peν) and exchange again γ, (eν)-quantum and ($e^-\nu$ - $e^+\nu$). This amounts to gauging the isotopic spin group SU(2) and is a realization of the original Yang-Mills idea of extending gauge group of electrodynamics to isospin.

Another example is a gauge theory of strong interactions where a number of real baryons (p, n, Λ, Σ ...) are the fermion particles and the real mesons (ρ, ω, A_2, f ...) are the exchanged vector mesons. The gauge principle and the assumed group structure determine the coupling.[13] It is a reasonable phenomenological model of strong interactions. Thus gauge symmetries are not necessarily fundamental symmetries of nature, but can also be a manifestation of composite structures.

The proliferation of the number of fundamental input particles and parameters in gauge theories is a point of great concern. This is, I think, generally recognized. Most of these assumed particles are unobserved and hypothetical. This is particularly true in supersymmetric gauge theories. Supersymmetry (symmetry between integer and half-integer spin levels) can be observed approximatively in some nuclear and particle levels,[6] and can be understood in terms of the constituents, and does not necessarily indicate a fundamental physical principle. At the fundamental level of today supersymmetry is not established. It has been introduced to have renormalizable gauge theories, i.e. to cancel some of the divergences.

We hold the renormalizability of field theories at such a high esteem that we are ready to pay the prize of introducing any number of new particles and fields in order to make the theory renormalizable in some orders of perturbation theory. Of course, the more particles the theory has the more possibilities can be arranged to cancel the divergent diagrams.

Gravitation and electromagnetism were the two main forces in physics until the beginning of particle physics around 1932. Since then two new interactions have been introduced, strong and weak. The phenomena associated with these four interactions have different symmetry properties, and different dynamics were assigned with each one. The attempts to unify these phenomena at the dynamical level tend to put all these particles and forces side by side and "unify" them by additional fields and particles as in unified gauge theories mentioned above. The different strengths of these interactions have to be put by hand. In that sense they are not really unified theories with a single parameter.

There is another possibility. This is to realize that a single theory, electromagnetism can have widely different manifestations as weak and strong forces depending on the composite structures.[14] It is possible to construct dynamical mechanisms based on electrodynamcis and based on the two absolutely stable constituents, electron and neutrino which can reproduce the complexity of particle physics without introducing any other new particles or fields.

At first sight it would seem to be impossible that the simple, humble electrodynamics can reproduce the multitudes of "fundamental particles" and their interactions. Firstly two absolutely stable particles, electron and neutrino, are sufficient to construct, group theoretically, all other particle states and quantum numbers by successive bound state construction. Secondly, the electromagnetic interactions for localized states at short distances (e.g. electron-positron system) can be so strong that new strongly bound states of matter can be formed. This is because the self-magnetic fields of such states are so large that they can sustain the stability of these states (self-focusing states). To understand these properties of electromagnetic interactions at short distances we need new nonperturbative techniques which reproduce the known excellent results of QED, but go beyond and can explain short distance behavior as well.[15]

What a Fundamental Theory Should Be?

A fundamental theory is one which starting from very few and very simple fundamental objects can in principle predict in composite structures all the other structures we observe. What are then the most fundamental objects? They must be simple. Elevating observed particles or a large number of particles to the level of fundamental interacting objects may lead to a false dynamics. It would seem that the most fundamental entity can only be an absolute stable object, unalterable, indivisible, undestructable, the original "atom" of Anaxegoras. All other objects would then be built as Newton has imagined: in order of decreasing stability. "Now the smallest particles of matter may adhere by the strongest attraction, and compose bigger particles of weaker virtue, and many of these adhere and compose bigger particles whose virtue is still weaker, and so on for diverse successions, until the progresson ends in the biggest particles on which the operations of chymistry, and the colours of natural bodies depend and which by cohering, compose bodies of a sensible magnitude."[16] Unstable objects cannot be considered to be elementary. The most satisfactory dynamical approach to end the chain of further and further subdivision, quarks, subquarks, preons, etc. seems to be to stop at the indivisible "atoms".

Conclusions

The examples given in this essay show, I think, that the theories based on symmetry and regularities on one side, and on the dynamics of more basic constituents on the other side, are both first of all interesting, useful and complementary. They have been part of the tradition and development of physics from its beginnings. Sometimes one is superior, sometimes the other. They cannot always be derived one from another. We have to be aware of the limitations of both approaches. The novel way in which symmetry ideas have influenced a particular line of development of dynamical models in particle physics has been discussed. It is argued that the direct translation of symmetry group of particle multiplets may lead to a unnecessarily complex and sometimes to false dynamics.

If perceiving the symmetry and regularities in phenomena may be called empiricism or positivism, the uncovering of the underlying simple dynamical ideas is the idealism. And we cannot be just content with empiricism, it is too complex; we need the idealism to achieve simplicity which is the very essence of science.

REFERENCES

1. M. M. Nieto, The letters between Titius and Bonnet and the Titius-Bode Law of Planetary distances, Am. J. Phys. 53, 22 (1985).
2. A. Buta, Toward an atomic model of the planetary system, Rev. Roumaine de Physique, 27, 321 (1982).
3. A. O. Barut, Beyond Symmetry: Homeometry and Homeometry Groups, in Topics in Mathematical Phsyics, Colorado Assoc. Univ. Press, 1977; and to be published.
4. E. Schrödinger, Measurement of Length and Angle in Quantum Mechanics, Nature, 173, 442 (1954).
5. A. O. Barut, Group Structure of the Periodic System, in The Structure of Matter, Rutherford Centennial Symposium (B. G. Wybourne, edit.), Univ. of Canterbury Press, 1972; p. 126-136.
6. See A. Böhm and D. H. Feng, these Proceedings.
7. A. O. Barut, On the Symmetry of Elementary Particles, I, II, Nuovo Cim. 10, 1146 (1958); Nuov. Cim. 27, 1267 (1963).
8. H. Bauer, Z. f. Naturforschung, 17a, 321 (1962), Z. f. Phys. 178, 390 (1964.
9. M. Gell-Mann, Symmetries of Baryons and Mesons, Phys. Rev. 125, 1067 (1962); Y. Ne'eman, Derivation of Strong Interactions from a Gauge Invariance, Nucl. Phys. 26, 222 (1961).
10. A. O. Barut, Connection between Electromagnetic and Dual Strings, Acta Physica Austriaca, Suppl. XI, 565 (1973); Appendix.
11. A. O. Barut, Description and Interpretation of Internal Symmetries of Hadrons as an Exchange Symmetry, Physica 114, 221 (1982).
12. A. O. Barut, Dynamical Groups for the Motion of Relativistic Composite Objects, chapter VI in Groups, Systems, and Many-Body Physics (edit. by P. Kramer), Vieweg, 1980, p. 285-317.
13. R. Raczka, A Spontaneously Broken Quintent Baryon Model for Strong Interactions, Trieste preprint (to be published).
14. A. O. Barut, Unification Based on Electrodynamics, Ann. der Physik, Jan. 1986; see also in Symmetry in Science, (B. Gruber and R. S. Millman, edits.) Plenum 1980, p. 39-54.
15. A. O. Barut and J. Kraus, Nonperturbative Quantumelectrodynamics Without Infinities, Univ. of Colorado preprint, 1986 (to be published).
16. Isaac Newton, Opticks, (ed. I. B. Cohen, Dover Publ. 1952), quest. 31, p. 394.

TOPOLOGICAL CONCEPTS IN NUCLEAR PHYSICS: THE DEUTERON AS A BI-SKYRMION*

L. C. Biedenharn

Duke University
Durham, NC, 27706, USA

E. Sorace and M. Tarlini

Università di Firenze
Firenze, I-50125, Italia

INTRODUCTION

This conference has been organized by Bruno Gruber around the theme of symmetry, and it is useful therefore to begin by recalling that symmetry groups appear in quantal physics in three distinct ways: (a) *linearly* in the Wigner mode, (b) *non-linearly* in the Heisenberg mode, and (c) *topologically* in what might well be called the Dirac mode, since the first application occurred in Dirac's famous magnetic monopole paper. The subject to be discussed here will actually involve all three of these modes.

Currently it is believed that quantum chromodynamics (QCD) is *the* fundamental theory of strong interactions, and accordingly all of nuclear physics must be a consequence. However, it will be extremely difficult to make this connection explicit, since nuclear phenomena are *low-energy* effects (on the scale of $\Lambda_{QCD} \sim 300$ MeV) exactly in the (so far unsolved) confinement regime of enormously strong coupling. Nonetheless nuclear physicists have fully accepted QCD and many expect explicit quarks must appear somehow in nuclear structure phenomena.

The conventional wisdom has always taken it for granted that composite systems which are spinorial can arise only from fields containing at least one elementary fermion (the first example being the Heisenberg-Pauli non-linear spinor theory). Long before quarks or QCD had been invented, however, Skyrme[1] showed that indeed *fermions can arise from purely bosonic fields as topologically stable excitations*. This is a startling result--at least until one realizes that Dirac gave the first example, thirty years before Skyrme, in the Dirac magnetic monopole problem, as we will discuss below.

It was the re-discovery of Skyrme's ideas, in the context of the large-N_c (N_c = number of colors) limit of QCD, that has created a new approach to nuclear physics. Put differently, one may say that non-perturbative topological symmetry concepts now offer hope for understanding QCD at its most complicated: the low-energy, long-distance, confinement regime that typifies nuclear structure. This is certainly true qualitatively, and even quantitatively at the 30% error level.

---

*Supported in part by the NSF.

Our aim will not be to survey the results achieved in this new approach--this is already too large a topic--but rather to discuss the underlying ideas and concepts that characterize it. In particular, we will focus on exactly how one is to obtain the observed quantum numbers of the nucleon and the deuteron from the Skyrmion approach to nuclear structure. Since the methods are inherently non-linear it is not at all clear, *a priori*, that one can hope to reproduce the spin and isospin quantum numbers defined in a nuclear structure based on linear group realizations.

THE MAGNETIC MONOPOLE EXAMPLE[2]

Let us consider the motion of a (spinless) charged particle (mass m, electric charge e), confined to move on the surface of a sphere of radius a, whose center contains a spinless, heavy, point magnetic monopole g. Thus there is a radial magnetic field $\vec{B} = g\,\hat{r}/r^2$.

The classical (non-relativistic) mechanics of this problem is not difficult; there are four integrals: the total energy $E = \frac{1}{2} m\vec{v}^2$ and the angular momentum vector $\vec{J} = m\vec{r} \times \vec{v} - (eg/c)\hat{r}$.

The unusual feature is that there is now a *radial* component of angular momentum, a fact which was explained (by Fierz) as the radial angular momentum of the associated electric and magnetic fields.

The quantal treatment of this problem is rather more difficult, since this requires a Hamiltonian and hence the use of electromagnetic potentials. But there exists *no* vector potential valid everywhere on the surface of the sphere, and one is forced to artifices[2]: either strings (Dirac), sections (Wu-Yang), or fiber bundles (Greub-Petry).

We may recover Dirac's famous charge quantization condition if we argue, heuristically, that all angular momentum is quantized in units of $\hbar/2$, so that the radial component must be $eg/c = N\hbar/2$, where N is an integer.

Let us emphasize two basic results, which are important for the generalizations to follow:

(a) For N an *odd* integer, this composite system is a *spinor*, yet *none of the constituents was taken to be spinorial!* (The half-integral angular momentum can be seen, from Fierz's result, to be contributed by the electromagnetic field, which is, however, itself *bosonic*.)

(b) The 'wave functions' for the quantal system are *monopolar harmonics*, which are described, topologically, as sections of a principal bundle (base space S^2 = SU2/U1), for which the monopolar interaction in the Lagrangian appears as an 'anomaly' (to use the generic term) contributing radial angular momentum.

[*Remark*: More precisely, we have two (local) sections, σ, which map points of the manifold, $\alpha\beta$, ($\in S^2$ = SU2/U1) into the U1 fiber ($\gamma = \pm\alpha$) lying over the point. In addition we have the spherical harmonics $\mathcal{D} \in \{D_{mm'}^{j*}(\alpha\,\beta\,\gamma)\}$ which map an SU2 group element $\alpha\,\beta\,\gamma$ into $\mathbb{R}$. The *monopolar harmonics* are the composite map: $\mathcal{D} \circ \sigma$. The radial angular momentum is $m'\hbar$ of the 'body-fixed' system of $D_{m,m'}^{j*}$.]

Results analogous to (a) and (b) will be shown to hold for the Skyrmions of the Skyrme-Witten model.

BRIEF RESUMÉ OF THE SKYRME MODEL[1,3,4]

The model is defined by the (Poincaré invariant) Lagrangian density:

$$L_{\text{Skyrme}} = \frac{F_\pi^2}{16} \operatorname{Tr}\left[\partial_\mu U \partial_\mu U^\dagger\right] + \frac{1}{32e^2} \operatorname{Tr}\left[(\partial_\mu U)U^\dagger, (\partial_\nu U)U^\dagger\right]^2 , \quad (1)$$

where $U \equiv \exp\left(\frac{2i}{F_\pi} \vec{\tau}\cdot\vec{\pi}(\vec{x},t)\right) \in SU2$, and $F_\pi \cong 186$ MeV; e is a dimensionless constant.

This Lagrangian shows that the model is a classical non-linear field theory of pions only, which has chiral $SU2 \times SU2$ symmetry, realized by the transformations: $U \to AUB^{-1}$, with A, B, U being SU2 matrices.

The first term in the Lagrangian is well-known from chiral dynamics. The second term, due to Skyrme, is *ad hoc* and chosen to ensure stability. (Under a scale transformation ($\vec{x}\to L\vec{x}$, $L > 0$) the first term scales as L, the second as L^{-1}, thus a stable static minimum exists.)

The energy corresponding to the Lagrangian density (1) is given by

$$E = \int d^3x \left\{ \frac{F_\pi^2}{16} \operatorname{Tr}\left(V^2{}_0 + \vec{V}^2\right) + \frac{1}{16e^2} \operatorname{Tr}\left(\vec{C}^2 + \vec{D}^2\right)\right\} , \quad (2a)$$

where

$$V_0 = iU^\dagger \partial_0 U, \quad \vec{V} = iU^\dagger \vec{\partial} U, \quad (2b)$$

$$C_i = \frac{1}{i}\left[V_0, V_i\right] , \quad D_i = \frac{1}{2i} \varepsilon_{ijk} \left[V_j, V_k\right] . \quad (2c)$$

To have finite energy solutions one must impose the conditions:

$$V_0(t,\vec{x}) \xrightarrow[|\vec{x}| \to \infty]{} 0, \qquad V(t,\vec{x}) \xrightarrow[|x| \to \infty]{} 0, \quad (3a)$$

which imply:

$$U(t,\vec{x}) \xrightarrow[|\vec{x}| \to \infty]{} U_0 \in SU(2), \quad (3b)$$

where U_0 is a space-time independent SU2 matrix. Utilizing the $SU2 \times SU2$ symmetry one can always choose $U_0 = \mathbb{1}$.

Topological considerations enter as follows: for a given time t, the matrix $U(t,\vec{x})$ is a mapping from three dimensional space $\mathbb{R}^3$ into SU2; the latter is topologically the three-sphere S^3. With the boundary conditions (3) this mapping includes the point at infinity. However, adding the point at infinity to three dimensional space compactifies it to another S^3. Thus at a given time $U(t,\vec{x})$ is a mapping from S^3 to S^3 where infinity is mapped onto $\mathbb{1} \in SU2$. All such maps are known to fall into equivalence classes under the homotopy group $\pi_3(SU2) = \mathbb{Z}$, which counts the number of times the image of the S^3 domain covers the S^3 range. Skyrme identified this integer as the baryon number, B_0, characterizing the mapping $U(t,\vec{x})$. One can calculate B_0, using the expression:

$$B_0 = \frac{\varepsilon_{ijk}}{24\pi^2} \int d^3x \operatorname{Tr}\left(V_i V_j V_k\right) . \quad (4)$$

This expression shows that B_0 is the charge corresponding to the topologically conserved current:

$$J_\alpha^B(t,\vec{x}) = \frac{\varepsilon_{\alpha\beta\gamma\delta}}{24\pi^2} \mathrm{Tr}\left(V^\beta V^\gamma V^\delta\right) \quad . \tag{5}$$

(This current is conserved independently of the equations of motion.)

[*Remark*: Although Skyrme's identification of B_0 as the baryon number was remarkably prescient, still this step requires proof that B_0 is the *same* quantum number B that is defined in the quark model. That this is indeed so follows from quantizing the zero modes of the Skyrme-Witten $SU3_f$ model, (see below). (Having noted this we drop the distinction between B_0 and B.)]

The sector $B=0$ describes the mesonic states in the theory. The state with lowest energy is a constant matrix U_0 which in view of our boundary condition we choose as $\mathbb{1}$. This is the classical vacuum state. The vacuum is not invariant under the full $SU2_L \times SU2_R$ group but only under its diagonal subgroup $SU2_f$. Space time dependent small oscillations about the vacuum give rise to pions.

To obtain a time-independent solution in the sector $B=1$, Skyrme introduced the ansatz (later called a "hedgehog"):

$$U_1 = \exp\left(\frac{2i}{F_\pi}\,\theta(r)\,\hat{r}\cdot\vec{\tau}\right) \tag{6}$$

and determined $\theta(r)$ by solving the Euler-Lagrange equation from (1). (The U_1 in (6) is called a "Skyrmion".)

Let us stress the group theoretic meaning of the ansatz (6). Under a spatial rotation $R: r_i \to R_{ij}r_j$, whereas an isospin transformation (the diagonal sub-group of chiral $SU2 \times SU2$) is realized by the adjoint action: $U_1 \to AU_1A^\dagger$. The ansatz (6) is *equivariant* under these two realizations, or, equivalently, (6) is invariant under combined spatial and isospin rotations, generated by $\vec{K} = \vec{J} + \vec{I}$, with $[\vec{K}, \hat{r}\cdot\vec{\tau}] = 0$. When the theory is quantized, this equivariance will lead to excitations having well defined $SU2_{spin} \times SU2_{isospin}$ transformation properties.

THE SKYRME-WITTEN MODEL

Witten made two basic contributions to the Skyrme model. The first concerns the relationship between QCD and the soliton model. Assuming confinement, QCD for any number of colors (N_c) is equivalent to a theory of mesons (and glueballs). This meson theory is almost unlimitedly complicated, but as 't Hooft[5] first pointed out, for N_c large the meson theory becomes weakly coupled since the coupling constant for meson-meson scattering, g, obeys the relation: $g \sim N_c^{-\frac{1}{2}}$. The baryons[6] in such a theory (composed of N_c quarks in a color singlet) have masses $\sim N_c$, size ~ 1, and meson-baryon interactions ~ 1. These features typify *solitons*, and in this way Witten re-obtained, from QCD for large-N_c, Skyrme's remarkable idea that baryons are topologically stable excitations of the pion field.

Witten's second contribution was to add to the Skyrme action *an anomaly term*, Γ. He observed[7] that there are symmetries of the Skyrme Lagrangian which are not symmetries of QCD. This Lagrangian, (1), is invariant under three discrete symmetries: (a) $U \to U^{tr}$, (b) $\vec{x} \to -\vec{x}$, $t \to t$, $U \to U$ and (c) $U \to U^{-1}$. The first symmetry is charge-conjugation and belongs to QCD, but the two symmetries (b) and (c) do not belong to QCD separately: only the product: $\vec{x} \to -\vec{x}$, $t \to t$, $U \to U^{-1}$ is a QCD symmetry (parity).

Witten showed that this unwanted extra symmetry is eliminated by a new term, the "anomaly". Adding this term to the action yields:

$$S = \int d^4x \; L_{\text{Skyrme}} + N_c \Gamma \quad , \tag{7}$$

where Γ is the anomaly and N_c can be shown to be an integer[7] which can be proved to be the number of colors. The anomaly cannot be written as an integral over space-time, but rather appears in the form:

$$\Gamma \equiv \frac{1}{240\pi^2} \int d\, \Sigma^{ijklm}\, \mathrm{Tr}(V_i V_j V_k V_l V_m) \quad , \tag{8}$$

with $V_i = -iU^{-1}\,\partial_i U$, $U \in SU3$, and $d\Sigma^{ijklm}$ a volume element in an extended five dimensional space; the boundary of the integration region is compactified space-time.

Since Γ and N_c are dimensionless quantities, the last term in (7) must be multiplied by $\hbar$ to be a term in the action. (This $\hbar$ is not exhibited explicitly because of the convention $\hbar = 1$.) It is important to be aware of this fact, since classically the anomaly disappears from the action.

The addition of the anomaly term, $N_c\Gamma$, to the Skyrme action is of crucial importance, since it is this term which is responsible for determining, uniquely, the angular momentum and isospin quantum numbers of the quantized Skyrmion. The anomaly, however, is non-vanishing only for three (or more) flavors (since for two flavors we have $SU2_f \approx S^3$ which is three dimensional and hence forces Γ to be zero.)

To extend the Skyrme model to chiral $SU3_f$ is, at first glance, not very difficult. One uses the same Lagrangian, but now takes U to be a 3×3 unitary unimodular matrix, an element of the group $SU3_f$. Left and right translations ($U \to AU$, $U \to UB^{-1}$) then realize the symmetries $SU3_L$ and $SU3_R$ of chiral SU3. The pion field of the Skyrme model is thus replaced by the octet of pseudoscalar meson fields for the Skyrme-Witten model.

There is, however, a problem in achieving equivariance. In Skyrme's case equivariance had the effect of identifying an orthogonal frame in the isospin Lie algebra with an orthogonal frame in physical space. Since the Lie algebra of SU3 is eight-dimensional while space is three-dimensional, equivariance can only be achieved by choosing a three-dimensional subalgebra for identifying the frames. Using the isospin sub-algebra we get the injection:

$$U_1^{SU2}(\vec{x}) \longrightarrow U_1^{SU3}(\vec{x}) \equiv \begin{pmatrix} U_1^{SU2}(\vec{x}) & \begin{matrix} 0 \\ 0 \end{matrix} \\ \begin{matrix} 0 & 0 \end{matrix} & 1 \end{pmatrix} \tag{9}$$

where U_1^{SU2} is the 2×2 matrix of the Skyrme ansatz (6). A theorem[8] guarantees that U_1^{SU3}, above, has $B = 1$ if U_1^{SU2} has $B = 1$.

QUALITATIVE CONSEQUENCES FOR QUANTIZED SKYRMIONS

Our purpose now is to illustrate the remarkable consequences[9,10] that flow from quantizing the Skyrmions having the chiral SU3 action given by (7). The anomaly is, in a sense, a link between the short distance behavior (quarks) and the long distance behavior (the meson octet field). We will see that the anomaly yields "quark results without quarks".

To discuss the quantization, we observe that, by construction, classical small oscillations about the (static) Skyrmion will have characteristic frequencies $\omega \geq 0$. The modes corresponding to $\omega \cong 0$ are termed "zero modes" and are a manifestation of the symmetry of the solution U_1. For the purpose of calculating the low lying excitations of the spectrum zero modes are to be quantized.

We therefore examine the invariance properties of U_1^{SU3}. Since we are discussing a baryon at rest, the relevant space-time symmetry is the quantal rotation group, $SU2_J$. An element $A \in SU2_J$ induces on U_1 a rotation:

$U_1(\vec{x}) \to U_1(R(A)\vec{x})$ where $R(A)$ is the image of A in SO3. The flavor symmetry group $SU3_f$ induces on U_1 the adjoint action: $U_1(\vec{x}) \to BU_1(x)B^\dagger$ where $B \in SU3_f$.

The effect of these transformations on the SU3 version of the equivariance ansatz (6), shows that (9) is invariant under the diagonal $SU2_K$ subgroup of $SU2_I \times SU2_J$, where $SU2_I$ is the isospin subgroup of $SU3_f$. Most importantly, we note that (9) is invariant under the hypercharge $U1_Y$ subgroup of $SU3_f$.

Quantization of the zero modes now proceeds by introducing time dependent spatial rotations and $SU3_f$ transformations:

$$U_1(\vec{x}) \to U_1(t,\vec{x}) \equiv A(t)U_1(R(t)\vec{x})A^\dagger(t), \tag{10}$$

and determining the quantal Hamiltonian from (1) and (7). This Hamiltonian can be shown to be the quantal Hamiltonian for an "axially symmetric rotator" in SU3.

Let us recall how the quantum numbers for the baryons are determined[11] in the Skyrme-Witten model. From (10) one sees that the manifold for the quantal solutions, A(t), is not SU3 but SU3/U1, because of the invariance of (9) to hypercharge rotations $U1_Y$. This seemingly minor technical result is actually very significant, as we can see from the magnetic monopole example. Noting that the manifold SU3/U1 here is the analog to the manifold SU2/U1 in that example, we see also that the anomaly $N_c\Gamma$ functions to yield "radial (= right action) hypercharge" (Y_R) exactly in analogy to the radial angular momentum (right action m') of the "magnetic monopole anomaly".

The left translations on this manifold are given by the flavor group $SU3_f$, and--because of equivariance--the right translations are the spin group $SU2_J$.

The anomaly implies[11] that the right hypercharge $Y_R = N_c B/3$.

Thus if we use the fact that the nucleon has hypercharge 1 (and strangeness 0), we can conclude* that $Y_R = 1$, so that (a) the number of colors N_c is three and (b) Skyrme's topological charge B_0 is indeed the baryon number B.

The lowest energy SU3 irrep then determines that the nucleon belongs to an octet, [210]. This, in turn, shows that the spin is 1/2, since the right translations on the SU3 irrep [210] having $Y_R = 1$, necessarily have $I_R = 1/2 = J$.

The energetically next higher SU3 irrep is the decuplet [300], whose spin is 3/2 (since the right translations on [300] with $Y_R = 1$ necessarily have $I_R = 3/2 = J$).

We have thereby determined the quantum numbers of the lowest energy $SU3_f$ baryon multiplets *uniquely from the Skyrme-Witten model.*

THE DEUTERON AS A BI-SKYRMION

Let us now consider the B = 2 sector of the Skyrme-Witten model. If we assume a spherical Skyrmion---spherical in the sense of equivariance, (6)--then we find from the Euler-Lagrange solution with B = 2 that this configuration has mass approximately *three* times that of the B = 1 Skyrmion. Thus the spherical B = 2 Skyrmion is unbound against decay into two B = 1 Skyrmions, and, roughly speaking, the nucleon-nucleon potential has a repulsive core $\sim$ 1 GeV. But a B = 2 spherical Skyrmion is quite unsatisfactory as a model of the deuteron, since (as the previous section shows) one necessarily has I = J, (contra to I = 0, J = 1 for the deuteron).

A more satisfactory model of the deuteron was given by Jackson, et al[12]. Using the fact that the homotopy product of two B = 1 Skyrmions has B = 2 and is equivalent to the group product, they constructed an approximation to a B = 2 Skyrmion (for $SU2_f$) as the product of two (undistorted) B = 1 Skyrmions located a distance R apart, one relatively rotated with respect to the other. They identified the classical energy of this system, after subtracting two Skyrmion masses, as the potential energy of the Skyrmion-Skyrmion interaction. In the absence of any relative rotation, the interaction was always *repulsive*, but if one Skyrmion was rotated by 180° about an axis perpendicular to the line joining the two Skyrmions, an attractive potential with a depth $\sim$40 MeV and a range $\sim$2 fm was obtained, and a bound classical solution. Asymptotically the interaction was similar to the tail of the one pion exchange potential.

Using these qualitatively satisfactory results, Braaten and Carson[13] analyzed the symmetry properties of this B = 2 configuration[12] and obtained the spin-isospin quantum numbers of the deuteron.

A solution to the Euler-Lagrange equations for B = 2 was obtained by Sorace and Tarlini[14] using a cylindrically symmetric bispherical coordinate system adapted to the two Skyrmion configuration. Their solution was restricted to have a very special form, the bispherical analog to the radial B = 1 Skyrmion, which automatically enforced cylindrically symmetric equivariance. This configuration is qualitatively similar to the *unrotated* two Skyrmion configuration of Jackson, et al[12]. This probably accounts for the

---

*This identifies the nucleon hypercharge with the 'body-fixed' hypercharge Y_R. Using other members of the baryon octet yields consistent results only if the body-fixed hypercharge is in fact the baryon number.

fact that ref. [14] found that the classical $B = 2$ energy always exceeded two $B = 1$ masses, that is, the interaction was repulsive.

These results pose an interesting question as to the role of equivariance in achieving minimal energy configurations. For $B = 1$ Skyrmions no one has questioned that spherically symmetric equivariant solutions are in fact the actual minimal energy configurations--but no proof that this is so exists to our knowledge. Already in the $B = 2$ case we observe that spherical equivariance cannot be minimal, since the Jackson, et al configuration[12]--which breaks *both* spherical *and* cylindrical equivariance--is certainly lower in energy. It would be surprising if the true minimal solution actually did break cylindrical symmetry (equivariance) but this cannot be ruled out.

We are at present calculating results for general configurations, but our work is not yet complete.

In this situation, it is fortunate that the extension to $SU3_f$ once again yields unique results for the $SU3_f$ quantum numbers of the deuteron (assuming a bound $B = 2$ classical Skyrmion state exists).

To demonstrate this, recall that the manifold for the $B = 2$ quantal solutions is necessarily SU3/U1, just as in the B=1 case. If follows from the anomaly, using $B = 2$ and $N_c = 3$, that the (right action) hypercharge $Y_R = 2$. Since the quantal Hamiltonian once again is a symmetric top in $SU3_f$, it follows that *the minimum energy* (for $Y_R = 2$) *is achieved by the* SU3 *irrep* [330]. This is precisely the desired $SU3_f$ irrep and, since the deuteron has hypercharge 2, it necessarily follows that the isospin is 0, that is to say, the $SU3_f$ state for the deuteron has the Gel'fand pattern $\begin{pmatrix} 3 & & 3 & & 0 \\ & 3 & & 3 & \\ & & 3 & & \end{pmatrix}$.

It is remarkable that these (flavor) quantum numbers are so directly obtained from the Skyrme-Witten model.

Any linking between the right action isospin group and the rotational (spin) quantum numbers will be determined by the equivariance symmetries, and this depends on determining the (so far unknown) classical minimal energy configuration. We can, however, by-pass this difficulty by noting that the quantal solutions will almost certainly contain (small) components where the two $B = 1$ Skyrmions are relatively far apart and undistorted. The known total quantum numbers from the linear realizations will then be found, and since these total quantum numbers are good quantum numbers of the non-linear system, they must be valid everywhere. This shows that the spin of the bi-Skyrmion must be 1, from the Pauli principle (using $I = 0$), and thus completes the determination of the quantum numbers of the deuteron. When the true minimal energy solution is known, it should be possible to determine the spin quantum numbers directly from the (non-linear) solution itself.

CONCLUDING REMARKS

When one considers the deuteron as a bi-Skyrmion, one is struck by the fact that this simple model clearly shows the deuteron to be crudely like a dumbbell, that is, to have a positive (prolate) quadrupole moment. (The numerical value predicted in the model is correct within the 30% accuracy level.) Taking the Skyrmion model quite literally, one sees that two nucleons to a considerable extent appear to preserve their integrity when binding. If this "stereotactic" view holds one might expect a triangular triton model and a tetrahedral alpha particle model to have some validity. Empirically the binding energy of nuclei has long been known to be given crudely by the number of alpha particles (Wefelmeier).

Let us pose an intriguing speculation: could the magic numbers of the nuclear shell model be explained by some kind of (spherical or tetrahedral) close packing of Skyrmions?

ACKNOWLEDGEMENTS

We would like to thank Prof. Yossef Dothan (Tel-Aviv) for discussions and Mr. Alec Schramm (Duke) for his help with the computations.

REFERENCES

1. T.H.R. Skyrme, Proc. Roy. Soc. A260 (1961) 127.

2. L.C. Biedenharn and J.D. Louck, "The Racah-Wigner Algebra in Quantum Theory", in Encl. of Math. and its Applications, Vol. 9, (G.-C. Rota, Ed.) (Addison-Wesley Publ. Co., Reading, MA 1981). A detailed review with citations of the original papers is given in Topic 2, "Monopolar Harmonics", p. 201 ff.

3. G. Adkins, C. Nappi and E. Witten, Nucl. Phys. B228 (1983) 552.

4. G. Holzwarth and B. Schwesinger, "Baryons in the Skyrme Model", Reports on Progress in Physics, (to be published).

5. G. 't Hooft, Nucl. Phys. B72 (1974) 461; ibid. B75 (1974) 461.

6. E. Witten, Nucl. Phys. B156 (1979) 269.

7. E. Witten, Nucl. Phys. B223 (1983) 422; ibid. 433.

8. R. Bott, Bull. Soc. Math. France, 84 (1956) 251.

9. E. Guadagnini, Nucl. Phys. B236 (1984) 35.

10. L.C. Biedenharn, Y. Dothan and A. Stern, Phys. Lett. 146D (1984) 289.

11. L.C. Biedenharn and Yossef Dothan, "Monopolar Harmonics in $SU3_f$ as Eigenstates of the Skyrme-Witten Model for Baryons", pps. 19-34 in "From SU3 to Gravity" (Ne'eman Festschrift), (Cambridge University Press, Cambridge, U.K.) 1986.

12. A. Jackson, A.D. Jackson and V. Pasquier, Nucl. Phys. A432 (1985) 567.

13. Eric Braaten and Larry Carson, "Nuclei in the Skyrme Model", preprint ANL-HEP-CP-85-67 presented at the LANL Workshop on Rel. Dyn. and Quarks in Nuclear Physics, June (1985), (to be published).

14. E. Sorace and M. Tarlini, Phys. Rev. D33 (1986) 253.

QCD AND NUCLEAR STRUCTURE

Konrad Bleuler

Institut für Theoretische Kernphysik der Universität Bonn

Nußallee 14-16, D-5300 Bonn, West-Germany

SUMMARY

QCD leads to a decisive renewal, i.e.redefinition of nuclear theory: an all over reconstruction starting directly from the basic quark-gluon level (thus replacing the conventional and half-phenomenological nucleon--boson scheme) is outlined within the framework of nuclear shell structure. Hereby essential - i.e. non-perturbative and group-theoretical - features of QCD play a decisive role. At the same time, a few characteristic inconsistencies - a.o. the solitonlike behaviour of the extended, in nuclear matter embedded, nucleons - of the conventional approach are overcome in a natural way.

I. INTRODUCTION

Theoretical physics is, by now, in a most critical and interesting stage: a decisive step to a deeper and unifying understanding of fundamental physical laws is foreshadowed by the advent of the recent ('heterotic') string approach[15]. An indication for this breath taking development constitutes, a.o. the well-known fact that - so far - the three fundamental interactions (gravitational, electroweak and strong) are all based on exactly the same geometric view-point, namely the gauge principle, as formulated by H. Weyl (the so-called second version) as early as 1929[1].

In view of the enormous difference in strength and physical interpretation of these three cases, it might be worthwhile to analize the characteristic formal relation between the conventional gauge theories and gravitation from an appropriate standpoint which yields more direct evidence for the common origin (as well as a typical difference) of these theories (see sect. II).

In this connection it should be stressed that the gauge theory of strong interaction, i.e. QCD leads a.o. to a perfectly new and deeper (i.e. parameter-free) understanding of the masses of all heavy systems, i.e. hadrons and nuclei all over the periodic table (see sect. III). In particular, it will be shown that this new QCD-theoretical basis points to a completely different, but, in a way much more systematic and natural description of nuclear structure which, in turn, exhibits an amazing similarity to atomic and molecular theory. A.o. it replaces the conventional and cumbersome N-nucleon system with its ill-defined expressions of so-called nuclear forces by a basic 'geometrical' law.

In sect. IV an example for the needed reformulation of nuclear theory is outlined, emphasizing the far-reaching group- and field theoretical methods connected to this new approach. From a more general view-point, it may also be realized that nuclear theory thus becomes a most important application of non-perturbative quantum field theoretical methods which, in turn, appear closely related to an interpretation of a few characteristic empirical nuclear properties which - within the conventional theory - were hard to be understood and lead in some cases to clearcut inconsistencies (see sect. V).

II. THE GEOMETRICAL BASIS OF GAUGE THEORIES

In order to define the general expression for a gauge field, we first consider a generalized 'parallel-transport' of an n-dimensional complex vector v for an infinitesimal displacement δx^μ on the Lorentz base space (x):

$$\delta v = \delta x^\mu A_\mu^{op}(x) v \tag{1}$$

where A^{op} represents an n-dimensional linear map. If, however, v is restricted by a fixed unitary norm A^{op} is given by the general element of the Lie-algebra of SU(n) and (1) takes the form

$$\delta v = i\delta x^\mu A_\mu^a(x) T_a^{op} v, \qquad a = 1\ldots n^2-1 \tag{1'}$$

where the T_a^{op} are the (n^2-1) elements of the Lie-algebra of SU(n). The n^2-1 component covariant vectorfield A thus represents the general expression for a gauge field of SU(n). Considering now a field $\psi(x)$ with the 'vector character' of v, the 'covariant' derivation D_μ reads (with coupling constant g and omitting now the operator sign):

$$D_\mu \psi \equiv (\partial_\mu + igA_\mu^a T_a)\psi \tag{2}$$

The corresponding 'curvature' R (the commutator represents geometrically the parallel transport along a closed infinitesimal circuit) is given by

$$R_{\mu\nu} = \frac{1}{ig}[D_\mu, D_\nu]_- \equiv F_{\mu\nu}^a T_a \tag{3}$$

thus leading to the well-known expression for the field strength (the c's represent the SU(n) structure constants):

$$F_{\mu\nu}^a = \frac{\partial A_\nu^a}{\partial x^\mu} - \frac{\partial A_\mu^a}{\partial x^\nu} - gc_{bc}^a A_\mu^b A_\nu^c \,. \tag{4}$$

As a consequence of this 'geometric' construction a local gauge transformation

$$\psi = S(x)\psi' \tag{5}$$

where S is an arbitrary space-time dependent unitary n-dimensional matrix (geometrically a general coordinate transformation on the fibre bundle)[9] yields automatically a'covariant' transformation law, i.e. the adjoint representation of SU(n) for F (in contrast to A where the term $S^{-1}\partial_\mu S$ appears) which enables us to introduce - enlarging the corresponding expression in the Maxwellian case n=1, i.e. U(1) - the totally invariant Lagrangian

(6) $$L = F^{a}_{\mu\nu}F^{\mu\nu}_{a}$$

leading, in turn, to the famous non-linear Yang-Mills field equations[8].

This geometric view-point suggests, of course, to introduce the interaction term in Dirac's equation by replacing ∂_μ by D_μ (according 2), i.e. writing (suppressing the spin-index):

(7) $$(\gamma^{\mu}D_{\mu})\psi + \kappa\psi = 0$$

thus obtaining (by the transformation (5)) and the Lagrangian (6) a perfectly gauge invariant theory in which the coupling term and the field equation are uniquely determined (apart from g which, however, represents just a scale parameter).

The main point of this section is to show, however, that our geometrical view-point leads in a natural way also to General Relativity which thus takes the form of an (enlarged) gauge theory: the form (5) of local gauge transformations suggests anyhow an enlargement to a more general, e.g. Riemannian ground space, i.e. a manyfold endowed with a metric tensor $g^{\mu\nu}(x)$. The Dirac anticommutation relations have then naturally to be enlarged to a condition for γ-fields, i.e.

(8) $$[\gamma^{\mu}(x)\gamma^{\nu}(x)]_{+} = 2\,Ig^{\mu\nu}(x)$$

As a decisive point one has now to introduce the parallel transport of a Dirac spinor ψ which has the general form (1):

(9) $$\delta\psi = \delta x^{\mu}\Gamma_{\mu}\psi$$

where the Γ_μ are 4 by 4 matrices acting now on the spinor index. They are, however, very much in contrast to the former case, to be expressed by the basic metric field g, i.e. the corresponding Christoffel symbol appearing naturally within the (formal) covariant derivation $\mathbb{D}_\mu$ of the vector field γ^ν: on one side $\delta\gamma^{\mu} = \delta x^{\sigma}\mathbb{D}_{\sigma}\gamma^{\mu}$ leaves (note $\mathbb{D}_{\sigma}g^{\mu\nu} = 0^{\mu}$) the commut. relat. (8) unchanged.

On the other hand Γ_μ is to be interpreted as the corresponding similarity transformation $S = I + \Gamma_{\mu}\delta x^{\mu}$ of the γ's thus leading to the commutator with the γ's and, therefore, to the geometrically evident defining relation:[16]

(10) $$\mathbb{D}_{\mu}\gamma^{\nu} = [\Gamma_{\mu}\gamma^{\nu}]_{-} \;.$$

Inserting - as before - the covariant derivation

(11) $$D_{\mu} = \partial_{\mu} + \Gamma_{\mu} \quad \text{i.e.} \quad \gamma^{\mu}(x)D_{\mu}\psi + \kappa\psi = 0$$

into the Dirac equation, one obtains - as the decisive point - the natural and needed invariance with respect to the general, i.e. space-time dependent similarity transformation S(x) of the γ-fields leaving the relation (8) invariant

(12) $$\gamma'^{\mu} = S^{-1}\gamma^{\mu}S \;.$$

The corresponding transformation of the Dirac field

(13) $$\psi = S(x)\psi'$$

has again the form of a gauge transformation (acting on spin), however, for the new gauge group SL(4). At first sight, again the expression $S^{-1}\partial_{\mu}S$ in (7) appears.

This characteristic(but disturbing) term will,in this case just be cancelled by the corresponding term appearing through the transformation of the Γ's. Inserting (12) into (10) one obtains, in fact,

(14) $$\Gamma'_{\mu} = S^{-1}\Gamma_{\mu}S - S^{-1}\partial_{\mu}S$$

This characteristic SL(4) gauge invariance of the general relativistic Dirac equation is, at the same time, equivalent to the (clumsy) construction based on the conventional 'Vierbeins':
(i) any explicit representation for the γ's satisfying the anticommutation relation (8) corresponds, in fact, to a choice of a 'Vierbein'-field to be determined by a special (vector) transformation a(x) of the γ's

(15) $$\gamma'^{\mu'} = a^{\mu'}_{\mu}(x)\gamma^{\mu}$$

and the corresponding 'diagonalizing' tensor transformation of $g^{\mu\nu}$ yielding

(16) $$[\gamma'^{\mu'}\gamma'^{\nu'}]_{+} = 2Ig^{o}$$

with the (constant) Lorentz metric g^{o}.
(ii) The general S-transformation (12) which leads to a new explicit solution thus corresponds (by the above construction (15) and (16))to a new (and general) 'Vierbein'-distribution leaving the general relativistic Dirac equations invariant.

Eventually, the invariance of the Dirac field with respect to general coordinate transformations x(x') follows easily through the vector transformations (in abbreviated notation)

(15) $$\gamma' = \frac{\partial x'}{\partial x}\gamma(x(x')), \qquad \psi' = \psi(x(x'))$$

and by verifying that Γ_{μ} (whose definition (10) contains the covariant derivation (!)) obeys the transformation law of a covariant vector.

In perfect analogy to eq. (10) the Riemannian curvature tensor R is obtained within this framework in spinorial form:

(16) $$R_{\mu\nu} = [D_{\mu}D_{\nu}]_{-}$$

but the Lagrangian will be defined differently, i.e. by the curvature scalar to be determined from $R_{\mu\nu}$ with the help of g and γ^{μ}. These considerations thus show that General Relativity may be based in a most natural way on the gauge group SL(4).

III. HADRONS AND NUCLEI IN THE LIGHT OF QCD

Considering the practically infinite system of hadrons with their mass values, inner excitations and mutual interactions (of about the same order of magnitude), it is clear (in analogy with the periodic table of atoms) that these particles can by no means assumed to be of elementary character: they behave, in fact, like systems built out of similar,more elementary particles bound, however, by a new type of basic interactions. It thus constitutes a true revelation that a far-reaching experimental research in connection with a most interesting group-theoretical analysis of the hadron mass spectra[10] showed that the (so far) elementary building blocks are the quarks, i.e. Dirac-fermions endowed with special inner degrees of freedom, in particular with the colour-index ranging from 1 to 3, and that the binding is due to the coupling to a gauge field (as described in sect. II) based on the colour index, i.e. on the gauge group $SU_{colour}(3)$ yielding thus (apart from the scale parameter g) a uniquely determined interaction.[17] In accordance to the general principles of field theory the transition to the corresponding quantized gauge field, i.e. to quantum chromodynamics (QCD) is[14] definitely needed; the corresponding far-reaching mathematical problems, e.g. a more direct proof of renormalizability still represents a major challenge for present-day theoretical research. On the other hand, the practically parameterfree determination of the relative masses of a few light hadrons through the so-called lattice-approach[11] to QCD represents a most impressive check of our basic assumptions and foreshadows that - in principle - the whole (infinite) hadronic mass spectrum is to be understood by our gauge-theoretical scheme, i.e. on a most simple and natural geometric view-point.

The main question to be discussed here deals, however, with the theoretical determination of the - empirically partly well-known - mutual interactions between the hadrons to be treated evidently on the same footing. In the special case of nucleons, these couplings are called (in an enlarged sense) nuclear forces and represent the very basis of nuclear physics, i.e. a most extended domain related to a nearly uncountable number of empirical data. These facts exhibit a striking analogy with atomic and molecular physics: here, the interaction between the elementary constitutes, i.e. the electrons, providing at the same time atomic structure and (according to present-day quantum chemistry) chemical binding is based on the Maxwellian field, i.e. the U(1) gauge theory whereas the inner structure of nucleons as well as nuclear forces (i.e. also nuclear structure) are provided by the SU(3) gauge field acting between the basic constituents, i.e. the quarks. In other words: the two gauge theories (U(1) and SU(3)) constitute, in fact, the basis for two really enormous realms: atoms and molecules on one side, nucleons (hyperons) and nuclei (hypernuclei) on the other.

In analogy to modern quantum chemistry it appears natural - and, as we will see, also suitable - to visualize the nucleus just as an overall system of quarks (and antiquarks) or, in other words, as a heavy (so to speak enlarged) hadron. As a consequence all of 'conventional' nuclear theory visualizing the nucleus as a system of (elementary) nucleons bound by (phenomenological) nuclear forces or (half-phenomenological) boson exchange is to be replaced - on the quark level - by the practically parameterfree gauge theory (or QCD). Specifying this statement explicitly: the reduction of the (most detailed and enormously extended) empirical nuclear data to a 'geometric' principle might appear unbelievable or at least - according to many scientists - unrealistic (reduction to one single scale parameter).

Our aim, however, is to make a very first attempt in this direction by considering, for the time being, the special (but important) case of medium heavy spherical nuclei using - in a new and different context - well-known but non-trivial approximation schemes. It turns out, as far as the first stages are concerned, that things work even better than to be anticipated; it will be seen, however, that the numerical results - as far as QCD is involved - were obtained through characteristic non-perturbative results which, in turn, are intimately related to the typical non-linearities of the underlying gauge theory.

IV. SHELL STRUCTURE BASED ON QCD

A. The Average Potential

Details of our explicit calculations of nuclear shell structure starting directly from quarks and QCD without using any additional parameters or interactions have been given elsewhere[2]. Here, I would just emphasize the arguments which are directly related to basic properties of QCD as well as those which are of special mathematical, i.e. group-theoretical interest.

(i) N-fermion systems occurring in physics (solid state, atoms, conventional nuclear structure) are usually treated by introducing a suitable average potential (e.g. Hartree-Fock). In the conventional nuclear (N--nucleon) problem this led, however, to enormous - to our experience practically unsurmountable - difficulties. In the case of the N-quark systems it appears, therefore, of greatest interest that QCD provides a perfectly new and basically different definition for a quantity which, so to speak, replaces the 'average potential': it is the so-called QCD-vacuum pressure as first introduced by the Russian school[3]. In fact, the energy density of the degenerate QCD vacuum is appreciably decreased in space domains occupied by 'valence' quarks, i.e. in the interior of the nucleus. Assuming a practically constant quark density, one might thus introduce (in the case of sufficiently large systems) a so-called surface pressure which had been used in high energy hadron reactions as well as in the MIT bag model[4] with roughly the same numerical value B. Assuming for the time being an ideal spherical surface and independent relativistic (Dirac-) quark states (i.e. orbitals) inside this spheres one obtains through a most elementary equilibrium condition (in connection with the value of B and an 'effective' mass for so-called 'dressed' quarks) the well-known radius law - const. $A^{1/3}$ - for spherical nuclei, thus replacing (in first approximation) on the quark level the (problematic) former expressions used on the nucleon level.

In addition, the quark orbitals exhibit automatically (as a natural relativistic effect) a spin-orbit splitting which will be - within the further steps of the calculations - taken up to the nucleon level where the conventional theory led to characteristic discrepancies.

(ii) The next and decisive step is, in fact, related to a 'preformation' of embedded nucleons within the apparently structureless quark sea: we have to satisfy the basic QCD-condition saying that all bound quark states (e.g. all hadrons) must have vanishing total colour (i.e. have to be 'white'). Writing in second quantization the (antisymmetric) N-quark state of independent orbitals characterized by spin and isospin in a given 'open shell':

(1) $$\psi = \prod_{r=1}^{N} a^{*}_{i_r \alpha_r} |0\rangle$$

where i=1...3 represents the colour-index and

$$\alpha \rightarrow {}^{-j\leq m\leq +j}_{\tau=-1/2,+1/2}$$

represents a combined index (assuming 2(2j+1) values) of the magnetic quantum number m (in the 'open' j-shell) and the 2-valued isospin τ. (The closed or filled shells are left out; see below). Out of these, in general, degenerate and 'coloured' states the 'white' ones are to be singled out by special linear combinations (corresponding to the 1-dimensional induced representations of SU(3)). Their explicit form reads (ε stands for the totally antisymmetric symbol) as will be shown:

(2) $$\psi^{o} = \prod_{r=1}^{N} \varepsilon^{ikl} a^{*}_{i\alpha_r} a^{*}_{k\beta_r} a^{*}_{l\gamma_r} |0\rangle$$

Note that, according to this form, the number of quarks has always to be a multiple of 3 and that the filled j-shells are automatically 'white' and may, therefore, be left out in expressions (1) and (2). In addition, the N 3-quark factors (or clusters) in expression (2) may naturally be visualized as (so far highly deformed and degenerate) embedded nucleons. In this connection it should be emphasized that QCD thus enforces a 'preformation' of a nucleon-structure (in fact, as a direct consequence of 3-valued colour) within nuclear matter (if antiquarks where admitted only $q\bar{q}$-pairs may be added) and that these 3 quark substructures should - in constrast to usual treatments - not be introduced 'by hand'.

(iii) In order to prove this really important fact, one just remarks that the system of all states according (2) yields automatically a (reducible) tensor representation of SU(3) and that the full reduction (into the irreducible representation) is to be obtained by the corresponding 3-line Young symmetrizers. The 1-dimensional ones are obtained by those with 3 lines of equal length and the corresponding symmetrization is just given by our expressions (2) in perfect analogy to Russel-Saunders.

B. The Direct Interaction

At this stage it has to be realized that QCD yields - in addition to the 'vacuum-pressure' - also 2-,3-body interactions between the embedded quarks which will automatically split the still highly degenerate states (2) in accordance with 'physical' expectation: the 3-quark clusters in (2) exhibit, in fact, a degeneracy given by independent values of α,β,γ whereas a nucleon in its ground state exhibits just a single α-value, i.e. m and τ. In order to simplify our calculations we assume in a first tentative stage a special 2-quark interaction P which (in second quantization) reads

(3)
$$P = g \sum_{i=1}^{3} A^i A^{i*} \quad \text{where}$$
$$A^i = \varepsilon^{ikl} a_{k\alpha} g^{\alpha\beta} a_{l\beta}$$

where the symmetric 'metric' $g^{\alpha\beta}$ is given explicitly through a generalized (i.e. SU(3) invariant) pairing-interaction scheme. (In analogy to superconductivity (i.e. BCS) and conventional nuclear physics; here g represents a so far unknown coupling strength).

A decisive mathematical part of one work consists in an exact diagonalization of the corresponding full Hamiltonian H:

(4) $$H = H_o + P$$

where H_o describes the independent obitals. The corresponding mathematical scheme is based on the well-known principle of symmetry breaking:
a) the 'white' eigenstates of H_o as given by (2) - again omitting the trivial closed shells - are uniquely characterized by the irreducible representations of the (maximal) invariance group:

(5) $$SU(n) \quad \text{with} \quad n = 2(2j+1)$$

i.e. a 'unitary rotation' of the n single quark states (within the 'open' j-shell) as functions of α for fixed values of the colour.

Introducing now the symmetry breaking term P this invariance group is reduced to the 'orthogonal' subgroup SO(n) of SU(n):

defined by the 'metric' $g^{\alpha\beta}$ within the expression (3) for P.
c) The states (2) generate first all induced but (in general) reducible representations of the subgroup SO(n) which,by now,have to be split (or decomposed) into the known irreducible ones. (According to Wigner, one has to form the 'right' linear combinations). If one assumes, however, that the eigenspaces of the full H (according (4)) do correspond to irreducible representations of SO(n) our 'right linear combinations' are, in fact, the eigenstates we were looking for[13]. (A detailed analysis has indeed shown that the above assumption is - apart from one exception - in fact, satisfied[5]).According Weyl[6], these linear combinations, i.e. the eigenstates of H can be written down explicitly: the lowest one exhibits a most natural form:

$$(7) \qquad \psi^{oo} = \prod_{r=1}^{n} \varepsilon^{ikl} a^{*}_{i\alpha_r} a^{*}_{k\beta} g^{\beta\gamma} a^{*}_{l\gamma} |o\rangle ,$$

i.e. there is a contraction with the help of the metric $g^{\beta\gamma}$ in each 3-quark factor, leaving only one single free value for α in accordance with the expected property of an embedded nucleon (which now consists of a diquark with zero spin and isospin and a 'valence' quark with the quantum number of the nucleon). All further states, i.e. ψ^{oR} are obtained by leaving out the contraction (stepwise)within these factors in connection with the well-known 'Schmidt' orthogonalization procedure. The corresponding eigenvalues were also obtained group-theoretically by expressing (as usual) the Hamiltonian H by the Casimir-operators of the two groups. It then turns out that the expression (7) is, in fact, (as assumed above) the lowest state representing N-independent (embedded) nucleons in accordance with conventional shell structure, whereas the states ψ^{oR} are naturally to be interpreted to represent inner excitations (i.e. so-called Δ-states) of R (embedded) nucleons.

In view of the fact that the states for R=1 are experimentally observed and measured, we may now determine the numerical value for the coupling strength in (3). The main point of our work which, however, needs further and more detailed investigations is this: a full-fledged QCD Feynmann path integration calculation[12] yields an expression which contains our special interaction (3) as the main part, i.e. it reproduces by and large its operational character as well as its strength g. This last statement is based on a comparison (using the so-called 'running coupling') to measured high energy processes. In this connection it was interesting to observe that the decisive contribution within the functional integration were due to special instanton solutions of the classical gauge equations. The corresponding topological quantum numbers are determined through the Atiyah-Singer index theorem.

A last - unfortunately uncompleted - calculation is related to the (definitely needed) introduction of a 3-body potential which acts only within colourfree 3-quark systems. An expression exhibiting this special property constitutes a characteristic consequence of QCD inasmuch it is related to a 'saturated' colour-flux; it is proportional to the shortest 'string' connection between the 3 quarks (so-called Mercedes-star) and is of similar strength as the well-known $q\bar{q}$-interaction. This term yields the needed spaceal correlation of the nucleon-clusters and explains, at the same time - in view of its typical saturation properties - the possibility of a free interpenetration of the rather extended embedded nucleons, definitely needed - without being understood - within conventional nuclear theory, i.e. a soliton-like behaviour.

Summarizing some results of this perfectly unconventional attempt, it should be emphasized that a few characteristic empirical nuclear properties appear intimately related to typical and non-trivial consequences of QCD:

(i) QCD-vacuum pressure related to the nuclear radius law.

(ii) The decisive importance of 3-valued colour in view of the formation of nucleonic substructures within nuclear matter.

(iii) Typical non-perturbative results of QCD in view of definitely needed properties of the quark-quark interactions.

As far as the mathematical methods (used here) are concerned, it appears appealing that group-theoretical (i.e. symmetry) arguments play a decisive role in nearly every step: the group SU(3), the 3-quark clusters and the symmetry breaking.

V. CONCLUSIONS

Looking over these calculations it should be stressed that they represent just a first attempt to derive nuclear structure - within the framework of a special case - from the outset on the basis of gauge theory, i.e. QCD. Our real motivation is, however, to reach the much more general aim of physical research, namely to try to interpret empirical data through laws which have a fundamental character inasmuch they contain the needed minimum number of basic constants and exhibit a mathematical structure which stands in its own right (cp. sect. II).

In the case of nuclear physics this abstract (or 'purist') standpoint has, however, also a rather important practical side: all conventional more or less phenomenological attempts lead - if followed systematically - to enormous computational difficulties and in certain cases to characteristic inconsistencies which, within the framework of special examples, are obviously overcome by a more basic approach; a few striking cases mentioned in sect. IV are:

(i) replacing the conventional nuclear potential by the QCD vacuum pressure;

(ii) a relativistic interpretation of spin-orbit splitting;

(iii) the new 'interpretation' of embedded nucleons determined by special properties of interactions based on QCD;

(iv) the interpretation of 'inner excitation' (i.e. Δ-states) of the embedded nucleons as a natural outcome of the quark approach; the conventional treatment leads in this case to severe inconsistencies[7] with respect to empirical properties related to shell structure.

In all these cases the - unfortunately to a large extent - suppressed difficulties of the conventional approach appear within our approach intimately related to characteristic consequences of QCD:

1. the most important properties of the QCD-vacuum;
2. the special behaviour (i.e. flux-conservation) of interactions due to QCD explaining the interpenetration of embedded nucleons.
3. the Pauli-principle to be applied directly to the quarks leads to a different interpretation of Δ-states within nuclear matter; conventionally these states had to be considered as new fundamental particles.

Surveying these facts, one might be tempted to state that a few special empirical nuclear properties lead - indirectly - to a verification of the validity of QCD, in particular with respect to certain non--perturbative calculations. The corresponding methods which are also of general interest should be - in any case - developed further: our present work thus rather represents an encouragement and motivation for a continuation making out of nuclear physics an important branch of general quantum field theory. At the same time, the so far foreshadowed possibility of a nearly parameterfree mainly group-theoretical interpretation of the enormous body of nuclear data represents a great challenge, especially in view of the fact that the basic (geometrical) assumptions may be seen as (according sect. I) a special part of a far reaching and universal theoretical view-point.

REFERENCES

1. H. Weyl, Z. Physik, 56 (1929) 330
2. H.R. Petry, Springer Lect.Not.Phys., Nr. 197, 236 (1984)
   K. Bleuler et al., Zeitschr.Naturf., 38a, 705 (1983)
   K. Bleuler, Proc. AMCO-7, 399 (Darmstadt 1984), Persp. Nucl.Phys. (World Scient.) 455 (1984)
   H.R. Petry, H. Bohr, K.S. Narain et al., 'An Application of QCD in Nuclear Structure', Phys.Lett. 159B, p. 365 (1985)
3. E.V. Shuryak, Nucl.Phys. B203 (1982) 93
   E.V. Shuryak, The QCD Vacuum, Hadrons and Superdense Matter, World Scientific, Singapore (1986), to appear
4. A. Chodos et al., P.R. D9 (1974) 3471 and
   E.V. Shuryak, CERN 83-01 (Febr. 1983)
5. H. Hofestädt and S. Merk, Diploma and Doctor Thesis, Bonn, Inst. f. Theor. Nucl. Phys., Nusallee 14-16, (unpublished)
6. H. Weyl, The Classical Groups, Princeton Math. Series (1946)
7. K. Bleuler, Nuclear Forces in the Light of Modern Gauge Theory, 6th Int. Conf. on Nuclear Reaction Mechanisms, ed. E. Gadioli, Univ. di Milano, in Ricerea Scientifica (1985)
8. C.N. Yang, R. Mills, Phys.Rev. 95, 631, 96, 191 (1954)
   R. Utiyamah, Phys.Rev. 101 (1956) 1597
9. W. Drechsler, M.E. Mayer, Fibre Bundle Techniques in Gauge Theories, Springer Lect. Notes in Physics, 67 (1977)
   A. Trautman, Reports on Math. Phys. 1 (1970) 29
   A. Held (ed.): General Relativity and Gravitation, Vol. 1, Plenum Press, New York (1980), 287
10. M. Gell'Mann, Y. Ne'eman, The Eightfold Way, W.A. Benjamin, New York (1964)
11. For example: C. Rebbi (ed.): Lattice Gauge Theories and Monte Carlo Simulations, World Scientific, Singapore (1983)
12. G. t'Hooft, Phys.Rev. D14 (1976) 3432
13. For very special cases see: H. Lipkin, Lie Groups for Pedestrians, North Holland (1967)
14. For a general introduction, see e.g. F.J. Yndurain, Quantum Chromodynamics, Springer 1983
15. See e.g. contribution by Y. Ne'eman in this volume
16. The explicit solution, i.e. the conventional (but less intuitive) expression reads (in our notation): $\Gamma_\mu = [\gamma^\sigma, \mathbf{D}_\mu \gamma_\sigma]_-$ and may be proved, a.o. by covariant derivation of the commutation relations (8).
17. Compare, e.g. F.J. Yndurain, Quantum Chromodynamics, Springer Monographs 1983, in particular the most extended list of literature therein. (Historically, the definite statement about introducing the quantized gauge field according SU(3), i.e. QCD, in relation to the colour degree of freedom of the quarks came only step by step).

QUANTUM MECHANICS AND SPECTRUM GENERATING GROUPS AND SUPERGROUPS[1]

A. Bohm

Center for Particle Theory, Deparment of Physics
The University of Texas at Austin
Austin, Texas 78712

I. INTRODUCTION: Collective Models are the Physical Basis for Dynamical Groups

In our quantum mechanics course we were taught that molecules consist of N electrons and M nuclei and that one has to solve a (N + M) body Schrödinger equation to understand their structure. But if one looks at the work of the practitioners in this area, e.g. the books of G. Herzberg,[1)] one sees that the practice is different: Low energy spectra and structure of molecules are analyzed in terms of rotators and oscillators (and at slightly higher energies in terms of Kepler systems (one electron outside a core)). This is shown in Figure 1a.

Figure 1b shows the energy levels and transitions of a rotating molecule. It is most economically described by associating to each level an irreducible representation space R^j of SO(3) in which the angular momenta J_i act. The transitions are performed by dipole operators Γ_i and the J_i and Γ_i together form the Lie algebra of $SO(3,1)_{\Gamma_i J_i} \supset SO(3)$ (or of $E(3)_{Q_i J_i} \supset SO(3)_{J_i}$ with commuting position operators Q_i which are obtained from the Γ_i by group contraction, or -- as is very fashionable these days in nuclear physics -- by $SO(4)_{\Gamma_i J_i} \supset SO(3)_{J_i}$ leading to finite multiplets). The irreducible representation (irrep) space

$$\mathcal{H}^{rot} \xrightarrow[SO(3)]{} \sum_j \oplus R^j \tag{1}$$

---

[1] Presented at: Symmetries in Science II, Carbondale, Illinois, March 1986.

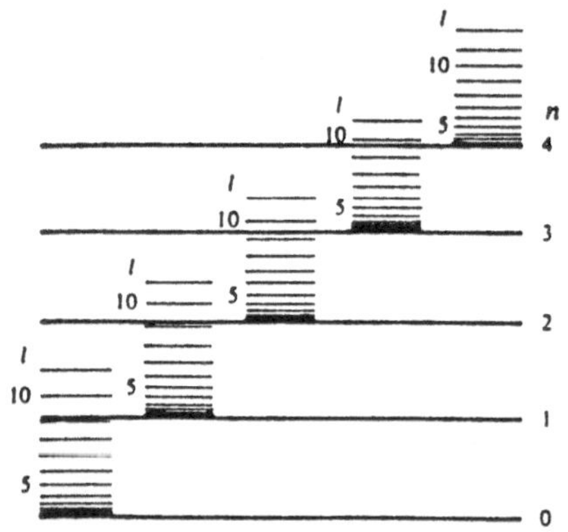

Fig. 1a. Energy levels of the vibrating rotator. For each of the first five vibrational levels, a number of rotational levels are drawn (short horizontal lines).

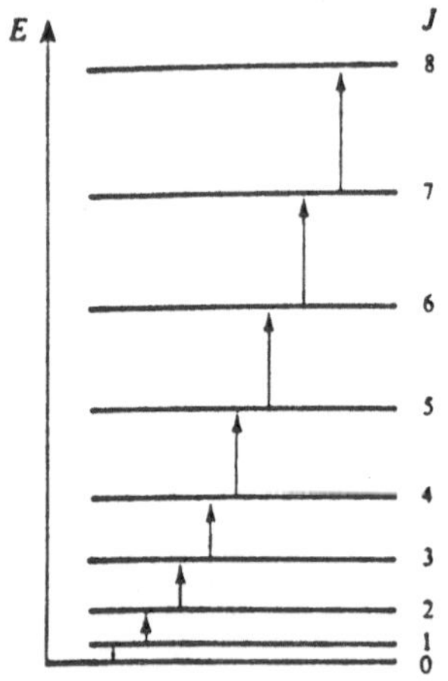

Fig. 1b. Energy levels and infrared transitions of a rigid rotator.

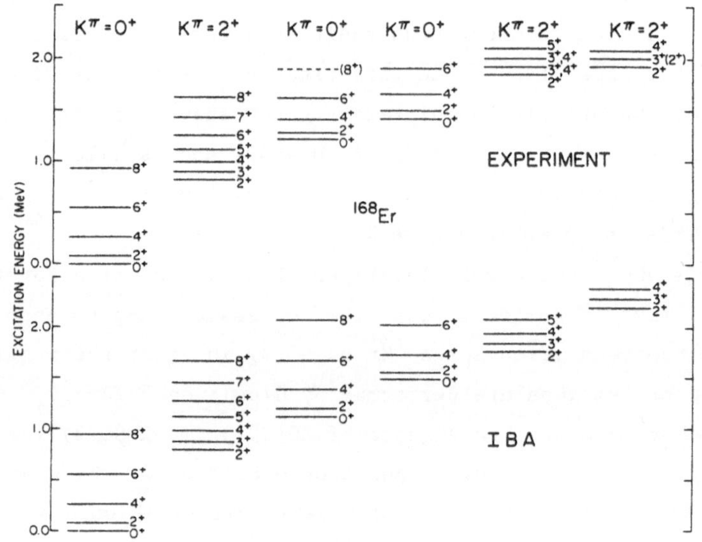

Fig. 2. Energy spectrum of ^{168}Er and its comparison with the IBA model (from D. D. Warner et al., Phys. Rev. Letters 22, 1761 (1980)). It consists of the ground state rotational band and two kinds of vibrational excitations with $K = 0^+$ and $K = 2^+$, also interpreted as β- and γ-vibrations, respectively.

describes the whole spectrum of the rotator. This SO(3,1) (or E_3, or SO(4)) is the simplest example of a spectrum generating group or dynamical group, which we shall denote by SG.[2] In this SG description the molecule is not analyzed in terms of constituents but in terms of its collective motions.

Like a molecule the nucleus can be considered as consisting of neutrons and protons but in nuclear physics the many-body problem is yet more complicated. Models built on this constituent picture, the so-called underlying microscopic theories, are also here not very useful for a fast interpretation of experimental data. Even more so than in molecular physics one resorts to "phenomenological models" of oscillators and rotators.[3] An example is shown in Fig. 2, where some energy levels of ^{168}Er are analyzed in terms of vibrational excitations (β- and γ-vibrations) with each vibrational level splitting into rotational bands. The explanation of this particular spectrum can be given by the Iachello Arima model which uses U(6) as the dynamical group.[4] The U(6) model has had enormous success in organizing the experimental data (low energy spectra, transitions) of a large class of nuclei.

The foundations for the spectrum generating group approach in nuclear physics were laid by the collective model of Bohr and Mottelson. It was also in connection with the collective motion of nuclei that a group was first used to describe the spectrum. Several years before the use of spectrum generating groups was made into a principle,[2] Goshen and Lipkin[5] used SO(2,1) ⊃ SO(2) to describe collective bands. An irrep space

$$\mathcal{H}^{osc} \Longrightarrow \sum_{\nu} \oplus \mathcal{H}^{\nu} \tag{2}$$

where $\mathcal{H}^{\nu}$ is an irrep space of SO(2) -- describes the whole spectrum.

Often one starts from individual nucleon variables and defines the group generators -- which represent the collective dynamical variables -- in terms of them.[6] Or one starts with six (more or less) boson creation and annihilation operators to define the group generators. But then all that can be essentially forgotten and the algebraic relations and the representations of the group become the only relevant properties. The constituents become unimportant. The physical states are described by the representation space of the SG and their properties are determined by the subgroup reduction chain of the dynamical group:

$$SG \supset SG^{(1)} \supset SG^{(2)} \ldots \supset G^{(n)} \tag{3}$$

The hamiltonian and transition operators are postulated in terms of group generators (collective variables) and e.g. given by

$$H = a_0 + \sum_i a_i^{(1)} C_i^{(1)} + \sum a_i^{(2)} C_i^{(2)} + \cdots \qquad (4)$$

where $a_i^{(m)}$ are numbers and $C_i^{(m)}$ are the Casimir operators of the m-th group in the subgroup chain (3). In the simple examples of (1) SG = SO(3,1), $G^{(n)}$ = SO(3) and of (2) SG = SO(2,1) and $G^{(n)}$ = SO(2).

In the constituent model one starts with the space of physical states as a direct product space of constituent spaces. E.g. for the diatomic molecule the space of physical states $\mathcal{H}$ is

$$\mathcal{H} = \mathcal{H}^{\mathrm{Nucl}_1} \otimes \mathcal{H}^{\mathrm{Nucl}_2} \otimes (\mathcal{H}^1 \wedge \mathcal{H}^2 \wedge \cdots \wedge \mathcal{H}^N) \qquad (5)$$

and the basis vectors are:

$$|\ \rangle = |\mathrm{Nucl}_1\rangle \otimes |\mathrm{Nucl}_2\rangle \otimes \text{antisym}\ |d_1\rangle \otimes |d_2\rangle \cdots \otimes |d_N\rangle \qquad (6)$$

where $d_1, \cdots, d_N$ stands for the electron quantum numbers of the N electrons and $\mathrm{Nucl}_1 \cdots$ stands for the quantum numbers of the nucleus

In the collective model one starts with the space of physical states as a direct product space of rotator spaces, oscillator spaces, Kepler system spaces (single constituent in the core potential) and the rest (core):

$$\mathcal{H} = \mathcal{H}^{\mathrm{rot}} \otimes \mathcal{H}^{\mathrm{osc}} \otimes \mathcal{H}^{\mathrm{Kepl}} \otimes \mathcal{H}^{\mathrm{core}} \qquad (7)$$

The basis vectors are

$$|\ \rangle = |j j_3 \cdots\rangle \otimes |\nu\rangle \otimes |n\rangle \otimes |\text{core q.n.}\rangle \qquad (8)$$

The (collective) motions are associated to groups so that $\mathcal{H}^{\mathrm{rot}}$ may be given by (1) and $\mathcal{H}^{\mathrm{osc}}$ by (2) and $G^{(n)}$ in (3) is given by SO(3) × SO(2) with the irrep space:

$$R^j \times \mathcal{H}^\nu \qquad (9)$$

The basis (8) is thus adapted to a subgroup chain (3).

The space of physical states $\mathcal{H}$ is the same whether one considers it from the collective model as in (7) and (8) or from the constituent model as in (5) and (6). (8) and (6) are two different basis systems of the same space $\mathcal{H}$ in the same way as the position eigenkets $|\vec{x}\rangle$ (particle) and the momentum eigenkets $|\vec{p}\rangle$ (wave) of a free quantum mechanical point particle are two different (generalized) basis systems for the same one-particle Hilbert space. (8) of the collective model are thus sort of complementary to (6) of the constituent model. The right hand sides of (7) and (5) are just two different ways of dissecting the space of physical states; (5) corresponds to the dissection of the physical system into its constituents, (7) corresponds to the dissection into its motion.

Though classically there may be a great difference between these two models, quantum mechanically they just correspond to two different choices of basis systems for the space of physical states (which may be complementary but which are at least incompatible to each other).

The second choice (7) is much simpler, especially if a large number of constituents is involved, because $\mathcal{H}^{core}$ is in practice just one dimensional, since the degrees of freedom corresponding to it are frozen (its set of quantum numbers take just one value). One often also need only take the "lowest" basis vector from $\mathcal{H}^{Kepl.}$ (below 1 eV) and under some circumstances (below 10^{-4} eV for most diatomic molecules) also only one basis vector from $\mathcal{H}^{osc}$; then one has the rigid rotator. Though the complete system of commuting operators has the same number of elements for (6) and (8), many of the observables for the basis vectors (8) take for all practical purposes only one eigenvalue. This is the reason why in the actual analysis of experiments in both molecular and nuclear physics the second description has proven itself of much greater value.

There is a quantitative difference though between molecules and nuclei: In molecular physics the three modes -- rotational, vibrational and particle excitations -- differ from each other in energy by factors of approximately 50. So rotations and vibrations are almost independent and they are very good, weakly interacting "parts" of the molecule. In nuclear physics the rotations and vibrations are not as well separated from each other and from the particle excitations, so that for the nuclear levels the rotation-vibration-particle interactions play a more important role. This could lead to a variety of subgroup reduction chains (3)[4),6)]; β- and γ-vibrations may not be the most practical parts of all nuclei.

To summarize the preceding discussions, we have seen that in molecular and nuclear physics we have a constituent -- motion dualism (reminiscent of the particle-wave dualism).[7)] On the one hand there "are" constituents and we understand by reduction to the simpler objects. On the other hand there are motions and we understand by reduction to the simpler motion. This latter is the understanding for which group theoretical tools are most suitable.

Does this dualism extend into the relativistic domain?

In hadron physics quantum chromodynamics is the "underlying theory." But as exact solutions or reliable approximations of QCD will be difficult to obtain, one must resort to phenomenological descriptions. On the one hand one has quarks and uses QCD corrections to the non-relativistic quark model. The complementary approach would be to understand hadrons as extended relativistic objects and analyze hadron structure and spectra in terms of relativistic motions. If the analogy between molecular and nuclear physics persists when we enter the relativistic domain then we would expect the lowest hadron states to be rotational and vibrational excitations with the only difference being that now the rotator and oscillator must be relativistic.

II. A SIMPLE RELATIVISTIC MODEL

a) $Osp(1,4) \supset SO(3,2)$ leads to a modified version of the lowest string mode

To show that collective models for extended relativistic objects can be constructed we will review now the results for the simplest such models, the quantal relativistic oscillator (QRO).[8] As dynamical (spectrum) supersymmetries have been successful in nuclear physics where they combine the spectrum of an even-even nucleus with the spectrum of an even-odd nucleus into a supermultiplet,[4] we will consider the supersymmetric QRO. It does not add anything essential to the QRO model: The SG of the QRO is $SO(4,2) \supset SO(3,2)$; adjoining to it Majorana spinor operators and closing to an $SU(2,2/1) \supset Osp(1,4)_{Q,\Gamma_\mu S_{\mu\nu}} \supset SO(3,2)_{\Gamma_\mu S_{\mu\nu}}$ gives the supersymmetric version. Whereas the vector operator Γ_i transforms from integer to integer spin or from half-integer to half-integer spin, the spinor operator Q_α transforms between integer and half-integer spin states.

Figure 3 shows how one can arrive at $Osp(1,4)$ and how the hamiltonian H is conjectured in analogy to non-relativistic supersymmetric quantum mechanics. Our choice of the hamiltonian is:

$$H = v(P_\mu P^\mu - \frac{1}{\alpha'} \sum \{Q_\beta, Q_\beta^+\}) \tag{10}$$

(where v = Lagrange multiplier of constraint Hamiltonian mechanics; P_μ = center of mass momentum, α'^{-1} = system constant of dimension energy per length connected with the string tension or spring constant). This, like every choice of a hamiltonian, cannot be derived, it can only be conjectured and can be justified only by success. For H of (10) one can show that it leads in the non-relativistic limit to the energy operator of the ordinary three-dimensional oscillator. Thus it is at least justified to call the physical system, whose relativistic hamiltonian is H, the quantal relativistic oscillator.

The irreps of $SO(4,2) \subset SU(2,2/1)$ which we use are very special ("remarkable") representations which remain irreducible under $SO(3,2) \subset SO(4,2)$. We will therefore continue our arguments just in terms of

non-relativistic
quantum mechanics
(one-dimensional or
three dimensional
oscillator)

S. Fubini, E. Rabinovici, CERN (1984)
E. Witten, N.P. B188 (1981);
F. Freedman, F. Cooper, Physica (1985)
A.B. Balantekin, Ann. Phys. (1985)
V.A. Kostellecky, M. Nieto, D.R. Truax,
Los Alamos (1984)
E. D'Hoker, L. Vinet, Comm. Math. Phys.
(1985)

supersymmetric
quantum mechanics
(oscillator with
spin-orbit coupling)

$H = H^{C.M.} + \frac{k}{2} \{Q, Q^+\}$

$SO(2,1) \supset SO(2)_{\Gamma_0}$

adjoining
Q, Q^+

$Osp(1,2)$

A.B., M. Loewe
P. Magnollay
L.C. Biedenharn
H. van Dam
M. Tarlini
P. Kielanowski
P.R.L. 53 (1984)
P.R. D31 (1985)

adjoining
S_{oi} (non-commuting
intrinsic position)
Γ_i (non-commuting momentum)

$SO(2,3)_{\Gamma_\mu S_{\mu\nu}}$

adjoining
$Q_\alpha, \bar{Q}_\alpha = - (Q^T C)$

$Osp(1,4)$

$H = v(P_\mu P^\mu - \frac{1}{\alpha^T} \sum_\alpha \{Q_\alpha, Q_\alpha^+\})$

relativistic
quantum mechanics
(quantal relativistic
oscillator)
(quantization of the
one-mode string with
non-commuting intrinsic
coordinates)

relativistic
supersymmetric
quantum
mechanics

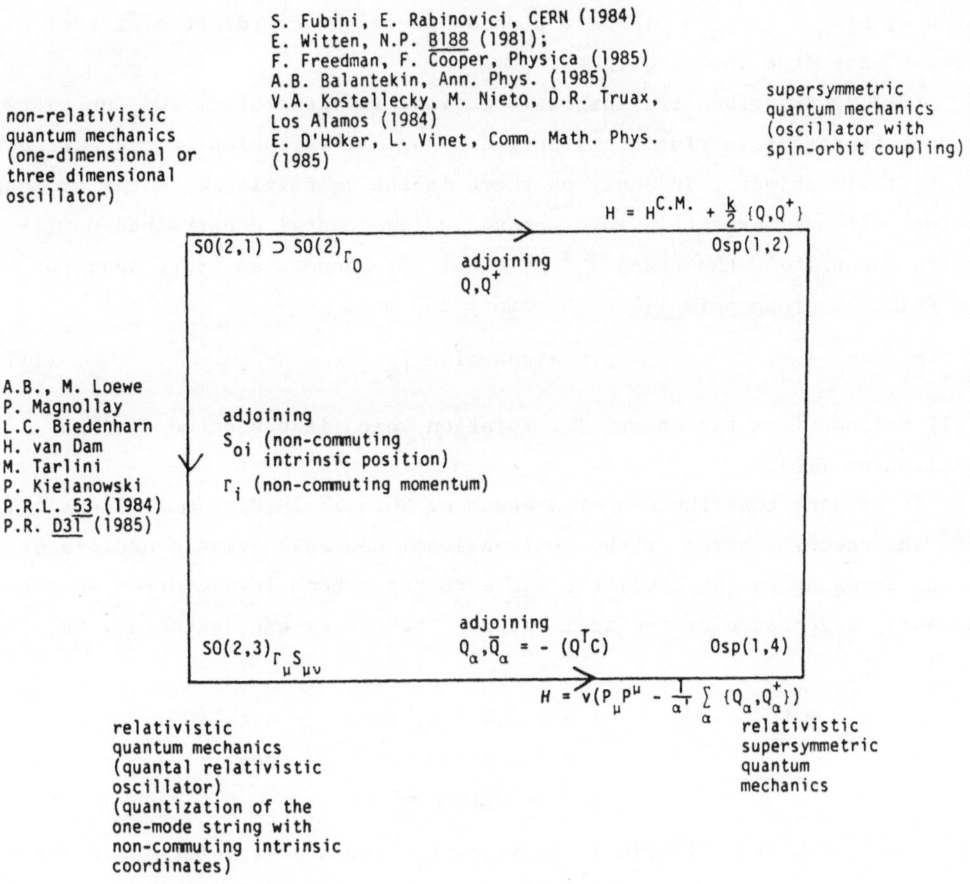

Fig. 3. Starting from non-relativistic quantum mechanics for the oscillator described by SO(2,1) one can adjoin Fermionic operators to obtain models with spectrum supersymmetry. The quantal relativistic oscillator is obtained if one adjoins the intrinsic position and momentum which leads to SO(2,3). To obtain its supersymmetric version one adjoins one (or two) Majorana spinor operators to obtain Osp(1,4) or SU(2,2/1).

SO(3,2). An idea of an infinite dimensional representation of a group, or supergroup like Osp(1,4), is provided by the weight diagram (K-type). Fig. 4a shows a typical weight diagram of $SO(3,2) \supset SO(3)_{S_{ij}} \times SO(2)_{\Gamma_0}$. In this diagram μ = eigenvalue Γ_0 is plotted versus j (where $j(j+1)$ = eigenvalue of $\frac{1}{2}S_{ij}S^{ij}$). Each dot displays an irrep space $\mathcal{R}^j \times \mathcal{H}^\mu$ of the maximal compact subgroup K = SO(3) × SO(2) which is the space (9) with $\mu = \mu_0 + \nu$. Fig. 4a is the weight diagram of the irrep conventionally denoted by $D(\mu_0 = 1, j_0 = 0)$ where μ_0 and j_0 are the coordinates of the "lowest" dot (the lowest weight).

SO(3,2) describes the intrinsic motion, its generators and functions thereof represent intrinsic collective dynamical variables of the extended relativistic object. In addition there is the relativistic center of mass motion. If one takes this into account using quantal constrained Hamiltonian mechanics[9] the space $\mathcal{R}^j \times \mathcal{H}^\mu$ of (9) becomes an irrep space of the Poincaré group $\mathcal{H}(m_\mu, j)$ with spin j and mass

$$m_\mu^2 = m_0^2 + \frac{1}{\alpha'}\mu \qquad \mu = \text{eigenvalue } \Gamma_0 \tag{11}$$

((11) follows from the constraint relation which is connected with the hamiltonian (10).

It happens that the weight diagram of SO(3,2) in Fig. 4a is identical with the energy diagram of the 3-dimensional non-relativistic oscillator, if one draws an energy level through each dot. Each level corresponds to an energy eigenspace of the oscillator. The energy and degeneracy is given by

$$E = E^{CM} + \hbar\omega(\nu + \tfrac{3}{2}) \qquad \begin{array}{ll} j = 0,2,4\cdots\nu & \text{for } \nu = \text{even } (\nu = \mu - 1) \\ j = 1,3,5\cdots\nu & \text{for } \nu = \text{odd} \\ E^{C.M} = \text{center of mass energy} & \end{array} \tag{12}$$

The observables are: intrinsic position ξ_i, momentum π_i, intrinsic angular momentum $S_{ij} = \xi_i \wedge \pi_j$ with the commutation relation

$$[\xi_i, \xi_j] = 0, \quad [\pi_i, \pi_j] = 0, \quad [\xi_i, \pi_j] = i\delta_{ij} \qquad i = 1,2,3 \tag{13}$$

The identity of the weight diagram of the irrep D(1,0) of SO(3,2) and the energy diagram is a reflection of the fact that D(1,0) of SO(3,2) goes by group contraction into the algebra of the three-dimensional oscillator. (This contraction describes the non-relativistic limit $c \to \infty$.)[8]

The usual relativistic generalization of the quantum oscillator, e.g. in (the one-mode version of) the relativistic string is

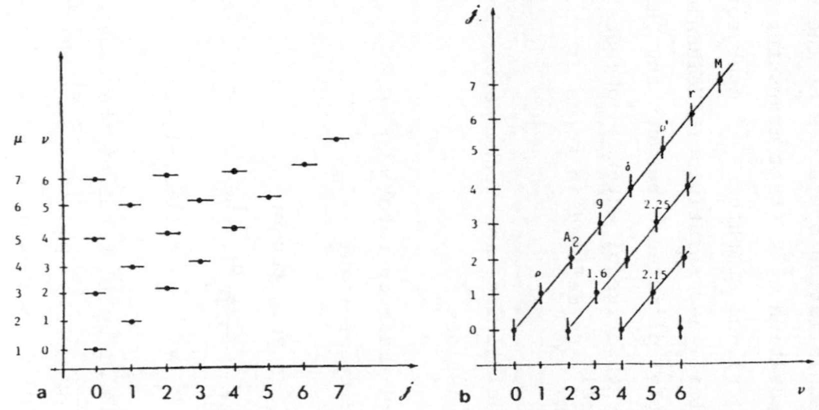

Fig. 4a. Weight diagram of the representation D(1,0) of SO(3,2). It is also the energy diagram of the three-dimensional oscillator. Fig. 4b is obtained by turning the picture of Fig. 4a around and drawing j versus $m_0^2 + \frac{1}{\alpha'} \nu = m^2$; $\nu = \mu - 1 = 0,1,2 \cdots$. With the free parameters m_0^2, α' fitted from the masses m of the meson resonances usually associated to the meson trajectories one obtains the picture of linearly rising Regge trajectories plus daughters.

$$S_{ij} \to S_{\mu\nu} \qquad \mu,\nu = 0,1,2,3$$
$$\pi^i \to \pi^\mu \qquad [\pi_\mu, \xi_\nu] = i\eta_{\mu\nu} \tag{14}$$
$$\xi^i \to \xi^\mu \qquad [\xi_\mu, \xi_\nu] = 0 = [\pi_\mu, \pi_\nu]$$

This leads to many difficulties well known from the old string theory.[10)]

Instead of using the ξ_μ and π_μ of (14) or their infinite mode generalizations

$$\xi^\mu(\alpha,\tau) = i\sqrt{2i\alpha'} \sum_{n=-\infty}^{+\infty} \frac{\alpha^\mu_n}{n} \cos n\sigma e^{-in\tau}$$

(where α^μ_{-n}, α^μ_n are the creation and annihilation operators of the n-th mode) we will use for the intrinsic position ξ^{rel}_μ (that means for the position relative to the center of mass) and momentum π^{rel}_μ operators which have non-commuting components and fulfill new relativistic Heisenberg commutation relations. This is the crucial difference between the QRO model and previous attempts of relativistic theories of extended objects. These new collective dynamical variables ξ^{rel}_μ are defined in terms of the group operators of $SO(3,2)_{\Gamma_\mu S_{\mu\nu}} \supset SO(3,1)_{S_{\mu\nu}}$.

We will now give a sequence of definitions and mathematic relations without much of an explanation: One defines

$$\xi^{rel}_\mu \equiv -S_{\mu\nu} \frac{P^\nu}{c^2M^2} ; \qquad M^2 = P_\mu P^\mu = (\text{mass operator})^2 \tag{15}$$
$$c = \text{velocity of light}$$

$$\Sigma_{\mu\nu} \equiv S_{\mu\nu} + \xi^{rel}_\mu \wedge P_\nu = \breve{g}_\mu{}^\rho \breve{g}_\nu{}^\sigma S_{\rho\sigma} = \text{spin tensor} \tag{16}$$
$$\pi^{rel}_\mu \equiv 2Mc\dot{\xi}^{rel}_\mu = 2Mc \frac{1}{i} [\xi^{rel}_\mu, H] = -\frac{1}{\alpha'} \breve{g}_\mu{}^\sigma \Gamma_\sigma \frac{1}{cM} \tag{17}$$
$$\text{where } \breve{g}_{\mu\nu} = \eta_{\mu\nu} - \hat{P}_\mu \hat{P}_\nu ; \quad \hat{P}_\mu = P_\mu/M, \quad \eta_{\mu\nu} = (1\text{-}1,-1,-1) \tag{18}$$

The definition of the momentum π^{rel}_μ is given in terms of the time derivative (proper time of the center of mass) of the position $\dot{\xi}^{rel}_\mu$ which then is calculated from the hamiltonian (10) using the Heisenberg equation of motion.

The intrinsic dynamics is spacelike ($g_{\mu\nu}$ projects into the plane perpendicular to P_μ). This excludes ghosts from the very beginning; $P^\mu \xi^{rel}_\mu = 0 = P^\mu \pi^{rel} = P^\mu \Sigma_{\mu\nu} = 0$. But more important is that these operators fulfill the c.r.

$$[\xi^{rel}_\mu, \xi^{rel}_\nu] = -\frac{1}{c^2M^2} \Sigma_{\mu\nu} ; \quad [\pi^{rel}_\mu, \pi^{rel}_\nu] = -\frac{i}{\alpha'^2c^2M^2} \Sigma_{\mu\nu} ,$$

and (19)

$$[\xi^{rel}_\mu, \pi^{rel}_\nu] = -i\breve{g}_{\mu\nu} \frac{1}{\alpha'} \frac{1}{P_\mu P^\mu} \hat{P}_\rho \Gamma^\rho \stackrel{C}{=} -i\breve{g}_{\mu\nu}$$

the last equality with the C follows from the constraint relation. (19)

means that the ξ_μ^{rel} and π_μ^{rel} are essentially SO(3,2) operators. In fact in the center of mass rest frame

$$\xi_i^{rel} = \frac{1}{cM} S_{oi} \qquad \pi_i^{rel} = -\frac{1}{\alpha' Mc} \Gamma_i \; ; \qquad i = 1,2,3 \tag{20}$$

Thus one has the fairly simple non-compact group SO(3,2) instead of the complicated string algebra, which has to be represented together with the center of mass Poincaré group. This makes a solution possible. Tachyons can still occur, but one can avoid them by the right choice of the representation, as will be done below in equation (23).

We have carried along the factor c to show that the non-commutativity of the intrinsic position is a relativistic effect. In the non-relativistic limit, $c \to \infty$, π_0, $\xi_0 \to 0$, the ξ_i, π_i become commuting, and fulfill the usual Heisenberg c.r. (13).[8]

The first of the equations (20) is a justification to call the ξ_i^{rel} position operators: $\frac{1}{cM} S_{oi}$ is the intrinsic Lorentz boost which in the Inönü-Wigner group contraction goes into the Galilean boost which is equal to the non-relativistic position operator . The second equation gives us a feeling for the physical meaning of the new operators Γ_i, it is reminiscent of the interpretation for the 4-dimensional Dirac γ-matrices γ_i in terms of the relativistic velocities. Another justification for interpreting the ξ_μ^{rel} as positions is that they perform a helical motion around the center-of-mass direction (Zitterbewegung), as expected of relativistic intrinsic positions. Cf. Fig. 5 where $-d_\mu$ = classical ξ_μ^{rel}).

b) An infinite supermultiplet describes Regge recurrences as yrast states and daughters as radial excitations

After this brief discussion concerning the interpretation of the operators (observables) we return to the energy diagram of Fig. 4.

Interpreting SO(3,2) as relativistic S.G. group the weights will represent hadron levels and the weight diagram leads to a very familiar picture: Turning Fig. 4a around and drawing j versus m_μ^2 one obtains the picture of linearly rising Regge trajectories plus daughters, as shown in Fig. 4b.

There are several things wrong with the irrep D(1,0) as SG for the hadron spectrum:

1) The lowest weight leads to a negative value of $m_{\nu=0}^2$ (tachyon with $j = 0$) if one determines the parameters m_0^2 and α' of (11) from the mass of the ρ-, A_2-, g- etc. mesons.

2) In the contraction limit D(1,0) leads to the spinless (non-relativistic) oscillator whereas the quark-antiquark pair of these mesons

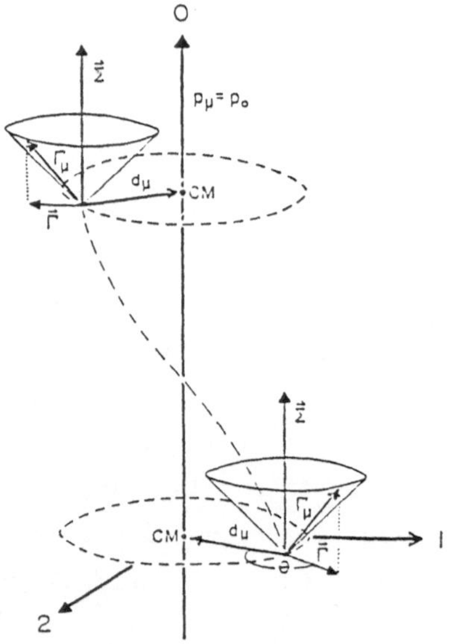

Fig. 5. Helical motion of the intrinsic position vector $\xi_\mu^{rel} = -d_\mu$ about the direction into which the center of mass (CM) of the extended object moves. This Zitterbewegung is derived for the classical analogue of ξ_μ^{rel} using (10) as the hamiltonian in the equations of motion (with commutators replaced by Poisson brackets). The expectation value of the operator ξ_μ^{rel} is zero in every hadron state.

has total quark spin equal to 1 and the triquarks of baryons have total quark spin equal to $\frac{1}{2}$.

There are other representations of SO(3,2) which are similar and can overcome these two difficulties. These are the irreps

$$D(\mu_0 = s + 1, j_0 = s) \qquad \text{for } s = \frac{1}{2}, 1, \frac{3}{2} \tag{21}$$

(s will represent the above mentioned total quark spin after contraction).

Further, two of these irreps of SO(3,2) (which are also irreps of SO(4,2)) combine into an irrep of the relativistic (spectrum) supersymmetry Osp(1,4):

$$D(s + 1, s) \oplus D(s + \frac{3}{2}, s = \frac{1}{2}) \qquad s = \frac{1}{2}, 1, \frac{3}{2}, 2 \tag{22}$$

We choose the representation

$$D(\frac{3}{2}, \frac{1}{2}) \oplus D(2,1) \tag{23}$$

Its weight diagram is shown in Fig. 6. It contains integer and halfinteger spin, but no $\mu = 1$, $j = 0$ state (tachyon).

When we assign meson and baryon resonances to it we will make use of the mass formula (11) that follows from the QRO hamiltonian (9). This will lead to a degeneracy in mass as for the three-dimensional non-relativistic oscillator, (12). To allow for minor variations of mass with spin (rotational bands for each vibrational excitation) and to adjust the lowest levels of mesons and baryons we have made fits of meson and baryon resonances using the formula

$$m^2 = \frac{1}{\alpha'} \mu + \lambda^2 j(j + 1) + \beta (2 - 2s^2) + \hat{m}_0^2 \tag{24}$$

($2 - 2s^2$ is the eigenvalue of the SO(3,2) Casimir operator). This is shown in Fig. 7 for the nuclear resonances and the normal j^P positive $C_n P$ non-strange meson resonances ($\rho - \omega$ - tower).

The particles on the Regge trajectories are reproduced as the yrast states (lowest mass for a given value of j), $\mu = j + 1$, of the relativistic collective model. In addition this model predicts daughters as radial excitations $\mu > j + 1$. The values of the parameters obtained from a fit to baryons and mesons are

$$\frac{1}{\alpha'} = (1.03 \pm 0.04)\ \text{GeV}^2 \qquad \lambda^2 = (0.015 \pm 0.008)\ \text{GeV}^2$$
$$\beta = (0.53 \pm 0.03)\ \text{GeV}^2 \tag{25}$$

Thus, the mass splitting is the same for mesons and baryons (this is the evidence for spectrum supersymmetries) and λ^2 is almost zero; $\lambda^2/(1/\alpha') \approx \frac{1}{100}$. This means that the pure oscillator hamiltonian (10) is a very

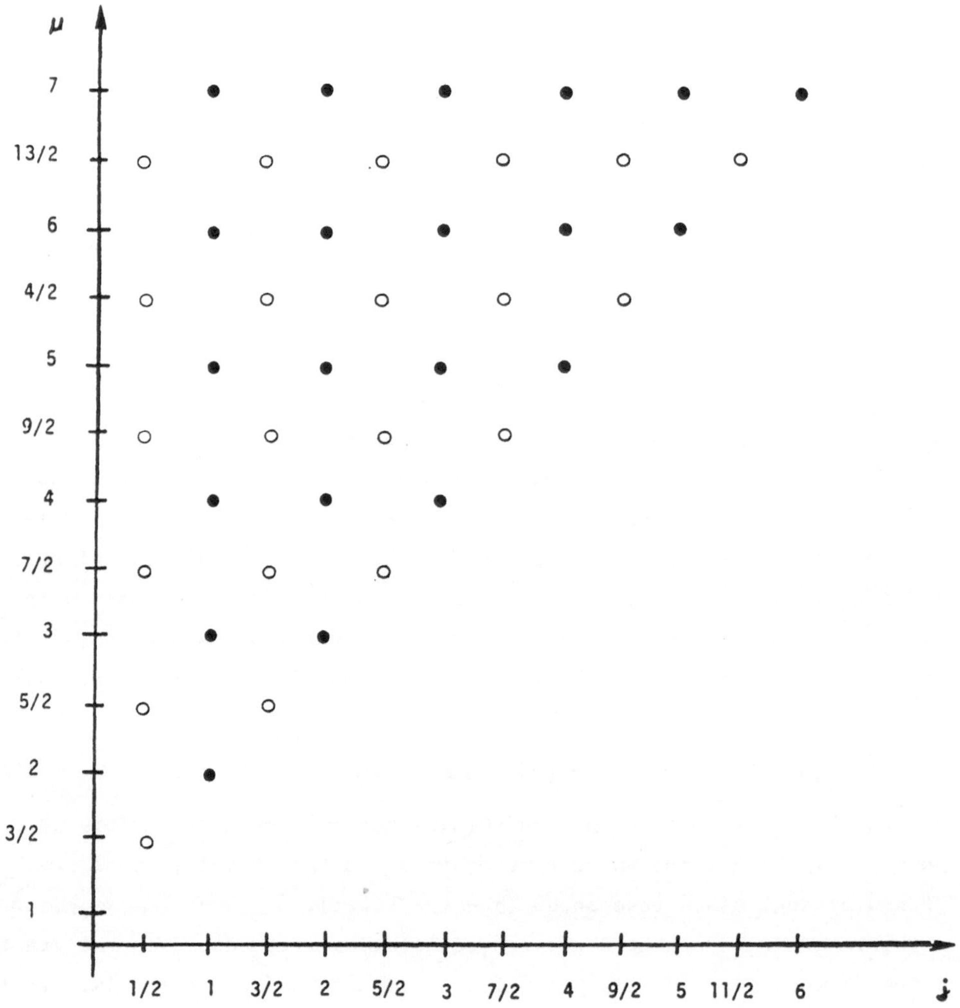

Fig. 6. Weight diagram for the representation D(3/2,1/2) ⊕ D(2,1) of Osp(1,4) ⊃ SO(3,2). The o make up the weight diagram for the representation D(3/2,1/2) of SO(3,2) and the • make up the weight diagram for the representation D(2,1).

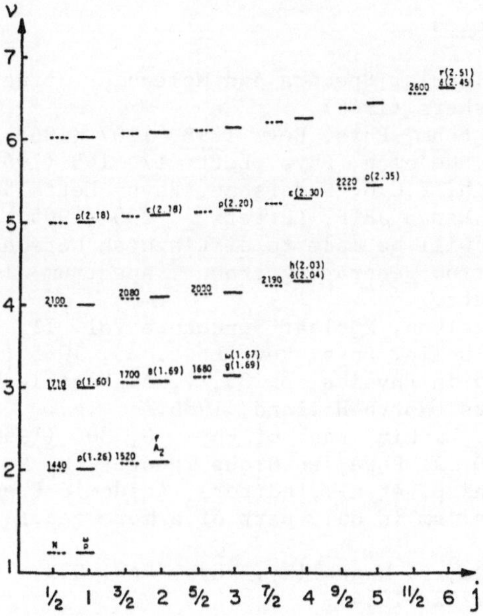

Fig. 7. The mass level diagram as obtained from a fit of the nucleon--and of the ($Y = 0$, $CP = +1$, j^P = normal) meson--resonances to the mass formula (24). On the horizontal axis is plotted the spin j of the resonance. Vertically is plotted $m^2 - \tilde{m}_0^2 - \beta(2 - s^2)$, where m^2 is the value calculated from (24) with the parameters (25), so that the baryon and meson ground state levels coincide.

good approximation, but there is also good evidence for the rotational bands (for which the hamiltonian will have to be amended).

Features which appeared already in molecular (10^{-13} GeV) and in nuclear (10^{-4} GeV) physics can be seen again in the relativistic domain (10^{+1} GeV), and one would of course expect them to reoccur again at the next level with many orders of magnitude higher energies. The reason for this is that not the constituents, but the motions determine these features and they have remained the same though the constituents have changed.

References and Footnotes

1) G. Herzberg, Molecular Spectra and Molecular Structure, D. van Norstrand Publishers (1966).

2) A. O. Barut, A. Bohm, Phys. Rev. 139B, 1107 (1965); Y. Dothan, M. Gell-Mann, Y. Ne'eman, Phys. Lett. 17, 145 (1965); N. Mukunda, L. O'Raifeartaigh, E.C.G. Sudarshan, Phys. Lett 15, 1041 (1965); I. A. Malkin V. Manko JETP, Letters 2, 230 (1965).
No attempt will be made to distinguish between the terms dynamical group, spectrum generating groups, spectrum algebras, spectrum supersymmetry, etc.

3) A. Bohr, B. Mottelson, Nuclear Structure Vol. II, Benjamin (1969).

4) A. Arima, F. Iachello, Phys. Rev. Lett. 35, 1065 (1975); F. Iachello, in Supersymmetry in Physics, p. 85, V. A. Kostelecky, David K. Campbell, editors, North-Holland, 1985.

5) S. Goshen, H. J. Lipkin, Ann. of Phys. 6, 301 (1959).

6) G. Rosensteel, D. J. Rowe, in Group Theoretical Methods in Physics, p. 115, R. T. Sharp, et al., editors, Academic Press, 1977.

7) "This latter dualism is only part of a more general pluralism" (E. P. Wigner).

8) A. Bohm, M. Loewe, P. Magnollay, Phys. Rev. D32, 791 (1985); Phys. Rev. Lett. 53, 2292 (1984).
A Bohm, M. Loewe, P. Magnollay, L. C. Biedenharn, H. van Dam, M. Tarlini, R. R. Aldinger, Phys. Rev. D32, 2828 (1985).

9) P. A. M. Dirac, Lectures on Quantum Mechanics, Yeshiva University Press (1984); N. Mukunda, H. van Dam, L. C. Biedenharn, Relativistic Models, Springer Verlag, N.Y. (1982), Chapter V.

10) E.g. J. Scherk, Rev. Mod. Phys. 47, 123 (1975).

SYMMETRIES IN HEAVY NUCLEI AND THE PROTON-NEUTRON INTERACTION

R.F. Casten

Brookhaven National Laboratory
Upton, New York, 11973

ABSTRACT

The Interacting Boson Approximation (IBA) nuclear structure model can be expressed in terms of the U(6) group, and thereby leads to three dynamical symmetries (or group chains) corresponding to different nuclear coupling schemes and geometrical shapes. The status of the empirical evidence for these three symmetries is reviewed, along with brief comments on the possible existance of supersymmetries in nuclei.

The relationships between these symmetries, the nuclear phase transitional regions linking them, and the residual proton-neutron interaction are discussed in terms of a particularly simple scheme for parameterizing the effects of that interaction.

INTRODUCTION

The Interacting Boson Approximation (IBA) model[1], originally proposed a decade ago, has led to a major resurgence of interest in nuclear structure, to the widespread use of algebraic techniques to elucidate that structure, and to an renewed appreciation of the role of symmetries in nuclei. The resulting systematic understanding of the broad scale evolution of nuclear structure throughout the Periodic Table has awakened a renewed awareness of the central importance of the proton-neutron interaction[2-4] in the development of collectivity in nuclei.

As summarized in ref. 1 and elsewhere[5] in this conference, the IBA model treats nuclei by assuming that their structure is dominated by the valence nucleons outside closed states and that the interactions of these fermions can be approximated, and simplified, by treating them in pairs as bosons. These bosons are allowed to occupy angular momentum 0(s) and 2(d) states. Their number, $N_B = n_s + n_d$, is equal to half the number of valence nucleons and is conserved for the low lying nuclear collective excitations. In the so-called IBA-1, no distinction is made between proton and neutron bosons. The group structure of the IBA-1 is that of U(6) and leads to three distinct non-trivial dynamical symmetries:

$$U(6) \rightarrow U(5) \supset O(5) \supset O(3) \quad (1a)$$
$$U(6) \rightarrow SU(3) \supset O(3) \quad (1b)$$
$$U(6) \rightarrow O(6) \supset O(5) \supset O(3) \quad (1c)$$

In these symmetries, denoted, for short, U(5), SU(3), and O(6), each successive step breaks a previous degeneracy, introduces one or more new quantum numbers to distinguish the now separated levels, adds a term to the eigenvalue equation, and provides new selection rules for transition rates. This process is illustrated for the O(6) limit in Fig. 1. Typical level patterns for each of these symmetries are shown in Fig. 2.

Each dynamical symmetry has a geometrical analogue. The U(5) symmetry corresponds to a very general[6] anharmonic vibrator, equivalent in many respects[7] to the fourth order anharmonic vibrator of Brink, de Toledo Piza, and Kerman[8]. If the anharmonicities are small (as in the example of Fig. 2), the spectrum of levels resembles the familiar vibrational spectrum with equally spaced, nearly degenerate multiplets, each corresponding to a particular phonon number.

The SU(3) symmetry is that of a deformed axially symmetric rotor. As will be seen below, while the U(5) limit permits a rich variety of spectra, SU(3) is a very specific type of symmetric rotor. It is characterized by sequences of rotational bands, with internal energy spacings proportional to J(J+1). The lowest of these is built on the ground state while the next two are analogous to the β and γ vibrations of the geometrical rotor. These are small amplitude quadrupole vibrations which preserve and destroy axial symmetry, respectively. In the

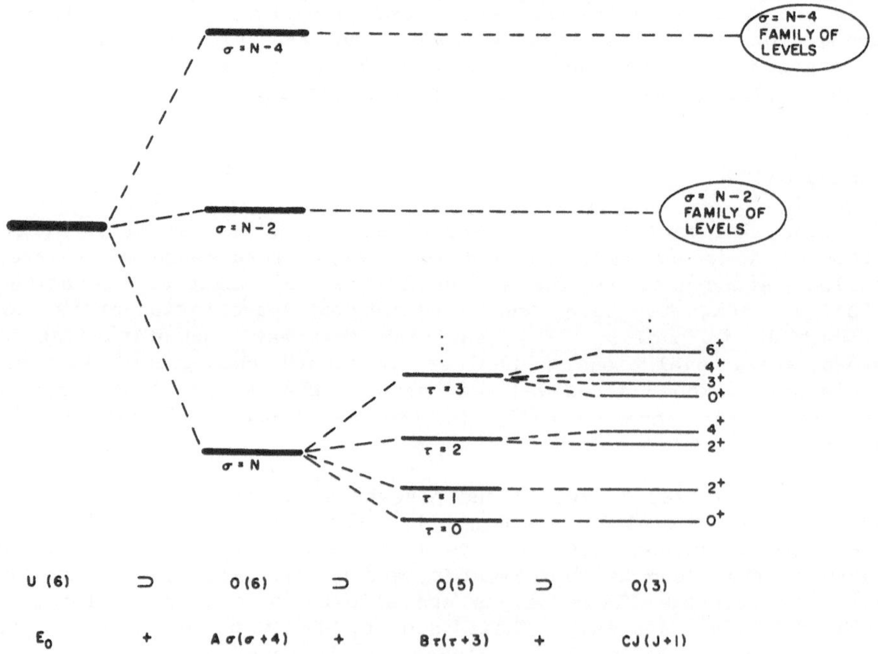

Fig. 1. Successive degeneracy breaking in the O(6) dynamical symmetry. The bottom line gives the eigenvalue expression corresponding to the group chain above.

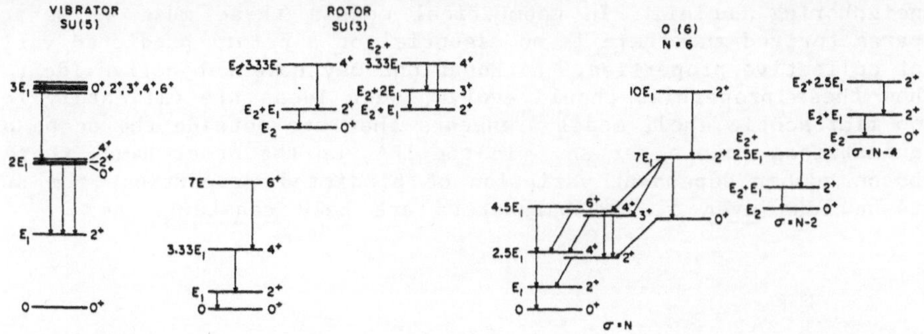

Fig. 2. Lowest levels and transitions of the three U(6) dynamical symmetries. For clarity the collective $\beta \rightarrow \gamma$ transitions of SU(3) are omitted.

SU(3) limit these two vibrations occur in the same representation and, thus, levels of equal spin are degenerate. In the standard geometrical models the β and γ vibrational energies are separately parameterized and need not be equal: empirically, in most deformed nuclei, $E_\beta > E_\gamma$.

The O(6) limit corresponds to an axially asymmetric rotor which is completely soft with respect to a degree of freedom, called γ, which specifies the degree of non-axiality. Here, the lowest levels are grouped according to a phonon-like quantum number τ but also display rotational band-like sequences, albeit with energy staggering (that gives large deviations from a J(J+1) spectrum) resulting from the non-axiality. Major families of levels are labelled by the quantum number σ, as evident in Fig. 1.

It is useful to depict the U(6) structure of the IBA with the symmetry triangle of Fig. 3. Here, each vertex corresponds to a symmetry while the legs describe the phase transitional sequences of nuclei whose properties vary smoothly from those of one symmetry to another. A particularly beautiful feature of the IBA (which will not be discussed at length here) is the simplicity of its treatment of transitional regions: as opposed to their complexity in most other models, they can be calculated in the IBA in terms of the smooth (indeed often linear) variation of a single parameter which, in effect, specifies the position of a nucleus along one of these transition legs.

Of course, the symmetries of the IBA are not identical to those of the corresponding geometrical models. The principal reason for this is that the boson number N_B appears explicitly and the predicted properties of the model depend critically on finite boson number. The collective models, by assuming a smooth nuclear surface, implicitly take the particle number to be infinite. Indeed, the IBA predictions go over into those of the geometrical model as $N \rightarrow \infty$. The explicit presence of finite boson number, N_B, in the IBA not only leads to predictions which deviate from those of the geometrical models, and which are verified empirically[9], but also lends the model a microscopic aspect. In geometrical model predictions, some of the observables for each nucleus are used to fix the free parameters. The same procedure is followed in the IBA. The difference lies in the predictions for neighboring nuclei. In geometrical models these must be separately parameterized and there is no essential or a priori predicted variation of collective properties. Although one may have subjective ideas as to how these properties should evolve, such ideas are ultimately founded on microscopic shell model arguments that are outside the scope of the geometrical models per se. In the IBA, on the other hand, there is a boson number dependent variation of predicted properties from nucleus to nucleus <u>even</u> if the parameters are held <u>constant</u>.

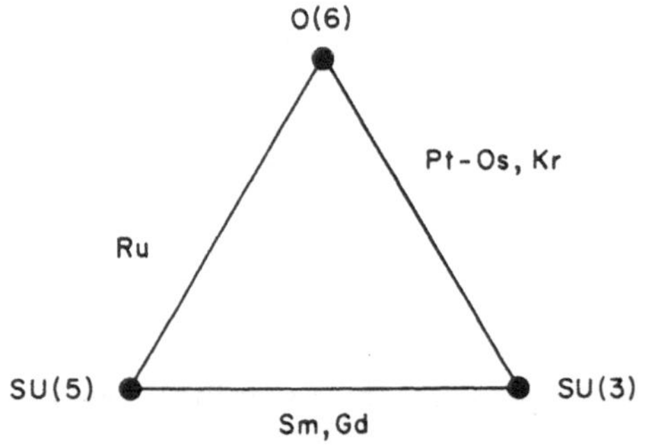

Fig. 3. A schematic symmetry triangle for the IBA-1.

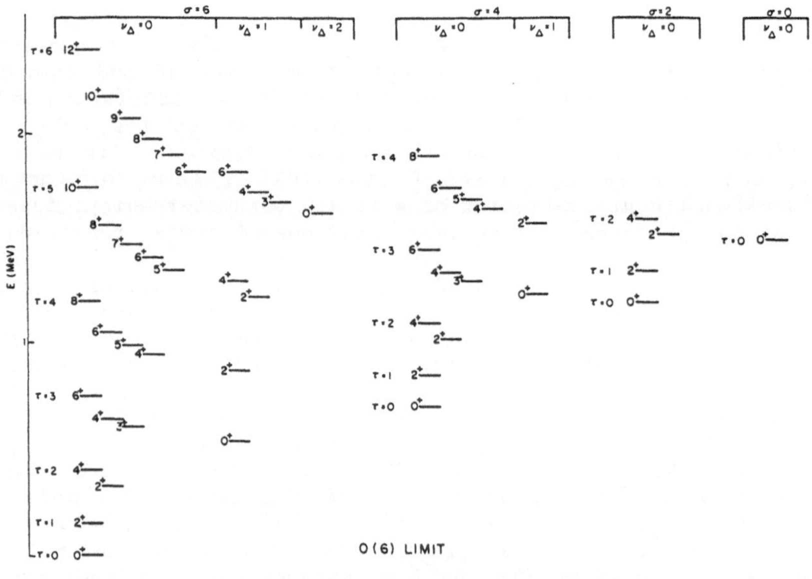

Fig. 4. Typical level scheme of O(6) for N_B = 6.

O(6)

In any case, the primary purpose of this review is not such points but rather the symmetries themselves. Owing to its historical importance, we first discuss the O(6) limit. The levels of the O(6) symmetry for $N_B = 6$ are shown in Fig. 4. The eigenvalue expression[1] is

$$E(N,\sigma,\tau,J) = A(\sigma-N_B)(\sigma+N_B+4) + B\tau(\tau+3) + CJ(J+1) \qquad (2)$$

where A, B, and C are parameters related to the scale of degeneracy breaking in the O(6), O(5), and O(3) steps in eq. 1c, respectively. Accordingly, major families of levels are distinguished by the quantum number σ, while states within a σ representation are labelled by τ and the total angular momentum J. The O(6) levels with $\sigma = N$ are characterized by a 0^+ ground state, a 2^+ first excited state and, then, by a doublet, 2^+-4^+, with $\tau=2$. This last feature clearly distinguishes the model from a typical vibrator (but not necessarily from U(5)--see below) which has a 2-phonon triplet 0^+-2^+-4^+. The higher $\sigma=N$ levels lie in τ multiplets whose degeneracy is broken by the J(J+1) term in eq. 2. Families of levels with $\sigma < N_B$ occur at higher energies and display exactly the same level sequences (albeit with different high τ and J cut-offs, since $\tau_{max} = 2\sigma$). The selection rules for electric quadrupole (or E2) γ-ray transitions are $\Delta\sigma=0$, $\Delta\tau=\pm1$.

Immediately after the O(6) limit was predicted it was discovered[10] empirically in ^{196}Pt. The essence of that discovery is summarized in Fig. 5, which shows the predicted and empirical energy levels as well as the relative E2 transition rates. These results can be summarized as follows. Despite some differences between predicted and observed energy levels, which have subsequently been understood[11] (see below), there is an excellent reproduction of the empirical level structure. There is a one-to-one correspondence between predicted and empirical levels up to nearly 2 MeV. The levels of the quasi-ground band ($0^+(\tau=0)$, $2^+(\tau=1)$, $4^+(\tau=2)$, $6^+(\tau=3)$) follow the $\tau(\tau+3)$ energy spacing predicted by eq. 2. Note also the absence of a two-phonon triplet but, instead, a doublet of levels with spin parity 2^+-4^+. Although, as noted above, the U(5) limit, in principle, allows large anharmonicities without symmetry breaking, and therefore can reproduce the low lying levels of an O(6)-like spectrum, the only nuclei that empirically exhibit such a level pattern are known from other data (as will be shown momentarily) to display the O(6) symmetry. Therefore it is reasonable to associate this "O(6)-like" level pattern with the true appearance of the O(6) limit rather than with a highly anharmonic U(5). A supporting argument comes from a recent study[12] of a Fermion model characterized by SO(8) and Sp(6) groups that can reproduce the level sequences of all three IBA Symmetries. In this shell model approach, the Hamiltonian is written in terms of multipole pairing operators. Extensive experience with shell model calculations dictates that monopole and quadrupole terms dominate the angular momentum (L=1) and octupole (L=3) terms that would lead to large J and τ splittings in the fermion analogs of both the vibrational and O(6) limits. This fermion model thus provides a microscopic argument for expecting a systematic decrease in the successive degeneracy splittings in Fig. 1.

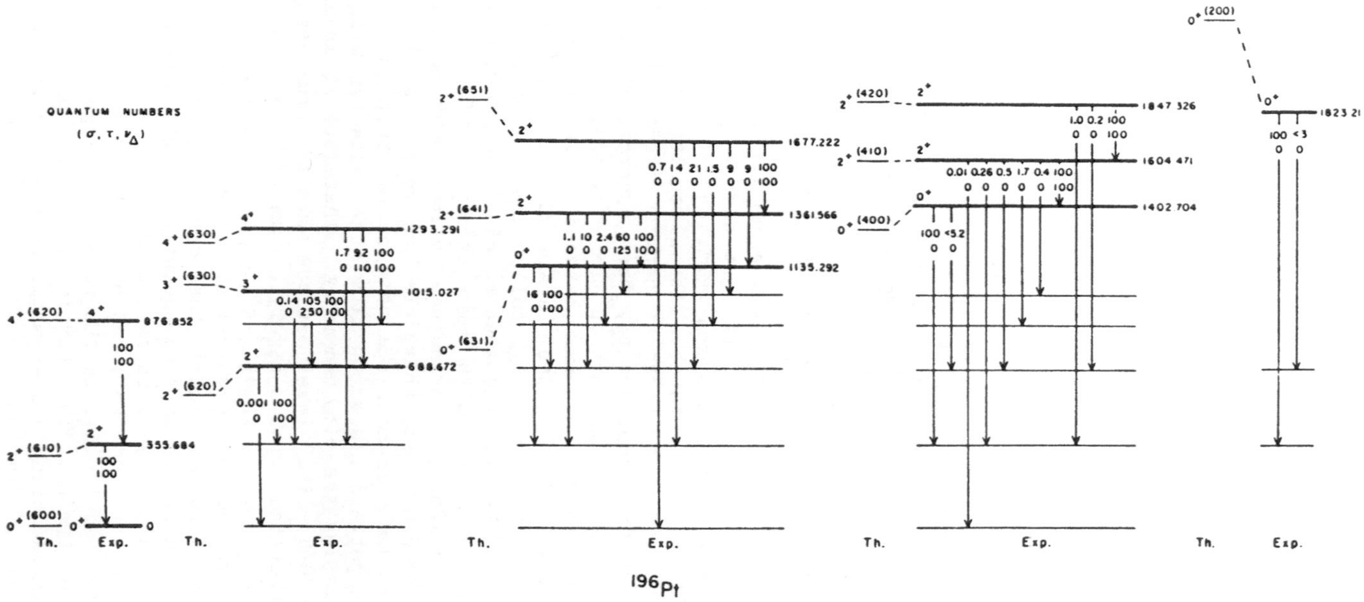

Fig. 5. Comparison of the empirical level scheme of ^{196}Pt with the 0(6) symmetry. The (σ, τ and n_Δ) 0(6) quantum numbers are indicated. The vertical transition arrows represent E2 transitions. The upper row of numbers gives the experimental relative B(E2) values, the lower row, the predicted set. From ref. 10.

In any case, more direct evidence for the O(6) symmetry comes from the $\sigma < N_B$ states which are also shown in Fig. 5. As noted above, the O(6) limit predicts that the level sequence of each successive σ group repeats that of the lowest group. Thus, in particular, one expects repeated sequences of levels with spin parity 0^+-2^+-2^+. The first of these is based on the ground state, the next on the $\tau=3$ 0^+ state of the $\sigma = N_B$ group. Further sequences are based on the $\sigma < N_B$ bandheads. Figure 5 shows three empirical 0^+-2^+-2^+ sequences. It also shows the 0^+ bandhead of the $\sigma = N-4$ group. This remains the only nucleus thus far in which such an extensive sequence of $\sigma<N_B$ levels has been identified.

In principle, the 0^+ bandheads of the $\sigma<N_b$ groups are forbidden to decay because of the $\Delta\sigma=0$ selection rule. Of course in practice they will decay, albeit with smaller matrix elements arising from perturbations to the strict O(6) limit. Nevertheless, one can make a qualitative prediction. Careful inspection of the origin of the $\Delta\sigma=0$ and $\Delta\tau=\pm1$ selection rules shows that the former is inherently weaker than the latter. Thus one expects the 0^+ bandheads of the $\sigma < N_B$ families to decay by breaking the σ rather than the τ selection rule, and hence to decay to the first excited 2^+ state (with $\tau=1$) rather than to the second. Figure 5 shows that this is indeed verified empirically.

A final and compelling set of evidence for the O(6) symmetry consists in the absolute B(E2) values (squares of intrinsic E2 matrix elements) which have been measured[13] for a number of states in ^{196}Pt. In the U(5) limit, for example, these are predicted to vary approximately with phonon number (except for an N_B dependence) such that the B(E2) value from the 2^+_2 level to the 2^+_1 should be just slightly less than twice that of the 2^+_1 level to the ground state. In the O(6) limit these B(E2) values grow much more slowly with increasing τ value. Table 1 summarizes the measured[13] and O(6) and U(5) B(E2) values for ^{196}Pt and shows remarkable agreement with the O(6) limit.

Even before the O(6) limit was recognized in the Pt region it was suspected that the nuclei near ^{134}Ba might also provide a realization of this symmetry. At the time, however, the data were rather sparce and this question remained ambiguous until recently. In the last few years data from the Koln group of Brentano and co-workers has led to the development of extensive level schemes in the A=130 region and it has recently been shown (see ref. 14 and references therein) that these

Table 1. Absolute $B(E2:J_i \rightarrow J_f)$ values in e^2b^2 in ^{196}Pt

	$2_1 \rightarrow 0_1$	$2_2 \rightarrow 2_1$	$2_2 \rightarrow 0_1$	$4_1 \rightarrow 2_1$	$0_2 \rightarrow 2_2$	$0_2 \rightarrow 2_1$	$4_2 \rightarrow 4_1$	$4_2 \rightarrow 2_2$	$4_2 \rightarrow 2_1$	$6_1 \rightarrow 4_1$
Exp a)	0.29	0.35	10^{-6}	0.40	0.14	0.022	0.19	0.18	0.003	0.42
O(6)	0.29 b)	0.38	0	0.38	0.38	0	0.19	0.20	0	0.38
U(5)	0.29 b)	0.48	0	0.48	0	0.48	0.28	0.30	0	0.58

a) From ref. 13.
b) Normalized to experiment.

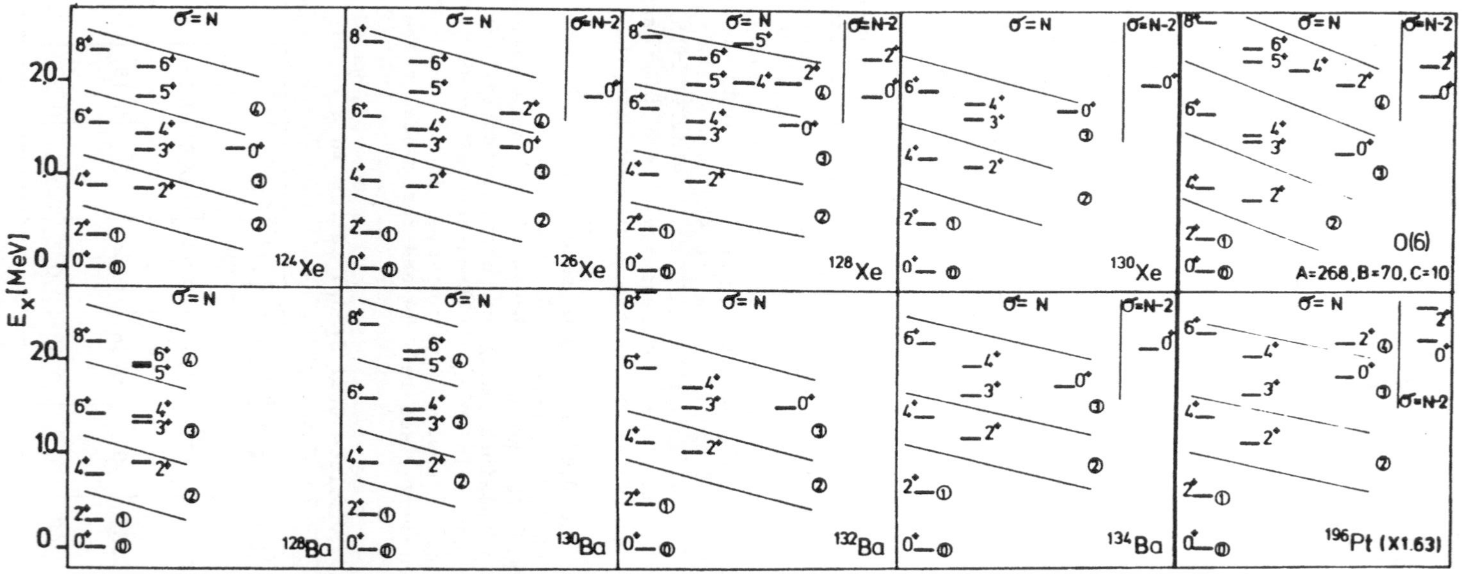

Fig. 6. Comparison of eight Xe and Ba nuclei with the O(6) limit shown with $N_B = 6$ and typical average parameters for the region) and with a scaled level scheme for ^{196}Pt. From ref. 14.

nuclei represent a very extensive O(6) region, more extensive in fact and, in many respects, better than the Pt nuclei. Results from this region are compared with the O(6) limit and with ^{196}Pt in Fig. 6. Absolute B(E2) values in the A=130 region, while less extensively known than in the Pt region, also support the O(6) symmetry.

A remarkable feature of the empirical realizations of the O(6) limit is that the parameters describing the symmetry in both the Pt and A=130 region are very similar (when scaled for a secular mass dependence). Moreover, even the discrepancies with the O(6) limit are the same in both regions. This suggests that a similar interpretation of these discrepancies might be forthcoming. The O(6) limit, as noted above, corresponds to a nuclear potential which is completely flat in the γ degree of freedom representing excursions from axial symmetry. Such a potential is, of course, inherently unstable and one expects it to be perturbed. Recent calculations[11] have introduced a very small perturbation to γ-independence which, however, maintains the same γ_{rms}. This perturbation, originally introduced to correct a problem with the energy staggering in the quasi γ vibrational band (that is, the level sequence $2^+_2(\tau=2)$, 3^+_1-$4^+_2(\tau=3)$, 5^+_1-$6^+_2(\tau=4)$,...) in fact corrects all the discrepancies with the O(6) limit that have been observed while preserving the good agreement for other observables. The interesting point is that the γ dependence in the potential thus introduced amounts only to a few percent of the potential and thus these nuclei, while not rigorously γ independent, are extremely γ soft and thus O(6)-like.

SU(3)

Turning now to the SU(3) symmetry, we recall that, while deformed nuclei abound, the special character of the SU(3) symmetry is not widespread in actual nuclei. There is of course no compelling reason why this or any other symmetry must appear in actual nuclear spectra. The usefulness and elegance of the symmetries is not inherently dependent upon their observation in nature. They retain a deep utility in providing benchmarks for the calculation of nuclei that are close in structure to these paradigms. Nevertheless, it cannot be denied that the discovery of examples of each of the three dynamical symmetries arising from the U(6) group would provide confidence that the IBA is not merely a convenient calculational strategy but rather that it, and the U(6) parent group, contain the essential ingredients describing the symmetry structure of medium and heavy even-even nuclei.

Therefore it is of interest that nuclei near neutron number N=104 (A≈170, especially the Hf and Yb nuclei) seem[15] to exhibit at least an underlying SU(3) symmetry even though noncollective excitations appear at energies comparable to some of the collective SU(3) vibrations and mix with the latter, thereby distorting and masking the observed level structure. The situation is seen most easily by considering Figs. 7-10. Figure 7 schematically depicts a typical rotational level scheme and lists the key SU(3) signatures. Figures 8-10 show the empirical evidence for each of these signatures.

The first, namely the degeneracy of the β and γ vibrations was mentioned earlier. A second is that, since these excitations occur in a different representation than the ground (g) band, γ→g and β→g E2 transitions are forbidden. Thirdly, for the same reason, there is no rotation-vibration coupling between β or γ vibrations and the ground state. Historically, such coupling has been introduced in geometrical

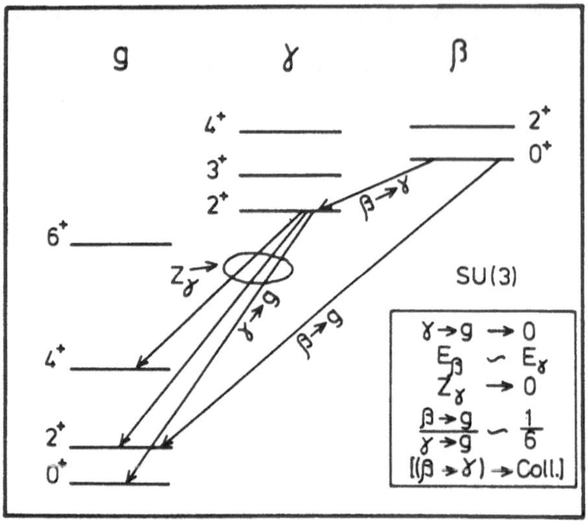

Fig. 7. A schematic illustration of the key signatures of SU(3). From ref. 15.

models and parameterized by a "bandmixing" coefficient Z_γ whose magnitude can be deduced from empirical E2 branching ratios. The SU(3) limit corresponds to $Z_\gamma=0$. Finally, while both $B(E2{:}2^+_\beta \to 2^+_g)$ and $B(E2{:}2^+_\gamma \to 2^+_g)$ values are zero in SU(3) it has been shown that their ratio is finite and equal to 1/6.

Figures 8-10 assemble the relevant data near A=170 and show that (see the dashed boxes which outline the key regions in each plot), near neutron number 104, all four of the characteristic SU(3) signatures are indeed empirically manifested. There is no single nucleus which exhibits all of them in pure form (although ^{174}Yb and ^{176}Hf come closest) but it seems highly unlikely that all four would appear fortuitously in one region. Nevertheless, it has been countered[16] that another interpretation is possible. The decrease in $\gamma \to g$ B(E2) values and the associated energy rise in the γ band could merely signal a decrease in collectivity. However, as just noted, the specific deformed rotor specified by the SU(3) limit also involves small (indeed vanishing) values for these B(E2) values and thus there is no necessary incompatability between these two viewpoints. The SU(3) limit is a specific case of relatively weak collective relationships between ground and vibrational degrees of freedom, in which the other characteristics summarized in Fig. 7 and evident empirically in Figs. 8-10 are also present.

A particular property of the SU(3) limit is worth noting here although, since it is preserved even with rather large perturbations to this limit, it cannot be used as a distinguishing signature of SU(3). It is the occurrence of allowed E2 transitions between β and γ vibrational bands. These are forbidden in traditional geometrical models unless β-γ mixing is introduced explicitly (in which case other difficulties arise) since they violate the selection rule limiting changes in phonon or vibration number to ±1. Such transitions were first observed[17] extensively in ^{168}Er, a well deformed nucleus which deviates substantially from the SU(3) limit. They also appear in Hf and Yb nuclei. These observations substantiate one of the key IBA predictions, and one which, incidentally, is a direct result of the explicit inclusion of finite boson number in the model.

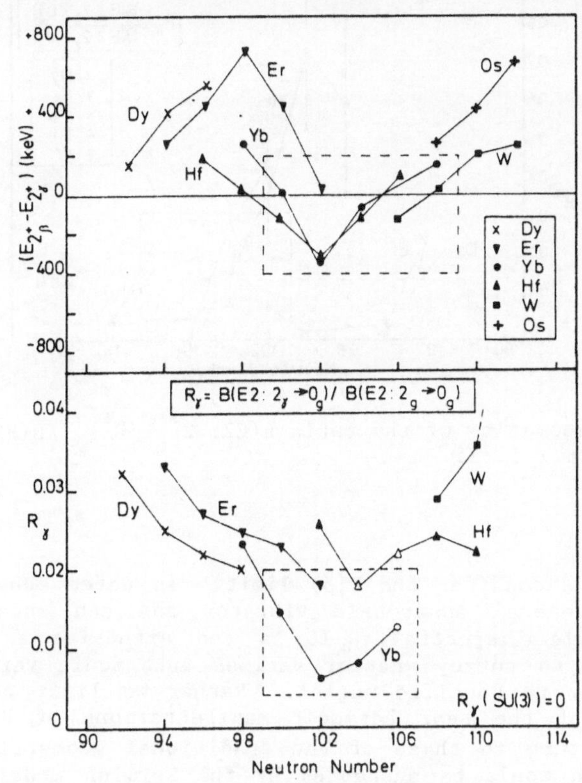

Fig. 8. Empirical systematics of the energy difference $E_{2^+_\beta}-E_{2^+_\gamma}$ (top) and of the $B(E2:2^+_\gamma \to 0^+_g)$ values (normalized by the intraband $B(E2:2^+_g \to 0^+_g)$ values (bottom)). The dashed boxes here and in figs. 9 and 10 indicate the region most closely approaching SU(3). From ref. 15.

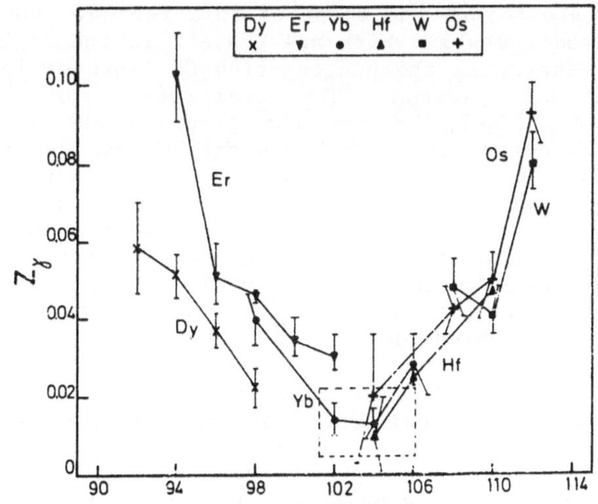

Fig. 9. Systematics of the parameter Z_γ. From ref. 15.

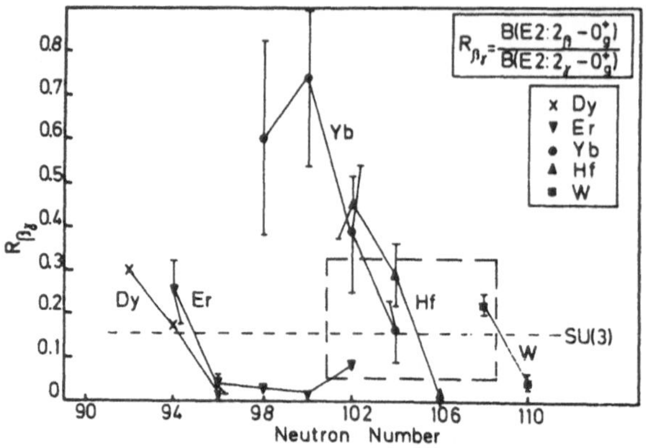

Fig. 10. Systematics of the ratio B(E2: $2^+_\beta \rightarrow 0^+)_g$)/B(E2: $2^+_\gamma \rightarrow 0^+_g$). From ref. 15.

U(5)

Finally, we consider the U(5) limit. As noted above, this limit is a rather general anharmonic vibrator and can encompass a wide variety of nuclear spectra[6]. It is too extensive a task for the present review to survey whether various anharmonic versions of this limit can apply to specific nuclei. Rather we limit ourselves to a consideration of the near harmonic manifestations of U(5) which are close in structure to those of the traditional geometric vibrational model and which would be suggested by the Fermion model of ref. 12. Until recently, all known nuclei which resemble vibrators also showed substantial deviations from such a scheme. These violations can be grouped into three categories: anharmonicities in the two-phonon or three-phonon multiplets in which the splitting of the degeneracy is comparable to the phonon energy itself, deviations from the E2 selection rules of the model, and the presence of extra or "intruder" states in the same energy region as the vibrational levels. In studies just completed by Aprahamian et al.[18], however, ^{118}Cd in fact is found to be free from such difficulties. The reasons why this nucleus reflects the U(5) symmetry are beyond the scope of this review: they relate to the fact that intruder levels which are present in the lighter Cd isotopes have risen in energy in the neutron-rich Cd isotopes leaving behind an intact vibrational spectrum. The level scheme for ^{118}Cd is shown in Fig. 11 and discloses, besides the ground state and one-phonon 2^+ state, a nearly degenerate two-phonon triplet and even a quintuplet of levels which can be assigned phonon number 3. Note that the energy splitting in the quintuplet is also much smaller than the phonon energy itself. (It will be noted that two levels in this quintuplet have two possible spin assignments. Only one is consistent with the U(5) symmetry: further measurements here would be most useful.) Moreover, there is a sequence of levels[18] (not shown) lying above the three phonon quintuplet which decays preferentially to this quintuplet. These levels may represent the remnants of a four-phonon group, although it would be difficult to imagine the Pauli principle permitting the undisturbed existence of an excitation of such high multiplicity.

A key concept of the vibrational picture of the nucleus is the selection rule which allows a change of one phonon in any E2 transition. Thus one expects the two-phonon triplet states to decay to the

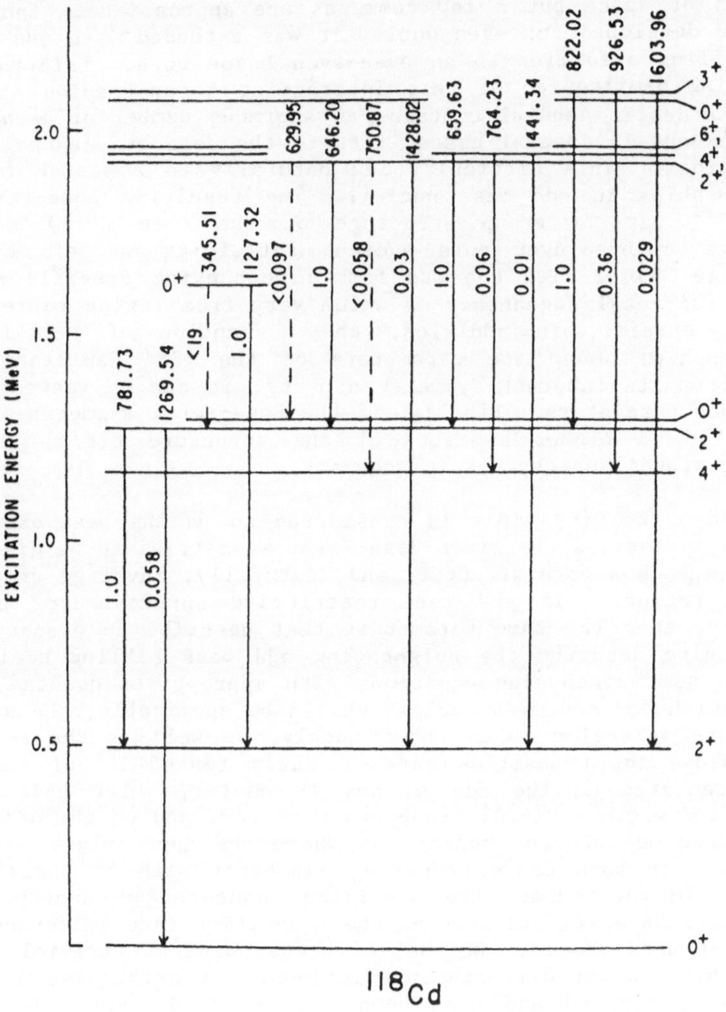

Fig. 11. Level scheme for ^{118}Cd from Aprahamian (ref. 18).

one-phonon 2^+ level and the three-phonon quintuplet states to decay to the two-phonon triplet states. A measure of the goodness of the vibrational symmetry is obtained by the relative B(E2) values of allowed and forbidden transitions. These are summarized in the figure and show that the the selection rule is rather strictly obeyed, with deviations at only the few percent level.

Symmetries in Odd Mass Nuclei

With these results in hand one sees that there are now examples of all three of the dynamical symmetries of the IBA. As pointed out, these symmetries are of interest in themselves and also as benchmarks for neighboring nuclei. They also serve the role of core symmetries in the development of the concept of bose-fermi symmetries for odd mass nuclei and of supersymmetries which link the properties of odd and even mass nuclei. This review will not treat this topic in any detail for reasons of space but a few comments are appropriate. Soon after the IBA was developed for even nuclei it was extended[19] to odd mass nuclei by coupling a fermion to an even-even boson core. Unfortunately, the resulting Hamiltonian of this Interacting Boson-Fermion Approximation (IBFA) model in general contains an enormous number of parameters such that, except in special cases, fits to the data by diagonalization are impractical. This difficulty led naturally to a search for symmetry relationships in odd mass nuclei. The resulting Bose-Fermi symmetries[20-21] have a group structure of the form $U^B(6)$ x $U^F(m)$ and envisage an even-even core nucleus satisfying one of the U(6) IBA symmetries along with the odd fermion occupying specific shell model orbits (of total degeneracy m) which vary from region to region. This concept entails, in addition, the assumption of certain specific relationships among the parameters of the IBFA Hamiltonian. Thus, aside from its inherent symmetry aspects, it can be viewed as a model for those parameters. Finally, if the Bose-Fermi symmetries themselves arise from a higher symmetry of the structure U(6/m) then one is dealing with a supersymmetric scheme.

Since the O(6) limit is considered to be the best established of the IBA symmetries the first Bose-Fermi symmetries to be discussed were based on such a core structure and, naturally, involved predictions in the Pt region. If the more restrictive supersymmetry concept has validity, then the same parameters that describe an even-even nucleus should also describe the neighboring odd mass sibling nucleus. That is, the same eigenvalue equation, with appropriate quantum numbers to distinguish odd and even nuclei, should be applicable, the same transition rate selection rules should apply, as well as the same analytic expressions for transition rates. Early tests[22-24] of supersymmetry ideas centered on the odd Au and Ir nuclei, which have odd proton number and where a U(6/4) group was utilized, and on the odd Pt nuclei, which have odd neutron number and where the appropriate supergroup is U(6/12). In both cases promising agreement with the predictions was found. In particular, the predicted sequences of energy levels and spins were observed and some of the transition rate selection rules and relations were obeyed. Neither of these points is trivial since other models have rather different predictions. Nevertheless, discrepancies were also observed and have been actively discussed. One of these, entailing the relative energies observed for the states of two different representations of the U(6/12) group in the odd Pt isotopes, led to the development[25] of an alternate supersymmetry group chain which is in better agreement with the data. These results encouraged an enormous burst of activity in the field of supersymmetries on both the theoretical and experimental sides. Recently, for example, analogous ideas have been applied[26] to nuclei in a near SU(3) region, namely the W isotopes. Here one would expect that, roughly speaking, the U(6/12) predictions for the odd nuclei W should resemble those of the standard Nilsson model which is the analog of the nuclear shell model for non-spherical shapes. Indeed, it was shown[26] that the wave functions resemble very closely those of the Nilsson model provided, however, that effects such as Coriolis coupling were included in the latter.

Very recently supersymmetry ideas have been extended even further to encompass odd-odd nuclei[27-29] as well. Here, one deals with a supergroup that encompasses the even core, an odd neutron nucleus, an odd proton nucleus, and a fourth member of a quartet with odd proton and odd neutron numbers. The existing data on odd-odd nuclei is very sparce but the initial comparisons with predictions look encouraging.

In concluding these remarks, it should be carefully stated that, to date, the evidence for true supersymmetry as opposed to Bose-Fermi symmetry is not resolved and this is an important question that remains to be elucidated. Indeed, the entire field of the application of supersymmetry ideas to nuclei is currently a highly active and particularly challenging one. It is also of broader interest outside the specialty of nuclear structure since supersymmetry ideas abound in other fields as well. Moreover, although the non-relativistic supersymmetry concepts applicable to nuclei differ from those pertaining to other fields, it is primarily in the realm of nuclear physics that such ideas can actually be subjected to empirical tests.

THE N_pN_n SCHEME: A UNIFIED INTERPRETATION OF HEAVY NUCLEI

Returning for the rest of this review to even-even nuclei, we turn now to the phase transitional regions linking the three IBA symmetries. Each of the latter corresponds to a particular choice of terms in the IBA Hamiltonian. As noted above, transition regions between two symmetries can be calculated in terms of the variation of a single parameter which is the ratio of the coefficients of those two terms characterizing the symmetries at either end of the transition region. All three transition legs of the triangle in Fig. 3 have now been explored[30-32]. More important than the quality of agreement or disagreement with the predictions of the IBA in phase transitional regions, for the purposes of this review, is the question of where one expects these symmetries to occur.

Nuclei near closed shells are generally spherical or vibrational in character and, as one goes away from closed shells, quadrupole deformation sets in. In the context of the IBA this signals an approach to SU(3) in midshell regions. However, inspection of nuclear data shows that, contrary to a common perception, the structure does not depend simply on the total number of valence nucleons, but has a complicated two-dimensional dependence on both N and Z. This is graphically illustrated in Fig. 12 which shows a typical nuclear observable (the energy of the first 2^+ state) which changes rapidly in the U(5)→rotor transition region around A=100.

It has been recognized for three decades[2-4] that the principal residual interaction which leads to shell model configuration mixing and deformation is the proton-neutron interaction among valence nucleons. This interaction is orbit dependent and is a maximum when the respective orbits are highly overlapping. One would expect that some parameterization which empahsizes the integrated strength of that interaction might provide a better systematizing parameter. It has recently been proposed[33] that an appropriate quantity is the product N_pN_n of the number of valence protons times the number of valence neutrons. The same data shown on the left in Fig. 12 are plotted on the right against this product: now one sees that the entire region can be described by a <u>single smooth curve</u>. Other examples of N_pN_n systematics, compared to traditional plots, are shown in Figs. 13-14 and reflect the same simplification. Moreover, different nuclear transition regions, which heretofore had been considered to be highly diverse in character, appear virtually identical in N_pN_n plots. This is shown, for another observable, in Fig. 15. The energy ratio shown ranges from approximately 2.0 for nuclei that are vibrational to values near 3.33 for the symmetric rotor. (Values for O(6)-like or γ-soft nuclei center on 2.5.) Whether the initial condition is that of U(5) or O(6), it is apparent that the crucial transition region corresponds to the range from ≈2.2 to ≈3 and that each region behaves similarly, except for displacements in N_pN_n. This suggests the possibility of a unified systematic interpretation of the structure of nearly all heavy nuclei. A simple explanation of the N_pN_n curves is that nuclei remain vibrational until a certain value of N_pN_n is obtained. This value corresponds to that position in the respective proton and neutron valence shells where the crucial, most highly overlapping, proton and neutron orbits are filling. At this point, which differs for each nuclear region (thus accounting for the horizontal displacements), the nuclei initiate a transition from a vibrator or γ-soft structure towards a rotor. For each region such a phase transition requires a change in N_pN_n of approximately 60-80 units. That is, once the crucial orbits in any given mass region begin to fill, deformation will ensue if additional valence protons and neutrons are added so that N_pN_n increases by roughly this amount.

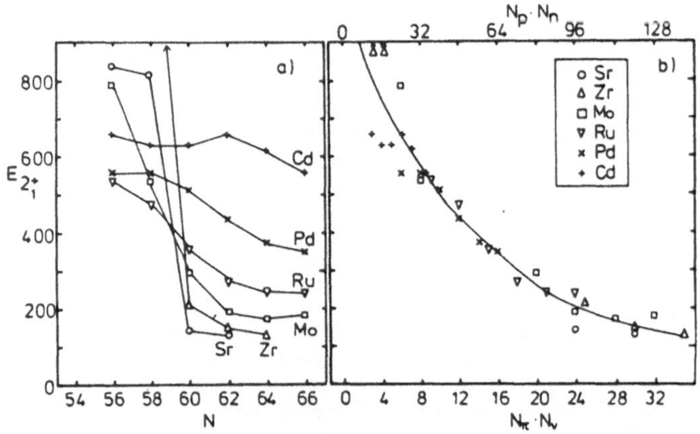

Fig. 12. Traditional and N_pN_n plots of the energy of the first 2^+ state in the A=100 region. $N_\pi N_\nu$ is the boson number product corresponding to (1/4) N_pN_n. From refs. 33.

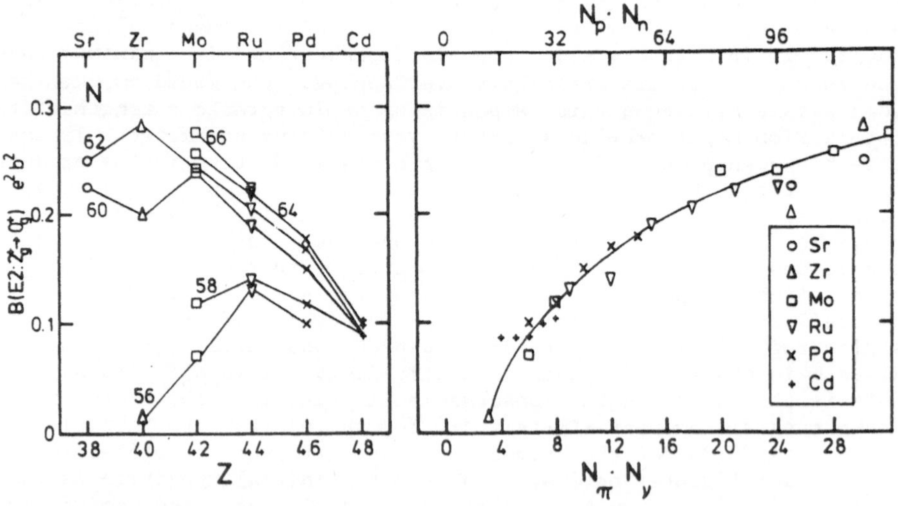

Fig. 13. Same as fig. 12 for the $B(E2{:}2^+{}_1 \rightarrow 0^+{}_1)$ values.

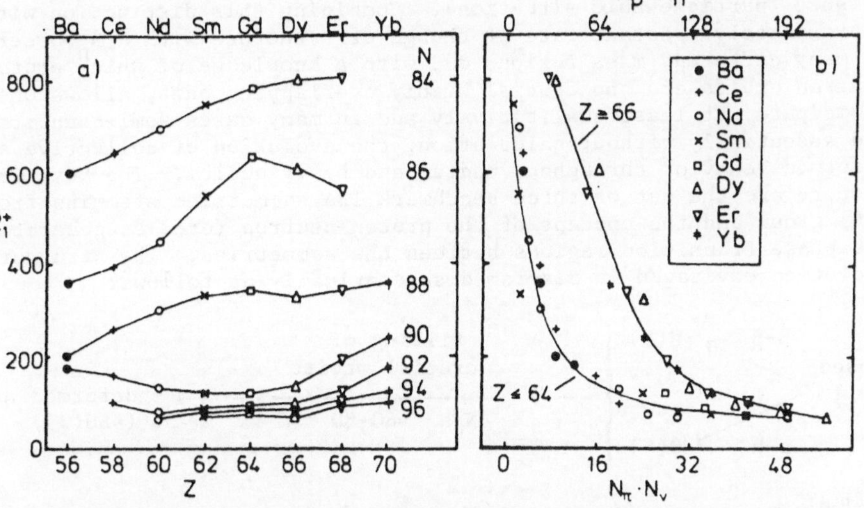

Fig. 14. Same as fig. 12 for $E_{2^+_1}$ values in the rare earth region. The two curves in the N_pN_n plot are for nuclei with protons filling below and above midshell (Z = 66) respectively.

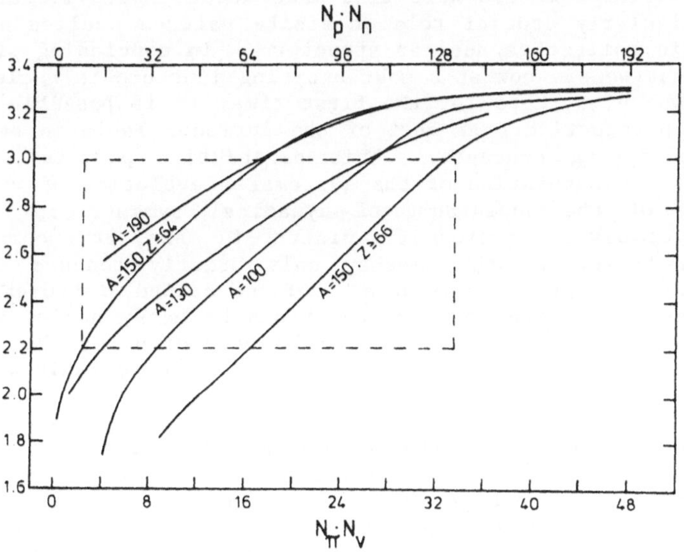

Fig. 15. Comparison of N_pN_n plots of the energy ratio $E_{4^+_1}/E_{2^+_1}$ for five different mass regions.

Adding in one more ingredient now allows a unified interpretation of collectivity in heavy nuclei. The U(5) symmetry tends to occur near closed shells when both protons and neutrons are particles or when both are holes. The O(6) symmetry, on the other hand, is preferred when one type of nucleon is just above and the other is just below a shell closure (particle-hole situation). Combining this distinction with the approximately constant rate of change of structure with N_pN_n observed in many different mass regions and with a knowledge of which proton and neutron orbits are the crucial highly overlapping ones, allows one to anticipate, at least qualitatively and in many cases semi-quantitatively, and essentially without calculation, the evolution of collective and deformed behavior throughout medium and heavy nuclei. The keys to this picture are the set of three benchmark IBA symmetries stemming from the U(6) group and the concept of the proton-neutron force as generating the phase transition regions between the symmetries. The structural evolution envisaged is diagramed schematically as follows:

$$
\text{Closed Shell} \left\{
\begin{array}{l}
\text{p-p} \rightarrow U(5) \\
\text{p-h} \rightarrow O(6)
\end{array}
\right\}
\longrightarrow
\begin{array}{c}
\text{filling of} \\
\text{crucial orbits} \\
\hline
\Delta N_pN_n \approx 60\text{--}80
\end{array}
\longrightarrow
\begin{array}{c}
\text{deformed nuclei} \\
(\approx SU(3))
\end{array}
$$

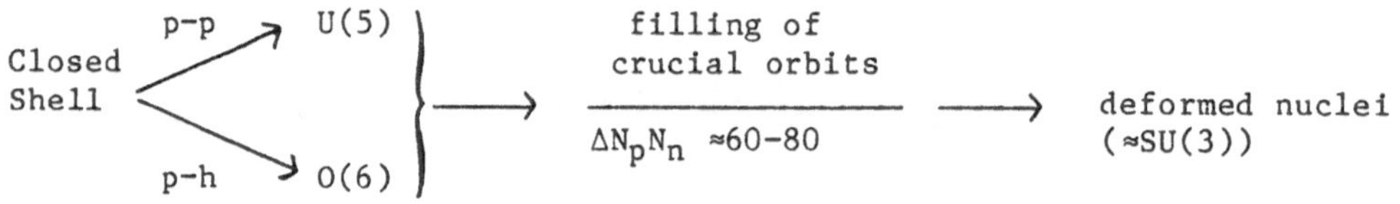

SUMMARY

The aim of this review has not been to exhaustively survey the existing literature but to scan the symmetry structure of medium and heavy nuclei through the perspective of the IBA and the U(6) group, to discuss the empirical evidence for these symmetries in nuclei, and then to delineate where these symmetries occur in terms of a simple picture of the neutron-proton force as the prime generator of nuclear collectivity. Brief comments were also made about supersymmetries and about the particularly crucial role of finite valence nucleon number in the IBA and in collective nuclear structure. In concluding, it seems fair to say that we are now at a most exciting juncture in nuclear structure physics where, perhaps for the first time, it is possible to glimpse a unifying perspective over most of the Periodic Table in terms of a few basic underlying concepts. This perspective permits a unified and universal interpretation of the systematic evolution of nuclear structure and of the appearance of dynamical symmetries, it links the behavior of diverse regions far distant in mass, it encompasses within the same framework other ideas, [34] only briefly touched upon in this review such as the occurrence of so-called intruder states, it simplifies collective model calculations of heavy nuclei and, finally, it provides a simple mechanism for the prediction of the properties of unknown nuclei far off stability in a much more reliable way.

The overview presented here brings together several strands in nuclear structure physics that have been developing and converging over the last several years. I am therefore grateful to numerous people working in many aspects of this field for inspiration and informative and enlightening discussions. It is impossible to mention all of these but I would particularly like to express my thanks to F. Iachello, A. Arima, I. Talmi, S. Pittel, D. D. Warner, A. Aprahamian, J. A. Cizewski, K. Heyde, A. Wolf, D. H. Feng, C. L. Wu, J. Ginocchio, O. Scholten, P. Van Isacker, A. E. L. Dieperink, B. R. Barrett, A. Frank, J. Wood, A. Gelberg, and P. von Brentano. Research has been performed under contract DE-AC02-76CH00016 with the United States Department of Energy.

References

1. A. Arima and F. Iachello, Phys. Rev. Lett. 25:1069 (1974), Ann. Phys. (N.Y) 99:253 (1976), 111:21 (1978), 123:468 (1979).
2. A. de Shalit and M. Goldhaber, Phys. Rev. 92:1211 (1953).
3. I. Talmi, Rev. Mod. Phys. 34:704 (1962).
4. P. Federman and S. Pittel, Phys. Lett. 69B:385 (1977), 77B:29 (1978), and Phys. Rev. C20:820 (1979).
5. D. H. Feng and F. Iachello, these proceedings.
6. A. Leviatan, A. Novoselsky, and I. Talmi, preprint, 1986).
7. A. Aprahamian, Ph.D. thesis, 1985.
8. D. M. Brink, A. F. R. de Toleda Piza, and A. K. Kerman, Phys. Lett. 19:413 (1965).
9. R. F. Casten, D. D. Warner, and A. Aprahamian, Phys. Rev. C28:894 (1983).
10. J. A. Cizewski et al., Phys. Rev. Lett. 40:167 (1978).
11. R. F. Casten et al., Nucl. Phys. A439:285 (1985).
12. J. N. Ginocchio, Ann. Phys. (N.Y.) 126:234 (1980). See also Wu et al., Phys. Lett. 168B:313 (1986).
13. H. Bolotin et al., Nucl. Phys. A370:146 (1981).
14. R. F. Casten and P. von Brentano, Phys. Lett. 152B:22 (1985).
15. R. F. Casten, P. von Brentano, and A. M. I. Hague, Phys. Rev. C31:1991 (1985).
16. W. Gelletly et al., to be published.
17. W. F. Davidson et al., J. of Phys. G7:455 (1981).
18. A. Aprahamian, in Recent Advances in the Study of Nuclei Off the Line of Stability, Chicago, Sept. 8-13, 1985, R. A. Meyer and D. S. Brenner, eds., to be published.
19. F. Iachello and O. Scholten, Phys. Rev. Lett. 44:679 (1979).
20. A. Arima and F. Iachello, Phys. Rev. Lett. 44:772 (1980).
21. A. B. Balantekin, I. Bars, and F. Iachello, Nucl. Phys. A370:284 (1981); Phys. Rev. Lett. 47:19 (1981).
22. J. A. Cizewski et al, Phys. Rev. Lett. 46:1264 (1981); J. Wood, Phys. Rev. C24:1788 (1981).
23. M. Vergnes et al., Phys. Rev. Lett. 46:584 (1981).
24. D. D. Warner et al., Phys. Rev. C26:1921 (1982).
25. H.-Z. Sun, A. Frank, and P. Van Isacker, Phys. Rev. C27:24320 (1983); Phys. Lett. 124B:275 (1983).
26. D. D. Warner and A. M. Bruce, Phys. Rev. C30:1066 (1984).
27. I. Bars, in Bosons in Nuclei, D. H. Feng, S. Pittel, and M. Vallieres, eds., World Scientific (1984), p. 155.
28. P. Van Isacker, J. Jolie, K. Heyde, and A. Frank, Phys. Rev. Lett. 54:653 (1985).
29. A. B. Balantekin and V. Paar, preprint.
30. R. F. Casten and J. A. Cizewski, Nucl. Phys. A309:477 (1978) and A425:653 (1984).
31. O. Scholten, F. Iachello, and A. Arima, Ann. Phys. (N.Y.) 115:366 (1978).
32. J. Stachel, P. Van Isacker, and K. Heyde, Phys. Rev. C25:650 (1982).
33. R. F. Casten, Phys. Lett. 152B:145 (1985), Phys. Rev. Lett. 54:1991 (1985), Nucl. Phys. A443:1 (1985).
34. K. Heyde et al., Phys. Lett. 155B:303 (1985).

TIME SYMMETRY AND THE THEORY OF MEASUREMENT

K. Datta

Department of Physics & Astrophysics
University of Delhi
Delhi 110007, India

INTRODUCTION

" Time present and time past
Are both perhaps present in time future,
And time future contained in time past.
If all time is eternally present
All time is unredeemable."

T.S.Eliot
Burnt Norton (1943)

I have chosen to speak on a problem which has been with us since the birth of quantum mechanics. On the one hand this has the advantage that much concerning this problem is part of the general education of a physicist and and it is possible to draw on a large body of earlier work; this report will, in large measure, be in the nature of a review. On the other hand, I shall have to conclude with the confession that inspite of some progress, the problem remains, in many important respects, unsolved.

The explicit statement of the absence of time symmetry in nature enters physical laws in two areas. First, in thermodynamics: the second law asserts that the entropy of an isolated system can only increase towards the future; second, in cosmology: if our universe is an open one, the universe only expands towards the future.

The laws of motion in quantum theory are, as in classical dynamics, time symmetric. There is, however, a point of view that asymmetry in time enters quantum theory in a fashion which has no classical counterpart viz. through the theory of measurement [1]. In the von Neumann theory of measurement[2], any mesurement performed on a quantum system changes its state discontinuously in a manner not to be described by the Schroedinger or Heisenberg equations of motion for an isolated system. A complete measurement is accompanied by a "collapse of the wave packet"; the density operator representing the state of the system-apparatus combine changes from a projection corresponding to a "pure" case to that associated with a "mixture". This change is accompanied, as was shown by von Neumann[2] and elucidated by London and Bauer[3] by an increase of entropy of the system-apparatus combination. It will be my effort to examine the nature of this transition and to ask if the rules of the quantum theory contain elements violating time symmetry or whether here, as in many other contexts, we are faced with the problem of explaining the irreversible behaviour of macroscopic systems even though the motion of microscopic constituents is governed by time reversible dynamical laws.

THE COLLAPSE POSTULATE

We adopt the "standard" view of the collapse postulate[1] ; other views are possible[4] but will remain unexamined here. In the measurement process, the measured system S and the measuring apparatus A are coupled for a duration of time τ , generally finite; the first step in the process is the establishment of a correlation between the states of S and of A (the spin state of the atom and its centre of mass in a Stern-Gerlach measurement of spin)

$$|\Phi_i\rangle = |A_0\rangle \otimes |\psi\rangle_S \longrightarrow |A^\psi\rangle \otimes |\psi\rangle_S \tag{1}$$

$|A^\psi\rangle$ is a state of the apparatus which bears a one-to-one correspondence with the system state $|\psi\rangle_S$. Consider first the case when $|\psi\rangle_S$ happens to be an eigenstate of a hermitian operator $\hat{S} = \sum \lambda_k |s_k\rangle\langle s_k|$. Then

$$|\Phi_i\rangle = |A_0\rangle \otimes |s_i\rangle \longrightarrow |A_i\rangle \otimes |s_i\rangle = |\Phi_f\rangle \tag{2}$$

This transition being one between "pure" states is achieved by an interaction operator $\exp(-iH_{AS}\tau/\hbar)$; our requirement that that the system state $|s_i\rangle$ remain unchanged after measurement is ensured if

$$[H_{AS}, \hat{S}] = 0 \tag{3}$$

The measurement then belongs to the category of "non-demolition" measurements[5] . From eq.(2) it follows that, on account of the linearity of the equations of motion, if

$$|\Phi_i\rangle = |A_0\rangle \otimes \sum c_k |s_k\rangle$$, then

$$|\Phi_f\rangle = \sum c_k |A_k\rangle \otimes |s_k\rangle \tag{4}$$

This development of the state in time is unitary and Hamiltonian; it can therefore be reversed. However, even though $|\Phi_f\rangle$ contains an essential element of the measurement process in the shape of correspondences between system and apparatus states, it cannot be said to represent the final stage in the measurement process. This is because $|\Phi_f\rangle$ in eq.(4) contains off diagonal correlations between different states of the apparatus. In an actual measurement, after the measurement has been completed(e.g. in a

Stern-Gerlach experiment after the deflected atom has triggered a counter on its path) no such off diagonal correlations between macroscopically distinguishable apparatus states persist. von Neumann therefore postulated a second stage in the measurement process in which the density operator of the system-apparatus combination collapses from that corresponding to the pure state given by eq. (4) viz. from

$$\rho_{pure} = \sum_{K,K'} C_K C_{K'}^* |A_K\rangle\langle A_{K'}| \otimes |S_K\rangle\langle S_{K'}| \tag{5}$$

to that of a mixture diagonal in the apparatus-system states i.e. to

$$\rho_{mix} = \sum_{K} |C_K|^2 |A_K\rangle\langle A_K| \otimes |S_K\rangle\langle S_K| \tag{6}$$

The correlations between apparatus-system states are preserved but all off-diagonal correlations are lost. The transition $\rho_{pure} \longrightarrow \rho_{mixture}$ is non-Hamiltonian and accompanied by an increase in the entropy $S = - kN\, tr(\rho \log \rho)$. The entropy of the apparatus-system combination in a pure state is zero; that of a mixture is necessarily positive.

What is the cause of this entropy increasing evolution of the density operator? von Neumann[2] , London and Bauer[3] and, in later years, Wigner[6] have ascribed this to the intervention of the consciousness of the observer. However, the possibility of a different kind of understanding of this collapse has been opened by the work of Zurek[7], Savage and Walls[8] and of Hillery and Scully[9]. These investigations ascribe the collapse to constant monitoring of the apparatus by an open environment. von Neumann had considered and rejected the notion that something is to be gained by such a coupling of the apparatus to the remaining part of the universe. Indeed, such is the case if the remaining part of the universe is merely another superapparatus[10] ;the mystery of the irreversible collapse is pushed into the third stage, not elucidated. We know, however, from the work of Caldeira and Leggett[11] that the coupling of a quantum system to an essentially open quantum reservoir can, under certain assumptions, lead to dissipative irreversible evolution of the quantum system over time scales large compared to the typical relaxation times of the reservoir: in this description, the phenomenon of dissipation consists of the transfer of energy from a single (or a few) degree(s) of freedom of the system + apparatus to the very complex degrees of freedom describing the environment. It is now possible to show, using simple couplings between the system and apparatus and between apparatus and environment that provided the environment has the characteristics of an open system, the off-diagonal correlations of the final system+apparatus density operator are rapidly quenched.

APPARATUS ENVIRONMENT COUPLING AND THE DISSIPATIVE FOKKER-PLANCK EQUATION

Consider a simple model of the apparatus as an oscillator with a single degree of freedom interacting with an environment consisting of a system of non-interacting oscillators at temperature T. The coupling between the systems is assumed to be linear in the coordinate of each oscillator.

$$H = H_A + H_E + H_{AE}$$

$$H_A = \hbar\omega\, b^+ b \quad ; H_E = \hbar \sum_K \omega_k\, d_k^+ d_k ; \; H_{AE} = x \sum_K g_k X_k \tag{7}$$

b and d_k are apparatus and environment operators; x and X_k are given by

$$x = (\hbar/2M\omega)^{\frac{1}{2}} (b^{+} + b)$$

$$X_k = (\hbar/2m\omega_k)^{\frac{1}{2}} (d_k^{+} + d_k)$$

The reduced density operator for the apparatus is obtained by the partial trace over environment variables of the final density operator; in the coordinate representation

$$\rho(xy;t) \equiv \rho_A(xy;t) = \int \prod dX_k \langle xX_k | \hat{\rho}(t) | yX_k \rangle$$

The problem of obtaining the propagation kernel for $\rho(xy;t)$ under the action of the Hamiltonian given by eq.(7) has been solved exactly by Feynman and Vernon[12] and shows no evidence irreversible evolution. However, if one now proceeds to the continuum limit for the frequency distribution of the environment oscillators and corresponding coupling strengths in a particular fashion[11] viz. if $\mu(\omega)$, the density of oscillators in frequency space, is given by

$$\mu(\omega)\, g^2(\omega) = 2m\omega^2\eta/\pi \qquad \omega \lessgtr \Omega$$
$$= 0 \qquad \omega > \Omega \qquad (8)$$

η being a constant and Ω a cutoff frequency, the equation for ρ, in the limit $\Omega \longrightarrow \infty$ is

$$\frac{\partial \rho}{\partial t} = -i\omega_R[b^{+}b, \rho] + \frac{i\gamma}{\hbar}[x, p\rho + \rho p]$$
$$- 2\gamma/\hbar \left(\bar{n} + \tfrac{1}{2}\right) m\omega_R [x, [x, \rho]] \qquad (9)$$

Here $\gamma = \eta / 2M$, p is the apparatus momentum operator, ω_R is a renormalized frequency

$$\omega_R^2 = \omega^2 - (4\gamma\Omega/\pi)$$

and $\bar{n}$ is the expected number of quanta in an oscillator of frequency ω at temperature T:

$$\bar{n} = 1/(\exp(\hbar\omega/k_B T) - 1)$$

The limit $\Omega \longrightarrow \infty$ has to be taken with care and considerable ingenuity is required in obtaining a finite theory. Eq.(9) is a dissipative Fokker-Planck equation with γ which is related to the square of the apparatus -environment coupling strengths as given by eq.(8) playing the role of a dissipation constant.

What is the effect of this environment induced damping on the off-diagonal correlations of the reduced apparatus density operator? This problem has been solved by Savage and Walls[8]. The main result may be simply stated. If the initial density operator is a superposition of coherent states:

$$\rho(0) = \sum_{\alpha,\beta} N_{\alpha\beta} |\alpha\rangle\langle\beta| \qquad (10)$$

where
$$\langle x\rangle_\alpha = (2\hbar/m\omega)^{\frac{1}{2}} \,\mathrm{Re}\,\alpha$$
$$\langle p\rangle_\alpha = (2\hbar m\omega)^{\frac{1}{2}} \,\mathrm{Im}\,\alpha$$
$$|\langle\alpha|\beta\rangle| = \exp(-\tfrac{1}{2}|\alpha - \beta|^2)$$

then at zero environment temperature

$$\rho(t) = \sum_{\alpha,\beta} N_{\alpha\beta} \langle\beta|\alpha\rangle^{(1 - \exp(-2\gamma t))} \cdot |(\exp-(\gamma+i\omega)t)\alpha\rangle\langle(\exp-(\gamma+i\omega)t)\beta| \qquad (11)$$

in the underdamped case ($\gamma \ll \omega$).

We see an exponential relaxation to the vacuum state; more important, there is a decay of the off diagonal parts of the density operator. For short times $2\gamma t \ll 1$, the decay occurs exponentially at the rate $-2\gamma \log |\langle\beta|\alpha\rangle|$, a quantity which increases with decreasing overlap of initial states. The results for the finite temperature case are of greater interest since it is in this case that the apparatus environment coupling of the form given in eq.(7) generates the dissipative mechanism responsible for the decay of off-diagonal coherences. If σ_x^2 and σ_y^2 are the variances of the diagonal and off-diagonal parts of in the coordinate basis i.e. of $\langle x-y|\rho|x+y\rangle$, we have

$$\sigma_x^2 = (\hbar/2m\omega)\,[2\bar{n}(1-\exp(-2\gamma t)) + 1]$$

$$\sigma_y^2 = (\hbar^2/2m\omega)^2[1/\sigma_x^2] \qquad (12)$$

Starting from the initial coherent state value, σ_x^2 increases in a time $\sim$ a few γ^{-1} to $(2\bar{n} + 1)(\hbar/2m\omega)$; the off- diagonal variance decreases to $(\hbar/2m\omega)(2\bar{n} + 1)^{-1}$. Even though the off-diagonal spreading increases with the temperature, so does the degree of elimination of off-diagonal coherences.

THE MEASUREMENT MODEL

A measurement model may now be constructed which takes advantage of the diagonalization of the density operator when coupled to a suitably modelled environment[13]. The system apparatus coupling is taken proportional to the number operator for system quanta. This allows an easy identification of the basis in which the system-apparatus density operator collapses to diagonal form.

$$H = H_S + H_A + H_E + H_{SA} + H_{AE}$$

$$H_S = \hbar\omega_o a^+ a \;;\; H_A = \hbar\omega b^+ b \;;\; H_E = \hbar\sum \omega_k d_k^+ d_k \qquad (13)$$

$$H_{SA} = \tfrac{1}{2}\hbar a^+ a\,(b\epsilon^* + b^+\epsilon) ;\; H_{AE} = b\Gamma^+ + b^+\Gamma ;\; \Gamma = \sum g_k d_k$$

a,b and d are system, apparatus and environment operators respectively; ϵ is a classical field coupling the system and apparatus. The density operator of interest is that for the system apparatus combination. Under the assumptions embodied in eq.(8) relating to the frequency distribution of environment oscillators, one has the equation

$$\frac{d\rho}{dt} = \tfrac{1}{2}[(\epsilon b^+ - \epsilon^* b)a^+a , \rho] + \tfrac{1}{2}\gamma(2b\rho b^+ - b^+b\rho - \rho b^+b) \qquad (14)$$

at zero environment temperature. If at t=0 we start with

$$\rho(0) = \sum_{n,m} \rho_{nm} (|n\rangle\langle m|)_S \otimes (|0\rangle\langle 0|)_A \qquad (15)$$

with $\rho_{nm} = \langle n|\rho_S(0)|m\rangle$, $|n\rangle$ being system number (and energy) eigenstates, the solution for $\rho(t)$ is[13]

$$\rho(t) = \sum_{n,m} \rho_{nm} \exp[(|\epsilon|^2/\gamma^2)(n-m)^2(1-\tfrac{1}{2}\gamma t - \exp(-\tfrac{1}{2}\gamma t))]$$
$$(|n\rangle\langle m|)_S \otimes \left[\frac{|\alpha_n(t)\rangle\langle\alpha_m(t)|}{\langle\alpha_m(t)|\alpha_n(t)\rangle}\right]_A \qquad (16)$$

$|\alpha_n(t)\rangle$ are coherent apparatus states with

$$\alpha_n(t) = (\epsilon n/\gamma)\,[1 - \exp(-\tfrac{1}{2}\gamma t)]$$

In the limit of large times

$$\rho(t) \longrightarrow \sum_n P_0(n)\,(|n\rangle\langle n|)_S \otimes (|\alpha_n\rangle\langle\alpha_n|)_A \qquad (17)$$

$P_0(n) = \rho_{nn}(0)$ is the initial number distribution of the system.

The correlations between the system and the apparatus states are induced by unitary evolution. The reduction of the density operator to a diagonal form is caused by the apparatus-environment coupling. The transfer of energy to the environmental degrees of freedom appears in the apparatus as a dissipation and causes the off-diagonal density matrix elements to decay as

$$\exp\,[\,(|\epsilon|^2/\gamma^2)(n-m)^2(1 - \tfrac{1}{2}\gamma t - \exp(-\tfrac{1}{2}\gamma t))]$$

The system-apparatus density is thereby reduced to a correlated diagonal form.A non-demolition measurement of system quanta has been performed;since $[H_{SA}, a^+a] = 0$, the system-apparatus density operator collapses to a form diagonal in the a^+a eigenstate basis. Zurek[14] has emphasised that the system-apparatus coupling will determine into which state the system-apparatus combination will collapse. This in turn determines the system observable measured; in the present case, this is $\sum \epsilon_n |n\rangle\langle n|$.

MEASUREMENT AND IRREVERSIBILITY

We will now examine the role of quantum measurements in irreversibility in the shape of a few observations[15].

(i) The model of measurements presented here does not in any way solve the longstanding problem of irreversible behaviour of a closed macroscopic system as arising from reversible microscopic equations of motion. The system+apparatus combination satisfies a dissipative Fokker-Planck equation and does not represent a closed system. Dissipation leads to irreversible behaviour, as is to be expected.

(ii) Is this irreversible development an essential element of a quantum measurement? The answer must be in the affirmative, for without it the density operator continues to be pure.

(iii) Fox[16] has shown that the entropy of an isolated perfect gas increases inspite of molecular motion being governed by time reversible equations. The solutions of a system of equations need not possess all the symmetries of the equations . In the present case all Hamiltonians are time reversible. Irreversible behaviour of the system+apparatus part results from the transfer of energy and information to the environment[7].

(iv) Has quantum theory introduced a new source of irreversibility? Not in this model, where irreversible collapse results from dissipative loss of coherences.

(v) It needs to be emphasized that complete loss of coherence occurs only for $t \longrightarrow \infty$. There cannot be any instantaneous collapse in any model of the measurement process in which collapse is viewed as a dynamical phenomenon and not as a result of re-interpretation of available information.[17]

REFERENCES

1. Yakir Aharanov, Peter G.Bergmann & Joel L.Lebowitz Phys.Rev. 134,B1410 (1964).

2. J.von Neumann, "Mathematical Foundations of Quantum Mechanics", Princeton University Press, Princeton(1955).
3. F.London and E.Bauer,The Theory of Observation in Quantum Mechanics, reprinted in: "Quantum Theory and Measurement", J.A.Wheeler & W.H.Zurek,ed.,Princeton University Press,Princeton(1983).
4. P.T.Landsberg and D.Home, Wavefunction collapse in the ensemble interpretation in : "Microphysical Reality and the Quantum Formalism", D.Reidel, Dordrecht(1986).
5. K.S.Thorne,R.P.Drever,C.M.Caves,M.Zimmermann & V.D.Sandberg, Rev.Mod.Phys. 52:667(1968).
6. E.P.Wigner, Am.J.Phys. 31:368(1963).
7. W.H.Zurek, Information Transfer in Quantum Measurements:Irreversibility and Amplification in : "Quantum Optics, Experimental Gravitation and Measurement Theory", P.Meystre & Marlan O.Scully, ed., NATO ASI Series B :Physics Vol. 94, Plenum Press, N.Y.(1983).
8. C.M.Savage & D.F.Walls, Phys.Rev.A32 : 2316(1985).
9. Mark Hillery & Marlan O.Scully, On State Reduction and Observation in Quantum Optics : Wigner's Friends and their Amnesia in : "Quantum Optics, Experimental Gravitation and Measurement Theory", P.Meystre & Marlan O.Scully,ed., NATO ASI Series B : Vol.94, Plenum Press,N.Y. (1983).
10. K.Gottfried, Quantum Mechanics : Vol.I, W.A.Benjamin, N.Y.(1966).
11. A.O.Caldeira & A.J.Leggett, Physica 121A : 587(1983).
12. R.P.Feynman & F.L.Vernon Ann.Phys. 24 : 118(1963).
13. D.F.Walls, M.J.Collet & G.J.Milburn, Phys.Rev. D32 : 3208(1985).
14. W.H.Zurek, Phys.Rev. D24 : 1516 (1981).
15. K.Datta, "Note on Irreversibility and Theory of Measurement", Univ. of Delhi preprint, Delhi, India (1986).
16. R.F.Fox, Am.J.Phys. 50 : 804(1982).
17. J.R.Fox, Am.J.Phys. 51 : 49(1983).

SYMMETRY AND TOPOLOGY OF THE CONFIGURATION SPACE AND QUANTIZATION

H. D. Doebner

A. Sommerfeld Institute for Mathematical Physics
Technical University of Clausthal
3392 Clausthal (F.R. Germany)

and

J. Tolar

Faculty of Nuclear Science and Physical Engineering
Czech Technical University
115 19 Prague, Břehová 7 (Czechoslovakia)

1. INTRODUCTION

In this contribution certain group-theoretical and topological aspects of quantum theory are reviewed. We consider systems localized on configuration spaces being homogeneous spaces or differentiable manifolds. Our approach to quantum kinematics is based on systems of imprimitivity in the case of homogeneous spaces, and their generalization to manifolds, which we call quantum Borel kinematics. We show that quantum kinematics can be classified in terms of global invariants - quantum numbers of group-theoretical or topological origin. Finally, quantum mechanics of a charged particle in the magnetic field of the Dirac monopole is presented as an example illustrating the interplay of group representation theory and non-trivial topology. The exposition is based on our joint work[1, 2, 3, 4] with B. Angermann and P. Šťovíček, and Ref. 5.

2. QUANTUM MECHANICS ON HOMOGENEOUS SPACES - SYSTEMS OF IMPRIMITIVITY

Given a system localized on a configuration space M, we say that the system has a geometrical symmetry group G if G is both the transformation group of space M and symmetry group of the system. For G acting transitively on M, space M is called a homogeneous psace of G. Every homogeneous space of group G is equivalent to a coset space G/K where K is the subgroup of G which leaves (an arbitrarily chosen but fixed) point of M invariant (isotropy subgroup).

There is an effective method[1, 6] for construction of quantum kinematics based on the theory of induced representations[7] provided the configuration space M is a homogeneous space of a Lie group G, M = G/K, and K is a closed subgroup of G.

In this approach[1, 6, 8] observables of generalized momenta are constructed from a projective (unitary) representation U of group G in a (separable) Hilbert space $\mathcal{H}$ as self-adjoint generators of its one-parameter subgroups. On the other hand, position operators are realized in terms of a projection-valued measure E: $B \mapsto E(B)$ mapping Borel subsets B of M ($B \in \mathcal{B}(M)$) into projection operators E(B) in $\mathcal{H}$ such that the axioms of localization hold:

$$\begin{aligned} E(B_1 \cap B_2) &= E(B_1).E(B_2), \\ E(B_1 \cup B_2) &= E(B_1) + E(B_2) - E(B_1 \cap B_2), \\ E\Big(\bigcup_{i=1}^{\infty} B_i\Big) &= \sum_{i=1}^{\infty} E(B_i) \quad \text{for mutually disjoint } B_i \in \mathcal{B}(M), \\ E(M) &= I \quad (\text{unit operator in } \mathcal{H}). \end{aligned} \tag{1}$$

For a given subset $B \in \mathcal{B}(M)$, the projection operator E(B) corresponds to a measurement (YES-NO experiment) which determines if the system is localized in B; its eigenvalues 1 (0) correspond to situations when the system is found completely inside (outside) B, respectively. For the system in a (normalized) state $\psi \in \mathcal{H}$, expectation value $(\psi, E(B)\psi)$ is equal to the probability to find the system in B. From the requirement that this probability be invariant under G-transformations one finds

$$U(g)\, E(B)\, U(g)^{-1} = E(g.B) \tag{2}$$

for all $g \in G$, $B \in \mathcal{B}(M)$; the left hand side represents the action of representation U on the family $\{E(B)\}$ of projection operators in Hilbert space $\mathcal{H}$ which corresponds to the transformation of subsets $B \mapsto g.B$ when G-action on M is denoted by $g: M \to M : m \mapsto g.m$. If $\psi \in \mathcal{H}$ is a state localized in B, then $U(g)\psi$ is a state localized in $g.B = \{m \in M;\ g^{-1}.m \in B\}$. For further details see[1, 4, 6, 8].

The pair (U, E) which satisfies (2) is called a transitive <u>system of imprimitivity</u> in $\mathcal{H}$, for group G, based on the homogeneous space M. Any system of imprimitivity is associated with a certain quantum kinematic on M. Their complete classification (up to unitary equivalence) is based on the

<u>Imprimitivity Theorem</u>[7]. Any transitive system of imprimitivity (U, E) is unitarily equivalent to a canonical system of imprimitivity (U^L, E^L).

For the construction of all canonical systems of imprimitivity (U^L, E^L) there is a standard general procedure[7]. We present it here in a slightly simplified form. For detailed exposition involving multiplier representations see[4, 7, 8]; we remind only that in the case when G is semi-simple, one should replace G by its universal covering group as the quantum mechanical symmetry group.

Let G be a connected Lie group (of finite dimension) and K be some closed subgroup of G. On the homogeneous space M = G/K we fix a quasi-invariant measure μ (i.e., for each $g \in G$, the measures $\mu: B \mapsto \mu(B)$ and $\mu \circ g$: $B \mapsto \mu(g.B)$ belong to the same measure class). Let L be a unitary representation of K in a separable Hilbert space H^L with inner product $\langle .,. \rangle$. The Hilbert space $\mathcal{H}$ is constructed as a completion of a linear space of vector-valued functions $\psi : G \to H^L$ which satisfy

$$\psi(gk) = L(k)^{-1}\, \psi(g), \quad k \in K, \tag{3}$$

and have finite norm induced by the inner product

$$(\varphi, \psi) = \int_{G/K} \langle \varphi(g), \psi(g) \rangle \, d\mu(m). \tag{4}$$

The representation U^L of G is then defined as a unitary induced representation

$$[U^L(g)\psi](a) = \sqrt{\frac{d\mu}{d(\mu \circ g)}} (g^{-1}aK) \, \psi(g^{-1}a), \tag{5}$$

where $d\mu/d(\mu \circ g)$ is the Radon-Nikodym derivative. The projection-valued measure E^L on G/K is defined as multiplication by characteristic functions of subsets B,

$$[E^L(B)\psi](a) = \tilde{\chi}_B(a) \, \psi(a), \tag{6}$$

where

$$\tilde{\chi}_B(a) = \begin{cases} 1 & \text{if } aK \in B \\ 0 & \text{if } aK \notin B \end{cases} \tag{7}$$

The pair (U^L, E^L) is the canonical system of imprimitivity in $\mathcal{H}$.

Thus we see that inequivalent (irreducible) systems of imprimitivity can be classified in terms of inequivalent (irreducible) unitary representations L of the isotropy subgroup K, i.e. in terms of quantum numbers labeling (irreducible) representations L (see also Sec. 5).

Given a transitive imprimitivity system (U, E) for G in $\mathcal{H}$ based on M, how quantum observables can be derived from it? The projection-valued measure E is simply related to a set of quantum position observables Q(f) corresponding to arbitrary classical smooth real functions f: M $\rightarrow$ R defined on configuration space M (e.g. coordinate functions, potentials, etc.). Operators Q(f) are uniquely determined by their spectral decompositions

$$Q(f) = \int_{-\infty}^{\infty} \lambda \, dE^f_\lambda, \tag{8}$$

where the spectral function E^f_λ is given by spectral measure $E^f(\Delta) = E(f^{-1}(\Delta))$ on subsets $\Delta = (-\infty, \lambda)$ of R. It can be shown that

$$[Q(f)\psi](m) = f(m).\psi(m) \tag{9}$$

for $\psi \in \mathcal{D}^\infty \subset \mathcal{H}$, where $\mathcal{D}^\infty$ will be given in (23). For instance, the operators $Q_j = Q(q_j) = q_j$ in $\mathcal{H} = L^2(R^n, d^nq)$ which quantize the Cartesian coordinate functions q_j in R^n are given by

$$Q_j = \int_{-\infty}^{\infty} \lambda \, dE^{q_j}_\lambda \tag{10}$$

with spectral functions

$$E^{q_j}_\lambda = E(\{ q \in R^n \, ; \, q_j \leq \lambda \}), \quad j = 1,\dots,n. \tag{11}$$

Unitary representation $U = U^L$ of Lie group G determines a comprehensive set of G-momenta uniquely associated with each generator of G. Now, a generator of G - an element $\underline{X}$ of the corresponding Lie algebra $\underline{G}$ - determines a one-parameter subgroup $a(t) = \exp(t\underline{X})$, $t \in R$, of G (and vice versa, $\underline{X} = a'(0)$). Any one-parameter group of G-transformations $\{a(t)\}$ of M determines a vector field X on M induced by $\underline{X} \in \underline{G}$:

$$X_m \psi(m) = \left[\frac{d}{dt} \psi(a(t).m) \right]_{t=0} . \qquad (12)$$

The representation of one-parameter group $\{a(t)\}$ by the (strongly continuous) one-parameter group of unitary operators $U(a(t))$ can be related, via

$$U(a(t)) = U(\exp(t\underline{X})) = \exp\left(-\frac{i}{\hbar} P(X)t\right), \qquad (13)$$

to the quantum operator $P(X)$ of G-momentum associated with a (complete) vector field X on configuration manifold M (by Stone's theorem, $P(X)$ is essentially self-adjoint). Let us note that in the classical limit $P(X)$ corresponds to classical generalized momentum $\sum_{i=1}^{n} X^i_m p_i$, where X^i_m are components of vector X_m in local coordinates q^i on M, and p_i are the conjugate canonical momenta.

Relation (2) can now be shown to yield a generalization of the Heisenberg commutation relations. Namely, for position operators (8), (9) we obtain

$$U(a(t))\, Q(f)\, U(a(-t)) = Q(f \circ \exp(-tX)) \qquad (14)$$

on $\mathcal{D}^\infty \subset \mathcal{H}$, which for infinitesimal t gives

$$[Q(f), P(X)] = i\hbar\, Q(X.f) \qquad (15)$$

on $\mathcal{D}^\infty \subset \mathcal{H}$ for all $f \in C^\infty(M)$ and all vector fields X generated by $\underline{X} \in \underline{G}$. Commutation relation (15) together with

$$[Q(f_1), Q(f_2)] = 0, \qquad (16)$$

$$[P(X), P(Y)] = -i\hbar\, P([X,Y]) \qquad (\text{on } \mathcal{D}^\infty \subset \mathcal{H}) \qquad (17)$$

generalize the Heisenberg commutation relations in terms of coordinate-independent global geometrical objects.

3. QUANTUM MECHANICS ON MANIFOLDS - QUANTUM BOREL KINEMATICS

A configuration manifold M need not have any geometrical symmetry in general. Even in such a case mathematical models of non-relativistic quantum mechanics on M can be constructed.

Localization of a quantum system on M is treated in the same way as in Sec. 2, i.e. in terms of a projection-valued measure E on $\mathcal{B}(M)$ satisfying (1). We assume that E is an elementary (i.e. multiplicity 1) spectral measure on $\mathcal{B}(M)$ in a separable Hilbert space $\mathcal{H}$. Equivalently, we may use the smooth position operators $Q(f)$ given by (8), (9).

To define generalized momenta, we use the set of all <u>complete</u> vector fields $X \in \mathcal{X}_c(M)$ on M, their basic useful property being a <u>one-to-one</u> correspondence with one-parameter groups of diffeomorphisms $\{\varphi^X_t;\ t \in R\}$ of M,

$$[X\psi](m) = \left[\frac{d}{dt} (\psi \circ \varphi^X_t)(m) \right]_{t=0} . \qquad (18)$$

Then the classical Borel kinematics[2, 3] on M is a pair $(\mathcal{B}(M), \mathcal{X}_c(M))$ with a flow model

$$B \longmapsto \varphi^X_t(B) = \{ m \in M;\ \varphi^X_{-t}(m) \in B \} \qquad (19)$$

of shifts along $X \in \mathfrak{X}_c(M)$ (Fig. 1).

Quantum Borel kinematics on M is then defined as a pair (E, P) where:

(1) E is an elementary projection-valued measure on $\mathcal{B}(M)$ acting in a separable Hilbert space $\mathcal{H}$;

(2) P associates to each complete vector field $X \in \mathfrak{X}_c(M)$ an infinitesimal generator P(X) of a unitary one-parameter group $V^X_t = \exp(-iP(X)t/\hbar)$ representing in $\mathcal{H}$ the flow model (19) of shifts along X:

$$V^X_t\, E(B)\, V^X_{-t} \;=\; E(\varphi^X_t(B)), \quad B \in \mathcal{B}(M),\ t \in R. \tag{20}$$

Hence a relation like (14) holds for Q(f), $f \in C^\infty(M)$, too.

$$V^X_t\, Q(f)\, V^X_{-t} \;=\; Q(f \circ \varphi^X_{-t}) \tag{21}$$

and implies commutation relation (15) on $\mathcal{D}^\infty \subset \mathcal{H}$,

$$[Q(f), P(X)] \;=\; i\hbar\, Q(X.f); \tag{15'}$$

we also have (16),

$$[Q(f_1), Q(f_2)] \;=\; 0. \tag{16'}$$

(3) P is a partial Lie algebra homomorphism, i.e.

$$P(X+aY) = P(X) + aP(Y) \quad \text{and} \quad [P(X), P(Y)] = -i\hbar\, P([X, Y]) \tag{17'}$$

for all $X, Y \in \mathfrak{X}_c(M)$, $a \in R$, for which $X+aY \in \mathfrak{X}_c(M)$ and $X, Y \in \mathfrak{X}_c(M)$, respectively;

(4) P is local* : if $\psi \in \mathcal{H}$, $\psi \neq 0$, is localized in $B \in \mathcal{B}(M)$, i.e. $(\psi, E(B)\psi) = 1$, and if $X \in \mathfrak{X}_c(M)$ vanishes in B, then

$$(\psi, P(X)\psi) = (\psi, P(0)\psi) \text{ whenever } \psi \in \mathcal{D}(P(0)) \cap \mathcal{D}(P(X)); \tag{22}$$

(5) $$\mathcal{D}^\infty = \bigcap_{f,X,\alpha_i,\beta_i,n} \mathcal{D}\Big(\prod_{i=1}^{n} Q(f)^{\alpha_i} P(X)^{\beta_i}\Big) \tag{23}$$

is dense in $\mathcal{H}$.

The most general form of quantum Borel kinematics was found[3, 10] under a more technical assumption concerning the domain $\mathcal{D}$ of smooth sections with compact supports of a complex line bundle η over M ($\mathcal{D} \subset \mathcal{D}^\infty$, $P(X)\mathcal{D} \subset \mathcal{D}$, $X \in \mathfrak{X}_c(M)$). It was proved[3, 10] that any (differentiable) quantum Borel kinematics is unitarily equivalent to a canonical one which depends on the choice of:

1/ smooth (Lebesgue) measure ν on M;

2/ fibre bundle $\eta = (\mathcal{E}, \pi, M; C)$ over M with fibres diffeomorphic to C

*Assumption (4) implies that P(X) are differential operators.

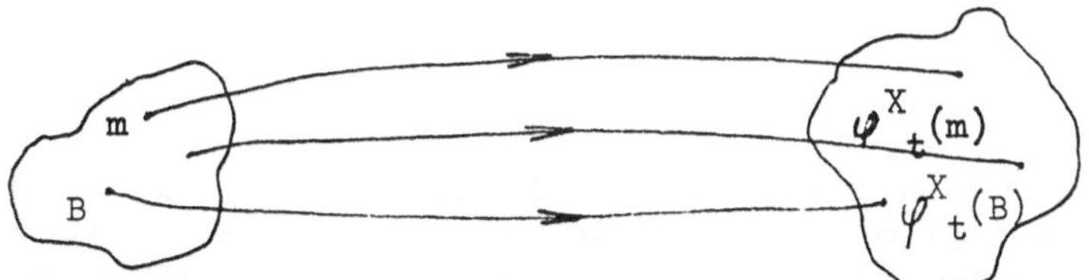

Fig. 1. Flow model for classical Borel kinematics.

(complex line bundle) equiped with Hermitean inner product $\langle z,z'\rangle = \bar{z}z'$;

3/ Hermitean flat connection ∇ on η, i.e. a connection compatible with inner product $\langle .,. \rangle$, $X\langle\sigma,\tau\rangle = \langle\nabla_X\sigma,\tau\rangle + \langle\sigma,\nabla_X\tau\rangle$ and with vanishing curvature;

4/ real number c.

The Hilbert space $\mathcal{H}$ can, in general, only be realized as a Hilbert space of sections of η, i.e. differentiable mappings $\sigma : M \to \mathcal{E}$ such that $\pi \circ \sigma = id_M$. Hilbert space of sections with finite norm induced by an inner product

$$(\sigma, \tau) = \int_M \langle\sigma(m), \tau(m)\rangle \, d\nu(m). \tag{24}$$

will be denoted $\mathcal{H} = L^2(\eta, \langle .,.\rangle, \nu)$. Position operators E(B) in $\mathcal{H}$ are uniquely given by

$$E(B)\sigma = \chi_B\sigma, \qquad \chi_B(m) = \begin{cases} 1 & \text{if } m\in B \\ 0 & \text{if } m\notin B \end{cases} \tag{25}$$

where $B \in \mathcal{B}(M)$, $\sigma\in\mathcal{H}$. Then also smooth position operators have the usual form of the Schrodinger representation (9),

$$Q(f)\sigma = f.\sigma, \quad f\in C^\infty(M), \quad \sigma\in\mathcal{D}^\infty. \tag{26}$$

Generalized momentum operators P(X) corresponding to complete vector fields $X \in \mathcal{X}_c(M)$ have the most general form

$$P(X)\sigma = -i\hbar\nabla_X\sigma + (-\frac{i\hbar}{2} + c)Q(\mathrm{div}_\nu X).\sigma, \tag{27}$$

where

$$\mathrm{div}_\nu X = \left[\frac{d\rho^X_t}{dt}\right]_{t=0}, \quad \rho^X_t.\nu = \nu\circ\varphi^X_t, \tag{28}$$

and the term with coefficient $-i\hbar/2$ makes P(X) a symmetric operator with respect to inner product (24); σ is a smooth section with compact support, $\sigma\in\mathcal{D}$.

We note that the dependence of quantum Borel kinematics on a smooth measure ν is inessential: smooth measures belong to the same measure class, so a Hilbert space $\mathcal{H}_1$ with ν_1 transforms by a simple unitary transformation $\sigma_1 \mapsto \sigma_2 = (d\nu_1/d\nu_2)^{1/2}.\sigma_1$ onto a Hilbert space $\mathcal{H}_2$ with ν_2.

We observe also that an arbitrary real constant c vanishes, if we require that "velocity be proportional to momentum"[3, 10]:

$$[H, Q(f)] = -i\hbar P(\mathrm{grad}_g f); \tag{29}$$

here $H = -(\Delta_g/2) + V$, Δ_g is the Laplace-Beltrami operator on M with Riemannian metric g, and grad_g is the vector field $g^{ik}\,\partial/\partial q^k$.

4. QUANTIZATION AND TOPOLOGY OF M

Let us consider the dependence of (differentiable) quantum Borel kinematics on the topology of M. The existence of a non-trivial complex line bundle η with connection ∇ (besides the trivial η_0 with total space $\mathcal{E} = M\times C$) is ensured if the second cohomology group $H^2(M, Z) \neq 0$. The complex line bundles with connections $(\eta, \langle .,.\rangle, \nabla)$ are classified by the elements of $H^2(M, Z)$ in the following sense[11]: if $R(X, Y) = \nabla_X \nabla_Y - \nabla_Y \nabla_X - \nabla_{[X, Y]}$ is a closed curvature 2-form of ∇, then bundles η exist with just those connections ∇ for which R represents an integer de Rham 2-cohomology class; i.e.,

$$\frac{1}{2\pi i} \int_{z_2} R = n, \quad n \in Z, \tag{30}$$

where z_2 is an arbitrary closed surface (2-cycle) in M. It follows from (27) that momenta P(X) fulfil

$$[P(X), P(Y)] = -i\hbar P([X, Y]) - \hbar^2 R(X, Y), \tag{31}$$

so our assumption (3) on quantum Borel kinematics, Eq. (17'), is equivalent with R = 0, i.e. flat connection ∇. Let us note however that quantum Borel kinematics (25) - (27) can be used[4,5] also for $R \neq 0$ satisfying (30) provided curvature R can be given a physical interpretation; see Sec. 5 for an example.

For a given bundle η possible inequivalent connections are further classified[11] in terms of the characters of the first homotopy group $\pi_1(M)$ (the Poincaré fundamental group of M). The set of all inequivalent characters (one-dimensional unitary representations $\chi : \pi_1(M) \to U(1)$) forms the character group $\pi_1(M)^*$. Using the Hurewicz isomorphism $\pi_1(M)/C(\pi_1(M)) \simeq H_1(M, Z)$ and the property that the commutator subgroup $C(\pi_1(M))$ (generated by elements of the form $aba^{-1}b^{-1}$ where $a, b \in \pi_1(M)$) is contained in the kernel of χ, we can equivalently classify quantizations involving inequivalent connections by the elements of $H^1(M, U(1)) = H_1(M, Z)^*$.

Let us describe the general structure of the abelian group $H_1(M, Z)$ for compact M. We have $H_1(M,Z) = F \oplus T$, where the free abelian group F is $F = Z \oplus \ldots \oplus Z$ (b_1 terms), with b_1 being the first Betti number of M, and the torsion abelian group is $T = Z_{\tau_1} \oplus \ldots \oplus Z_{\tau_k}$ with Z_{τ_i} being cyclic groups of orders τ_i (torsion coefficients) such that τ_{i+1}/τ_i = positive integer. Thus the elements of $H^1(M, U(1))$ can be parametrized by (b_1+k)-tuples

$$(\exp(2\pi i\, \theta_1), \ldots, \exp(2\pi i\, \theta_{b_1});\ \exp(2\pi i \tfrac{m_1}{\tau_1}), \ldots, \exp(2\pi i \tfrac{m_k}{\tau_k}) \tag{32}$$

with the numbers $\theta_l \in [0, 1)$, $l = 1, \ldots, b_1$, and $m_i = 0, 1, \ldots, \tau_i - 1$, $i = 1, \ldots, k$, classifying inequivalent quantum Borel kinematics (flat connections). For a short list of examples see Table 1.

In a local trivialization of η (and, of course, in a trivial bundle) sections $\sigma : M \to \mathcal{E}$ can be replaced by usual wave functions $\psi : M \to C$ via

$$\sigma(m) = (m, \psi(m)) \tag{33}$$

and $\mathcal{H} = L^2(M, \nu)$. Then quantum Borel kinematics (25) - (27) takes the form

$$\begin{aligned} E(B)\psi &= \chi_B \cdot \psi , \qquad Q(f)\psi = f.\psi , \\ P(X)\psi &= -i\hbar\,(X + \tfrac{1}{2}\mathrm{div}_\nu X + \omega(X))\psi + c.\mathrm{div}_\nu X.\,\psi , \end{aligned} \tag{34}$$

where ω is a connection 1-form on M with curvature $R = d\omega = 0$ (or integral class). For trivial η_0 inequivalent quantum Borel kinematics correspond to the classes of closed 1-forms ω (R = 0) modulo logarithmically exact 1-forms dF/F, i.e. up to a gauge transformation $\omega \mapsto \omega + (dF/F)$ (Refs. 2, 12).

A physical meaning of ω can be obtained by comparison with quantum mechanics of a non-relativistic particle of charge e in $M = R^3$ interacting with an external stationary magnetic field $\vec{B} = \mathrm{rot}\,\vec{A}$. Here the canonical momenta p_i are quantized as $p_i = -i\hbar\,\partial/\partial q^i$ in $L^2(R^3, d^3q)$ but the Hamiltonian involves kinetic momenta

$$p_i - eA_i = -i\hbar\left(\frac{\partial}{\partial q^i} - i\frac{e}{\hbar}A_i\right) = -i\hbar\,\nabla_{\partial/\partial q^i} = P\left(\frac{\partial}{\partial q^i}\right) \tag{35}$$

quantized according to (34) with c = 0, $\omega = -i\frac{e}{\hbar}A_i\,dq^i$ and $R = d\omega = -i\frac{e}{\hbar}d(A_i\,dq^i) = -i\frac{e}{\hbar}B$.

5. QUANTUM MECHANICS OF A CHARGED PARTICLE IN THE MAGNETIC FIELD OF THE DIRAC MONOPOLE

Maxwell's equations of magnetostatics without currents

$$\mathrm{div}\,\vec{B} = 0, \quad \mathrm{rot}\,\vec{B} = 0, \tag{36}$$

admit in the region $M = R^3 \setminus \{(0, 0, 0)\}$ a family of static spherically symmetric solutions

$$\vec{B} = \frac{g}{4\pi}\frac{\vec{r}}{r^3}, \quad r \neq 0,\ g \in R,\ g \neq 0, \tag{37}$$

which can be interpreted as the field of a point-like magnetic monopole of "magnetic charge" g placed at the origin. Magnetic field (37) of a hypothetical magnetic monopole has an unusual global property: for any closed surface S containing the origin, the magnetic flux through S does not vanish,

$$\Phi = \int_S \vec{B}.d\vec{S} = g \neq 0, \tag{38}$$

hence field (37) does not possess in M any vector potential $\vec{A}$ without singularities ($\vec{A}$ defined by $\vec{B} = \mathrm{rot}\,\vec{A}$). Namely, we observe that according to Stokes' theorem (∂S = boundary of S)

$$g = \int_S \vec{B}.d\vec{S} = \int_{\partial S} \vec{A}.d\vec{r} = 0. \tag{39}$$

Thus for $g \neq 0$ a singularity-free potential $\vec{A}$ does not exist. In the language of differential geometry, magnetic field (37) is described by a 2-form

$$R = -i\frac{e}{\hbar}B = -i\frac{e}{\hbar}\frac{g}{4\pi r^3}(q^1\,dq^2 \wedge dq^3 + \mathrm{cycl.}) \tag{40}$$

which is closed

$$dR = 0 \iff \mathrm{div}\,\vec{B} = 0 \tag{41}$$

but not exact, i.e. a 1-form ω such that $R = d\omega$ does not exist.

Classical motion of a particle with charge e in the spherically symmetric field (37) was investigated already by H. Poincaré[13] who showed that the equations of motion

Table 1. Topology of configuration space M and inequivalent quantizations

Quantum System	M	$H^2(M, Z)$	$H_1(M, Z)$	Parameters of Quantizations
Spinless particle	R^3	0	0	-
Rotator with fixed axis	S^1	0	Z	$\theta \in [0, 1)$
Aharonov-Bohm effect[19]	$R^3 \setminus R$	0	Z	$\theta \in [0, 1)$
Rotator (two-dimensional)	S^2	Z	0	$n \in Z$
Dirac's monopole	$R^3 \setminus 0$	Z	0	$n \in Z$
Top	SO(3)	Z_2	Z_2	m=0, 1
2 identical particles[20]	RP^2	Z_2	Z_2	m=0, 1

$$m\ddot{\vec{r}} = e \frac{g}{4\pi} \frac{\vec{r}}{r^3} \tag{42}$$

possessed conserved total angular momentum

$$\vec{J} = m \vec{r} \times \dot{\vec{r}} - \frac{eg}{4\pi} \frac{\vec{r}}{r} . \tag{43}$$

The additional term in (43) represents[14] the angular momentum of the electromagnetic field produced by the particle-monopole system.

Quantum theory of a charged particle in field (37) was first formulated by Dirac[15] with a rather unexpected result. According to Dirac, a quantal description exists only under the condition that the dimensionless quantity $e\phi/2\pi\hbar$ is an integer, i.e.

$$\frac{eg}{2\pi\hbar} = n \in Z. \tag{44}$$

In the standard quantum mechanical formalism Dirac was forced to introduce special rules for working with unphysical singularities of the vector potential $\vec{A}$ distributed along Dirac's string. It was recognized only much later[16,17] which global geometrical objects are needed for the model.

In using quantum Borel kinematics (25) - (27) on $M = R^3 \setminus \{(0, 0, 0)\}$ with c = 0 and integral R one actually works with those global geometrical objects and the singularities of $\vec{A}$ do not appear. The bundle η with connection ∇ exists for quantum Borel kinematics on M, if and only if curvature R is integral, Eq. (30). For the Dirac monopole field (40) this leads to

$$\frac{1}{2\pi i} \int_{S^2} R = \frac{1}{2\pi i} \left(-i \frac{e}{\hbar}\right) \int_{S^2} \vec{B}.d\vec{S} = \frac{-e}{2\pi\hbar} g = -n \in Z. \tag{45}$$

Thus Dirac's condition (44) is exactly the condition for the existence of a (non-trivial) complex line bundle over M with connection ∇ and curvature $R \neq 0$.

In order to write explicit quantum mechanical relations including the Schrödinger equation, the method of local trivializations of η is useful (Refs. 4, 5, 17). There the manifold M is covered with two patches $U_\pm = M \setminus \{\mp z\text{-axis}\}$, and two reference sections $\rho_\pm$ of the corresponding principal bundle (Hopf's fibration) are chosen[4, 5]. Then connection ∇ is determined by a pair of 1-forms

$$\omega_\pm = -i\frac{e}{\hbar}\left(\pm\frac{g}{4\pi}\right)(1 \mp \cos\vartheta)\, d\varphi \tag{46}$$

defined on $U_\pm$ and linked on $U_+ \cap U_-$ by a gauge transformation

$$\omega_+ = \omega_- + (dF/F), \qquad F = \exp\left(-i\frac{e}{\hbar}\frac{g}{2\pi}\varphi\right). \tag{47}$$

The corresponding vector potentials

$$\vec{A}_\pm = \pm\frac{g}{4\pi}\frac{\vec{e}_z \times \vec{r}}{r \pm z}, \qquad \vec{e}_z = (0, 0, 1), \tag{48}$$

are singular on the $\mp z$-axes excluded from the patches $U_\pm$. Their gauge transformation on $U_+ \cap U_-$ can be written $\vec{A}_+ = \vec{A}_- + \overrightarrow{\mathrm{grad}}(g\varphi/2\pi)$ but the function $\lambda = g\varphi/2\pi$ is not smooth on $U_+ \cap U_-$; therefore the notation (47) is preferable.

Both 1-forms $\omega_\pm$ yield the curvature 2-form

$$R = d\omega_+ = d\omega_- = -i\frac{e}{\hbar}\frac{g}{4\pi}\sin\vartheta\, d\vartheta \wedge d\varphi, \tag{49}$$

which, according to (40), belongs to the magnetic field (37). Each section is replaced by a pair of complex wave functions $\psi_\pm$ defined on patches U_+, U_- , their gauge transformation on $U_+ \cap U_-$ being (see (47))

$$\psi_+ = F.\,\psi_- . \tag{50}$$

The resulting quantum Borel kinematics (E, P) on M involves the projection-valued measure on subsets $B \in \mathcal{B}(M)$,

$$E(B) : (\psi_+, \psi_-) \longrightarrow (\chi_{B+}\cdot\psi_+, \chi_{B-}\cdot\psi_-), \quad B_\pm = B \cap U_\pm, \tag{51}$$

and pairs of operators of generalized momenta $P_\pm(X)$ given by (27), (28) with c = 0 and 1-forms $\omega_\pm$:

$$P_\pm(X) = -i\hbar\left(X + \omega_\pm(X) + \frac{1}{2}\,\mathrm{div}\, X\right) \tag{52}$$

Because of spherical symmetry, the system can also be quantized with the help of systems of imprimitivity considered in Sec. 2. We introduce Mackey's transitive systems of imprimitivity on spheres $M = S^2_r$ for quantum mechanical symmetry group G = SU(2). The induced representations U^L are classified by (irreducible) representations L_n of the isotropy subgroup U(1),

$$L_n : U(1) \longrightarrow U(1) : \tau \longmapsto \tau^{-n}, \quad n \in Z. \tag{53}$$

Explicit calculation of generators J_k of the induced representation U^{L_n} defined by

$$U^{L_n}\left(e^{-it\frac{\sigma_k}{2}}\right) = \exp\left(-\frac{i}{\hbar}tJ_k\right), \quad t \in R, \ k = 1, 2, 3, \tag{54}$$

yields the components of angular momentum operators in the two patches $(J_k : (\psi_+, \psi_-) \rightarrow (J_{k+}\psi_+, J_{k-}\psi_-),\ u^k = q^k/r)$:

$$J_{k\pm}\psi_{\pm} = -i\hbar(L_k + \omega_{\pm}(L_k))\psi_{\pm} - \frac{1}{2}n\hbar u^k \psi_{\pm} . \quad (55)$$

The result agrees with classical formula (43) since $L_k = \varepsilon_{k\ell m} q^{\ell}\partial/\partial q^m$ is the vector field on S^2_r induced by the action of the one-parameter subgroup of SU(2) generated by the element $\sigma_k/2$ of Lie algebra su(2). We observe that Mackey's method leads in this case to angular momentum operators (55) differring from operators (52) for $X = L_k$,

$$J_{k\pm} = P_{\pm}(L_k) - \frac{1}{2}n\hbar u^k . \quad (56)$$

Operators J_k generating representation U^{L_n} of the symmetry group SU(2) contain an additional term so that su(2) commutation relations are valid:

$$[J_k, J_\ell] = i\hbar\, \varepsilon_{k\ell m} J_m . \quad (57)$$

On the other hand, the commutator

$$[P(L_k), P(L_\ell)] = -i\hbar\,\varepsilon_{k\ell m}P(L_m) + i\hbar^2 \frac{n}{2}\varepsilon_{k\ell m}u^m \quad (58)$$

agrees with (31) for $R \neq 0$.

Stationary states of an electron in the Dirac monopole field are solutions of the Schrödinger equation which can be written as two equations

$$\left[\frac{1}{2m}(\vec{p} - e\vec{A}_{\pm})^2 + (V - E)\right]\psi_{\pm} = 0 \quad (59)$$

in two local trivializations over the patches $U_{\pm}$; in $U_+ \cap U_-$ condition (50) must be fulfilled. If an additional spherically symmetric potential is present, angular dependence of solutions of Eq. (59) can be separated out. Let us only note that the monopole harmonics[18] can be simply obtained as bases in $\mathcal{H}$ for irreducible components of the reducible unitary representation U^{L_n} of SU(2) into irreducibles by application of the Peter-Weyl theorem:

$$Y_{\frac{n}{2}jm}(\vartheta,\varphi) = \mathcal{D}^{j}_{m,-\frac{n}{2}}(0,\vartheta,\varphi) =$$

$$= \sqrt{\frac{2j+1}{4\pi}}\,(-1)^{m+\frac{n}{2}}\, d^{j}_{m,-\frac{n}{2}}(\vartheta)\, e^{i(m+\frac{n}{2})\varphi} ; \quad (60)$$

here $n = eg/2\pi\hbar$, $j = |n/2|, |n/2|+1, \ldots,$ $m = -j, \ldots, j$.

REFERENCES

1. H. D. Doebner, J. Tolar, Quantum mechanics on homogeneous spaces, J. Math. Phys. 16 (1975), 975-984.
2. H. D. Doebner, J. Tolar, On global properties of quantum systems, in: "Symmetries in Science", Proc. Einstein Centennial Celebration Symposium at Carbondale, Ill., B. Gruber and R. S. Millman, eds., Plenum Publ. Co., New York (1980), 475-486.
3. B. Angermann, H. D. Doebner, J. Tolar, Quantum kinematics on smooth manifolds, in: "Non-Linear Partial Differential Operators and Quantization Procedures", Proceedings of a workshop held at Clausthal, 1981, S. I. Andersson and H. D. Doebner, eds., Lecture Notes in Mathematics, Vol. 1037, Springer-Verlag, Berlin (1983), 171-208.
4. P. Šťovíček, J. Tolar, Topology of configuration manifold and quantum mechanics (in Czech), Acta Polytechnica - Práce ČVUT v Praze, No. 6 (IV, 1), (1984), 37-75.
5. P. Šťovíček, Dirac monopole derived from representation theory, Suppl.

Rend. Circ. Mat. Palermo, Ser. II, No. 3, (1984), 301-306.
6. G. W. Mackey, "Induced Representations and Quantum Mechanics", Benjamin, New York (1968).
7. G. W. Mackey, "The Theory of Induced Representations", mimeographed lecture notes, University of Chicago Press, Chicago (1955).
8. V. S. Varadarajan, "Geometry of Quantum Theory, Vol. 2 - Quantum Theory of Covariant Systems", Van Nostrand Reinhold Co., New York (1970).
9. J. von Neumann, Die Eindeutigkeit der Schrödingerschen Operatoren, Math. Ann. 104 (1931), 570-588.
10. B. Angermann, Über Quantisierungen lokalisierter Systeme: Physikalisch interpretierbare mathematische Modelle, Ph. D. Thesis, Technische Universität, Clausthal (1983).
11. B. Kostant, Quantization and unitary representations: Part I. Prequantization, in: Lecture Notes in Mathematics, Vol. 170, Springer-Verlag, New York (1970), 87-208.
12. I. E. Segal, Quantization of nonlinear systems, J. Math. Phys. 1 (1960), 468-488.
13. H. Poincaré, Comptes Rendus 125 (1896), 530.
14. M. Fierz, Helv. Phys. Acta 17 (1944), 27.
15. P. A. M. Dirac, Quantized singularities in the electromagnetic field, Proc. Roy. Soc. (London) A133 (1931), 60.
16. W. Greub, H. R. Petry, Minimal coupling and complex line bundles, J. Math. Phys. 16 (1975), 1347-1351.
17. T. T. Wu, C. N. Yang, Concept of nonintegrable phase factors and global formulation of gauge fields, Phys. Rev. D12 (1975), 3845-3857.
18. T. T. Wu, C. N. Yang, Dirac monopole without strings: monopole harmonics, Nucl. Phys B107 (1976), 365-380.
19. C. Martin, A mathematical model for the Aharonov-Bohm effect, Lett. Math. Phys. 1 (1976), 155-163.
20. H. D. Doebner, P. Šťovíček, J. Tolar, Quantizations of the system of two indistinguishable particles, Czech. J. Phys. B31 (1981), 1240-1248.

ALGEBRAIC REALIZATION OF THE ROTOR⁺

J. P. Draayer and Yorck Leschber

Department of Physics and Astronomy
Louisiana State University
Baton Rouge, LA 70803-4001

INTRODUCTION

The rotor enjoys a prominent place in physics. In classical mechanics it is usually offered as the most challenging example of rigid-body motion.[1] Applications extend from the simple symmetrical top to the dynamics of mechanical gyros, satellite behavior, and even planetary motion. All physicists learn early in their career that "the polhode rolls on the herpolhode..."!

One of the first applications of the new quantum mechanics was to the rotor. The symmetric case was considered first.[2] Its eigenstates are known to have a relatively simple representation in terms of Euler angles and rotation matrices.[3] Because of the symmetry, the projection of the angular momentum on the body-fixed symmetry axis is a good quantum number. The asymmetric case is a more challenging problem.[4] Its eigenstates are Lamé functions. These can be expressed as a linear combination of the eigenfunction of the symmetric rotor. A major application is found in molecular physics.[5] The simplest structure possessing an asymmetric rotor geometry is H_2O, the ordinary water molecule. Some excellent reviews of rotor dynamics as it applies in molecular physics are available.[6]

It is for the rotor that algebraic methods, as understood and used today, were first introduced in physics. The thesis of H.G.B. Casimir[7] is a treatise on the "Rotation of a Rigid Body in Quantum Mechanics". Following Klein, Casimir used Heisenberg's matrix mechanics in an analysis of the dynamics of the rotor. This is in contrast, for example, with the work of Kramers and Ittman which was based on Schrodinger's wave mechanics. Casimir established the connection between eigenfunction of the rotor and irreducible representations of the rotation group in three dimensions. Indeed, in his thesis one finds an outline of the general theory of continuous groups. The rotor example clearly illustrates the advantages of using algebraic over analytic methods when the system hamiltonian possesses a high degree of symmetry.

⁺Supported in part by the National Science Foundation.

Two decades after these initial efforts, a major new application of rotor dynamics emerged for the spectra of many nuclei were found to display rotational characteristics.[8] In contrast with molecular applications where the electronic ($\sim 10^1$eV), vibrational ($\sim 10^{-1}$eV), and rotational ($\sim 10^{-3}$eV) modes are well-separated in energy, in nuclei the intrinsic nucleonic modes are energetically comparable to vibrational excitations ($\sim$1 MeV) while rotations, when they occur, are again smaller ($\sim$0.1 MeV) but only by about one order of magnitude.[9] Since rotational energies depend quadratically on the angular momentum [$E_I \sim I(I+1)$], except for the very lowest spins [I=0,2,(4,6)] rotational excitations in nuclei are not well-separated from the other modes. Nonetheless, the success of the collective model in describing the properties of nuclei is now well-established and in this rotations play an important, often dominant role.[10]

Although the idea of collective rotations seemed to be in direct conflict with parallel developments supporting a shell-model interpretation of nuclear structure,[11] as evidence was gathered it became clear that both pictures had validity.[12] The challenge of understanding this duality continues to be a driving force stimulating new work in nuclear theory; the question, How does a strongly interacting many-body system support rotational behavior?, lives on. Elliott made a giant step forward towards resolving the paradox when he demonstrated that a quadrupole-quadrupole interaction operating in a space partitioned into irreducible representation of SU(3), the symmetry group of the oscillator, gives rise in a very natural and simple way to I(I+1) rotational spectra.[13]

In the last decade other algebraic theories have been introduced that provide an even simpler shell-model interpretation of nuclear structure. The most popular of these is the Interacting Boson Model (IBM).[14] The building blocks of the theory are s(L=0) and d(L=2) bosons which represent coupled pairs of fermions. The largest symmetry group of the model is U(6) with its thirty-six dimensional Lie algebra generated by bilinear combinations of the boson creation and annihilation operators. It supports three subgroup chains which have been associated with vibrational, rotational, and so-called gamma unstable nuclear configurations. A complementary theory is the Microscopic Collective Model (MCM).[15] The largest symmetry group in this case is Sp(3,R), the dynamical group of the oscillator with a twenty-one dimensional Lie algebra built out of bilinear combinations of the fermion coordinate and momentum operators. The theory is a multishell generalization of the simplest Elliott model and contains the rotor angular momentum and quadrupole operators as a subalgebra.

Both the IBM and MCM models, as well as Elliott's SU(3) model which works for light nuclei,[16] and its extension, the so-called pseudo SU(3) model, which applies for heavier deformed systems,[17] contain a root SU(3) → SO(3) group structure, see Table 1. In each case that root structure has been associated with rotational motion. And for each there is a prescription for determining the SU(3) content of the model space and the representations of SU(3) that are expected to be the dominant ones in a description of the low-lying states of a system. From this, and certain complementary developments in group theory and statistical spectroscopy, we were convinced that the SU(3) → SO(3) algebra truly embodies the dynamics of rotational motion. Could it be, even though its been a quarter of a century since the pioneering work of Elliott first appeared, that there is yet more to be learned about SU(3)? We believe the answer to that question is an emphatic yes! What follows is a report on our findings.

Table 1. The SU(3) → SO(3) root structure in shell-model theories.

BOSON	FERMION		
IBM	MCM	PSEUDO-SU(3)	ELLIOTT
U(6)	$S_p(3,R)$	$SU(N_p) \times SU(N_n)$	U(6)
↓	↓	↓	↓
SU(3)	**SU(3)**	**SU(3)**	**SU(3)**
↓	↓	↓	↓
SO(3)	**SO(3)**	**SO(3)**	**SO(3)**

ASYMMETRIC ROTOR REVISITED

The hamiltonian of the asymmetric rotor assumes a very simple form when written in terms of projections of the angular momentum operator I onto the body-fixed symmetry axes of the system,

$$H_{ASR} = \sum_{\alpha=1}^{3} A_\alpha I_\alpha^2 . \tag{1}$$

In (1), the A_α are inertia parameters. For rigid-body motion these are given by $A_\alpha = 1/2\zeta_\alpha$ where the ζ_α are eigenvalues of the inertia tensor,

$$I_{\alpha\beta} = \int \rho(\overline{r})\ (r^2\delta_{\alpha\beta} - \chi_\alpha\chi_\beta)\ d^3r. \tag{2}$$

It will prove to be convenient to have everything expressed in terms of the eigenvalues, λ_α, of the traceless quadrupole operator,

$$Q^c_{\alpha\beta} = \int \rho(\overline{r})\ (3\chi_\alpha\chi_\beta - r^2\delta_{\alpha\beta})\ d^3r. \tag{3}$$

The superscript "c" will be appended as necessary to denote collective model operators. The inertia and quadrupole tensors are related in a very simple way:

$$I_{\alpha\beta} = \frac{1}{3}\ (\zeta\ \delta_{\alpha\beta} - Q^c_{\alpha\beta}), \quad \text{where } \zeta = 2 \int \rho(\overline{r})\ r^2 d^3r. \tag{4}$$

It follows from this that $\zeta_\alpha = (\zeta - \lambda_\alpha)/3$, which is a result that can be used to define the moments of inertia of a system even if the motion is not rigid-rotor-like.

It should be clear that the asymmetric rotor hamiltonian is invariant under π rotations about the principal axes; that is, H_{ASR} commutes with the transformation operators $T_\alpha = \exp(i\pi I_\alpha)$. The set of operators $\{E, T_1, T_2, T_3\}$, where E is the identity, forms a realization of the Vierergruppe (D_2). As a consequence, eigenstates of the rotor can be classified according to their transformation properties under this four-element symmetry group. The symmetrized eigenstates can be expressed in terms of eigenstates of the symmetric rotor for which $A_1 = A_2 \neq A_3$ and, as a consequence, the magnitude of K, the eigenvalue of I_3, is a good quantum number,[18]

$$\Psi^{(\lambda\mu)|K|I}_{SYM\ \ M} \equiv \frac{1}{\sqrt{2(1+\delta_{K0})}} (D^I_{KM} + (-1)^{\lambda+\mu+I} D^I_{-KM}),$$

$$\Psi^{(\lambda\mu)\nu I}_{ASR\ \ M} = \sum_{K>0}{}' C^{(\lambda\mu)\nu I}_{K} \Psi^{(\lambda\mu)KI}_{SYM\ \ M} \qquad (5)$$

In (5), the D^I_{KM} are rotation matrices and the prime on the summation for the asymmetric wavefunction means that only even or odd K values are to be included. The quantity λ can be either an even (0) or odd (1) integer while μ must be even for K even and odd for K odd. The transformation properties of these functions under D_2 are given in Table 2. The quantity ν is a running index used to give a unique labelling to the eigenstates since in general there are more than one of symmetry type $(\lambda\mu)$ and spin I.

Table 2. Properties of eigenstates of the asymmetric rotor. The symmetry types are classes of the Vierergruppe (D_2) with elements $T_\alpha = \exp(i\pi I_\alpha)$ that generate rotations by π about the principal symmetry axes of the system.

Symmetry* Type	Transformation E	T_1	T_2	T_3	Index+ λ	μ	Dimension± I(even)	I(odd)
A	1	1	1	1	e	e	(I+2)/2	(I-1)/2
$B_{1:3} \rightarrow B_3$	1	-1	-1	1	o	e	I/2	(I+1)/2
$B_{2:2} \rightarrow B_2$	1	-1	1	-1	o	o	I/2	(I+1)/2
$B_{3:1} \rightarrow B_1$	1	1	-1	-1	e	o	I/2	(I+1)/2

*The first of the two indices on the B-type symmetry label is the usual one found in character tables for the crystallographic point groups while the second one indicates directly the axes of the rotation. In what follows, the first index will be suppress in favor of the second to gain the added simplicity of a direct relationship between the B_α and T_α.

+The symbols e and o refer to even and odd character of the integer indices λ and μ that are used to specify the symmetry type.

±The number of eigenstates with total angular momentum I of a specific D_2 symmetry type. Note that the sum of each column is 2I+1. For I=0 there is only A-type symmetry and for I=1 there are only B-types, each one occuring once.

The invariance of the hamiltonian under D_2 means that for angular momentum I the (2I+1)-dimensional matrix equation (K=-I, -I+1,...,I) breaks up into block diagonal form. The dimension of each of these submatrices, which can be labelled by its D_2 symmetry type or, equivalently, by the odd or even character of the integer indices λ and μ, is also given in Table 2. Since most even-even rotational nuclei have a positive parity I=0,2,4,... ground state band rotational sequence, it seems that type A symmetry applies.[19] However, there is no apriori reason to exclude the others. Indeed, as will be shown below, the shell model supports all four symmetry types.

Because of the invariance of H_{ASR} under D_2, the eigenenergies of the rotor satisfy simple sum-rule relations. These form a powerful tool to use in checking whether or not a spectrum is rotational for they apply independent of the choice of the rotor's inertia parameters. Probably the best known example, which applies for A-type symmetry, is $E_{12}+ E_{22}= E_3$; that is, the sum of the eigenenergies for I=2 is equal to the sum of the eigenenergies for I=3. Indeed, for A-type symmetry it can be shown that the trace of the submatrix of H_{ASR} for spin I is equal to the trace of the submatrix of H_{ASR} for spin (I+1),[20]

$$\mathrm{Tr}(\mathbf{H}_{ASR})_I = \mathrm{Tr}(\mathbf{H}_{ASR})_{I+1}. \tag{6}$$

A summary of analytic results for traces of H_{ASR}, from which results such as (6) can be deduced, are given in Table 3. Analytic results for traces of the square of H_{ASR} can also be written down. By combining the two, analytic results for the centroids $[\varepsilon = d^{-1}\mathrm{Tr}(\mathbf{H})]$ and variances $[\sigma^2 = d^{-1}\mathrm{Tr}(\mathbf{H} - \varepsilon)^2]$ of rotor spectra can be given:

$$\varepsilon_{ASR}/S_\varepsilon = I(I+1)/3$$

$$\sigma^2_{ASR}/S_\sigma = I(I+1)/45 \left\{ \begin{matrix} (4I-3)(I+4) & I(\text{even}) \\ (I-3)(4I+7) & I(\text{odd}) \end{matrix} \right\}. \tag{7}$$

In (7), $S_\varepsilon = A_1+A_2+A_3$ and $S_\sigma = A_1^2+A_2^2+A_3^2-A_1A_2-A_1A_3-A_2A_3$. Centroid and variance measures will be used below in a comparison of rotor and SU(3) → SO(3) shell-model results.

Before proceeding to a consideration of the SU(3) → SO(3) algebra, it is important to further prepare the way by giving a frame independent representation for the rotor hamiltonian. To achieve this consider the following three rotational scalars,

$$\mathbf{I}^2 = \sum_\alpha \mathbf{I}_\alpha^2 = \mathbf{I}_1^2+ \mathbf{I}_2^2+ \mathbf{I}_3^2$$

$$\mathbf{X}_3^c = \sum_{\alpha,\beta} \mathbf{I}_\alpha \mathbf{Q}^c_{\alpha\beta} \mathbf{I}_\beta = \lambda_1 \mathbf{I}_1^2+ \lambda_2 \mathbf{I}_2^2+ \lambda_3 \mathbf{I}_3^2$$

$$\mathbf{X}_4^c = \sum_{\alpha,\beta,\gamma} \mathbf{I}_\alpha \mathbf{Q}^c_{\alpha\beta} \mathbf{Q}^c_{\beta\gamma} \mathbf{I}_\gamma = \lambda_1^2 \mathbf{I}_1^2+ \lambda_2^2 \mathbf{I}_2^2+ \lambda_3^2 \mathbf{I}_3^2. \tag{8}$$

In (8), $Q^c_{\alpha\beta}$ is the quadrupole operator defined by (3). The last forms given for the X^c_3 and X^c_4 operators follow because they are rotational scalars and can therefore be evaluated without loss of generality in the principal axis system where $Q^c_{\alpha\beta} = \lambda_\alpha\delta_{\alpha\beta}$. Equations (8) can be inverted to yield expressions for the I^2_α in terms of I^2, X^c_3, X^c_4:

$$I^2_\alpha = (\lambda_\beta\lambda_\gamma I^2 + \lambda_\alpha X^c_3 + X^c_4)/(2\lambda^2_\alpha + \lambda_\beta\lambda_\gamma),$$

$$(\alpha,\beta,\gamma) \xleftrightarrow[\text{permutation}]{\text{cyclic}} (1,2,3). \qquad (9)$$

It follows that

$$H_{ASR} = aI^2 + bX^c_3 + cX^c_4;$$

$$a = \sum_\alpha a_\alpha A_\alpha, \quad a_\alpha = \lambda_\beta\lambda_\gamma/(2\lambda^2_\alpha + \lambda_\beta\lambda_\gamma)$$
$$b = \sum_\alpha b_\alpha A_\alpha, \quad b_\alpha = \lambda_\alpha/(2\lambda^2_\alpha + \lambda_\beta\lambda_\gamma)$$
$$c = \sum_\alpha c_\alpha A_\alpha, \quad c_\alpha = 1/(2\lambda^2_\alpha + \lambda_\beta\lambda_\gamma). \qquad (10)$$

Since this new form for H_{ASR} does not explicitly display the D_2 symmetry of the rotor, it might appear to represent a step backward rather than forward. While that may be so if the objective is only to solve the rotor problem, it is not so if the objective is to explore the underpinnnings of rotational behavior in microscopic systems. We will now show that the SU(3) → SO(3) group structure supports a faithful realization of the rotor dynamics; that is, there is an SU(3) → SO(3) hamiltonian that is the microscopic image of (10) and its eigenstates display all the features of eigenstates of the rotor, including the D_2 invariance!

Table 3. Sum rules for eigenenergies of the rotor. Traces of H_{ASR} are given in terms of $S=(A_1+A_2+A_3)/6$ and $S_\alpha = A_\alpha/2$ where the A_α are the inertia parameters of the rotor.

Symmetry Type	$Tr(H_{ASR})_I$ I(even)	$Tr(H_{ASR})_I$ I(odd)
A	$I(I+1)(I+2)S$	$I(I-1)(I+1)S$
B_1	$I(I-1)(I+1)S + I(I+1)S_1$	$I(I+1)(I+2)S - I(I+1)S_1$
B_2	$I(I-1)(I+1)S + I(I+1)S_2$	$I(I+1)(I+2)S - I(I+1)S_2$
B_3	$I(I-1)(I+1)S + I(I+1)S_3$	$I(I+1)(I+2)S - I(I+1)S_3$
Σ	$2I(I+1)(2I+1)S$	$2I(I+1)(2I+1)S$

SU(3) → SO(3) ALGEBRA

The use of SU(3) in physics can be traced back to the pioneering work of Racah whose quest to provide a simple interpretation of regularities in atomic spectra led him to explore various group theoretical coupling schemes.[21] The SU(3) → SO(3) structure enters for p-shell applications. The objective was to find a set of operators with simple eigenvalues that could be used to form a complete orthonormal labelling scheme for basis states of multiparticle configurations. The hope was that the eigenvalues of these operators would be associated with important physics.

The SU(3) algebra is spanned by eight generators, three angular momentum operators, $\mathbf{L}_\mu$; $\mu=0,\pm1$, which are associated with its SO(3) subgroup, and five quadrupole operators, $\mathbf{Q}_\mu$; $\mu=0,\pm1,\pm2$. (It is important to note that whereas the angular momentum operator $\mathbf{L}$ is the microscopic equivalent of the collective model $\mathbf{I}$, the spherical tensor operators $\mathbf{Q}_\mu$ of the SU(3) algebra are not the microscopic equivalent of the collective model cartesian operators $\mathbf{Q}^c_{\alpha\beta}$. They are the combination of $r^2Y_{2\mu}(\theta_r,\phi_r)$ and $p^2Y_{2\mu}(\theta_p,\phi_p)$ that in a harmonic oscillator basis have no matrix elements coupling different major shells. However, within a major shell equivalence holds.) Since SU(3) is a rank 2 group it has two Casimir invariants, one of degree two in the generators, which we label $\mathbf{C}_2$, and one of degree three which we label $\mathbf{C}_3$. The eigenvalues of these operators are related to the irreducible representation labels $(\lambda\mu)$ in a simple way,

$$\langle \mathbf{C}_2\rangle^{(\lambda\mu)} = \lambda^2+\lambda\mu+\mu^2+3\lambda+3\mu$$

$$\langle \mathbf{C}_3\rangle^{(\lambda\mu)} = 2\lambda^3+3\lambda^2\mu-3\lambda\mu^2-2\mu^3+9\lambda^2-9\mu^2+9\lambda-9\mu \tag{11}$$

Three labels are required to specify states within an irreducible representation. When SU(3) is reduced with respect to SO(3), the angular momentum and its projection provide two of these three labels. One additional state-labelling operator is necessary.

Racah recognized that there are two candidates for resolving the SU(3) → SO(3) state-labelling problem, $\mathbf{X}_3=(\mathbf{LQL})^0$ and $\mathbf{X}_4=(\mathbf{LQQL})^0$. But neither he nor his students were able to find a linear combination of these operators that has simple eigenvalues and yields an orthonormal resolution of multiply occurring L values.[22] In addition to this state-labelling problem, SU(3) is not simply reducible; that is, in the reduction of a product of two SU(3) representations a given representation may occur more that once. These two features make it the simplest example of the situation that is usually encountered in algebraic theories. For these technical reasons and other physical ones, there has been an abundance of theoretical work on the SU(3) → SO(3) group structure.[23] Only SO(3), which is simply reducible and multiplicity free, has been better studied.[24]

Perhaps the most significant of the theoretical developments relates to the introduction and use of the integrity basis concept. It provides a simple answer to the question, What is the complete set of state-labelling operators? Or, stated differently and in a way that elevates the question beyond the state-labelling problem, What SO(3) scalars can be built out of the SU(3) generators? Each such scalar or combination of scalars is a candidate for a state-labelling operator. In this form the question can be identified as a classic problem of group theory that is addressed, for example, in the work of Molien, Noether, and Weyl.[25] The solution is simple: all the SO(3) scalar operators that can be built out of the generators of SU(3) can be expressed as polynomial functions of a finite subset of the SU(3) → SO(3) scalars. That set is called the SU(3) → SO(3) integrity basis.

The SU(3) → SO(3) integrity basis consist of six operators.[26] These can be chosen to be the second and third order Casimir invariants of SU(3), C_2 and C_3, the Casimir invariant of SO(3), L^2, and three non-Casimir invariant scalar operators, one of degree three in the generators of SU(3), $\mathbf{X}_3 \equiv (\mathbf{L}\times\mathbf{Q}\times\mathbf{L})^0 \equiv (\mathbf{LQL})^0$, one of degree four, $\mathbf{X}_4(k) \equiv [(\mathbf{L}\times\mathbf{Q})^k \times (\mathbf{Q}\times\mathbf{L})^k]^0$ with $\mathbf{X}_4(k=1) \equiv \mathbf{X}_4 \equiv (\mathbf{LQQL})^0$, and an antihermitian one of degree six that can be represented by the commutator of X_3 and X_4, $X_6 \equiv [X_3, X_4]$. The square of X_6, which is necessarily hermitian, has a polynomial representation in terms of the other integrity basis operators so in any polynomial expansion $\mathbf{X}_6$ should only enter linearly.

In probing how rotations might emerge in an SU(3) → SO(3) algebraic theory, we had to ask and answer the question, What is the most general SO(3) scalar interaction the SU(3) algebra supports?[27] Clearly this is the same question as the one raised in seeking a resolution of the state-labelling problem. The answer is a polynomial function of the integrity basis operators,

$$H = \sum_{abcde} h_{abcde} \, (L^2)^a \, \mathbf{X}_3^b \, \mathbf{X}_4^c \, \mathbf{C}_2^d \, \mathbf{C}_3^e. \qquad (12)$$

Note that the antihermitian operator $\mathbf{X}_6$ is not included because we have assumed time reversal invariance applies. Now within a single representation of SU(3) the Casimir invariants C_2 and C_3 act as a simple multiple of the identity; they only contribute to a shift in the centroid energy of the representation. The effective interaction can therefore be written as a polynomial function of just three operators, $\{L^2, \mathbf{X}_3, \mathbf{X}_4\}$. For an $\mathbf{H}$ of maximum degree k in the generators of SU(3) we have

$$\mathbf{H}(k) = \sum_{2a+3b+4c\leq k} h_{abc} \, (L^2)^a \, \mathbf{X}_3^b \, \mathbf{X}_4^c. \qquad (13)$$

We have published results which show that there exists a fourth order SU(3) → SO(3) integrity basis interaction that reproduces the eigenvalue spectrum of an axially symmetric rotor hamiltonian,[28]

$$\frac{1}{2\partial}\, \mathbf{I}^2 + \left(\frac{1}{2\partial_3} - \frac{1}{2\partial}\right) \mathbf{I}_3^2 \equiv H_{SYM} \longleftrightarrow H_{SU3} \equiv a\mathbf{L}^2 + b\mathbf{X}_3 + c\mathbf{X}_4 + d\mathbf{L}^4. \qquad (14)$$

Actually, the mapping can be established without including the L^4 term in H_{SU3}. It represents a centrifugal stretching or antistretching "correction" that is only present in a higher-order collective model theory.

Before proceeding further with the derivation of analytic results for mapping between collective (macroscopic) and algebraic (microscopic) rotor hamiltonians, it seems appropriate to give explicit results for matrix elements of the integrity basis operators for then regardless of the form chosen for H, because of (12), its matrix elements can be evaluated,[29]

$$<(\lambda\mu)\kappa'LM|\mathbf{L}^2|(\lambda\mu)\kappa LM> = L(L+1)\delta_{\kappa',\kappa}$$

$$<(\lambda\mu)\kappa'LM|\mathbf{X}_3|(\lambda\mu)\kappa LM> = L(L+1)\sqrt{3(2L+1)}W(L,1,L,1;L,2) \times <(\lambda\mu)\kappa'L\|\mathbf{Q}\|(\lambda\mu)\kappa L>$$

$$<(\lambda\mu)\kappa'LM|\mathbf{X}_4|(\lambda\mu)\kappa LM> = L(L+1)\sqrt{(2L+1)} \times \sum_{\kappa''L''}(-1)^{L+L''+1}\sqrt{(2L''+1)}[W(1,L,2,L'';L,2)]^2 \times <(\lambda\mu)\kappa'L\|\mathbf{Q}\|(\lambda\mu)\kappa''L''><(\lambda\mu)\kappa''L''\|\mathbf{Q}\|(\lambda\mu)\kappa L>. \quad (15)$$

In (15), W denotes an SO(3) Racah coefficient and the double-barred matrix elements of the quadrupole operator are given by

$$<(\lambda\mu)\kappa'L\|\mathbf{Q}\|(\lambda\mu)\kappa L>=(-1)^{\phi}2<\mathbf{C}_2>^{1/2}<(\lambda\mu)\kappa L;(11)12\|(\lambda\mu)\kappa'L>_{\rho=1}. \quad (16)$$

The phase factor ϕ is 1 if $\mu\neq 0$ and 0 if $\mu=0$. As in (11) above, the factor $<\mathbf{C}_2>$ represents the expectation value of the second order Casimir invariant of SU(3) and $<(\lambda\mu)\kappa L;(11)12\|(\lambda\mu)\kappa'L>_{\rho=1}$ is a reduced SU(3) $\rightarrow$ SO(3) Wigner coefficient.[30] The operator $\mathbf{Q}$ is normalized so that $\mathbf{Q}\cdot\mathbf{Q} = 4\mathbf{C}_2 - 3\mathbf{L}^2$. The symbol κ is used in place of the Elliott K label to denote the fact that an orthogonal labelling of the multiple occurrences of L values in the $(\lambda\mu)$ representation of SU(3) is being employed. The κ-scheme preserves as best possible the physical significance of the Elliott state-labelling prescription. In particular, all κ=0 states are pure Elliott K=0 states, a κ=2 state is an Elliott K=2 state with sufficient K=0 admixture to make it orthogonal to the κ=0 state, etc.

There is an obvious choice for the SU(3) $\rightarrow$ SO(3) image of an asymmetric rotor hamiltonian,

$$H_{SU3} = a\mathbf{L}^2 + b\mathbf{X}_3 + c\mathbf{X}_4. \quad (17)$$

This follows from (10) under the substitution $\mathbf{I} \longleftrightarrow \mathbf{L}$ and $\mathbf{Q}^c \longleftrightarrow \mathbf{Q}$. But what about the values of the a,b and c parameters? In the collective model they are given in terms of the inertia parameters of the system. In the SU(3) $\rightarrow$ SO(3) model we have, until now, taken them as parameters to be determined in any particular case by a least squares fit to the data.[31] However, one can do better than that. Analytic expressions for the a,b and c paramters of the SU(3) $\rightarrow$ SO(3) theory can be obtained by establishing a relationship between the λ_α and the representation labels λ and μ of SU(3).

Under the Elliott prescription for projecting states of good angular momentum from the intrinsic highest weight SU(3) state,

$$<\mathbf{Q}_0> = <2\mathbf{N}_3 - \mathbf{N}_1 - \mathbf{N}_2> \sim 2\lambda+\mu$$

$$<2\mathbf{\Lambda}_0> = <\mathbf{N}_1 - \mathbf{N}_2> \sim \mu. \quad (18)$$

In (18), $\mathbf{N}_\alpha$ counts the number of oscillator quanta in the α-th direction. Since $\langle x_\alpha^2\rangle \sim \langle \mathbf{N}_\alpha\rangle$ for the oscillator, it follows that $\langle \mathbf{N}_\alpha\rangle \sim \lambda_\alpha$ and, accordingly,

$$\lambda_1 \sim (-\lambda + \mu)/3 + \sigma_1$$

$$\lambda_2 \sim (-\lambda - 2\mu)/3 + \sigma_2$$

$$\lambda_3 \sim (2\lambda + \mu)/3 + \sigma_3. \qquad (19)$$

In (19) the σ_α are constants that satisfy $\sigma_1+\sigma_2+\sigma_3=0$. To fix the σ_α and hence uniquely specify the relationship between the λ_α and λ and μ, we require that the collective model invariants $\mathrm{Tr}[(\mathbf{Q}^c)^2]$ and $\mathrm{Tr}[(\mathbf{Q}^c)^3]$ go over into invariants of the SU(3) algebra. For this $\sigma_1=0$, $\sigma_2=-1$, $\sigma_3=+1$, in which case

$$\mathrm{Tr}[(\mathbf{Q}^c)^2]=\lambda_1^2+\lambda_2^2+\lambda_3^2 \sim \frac{2}{3}(\lambda^2+\lambda\mu+\mu^2+3\lambda+3\mu+3) = \frac{2}{3}\langle \mathbf{C}_2\rangle + 2 \qquad (20)$$

$$\mathrm{Tr}[(\mathbf{Q}^c)^3]=\lambda_1^2+\lambda_2^2+\lambda_3^2 \sim \frac{1}{9}(2\lambda^3+3\lambda^2\mu-3\lambda\mu^2-2\mu^2+9\lambda^2-9\mu^2+9\lambda-9\mu) = \frac{1}{9}\langle \mathbf{C}_3\rangle.$$

The microscopic equivalent of the rotor hamiltonian can now be given. Care must be exercised in associating the cartesian operators of the rotational model to the spherically coupled tensor operators of the integrity basis: $\mathbf{X}_3 \longleftrightarrow (6/\sqrt{10})\mathbf{X}_3^c$ and $\mathbf{X}_4 \longleftrightarrow -(18/5)\,\mathbf{X}_4^c$. Taking these differences into account we finally have

$$AI_1^2 + BI_2^2 + CI_3^2 \equiv H_{ASR} \longleftrightarrow H_{SU3} \equiv aL^2 + b\mathbf{X}_3 + c\mathbf{X}_4$$

$$A = a - \frac{2}{\sqrt{10}} f_1 b + \frac{2}{15} f_1 \frac{(f_2^2 - f_3^2)}{g_1} c$$

$$B = a - \frac{2}{\sqrt{10}} f_2 b + \frac{2}{15} f_2 \frac{(f_3^2 - f_1^2)}{g_2} c$$

$$C = a - \frac{2}{\sqrt{10}} f_3 b + \frac{2}{15} f_3 \frac{(f_1^2 - f_2^2)}{g_3} c$$

$$a = -\frac{1}{9}\left(\frac{f_2 f_3}{g_2 g_3} A + \frac{f_3 f_1}{g_3 g_1} B + \frac{f_1 f_2}{g_1 g_2} C\right)$$

$$b = \frac{\sqrt{10}}{18}\left(\frac{f_1}{g_2 g_3} A + \frac{f_2}{g_3 g_1} B + \frac{f_3}{g_1 g_2} C\right)$$

$$c = \frac{5}{18}\left(\frac{1}{g_2 g_3} A + \frac{1}{g_3 g_1} B + \frac{1}{g_1 g_2} C\right)$$

where $f_1 = \lambda-\mu$ $\quad g_1 = \lambda+\mu+2$

$$f_2 = \lambda+2\mu+3 \qquad g_2 = -(\lambda+1)$$

$$f_3 = -(2\lambda+\mu+3) \qquad g_3 = -(\mu+1) \tag{21}$$

SOME EXAMPLES

So what? By some formal manipulations we have arrived at what appears to be an SU(3) → SO(3) integrity basis interaction that is the image of the asymmetric rotor hamiltonian. But does the mapping reproduce rotor dynamics? In particular, do eigenstates of H_{SU3} yield rotor values for energies, electromagnetic transition rates, etc.? And what about the D_2 invariance; can the eigenstates of H_{SU3} be classified according to symmetry classes of the Vierergruppe? This seems quite unlikely for even simpler things do not seem to match up. For example, the $(\lambda\mu)$ representation of SU(3) is finite dimensional with the number of occurrences of a specific L value given by the rather awkward formula[32]

$$d(\lambda\mu,L) = [(\lambda+\mu+2-L)/2] - [(\lambda+1-L)/2] - [(\mu+1-L)/2]. \tag{22}$$

In contrast, for the rotor there are 2I+1 states of spin I. The heavy brackets in (22), [ ], denote the greatest integer function. Furthermore, there is no angular momentum cutoff for the rotor whereas for SU(3) the maximum L value in the $(\lambda\mu)$ representation is $\lambda+\mu$. Despite all this, we will now show that H_{SU3} is indeed a true image of H_{ASR}, with eigenstates that reproduce rotor values for observables and even possess D_2 invariance!

First consider the dimension question. It is a simple exercise to verify, using (22), that for both λ and μ even and the angular momentum L less than or equal to the $\min(\lambda,\mu)+1$, the multiplicity of the L value in the $(\lambda\mu)$ representation of SU(3) is (L+2)/2 for even L and (L-1)/2 for odd L. This is the same result as the one that applies for the A-type, D_2 symmetry subspace of the rotor, see Table 1. Similarly, for either λ or μ odd or both λ and μ odd, the dimensions agree with those for the B_α-type, D_2 symmetries. For L values greater than $\min(\lambda,\mu)+1$, the $(\lambda\mu)$ representation produces less states than exist for the rotor. This is a direct consequence of the microscopic underpinnings of the algebraic theory. For example, if the low-lying states of ^{24}Mg are taken to be $(s)^4(p)^{12}(ds)^8$ shell-model configurations, there is only one way to make an L=12, S=0, T=0 state. It is found in the leading $(\lambda\mu)=(8,4)$ representation of SU(3). For the rotor, on the other hand, there are seven I=12 states of A-type symmetry. Six of those seven are blocked by the finite-space/fermion-statistics constraint of the shell-model picture.

In Figure 1, six sets of spectra are given, three for H_{ASR} and three corresponding ones for H_{SU3}. The three sets for the asymmetric rotor are labelled by the asymmetry parameter,

$$\kappa = (2A_1 - A_3 - A_2)/(A_3 - A_2). \tag{23}$$

By convention, $A_2 < A_1 < A_3$. Accordingly, the asymmetry parameter has

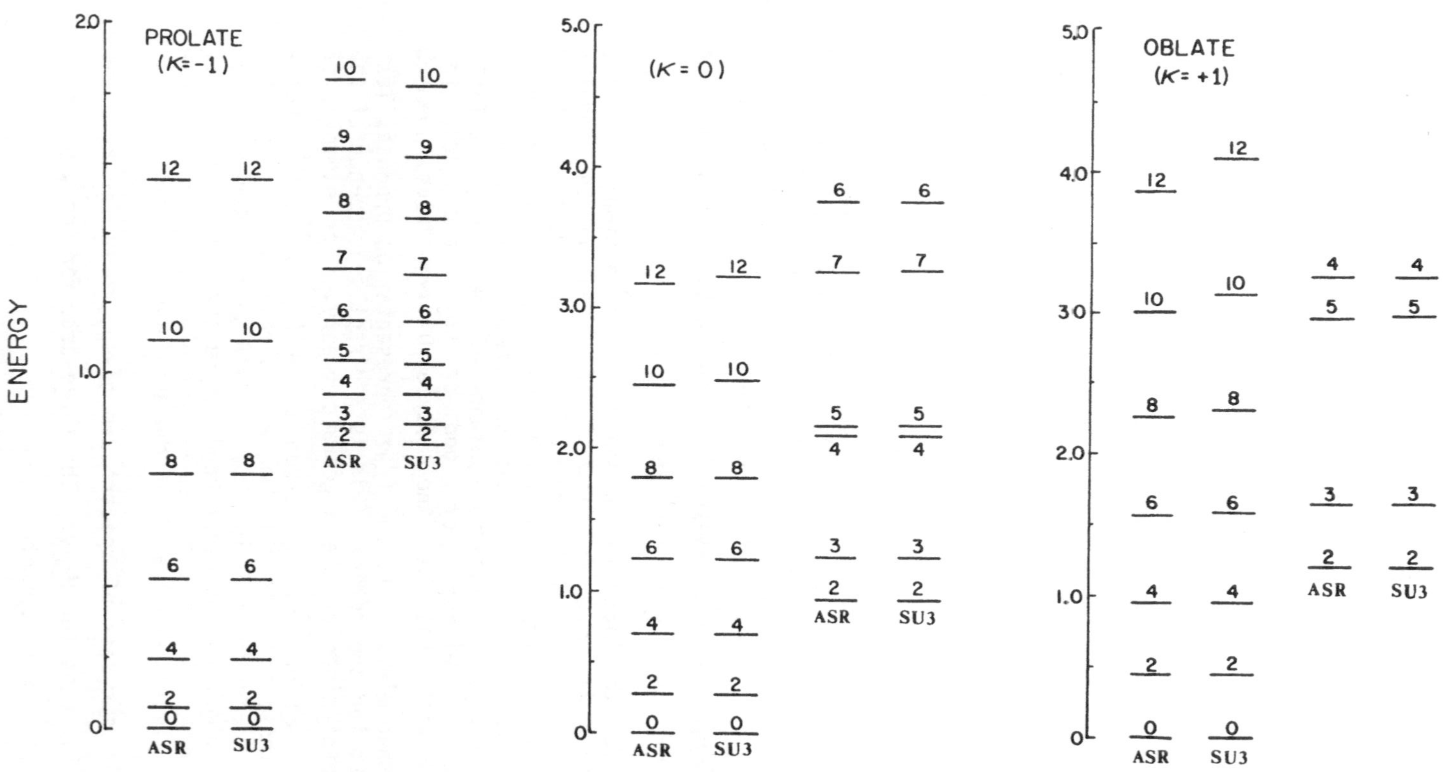

Figure 1. A comparison of the spectra of H_{ASR} and H_{SU3} for prolate, most asymmetric, and oblate rotor geometries: A_2=10 and A_3=200 with A_1=10, 105, 200 corresponding to values κ = -1, 0, +1 for the asymmetry parameter which is defined by $\kappa = (2A_1-A_3-A_2)/(A_3-A_2)$. The algebraic results are for the $(\lambda\mu)$ = (40,8) representation with the interaction parameters determined using the connection formulae. Only for the prolate case is K a good quantum number; the separation of the spectra into bands for the other two cases is only to aid in making a comparison of the spectra.

a value -1 for a prolate shape ($A_2 = A_1 < A_3$), 0 for what can be called the most asymmetric shape ($A_2 < A_1 = (A_2 + A_3)/2 < A_3$), and +1 for an oblate shape ($A_2 < A_1 = A_3$). In the Figures, $A_2=10$ and $A_3=200$ so $A_1=10,105,200$ for the $\kappa = -1,0,+1$ results, respectively. Only for the prolate case ($\kappa=-1$) is K a good quantum number; for the most asymmetric ($\kappa=0$) and oblate ($\kappa=+1$) cases K is not a good quantum number. So the spectra, as shown, are somewhat misleading because the groupings suggest a K=0 and K=2 band structure in all three cases whereas in reality that labelling only applies to the first.

The SU(3) results shown in Figure 1 are for $(\lambda\mu)=(40,8)$ with the parameters a,b,c of H_{SU3} set by the connection formula, (21). For example, for the $\kappa=0$ case with $A=A_1=105$, $B=A_2=10$, $C=A_3=200$ one finds that a=183.03, b=2.9291, c=0.045769. Notice how closely the SU(3) eigenenergies track the rotor results. Whereas one might be tempted to dismiss the prolate results as elementary and fortuitous for there are only really two important features to the spectra, the I(I+1) spacing and the K-band splitting, in the most asymmetric and oblate cases one cannot for the rotor spectra are quite complex and show very little regularity. Nonetheless, the algebraic results reproduce the rotor results with comparable quality in all three cases. In particular, note how accurately the 7-6 and 5-4 "inversions" are reproduced in the $\kappa=0$ and $\kappa=+1$ cases, respectively.

The dimensionality arguments given above suggest, at least for L values less than $\min(\lambda,\mu)+1$, that the λ and μ labels might provide a key to gaining an understanding of the transformation properties of SU(3) $\rightarrow$ SO(3) basis states under the action of generators of the Vierergruppe. To see how this comes about, we draw on known symmetry relations for the transformation coefficients between SU(3) $\rightarrow$ SU(2) x U(1) and SU(3) $\rightarrow$ SO(3) basis states.[33] For SO(3) projection from an intrinsic highest weight SU(3) state one has that

$$\langle(\lambda\mu)\alpha|(\lambda\mu)\text{-}KLM\rangle = (-1)^{\lambda+\mu+L}\langle(\lambda\mu)\alpha|(\lambda\mu)KLM\rangle. \tag{24}$$

Since the $|(\lambda\mu)\alpha\rangle$ form a complete set and the phase factor in (24) is independent of α, it follows that

$$|(\lambda\mu)\text{-}KLM\rangle = (-1)^{\lambda+\mu+L}|(\lambda\mu)KLM\rangle. \tag{25}$$

The SU(3) $\rightarrow$ SO(3) states can therefore be reorganized to have the same transformation properties under the action of the generators of D_2 as eigenstates of the symmetric rotor,

$$|(\lambda\mu)|K|LM\rangle \equiv \frac{1}{\sqrt{2(1+\delta_{K0})}}\left(|(\lambda\mu)KLM\rangle+(-1)^{\lambda+\mu+L}|(\lambda\mu)\text{-}KLM\rangle\right). \tag{26}$$

And just as for the rotor, the odd-even character of λ and μ then dictate the D_2 symmetry class to which the SU(3) $\rightarrow$ SO(3) eigenstates belong, see Table 2.

To provide further support for this association between the odd-even character of the λ and μ values and D_2 rotor symmetry types, we calculated and compared sum-rule measures for eigenenergies of H_{SU3} and H_{ASR}. The results are given in Figure 2(a,b). The solid curves are $Tr(H_{ASR})_I/I(I+1)$ which, according to the results given in Table

3, depend linearly on I. The symbols label the corresponding H_{SU3} measures for selected $(\lambda\mu)$ values. (The $(\lambda\mu)=(30,8)$ representation is the leading one in a pseudo SU(3) description of the rare earth nucleus ^{168}Er. The other $(\lambda\mu)$ values are its even-odd, odd-even, and odd-odd neighbors.) In each case the results shown are for the most asymmetric rotor geometry ($\kappa=0$) with a,b,c determined by the mapping formulae, (21). The agreement is remarkable; for $L<\min(\lambda,\mu)+1$ the deviations fall below the 1% level. This can also be seen in Figure 3(a,b) where results for centroids and variances of the rotor and algebraic, $(\lambda\mu)=(30,8)$, theories are compared. For $L<\min(\lambda,\mu)+1=9$, the agreement is again excellent, better for the centroids than for the variances. This is to be expected for the variance is a measure of the spread in the eigenvalues of a spectrum about the centroid which serves to fix its location. Note that $\sigma^2_{SU3} \to 0$ for L=37 and L=38 as is required for each of these L values occurs only once in the $(\lambda\mu)=(30,8)$ representation.

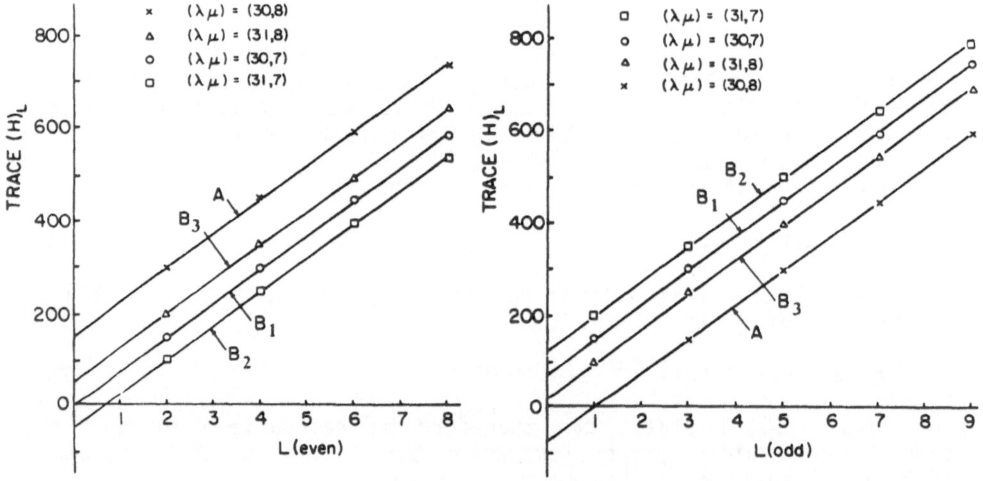

Figure 2. A comparison of rotor and algebraic results for traces [divided by L(L+1)] of the interaction H_{ASR} and H_{SU3} in the A,B_α-type symmetry subspaces of D_2. The same set of inertia parameters were used in all four cases with $(\lambda\mu)$ values chosen as indicated. The parameters of H_{SU3} were fixed from those of H_{ASR} by using the connection formulae. Whereas algebraic results exist for H_{ASR} (Table 3) the H_{SU3} numbers were determine by first calculating the various L-submatrices and explicitly evaluating their traces.

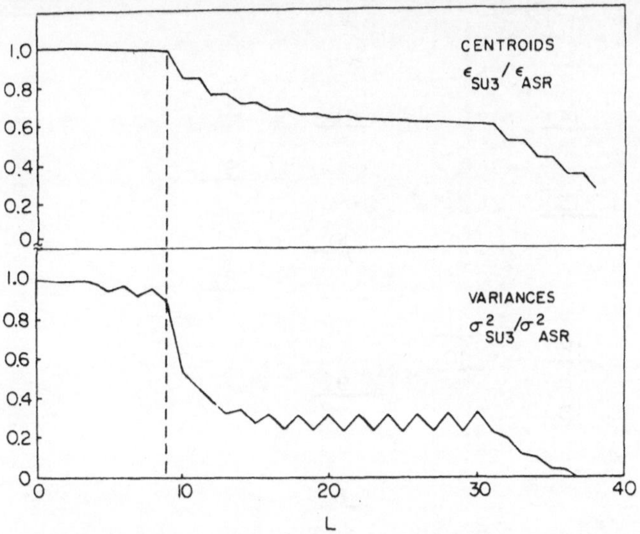

Figure 3. A comparison of centroid and variance measures for the asymmetric rotor (κ=0) and its algebric [($\lambda\mu$) = (30,8)] image. Up to L=min(λ,μ)+1=9, every state of the rotor has an algebraic image. The fall-off for L>9 is a direct consequence of "missing" states in the algebraic theory. Note that for L=37 and L=38 the variance, which is a measure of the spread in eigenvalues, goes to zero as it must for there is but one of each of those states in the (30,8) representation.

In Figure 4 a typical spectrum for $H_{ASR}(\kappa=0)$ is shown. The corresponding results for H_{SU3}, with ($\lambda\mu$)=(8,4) and the a,b,c parameters determined using the connection formulae (21), are also given. A comparison of the two gives a good indication of both the accuracy and limitations of the mapping procedure. All states of the (8,4) representation were included in generating the SU(3) eigenvalue spectrum: $L=0^1,2^2,3^1,4^3,5^2,6^3,7^2,8^3,9^2,10^2,11^1,12^1$. These states are a shell-model image of those associated with the eigenvalues enclosed in the box on the rotor spectrum. Note that for the ground state band, which terminates at L=12 in the SU(3) theory, the agreement between eigenenergies is almost perfect. The agreement is also excellent for the lowest members of the second band (L $\lesssim$ 7). Bigger differences are found between eigenenergies of excited states of the third band. Note also that it is only for L value up to five

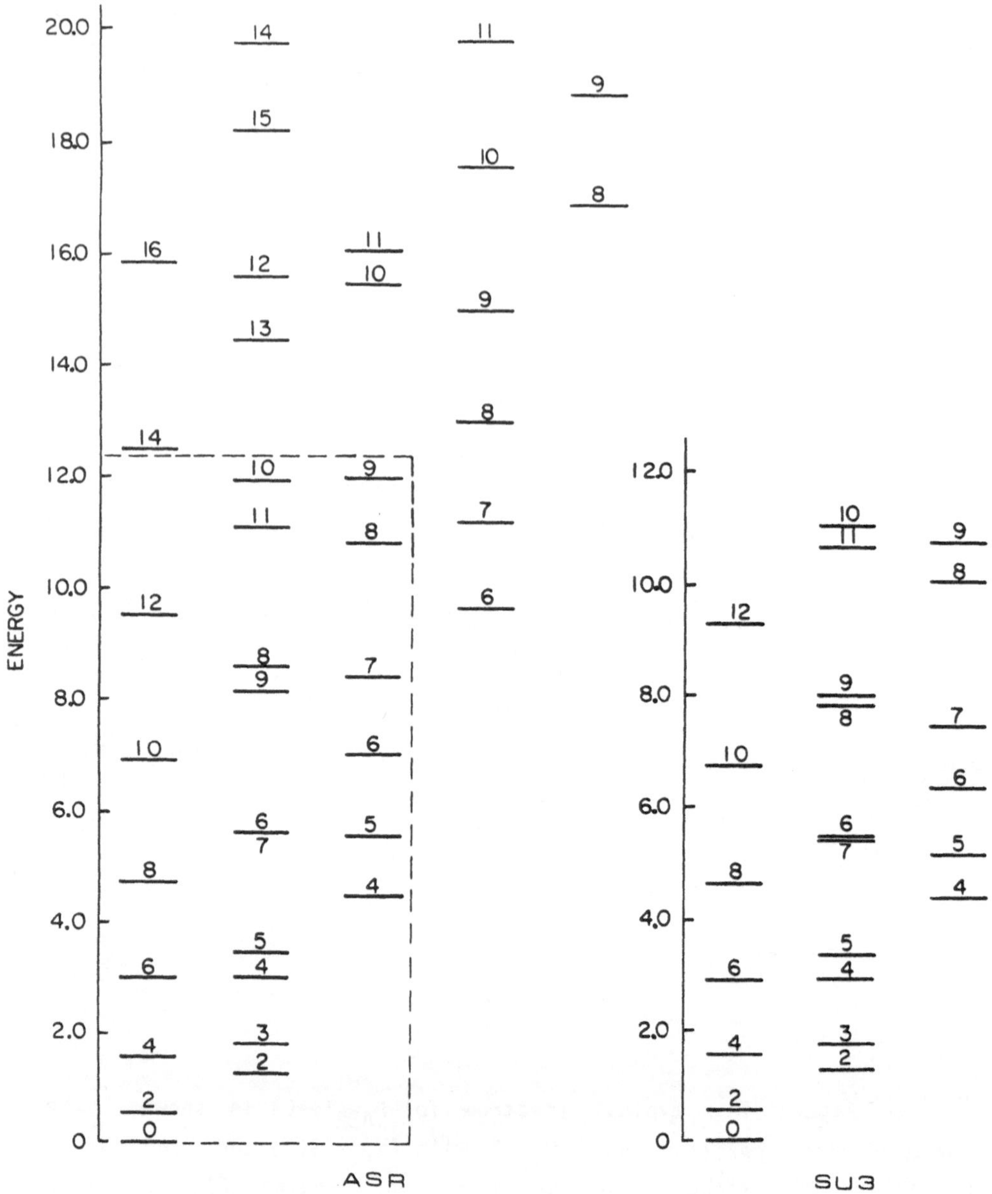

Figure 4. Asymmetric rotor (κ=0) excitation spectra and its algebraic image for the $(\lambda\mu)$=(8,4) representation of SU(3). Only rotor levels within the broken-boxed area have an SU(3) image in the (8,4) representation. Note that the agreement is excellent for ground-band states. It deteriorates slightly with increasing angular momentum in the excited bands with the onset of significant deviations occurring for lower angular momenta in the higher bands.

that the (8,4) representation accounts for all the L values of the rotor; the rotor has four L=6 states whereas in the (8,4) representation there are only three, etc. These results are typical: the finite space constraint of an SU(3) theory results in deviations from rotor values for observables that increase with increasing L within each band with the onset occuring at lower L values in the higher bands. But even for a relatively small representation like the (8,4), which is the leading one in a $(s)^4(p)^{12}(ds)^8$ shell-model description of ^{24}Mg, the algebraic and rotor results agree amazingly well!

As the final example, we now give some results for ^{24}Mg. The experimental and calculated spectra are shown in Figure 5. Using a nonlinear-least-squares procedure we first determined the parameters of the asymmetric rotor that yielded eigenenergies that best reproduce the experimental excitation energies: A_1=195.79, A_2=195.30, A_3=848.50 (keV). This implies a symmetric (κ=-0.998) rotor configuration for ^{24}Mg. We then used the mapping formulae to determine the parameters of the SU(3) hamiltonian under the assumption that the leading representation is $(\lambda\mu)$=(8,4): a=239.61, b=20.955, c=-1.4370. The results shown are for this SU(3) → SO(3) integrity basis interaction. As already seen in Figure 3 for the most asymmetric case, the H_{SU3} and H_{ASR} results differ very little for states of the (8,4) representation with L≤7. It is for that reason the rotor results are not shown. The fit to the data, determined in this way, is quite reasonable. However, one can do better by applying the

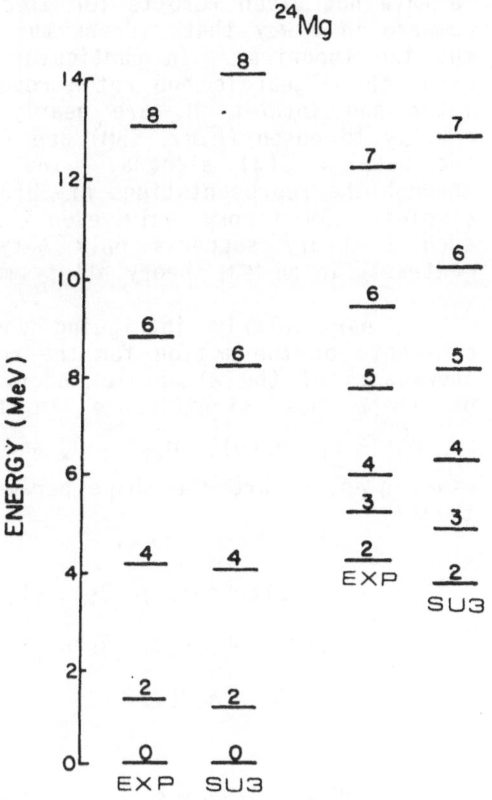

Figure 5. A comparison of experimental (EXP) and theoretical (SU3) energy spectra for the ground band and gamma band states of ^{24}Mg. The theoretical spectra is the result of applying the connection formulae for the $(\lambda\mu)$ = (8,4) representation to inertia parameters of the rotor where the latter were determined by a best fit to the data.

least-square procedure directly to the SU(3) theory without going through the intermediate steps of fitting to the rotor and using the connection formulae.[34] This suggests that the SU(3) picture, while it can be used to reproduce rotor results, actually goes beyond it!

CONCLUSION

We have shown that the SU(3) → SO(3) group structure provides for an algebraic realization of the rotor. Given the three inertia parameters of the rotor and the representation labels λ and μ of SU(3), the connection formulae, (21), can be used to determine the parameters of an SU(3) → SO(3) integrity basis interaction that "reproduces" rotor results when acting in the $(\lambda\mu)$ space. (Although we have not given results for electromagnetic transition rates, they compare in a way that mirrors the agreement between eigenenergies of the two theories. In particular, for low L values of the lowest bands the algebraic and rotor results for E2 transition rates, both intra and inter-band, are nearly identical.) The theory applies equally to boson (e.g., IBM) and fermion (e.g., MCM) realizations of the SU(3) → SO(3) algebra. The difference in the physics enters through the representations of SU(3) that occur. For example, in the simplest IBM theory only even λ and μ values enter. Accordingly, such a theory supports only A-type, D_2 rotor configurations. In contrast, in an MCM theory all symmetry types can be realized.

A particularly intriguing and important feature is that the constants of the motion for the rotor are carried over into Casimir invariants of the algebraic theory. This suggests that the concept of shape has significance in both pictures. In particular, if $\langle Q_0^c\rangle \sim k\beta\cos(\gamma)$, $\langle Q_{\pm1}^c\rangle \sim 0$, and $\langle Q_{\pm2}^c\rangle \sim 1/\sqrt{2}\,k\beta\sin(\gamma)$, where as usual β and γ are the shape parameters of the hydrodynamical model, then[35]

$$
\begin{aligned}
\lambda_1 &\sim -(1/2)k\beta[\cos(\gamma)-\sqrt{3}\sin(\gamma)] \sim -(\lambda-\mu)/3\\
\lambda_2 &\sim -(1/2)k\beta[\cos(\gamma)+\sqrt{3}\sin(\gamma)] \sim -(\lambda+2\mu+3)/3\\
\lambda_3 &\sim \quad k\beta\cos(\gamma) \qquad\qquad\qquad \sim +(2\lambda+\mu+3)/3
\end{aligned}
\tag{27}
$$

and

$$
\begin{aligned}
\mathrm{Tr}[(Q^c)^2] &\sim (3/2)k^2\beta^2 \sim (2/3)\langle C_2\rangle + 2\\
\mathrm{Tr}[(Q^c)^3] &\sim (3/4)k^3\beta^3\cos(3\gamma) \sim (1/9)\langle C_3\rangle.
\end{aligned}
\tag{28}
$$

In this way β and γ are seen to be determined by $\langle C_2\rangle$ and $\langle C_3\rangle$ and are therefore the same for all members of a representation.

We have said very little about how this fits into a shell-model picture of nuclear structure. To see how this goes, consider the Elliott model. The full shell-model space is first partitioned into it space [irreducible representations of U(N(N+1)/2) for the N-th oscillator shell] and supermultiplet [irreducible representations of U(4)] parts. The space part is then further partitioned into irreducible representations of SU(3) and its SO(3) subgroup to gain the $(\lambda\mu)$ and L state labels. The space geometry is therefore given by the group chain U(N(N+1)/2) → SU(3) → SO(3) with the corresponding state labels $[f]\alpha(\lambda\mu)\kappa L$. Likewise, the supermultiplet symmetry must be reduced with respect to its spin-isospin subgroup to gain the S and T state labels, $U(4) \rightarrow SU_S(2) \times SU_T(2)$ which yields the state labels $[\tilde{f}]\beta ST$. Antisymmetric states of good total angular momentum are formed by coupling [f] and $[\tilde{f}]$ to $[1^n]$ and L and S to J,

$$|\Psi\rangle \longrightarrow |[1^n];[f]\alpha(\lambda\mu)\kappa L;[\tilde{f}]\beta STM_T;JM_J\rangle. \tag{29}$$

In this way the full space is partitioned into representations of SU(3). If the true hamiltonian for the many-body system has the form

$$H = H_{SU3} + H' \tag{30}$$

with the model space norm of H' small compared to that of H_{SU3}, the shell-model dynamics will appear to be that of a collection of rotors interacting through the H' term.

An interesting feature of this scenario is that if H is truly a microscopic interaction, the H_{SU3} part of H will have the same form in all representations of SU(3). The connection formulae can then be used to determine the effective inertia parameters for the various $(\lambda\mu)$ that occur. Suppose, as an example, that in the ^{24}Mg case H gives rise to an H_{SU3} that corresponds to the most asymmetric shape for the leading $(\lambda\mu)$=(8,4) representation: a=49.400, b=0.82264, c=-0.41887 for which A=50, B=100, and C=150 so κ=0. That same interaction will produce different asymmetry measures in other SU(3) representations. Examples are given in Table 4. The end result is that if H can be written in the form suggested by (30), then the shell model can be viewed as a collection of interacting rotors of mixed symmetry types and different inertia parameters. Of course, if H' dominates, the rotor features will be lost. But there is good evidence suggesting that it does not, and in that case the next task is to discover how to extract H_{SU3} and H' from H and give appropriate measures for their relative importance. In this the methods of statistical spectroscopy can be brought to bear.[35]

Table 4. Inertia and asymmetry parameters of an SU(3) → SO(3) integrity basis interaction for which κ=0 in the $(\lambda\mu)$=(8,4) representation. The representations given are important ones in the leading space symmetry of ^{24}Mg.

(λ,μ)	A_1	A_2	A_3	κ	D_2 Symmetry
(8,4)	100	50	150	0.00	A
(7,3)	84	50	127	-0.12	B_1
(8,1)	71	54	127	-0.53	B_2
(4,6)	100	51	107	+0.75	A
(5,4)	84	49	107	+0.21	B_3

Of course there are also other open questions. For example, the simplicity of our results suggest that it might be possible to find analytic expressions for matrix elements of the integrity basis operators.[36] Also, it should be clear that the SU(3) → SO(3) algebra offers more than just rotor dynamics; we have not explored features that may enter with the inclusion of higher-order terms in H_{SU3}. Obviously the L^4 term that is usually associated with centrifugal distortions in the collective model is there, but what about terms like $L^2\mathbf{X}_3$ or $\mathbf{X}_3\mathbf{X}_4$, what do they do? And since the eigenfunctions of the asymmetric rotor are Lamé functions, can more be said about analytic forms for basis states of SU(3)? In short, it seems that that SU(3) remains a fertile field for future farming!

REFERENCES

1. F. Klein and A Sommerfeld, "Theorie des Kreisels," Volumes 1-4, Teubner, Leipzig (1897-1910);
   H. Goldstein, "Classical Mechanics," Addison-Wesley Reading, Massachusetts (1980).
2. D. M. Dennison, Phys. Rev. 28:318 (1926);
   F. Reiche and H. Rademacher, Zeits. f. Physik 39:444 (1926); 41:453 (1927);
   R. de L. Kronig and I. I. Rabe, Phys. Rev. 29:262 (1927);
   C. Manneback, Phys. Zeits. 28:72 (1927).
3. A. R. Edmonds, "Angular Momentum in Quantum Mechanics," Princeton, New Jersey (1957).
4. E. Witmer, Proc. Nat. Acad. 13:60 (1927);
   S. C. Wang, Phys. Rev. 34:243 (1929);
   H. A. Kramers and G. P. Ittmann, Zeits. f. Physik 53:553 (1929); 58:217 (1929); 60:663 (1930);
   O. Klein, Zeits. f. Physik 58:730 (1929).
5. R. S. Mulliken, Rev. Mod. Phys. 3:89 (1931);
   D. M. Dennison, Rev. Mod. Phys. 3:280 (1930).
6. R. S. Mulliken, Phys. Rev. 59:873 (1941);
   G. W. King, R. M. Hainer, and P. C. Cross, J. Chem. Phys. 11:27 (1943);
   C. H. Townes and A. L. Schawlow, "Microwave Spectroscopy," McGraw-Hill, New York (1955).
7. H. B. G. Casimir, "Rotation of a Rigid Body in Quantum Mechanics," J. B. Wolter's, The Hague (1931).
8. A. Bohr and B. R. Mottelson, Phys. Rev. 89:316 (1953); 90:717 (1953).
9. A. Faessler, W. Greiner, and R. K. Sheline, Nucl. Phys. 70:33 (1965).
10. J. M. Eisenberg and W. Greiner, "Nuclear Theory," Volumes 1, 2, & 3, North-Holland, Amsterdam (1975-1976);
    A. Bohr and B. Mottelson, "Nuclear Structure," Volumes I & II, Benjamin, Reading, Massachusetts (1969-1975).
11. M. G. Mayer, Phys. Rev. 75:1969 (1949); 78:16 (1950);
    O. Haxel, J. H. D. Jensen, and H. Suess, Phys. Rev. 75:1766 (1949); Zeits. f. Physik 128:295 (1950).
12. D. L. Hill and J. A. Wheeler, Phys. Rev. 89:1102 (1953).
13. J. P. Elliott, Proc. Roy. Soc. A245:128,562 (1958).
14. A. Arima and F. Iachello, Ann. of Phys. 99:253 (1976); 111:201 (1978); 123:468 (1979).
15. G. Rosensteel and D. J. Rowe, Ann. of Phys. 123:36 (1979); 126:198,343 (1980).
16. J. P. Elliott and M. Harvey, Proc. Roy. Soc. A272:557 (1963);

J. P. Elliott and C. E. Wilsdon, Proc. Roy. Soc. A302:509 (1968).
17. R. D. Ratna Raju, J. P. Draayer and K. T. Hecht, Nucl. A202:433 (1973).
18. A. S. Davydov and G. F. Fillippov, Nucl. Phys. 8:237 (1958);
A. S. Davydov and A. A. Chaban, Nucl. Phys. 20:499 (1960);
J. P. Davidson, Nucl. Phys. 33:664 (1962);
J. P. Davidson, Rev. Mod. Phys. 37:105 (1965);
A. S. Davydov, "Quantum Mechanics," NEO Press, Ann Arbor, Michigan (1966).
19. A. Bohr, Mat. Fys. Medd. Dan. Vid. Selsk. 26:No.14 (1952);
A. Bohr and B. R. Mottelson, Mat. Fys. Medd. Dan. Vid. Selsk. 27: No.16 (1953);
C. Marty, Nucl. Phys. 1:58 (1956); 3:193 (1957).
20. S. A. Williams, Nucl. Phys. 63:581 (1965).
21. G. Racah, Group Theory and Spectroscopy, in: CERN reprint of lectures delivered at the Institue for Advanced Study, Princeton, New Jersey (1951);
G. Racah, Lectures on Lie Groups, in: "Farkas Memorial Volume," A. Farkas and E. P. Wigner, eds., Research Council of Israel, Jerusalem (1952);
G. Racah and I. Talmi, Phys. Rev. 89:913 (1953).
22. R. N. Sen, "Construction of Irreducible Representations of SU(3)," Thesis, Hebrew University, Jerusalem (1963).
23. L. C. Biedenharn, Phys. Lett. 28B:537 (1969);
J. A. Castilho Alcaras, L. C. Biedenharn, K. T. Hecht, and G. Neely, Ann. Phys. 60:85 (1970);
J. W. B. Hughes, J. Phys. A6:48,281 (1973);
M. Moshinsky, J. Patera, R. T. Sharp, and P. Winternitz, Ann. Phys. 95:139 (1975);
H. E. DeMeyer, G. Vanden Berghe, J. W. B. Hughes, J. Math. Phys. 22:2360,2366 (1981); 24:1025(1983).
24. D. M. Brink and G. R. Satchler, "Angular Momentum," Clarendon Press, Oxford (1962).
25. T. Molien, Preuss. Akad. Wiss. Sitzungberichte 1152 (1897);
E. Noether, Math. Ann. 77:89 (1916);
H. Weyl, "The Classical Groups," Princeton, New Jersey (1946).
26. B. R. Judd, W. Miller Jr., J. Patera, and P. Winternitz, J. Math. Phys. 15:1787 (1974).
27. J. P. Draayer, Spectroscopy from the Bottom Up, in: "VIII Symposium on Nuclear Physics," Proceedings, Institute de Fisica, U.N.A.M., Oaxtepec, Morelos, Mexico (1985).
28. Yorck Leschber and J. P. Draayer, Phys. Rev. C33:749 (1986).
29. J. P. Draayer and G. Rosensteel, Nucl. Phys. A439:61 (1985).
30. Y. Akiyama and J. P. Draayer, Comp. Phys. Comm. 5:405 (1973).
31. J. P. Draayer and K. J. Weeks, Ann. Phys. 156:41 (1984).
32. G. Racah, Rev. Mod. Phys. 21:494 (1949);
J. P. Draayer and S. A. Williams, Nucl. Phys. A119:577 (1968).
33. J. P. Draayer and Y. Akiyama, J. Math. Phys. 14:1904 (1973).
34. J. P. Draayer, Algebraic Methods and the Microscopic/Macroscopic Connection for Nuclear Rotational Motion, in: "IX Symposium on Nuclear Physics," Proceedings, Institute de Fisica, U.N.A.M., Oaxtepes, Morelos,Mexico (1986).
35. J. B. French, Phys. Lett. B26:75 (1967);
F. S. Chang, J B. French, and T. H. Thio, Ann. of Phys. Ann. Phys. 66:137 (1971).
36. H. DeMeyer, G. Vanden Berghe, and J. Van der Jeugt, J. Math. Phys. 26:3109 (1985).

SPECIAL FUNCTIONS AND REPRESENTATIONS OF su(2)

Philip Feinsilver
Mathematics Department
Southern Illinois University
Carbondale, IL 62901

INTRODUCTION

Here we present a general approach to integrating representations of a Lie algebra that are explicitly presented acting on quotients of the universal enveloping algebra (or as equivalent induced representations). The idea is to think of "group elements" as "non-commutative generating functions" for the representations. One can then shift from non-commutative elements acting on non-commutative elements to non-commutative operators acting on (functions of) commutative variables. The representation so transformed is thus typically presented in terms of differential operators and the usual boson operator calculus can be applied for performing computations.

We focus in this presentation on some examples drawn from work of Gruber and Klimyk [2]. Initially, however, the universal (or "master") representation, that is, the representation on the universal enveloping algebra, $\mathcal{U}$, will be presented, using a slightly different presentation of the Lie algebra. We then illustrate the quotient representation corresponding to a usual irreducible representation. Then two examples from [2] are presented to illustrate further our approach and to show some of the interesting phenomena in even some of the simplest situations. We remark that the representations that arise are in general "indecomposable representations" not necessarily (though not necessarily not) irreducible.

METHOD: General Approach

We briefly outline the main steps in our procedure. The initial information consists of the commutation rules for the Lie algebra in terms of a chosen (thence fixed) basis. A basis for the universal enveloping algebra $\mathcal{U}$ is thus the set of ordered monomials in the given basis (après Poincaré-Birkhoff-Witt).

Note: Examples are subsequent. The outline to follow is not as obscure as it may initially seem.

Step 1. Compute the universal representation. This is the action of the basis elements on monomials in $\mathcal{U}$. For a quotient representation, one reduces to the appropriate basis first.

Step 2. Convert to the dual representation (or "Fourier transform") in terms of (generally) boson operators. This representation is contravariant, as Lie brackets are reversed.

Step 3. Compute exponentials for the dual representation using Heisenberg/Hamilton's equations.

Step 4. Interpret the group law (or representation) as a group action on functions of commuting variables. Products are computed in reverse order (cf. Step 2) to compute the group law or group action.

Step 5. Expanding in series and collecting terms one finds the matrix elements for the group action in terms of special functions.

We proceed to illustrate this approach.

Notation. The operators ℓ_+, ℓ_-, ℓ_3 of [2] satisfy $[\ell_+,\ell_-] = \ell_3$, $[\ell_3,\ell_\pm] = \pm\ell_\pm$. Δ, R, ρ used below may be expressed as $R = \sqrt{2}\ell_+$, $\Delta = -\sqrt{2}\ell_-$, $\rho = 2\ell_3$.

UNIVERSAL REPRESENTATION

We choose the basis Δ, R, ρ satisfying $[\Delta,R] = \rho$, $[\rho,R] = 2R$, $[\Delta,\rho] = 2\Delta$. As a preliminary step it is convenient to compute the action on one-variable polynomials.

Proposition 1. For any polynomial f:

1. $[\Delta,f(R)] = f'(R)\rho + Rf''(R)$
2. $[f(\Delta),\rho] = 2\Delta f'(\Delta)$
3. $[\rho,f(R)] = 2Rf'(R)$.

Proof: 2 and 3 follow readily: e.g., $\rho R = R(\rho+2)$ implies $\rho R^n = R^n(\rho+2n)$, i.e., $[\rho,R^n] = 2R(R^n)'$. Then 1 follows (from 3) using the general formula (in any associative algebra) $[a,b^n] = \sum_{j=0}^{n-1} b^{n-1-j}[a,b]b^j$ with $a = \Delta$, $b = R$. □

As a basis for $\mathcal{U}$ it is convenient to choose, $\ell,m,n \geq 0$, $X(\ell mn) = R^\ell \rho^{(m)}\Delta^n$, where $\rho^{(m)} = \rho(\rho-1)\dots(\rho-m+1)$. Thus the generating function

$$g(abc) = \sum_{\ell,m,n} a^\ell(b-1)^m c^n X(\ell mn)/\ell!m!n!$$

$$= e^{aR}b^\rho e^{c\Delta} \quad \text{(using the binomial theorem for the m-sum)}$$

This is a group element $\exp(aR)\exp(\beta\rho)\exp(c\Delta)$, with $\beta = \log b$. These formulas are most conveniently interpreted in the sense of formal power series.

The dual representation in general is obtained by considering g(abc) as the bilinear pairing $\langle Y,X\rangle$ of two (formal) infinite vectors with components $Y(\ell mn) = a^\ell(b-1)^m c^n$ and $X(\ell mn)$. The action of a typical basis element has an adjoint action $\hat{\xi}$ defined via $\langle Y,\xi X\rangle = \langle\hat{\xi}Y,X\rangle$; $\hat{\xi}$ is generally a differential operator. In our present case we find, using Proposition 1, (∂_x denoting differentiation with respect to x)

$$\hat{R} = \partial_a,\quad \hat{\rho} = 2a\partial_a + b\partial_b,\quad \hat{\Delta} = a^2\partial_a + ab\partial_b + b^2\partial_c.$$

E.g., to compute $\hat{\rho}$:

$$\rho e^{aR_b}\rho_e^{c\Delta} = e^{aR}\rho b^{\rho} e^{c\Delta} + 2aRe^{aR_b}\rho_e^{c\Delta}$$

$$= (b\partial_b + 2a\partial_a)g(abc).$$

The next step is to compute exponentials in the dual representation. We use Heisenberg/Hamilton's equations. Namely, if $H = H(x,p) = H(x,\partial_x)$, then $x(t) = e^{tH}xe^{-tH}$ and $p(t) = e^{tH}pe^{-tH}$ satisfy $\dot{x} = \frac{\partial H}{\partial p}$, $\dot{p} = -\frac{\partial H}{\partial x}$, where in H we put all derivatives, p's, to the right of all x's. In the present case only $\hat{\Delta}$ is of any difficulty. To compute $e^{t\hat{\Delta}}$ we solve $\dot{a}(t) = a(t)^2$, $\dot{b}(t) = a(t)b(t)$, $\dot{c}(t) = b(t)^2$ with $a(0) = a$, $b(0) = b$, $c(0) = c$. Thus $a(t) = a(1-at)^{-1}$, $b(t) = b(1-at)^{-1}$, $c(t) = c+b^2t(1-at)^{-1}$. Note that $\hat{\rho}$ simply generates scaling transformations: $\lambda^{\rho}g(abc) = g(a\lambda^2, b\lambda, c)$.

We can now easily compute the group law via the action of the dual representation (excuse the notational changes):

$$g(b\mu s)g(a\lambda t) = e^{s\hat{\Delta}}\mu^{\hat{\rho}}e^{b\hat{R}}g(a\lambda t)$$

$$= e^{s\hat{\Delta}}g(a\mu^2+b, \lambda\mu, t)$$

$$= g(b+a\mu^2(1-as)^{-1}, \lambda\mu(1-as)^{-1}, t+\lambda^2 s(1-as)^{-1}).$$

Expanding in series we find the matrix elements of the action of g on the basis $X(\ell mn)$:

$$g(b\mu s) \sum_{j,k,\ell} a^j(\lambda-1)^k t^\ell X(jk\ell)/j!k!\ell!$$

$$= \sum_{j,k,\ell} \sum_{J,K,L} (a^j(\lambda-1)^k t^\ell/j!k!\ell!)\ M(b\mu s)^{JKL}_{jk\ell}\ X(j+J, k+K, \ell+L)$$

where the "shift indices" J,K,L run through $\mathbb{Z}$ as long as the resulting indices of X are nonnegative. $M(b\mu s)$ denotes the matrix corresponding to the group element $g(b\mu s)$. Let me indicate the main steps:

1. Expand

$$\exp((b+a\mu^2(1-as)^{-1})R)\cdot(1+(\lambda\mu(1-as)^{-1}-1))^{\rho}\cdot\exp((t+\lambda^2 s(1-as)^{-1})\Delta).$$

2. Re-expand λ^p, say, via $(\lambda-1+1)^p$.

3. Label exponents: $a^j(\lambda-1)^k t^\ell$ and relabel indices, identifying $R^\ell\rho^{(m)}\Delta^n$ with $R^{j+J}\rho^{(k+K)}\Delta^{\ell+L}$.

4. Finally, collect terms to yield (not quite immediately):

$$M(b\mu s)^{JKL}_{jk\ell} =$$

$$\frac{b^J\mu^{2j}s^L}{J!K!L!}\ \frac{\binom{2L}{k}}{\binom{k+K}{k}}\ (-1)^{k+K}\sum_\beta \frac{(2L+1)_\beta(-k-K)_\beta}{(2L-k+1)_\beta\beta!}\ \mu^\beta\ {}_2F_1\left(\begin{matrix}-j, -j-\beta-L+1\\ J+1\end{matrix}; \frac{bs}{\mu^2}\right)$$

for $J,K,L \geq 0$.

5. To handle negative J,K,L that may arise, we note that it is required that $j+J$, $k+K$, $\ell+L \geq 0$. For hypergeometric functions observe, e.g., that for $C < 0$, integral,

$$\frac{1}{C!}\,{}_2F_1\left({A,B \atop C+1};\, z\right) \to \frac{z^{-C}}{(-C)!}(A)_{-C}(B)_{-C}\,{}_2F_1\left({A-C,B-C \atop -C+1};\, z\right)$$

where, as usual, $(A)_n = A(A+1)\ldots(A+n-1)$, e.g.

Let us illustrate the calculations for the representation induced from $\Delta\Omega = 0$, $\rho\Omega = c\Omega$. That is, the elements Δ, R, ρ act on $\mathcal{U} \otimes_{\mathcal{B}} \Omega$ where $\mathcal{B}$ is generated by ρ, Δ. The basis reduces to $R^n \otimes \Omega$ which we simply write as R^n, noting that the representation becomes $RR^n = R^{n+1}$, $\rho R^n = (c+2n)R^n$, $\Delta R^n = n(c+n-1)R^{n-1}$. To find the action $g(\lambda\mu\nu)R^n$ we employ the function $I_c(aR) = \sum_{n=0}^{\infty} a^n R^n/n!(c)_n$ as generating function since, as is easily checked, the action of Δ on $I_c(aR)$ will simply be multiplication by a. We have:

$$e^{\lambda R}\mu^{\rho}e^{\nu\Delta}I_c(aR) = e^{\lambda R}\mu^{\rho}e^{\nu a}I_c(aR)$$

$$= \sum_{q,r,s} \frac{\lambda^q R^q}{q!}\,\mu^{c+2s}\,\frac{\nu^r a^r}{r!}\,\frac{a^s R^s}{s!(c)_s}\,.$$

Put $r+s = \ell$, $q+s = \ell+L$ so that $q = r+L$. We have terms of the form:

$$\frac{a^{\ell}}{\ell!(c)_{\ell}}\,R^{\ell+L}\cdot\frac{\ell!(c)_{\ell}\lambda^{r+L}}{(r+L)!}\,\mu^{c+2(\ell-r)}\,\frac{\nu^r}{r!}\,\frac{1}{(c)_{\ell-r}(\ell-r)!}$$

The matrix elements are thus

$$\frac{\lambda^L\mu^{c+2\ell}}{L!}\,{}_2F_1\left({-\ell,\, 1-c-\ell \atop L+1};\, \frac{\lambda\nu}{\mu^2}\right)$$

which are essentially Jacobi polynomials.

We continue with two further examples (that will conclude our presentation).

EXAMPLE (see [2], eqs. 6.1, 6.2)

In this section and the next we use the basis $\ell_{\pm}, \ell_3$ with $[\ell_+, \ell_-] = \ell_3$, $[\ell_3, \ell_{\pm}] = \pm\ell_{\pm}$. Here we consider the representation induced from $\ell_3\Omega = c\Omega$, c a complex number. We thus have, on the basis $X(nm) = \ell_+^n\ell_-^m$:

$$\ell_3 X(nm) = (c+n-m)X(nm), \quad \ell_+ X(nm) = X(n+1,m),$$

$$\ell_- X(nm) = X(n,m+1) - n(c-m+(n-1)/2)X(n-1,m).$$

Using exponential generating functions, i.e., $e^{A\ell_+}e^{B\ell_-}$, we have the dual representation:

$$\hat{\ell}_3 = c + A\partial_A - B\partial_B,\quad \hat{\ell}_+ = \partial_A,\quad \hat{\ell}_- = (1+AB)\partial_B - cA - \tfrac{1}{2}A^2\partial_A.$$

It is clear how to exponentiate $\hat{\ell}_3$ and $\hat{\ell}_+$. Observe the following:

Proposition 2. Let $H\psi(x) = 0$, where $H = H(x, \partial_x)$. Then $e^{tH}f(x) = \frac{f(x(t))}{\psi(x(t))}\psi(x)$, where $x(t) = e^{tH}xe^{-tH}$.

Proof: $f(x(t))/\psi(x(t)) = e^{tH}f\psi^{-1}(x)e^{-tH}$. Applying to $\psi(x)$, $e^{-tH}\psi = \psi$, so the result is $e^{tH}f$. □

Thus, to exponentiate $\hat{\ell}_-$ we proceed as follows. Let $H = \hat{\ell}_-$. Then Hamilton's equations are:

$$\dot{A}(t) = -A(t)^2/2, \quad \dot{B}(t) = 1+A(t)B(t)$$

with solution $A(t) = A(1+At/2)^{-1}$, $B(t) = (1+At/2)(t+B(1+At/2))$. Observe that $\psi(A) = A^{-2c}$ satisfies $H\psi = 0$. Thus, $e^{tH}f(A,B) = f(A(t),B(t))\cdot(1+At/2)^{-2c}$. We can thus compute the group action in the dual representation:

$$\exp(u\hat{\ell}_3)\exp(t\hat{\ell}_-)\exp(s\hat{\ell}_+)f(A,B)$$

$$= e^{uc}(1+Ae^u t/2)^{-2c} f(s+Ae^u(1+Ate^u/2)^{-1}, (1+tAe^u/2)(t+Be^{-u}+ABt/2)).$$

One can compute the group action $\exp(s\ell_+)\exp(t\ell_-)\exp(u\ell_3)$ on $X(nm)$ by choosing $f(A,B) = e^{A\ell_+ + B\ell_-}$, expanding the above result, and collecting terms to find the matrix elements as in the previous section.

ANOTHER EXAMPLE (see [2], eqs. 7.8-7.12)

For notational clarity we let $A = \ell_+$, $B = \ell_-$, $C = \ell_3$ acting on a basis X_r^+, X_n^-, $r,n \geq 0$, as follows:

$$AX_r^+ = -r(c+(r+1)/2)X_{r+1}^+, \quad BX_r^+ = X_{r-1}^+, \quad CX_r^+ = (c+r)X_r^+$$

$$AX_n^- = X_{n-1}^-, \quad BX_n^- = (n+1)(c-n/2)X_{n+1}^-, \quad CX_n^- = (c-n)X_n^-.$$

(Here c denotes any complex scalar, Λ in ref. [2]).

Consider the action on $\sum_{n=0}^{\infty} \beta^n X_n^-/n!$. The dual representation is:

$$\hat{A} = I_\beta, \quad \hat{B} = c\partial_\beta\beta\partial_\beta - \beta(1+\beta\partial_\beta/2)\partial_\beta^2, \quad \hat{C} = c - \beta\partial_\beta$$

where I_β denotes integration: $I_\beta\beta^k = \beta^{k+1}/(k+1)$. This is not a very lucid representation. We can transform to a new set of boson operators. First, observe that for boson operators x, D, with $[D,x] = 1$, we can introduce $J = D^{-1}$ on an appropriate module. J is essentially the operator of integration, and a representation of the algebra generated by x, D, J, with $JD = DJ = 1$ and $[J,x] = -J^2$, can be induced from $J\Omega = a\Omega$, $D\Omega = a^{-1}\Omega$, for $a \neq 0$ (think of $\Omega = e^{x/a}$). It is readily checked that xD^2 and J form a pair of boson operators: $[xD^2,J] = 1$. The above representation in terms of the boson operators $\delta = \beta\partial_\beta^2$, $\xi = I_\beta$ becomes $\hat{A} = \xi$, $\hat{B} = \delta(c - \delta\xi/2) + c\xi^{-1}$, $\hat{C} = c-\delta\xi$ which can directly be seen to satisfy the dual commutation relations. We easily see that on ξ^n:

$$\hat{A}\xi^n = \xi^{n+1}, \quad \hat{B}\xi^n = (n+1)(c-n/2)\xi^{n-1}, \quad \hat{C}\xi^n = (c-n-1)\xi^n.$$

In this case, for $c = m/2$ we have $\hat{B}\xi^m = 0$ so that ξ^m becomes an extremal vector and with $\xi^m = \Omega$ we have a representation with basis $\xi^k\Omega$, $k \geq 0$ (cf. [2], pp. 763-764).

Similarly, the action on $\sum_{n=0}^{\infty}\alpha^n X_n^+/n!$ yields the dual representation

$$\hat{A} = -\alpha(c+1+\alpha\partial_\alpha/2)\partial_\alpha^2, \quad \hat{B} = I_\alpha, \quad \hat{C} = c + \alpha\partial_\alpha.$$

In terms of boson operators $\xi = I_\alpha$, $\delta = \alpha\partial_\alpha^2$:

$$\hat{A} = -\delta(c + \frac{1}{2} + \xi\delta),\ \hat{B} = \xi,\ \hat{C} = c + \delta\xi.$$

On ξ^n:

$$\hat{A}\xi^n = -n(n + c + \frac{1}{2})\xi^{n-1},\ \hat{B}\xi^n = \xi^{n+1},\ \hat{C}\xi^n = (c + n + 1)\xi^n.$$

This last representation is virtually the same as that discussed above in the latter part of the section on the universal representation and the group action may be similarly found.

REFERENCES

1. B. Gruber, H. D. Doebner, P. J. Feinsilver, Representations of the Heisenberg-Weyl algebra and group, Kinam, 4:241 (1982).
2. B. Gruber, A. U. Klimyk, Matrix elements for indecomposable representations of complex su(2), JMP 25, 4:755 (1984).
3. Also see the article by Moshinsky in this volume for boson operator techniques.

THE FDS (FERMION DYNAMICAL SYMMETRY) MODEL, NUCLEAR SHELL MODEL AND COLLECTIVE NUCLEAR STRUCTURE PHYSICS*

Da Hsuan Feng

Department of Physics and Atmospheric Science
Drexel University,
Philadelphia, Pennsylvania, 19104

INTRODUCTION

We often hear nowadays that the "fundamental theory of strong interaction physics is QCD". All the models and theories in that field are motivated by it. Nevertheless, due to the enormous complexities of QCD, one still awaits its precise usage in practical calculations. In nuclear structure physics, there is a parallel dilemma. The fundamental theory of nuclear structure is the **shell model.** Let me be precise about this. By the term shell model, I mean the **diagonalization of a nuclear hamiltonian with certain residual "effective" interactions in a truncated many body basis which is constructed from the spherical single particle orbits truncated within, say, one major physical shell.** This is nuclear structure physics' utopia. Unfortunately, while the shell model has proven to be extremely useful for studying properties of nuclei near closed shells, its general practice for systems with large number of valence nucleons has been the major challenge in the nuclear structure physics arena for the last thirty years. In order to use the shell model to study such systems where very " rich physics" presides, further drastic truncation of the space is a must. Therefore, how to further truncate the fermion space is perhaps the central question. In this talk, I hope to convince you that, through twists and turns, the idea of dynamical symmetry has played a pivotal role in this entire effort and that we may now have a glimpse as to how this further truncation of the shell model space can be achieved. We may be at the verge of entering a new stage of nuclear structure physics where one is able to reconcile on the one hand the collective nuclear structure physics with, on the other the spherical shell model.

For nuclei far away from closed shells, they generally exhibit rich collective behaviors. These behaviors all seem to have simple geometrical interpretations (Bohr and Mottelson, 1955; 1975). By exploring these geometrical pictures, very elegant **semi-quantal models** were developed over the last thirty years. The successes of these models are indeed very well documented and therefore will not be repeated here. However, a persistent question of the relationship between the shell model as I have outlined previously and these "geometrical models" remains.

Around the late 50's, Elliott proposed the famous SU(3) model. The model is based on shell structure (s-d shell) of nuclei with certain effective interactions and although its applications in nuclear structure are very much restricted in both scope and region of the periodic table, it is nevertheless the first model to make a connection between collectivity and microscopy. The model was later extended , known as the pseudo-SU(3) model (Arima, Harvey and Shimizu, 1969; Hecht and Adler, 1969) to the higher physical shells. Again, just as the SU(3) model, the pseudo SU(3) model also is primarily applicable to the region where strong deformation prevails.

The study of low energy nuclear structure physics took a major turn about a decade ago. This turn was brought about by the introduction of an elegant and useful phenomenological model known as the Interacting Boson Model (IBM) (Arima and Iachello (1975)). The physics input to this model is remarkably simple. It is based on the knowledge that nucleons in the valence shells tend to couple to coherent S (l=0) and D (l=2) pairs (True and Schiffer, 1976). Thus, the cardinal mathematical structure of the IBM is that by simulating the valence nucleon pairs as s (l=0) and d (l=2) bosons, it envisions the U(6) as the highest symmetry and possesses three group chains:

$$U^B(6) \supset SU^B(3) \supset O^B(3) \quad (1a)$$

$$U^B(6) \supset U^B(5) \supset O^B(5) \supset O^B(3) \quad (1b)$$

$$U^B(6) \supset O^B(6) \supset O^B(5) \supset O^B(3) \quad (1c)$$

As I have mentioned earlier, before the IBM, the study of nuclear structure physics must rely on either the use of intuitive geometrical ideas (for example Bohr-Mottelson(1955), Wilet-Jean (1954)) or the use of more microscopic group theoretical methods like the SU(3) or pseudo-SU(3) models (Elliott,1958; Arima, Harvey and Shimizu,1969; Hecht and Adler,1969; Draayer,1985) which are designed primarily to describe one type of collective behaviors in nuclei: rotational.

There are two fundamental differences between these models and the IBM. The differences are:

(a) All of these models treat, if at all, the different colleactive motions as separate entities. The IBM, on the other hand, describes (phenomenologically at least) all the low energy collective behavior in nuclei via the concept of multichain as in eq.(1), and

(b) None of the above mentioned models, even the group theoretical ones, propose that the dynamical symmetries could be interpreted as "bench-marks" of collective motions which are susceptible to experimental detection .

So, prior to the introduction of the IBM, there was no comprehensive description of the various collective degrees of freedom exhibited by nuclei. In the language of particle physics, there was no nuclear "GUT" to unify the various collective behavior of nuclei with the shell model theory as the underpinning. With the phenomenological IBM, which is intuitively connected to the shell model, we are now one step closer towards having a **nuclear GUT**, although the step connecting the IBM with the underlying shell model remains not completely understood. Furthermore, thanks to the hard work of many outstanding experimentalists for the past decade, we are now at the fortunate stage where we can truly say that these IBM dynamical symmetries are in fact experimentally detectable. Hence, the IBM seems to point to the fact that not only may these chains providing us with clues as to how to truncate the fermion space , but also point to the fact that these dynamical symmetries, to a very large extend, are realizable in nature. For details of this aspect of the IBM, I shall refer the interested readers to the excellent article by Rick Casten (1986) in these proceedings.

Having located these "bench mark" dynamical symmetries, one must seek their microscopic origin. Of course, at the phenomenological level, the IBM has only a tenuous link to the fundamental shell structure (through the finite boson number). Therefore, although it provides an important unified view for various collective modes, it is beyond the scope of the model to provide a deeper understanding of these collective modes and the related dynamical symmetries from the fermion point of view. Although there have been intense efforts in recent years (Otsuka, 1984, Barrett, 1984, Pittel, 1984, Moszkowski, 1984, Yang et. al., 1984, Arima, 1984, P. Ring, 1984, Wu and Feng,1984, Xu, 1984) to provide a better way of mapping a fermion system to a boson system so that a microscopic understanding of the IBM may emerge, such efforts have so far not bear fruit in the understanding of the origins of these dynamical symmetries. **In our view, the dynamical symmetries unraveled by the IBM constitute the fundamental behavior of nuclei and therefore should, and must, manifest themselves directly from the fermion degrees of freedom without introducing bosons.**Therefore it seems to us that the question one needs to ask at this point is: Suppose one begins with the same input physics as the IBM, namely nucleons in the valence shells tend to couple to the coherent S (l=0) and D (l=2) pairs, can one still find the dynamical symmetries which seem to manifest in the IBM. This immediately leads to perhaps a profound question of whether there exist dynamical symmetries rooted in the fermionic nature of the systems. The next paper in these proceedings (Wu, in these proceedings) is precisely devoted to answer this question.

THE GINOCCHIO SO(8) MODEL

An important milestone in the search for fermion dynamical symmetries in nuclear structure occured in the later 70's when Ginocchio (1978, 1980) proposed a fermion model and investigated its possible dynamical symmetries. In his model, the single-nucleon angular momentum **j** is separated into a pseudo-orbital angular momentum **k** and pseudo-spin **i**, where **j = k + i**. The single nucleon creation operator in the k-i scheme is related to the original fermion creation operator by just a Clebsch-Gordon (CG) coefficient:

$$\mathbf{a}^{\dagger}_{jm} = \sum (k\,\alpha\;\; i\;\; \beta \mid j\; m)\; \mathbf{b}^{\dagger}_{k\alpha i\beta} \qquad (2)$$

Ginocchio pointed out that in the language of the k-i basis, there are two alternatives to construct the S and D fermion pair operators as well as the multipole operators denoted as P^r so that the (S,D) fermion subspace is automatically decoupled from the other fermion states. The first is to take k=1 and couple i to zero for the fermion pair. The second is to take i=3/2 and couple k to zero. The first scheme gives rise to the Sp(6) Lie algebra while the latter the SO(8) algebra. Thus, by assuming that the nuclear Hamiltonian is rotationally invariant and can be constructed from of the generators of Sp(6) and SO(8), Ginocchio further showed that the SO(8) and Sp(6) have the following (multi-chain) **fermion dynamical symmetries** :

$$SO^F(8) \supset SO^F(5) \times SU^F(2) \supset SO^F(5) \supset SO^F(3) \qquad (3a)$$

$$SO^F(8) \supset SO^F(6) \qquad \supset SO^F(5) \supset SO^F(3) \qquad (3b)$$

$$SO^F(8) \supset SO^F(7) \qquad \supset SO^F(5) \supset SO^F(3) \qquad (3c)$$

and

$$Sp^F(6) \supset SU^F(3) \qquad \supset SO^F(3) \qquad (4a)$$

$$Sp^F(6) \supset SU^F(2) \times SO^F(3) \supset SO^F(3) \qquad (4a)$$

In eqs.(3) and (4), I have used the clumsy supercript F in all the groups to emphasize that the irreps of these groups must be fully antisymmetric. From now on, unless otherwise stated, all the groups will denote fermion symmetries and the superscript will not be included for simplicity. Just as the familiar dynamical symmetries of the IBM, in constructing the SO(8) dynamical symmetries, one can write the nuclear Hamiltonian as a linear combination of the first and second order Casimir operators of all the groups in eq. (3). Such a Hamiltonian can be transformed into a general Hamiltonian with the residual interactions equal to the monopole and quadrupole pairing interactions and the multipole pairing interactions $\mathbf{P}^r \cdot \mathbf{P}^r$ with $r \leq 3$. Likewise for the Sp(6) dynamical symmetries. Obviously, for the SO(8), with the suitable choices of the parameters in the Hamiltonian, one may be in any of the three limiting symmetry situations. For theSO(5)xSU(2) limit, the energy apectra are identical, while the transition matrix element are identical to within a Pauli factor to withinto those of the U(5) vibrational limit of the IBM. For the SO(6) limit, they are identical to the O(6) γ-soft limit of the IBM. It is most intriguing that the SO(7) limit has no obvious counterpart in the IBM, which means that if one could in fact find the experimental manifestation of this dynamical symmetry, then it may be the "smoking gun" of the fermion-based dynamical symmetry and thereby further provide evidence that **the dynamical symmetries observed via the phenomenological IBM are not fortuitous but in fact are deeply rooted in the fermion structure of nuclei.** Discussions about the exciting empirical evidence of the SO(7) symmetry will be given in the next article by Professor Cheng-Li Wu in these proceedings. Also, a preprint of this work is now available (Casten et. al., 1986).

Notice that none of the SO(8) limiting dynamical symmetries is rotational-like. On the other hand, the Sp(6) dynamical symmetry of eq.(4) has two chains, one of which is the SU(3) limit and the other is another vibrational-like chain SU(2)xSO(3). This SU(3) has a great deal of similarity with the IBM's SU(3) (see eq.(1)). First of all, just as the IBM SU(3), the ground rotational band of a well deformed nucleus is the representation $(\lambda,\mu) = (2N,0)$ where N is the valence pair number while the two important side bands, β- and γ- bands naturally belong to the (2N-2,4) representation. Although the Sp(6) symmetry contains the important SU(3) chain, Ginocchio noticed that it suffers from a "fatal flaw"; namely Pauli principle forbids the occurence of the highest SU(3) (2N,0) representation when the nucleon pair number $N \geq \Omega/3$ where $\Omega = \sum (2j+1)/2$, the total pair degeneracy of a major shell. Thus, at first sight, the Ginocchio model seems to forbid the use (2N,0) to describe the ground band of most rotational nucleus since these nuclei generally exist in or near midshell ($\Omega/2$) and thus cannot reconcile with the successful IBM's SU(3). Because of this reason, Ginocchio abandoned the Sp(6) model (along with its important rotational limit). In fact, in the literature, the Ginocchio model is synoymous with the SO(8) symmetry of eq. (3).

Despite the apparently serious difficulty associated with the SU(3) chain, and the ill-defined relation to the shell model, the Ginocchio effort is, in our opinion, the first attempt in nuclear structure physics to seek a variety of dynamical symmetries (representing various collective motions) from a fermion point of view. Some of these dynamical symmetries could only previously be studied via the phenomenological IBM.

THE FDS MODEL

It should be mentioned at this point that the shell structure of nuclei is one of the most **profound** aspect of nuclear structure and perhaps the most difficult to properly handle. For one thing, except for the light s-d shell ($A \approx 24$),each physical major shell has a common feature, namely due to the strong spin-orbit force, the largest angular momentum normal parity single particle orbit is replaced by an "intruding" abnornal parity single particle orbit. For example, if the spin-orbit interaction were weak, the second s-d-g shell should be

$s_{1/2}$, $d_{3/2}$, $d_{5/2}$, $g_{7/2}$ and $g_{9/2}$.

Yet, in the presence of a strong spin-orbit interaction, the single particle levels are

$s_{1/2}$, $d_{3/2}$, $d_{5/2}$, $g_{7/2}$ and $h_{11/2}$.

Both the $h_{11/2}$ and $g_{9/2}$ are "kicked downstairs", so-to-speak.Obviously, any realistic fermion model of nuclei must reflect this physical reality. For example, in the pseudo-SU(3) model of Arima et. al (1969) and Hecht and Adler (1969) , the pseudo-orbital symmetry of SU(3) are constructed from the normal parity levels $s_{1/2}$, $d_{3/2}$, $d_{5/2}$, $g_{7/2}$ by the reclassification of these four levels as a coupling between the pseudo f-p orbit with a pseudo-spin 1/2. In the same vein, in its "raw " form, the Ginocchio SO(8) model cannot be used to study real nuclei because the connection between the k-i single particle basis and the shell structure as we have just discussed was not made. Also, it is difficult to envision that a model without a rotational limit which is clearly one of the most common collective behaviors of nuclei, can in fact be a serious model for nuclear structure.

Recently, we (Wu et. al., 1986) proposed a fermion dynamical symmetry (FDS) model. The FDS model begins with the same physics input as the IBM: namely (no matter how large individually the single particle angular momenta j_1 and j_2 are) nucleons in the valence shells tend to couple to coherent S (l=0) and D (l=2) pairs. Also, it is suitable to describe the physics which is within the context of one major shell for neutrons and protons. With the aformentioned truncation as the physics input, the theory has three important components:

(a) **A connection is made between the k-i basis and the physical single particle shell basis. It is noticed that each physical shell has a *unique* k-i scheme.**

(b) **By identifying that the abnormal parity single particle orbit as a particular k-i decomposition in all major shells, this orbit is easily incooperated into the theory. As a "bonus", the incooperation of the abnormal orbit is the most natural way to remove the "fatal flaw".**

(c) **The model's dynamics is directly obtained from the nuclear many body hamiltonian. This means that under certain assumptions, one can derive a model nuclear Hamiltonian from the original nuclear Hamiltonian which possesses either SO(8) or Sp(6) symmetry for each major shell in accordance to the particular k-i scheme.**

Since points (a), (b) and (c) have been discussed in great details in a series of articles(Wu et. al., 1986; Guidry et. al., 1986), I shall only discuss the essence of each here.

Point (a) constitutes the selection of the kinematics of the problem, i.e. the selection of the necessary Hilbert space, for our problem. Obviously, the central problem of any quantum mechanical problem is the selection of the most appropriate Hilbert space. In selecting the single particle basis suitable for the physics in low energy region, we are motivated by the fact that the total angular momentum of a valence pair must either be coupled to the relatively small number of 0 or 2. This corresponds physically to the fact that somehow the short-range nucleon-nucleon interaction in the low energy region tends to "freeze" , i.e. couple to zero, most of the single particle angular momenta j_1 and j_2 of any two valence nucleons, leaving the remaining angular momenta coupled to a final value of 0 and 2. We shall refer to the part of the angular momentum j which will, upon coupling, produce the pair J=0,2 as the "active" part. Furthermore, to build a complete basis for a many body system, we need only to "unfreeze" the inactive parts, i.e. allow it to couple to non-zero values of angular momentum. The above mentioned schemes are depicted in Figs. (1a) through (1c). With this as preamble, I want to show you that if $\mathbf{j} = \mathbf{k} + \mathbf{i}$, and i (k) is the frozen part, then the active part k(i) must uniquely be 1(3/2). Also, in order to reproduce exactly (no more, no less) the multi- j values corresponding to the normal levels of a physical major shell, it must either uniquely be k-active or i-active. On the other hand, for the abnormal orbit within the shell, it must always be k-active with k=0 (therefore j = i). This latter statement constitutes a modification of the Ginocchio basis. We see clearly from Table 1 that for either k-active or i-active scheme , we must have k =1 or i = 3/2 to ensure that only 0 and 2 total J is allowed. Any other possibilities will give a range of J which does not have, or is beyond 0 and 2.

Now I want to discuss about point (b). It is well known that, reclassification of the shell structure within a major shell, must imply that the same levels are merely "relabeled", so-to-speak. For example, the SU(3) model of Elliott is built on the fact that within the s-d shell, all the j levels of s and

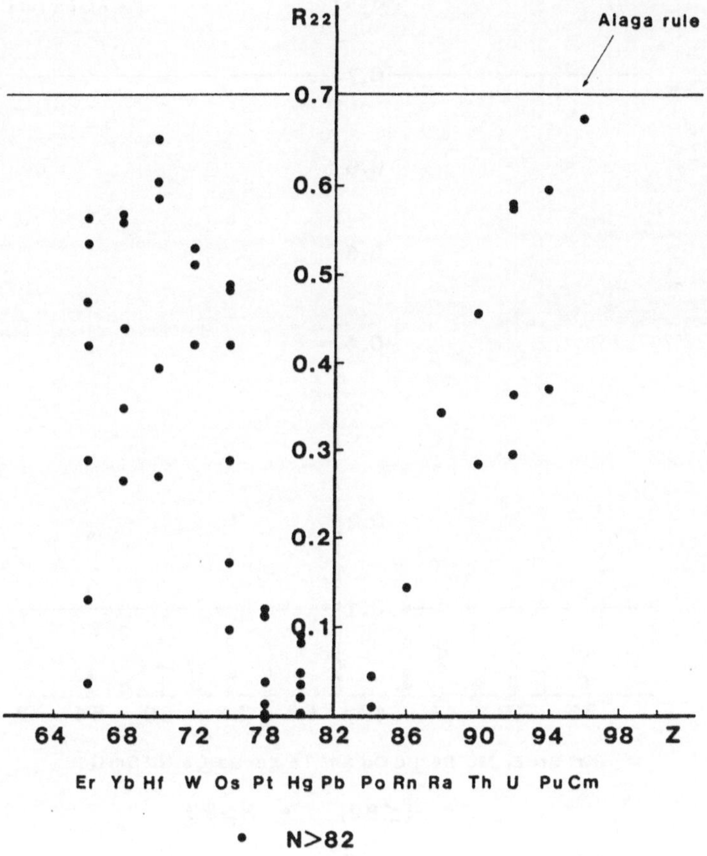

Figure 1. The B(E2) Ratio R_{22} of Even-Even Nuclei Data are taken from Ref. 7 and 8.

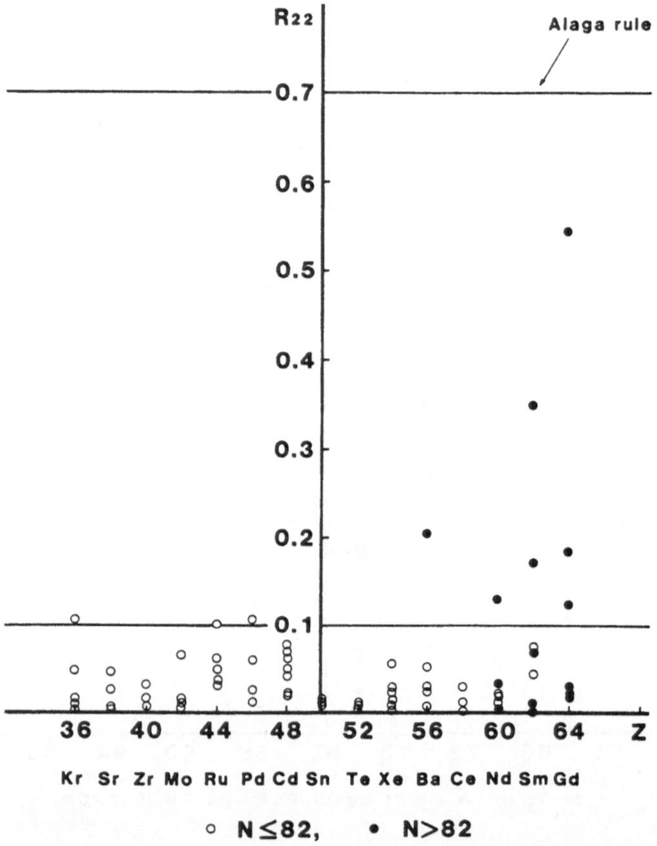

Figure 2. The B(E2) Ratio R_{22} of Even-Even Nuclei Data are taken from Ref. 7 and 8.

There is one additional subtlety associated with Table 2 which must be emphasized. This is the question about the abnormal parity orbit. The one-to-one correspondence I stated before actually forces this orbit, in terms of k-i, to have k=0 and j = i. All in all, the rather remarkable unique result we have obtained here is motivated by the basic assumption that the short range nuclear force (the ultimate **dynamics** of the system) is responsible for the formation of pairs of valence nucleons in the low energy region to be S and D. Once each physical shell is associated with a definite k-i decomposition, we may then use the results as shown by Ginocchio, to directly link the shell structure to the symmetry of either Sp(6) or SO(8). Hence, Table 2 conveys two very important points: first, it shows the the unique reclassification scheme according to k-i for all the major single particle shell levels and second, it shows that for each major shell, the abnormal parity orbit possesses a definite symmetry. For example, for shells 7 and 8, the normal parity orbits' symmetries are Sp(6). Similarly, the normal parity orbits' symmetry is SO(8) for shell 6. For the abnormal parity orbit in each major shell, it must be the quasi-spin symmetry $SU(2)$.

The importance of the **abnormal parity orbit** in the FDS model cannot be under emphasized. Although it exists in every major physical shell (shell 5:$g_{9/2}$, shell 6 $h_{11/2}$, shell 7: $i_{13/2}$ and shell 8: $j_{15/2}$) its importance can be seen most vividly from the shells which possess Sp(6) x$SU(2)$ symmetry. As I have mentioned earlier, the Sp(6) symmetry was abandoned because of a certain subtle Pauli principle. In the Sp(6) x$SU(2)$ symmetry, however, the reason for abandonment vanishes rather naturally. Roughly it is as follows: For a nucleus with N_1 and N_0 valence nucleon pairs in the normal and abnormal parity orbits respectively, the highest SU(3) irrep ($2N_1$, 0) is also not allowed when $N_1 > \Omega_1/3$ (where Ω_1 is the pair number in the normal parity levels), instead of

Table 1

k-active			i-active		
k_1	k_2	K = J	i_1	i_2	I = J
0	0	0	1/2	1/2	0
0	1	1	1/2	3/2	1,2
1	**1**	**0,2**	**3/2**	**3/2**	**0,2**
1	2	1,2,3	3/2	5/2	1,2,3,4
2	2	0,2,4	5/2	5/2	0,2,4

Allowed values of the total angular momentum J for different choices of k and i d (1/2, 3/2 and 5/2) are present. In this way, one may use either the L-S coupling scheme or the j-j coupling scheme, in accordence to physics or convenience or both. However, for heavier systems, the decomposition of j into l and s is obviously not a desirable thing to do since there always will be an additional state occuring in the l-s scheme. The correspondence principle between j's and any other coupling scheme, L-S (or k-i), is that the **number of levels must be the same**. Once this principle is imposed, as it must be, the k-i scheme for the entire shell levels becomes unique. The result of the entire single particle shell model basis (j) in terms of the k-i basis are given in Table 2.

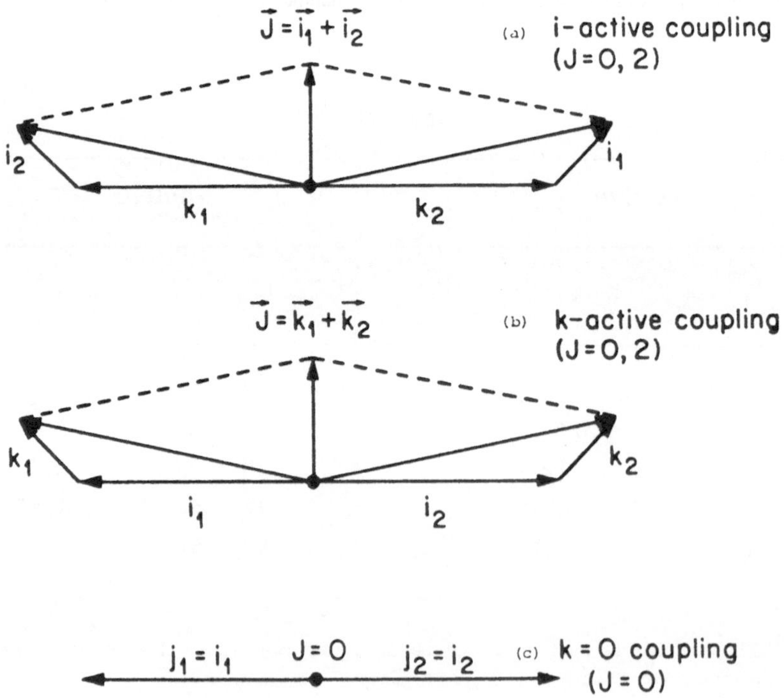

Figure 3

Table 2

No	1	2	3	4	5		6		7			8		
n	0	1	2	3	3	4	4	5	5	5	6	6	6	7
k	0	1	1	0	1	0	2	0	1	1	0	1	1	0
i	1/2	1/2	3/2	7/2	3/2	9/2	3/2	11/2	1/2	7/2	13/2	3/2	9/2	15/2
CONFIGURATION	$s_{1/2}$	$p_{1/2}$ $p_{3/2}$	$s_{1/2}$ $d_{3/2}$ $d_{5/2}$	$f_{7/2}$	$p_{1/2}$ $p_{3/2}$ $f_{5/2}$	$g_{9/2}$	$s_{1/2}$ $d_{3/2}$ $d_{5/2}$ $g_{7/2}$	$h_{11/2}$	$p_{1/2}$ $p_{3/2}$	$f_{5/2}$ $f_{7/2}$ $h_{9/2}$	$i_{13/2}$	$s_{1/2}$ $d_{3/2}$ $d_{5/2}$	$g_{7/2}$ $g_{9/2}$ $i_{11/2}$	$j_{15/2}$
SYM			$G_6\ G_8\ G_3$		$G_6\ G_8\ G_3$		G_8		G_6			G_6		
Ω_0	0	0	0	0	5		6		7			8		
Ω_1	1	3	6	4	6		10		15			21		
Ω	1	3	6	4	11		16		22			29		
$\mathcal{N}$	2	8	20	28	50		82		126			184		

$$G_6 = (Sp_6^k \times SO_3^i) \times (S\mathcal{U}_2 \times S\mathcal{O}_3)$$

$$G_3 = (SU_3^k \times SO_6^i) \times (S\mathcal{U}_2 \times S\mathcal{O}_3)$$

$$G_8 = (SO_8^i \times SO_3^k) \times (S\mathcal{U}_2 \times S\mathcal{O}_3)$$

No labels the ordering of each shell. n,k,l label the principle, pseudo-orbit and pseudo-spin quantum numbers.

$\Omega/3$. On the other hand, since $N = N_1 + N_0$, therefore even when N is roughly $\Omega/2$, it is still possible that $N_1 \leq \Omega_1/3$, thus realizing the $(2N_1, 0)$ ground band. Physically, this means that the abnormal parity orbits can be regarded as a "sink" to absorb the Pauli driven out nucleons from the normal parity orbits. Therefore, we see that **it is the separation of the normal and abnormal parity orbits in the FDS model is responsible for the "revival" of the SU(3) symmetry in the fermion model.** For details of this very important point, I will have to refer the reader to our paper (Wu et. al., 1986). I should add a comment here aboiut the two Fermion SU(3) models which I have discussed here. The first one, which is the pseudo-SU(3) model of Arima et. al. (1969) and Hecht et. al. (1969) is a direct descent of the Elliott's SU(3) model. As such, it is only purported to incooperate one type of collective motion. The FDS' SU(3), on the other hand, is a direct descent of the same IBM input physics and is just one of the several manifestations of collective behavior in nuclei.

I now turn to point (c). Up to now, I have said nothing about the dynamics of the model. It is important, before I discuss the dynamics, to see things in perspective at this stage. The IBM is a phenomenolgical model and has bosons as the building blocks. Perhaps the most important lesson one learnes from the IBM is that the "bench-mark" dynamical symmetries are close, if not always totally accurate, descriptions of certain collective modes in nuclei. Having said this, it seems clear that there are two separate questions one needs to ask.

(i) Can one construct as well from the "same input physics" a similar phenomenological model which must have at least the same sort of "bench-mark" symmetries as the IBM because of its proven successes, but is constructed out of fermions.

(ii) Can one find the dynamics of such a model from the known nuclear Hamiltonian?

The reason why question (i) was raised is obvious. We see that having made the s, d boson assumption, the IBM immediately gives rise to the multi-chain dynamical symmetry structure (see eq.(1)). Therefore, a nagging question is that whether these dynamical symmetries could be just the results of the boson assumption? If in fact that it is just that, than the resulting dynamical symmetries of the boson model may just be fortuitous. Therefore, it is imperative to inquire as to whether such dynamical symmetries can arise from the more fundamental constitutents of nuclei: nucleons. To find these dynamical symmetries in a fermion phenomenological model would constitute, in my opinion, an **"existence proof"** of such dynamical symmetries. The FDS model, prior to any dynamics, may already be very useful in providing such "proofs". Furthermore, it has all the IBM's phenomenology (and more, see Professor Wu's contribution in these proceedings) and therefore immediately shares with the IBM the same level of successes. Clearly a phenomenological model which is built upon the fermion foundation is a **major** step forward in the understanding of the boson phenomenology.

Having asked the first question, we obviously need to follow it by the second question. It is well known that the fundamental question researchers have asked in this area of physics ever since the inception of the IBM is whether the model can be **microscopically justified**. As I have mentioned earlier, there have been rather extensive and exhaustive research carried out in this direction. The main difficulties of all such efforts are the structure of a "single" boson (in nuclei) and the boson to fermion mapping (or vice versa) procedure. On the other hand, starting with the same physics as the IBM, the FDS model is able to encompass all the phenomenology of the IBM. Yet, because the FDS model resides in the fermion space, to find its "microscopy" requires no such mapping procedures.

It turns out that the dynamics of the model can be derived, under certain assumptions, no doubt, from the shell model Hamiltonian. This derivation is rather lengthy and is already discussed in great depth in Wu et. al. (1986) and therefore will not be repeated here. I shall nevertheless give the essence in this talk.

Basically, three assumptions are needed to simplify the nuclear Hamiltonian.They are

(i) The residual interaction is dominated by the monopole and quadrupole pairing interactions.

(ii) Parameterize the two body nuclear matrix elements in accordance to the shell degeneracy.

(iii) Discard terms which do not give rise to a close algebra.

Under these assumptions, the original nuclear Hamiltonian can be shown to possess the following form:

$$H_{FDS} = e_0 n_0 + e_1 n_1 + \sum_{kk'} G_0^{kk'} \mathbf{S}^\dagger(k)\cdot\mathbf{S}(k') + G_2 \mathbf{D}^\dagger . \mathbf{D}$$

$$+ \sum_{r,\alpha\alpha'} B_r^{\alpha\alpha'} \mathbf{P}^r(\alpha)\cdot\mathbf{P}^r(\alpha') \qquad (5)$$

In eq.(5), the operators $\mathbf{n}_0$ and $\mathbf{n}_1$ are the number operators in the abnormal and normal parity orbitals respectively while e_0 and e_1 are the energy of the abnormal parity orbital and average single particle orbitals of the normal parity orbitals respectively. The operators $\mathbf{S}^\dagger(k)$ and $\mathbf{S}(k')$ are the monopole pairing operators which is defined as

$$\mathbf{S}^\dagger(k) = \sum_j \sqrt{\Omega_j}/2 \; [\mathbf{a}^\dagger_j \, \mathbf{a}^\dagger_j]^0 \qquad (6)$$

$k = 0$ ($\neq 0$) in eq. (6) denotes the monopole pair in the abnormal (normal) parity orbital(s). $G_0^{kk'}$ is the strength of the monopole monopole interactions. The fourth term in the Hamiltonian is the quadrupole pairing interaction. Depending on whether the shell is k or i active, the quadrupole pairing operator $\mathbf{D}^\dagger$ is defined as

$$\mathbf{D}^\dagger = \mathbf{D}^\dagger(k) = \sum_i \sqrt{\Omega_{1i}}/2 \; [\mathbf{b}^\dagger_{1i} \, \mathbf{b}^\dagger_{1i}]^{20} \qquad \text{for k active} \qquad (7a)$$

$$= \mathbf{D}^\dagger(i) = \sqrt{\Omega_{k3/2}}/2 \; [\mathbf{b}^\dagger_{k3/2} \, \mathbf{b}^\dagger_{k3/2}]^{02} \qquad \text{for i active} \qquad (7b)$$

In eq. (7), Ω_{ki} = (2 k + 1)(2 i + 1)/2. For the abnormal parity orbit, there are only monopole pairs. Again, G_2 is the quadrupole pairing strength. Finally, the operator $\mathbf{P}^r(\alpha)$ is the multipole operator which is defined as

$$\mathbf{P}^r(k) = \sum_i \sqrt{\Omega_{ki}}/2\ [\mathbf{b}^\dagger_{ki}\tilde{\mathbf{b}}_{ki}]^{r0} \qquad \text{for } k \text{ active} \qquad (8a)$$

$$\mathbf{P}^r(i) = \sum_k \sqrt{\Omega_{ki}}/2\ [\mathbf{b}^\dagger_{ki}\tilde{\mathbf{b}}_{ki}]^{0r} \qquad \text{for } i \text{ active} \qquad (8a)$$

In (8a), when k = 0, it corresponds to the abnormal orbital's multipole operators. $B_r^{\alpha\alpha'}$ corresponds to the multipole interaction strengths. This model Hamiltonian possesses dynamical symmetries according to Fig. 2. For the interested readers about the detail implications, may I again refer you to the forthcoming article (Wu et. al., 1986).

EFFECTIVE INTERACTIONS OF THE FDS MODEL

As I have emphasized from the outset in this talk that some sort of **truncation** procedure is a must in nuclear structure physics since solving the full shell model Hamiltonian (infinite Hilbert space with bare residual interactions) is impossible. Such efforts will necessarily lead to "effective interactions" (hereafter known as EI) in nuclei. It is of course very well known to nuclear physicists and a beautiful summary was recently given by Kirson (1985), that the EI is intimately connected with the truncation of the Hilbert space. In the remaining time, I would like to discuss the meaning of the FDS model from the EI point of view. Also, we shall compare the FDS-EI with the more traditional ones, such as the pairing plus quadrupole (P+Q)-EI and the shell model(SM)-EI.

In principle, given a truncated basis, the effective interaction can be calculated from the bare nucleon-nucleon interaction via the many-body theory (Fetter and Walecka, 1969, Kuo, 1985). To date, there is still great difficulties to obtain effective interactions which can accurately reproduce data, especially in the medium to heavy weight nuclei. Therefore, one has to either resort to introduce, by means of physical considerations, a simple model for the effective interaction, or simply treat the two-body matrix elements of the effective interaction as parameters and then determine them empirically. The concept of the latter scheme, pioneered immediately after the inception of the shell model theory by Talmi in the early 50's (Talmi, 1952), is now the standard procedure to obtain the shell model effective interaction (SM-EI) (Wildenthal, 1985). On the other hand, both the FDS and the P+Q interactions are models of effective interactions which may be regarded as different simplifications of the "real" effective interactions.

First we shall discuss the FDS-EI and P+Q-EI. There are some similarities between the two models:

(i) Both are based on multipole expansion of the nuclear force and the EIs are expressed in terms of the pairing plus multipole interactions;

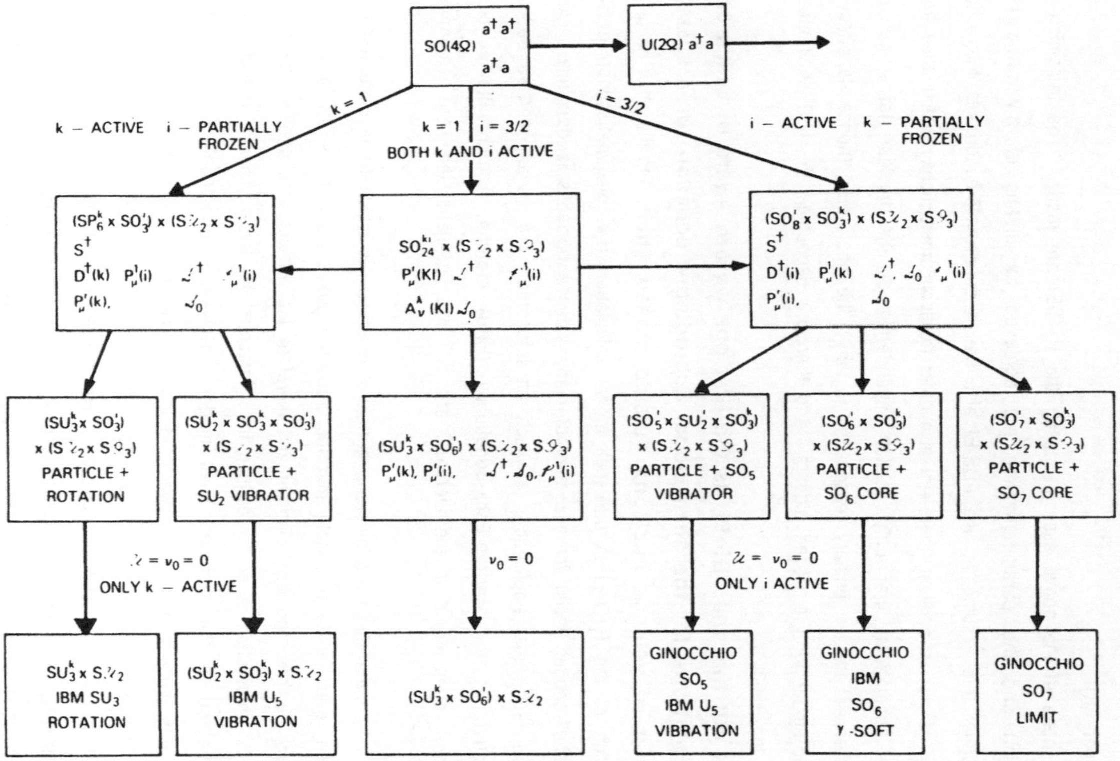

$S^{\dagger} = \sum_i A_0^{0\dagger}(k0, i0)$, $D_\mu^{\dagger}(k) = \sum_i A_\mu^{2\dagger}(k2, i0)$, $\mathcal{D}_\mu(i) = A_\mu^{2\dagger}(k0, i2)$, $\mathcal{S}^{\dagger} = A_0^{0\dagger}(00, i0)$, $\mathcal{S}_0 = (n_0 - \Omega_{i_0})/2$

$P_\mu^r(k) = \sum_i F_\mu^r(kr, i0)$, $P_\mu^r(i) = P_\mu^r(k0, ir)$, $\mathcal{P}_\mu^r(i) = P_\mu^r(00, ir)$, $P_\mu^r(KI) = P_\mu^r(1K, 3/2\, I)$, $A_\nu^{\lambda}(KI) = A_\nu^{\lambda}(1K, 3/2\, I)$

Figure 4

(ii) Both have the same pairing interaction and assume that it is the most important part of the EI;

(iii) Both truncate the multipole expansion and parametrized the matrix elements.

The main difference between these two models is the manner of truncation and parameterization of the reduced matrix elements for the , say quadrupole operator. In the P+Q model, the assumption of **spatial dependence, separability** and a **spatial form of the radial part** of the interaction are introduced in order to obtain a simple Q·Q interaction (Q is the quadrupole moment operator).

In the FDS model, none of the above assumptions for the P+Q-EI are made. The truncation of the multipole expansion and the parameterization are achieved by means of **symmetry consideration**. Under this assumption, the FDS-EI also has a quadrupole-quadrupole interaction which is denoted as $\mathbf{P}^2 \cdot \mathbf{P}^2$. In addition, there are also other multipole interactions which arise from the relaxation of the assumption that only spatial dependence of the nuclear force will be considered. These differences are the primary reasons as to why the operators (hence its effective interactions) in the FDS-model form a closed algebra and therefore decouple the S-D shell model space from the rest of the space.

We should point out that the FDS-EI is the first one we are aware of which can provide <u>analytic solutions</u> for the shell model Schrodinger equation for a system with a large number of valence nucleons outside closed shell. On the other hand, a straightforward application of the P+Q-EI in the shell model calculation (i.e. diagonalizatioon of the EI within a spherical shell model basis) for a system of many valence nucleons is definitely beyond reach at this stage. In fact, most applications of such an interaction are usually carried out via theoretical schemes such as the Hartree-Fock-Bogoliubov (HFB) method. The results of such approaches are very well documented in the literature and will not be elaborated here (Goodman, 1973).

Since the FDS-EI and the P+Q-EI are different, a natural question is which EI is more realistic? To answer this question, one should first note that the truncated space as well as the methods of application for these two EIs are also very different and therefore the answer can, at best, be only a nebulous one. The FDS interaction is an EI defined in a truncated k-i basis within one major shell and the theoretical method is the standard shell model. However, the definition (or physical meaning) of the P+Q-EI is much less straightforward. In fact, its usage permeates in several different theoretical schemes. One such application is to regard the P+Q-EI as a phenomenological "bare" nucleon-nucleon interaction (e.g. the HFB calculations). **Because the EI is directly a result of the truncated space, therefore different truncated spaces can, in fact, result in different wavefunctions as well as effective interactions.** Consequently, as emphasized by Kirson (1985), it is not always meaningful to carry out a comparison of the wavefunctions and the EIs (when these EIs exist in different truncated Hilbert spaces). In this context, it is not surprizing that the P+Q-EI and the FDS-EI are different. What is important is that the **respective physics outcome must be compatible with empirical data** since they purport to describe the same physics. As we have seen in the previous sections, the FDS-EI within the k-i truncated basis does contain all the well known low energy collective modes of even-even nuclei, therefore it is very suggestive that such an EI is a good caricature of the realistic one, even though in form it is different from the more conventional EIs.

We now turn to the question of the relations between the SM-EI and the FDS-EI. It would still be most interesting, of course, to compare the "standard" SM-EI with the FDS-EI. There are several reasons why this would be the case. Firstly, unlike the P+Q-EI, the corresponding truncated space for the SM-EI is well defined (i.e. one major shell); Secondly, it is strictly empirically obtained and probably is more close to the realistic effective interactions (certainly true in the s-d shell and at least for the beginning of the shell for heavy weight nuclei); Thirdly, the FDS model is also a shell model in a (further) k-i truncated basis in one major shell. Thus the FDS model and the effective SM-EI share more common foundation to render such a comparison meaningful. It should be pointed out, however, that if the two interactions appear to be similar in characteristics, it implies that the k-i truncated basis may in fact be a good approximation of the entire one major shell space. On the other hand, if the EIs appear different, it merely suggests that the space left out by the FDS model is not negligible and therefore, as emphasized by Kirson, will either cause the EI to vary or the EI in the mid-shell may, in fact, be quite different from that which is appropriate for the beginning of the shell, or both. After all, the shell model EIs in the heavy weight nuclei are generally obtained empirically only near close shell. Contrary to the shell model-EI, the FDS-EI is mainly designed to describe low lying collective states of nuclei where, in general, there are many valence nucleons. Hence, there is no obvious way to ascertain its validity in mid-shell (except for the s-d shell where a smooth change of EI is demanded by the data).

Of course, the FDS-EI may still be oversimplified as it stands. Therefore, by studying the connection between the shell model-EI and the FDS-EI, a better EI within the k-i basis for heavy nuclei in midshell may emerge. Finally, we like to point out that the most promising feature of the FDS model is that it provides a tractable truncated k-i basis for nuclei with many valence nucleons. The validity of this truncation basis is demonstrated by the fact that in all the previously discussed limiting cases, the pure configuration withou mixing can describe various collective motions well, at least qualitatively. Therefore, if indeed we have at our disposal all the necessary symmetry breaking terms (e.g. single-particle energy splitting, the quadrupole interaction between the "core" and decoupled particles, $\mathbf{P}^2(k).\mathbf{P}^2(i)$) etc, and then diagonalize these symmetry breaking terms in the k-i space, it is possible to study the various intermediate symmetry situation.

SUMMARY

In this paper, we have presented an effective nuclear model called the Fermion Dynamical Symmetry (FDS) model. The model is based on the philosophy of the Interacting Boson Model and generalized the Ginocchio model. It has all the low energy phenomenology of collective nuclear structure physics. Before I conclude this talk, I would just like to point out that the model include not only low energy collective phenomenology of even-even nuclei at this stage, but automatically encompass the even-odd as well as odd-odd systems as well. This is actually very easy to see because what we have discussed so far are the irreps of Sp(6) or SO(8) with u = 0 where u is the so-called S-Dseniority (number of nucleon pairs not coupled to 0(S) and 2(D)). Merely by allowing u ≠ 0 (e.g. 1 for odd), we immediately bring into the theory the other systems. Furthermore, by studying the interactions between u = 0 and u = 2, we can study the effect of high spin physics. I would like to point out that a preliminary account of this paper has already been written up and submitted to Phys. Lett. for publications (Guidry et. al. , 1986).

It seems to me that what we now possess is a theory which provides us with (a) a **further truncation** of the shell model space and (b) the **necessary effective interactions** to describe the physics, for both the low as well as high spin physics. Obviously, the next exciting step in this whole effort is clearly tailored for the standard shell model computer codes which several groups have developed over the last twenty years. The implimentation of this scheme is now being carried out by our group and are spear-headed by **Xiang-Dong Ji**, **Micheal Vallieres** and **Phil Halse**, on the standard shell model codes. Needless to say, we are anxiously awaiting the day when one can study the collective nuclear structure physics from the spherical shell model codes. I expect that this anticipated exciting outcome is not far in the future.

ACKNOWLEDGEMENTS

I must thank my collaborators in this research effort. One of them, **Cheng-Li Wu**, is the author of the following article in these proceedings. In the early stages of this work, Wu and I have collaborated closely with **Xuan-Gen Chen** and **Jin-Quan Chen**, both from Nanjing University and **Michael W. Guidry** of the University of Tennessee. More recently, we are fortunate to have the additional outstanding collaboration from **J.N.Ginocchio** and **R.F.Casten**. Also discussions with as well as critical communications from many people, especially Bruce Barrett, Larry Biedenharn, Xiao-Ling Han, Ted Hecht, Franco Iachello, Xiang-Dong Ji, Z.-P. Li, T. Otsuka, Lee L. Riedinger, Michel Vallieres, D.-D. Warner, B. Hobson Wildenthal and Hua Wu are obviously most useful in formulating many of the ideas presented in this paper. Finally, I cannot thank enough Professor Bruno Gruber who has performed a yeoman's work in organizing this very interesting and scientifically beneficial conference. The work is supported by the **National Science Foundation.**

REFERENCES

Arima, A., Harvey, M., and Shimizu, K.,1969, "Pseudo LS Coupling and Pseudo SU_3 Coupling Schemes," Phys. Lett., 30B: 8.

Arima, A., and Iachello, F., 1975, "Collective Nuclear States as Representations of a SU(6) Group, " Phys. Rev. Lett., 35:1069.

Arima, A., 1984, Foundation and Extension of Interacting Boson Model, in: "Proceedings of the International Workshop on Collective States in Nuclei," L.-M. Yang, X.-Q. Zhou, G.-O. Xu, S.-S. Wu and D. H. Feng ed., North Holland, Amsterdam.

Barrett, B. R., 1984, The IBM-2 Parameters and their Microscopic Interpretations, in: "Interacting Boson-Boson and Boson-Fermion Systems," O. Scholten ed., World Scientific, Singapore.

Bohr, A., and Mottelson, B. R., 1955, "Moments of Inertia of Rotating Nuclei", Dan. Mat. Fys. Medd., 30:1.

Bohr, A., and Mottelson, B. R., 1975, "Nuclear Structure I and II", Benjamin, New York and references therein.

Casten, R. F., 1986, in these Proceedings.

Casten, R. F., Wu , C.-L., Feng, D. H. , Ginocchio J. N., and Han, X.-L., 1986 "Empirical Evidence for an SO(7) Fermion Dynamical Symmetry in Nuclei", Phys. Rev. Lett. (in press).

Draayer, J. P. and Weeks, K., J., 1983, "Shell-Model Description of the Low-Energy Structure of Strongly Deformed Nuclei," Phys. Rev. Lett. 51:1422.

Elliott, J. P., Collective Motion in the Nuclear Shell Model, I. Classification Schemes for States of Mixed Configurations, Proc. Roy. Soc. of London, Ser. A 245: 128; ibid. 562.

Fetter, A. L., and Walecka, J. D., 1971, Quantum Theory of Many-Body-Particle Systems, McGraw-Hill, NY.

Ginocchio, J. N., 1978, "An Exact Fermion Model with Monopole and Quadrupole Pairing", Phys. Lett. 79B:173.

Ginocchio, J. N., 1980, "A Schematic Model for Monopole and Quadrupole Pairing in Nuclei", Ann. Phys (N.Y.), 126:234.

Goodman, A. L., 1979, The Hartree-Fock-Bogoliubov Theory with Applications to Nuclei, in: "Advances in Nuclear Physics, Vol. 11," J. W. Negele and E. Vogt ed., Plenum, New York and references therein.

Guidry, M. W., Wu, C.-L., Feng, D. H., Ginocchio, J. N, Chen, X.-G. and Chen, J.-Q., "The Fermion Dynamical Symmetry Model and High Spin Physics", Phys. Lett. (submitted).

Hecht, K., and Adler, A., 1969, Generalized Seniority for Favored J≠0 Pairs in Mixed Configurations, Nucl. Phys. A137:129.

Kirson, M. W., 1985, The Shell Model and Effective Interactions, in: "International Symposium on Nuclear Shell Models," M. Vallieres and B. H. Wildenthal ed., World Scientific, Singapore.

Kuo, T.T.S., 1985, Methods for Deriving Nuclear Single Particle Potentials, in: "International Symposium on Nuclear Shell Models," M. Vallieres and B. H. Wildenthal ed., World Scientific, Singapore.

Moszkowski, S., 1984, Toward a Microscopic Theory of the Interacting Boson Model, in: "Interacting Boson-Boson and Boson-Fermion Systems," O. Scholten ed., World Scientific, Singapore.

Otsuka, T.,1984, Microscopic Calculation for Deformed Nuclei, in: "Interacting Boson-Boson and Boson-Fermion Systems," O. Scholten ed., World Scientific, Singapore, and ref. therein.

Pittel, S.,1984, Mean-Field Approximations for Deformed Boson Systems, in: "Interacting Boson-Boson and Boson-Fermion Systems," O. Scholten ed., World Scientific, Singapore, and ref. therein.

Ring, P.,1984, Microscopic Description of Nuclear Collective States by Symmetry Conserving Mean Field Theory, in: "Proceedings of the International Workshop on Collective States in Nuclei," L.-M. Yang, X.-Q. Zhou, G.-O. Xu, S.-S. Wu and D. H. Feng ed., North Holland, Amsterdam.

Schiffer, J. P., and True, W., 1976, "The Effective Interaction between Nucleons Deduced from Nuclear Spectra", Rev. Mod. Phys. 48:191.

Wildenthal, B. H., 1985, The Common Genesis of Energy Eigenstates in the Nuclear 1s,0d Shell, in: "International Symposium on Nuclear Shell Models," M. Vallieres and B. H. Wildenthal ed., World Scientific, Singapore, and references therein.

Wilets L., and Jean, M. , 1956, Surface Oscillations in Even-Even Nuclei, Phys. Rev. 102: 788.

Wu, C.-L. and Feng, D. H., 1984, The Composite Particle Representation Theory and the Effective Boson-Boson and Boson-Fermion Interactions in Nuclei, in: "Proceedings of the International Workshop on Collective States in Nuclei," L.-M. Yang, X.-Q. Zhou, G.-O. Xu, S.-S. Wu and D. H. Feng ed., North Holland, Amsterdam and references therein.

Wu. C.-L., Feng, D. H., Chen, X.-G., Chen, J.-Q. and Guidry, M. W., 1986, "Fermion Dynamical Symmetry Model and the Nuclear Shell Model", Phys. Lett. 168B, 1986.

Wu. C.-L., Feng, D. H. , Chen, X.-G., Chen, J.-Q. and Guidry, M. W., 1986, "A Fermion Dynamical Symmetry Model of Nuclei (I) Basis, Effective Interactions and Symmetries", Phys. Rev. C (submitted).

Wu. C.-L., 1986, in these proceedings.

Xu, G.-O, 1984, Microscopic Investigation of Nuclear Collective Models, in: "Proceedings of the International Workshop on Collective States in Nuclei," L.-M. Yang, X.-Q. Zhou, G.-O. Xu, S.-S. Wu and D. H. Feng ed., North Holland, Amsterdam.

Yang, L.-M., Lu, D.-H., and Zhou, Z.-N., 1984, Microscopic Investigation of IBM and IBFM, in: "Proceedings of the International Workshop on Collective States in Nuclei," L.-M. Yang, X.-Q. Zhou, G.-O. Xu, S.-S. Wu and D. H. Feng ed., North Holland, Amsterdam.

DYNAMICAL SUPERSYMMETRIC DIRAC HAMILTONIANS

Joseph N. Ginocchio

Theoretical Division
Los Alamos National Laboratory
Los Alamos, NM 87545

INTRODUCTION

Most recently a relativistic quantum mechanical approach to nuclear physics has proven promising.[1] However we know from the interacting boson model of nuclei[2] and the fermion dynamical symmetry model of nuclei[3-5] that nuclear spectra exhibit dynamical symmetries.* Perhaps these symmetries have their basis in a realtivistic theory, particularly since the spin-orbit potential is a relativistic effect. For this reason I would like to explore in this Symposium certain Dirac Hamiltonians, although I don't pretend to come anywhere near the above-stated goal in this talk. The Dirac Hamiltonian I would like to consider is that of a neutral fermion interacting with a tensor field, say for example the electromagnetic field tensor. In this paper I shall use the language of QED since it is our best known realtivistic quantum theory. However, the Hamiltonians can have a more general applicability for example to QHD[1] or QCD[6].

First we shall discuss a supersymmetry found for a general Dirac Hamiltonian of this type. Then we shall discuss a special case of this type of Dirac Hamiltonian.

THE DIRAC HAMILTONIAN

A neutral fermion can interact with an external electormagnetic field through its "anomalous" magnetic moment μ', i.e., the difference between its measured magnetic moment and the Dirac magnetic moment. For example the anomalous magnetic moment of the neutron could be due to its meson

---

*See also talks by R. Casten, D. H. Feng, and C. L. Wu in this Symposium.

cloud. In any case the Dirac equation is

$$H\Psi_s = \beta\left[\vec{\gamma}\cdot\vec{p} + \frac{i\mu'}{c}\gamma_\mu\gamma_\nu F^{\mu\nu} + mc\right]\psi_s = \epsilon_s\psi_s \tag{1}$$

We use the usual notation for the 4 × 4 Dirac matrices,

$$\gamma_0 = \begin{pmatrix} 1 & 0 \\ 0 & -1 \end{pmatrix} = \beta \tag{2a}$$

$$\gamma_k = \begin{pmatrix} 0 & \sigma_k \\ -\sigma_k & 0 \end{pmatrix} = \beta\,\alpha_k, \quad k = 1, 2, 3 \tag{2b}$$

The bound states are denoted by the quantum number s and σ_k are the 2 × 2 Pauli matrices.

The electromagnetic field tensor is

$$F^{\mu\nu} = \frac{\partial}{\partial x_\nu} A^\mu - \frac{\partial}{\partial x_\mu} A^\nu \tag{3a}$$

where the four-vector potential, electric field and magnetic field are

$$A_\mu = (\phi, \vec{A}) \tag{3b}$$

$$E_i = F^{oi}, \quad i = 1, 2, 3 \tag{3c}$$

$$B_i = \epsilon_{ijk}F^{jk} \tag{3d}$$

Furthermore we assume that B = 0, and the electric field is centrally symmetric $\vec{E} = \hat{r}\,\frac{\partial\phi(r)}{\partial r}$. In this case the Dirac Hamiltonian is spherically symmetric and conserves the total anagular momentum j and projection μ, helicity ν,

$$\nu = \left[\frac{\vec{\sigma}\cdot\vec{j}+1}{j+\frac{1}{2}}\right]\beta \tag{4a}$$

which has eigenvalues ±1, and parity π,

$$\pi = \nu(-1)^{j-\frac{1}{2}} \tag{4b}$$

Thus we shall have

$$s = (n\nu j\mu) \tag{4c}$$

where n is the radial quantum number.

We write the Dirac wavefunction as $(\psi_s^{(+)}, \psi_s^{(-)})$ where $\psi_s^{(+)}$ is the upper component, and $\psi_s^{(-)}$ the lower component. Then the Dirac Hamiltonian (1) reduces to

$$H = \begin{pmatrix} mc & h^\dagger \\ h & -mc \end{pmatrix} \tag{5a}$$

where

$$h^\dagger = \vec{\sigma}\cdot(\vec{p} + i\,\mu'\vec{E}) \tag{5b}$$

If we square H, we get a diagonal Hamiltonian $\bar{H}$

$$\bar{H} = H^2 = \begin{pmatrix} \bar{H}_+ & 0 \\ 0 & \bar{H}_- \end{pmatrix} \tag{6a}$$

where

$$\bar{H}_+ = (mc)^2 + h^\dagger h = p^2 + V - W \tag{6b}$$

$$\bar{H}_- = (mc)^2 + hh^\dagger = p^2 + V + W \tag{6c}$$

and the potentials V and W are

$$V = (mc)^2 + \left[\frac{\mu'}{c}\frac{d\phi}{dr}\right]^2 \tag{6d}$$

$$W = \frac{\hbar\mu'}{c}\left[\frac{1}{r^2}\left(\frac{d}{dr} r^2 \frac{d\phi}{dr}\right) + \frac{2}{r}\frac{d\phi}{dr}\,\vec{\sigma}\cdot\vec{L}\right] \tag{6e}$$

Clearly the eigenvalues of $\bar{H}$ are ϵ_s^2. Furthermore $\bar{H}_+$ and $\bar{H}_-$ are each Schrodinger equations which have <u>different</u> potentials, but the <u>same</u> eigenvalues:

$$\bar{H}_+\psi_s^{(+)} = \epsilon_s^2\psi^{(+)} \tag{7a}$$

$$\bar{H}_-\psi_s^{(-)} = \epsilon_s^2\psi^{(-)} \tag{7b}$$

This means that the square of the Dirac Hamiltonian is a supersymmetric Schrodinger Hamiltonian[7,8] in <u>three-dimensional</u> space[9]. The supersymmetric generators are

$$Q^\dagger = \begin{pmatrix} 0 & h^\dagger \\ 0 & 0 \end{pmatrix} \tag{8a}$$

$$Q = \begin{pmatrix} 0 & 0 \\ h & 0 \end{pmatrix} \tag{8b}$$

and the supersymmetric Hamiltonian is

$$\bar{H} = (mc)^2 + \{Q^\dagger, Q\} \tag{9}$$

and clearly

$$[Q^\dagger, \bar{H}] = [Q, \bar{H}] = 0 \tag{10}$$

Thus $Q^\dagger, Q, \bar{H}$ generate an sl(1/1) supersymmetry.

For a singlet representation of this algebra we would have

$$Q\psi_s = Q^\dagger \psi_s = 0 \tag{11}$$

which implies that

$$h^\dagger \psi_s^{(-)} = 0 \tag{12a}$$

$$h\psi_s^{(+)} = 0 \tag{12b}$$

From (5) this leads to

$$\psi_s^{(\pm)} = A_{\nu j}^{(\pm)} e^{\frac{\mp\mu'}{\hbar c}\phi(r)} r^t \tag{13a}$$

where

$$t = \pm\nu(j + \tfrac{1}{2}) - 1 \tag{13b}$$

and $A_{\nu j}^{(\pm)}$ is the normalization.

From (9) we see that these singlet eigenfunctions will be eigenfunctions of the supersymmetric Hamiltonian with $\epsilon_s^2 = (mc)^2$. Of course whether or not they are well-behaved eigenfunctions, i.e., regular at the origin and normalizable depends on the details of $\phi(r)$. However it is clear from (13) that there may be more than one of these eigenfunctions with the same eigenenergy.

Furthermore the Dirac eigenfunctions will be $\psi_s = \left[\psi_s^{(+)},0\right]$ $\psi'_s = \left[0,\psi_s^{(-)}\right]$ with $\epsilon_s = + mc$ and $\epsilon_s = - mc$ respectively.

In particular if $\phi(r) \overset{r\to\infty}{\to} r^P$, where P>0, then clearly $\psi_s^{(+)}$ will be normalizable and well-behaved for positive helicity states, $\nu = 1$. But $\psi_s^{(-)}$ will not be normalizable. Hence there will exist an infinite multiplet of states ($j = \frac{1}{2}, \frac{3}{2}, \ldots$) with the same energy $\epsilon_s = mc$, and which do not have a lower component.

For a doublet representation, the algebra gives for each eigenfunction with $\epsilon_s^2 \neq mc^2$,

$$Q^{\dagger}\Psi_s^{(-)} = (\epsilon_s - mc)\Psi_s^{(+)} \tag{14a}$$

$$Q\Psi_s^{(+)} = (\epsilon_s + mc)\Psi_s^{(-)} \tag{14b}$$

$$Q^{\dagger}\Psi_s^{(+)} = Q\Psi_s^{(-)} = 0 \tag{14c}$$

The covariant Dirac Hamiltonian will then have a dynamic supersymmetry,

$$H_c = \beta H = m + Q^{\dagger} - Q$$

since it is a linear combination of the generators. The eigenfunctions will be either supersymmetric doublets or singlets. In fact we could have used H_c and $H_c^{\dagger}$ as the supersymmetry generators since

$$\bar{H} = \frac{1}{2}\{H_c, H_c^{\dagger}\} \tag{15}$$

In this Symposium, a similar supersymemtry has been found for the Dirac equation with minimal coupling to the external four-vector potential*. However this supersymmetry differs from the one we have discussed in that γ_5 is diagonal rather than γ_0. Hence chiral symmetry is preserved rather than parity. We can call the supersymmetry discussed here γ_0-supersymmetry, and the other γ_5-supersymmetry.

---

*See talk by L. O'Raifeartaigh in this Symposium.

Let us consider the special case of a harmonic electric potential $\phi = \phi_0 r^2$. Then the potential V is also harmonic $V = (mc)^2 + 4\left[\frac{\mu'}{c}\phi_0\right]^2 r^2$, and the potential W has no radial dependence, $W = W_0(\vec{\sigma}\cdot\vec{L} + \frac{3}{2})$, where $W_0 = (2\hbar\omega m)$ and $\omega = 2\mu'\phi_0/mc$. Then the supersymmetric Hamiltonian can be written in terms of the harmonic oscillator quantum number operator,

$$\bar{H}_+ = (mc)^2 + 2\, m\hbar\omega[\hat{N} - \vec{\sigma}\cdot\vec{L}] \tag{16a}$$

$$\bar{H}_- = (mc)^2 + 2\, m\hbar\omega[\hat{N} + 3 + \vec{\sigma}\cdot\vec{L}] \tag{16b}$$

where $\hat{N} = a^\dagger \cdot \tilde{a}$ and counts the number of oscillator quanta

$$a_q^\dagger = \frac{1}{\sqrt{2m\hbar\omega}}\left[p_q + im\omega r_q\right] \tag{17a}$$

and

$$h^\dagger = \sigma\cdot a^\dagger \tag{17b}$$

$$h = \sigma\cdot a \tag{17c}$$

This special Dirac Hamiltonian[10] has been considered as a model for spinor quarks[11], before supersymmetric quantum mechanics. Later the supersymmetric Hamiltonian (16) was considered without (apparently) recognizing its relationship to the Dirac Hamiltonian.[9,12,13]

The radial eigenfunctions are clearly given in terms of the harmonic oscillator eigenfunctions with radial quantum number n = 0, 1, ...,. From (13), and as noted in the last section, there are normalizable and well-behaved eigenfunctions with $\epsilon_s = mc$ which correspond to $n = 0$ and $\nu = 1$,

$$\Psi^{(+)}_{n=0,\ \nu=1,j} = A_j^{(+)} r^{j-\frac{1}{2}} e^{-\frac{\mu'}{\hbar c}\phi_0 r^2} \tag{18}$$

However the lower component $\Psi^{(-)}_{n=0,\nu=1,j}$ in (13) is not normalizable. This corresponds to other results[7,8,9,12,13,14] that the ground state of a supersymmetric Hamiltonian has only a "boson" normalizable eigenfunction.

We also note that the ground state exists for an infinite number of j, $j = \frac{1}{2}, \frac{3}{2}, \ldots$,as noted in the last section. In fact this infinite degeneracy is even more widespread. If we look at the exact eigenergies for all states[11], which we can find easily from (16),

$\nu=1$

$$\epsilon_{n\nu=1j} = \left[(mc)^2 + 4\, m\hbar\omega n \right]^{\frac{1}{2}} \tag{19a}$$

$\nu=-1$

$$\epsilon_{n\nu=-1j} = \left[(mc)^2 + 4\, m\hbar\omega(n + j + 1) \right]^{\frac{1}{2}} \tag{19b}$$

Hence we see that, for the states with positive helicity, the energy depends only on the radial quantum number n and not on j, and hence these states are infinitely degenerate for each radial quantum number, not only for n=0. The states with negative helicity have finite degeneracies. For example for $\nu=-1$, n=0, $j=\frac{3}{2}$ and n=1, $j=\frac{1}{2}$ are degenerate. Because of these degenerate bands, it is convenient to define a new quantum number in place of the radial quantum number

$$k = 2n + (1 - \nu)(j+1) \tag{20}$$

Then all the states in a degenerate band have the same value of j. Also the bands with $\nu=1$ have k even, k=0, 2, ... and those with $\nu=-1$ have k odd, k = 3, 5, ...

The eigenenergies will then depend only on the integer k,

$$\epsilon_k = [(mc)^2 + 2mh\omega k]^{\frac{1}{2}} \tag{21a}$$

and the allowed values of k are

$$k = 0,\ 2,\ 3,\ 4,\ 5,\ \ldots \tag{21b}$$

and the allowed values of j are

$$j = \frac{1}{2}, \frac{3}{2}, \ldots, \frac{k}{1-\nu} - 1 \tag{21c}$$

Because of these additional degeneracies we may ask whether or not there is a higher symmetry in this special case similar to the higher (SU_3) symmetry that occurs in the non-relativistic harmonic oscillator. This does not seem to be the case[14,15]. However we have found the ladder operator which steps up from one Dirac state in the degenerate band to the next Dirac state. This operator is

$$\vec{A}^\dagger = \begin{pmatrix} (\vec{\sigma}\times\vec{j})h^\dagger & 0 \\ 0 & h^\dagger(\vec{\sigma}\times\vec{j}) \end{pmatrix} \tag{22a}$$

and it has the ladder property

$$\vec{A}^\dagger \psi_{kj} = c_{kj}\psi_{k,j+\nu} , \tag{22b}$$

where c_{kj} is some constant. That is, it leaves k constant but increases (decreases) the angular momentum for positive (negative) helicity states thereby generating all the states in a band. The $\vec{A}^\dagger$ and $\vec{A}$ commute with the Hamiltonians,

$$[\vec{A}^\dagger,\bar{H}] = [\vec{A},\bar{H}] = [\vec{A}^\dagger,H] = [\vec{A}, H] = 0, \tag{23}$$

but $\vec{A},\vec{A}^\dagger$ do not seem to form a closed algebra.

CONCLUSIONS

We have shown that the square of the Dirac Hamiltonian of a neutral fermion interacting via an anomolous magnetic moment in an electric potential is equivalent to a three-dimensional supersymmetric Schrodinger equation. If the potential grows as a power of r as r increases, $\phi \overset{r\to\infty}{\to} \phi_0 r^P$, P>0, then the lowest energy of the Hamiltonian equals the rest mass of the fermion, and the Dirac eigenfunction has only an upper component which is normalizable. Furthermore there will be an infinite number of states with the same energy but different angular momenta, $j = \frac{1}{2}, \frac{3}{2}, \ldots$. Also, the higher energy states have upper and lower components which form a supersymmetric doublet, and each separately are eigenfunctions of the supersymmetric Hamiltonian with the same eigenvalue.

REFERENCES

1. B. D. Serot and J. D. Walecka, The Relativistic Nuclear Many-Body Problem, Adv. in Nuclear Physics, 16:1 (1986).
2. A. Arima and F. Iachello, The Interacting Boson Model, Adv. in Nuclear Physics, 13:139 (1984).
3. J. N. Ginocchio, A Schematic Model for Monopole and Quadrupole Pairing in Nuclei, Ann. Phys., 126:234 (1980).
4. C. L. Wu, D. H. Feng, X. G. Chen, J. Q. Chen, and M. W. Guidry, A Fermion Dynamical Symmetry Model and the Nuclear Shell Model, Phys. Lett. 168B:313 (1986).

5. M. W. Guidry, C. L. Wu, D. H. Feng, and J. N. Ginocchio, A Fermion Dynamical Symmetry Model for High Spin Physics, Los Alamos preprint LA-UR-86-407, to be published in Phys. Lett. (1986).
6. I. J. R. Aitchison, An Informal Introduction to Gauge Field Theories, Cambridge University Press, New York (1982).
7. E. Witten, Constraints on Supersymmetry Breaking, Nucl.Phys. B202:253 (1982).
8. F. Cooper and B. Freedman, Aspects of Supersymmetric Quantum Mechanics, Ann. Phys., 146:262 (1983).
9. H. Ui, Supersymmetric Quantum Mechanics in Three-Dimensional Space. I, Prog. Theor. Phys. 72:813 (1984).
10. N. V. V. J. Swamy, Exact Solution of the Dirac Equation with an Equivalent Oscillator Potential, Phys. Rev. 180:1225 (1969).
11. Y. M. Cho, A Potential Approach to Spinor Quarks, Il Nuovo Cimento 23A:550 (1974).
12. A. B. Balantekin, Accidental Degeneracies and Supersymmetric Quantum Mechanics, Ann. Phys. 164:277 (1986).
13. V. A. Kostelecky, M. M. Nieto, and D. R. Truax, Supersymmetry and the Relationship Between Coulomb and Oscillator Problems in Arbitrary Dimensions, to be published in Phys. Rev. D.
14. H. Ui and G. Takeda, Does Accidential Degeneracy Imply a Symmetry Group?, Prog. Theor. Phys. 72:266 (1984).
15. M. Moshinsky and C. Quesne, Does Accidental Degeneracy Imply a Symmetric Group?, Ann. Phys. 148:462 (1983).

FINITE DIMENSIONAL INDECOMPOSABLE REPRESENTATIONS OF THE POINCARÉ ALGEBRA

Bruno Gruber* and Romuald Lenczewski**

*Physics Department, Southern Illinois University
Carbondale, IL 62901

**Mathematics Department, Southern Illinois University
Carbondale, IL 62901

1. INTRODUCTION

The Poincaré algebra $P = \text{iso}(3,1)$ is an example of a non-semi-simple Lie algebra. More precisely, it is a semi-direct product of the Lorentz algebra $L = \text{so}(3,1)$ and the algebra of translations K in the 4-dimensional Minkowski's space-time.

K forms not only a subalgebra of P but also an ideal i.e. $[K,P] = K$, where $[\ ,\ ]$ denotes the Lie product in P. This is the reason why finite dimensional indecomposable representations occur in the case of the Poincaré algebra in opposition to semi-simple Lie algebras where all finite-dimensional representations are irreducible or completely reducible.

It is of considerable interest to find as many finite dimensional indecomposable representations of P as possible, the motivation being provided by physics of unstable particles ([1],[2],[3]). Paneitz [4] has defined indecomposable representations of P up to dimension 8. In our work [5] we obtained chains of new representations with no dimension bound, although not all of those with dimensions less than or equal to 8 were recovered. Nevertheless, the generality of the algebraic approach we used seems to throw more light onto the subject and may lead to its complete treatment.

A detailed presentation of the method used can be found in our previous work ([5],[6],[7]). Let us only recall that the representations are obtained on the universal enveloping algebra Ω of P through the left multiplication of Ω by P. This makes Ω a left P-module. The corresponding representation is denoted by ρ. If we take Ω/I, where I is an ideal, then we obtain a quotient module. The corresponding representation can be referred to as the quotient representation. Of particular interest will be Ω_+ generated by raising operators, Ω_- generated by lowering operators and Ω_k generated by K.

As it will become apparent later, the embedding $R \subset L \subset P$ (where $R = \text{so}(3)$) is of crucial importance. On one hand it leads to the angular momentum basis and on the other to the identification of interacting [4] L-representations in the indecomposable P-representations.

All finite dimensional indecomposable representations obtained are of the following block-form:

$$\begin{pmatrix} [\sigma_1] & & & & \\ [\xi_{12}] & [\sigma_2] & & & \\ & [\xi_{23}] & [\sigma_3] & & \\ & & & \ddots & \\ & & & [\xi_{k-1,k}] & [\sigma_k] \end{pmatrix}$$

where each $[\sigma_i]$ stands for an irreducible L-representation (finite dimensional), each $[\xi_{ij}]$ represents the interaction between $[\sigma_i]$ and $[\sigma_j]$ and the dimensions of $[\sigma_i]$'s are strictly decreasing. Bases for these representations are defined and dimension formulae are introduced.

More precise definitions of the terms used can be found in general references ([8],[9],[10]).

2. COMMUTATION RELATIONS

The basis for the Poincaré algebra P will consist of the following infinitesimal generators:

$\{h_3,h_+,h_-\}$ - rotations R in the 3-dimensional space

$\{p_3,p_+,p_-\}$ - Lorentz boosts

$\{k_0,k_3,k_+,k_-\}$ - translations in Minkowski's space-time

with the following commutation relations for L:

$$[L,L]\begin{cases} [h_3,h_\pm] = \pm h_\pm \quad [h_+,h_-] = 2h_3 \\ [h_3,p_\pm] = \pm p_\pm \quad [h_+,p_-] = [p_+,h_-] = 2p_3 \\ [p_3,p_\pm] = \mp h_\pm \quad [p_3,h_\pm] = \pm p_\pm \quad [p_+,p_-] = -2h_3 \end{cases}$$

and the interaction between K and L:

$$[K,L]\begin{cases} [k_3,h_\pm] = \pm k_\pm \quad [h_3,k_\pm] = \pm k_\pm \\ [k_0,p_\pm] = -k_\pm \quad [p_3,k_3] = -k_0 \\ [p_3,k_0] = k_3 \quad [h_\pm,k_\mp] = \pm 2k_3 \\ \qquad [p_\pm,k_\mp] = -2k_0 \end{cases}$$

In the above it is understood that only nonvanishing Lie products are listed and all upper signs or all lower signs are taken simultaneously.

For the Poincaré algebra extended by dilations D, we have to add the following nonvanishing Lie products:

$$[K,D]\ \{\ [k_\pm,d] = k_\pm \qquad [k_3,d] = k_3 \qquad [k_0,d] = k_0$$

where d is the infinitesimal generator of D. The extended Poincaré algebra will be denoted by P'.

3. REPRESENTATIONS ON Ω_k

The left multiplication of Ω by P induces a representation on Ω_k provided that $\rho(h_3)\mathbf{1} = \rho(p_3)\mathbf{1} = 0$. In other words, Ω becomes then a left P-module. An infinite dimensional indecomposable representation is obtained:

$$\Omega_k: \{X(p,q,z,w) = k_-^p k_+^q k_3^z k_0^w, \qquad p,q,z,w \in \mathbb{N}\}$$

(where $\mathbb{N}$ = nonnegative integers)

$$\rho(L)\begin{cases} \rho(h_3)X = (q-p)X \\ \rho(h_-)X = zX(p+1,z-1)-2qX(q-1,z+1) \\ \rho(h_+)X = -zX(q+1,z-1)+2pX(p-1,z+1) \\ \rho(p_3)X = wX(z+1,w-1)-zX(z-1,w+1) \\ \rho(p_-)X = wX(p+1,w-1)-2qX(q-1,w+1) \\ \rho(p_+)X = wX(q+1,w-1)-2pX(p-1,w+1) \end{cases}$$

$$\rho(K)\begin{cases} \rho(k_3)X = X(z+1) \\ \rho(k_0)X = X(w+1) \\ \rho(k_-)X = X(p+1) \\ \rho(k_+)X = X(q+1) \end{cases}$$

One can notice that $\rho(L)$ does not change $N = p+q+z+w$ whereas $\rho(K)$ increases N by one. This gives rise to indecomposable representations. On the quotient spaces, ρ induces finite dimensional representations. If they are indecomposable, then they consist of several interacting Lorentz representations. For example, representations of dimension < 20 have the following $R \subset L$ decomposition:

$$5 = (3+1) + 1$$
$$14 = (5+1) + (3+1) + 4$$
$$15 = (5+1) + (3+1) + 4 + 1$$

i.e. the sums give the dimensions of interacting L-representations with sums inside the parentheses giving the R-content.

A general formula for the dimensions of finite dimensional indecomposable representations induced on quotient spaces of Ω_k can be obtained:

$$\dim \rho(N,M) = \{(N+1)(N+2)(N+3)(N+4) - M(M+1)(M+2)(M+3)\}$$

$$N = 1,2,3,\ldots, \qquad M = 0,1,2,\ldots,N-1.$$

The representation of P on Ω_k can be extended to a representation of P' on Ω'_k. The latter is defined as a natural extension of Ω_k with one more condition: $\rho(d)\mathbf{1} = \Lambda_3\,\mathbf{1}$ where $\Lambda_3 \in$ ¢, the corresponding induced representation is the extension of the above representation in the sense that the formulae for $\rho(P)$ are the same and moreover

$$\rho(d)X = (\Lambda_3-p-q-z-w)X.$$

Clearly, the indecomposability of the representation ρ and subsequently, of finite dimensional quotient representations is not affected. All extensions are parametrized by Λ_3.

4. REPRESENTATIONS ON Ω_+

The representations induced on the quotient modules Ω_+ (sometimes referred to as Verma modules) will be discussed in this section. The module Ω_+ is defined as a quotient module Ω/I, where the ideal I is generated by: $h_3 - \Lambda_1$, $p_3 - \Lambda_2$, h_-, p_-, k_-, k_0, k_3, where Λ_1 and Λ_2 are arbitrary complex numbers.

For fixed values of Λ_1 and Λ_2 one obtains the following induced representation:

$$\rho(L)\begin{cases} \rho(h_3)X = (\Lambda_1+n+s+q)X \\ \rho(h_+)X = X(n+1) \\ \rho(h_-)X = n(-2\Lambda_1-2s-2q-n+1)X(n-1)-2s\Lambda_2X(s-1)+s(s-1)X(s-2,n+1) \\ \rho(p_3)X = \Lambda_2X+nX(s+1,n-1)-sX(n+1,s-1) \\ \rho(p_+)X = X(s+1) \\ \rho(p_-)X = s(2\Lambda_1+2n+2q+s-1)X(s-1)-2n\Lambda_2X(n-1)-n(n-1)X(s+1,n-2) \end{cases}$$

$$\rho(K)\begin{cases} \rho(k_3)X = nX(n-1,q+1) \\ \rho(k_0)X = -sX(s-1,q+1) \\ \rho(k_+)X = X(q+1) \\ \rho(k_-)X = -n(n-1)X(n-2,q+1)-s(s-1)X(s-2,q+1) \end{cases}$$

In order to carry out a change of basis from the basis of standard monomials X(s,n,q) to the angular momentum basis it is necessary to find all $\rho(k_-)$-extremal vectors. These are defined as follows:

$$\rho(h_-)y_{Nq} = 0$$

$$y_{Nq} = \sum_{\substack{k\geq 0\\ k+q\leq N}} c_{kq}\, X(N-k-q,k,q)\ .$$

The new basis for the Verma module Ω_+ is obtained by acting on extremal vectors with the raising operator h_+, namely,

$$\{y^n_{Nq} = h^n_+\, y_{Nq},\qquad n,N \in \mathbb{N}\ \}.$$

The representation ρ assumes the following form in the new (angular momentum) basis:

$$\rho(h_3)y^n_{Nq} = G^n_N y^n_{Nq} \qquad \rho(h_+)y^n_{Nq} = y^{n+1}_{Nq}$$

$$\rho(h_-)y^n_{Nq} = nF^n_N y^{n-1}_{Nq}$$

$$\rho(p_3)y^n_{Nq} = -\alpha_{Nq}F^n_N y^{n+1}_{N-1,q} + \beta_{Nq}G^n_N y^n_{Nq} + ny^{n-1}_{N+1,q}$$

$$\rho(p_+)y^n_{Nq} = \alpha_{Nq}y^{n+2}_{N-1,q} + \beta_{Nq}y^{n+1}_{Nq} + y^n_{N+1,q}$$

$$\rho(p_-)y^n_{Nq} = -\alpha_{Nq}F^n_N(F^n_N+1)y^n_{N-1,q} + \beta_{Nq}\, n\, F^n_N\, y^{n-1}_{Nq} - n(n-1)y^{n-2}_{N+1,q}$$

$$\rho(k_-)y^n_{Nq} = -\delta_{Nq}F^n_N(F^n_N+1)y^n_{N-1,q+1} + \gamma_{Nq}\, n\, F^n_N\, y^{n-1}_{N,q+1} - n(n-1)y^{n-2}_{N+1,q+1}$$

$$\rho(k_+)y^n_{Nq} = \delta_{Nq}y^{n-2}_{N-1,q+1} + \gamma_{Nq}y^{n+1}_{N,q+1} + y^n_{N+1,q+1}$$

$$\rho(k_3)y^n_{Nq} = -\delta_{Nq}F^n_N y^{n+1}_{N-1,q+1} + \gamma_{Nq}\, G^n_N\, y^n_{N,q+1} + n y^{n-1}_{N+1,q+1}$$

$$\rho(k_0)y^n_{Nq} = -(N-q)y^n_{N,q+1}$$

where:

$$F^n_N = -2\Lambda_1 - 2N - n + 1, \qquad G^n_N = N + n + \Lambda_1$$

$$\alpha_{Nq} = \frac{[\Lambda_2^2+(1-\Lambda_1-N)^2](N-q)(-2\Lambda_1+2-N-q)}{(-\Lambda_1+1-N)^2(-2\Lambda_1-2N+3)(-2\Lambda_1-2N+1)}$$

$$\delta_{Nq} = \frac{(N-q)(N-q-1)[\Lambda_2^2+(-\Lambda_1-N+1)^2]}{(-2\Lambda_1-2N+1)(-2\Lambda_1-2N+3)(-\Lambda_1-N+1)^2}$$

$$\beta_{Nq} = \frac{\Lambda_2(-\Lambda_1+1-q)}{(\Lambda_1+N)(-\Lambda_1+1-N)}, \qquad \gamma_{Nq} = \frac{\Lambda_2(N-q)}{(\Lambda_1+N)(-\Lambda_1+1-N)}$$

When restricted to *L*, one obtains representations of *L* discussed in our previous articles ([6],[7]).

The above formulae derived on the angular momentum basis y^n_{Nq} for $n,q,N \in \mathbb{N}$ and $N \geq q$ can be extended to all integral values of n,q,N. The original space Ω_+ then can be viewed as an invariant subspace of an extended space Ω^e_+. The latter has a basis consisting of abstract elements y^n_{Nq} for $n,q,N \in \mathbb{Z}$.

On the N-n plane, the action of $\rho(L)$ is as follows:

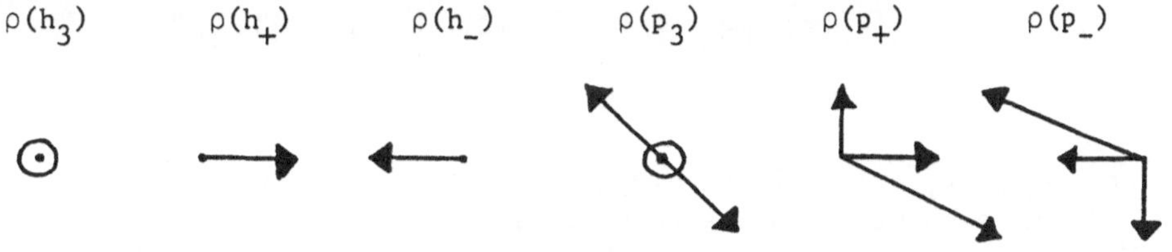

where the vertical axis represents N and the horizontal axis represents n. If the parameter q is added to the geometrical picture, the action of $\rho(K)$ can be identified. Namely, $\rho(k_3)$, $\rho(k_+)$ and $\rho(k_-)$ act as $\rho(p_3)$, $\rho(p_+)$ and $\rho(p_-)$, respectively, composed with a translation by 1 in the direction of increasing q. Moreover, $\rho(k_0)$ is precisely a translation just mentioned.

In general, there are 3 sources of indecomposability of the above representations for certain values of Λ_1,Λ_2:

(a) vanishing of F_N^n

(b) vanishing of α_{Nq}, δ_{Nq}

(c) q-indecomposability due to the fact that $\rho(K)$ increases the value of q by 1.

On each q-plane, the geometric picture coincides with the Lorentz algebra case and is shown in Figure 1.

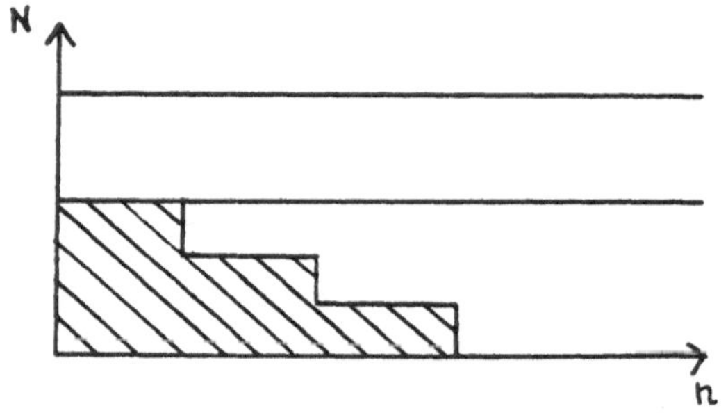

FIGURE 1

The picture should be interpreted in such a way, that the elements lying above the lines of separation form invariant subspaces. For a more detailed presentation the reader is referred to our previous work [5].

The q-indecomposability is easily seen in 3-dimensions and it is a characteristic feature of P-representations. It gives rise to finite dimensional indecomposable representations realized as towers supported by a finite number of q-planes. The latter gives the number of interacting L-representations comprising an indecomposable P-representation.

It is not difficult to notice that integral and half-integral values of Λ_1, $-i\Lambda_2$ give indecomposability of types (a) and (b) listed above. Thus, it is also the only way to obtain finite dimensional indecomposable representations. Those are obtained on quotient spaces.

One obtains the following bases for finite dimensional quotient spaces:

(i) if $\Lambda_1 = -\frac{M}{2}$, $\Lambda_2 = -i\,\frac{n}{2}$ $\quad$ $(M,n = 3,5,7,\ldots,\quad n < M)$

or $\Lambda_1 = -M$, $\Lambda_2 = -in$ $\quad$ $(M,n = 1,2,3,\ldots,\quad n < M)$

$$T_{(\Lambda_1,-i\Lambda_2,q_c)} = \{y^n_{Nq}\mid q \le N \le -\Lambda_1 - i\Lambda_2,\ 0 \le n \le -2\Lambda_1 - 2N,\ 0 \le q \le q_c\}$$

where $0 \le q_c \le -\Lambda_1 - i\Lambda_2$ represents the number of interacting L-representations. Thus, each of the above representations can be labelled with a triplet of numbers $(\Lambda_1,-i\Lambda_2,q_c)$ (integers or half-integers). The $R \subset L$ decomposition of those with dimensions ≤ 20 is given below:

$(-1,0,1)$	$5 = 4 + 1 = (3 + 1) + 1$
$(-\frac{3}{2},-\frac{1}{2},1)$	$8 = 6 + 2 = (4 + 2) + 2$
$(-2,-1,1)$	$13 = 9 + 4 = (5 + 3 + 1) + (3 + 1)$
$(-2,0,2)$	$14 = 9 + 4 + 1 = (5 + 3 + 1) + (3 + 1) + 1$

$(-\frac{5}{2},-\frac{3}{2},1)$ $\quad 14 = 10 + 4 = (6 + 4) + 4$

$(-3,-2,1)$ $\quad 17 = 12 + 5 = (7 + 5) + 5$

$(-\frac{5}{2},-\frac{1}{2},2)$ $\quad 20 = 12 + 6 + 2 = (6+4+2) + (4+2) + 2$

$(-\frac{7}{2},-\frac{5}{2},1)$ $\quad 20 = 14 + 6 = (8 + 6) + 6$

etc., ...

(ii) if $\Lambda_1 = \frac{M}{2}$, $\Lambda_2 = i\,\frac{n}{2}$ $\qquad (M,n = 3,5,7,\ldots, \quad n > M)$

or $\Lambda_1 = M$, $\Lambda_2 = in$ $\qquad (M,n = 1,2,3,\ldots, \quad n > M)$

$$T_{(\Lambda_1,-i\Lambda_2,q_c)} = \{y^n_{Nq} \mid q \leq N \leq -\Lambda_1 - i\Lambda_2,\ \ 2\Lambda_1 - 2N+1 \leq n < 0$$
$$0 \leq q \leq q_c\}$$

where $0 \leq q_c \leq -\Lambda_1 - i\Lambda_2$ has the same meaning as above. The $R \subset L$ decomposition of the representations with dimensions ≤ 20 is given below:

$(1,2,1)$ $\quad 7 = 4 + 3 = (1 + 3) + 3$

$(\frac{3}{2},\frac{5}{2},1)$ $\quad 10 = 6 + 4 = (2 + 4) + 4$

$(2,3,1)$ $\quad 13 = 8 + 5 = (5 + 3) + 5$

$(\frac{5}{2},\frac{7}{2},1)$ $\quad 16 = 10 + 6 = (4 + 6) + 6$

$(1,3,1)$ $\quad 17 = 9 + 8 = (1 + 3 + 5) + (3 + 5)$

$(3,4,1)$ $\quad 19 = 12 + 7 = (5 + 7) + 7$

etc., ...

A general formula for the dimensions of the finite dimensional indecomposable P-representations on Ω^e_+ can be obtained in the following form:

$$d_{(\Lambda_1,-i\Lambda_2,q_c)} = \left|\frac{1}{6}(q_c+1)(q_c+2)(-6\Lambda_1-4q_c+3)+(q_c+1)[(-\Lambda_1-q_c)^2+\Lambda_2^2]\right|$$

Let us consider 2 examples. Namely, we are going to assign matrices to representations $(-1,0,1)$ and $(-\frac{3}{2},\frac{1}{2},1)$, i.e., a 5-dimensional and an 8-dimensional P-representation, respectively. Both of them incorporate 2 interacting L-representations.

<u>Example 1</u>. (See Fig. 2)

$(-1,0,1)$ $\quad 5 = 4 + 1 = (3 + 1) + 1$

(i) $\sigma_1(h_3) = \begin{pmatrix} -1 & & & \\ & 0 & & \\ & & 1 & \\ & & & 0 \end{pmatrix}$ $\qquad \sigma_1(h_+) = \begin{pmatrix} 0&0&0&0 \\ 1&0&0&0 \\ 0&1&0&0 \\ 0&0&0&0 \end{pmatrix}$

$$\sigma_1(h_-) = \begin{pmatrix} 0&2&0&0\\ 0&0&2&0\\ 0&0&0&0\\ 0&0&0&0 \end{pmatrix} \qquad \sigma_1(p_3) = \begin{pmatrix} 0&0&0&0\\ 0&0&0&-1\\ 0&0&0&0\\ 0&1&0&0 \end{pmatrix}$$

$$\sigma_1(p_+) = \begin{pmatrix} 0&0&0&0\\ 0&0&0&0\\ 0&0&0&1\\ 1&0&0&0 \end{pmatrix} \qquad \sigma_1(p_-) = \begin{pmatrix} 0&0&0&-2\\ 0&0&0&0\\ 0&0&0&0\\ 0&0&-2&0 \end{pmatrix}$$

$$\sigma_1(k_0) = \sigma_1(k_3) = \sigma_1(k_-) = \sigma_1(k_+) = 0$$

(ii) σ_2 is a trivial representation

(iii) $\xi_{12}(h_3) = \xi_{12}(h_+) = \xi_{12}(h_-) = \xi_{12}(p_3) = \xi_{12}(p_+) = \xi_{12}(p_-) = 0$

$\xi_{12}(k_+) = (1,0,0,0) \qquad \xi_{12}(k_-) = (0,0,-2,0)$

$\xi_{12}(k_0) = (0,0,0,-1) \qquad \xi_{12}(k_3) = (0,1,0,0)$

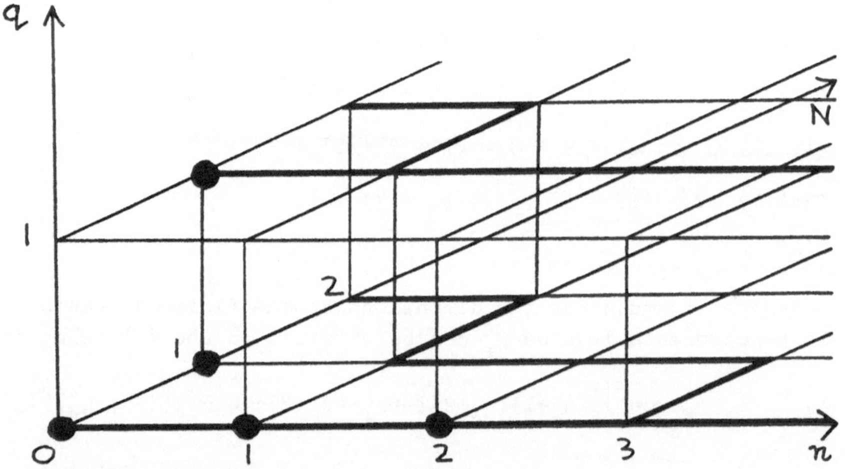

FIGURE 2. 5-dimensional representation

<u>Example 2</u>. (See Fig. 3)

$(-\frac{3}{2},-\frac{1}{2},1) \qquad 8 = 6 + 2 = (4 + 2) + 2$

(i) $\sigma_1(h_3) = \begin{pmatrix} -\frac{3}{2} &&&&& \\ & -\frac{1}{2} &&&& \\ && \frac{1}{2} &&& \\ &&& \frac{3}{2} && \\ &&&& -\frac{1}{2} & \\ &&&&& \frac{1}{2} \end{pmatrix} \quad \sigma_1(h_+) = \begin{pmatrix} 0&0&0&0&0&0\\ 1&0&0&0&0&0\\ 0&1&0&0&0&0\\ 0&0&1&0&0&0\\ 0&0&0&0&0&0\\ 0&0&0&0&1&0 \end{pmatrix}$

$$\sigma_1(h_-) = \begin{pmatrix} 0 & 3 & 0 & 0 & 0 & 0 \\ 0 & 0 & 4 & 0 & 0 & 0 \\ 0 & 0 & 0 & 3 & 0 & 0 \\ 0 & 0 & 0 & 0 & 0 & 0 \\ 0 & 0 & 0 & 0 & 0 & 1 \\ 0 & 0 & 0 & 0 & 0 & 0 \end{pmatrix} \quad \sigma_1(p_3) = \begin{pmatrix} -\frac{i}{2} & 0 & 0 & 0 & 0 & 0 \\ 0 & -\frac{i}{6} & 0 & 0 & -\frac{8}{9} & 0 \\ 0 & 0 & \frac{i}{6} & 0 & 0 & \frac{4}{9} \\ 0 & 0 & 0 & \frac{i}{2} & 0 & 0 \\ 0 & 1 & 0 & 0 & -\frac{5i}{6} & 0 \\ 0 & 0 & 2 & 0 & 0 & \frac{5i}{6} \end{pmatrix}$$

$$\sigma_1(p_+) = \begin{pmatrix} 0 & 0 & 0 & 0 & 0 & 0 \\ \frac{i}{3} & 0 & 0 & 0 & 0 & 0 \\ 0 & \frac{i}{3} & 0 & 0 & \frac{4}{9} & 0 \\ 0 & 0 & \frac{i}{3} & 0 & 0 & \frac{4}{9} \\ 1 & 0 & 0 & 0 & 0 & 0 \\ 0 & 1 & 0 & 0 & \frac{5i}{3} & 0 \end{pmatrix} \quad \sigma_1(p_-) = \begin{pmatrix} 0 & i & 0 & 0 & -\frac{8}{3} & 0 \\ 0 & 0 & \frac{4i}{3} & 0 & 0 & -\frac{8}{9} \\ 0 & 0 & 0 & i & 0 & 0 \\ 0 & 0 & 0 & 0 & 0 & 0 \\ 0 & 0 & -2 & 0 & 0 & \frac{5i}{3} \\ 0 & 0 & 0 & -6 & 0 & 0 \end{pmatrix}$$

$$\sigma_1(k_0) = \sigma_1(k_3) = \sigma_1(k_+) = \sigma_1(k_-) = 0$$

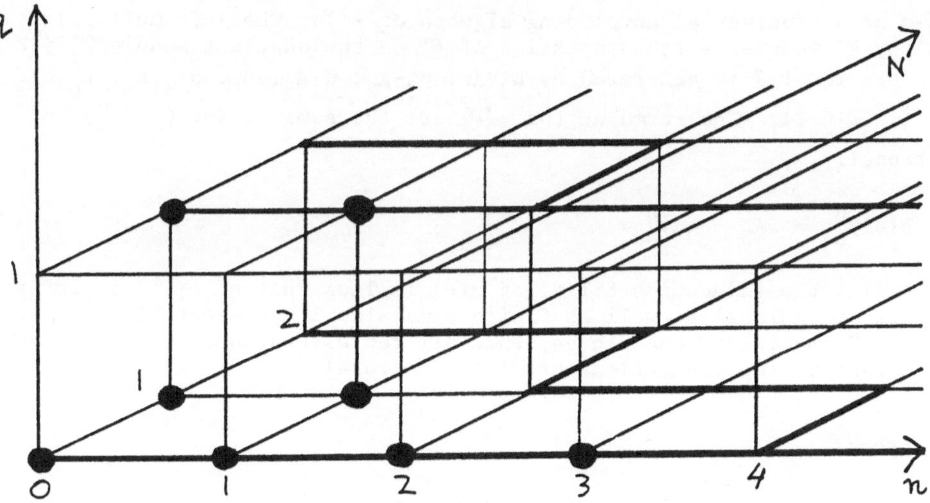

FIGURE 3. 8-dimensional representation

(ii) $\sigma_2(h_3) = \begin{pmatrix} -(1/2) & \\ & 1/2 \end{pmatrix} \quad \sigma_2(h_+) = \begin{pmatrix} 0 & 0 \\ 1 & 0 \end{pmatrix}$

$$\sigma_2(h_-) = \begin{pmatrix} 0 & 1 \\ 0 & 0 \end{pmatrix} \qquad \sigma_2(p_3) = \begin{pmatrix} -(i/2) & \\ & i/2 \end{pmatrix}$$

$$\sigma_2(p_+) = \begin{pmatrix} 0 & 0 \\ i & 0 \end{pmatrix} \qquad \sigma_2(p_-) = \begin{pmatrix} 0 & i \\ 0 & 0 \end{pmatrix}$$

$$\sigma_2(k_0) = \sigma_2(k_3) = \sigma_2(k_+) = \sigma_2(k_-) = 0$$

(iii) $\xi_{12}(h_3) = \xi_{12}(h_+) = \xi_{12}(h_-) = \xi_{12}(p_+) = \xi_{12}(p_-) =$

$$\xi_{12}(k_0) = \begin{pmatrix} 0 & 0 & 0 & 0 & -1 & 0 \\ 0 & 0 & 0 & 0 & 0 & -1 \end{pmatrix} \quad \xi_{12}(k_+) = \begin{pmatrix} 1 & 0 & 0 & 0 & 0 & 0 \\ 0 & 1 & 0 & 0 & \frac{2i}{3} & 0 \end{pmatrix}$$

$$\xi_{12}(k_-) = \begin{pmatrix} 0 & 0 & -2 & 0 & 0 & \frac{2i}{3} \\ 0 & 0 & 0 & -6 & 0 & 0 \end{pmatrix} \quad \xi_{12}(k_3) = \begin{pmatrix} 0 & 1 & 0 & 0 & \frac{i}{3} & 0 \\ 0 & 0 & 2 & 0 & 0 & -\frac{i}{3} \end{pmatrix}$$

All the above representations were obtained on Ω_+^{ℓ}, in opposition to our previous analysis [5], where some of them were derived on Ω_- and some on Ω_+^{ℓ}. It presents no major difficulty to go from one version to the other, since all representations induced on Ω_- can be obtained from those induced on Ω_+ through a Lie algebra automorphism [5].

The indecomposable representations obtained on Ω_+ (Ω_+^{ℓ}) can also be extended to P' as in the case of Ω_k. The carrier space is then Ω_+' ($\Omega_+'^{\ell}$), the natural extension of Ω_+ (Ω_+^{ℓ}). Strictly speaking, Ω_+' is the quotient module of Ω' (universal enveloping algebra of P'). The left multiplication of Ω' by P' induces a representation of P' on the quotient module $\Omega_+' = \Omega'/J$, where the ideal J is generated by $h_3-\Lambda_1$, $p_3-\Lambda_2$, $d-\Lambda_3$, h_-, p_-, k_-, k_0, k_3 ($\Lambda_1,\Lambda_2,\Lambda_3, \in ¢$). The formulae for $\rho(P)$ are the same as for Ω_+ (Ω_+^{ℓ}) and additionally,

$$\rho(d)y^n_{Nq} = (\Lambda_3 - q)y^n_{Nq}.$$

Again, the indecomposability of $\rho(P)$ implies that of $\rho(P')$ as far as infinite dimensional as well as finite dimensional representations are concerned. Summarizing, one can say that all representations of P obtained here extend to representations of P' in a natural way.

REFERENCES

1. P. A. M. Dirac, "The Future of Atomic Physics", Int. J. Th. Phys., 23, No. 8, 677-681 (1984).
2. A. O. Barut and R. Raczka, "Theory of Group Representations and Applications", Polish Scientific Publishers, Warsaw (1977).
3. L. Hlavaty and J. Niederle, "Relativistic Equations and Indecomposable Representations of the Lorentz Group SL(2,C)", Czech. J. Phys. B, 29, No. 3, 283-288 (1979).
4. S. M. Paneitz, "All linear Representations of the Poincaré Group up to Dimension 8", Ann. Inst. H. Poincaré, 40, No. 1, 35-57 (1984).

5. R. Lenczewski and B. Gruber, "Indecomposable Representations of the Poincaré Algebra", J. Phys. A, 19, 1-20 (1986).
6. B. Gruber and R. Lenczewski, "Indecomposable Representations of the Lorentz Algebra in an Angular Momentum Basis", J. Phys. A, 16, 3703-3722 (1983).
7. R. Lenczewski and B. Gruber, "Representations of the Lorentz Algebra on the Space of its Universal Enveloping Algebra", Springer Lecture Notes (to appear).
8. N. Jacobson, "Lie Algebras", Interscience Publishers, New York (1962).
9. J. E. Humphreys, "Introduction to Lie Algebras and Representation Theory", Springer-Verlag, Berlin (1972).
10. J. Dixmier, "Enveloping Algebras", North-Holland Publishers, Amsterdam (1978).

ON THE BEHAVIOUR OF SOLUTIONS OF THE LAPLACE-BELTRAMI EQUATIONS*

A.M. Grundland**

Department of Mathematics and Statistics
Memorial University
St. John's, Newfoundland, Canada, A1C 5S7

1. INTRODUCTION

This paper presents an approach to the nonlinar second order P.D.E.'s based on group theoretical and differential geometric methods - namely symmetry reduction method and Riemann invariants method which provide us with complementary results. The presentation here is based on the results of the previous papers[1,2] (with some modifications) and is their continuation.

We consider the second order P.D.E. $F(\lambda u_{x^0} + \nabla^2 u, (\nabla u|\nabla u), u) = 0$ in an n-dimensional vector space with a scalar product having arbitrary signature $(p,n-p)$ (this type of equation appears in various branches of mathematical physics). We look for certain classes of solutions satisfying the condition of constancy $(\nabla u|\nabla u)$ on the levels of the function u. We prove that the solutions with this property exist and are functionally invariant. For the purpose of further analysis we distinguish two types of these solutions - isotropic and nonisotropic ones.

Using the Riemann invariants method we show (section 3), that all isotropic solutions in Minkowski space $M(n,1)$ are simple waves. In general, for the space with arbitrary signature $(p,n-p)$, $1 < p < n-1$ a certain family of solutions, so called nonplanar simple waves, is discussed.

In section 4 we present nonisotropic solutions u of the postulated class, obtained by the symmetry reduction method according to[1]. We introduce a set of independent variables called symmetry variables which are invariants of the assumed subgroup G_i of the Poincaré group $P(n,1)$ having generic orbits of codimension 1 in Minkowski space $M(n,1)$. This method permits us to obtain in a systematic way all the classifications of the symmetry group of the P.D.E. and enables us to reduce the number of independent variables for some problems, so the solution of the basic system of equations is simplified.

The new elements in this paper are a geometrical interpretation and a systematic analysis of the symmetry groups of certain overdetermined auxiliary systems (2.1) which define symmetry variables ξ such that after the transformation $u = Y(\xi)$, the basic P.D.E. reduces to a second order O.D.E. We demonstrate that the symmetry groups of these auxiliary systems for the cases of the isotropic and nonisotropic solution are infinite dimensional and the 16 parameter Lie group respectively. In the second case this Lie group contains the similitude group Sim(n,1) (i.e. the Poincaré group P(n,1) extended by dilation) which is a more general case than the one considered in previous works[1,3].

2. FORMULATION OF THE PROBLEM

Let E be a $(n + 1)$-dimensional $(n \geq 1)$ real vector space with nondegenerated symmetric scalar product $(\cdot|\cdot):E \times E \to R^1$ with arbitrary signature $(p, n - p)$ (not necessarily positive definite). Let a function $\xi : \Theta \subset E \to R^m$ satisfy the following over-determined system of P.D.E.'s.

$$(\nabla\xi^j|\nabla\xi^k) = \alpha^{jk}(\xi^1, \ldots, \xi^m) \qquad 1 \leq j, k \leq m \leq n + 1$$
$$\lambda\xi^j_{x^o} + \nabla^2\xi^j = \beta^j(\xi^1, \ldots, \xi^m) \qquad \lambda \in R^1 \tag{2.1}$$

where ∇ and ∇^2 denote the gradient and Laplace-Beltrami operator respectively. We assume that $(\nabla\xi^j|\nabla\xi^k)$ and $\lambda\xi^j_{x^o} + \nabla^2\xi^j$ can be expressed in terms of $\xi^1, \ldots, \xi^m$. Then the second order partial differential equation (P.D.E.) of the form

$$F(\lambda u_{x^o} + \nabla^2 u, (\nabla u|\nabla u), u) = 0 \tag{2.2}$$

where $u : \Theta \subset E \to R^1$, after the substitution

$$u = Y(\xi^1, \ldots, \xi^m), \tag{2.3}$$

reduces to a second order P.D.E. for the function Y in m-independent variables

$$F(\alpha^{jk}(\xi) \frac{\partial^2 Y}{\partial\xi^j\partial\xi^k} + \beta^j(\xi) \frac{\partial Y}{\partial\xi^j}, \alpha^{jk}(\xi) \frac{\partial Y}{\partial\xi^j} \frac{\partial Y}{\partial\xi^k}, Y) = 0 \tag{2.4}$$

This follows since from equation (2.3) we have

$$u_{x^o} = \frac{\partial Y}{\partial\xi^j} \xi^j_{x^o}, \quad \nabla u = \frac{\partial Y}{\partial\xi^j} \nabla\xi^j$$

and by virtue of (2.1) we obtain

$$(\nabla u|\nabla u) = \frac{\partial Y}{\partial\xi^j} \frac{\partial Y}{\partial\xi^k} (\nabla\xi^j|\nabla\xi^k) = \alpha^{jk} \frac{\partial Y}{\partial\xi^j} \frac{\partial Y}{\partial\xi^k}$$

$$\lambda u_{x^0} + \nabla^2 u = \lambda \frac{\partial Y}{\partial \xi^j} \xi_{x^0} + \frac{\partial^2 Y}{\partial \xi^j \partial \xi^k} (\nabla \xi^j | \nabla \xi^k) + \frac{\partial Y}{\partial \xi^j} \nabla^2 \xi^j =$$

$$= \alpha^{jk} \frac{\partial^2 Y}{\partial \xi^j \partial \xi^k} + \beta^j \frac{\partial Y}{\partial \xi^j} .$$

System (2.1), where α^{jk} and β^j are treated as arbitrary (not fixed) functions, characterizes the congruence of the levels of functions $\xi^1, \ldots, \xi^m$. Thus from every solution ξ of the system (2.1) we can generate (by integrating the P.D.E. (2.4)) an $\frac{m}{2}(m + 3)$ - parameter family of solutions of the equation (2.2).

Let us consider now the special case when $m = 1$. Then equations (2.1) and (2.4) reduce to a second order O.D.E. for unknown function Y, i.e.

$$F(\alpha(\xi) Y_{\xi\xi} + \beta(\xi) Y_\xi, \alpha(\xi) Y_\xi^2, Y) = 0 \tag{2.5}$$

and to the following system of equations

$$\text{(i)} \quad (\nabla\xi | \nabla\xi) = \alpha(\xi), \qquad \text{(ii)} \quad \lambda \xi_{x^0} + \nabla^2 \xi = \beta(\xi) \tag{2.6}$$

respectively, where α and β are certain functions of one variable ξ. So every solution of the system (2.6) will provide a variable ξ of x in which (2.2) reduces to the O.D.E. (2.5). With no loss of generality, if $\alpha(\xi) \not\equiv 0$ then (as is possible after the transformation $\xi \to Y = \Psi(\xi)$ with $\Psi_\xi = |\alpha(\xi)|^{-\frac{1}{2}}$) we may assume that $\alpha(\xi) = \pm 1$. Hence, instead of (2.6) we obtain

$$(\nabla\xi | \nabla\xi) = \kappa, \quad \lambda \xi_{x^0} + \nabla^2 \xi = \beta(\xi), \quad \kappa = 0, \pm 1. \tag{2.7}$$

So from (2.5) and (2.7) we find

$$F(\kappa Y_{\xi\xi} + \beta(\xi) Y_\xi, \kappa Y_\xi^2, Y) = 0 \tag{2.8}$$

Thus from each solution of the system (2.7) we can generate with the help of the solution of equation (2.8) the two-parameter family of solutions of equation (2.2).

Let us introduce now the following definition. A function ξ defined on an open subset $\Theta \subset E$ is called an isotropic function at the point $x_o \in \Theta$ if $(\nabla\xi | \nabla\xi) = 0$ in a certain neighborhood $\Theta_{x_o} \subset \Theta$ of the point x_o. Otherwise ξ is called a nonisotropic function at the point $x_o \in \Theta$.

Let n-dimensional hypersurface $\Sigma \subset E$ be one of the levels of the function ξ. The hypersurface Σ is called the nonisotropic one if the transversality condition for the equation

$$(\nabla\xi | \nabla\xi) = \pm 1 \tag{2.9}$$

is satisfied (which guarantees the local existence and uniqueness of the solution of (2.9)). It means, that at each point x of the surface Σ there exists the nonisotropic normal unit vector $n(x)$ such that $(n(x)|n(x)) = \pm 1$. One can prove in the same way as that given in the paper[2], that the solution of (2.9) with the boundary condition $\xi|_{\Sigma} = 0$ satisfies the equation (2.7) iff a nonisotropic hypersurface Σ has all principal curvatures constant.

In this paper we discuss in detail the general form of isotropic and nonisotropic solutions of the nonlinear Laplace-Beltrami equation (2.2) with the property (2.7) for the case when $\lambda = 0$. The case of $\lambda \neq 0$ will be the subject of a future investigation. Examples (solutions of nonlinear field equations) illustrating the described procedures can be found in[2,3,4].

3. ISOTROPIC SOLUTIONS

The procedure of constructing the isotropic solutions of the system (2.1) demonstrated below is based on the Riemann invariants method, modified for the second order P.D.E.'s (according to[2]).

Let M be an $n = m + 1$ dimensional $(n \geq 2)$ Minkowski space and let $g := (\cdot|\cdot) : M \times M \to \mathbb{R}^1$ be a degenerate symmetric scalar product wtih signature $(1,m)$. By $x = (x^o, \ldots, x^m)$ we denote Cartesian coordinates in M so that the scalar product is defined by

$$(x|y) = x^{\mu}y_{\mu} = x_o y_o - \sum_{i=1}^{m} x_i y_i$$

We use also the following standard notation

$$\partial_{\mu} := \frac{\partial}{\partial x^{\mu}},\ \xi_{\mu} := \frac{\partial \xi}{\partial x^{\mu}},\ \xi_{\mu\nu} := \frac{\partial^2 \xi}{\partial x^{\mu}\partial x^{\nu}},\ \xi^o := \xi_o,\ \xi^i := -\xi_i,$$

$$\xi^i_{\mu} := -\xi_{i\mu},\ \text{etc.}\quad F' := \frac{dF(\xi)}{d\xi},\ F'' := \frac{d^2F(\xi)}{d\xi^2},\ \text{etc.}$$

Now we prove the following statement

<u>Lemma 1.</u> If the function ξ defined on some neighborhood of the point x_o of Minkowski space M satisfies

$$(\nabla\xi|\nabla\xi) = 0 \tag{3.1}$$

then for any $a \in R^1$ and

$$z_i = \sum_{j=1}^{m} \xi_{ij}\, n_j - a n_i, \quad \text{where} \quad n_i := \frac{\xi_i}{\xi_o} \tag{3.2}$$

the following identity holds:

$$\sum_{i,j=1}^{m} (\xi_{ij} - z_i n_j - z_j n_i)^2 = 4a^2 - 4a \sum_{i,j=1}^{m} \xi_{ij} n_i n_j + \sum_{i,j=1}^{m} \xi_{ij} \xi_{ij}$$

$$2 \sum_{i,j,k,\ell=1}^{m} \xi_{ik} \xi_{j\ell} n_k n_j n_\ell n_i - 2 \sum_{i,j,k=1}^{m} \xi_{ij} \xi_{ik} n_j n_k \tag{3.3}$$

Proof. From (3.1) and (3.2) we obtain the following formulae

$$n_i^2 = \xi_o^{-2} \sum_{i=1}^{m} \xi_i^2 = 1, \quad z_i n_j = \sum_{k=1}^{m} \xi_{ik} n_k n_j - a n_i n_j \tag{3.4}$$

Hence

$$\sum_{i,j=1}^{m} (\xi_{ij} - z_i n_j - z_j n_i)^2 = \sum_{i,j=1}^{m} \{[\xi_{ij} - \sum_{k=1}^{m} (\xi_{ik} n_k n_j - \xi_{jk} n_k n_i)]$$

$$+ 2a n_i n_j\}^2 = \sum_{i,j=1}^{m} \xi_{ij} \xi_{ij} + 2 \sum_{i,k,\ell=1}^{m} \xi_{ik} \xi_{i\ell} n_k n_\ell$$

$$- 4 \sum_{i,j,k=1}^{m} \xi_{ij} \xi_{ik} n_k n_j + 2 \sum_{i,j,k,\ell=1}^{m} \xi_{ik} \xi_{j\ell} n_k n_j n_\ell n_i + 4a^2 n_i^2 n_j^2$$

$$+ 4a\{ \sum_{i,j=1}^{m} \xi_{ij} n_i n_j - 2 \sum_{k,i=1}^{m} \xi_{ik} n_k n_i\} = 4a^2 - 4a \sum_{i,j=1}^{m} \xi_{ij} n_i n_j$$

$$+ 2 \sum_{i,j,k,\ell=1}^{m} \xi_{ik} \xi_{j\ell} n_k n_j n_\ell n_i - 2 \sum_{i,k,\ell=1}^{m} \xi_{ik} \xi_{i\ell} n_k n_\ell$$

$$+ \sum_{i,j=1}^{m} \xi_{ij} \xi_{ij} .$$

Q.E.D.

Therefore the following theorem[2] holds.

Theorem 1. For a function $\xi : \Theta \subset M \to R^1$, of the clsss C^3, defined on some neighborhood of the point $x_o \in \Theta$ of Minkowski space M the following conditions are equivalent:

(i) $(\nabla\xi|\nabla\xi) = 0, \quad \nabla^2\xi = 0$ (3.5)

(ii) $(\nabla\xi|\nabla\xi) = 0, \quad \nabla^2\xi = \beta(\xi)$ (3.6)

for an arbitrary function β.

(iii) $(\nabla\xi|\nabla\xi) = 0, \quad (\nabla(\nabla^2\xi)|\nabla\xi) = 0$ (3.7)

(iv) $\Theta \ni x \to \xi(x)$ may be defined in implicit form by the equation

$$\xi(x) = F((q(\xi)|x - x_o)) \tag{3.8}$$

where F is an arbitrary function of one variable and $\xi \to q(\xi) \in M$ is a one parameter family of non-zero isotropic vectors, i.e. such that $(q(\xi)|q(\xi)) = 0$. The function ξ is called Riemann invariant.

Proof. Obviously (i) implies (ii) and since $(\nabla\beta(\xi)|\nabla\xi) = \beta'(\xi)(\nabla\xi|\nabla\xi)$, (ii) implies (iii). The compatibility conditions for the systems (3.6) require that $\beta = 0$, so (ii) implies (i). Furthermore, differentiating (3.8) we have

$$\nabla\xi = \frac{F'((q|x - x_o))}{1 - F'((q|x - x_o))(q'|x - x_o)} \, q$$

So $\nabla\xi$ is isotropic vector where, $F'((q|x - x_o))(q'|x - x_o) \neq 1$, which is just the condition of a local solvability of the system (3.8). By successive differentiation we obtain

$$\nabla^2\xi = \frac{2F'((q|x - x_o))(q'|\nabla\xi)}{1 - F'((q|x - x_o))(q'|x - x_o)} = \frac{2F'^2((q|x - x_o)) \cdot (q'|q)}{\{1 - F'((q|x - x_o))(q'|x - x_o)\}^2} = 0,$$

since $(q(\xi)|q(\xi)) = 0$ implies $(q'(\xi)|q(\xi)) = 0$. Then the function ξ is a general integral of system (3.5). Therefore (iv) implies (i) and it remains to prove that (iii) implies (iv).

Let us assume that ξ satisfies condition (iii). Differentiating the isotropy condition (3.7) we have

$$\sum_{\lambda=0}^{m} \xi^\lambda \xi_{\lambda\mu} = 0, \quad \sum_{\lambda=0}^{m} \xi^\lambda \xi_{\lambda\mu\nu} + \xi^\lambda{}_\mu \xi_{\lambda\nu} = 0 \tag{3.9}$$

Thus the second equation of the system (3.7) may be written in the form

$$\sum_{\lambda,\mu=0}^{m} \xi^\lambda{}_\mu \, \xi^\mu{}_\lambda = 0 \tag{3.10}$$

If $\nabla\xi \neq 0$ i.e. $\xi_o = \varepsilon\, (\sum_{i=1}^{m} \xi_i\, \xi_i)^{\frac{1}{2}} \neq 0$, $\varepsilon = \pm 1$ then from (3.9) we obtain

$$\xi_{oi} = \xi_o^{-1} \sum_{j=1}^{m} \xi_{ij}\, \xi_j, \quad \xi_{oo} = \xi_o^{-2} \sum_{i,j=1}^{m} \xi_{ij}\, \xi_i\, \xi_j \tag{3.11}$$

So (3.10), by virtue of (3.2), may be expressed as

$$0 = (\xi_{oo})^2 - 2 \sum_{i=1}^{m} \xi_{oi}\, \xi_{oi} + \sum_{i,j=1}^{m} \xi_{ij}\, \xi_{ij} +$$

$$= \sum_{i,j,k,\ell=1}^{m} \xi_{ij}\, \xi_{k\ell}\, n_i n_j n_k n_\ell - 2 \sum_{i,j,k=1}^{m} \xi_{ij}\, \xi_{ik}\, n_j n_k +$$

$$+ \sum_{i,j=1}^{m} \xi_{ij}\, \xi_{ij} \tag{3.12}$$

Taking into account Lemma 1 one can see that equation (3.12) describes the fact that all levels of the function ξ are hyperplanes in M. In fact, substituting (3.11) into (3.12) we get

$$\sum_{i,j=1}^{m} (\xi_{ij} - z_i n_j - z_j n_i)^2 = (2a - \sum_{i,j=1}^{m} \xi_{ij}\, n_i n_j)^2 ,$$

which means that

$$\xi_{ij} = z_i n_j + z_j n_i \tag{3.13}$$

for z_i given by (3.2), with $a = \frac{1}{2} \sum_{i,j=1}^{m} \xi_{ij}\, n_i n_j$. Thus using (3.11) we obtain

$$\xi_{\mu\nu} = \beta_\mu\, \xi_\nu + \beta_\nu\, \xi_\mu \quad \text{where} \quad \beta_o = \frac{1}{2}\, \xi_o^{-1} \sum_{i,j=1}^{m} \xi_{ij}\, n_i n_j ,$$

$$\beta_i = z_i\, \xi_o^{-1} \tag{3.14}$$

If X is a vector field in M tangent to the levels of the function ξ, i.e. $\sum_{\mu=0}^{m} X^\mu\, \xi_\mu = 0$, then (3.14) gives

$$\nabla_X\, \xi_\mu = \sum_{\nu=0}^{m} (\beta_\nu\, X^\nu)\, \xi_\mu .$$

Hence $\nabla\xi$ has constant direction on each level of the function ξ. So $\nabla\xi = \phi \cdot q(\xi)$ where $q(\xi)$ is an isotropic vector and ϕ is a function defined in a neighborhood of x_o. Therefore

$$\nabla(q(\xi)|x - x_o) = \{(q'(\xi)|x - x_o) + \phi^{-1}\}\, \nabla\xi,$$

and thus the function ξ may be expressed as a function on $(q(\xi)|x - x_o)$. So (3.8) holds. Q.E.D.

Remark 1. Let $R^1 \ni \xi \to q(\xi)$ be an isotropic vector in Minkowski space $M = E_1 \ominus E_m$, $m := n - 1$ where $E_1 \cong R^1$ and E_m are Euclidean spaces of dimension 1 and m respectively. The symbol $\ominus$ denotes, that M is a Cartesian product $E_1 \times E_m$ with signature $(1,m)$. After suitable normalization the isotropic vector $q(\xi)$ can be written in the form $q(\xi) = (1, \bar{e}(\xi))$, where $\bar{e}(\xi) \in E_m$ is a unit vector. So (3.8) takes the form

$$\xi = F(x^o - \bar{e}(\xi) \cdot \bar{x}), \quad (\bar{e}(\xi)|\bar{e}(\xi)) = 1 \tag{3.14}$$

The form of solution (3.14) suggests that the vector q should be treated as an analogue of the wave vector $(\omega,\bar{k})$, which determines the velocity and direction of the propagation of the wave. Here $(\omega,\bar{k})$ depends also on the value of the solution, therefore the profile of the wave changes during propagation. This is due to the implicit form of the expression (3.8).

Remark 2. In general for arbitrary signature $(p, n - p)$ of the metric g the solution of (3.5) may be written in implicit form

$$\xi = F((\overset{1}{q}(\xi)|x), \ldots, (\overset{r}{q}(\xi)|(x)), \quad r \leq \min (p, n - p) \qquad (3.15)$$

where $(\overset{1}{q}(\xi), \ldots, \overset{r}{q}(\xi))$ is a basis of an isotropic space $I(\xi) \subset M$ of the dimension r, according to the Witt lemma (i.e. maximal dimension of isotropic subspaces $I \subset E$ is a $\min (p, n - p) =: r$). F is an arbitrary real function of r variables. The solution (3.15) is a generalization of the equation (3.8) and is well known[5] as simple or Riemann waves (nonplanar simple waves when $r > 1$). Note, that each level of the function ξ given by (3.15) is the cylinder Σ_c, i.e.

$$F((\overset{1}{q}(c)|x), \ldots, (\overset{r}{q}(c)|x)) = c$$

profiled by the level of F and with $(\overset{1}{q}(c), \ldots, \overset{r}{q}(c))^{\perp}$ as the generating subspace. In the simplest case when $r = 1$, equation (3.15) reduces to (3.8) and the cylinders become the hyperplanes P_c, orthogonal to the isotropic vector q. From (3.8) it follows that on the hypersurface given by the relations

$$\xi = F((q(\xi)|x)), \quad F'((q(\xi)|x))(q'(\xi)|x) = 1$$

the gradient of the function ξ becomes infinite; this situation is called the gradient catastrophe. In this case, certain discontinuities can arise - e.g. shock waves. In general (except for a few cases - e.g. if planes P_c are parallel) there exists a developable surface, which is an envelope of this family of planes P_c. This surface is exactly the place of the gradient catastrophe.

Using the notations from Remark 1 we have

Theorem 2. Let the mapping $\xi : M \supset \Theta \to R$ (of the class c^2) be given in implicit form

$$x^o = \tau(\xi(x^o,\bar{x}),\bar{x}), \quad \text{where} \quad (x^o,\bar{x}) \in E_1 \times E_m \qquad (3.16)$$

then the following conditions are equivalent

$$\text{(i)} \quad (\nabla\xi|\nabla\xi) = 0, \ \nabla^2\xi = 0 \ \text{ in } n\text{-dimensional Minkowski space } M \qquad (3.17)$$

$$\text{(ii)} \quad (\bar{\nabla}\tau|\bar{\nabla}\tau) = 1, \ \bar{\nabla}^2\tau = 0 \ \text{ in } m\text{-dimensional Euclidean space } E_m \qquad (3.18)$$

where the function τ depends on ξ in parametric form. Furthermore, the general solution of (i) is given by (3.8) and the general solution of (ii) is given by

$$\tau(\xi,\bar{x}) = (\bar{w}(\xi)|\bar{x}) + \phi(u), \quad \text{where} \quad (\bar{w}(\xi)|\bar{w}(\xi)) = 1 . \qquad (3.19)$$

Proof. Differentiating (3.16) we get

$$\xi_{x^o} = (\tau_\xi)^{-1}, \ \xi_{x^i} = -(\tau_\xi)^{-1} \tau_{x^i} ,$$

$$\xi_{x^o x^o} = -(\tau_\xi)^{-3} \tau_{\xi\xi}, \ \xi_{x^o x^i} = (\tau_\xi)^{-3}[\tau_{\xi\xi} \tau_{x^i} - \tau_{\xi x^i} \tau_\xi] ,$$

$$\xi_{x^i x^j} = (\tau_\xi)^{-3}[(\tau_{\xi x^i} \tau_{x^j} + \tau_{\xi x^j} \tau_{x^i})\tau_\xi - \tau_{\xi\xi} \tau_{x^i} \tau_{x^j} - \tau_{x^i x^j} \tau_\xi^2] .$$

So we have

$$(\nabla\xi|\nabla\xi) = \xi_o^2 - (\overline{\nabla}\xi|\overline{\nabla}\xi) = (\tau_\xi)^{-2} [1 - (\overline{\nabla}\tau|\overline{\nabla}\tau)] ,$$

$$\nabla^2\xi = \xi_{oo} - \overline{\nabla}^2\xi = (\tau_\xi)^{-3}[-\tau_{\xi\xi} - \tau_\xi \partial_\xi(\overline{\nabla}\tau|\overline{\nabla}\tau) + \tau_{\xi\xi} (\overline{\nabla}\tau|\overline{\nabla}\tau) + \tau_\xi^2 \overline{\nabla}^2\tau] .$$

Hence (i) is equivalent to (ii), and from Theorem 1 the general integral is given by (3.8) and (3.19) respectively. Q.E.D.

Now using group properties we show (according to the procedure proposed by P. J. Olver[6]) how to find the Lie algebra L of the infinitesimal symmetries of the system (3.5) and the one-parameter groups G_i of symmetry group G of this system.

Theorem 3. The Lie algebra of infinitesimal symmetries of the system (3.5) is spanned by the generators

$$\alpha_\mu = P_\mu = \partial_{x^\mu}, \quad \mu = 0, 1, 2, 3 . \qquad (3.20\text{-i})$$

and the infinite-dimensional subalgebra

$$\alpha_4 = Q(A) = A(\xi)\partial_\xi \qquad (3.20\text{-ii})$$

$$\alpha_5 = D(G) = G(\xi)x^\mu\partial_{x^\mu}$$

where A and G are arbitrary functions of ξ satisfying the commutator relations given in Table 1.

Proof. (following P. J. Olver[6]). Note that in this case $p = 4$, $q= 1$ and $k = 2$. The system (3.5) is the linear subvariaty of $X \times U^{(2)}$ given by

$$\Delta^1(x,\xi^{(1)}) := \xi^\mu\xi_\mu = 0, \ \Delta^2(x,\xi^{(2)}) := \xi^\mu_{\ \mu} = 0 \qquad (3.21)$$

The infinitesimal generator α is a smooth vector field on $X \times U \cong R^4 \times R$ given by

$$\alpha = \eta^\mu(x,\xi)\partial_\mu + \phi(x,\xi)\partial_\xi \qquad (3.22)$$

Table 1. The commutation relations.

	$D(G_2)$	$Q(A_2)$	P_0	P_1	P_2	P_3
$D(G_1)$	0	$-D(A_2G_1')$	$-P_0(G_1)$	$-P_1(G_1)$	$-P_2(G_1)$	$-P_3(G_1)$
$Q(A_1)$	$D(A_1G_2')$	$Q(A_1A_2'-A_1'A_2)$	0	0	0	0
P_0	$P_0(G_2)$	0	0	0	0	0
P_1	$P_1(G_2)$	0	0	0	0	0
P_2	$P_2(G_2)$	0	0	0	0	0
P_3	$P_3(G_2)$	0	0	0	0	0

Note that in the above algorithm the variables x^μ , ξ , are treated as independent variables, the dependent ones are η^μ , ϕ and Δ^i , $i = 1,2$. According to theoretical considerations, we construct the second prolongation of the vector field α given by

$$pr^{(2)}\alpha = \eta^\mu\partial_\mu + \phi\partial_\xi + \left(\frac{d\phi}{dx^\mu} - \xi_\nu \frac{d\xi^\nu}{dx^\mu}\right)\partial_{\xi_\mu} + \tag{3.23}$$

$$+ \left\{\frac{d^2\phi}{dx^\mu dx^\nu} - \xi_\rho \frac{d^2\eta^\rho}{dx^\mu dx^\nu} - \xi_{\mu\rho}\frac{d\eta^\rho}{dx^\nu} - \xi_{\nu\rho}\frac{d\eta^\rho}{dx^\mu}\right\}\partial_{\xi_{\mu\nu}}$$

where the total derivative $\frac{d}{dx^\mu}$ is defined as follows

$$\frac{d}{dx^\mu} = \partial_\mu + \xi_\mu\,\partial_\xi + \xi_{\mu\nu}\,\partial_{\xi_\nu} \tag{3.24}$$

The infinitesimal symmetry criterion for (3.5) is

$$pr^{(i)}\alpha[\Delta^i(x,\xi^{(i)})] = 0 \qquad i = 1,\ 2 \tag{3.25}$$

which must be satisfied whenever (3.5) is satisfied. Substituting equations (3.23) and (3.24) into (3.25) replacing ξ_{oo} by $\sum_{i=1}^{m}\xi_{ii}$ and equating the coefficients of various partial derivatives of ξ yields the following system of 23 determining equations.

$$\phi_{x^\mu} = 0\ ,\ \eta^\mu_{x^\mu} - \eta^\nu_{x^\nu} = 0,\ \mu \neq \nu = 0,\ 1,\ 2,\ 3$$

$$\eta^o_{x^k} - \eta^k_{x^o} = 0\ ,\ \eta^\ell_{x^k} + \eta^k_{x^\ell} = 0,\ k \neq \ell = 1,\ 2,\ 3$$

$$\nabla^2\phi = 0\ ,\ \nabla^2\eta^k + 2\phi_{\xi x^k} = 0\ ,\ \eta^k_{\xi x^k} = 0\ ,\ k = 1,\ 2,\ 3$$

$$\nabla^2\eta^o - 2(\Delta\eta^o_{\xi x^o} - \phi_{\xi x^o}) = 0\ ,\ \text{where}\ \Delta := (\xi_x{}^2 + \xi_y{}^2 + \xi_z{}^2)^{\frac{1}{2}} \neq 0$$

Integrating the first order equations we obtain results in many arbitrary functions in terms of x^μ and ξ some of which are eventually eliminated through compatibility conditions and the remaining initial second order equations.

The most general solution of the symmetry equation is

$$\eta^\mu = G(\xi)x^\mu + c^\mu, \quad \phi = A(\xi) \tag{3.26}$$

where G and A are arbitrary function of ξ and c^μ are arbitrary constants. Substitution (3.26) into (3.22) and collecting coefficients of $G(\xi)$ and $A(\xi)$ gives

$$\alpha = G(\xi)\, x^\mu \partial_\mu + A(\xi)\partial_\xi + c^\mu \partial_\mu \tag{3.27}$$

Setting successively each of the arbitrary functions G, A and next the constants c^μ equal to one with the rest zeroed, into (3.27) we obtain the Lie algebra of infintesimal symmetries of the system (3.5). The Lie algebra is spanned by the four vector fields $\alpha_\mu = \partial_\mu$, $\mu = 0, 1, 2, 3$ and the infinite-dimensional subalgebras $\alpha_4 = A(\xi)\partial_\xi$, $\alpha_5 = G(\xi)x^\mu\partial_\mu$. The commutator relations of these vector fields are presented in Table 1.

Q.E.D.

Now we proceed to find the one-parameter groups $G_i = \exp(\lambda\alpha_i)$ generated by the generators defined by (3.20).

$G_0 : (x^o + \lambda, x^1, x^2, x^3, \xi)$, similarly for G_1, G_2 and G_3,

$G_4 : (x^o, x^1, x^2, x^3, B^{-1}(\lambda + B(\xi))$ where $B(\xi) = \displaystyle\int \frac{d\xi'}{A(\xi')}$

$G_5 : (x^o e^{\lambda G(\xi)}, x^1 e^{\lambda G(\xi)}, x^2 e^{\lambda G(\xi)}, x^3 e^{\lambda G(\xi)}, \xi)$

Hence $G_0, \ldots, G_3$ reflect the fact that system (3.5) has constant coefficients, so the generators $\alpha_0, \ldots, \alpha_3$ represent translations in time and space directions, respectively, denoted by $P_0, \ldots, P_3$. G_4 is a consequence of the fact that the system (3.5) prossesses functionally invariant solutions in a certain domain of independent variables x^μ. This means that if $\xi_1(x)$ is a solution of the system (3.5) and $F(\cdot)$ is an abritrary function of one variable, then $\xi_2(x) := F(\xi_1(x))$ again satisfies the system (3.5). G_5 is the well known scale symmetry, so α_5 generates a dilation (denoted D), which is related with Riemann waves.

4. NONISOTROPIC SOLUTIONS

In the literature[7-12] various methods of solving the system of equations (2.6) for the case when $\lambda = 0$ have been presented. In this section we describe a systematic computational method for finding solutions, exact and implicit, of (2.6) when $\lambda = 0$ by the method of symmetry group analysis of P.D.E.'s.

We are interested here in the case when the P.D.E. is reduced to an O.D.E. So, after finding the symmetry group G of the system (2.6) and it's Lie algebra L, which is the starting point in the symmetry reduction method, we have to classify all subalgebras L_i and all subgroup $G_i \subset G$ having generic orbits of codimension 1 in the space of independent and dependent variables. This allows us te determine so called symmetry variables ξ being the solutions of the system of O.D.E.'s

$$\frac{dx^\mu}{\eta^\mu(x, u)} = \frac{du}{\phi(x, u)}, \quad \mu = 0, 1, \ldots, n \tag{4.1}$$

where η^μ and ϕ are given from the generators α_i of the Lie subalgebras L_i. From the first integrals $I_i(x, u) = c_i$, $i = 1, 2$ of (4.1) one can define a new independent variable ξ of x such that it is an invariant of the assumed subgroup G_i of the Lie group G. Then solving the system of equations $I_i(x, u) = c_i$ with respect to u we obtain the following relation

$$u(x) = \alpha(x)F(\xi(x)) + \beta(x) \tag{4.2}$$

where α, β and ξ are given functions of x and F is an arbitrary function of ξ only. Substituting (4.2) into basic P.D.E. (2.2) we obtain an O.D.E. for the function F of ξ. Each variable ξ being the invariant of a given subgroup G_i corresponds to a different solution of the basic P.D.E. (2.2). Let us present now the result as a theorem, for which the proof can be found in[3].

Theorem 4: Let us assume that

(i) G is a Lie group of the Poincare group $P(n, 1)$ having generic orbits of codimension 1 in $n + 1$ - dimensional Minkowski space M.

(ii) ξ_i, $1 \le i \le k$ is a set of functionally independent invariants of a subgroup $G \subset P(n, 1)$.

(iii) $\xi_i : M \supset \Theta \to R^1$ is a nonisotropic mapping at the point $x_o \in \Theta$. Then in a certain neighbourhood $\Theta_{x_o} \subset \Theta$ of the point x_o the functions ξ_i are solutions of the system

$$(\nabla\xi|\nabla\xi) = \varepsilon, \quad \nabla^2\xi = \beta(\xi), \quad \varepsilon = \pm 1. \tag{4.3}$$

iff ξ_i is of one of the following forms given in Table 2.

Classes of solutions described by this Theorem contain in particular plane waves $\xi(x) = (q|x)$ and spherically symmetric solutions (with respect to the group $O(1, n - 1)$) $\xi(x) = \sqrt{\varepsilon(x|x)}$, $\varepsilon = \pm 1$.

The symmetry variables ξ obtained in the Table 2 satisfy the equation (2.7) with $\kappa = \pm 1$, $\lambda = 0$ and $\beta(\xi) = \frac{k}{\xi}$. Thus equation (2.8) has the form

Table 2. Subgroups of $P(n,1)$ with orbits of codimension 1 and their nonisotropic invariants[3].

No	Algebra	Variables ξ_i	$(\nabla\xi\|\nabla\xi)$	$\nabla^2\xi$
1.	M_{ik}, P_i $1<i,k\le n$	x^o	1	0
2.	$M_{\mu\nu}, P_\mu$ $\mu,\nu=0,2,\dots,n$	x^1	-1	0
3.	$M_{ab}, M_{oa}-M_{1a}, P_a$, $M_{o1}+\lambda P_2, P_o-P_1$ $3\le a\le n,\ \lambda\neq 0$	$x^2+\lambda \ln(x^o+x^1)$	-1	0
4.	$M_{ab}, M_{oa}-M_{1a}, P_a$ $M_{o2}-M_{12}+P_0+P_1, P_o-P_1$ $3\le a\le n$	$x^2+\frac{1}{4}(x^o+x^1)^2$	-1	0
5.	$M_{\alpha\beta}, P_a, M_{ab}$ $\alpha,\beta=o,k+1,\dots,n$ $1\le a,b\le k$	$[(x^1)^2+\dots+(x^k)^2]^{\frac{1}{2}}$	-1	$\frac{k-1}{\xi}$
6.	$M_{ab}, P_a, M_{\alpha\beta}$ $k+1\le a,b\le n$ $o\le\alpha,\beta\le k$	$[(x^o)^2-(x^1)^2-\dots-(x^k)^2]^{\frac{1}{2}}$	1	$\frac{k}{\xi}$

$$F(\kappa\, y_{\xi\xi}+\frac{k}{\xi}y_\xi,\ \kappa\, y_\xi^{\,2},\ y)=0\ ,\ k=0,1,\dots,n,\ \kappa=\pm 1 \qquad (4.4)$$

Such O.D.E.'s were investigated by P. Painlevé and R. Gambier[17,18] and others[19,20]. In particular for some special class of functions F these equations are called Enden type equations[21].

Furthermore, the degenerated symmetry variables have also been studied in the past[1,3] and are listed below.

(i) For $(\nabla\xi_1|\nabla\xi_1)=-1$ and $\nabla^2\xi_1=0$ equation (2.2) is reduced to the O.D.E. given by equation (4.4) with $\kappa=-1$ and $k=0$. The symmetry variable is

$$\xi_1=(\vec{B}(\rho)|\vec{x})+\psi(\rho) \qquad (4.5)$$

where $(\overline{B}_i(\rho)|\overline{B}_j(\rho)) = -1$, $(\overline{A}|\overline{B}) = 0$, $\rho = (\overline{x}|\overline{A})$ and $\overline{A}$ is a constant isotropic vector in $M(n-1,1)$, i.e. $(\overline{A}|\overline{A}) = 0$.

(ii) For $(\nabla\xi_k|\nabla\xi_k) = -1$ and $\nabla^2\xi_k = -\frac{k}{\xi_k}$ equation (2.2) is reduced to the O.D.E. given by equation (4.4) with $\kappa = -1$ and the value of k in the numerator coincides with the subscript of ξ_k. The degenerate symmetry variable is

$$\xi_k = \left[\sum_{i=1}^{k+1} (\overline{x} + \overline{c}(\rho)|\overline{B}_i(\rho))^2\right]^{\frac{1}{2}} \tag{4.6}$$

where $(\overline{B}_i(\rho)|\overline{B}_j(\rho)) = -\delta_{ij}$, $(\overline{B}_i(\rho)|\overline{A}) = 0$, $(\vec{A}|\vec{A}) = 0$ and $\rho = (\overline{x}|A)$

(iii) For $(\nabla\tau|\nabla\tau) = -1$ and $\nabla^2\tau = 0$ equation (2.2) is reduced to the O.D.E. given by equation (4.4) with $\kappa = -1$ and $k = 0$. The degenerate symmetry variable is

$$\tau = \frac{x^2 + \epsilon(x^o + x^1)x^3}{[1 + (x^o + x^1)^2]^{\frac{1}{2}}} + \psi(x^o + x^1) \tag{4.7}$$

where ψ is an arbitrary function and $\epsilon = \pm 1$.

Now we study the symmetry group of these particular forms of the system (2.7).

Theorem 5. The system of equations

$$(\nabla\xi|\nabla\xi) = \varepsilon\,,\ \xi\nabla^2\xi = K,\ \varepsilon = \pm 1,\ K = 0, 1, \ldots, 3 \tag{4.8}$$

in 4-dimensional Minkowski space M is invariant with respect to the 16 and 13 dimensional Lie algebras spanned by the generators

$$\alpha_\mu := P_\mu = \partial_{x^\mu},\ \alpha_{3+\ell} = M_{o\ell} = x^o\partial_{x^\ell} + x^\ell\partial_{x^o},\ \alpha_{6+\ell} := M_{i\ell} = x^\ell\partial_{x^i}$$
$$- x^i\partial_{x^\ell},\ \alpha_{10} := D = \sum_{\mu=0}^{3} x^\mu\partial_{x^\mu} + \xi\partial_\xi,\ 0 \le \mu \le 3,\ 1 \le i, \ell \le 3, \tag{4.9}$$

and

(i) when $K = 0$

$$\alpha_{11} := Q = \partial_\xi,\ \alpha_{11+\ell} := R_\ell = \xi\partial_{x^\ell} - \varepsilon x^\ell\partial_\xi,$$
$$\alpha_{15} = B = \xi\partial_{x^o} + \varepsilon x^o\partial_\xi, \tag{4.10-i}$$

(ii) when $K \neq 0$

$$\alpha_{11} = G_1\ \xi^{\gamma_1}\partial_\xi,\ \alpha_{12} = G_2 = \xi^{\gamma_2}\partial_\xi,\ \text{where}$$
$$\gamma_{\frac{1}{2}} = \tfrac{1}{2}(1 - \varepsilon k \pm \sqrt{1 + k^2 - 6\varepsilon k}) \neq 1, \tag{4.10-ii}$$

respectively. The commutation relations are given in Tables 3 and 4 respectively.

Proof. (following P. J. Olver[6]). Note that in this case $p = 4$, $q = 1$ and $k = 2$. The infinitesimal generator α is a vector field on $X \times U \cong R^4 \times R^1$ given by (3.22). The infinitesimal symmetry criterion for (4.8) must be satisfied whenever (4.8) is satisfied. This procedure leads us to the following systems of determining equations.

$$\eta^\mu_{x^\mu} - \eta^\nu_{x^\nu} = 0, \quad \eta^o_{x^\mu} - \eta^\mu_{x^o} = 0, \quad \eta^k_{x^\ell} + \eta^\ell_{x^k} = 0, \quad 0 \le \mu, \nu \le 3$$

$$1 \le k \ne \ell \le 3$$

and (i) when $K = 0$

$$\varepsilon\, \eta^k_\xi + \phi_{x^k} = 0, \quad (\varepsilon + \sum_{k=1}^{3} \xi_{x^k}^2)^{\frac{1}{2}}(\varepsilon\eta^o_\xi - \phi_{x^o}) + \varepsilon(\phi_\xi - \eta^o_{x^o}) = 0$$

$$\varepsilon\eta^o_{\xi\xi} + \nabla^2\eta^o - 2\phi_{\xi x^o} = 0, \quad \varepsilon\eta^k_{\xi\xi} + \nabla^2\eta^k + 2\phi_{\xi x^k} = 0, \tag{4.11-i}$$

$$\varepsilon(2\eta^o_{\xi x^o} - \phi_{\xi\xi}) - \nabla^2\phi = 0$$

(ii) when $K \ne 0$

$$\eta^o_\xi = 0, \quad \varepsilon(\phi_\xi - \eta^o_{x^o}) - (\varepsilon + \sum_{k=1}^{3} \xi_{x^k}^2)^{\frac{1}{2}}\phi_{x^o} = 0, \quad \nabla^2\eta^o - 2\phi_{\xi x^o} = 0,$$

$$\varepsilon\eta^k_\xi + \phi_{x^k} = 0, \quad \eta^k_{x^k} + \nabla^2\phi = 0, \quad K\eta^k_\xi + \varepsilon\xi\eta^k_{\xi\xi} + \xi\nabla^2\eta^k + 2\xi\phi_{\xi x^k} = 0 \tag{4.11-ii}$$

In both cases integrating the first order equations we obtain results in many arbitrary functions in terms of x^μ and ξ, which are eventually eliminated through compatibility conditions and the remaining initial second order equations. Therefore the most general solutions of the symmetry equations (4.11-i) and (4.11-ii) are

(i) when $K = 0$

$$\eta^o = B^o x^o + C^1 x^1 + C^2 x^2 + C^3 x^3 + A^o \xi + D^o$$

$$\eta^1 = C^1 x^o + B^o x^1 + B^1 x^2 + B^2 x^3 + A^1 \xi + D^1$$

$$\eta^2 = C^2 x^o - B^1 x^1 + B^o x^2 + B^3 x^3 + A^2 \xi + D^2 \tag{4.12-i}$$

$$\eta^3 = C^3 x^o - B^2 x^1 - B^3 x^2 + B^o x^3 + A^3 \xi + D^3$$

$$\phi = \varepsilon(A^o x^o - A^1 x^1 - A^2 x^2 - A^3 x^3) + B^o \xi + D^4$$

where A^μ, B^μ, C^k, D^μ, D^4 are arbitrary constants.

Table 3. The commutation relations.

	D	M_{01}	M_{02}	M_{03}	M_{12}	M_{13}	M_{23}	R_1	R_2	R_3	B	P_1	P_2	P_3	P_0	Q
D	0	0	0	0	0	0	0	0	0	0	0	$-P_1$	$-P_2$	$-P_3$	$-P_0$	$-Q$
M_{01}	0	0	$-M_{12}$	$-M_1$	$-M_{02}$	$-M_{03}$	0	$-B$	0	0	$-R_1$	$-P_0$	0	0	$-P_1$	0
M_{02}	0	M_{12}	0	$-M_{23}$	M_{01}	0	M_{03}	0	$-B$	0	$-R_2$	0	$-P_0$	0	$-P_2$	0
M_{03}	0	M_{13}	M_{23}	0	0	M_{01}	0	0	0	$-B$	$-R_3$	0	0	$-P_0$	$-P_3$	0
M_{12}	0	M_{02}	$-M_{01}$	0	0	M_{23}	$-M_{13}$	R_2	$-R_1$	0	0	P_2	$-P_1$	0	0	0
M_{13}	0	M_{03}	0	$-M_{01}$	$-M_{23}$	0	M_{12}	R_3	0	$-R_1$	0	P_3	0	$-P_1$	0	0
M_{23}	0	0	$-M_{03}$	0	M_{13}	$-M_{12}$	0	0	R_3	$-R_2$	0	0	P_3	$-P_2$	0	0
R_1	0	B	0	0	$-R_2$	$-R_3$	0	0	εM_{12}	εM_{13}	$-\varepsilon M_{01}$	εQ	0	0	0	$-P_1$
R_2	0	0	B	0	R_1	0	$-R_3$	$-\varepsilon M_{12}$	0	εM_{23}	$-\varepsilon M_{02}$	0	εQ	0	0	$-P_2$
R_3	0	0	0	B	0	R_1	R_2	$-\varepsilon M_{13}$	$-\varepsilon M_{23}$	0	$-\varepsilon M_{03}$	0	0	εQ	0	$-P_3$
B	0	R_1	R_2	R_3	0	0	0	εM_{01}	εM_{02}	εM_{03}	0	0	0	0	εQ	$-P_0$
P_1	P_1	P_0	0	0	$-P_2$	$-P_3$	0	$-\varepsilon Q$	0	0	0	0	0	0	0	0
P_2	P_2	0	P_0	0	P_1	0	$-P_3$	0	$-\varepsilon Q$	0	0	0	0	0	0	0
P_3	P_3	0	0	P_0	0	P_1	P_2	0	0	$-\varepsilon Q$	0	0	0	0	0	0
P_0	P_0	P_1	P_2	P_3	0	0	0	0	0	0	$-\varepsilon Q$	0	0	0	0	0
Q	Q	0	0	0	0	0	0	P_1	P_2	P_3	P_0	0	0	0	0	0

Table 4. The commutation relations.

	D	M_{12}	M_{13}	M_{23}	M_{o1}	M_{o2}	M_{o3}	P_1	P_2	P_3	P_0	G_1	G_2
D	0	0	0	0	0	0	0	$-P_1$	$-P_2$	$-P_3$	$-P_0$	$(\gamma_1-1)G_1$	$(\gamma_2-1)G_1$
M_{12}	0	0	M_{23}	$-M_{12}$	M_{02}	$-M_{01}$	0	P_2	$-P_1$	0	0	0	0
M_{13}	0	$-M_{23}$	0	M_{12}	M_{03}	0	$-M_{01}$	P_3	0	$-P_1$	0	0	0
M_{23}	0	M_{12}	$-M_{12}$	0	0	M_{03}	$-M_{02}$	0	P_3	$-P_2$	0	0	0
M_{o1}	0	$-M_{02}$	$-M_{03}$	0	0	$-M_{12}$	$-M_{13}$	$-P_0$	0	0	$-P_1$	0	0
M_{o2}	0	M_{01}	0	$-M_{03}$	M_{12}	0	$-M_{23}$	0	$-P_0$	0	P_2	0	0
M_{o3}	0	0	M_{01}	M_{02}	M_{13}	M_{23}	0	0	0	$-P_0$	$-P_3$	0	0
P_1	P_1	$-P_2$	$-P_3$	0	P_0	0	0	0	0	0	0	0	0
P_2	P_2	P_1	0	$-P_3$	0	P_0	0	0	0	0	0	0	0
P_3	P_3	0	P_1	P_2	0	0	P_0	0	0	0	0	0	0
P_0	P_0	0	0	0	P_1	$-P_2$	P_3	0	0	0	0	0	0
G_1	$(1-\gamma_1)G_1$	0	0	0	0	0	0	0	0	0	0	0	0
G_2	$(1-\gamma_2)G_2$	0	0	0	0	0	0	0	0	0	0	0	0

(ii) when $K \neq 0$

$$\eta^o = C^o x^o + B^1 x^1 + B^2 x^2 + B^3 x^3 + D^o,$$
$$\eta^1 = B^1 x^o + C^o x^1 + C^1 x^2 + C^2 x^3 + D^1,$$
$$\eta^2 = B^2 x^o - C^1 x^1 + C^o x^2 - C^3 x^3 + D^2, \qquad (4.12\text{-ii})$$
$$\eta^3 = B^3 x^o - C^2 x^1 - C^3 x^2 + C^o x^3 + D^3,$$
$$\phi = C^o \xi + E^1 \xi^{\gamma_1} + E^2 \xi^{\gamma_2},$$

where C^μ, D^μ, B^k, E^i are arbitrary constants.

In both cases setting successively each of the constants equal to one, with the rest zeroed, into (4.12-i) and (4.12-ii) and then those results into (3.22) successively, we obtain the 16 and 13 vector fields (4.9), (4.10-i) and (4.10-ii) respectively that span the Lie algebra of the infinitesimal symmetries of the system (4.8). The commutator relations of these vector fields are in Table 3 respectively. Q.E.D.

The one-parameter groups $G_i = \exp(\lambda\alpha_i)$ are found to be:

G_o : $(x^o + \lambda, x^1, x^2, x^3, \xi)$, similarly for G_1, G_2 and G_3,

G_4 : $(x^o, x^1, x^3 \sin\lambda + x^2 \cos\lambda, -x^2 \sin\lambda + x^3 \cos\lambda, \xi)$,
similarly for G_5 and G_6,

G_7 : $\left(\frac{x^1 + x^o}{2} e^{\lambda} - \frac{x^1 - x^o}{2} e^{-\lambda}, \frac{x^1 + x^o}{2} e^{\lambda} + \frac{x^1 - x^o}{2} e^{-\lambda}, x^2, x^3, \xi\right)$,
similarly for G_8 and G_9,

G_{10} : $(x^o e^\lambda, x^1 e^\lambda, x^2 e^\lambda, x^3 e^\lambda, \xi e^\lambda)$

and (i) when $K = 0$

G_{11} : $(x^o, x^1, x^2, x^3, \xi + \lambda)$

G_{12} : $\left(x^o, \frac{\xi}{\sqrt{\varepsilon}} \sin(\sqrt{\varepsilon}\lambda) + \frac{x^1}{\sqrt{\varepsilon}} \cos(\sqrt{\varepsilon}\lambda), x^2, x^3, -x\sqrt{\varepsilon} \sin(\sqrt{\varepsilon}\lambda) + \xi\sqrt{\varepsilon} \cos(\sqrt{\varepsilon}\lambda)\right)$,
similarly for G_{13} and G_{14}

G_{15} : $\left(\frac{\xi + \sqrt{\varepsilon}x^o}{2\sqrt{\varepsilon}} e^{\sqrt{\varepsilon}\lambda} - \frac{\xi - \sqrt{\varepsilon}x^o}{2\sqrt{\varepsilon}} e^{-\sqrt{\varepsilon}\lambda}, x^1, x^2, x^3, \frac{\xi + \sqrt{\varepsilon}\, x^o}{2} e^{\sqrt{\varepsilon}\lambda} + \frac{\xi - \sqrt{\varepsilon}x^o}{2} e^{-\sqrt{\varepsilon}\lambda}\right)$,

(ii) when $K \neq 0$

G_{11} : $\left(x^o, x^1, x^2, x^3, (\lambda(1-\gamma_1) + \xi^{1-\gamma_1})^{1/1-\gamma_1}\right)$, similarly for G_{12}.

Hence, the generators P_μ and Q represent translations, $M_{i\ell}$ and R_ℓ represent rotations, $M_{o\ell}$ represents Lorentz boosts, D represents a dilation, B represents a Gallilean boost, and G_i is a consequence of the fact that the system (4.8) possesses functionally invariant solutions.

Let us note, that the symmetry Lie group of the equation (4.8) contains the similitude group Sim (3,1) given by (4.9) which is the more general case than previously described by the Theorem 4. Since all subgroups G_i of the similitude group S(3,1) are known[22,23], it is possible to construct a larger class of solutions of (4.8) than the one given in the Table 2. For each of the symmetry variables ξ_i being the invariants of a given subgroup $G_i \subset$ Sim (3,1) the basic P.D.E. (2.2) will reduce to a different O.D.E. The systematic classification of the solutions obtained for this case will be the subject of further investigation.

REFERENCES

1. A. M. Grundland, J. Harnad, P. Winternitz, Symmetry reduction for nonlinear relativistically invariant equations, J. Math. Phys. 25, 4, 791-806, (1984).

2. G. Cieciura, A. M. Grundland, A certain class of solutions of the nonlinear wave equation, J. Math. Phys. 25, 12, 3460-3469, (1984).

3. A. M. Grundland, J. Harnad, P. Winternitz, Solutions of the multidimensional Sine Gordon equation obtained by symmetry reduction, KINAM Rev. Fis. 4, 333-344, (1982).

4. A. M. Grundland, J. A. Tuszyński, P. Winternitz, Solutions of the multidimensional classical ϕ^6 field equations, (to be published).

5. Z. Peradzynski, Riemann invariants for the nonplanar k-waves, Bull. Acad. Pol. Sci. 29, 10, 67-74, (1971).

6. P. J. Olver, Symmetry groups and group invariant solutions of partial differential equations, J. Diff. Geom. 14, 497-542, (1979).

7. C. B. Collins, All solutions to a nonlinear system of complex potential equations, J. Math. Phys. 21, 2, 240-248, (1980).

8. C. B. Collins, Complex potential equations, special relativity and complexified Minkowski space-time, J. Math. Phys. 21, 2, 249-255, (1980).

9. F. G. Friedlander, The wave equation on a curved space, Cambridge Univ. Press, 1975.

10. S. Helgason, A formula for the radial part of the Laplace-Beltrami operator, J. Diff. Geom. 6, 411-419, (1972).

11. S. Helgason, Wave equations on homogeneous space, Acta Math. 9, 1-33, (1984).

12. H. Urbantke, On complex-valued scalar waves of the simply-progressive type, Report from the Institute of Theoretical Physics, Vienna University, 15, 1-11, (1981).

13. G. W. Bluman, J. D. Cole, Similarity methods for differential equations, Springer-Verlag, New York, 1974.

14. L. V. Ovsiannikov, Group analysis of differential equations, Academic Press, New York, 1982.

15. N. H. Ibragimov, Transformation groups applied to mathematical physics, D. Reidel, Boston, 1984.

16. P. Winternitz, Lie groups and solutions of nonlinear differential equations in Nonlinear Phenomena, Lecture notes in Physics, Springer-Verlag, New York, 263, 1982.

17. P. Painlevé, Sur les équations différentielles du second ordre et d'ordre supérieur dont l'intégrale générale est uniforme, Acta Math. 25, 1, 1-85, (1902).

18. B. Gambier, Sur les équations différentielles du second ordre et du premier degré dont l'intégrale générale est a points critiques fixes, Acta Math. 33, 1, 1-55, (1910).

19. M. J. Ablowitz, A. Ramani, H. Segur, A connection between nonlinear evolution equations and ordinary differential equations of P-type, J. Math. Phys. 21, 4, 715-721 and 1006-1021, (1980).

20. V. Golubev, Lectures on integrals of equations of motion, State Publ. House, Moscow, 1953.

21. H. Davis, Introduction to nonlinear differential and integral equations, Dover, New York, 1962.

22. J. Patera, P. Winternitz, R. Sharp, H. Zassenhaus, Subgroups of the similitude group of three-dimensional Minkowski space, Can. J. Phys., 54, 9, 950-961, (1976).

23. J. Patera, P. Winternitz, H. Zassenhaus, Continuous subgroups of the fundamental groups of physics II. The similitude groups, J. Math. Phys. 16, 8, 1615-1623, (1975).

* Lectures presented at the Symposium on Symmetry of Science, Carbondale, U.S.A., March 24-26, 1986.

**Supported by the Natural Sciences and Engineering Research Council of Canada.

ALGEBRAIC STRUCTURES OF DEGENERATE SYSTEMS, PHYSICAL REQUIREMENTS AND THE INDEFINITE METRIC

Hendrik Grundling and C. A. Hurst

University of Adelaide

ABSTRACT. The basic algebraic structures of systems with a gauge degeneracy have been developed in a C^*- algebra framework in [1], of which the present work is a natural extension. We consider the treatment of algebraic conditions in more detail, as well as the compatibility conditions between a given constraint set and a given set of physical transformations. A list of reasonable physical requirements on the algebraic structure is given. We also consider the existence and structure of indefinite inner product representations and obtain various results for obtaining an indefinite inner product representation.

1. INTRODUCTION

In a previous paper [1] we studied the abstract algebraic structure of degenerate systems in a C^* - algebra. Here we look at two further aspects - how physical requirements can constrain automorphisms of the physical algebra and how indefinite inner product representations can be constructed.

The term "degenerate systems" means systems which have mathematical degrees of freedom without any physical counterparts. In practice, degenerate systems are characterized by supplementary conditions which may be ad hoc, or may be canonical in the sense of Dirac [2], or be generators of non-physical transformations. The imposition of supplementary conditions is intended to select the physical theory.

Quantized systems consist of an algebra of operators acting on a Hilbert space (or a rigged Hilbert space), and supplementary conditions may be imposed as algebraic conditions, written $\underline{A} = 0$, or state conditions, $\underline{A}\,|\psi\rangle = 0$. In order to avoid the complication of unbounded operators, it is convenient to express such constraints as $\underline{U}\lambda = 1$ or $\underline{U}\lambda|\psi\rangle = |\psi\rangle$ where $\underline{U}\lambda = \exp i\lambda\underline{A}$. If $\underline{A}$ is hermitian, $\underline{U}\lambda$ is unitary and it may then be possible to define abstract elements in a C^* - algebra corresponding to them. If $\underline{A}$ is non hermitian, but $\underline{A}^*$ is also a constraint, this pair can be replaced by an equivalent hermitian pair, $\underline{A} + \underline{A}^*$ and $i(\underline{A} - \underline{A}^*)$. If $\underline{A}^*$ is not a constraint then it is necessary to use the product $\underline{A}^*\,\underline{A}$ in order to define $\underline{U}\lambda$, which would take us outside the framework of linear field theories, such as developed by Segal [3], and quantization then becomes more difficult.

In the next section, we give a brief summary of our previous work. In the third section we shall discuss the requirements that physical aspects impose on the automorphisms of the physical algebra. In the final section we shall discuss the formalism of Strocchi and Wightman [4] for constructing indefinite inner product representation within the algebraic framework.

2. BASIC STRUCTURE OF DEGENERATE SYSTEMS

This section contain a brief summary of the results presented in [1], which should be consulted for proofs and other details. Our approach is centred about Dirac's method of constraint, with the intention of presenting that in a suitable mathematical form. The two assumptions that form the foundation of our approach are the following:

(a) $\exists$ a unital C^* - algebra $\mathfrak{F}$, called the field algebra, and a set of states $\mathcal{S}$ on it. All physical information is contained in this pair.

(b) $\exists$ a family of one parameter groups $\{U_i(\lambda) \mid \lambda \in \mathbb{R}, i \in I\}$ in $\mathfrak{F}$ called the state conditions, where I is a (not necessarily finite) index set.

These assumptions need to be verified for particular models, and so far we have only done this for linear field theories.

The Dirac states, which carry the physical information on $\mathfrak{F}$, are defined by:

$$\mathcal{S}_D : \{\omega \in \mathcal{S} \mid \langle \omega ; U_i(\lambda) \rangle = 1, \ \forall i \in I, \lambda \in \mathbb{R}\}.$$

In [1] we showed that this implies that:

$$\langle \omega ; U_i(\lambda) \rangle = 1, \ \forall i \in I, \lambda \in \mathbb{R} \quad \text{iff} \quad \langle \omega ; A U_i(\lambda) \rangle = \langle \omega ; A \rangle = \langle \omega ; U_i(\lambda) A \rangle$$
$$\forall A \in \mathfrak{F}, \ i \in I, \lambda \in \mathbb{R}.$$

Instead of the $U_i(\lambda)$, it is convenient to use instead the quantities $L_i(\lambda) = U_i(\lambda) - 1$, and then:

$$\omega \in \mathcal{S}_D \quad \text{iff} \quad \{L_i(\lambda)\} \subset \operatorname{Ker} \omega \quad \text{iff} \quad \mathfrak{F}\{L_i(\lambda)\} \cup \{L_i(\lambda)\}\mathfrak{F} \subset \operatorname{Ker} \omega.$$

The first result extablished was the condition that $\mathcal{S}_D$ is non-empty. If we denote by $\mathcal{A}(L)$ the C^* - algebra generated by $\{L_i(\lambda)\}$ then:

Theorem: $\mathcal{S}_D \neq \emptyset$ iff $1 \notin \{\mathfrak{F}\mathcal{A}(L)\} \cup \{\mathcal{A}(L)\mathfrak{F}\} \Rightarrow \mathcal{S}_D$ contains pure states.

The next step is to specify the C^* - subalgebra of $\mathfrak{F}$ which incorporates the notion of the supplementary conditions of Dirac. This subalgebra is denoted by $\mathcal{B}$, and is called the constraint algebra. It is the largest C^* - algebra annihilated by all Dirac states. In general it is larger than $\mathcal{A}(L)$. It is defined by:

$$\mathcal{B} = \mathcal{N} \cap \mathcal{N}^*,$$

where $\mathcal{N} = [\mathfrak{F}\mathcal{A}(L)]$, the closed linear space generated by $\mathfrak{F}\mathcal{A}(L)$. An alternative definition, which can be shown to be equivalent, and which is often more convenient in practice is:

$$\mathcal{B} = \mathcal{S}^* \mathcal{A}(L) \mathcal{S},$$

where $\mathcal{S} \subset \mathfrak{F}$ is the largest set for which

$$\mathcal{A}(L)\mathcal{S} \subset [\mathfrak{F}\mathcal{A}(L)].$$

The next step is to define the observable algebra, $\mathcal{O}$, which must be distinguished from the physical algebra $\mathcal{R}$, which will be defined shortly. $\mathcal{O}$ is the multiplier algebra of $\mathcal{D}$:

$$\mathcal{O} = \{ F \in \mathcal{F} \mid FD, DF \in \mathcal{D}, \forall D \in \mathcal{D} \}.$$

Alternatively it is given by:

$$\mathcal{O} = \mathcal{S} \cap \mathcal{S}^*.$$

It follows easily that the commutator algebra $\mathcal{A}(L)'$ of $\mathcal{A}(L)$ is contained in $\mathcal{O}$. This algebra is what one might expect to be the observable algebra as it commutes with all constraints, but it is only suitable when $\mathcal{A}(L)$ is obelian. $\mathcal{O}$ is the largest C^* - subalgebra of $\mathcal{F}$ to which the constraints may be consistently applied.

The rigorous statement of Dirac's heuristic state constraints is:

Theorem: $\omega \in \mathcal{S}_D$ iff $\pi_\omega(\mathcal{D})\Omega_\omega = 0$ where π_ω, Ω_ω are respectively the GNS representation associated with ω and the cyclic vector.

It follows immediately that $\pi_\omega(\mathcal{D})$ not only annihilates Ω_ω but also all vector states in the representation π_ω, and that is the full statement of Dirac's requirement.

The C^* - algebra of physical observables is then defined by:

$$\mathcal{R} = \mathcal{O}/\mathcal{D},$$

and the pure states on $\mathcal{R}$ may be obtained by bijection from pure Dirac states on $\mathcal{O}$.

Automorphisms α on $\mathcal{F}$ can be translated canonically to automorphisms α' on $\mathcal{R}$ if $\alpha(\mathcal{D}) = \mathcal{D}$ i.e. they leave the set of constraints invariant. In the next section we shall study the automorphisms α' more closely.

The preceding discussion has been of state conditions. Algebraic conditions may be treated in an analogous fashion. Indeed, if they are not then ambiguities may arise which are difficult to interpret. A fuller discussion is contained in a forthcoming publication [5].

The assumption that the constraints may be consistently and simultaneously imposed implies that we are dealing with what Dirac calls "first class constraint". At present the only consistent way to handle second class constraint is at the classical level before quantizing. Because of well known ambiguities in the passage from classical to quantum theory it is not clear whether the consistent incorporation of second class constraints into a quantum formulation would be the same as imposing them before quantization. This question requires further study.

3. PHYSICAL REQUIREMENTS FOR DEGENERATE SYSTEMS

Reasonable requirements on the physical and constraint algebras $\mathcal{R}$ and $\mathcal{D}$ are the following:

(i) $1 \notin \mathcal{D}$, (ii) $\mathcal{R}$ is non-trivial, (iii) $\mathcal{R}$ is simple,
(iv) α' is inner on $\mathcal{R}$.

Requirement (i) ensures that $\mathcal{R} \neq \emptyset$, whilst (ii) means that $\mathcal{R} \cong \mathbb{C}$ is excluded. This will be so if $\mathcal{A}'(L)$ is non trivially larger than $\mathcal{D} \oplus 1\mathbb{C}$.

Requirement (iii) is imposed for the physical reason that no physical algebra should have ideals, or, in other words, there are no states which can fail to distinguish between distinct elements of $\mathcal{R}$. This last requirement is additional to the assumptions already made. The last requirement (iv) is designed to exclude operations which lie outside $\mathcal{R}$. The simplest example of this is the time translations of the electromagnetic field. Although the Hamiltonian is initially expressed as a functional of the four vector potential, so far as the electromagnetic fields alone are concerned, it may be taken as having the usual form:

$$H_{physical} = \frac{1}{2}\int d^3x \, (\underline{E}^2 + \underline{B}^2).$$

In this section we shall study the automorphism group Υ of $\mathcal{F}$ for which $\alpha(\mathcal{B}) = \mathcal{B}$, $\forall \alpha \in \Upsilon$. The first result is straightforward.

Theorem: If $\alpha \in \mathrm{Inn}\,\mathcal{F} \cap \Upsilon$, then $\alpha' \in \mathrm{Inn}\,\mathcal{R}$.

If however $\alpha \notin \mathrm{Inn}\,\mathcal{F}$ a more involved discussion is required. One such statement is the following, for the case where α is an element of a one parameter subgroup of Υ. A typical example is the time translation group.

Theorem: $\alpha' \in \mathrm{Inn}\,\mathcal{R}$ iff $\mathrm{Ker}\,\hat{f} = \emptyset, \forall f \in \{f \in L^1(\mathbb{R}) \mid \mathcal{B} \neq \alpha_f(\mathcal{B}) \subset \mathcal{B} \;\forall \mathcal{B} \in \mathcal{H}^{\alpha}_{\mathcal{B}}(\mathcal{O})\}$

Some explanation of this statement is necessary. $\mathcal{H}^{\alpha}(\mathcal{Q})$ is the set of α-invariant hereditary C^* - subalgebras of $\mathcal{Q}$. (A C^* - algebra is hereditary if $x \in \mathcal{B}_+ \subset \mathcal{B} \subset \mathcal{Q} \Rightarrow y \in \mathcal{B}_+ \,\forall y \mid 0 \leq y \leq x$). The subscript $\mathcal{B}$ on $\mathcal{H}^{\alpha}_{\mathcal{B}}(\mathcal{O})$ means that $\mathcal{B} \subset \mathcal{B} \neq \mathcal{B}$ for $\mathcal{B} \in \mathcal{H}^{\alpha}(\mathcal{O})$. $\hat{f}$ is the Fourier transform of f and $\alpha_f(\mathcal{B})$ is the set:

$$\alpha_f(x) = \int_{-\infty}^{\infty} \alpha_t(x) f(t)\, dt.$$

A more constructive result is:

Theorem: $\alpha' \in \mathrm{Inn}\,\mathcal{R}$ iff $\exists y \in \mathcal{O}$ and a linear involutive map $D: \mathcal{O} \mapsto \mathcal{B}$ such that:

(i) $\alpha(x) = y x y^* + D(x), \; \forall x \in \mathcal{O}$,
(ii) $y y^* - 1 = D_2, \; y^* y - 1 = D_1, \; D_1, D_2 \in \mathcal{B}$,
(iii) $D(1) = -D_2$, (iv) $D(xz) = y x D_1 z y^* + D(x) y z y^* + y x y^* D(z) + D(x) D(z), \; \forall x, z \in \mathcal{O}$.

In more heuristic terms, if α^{π}_t is generated by a Hamilton H^{π} in a π (so that it is inner) then if $[H^{\pi}, \pi(\mathcal{B})] \subset \pi(\mathcal{B})$, α'^{π}_t is also inner. This is the situation in electromagnetism.

4. SYSTEMS WITH INDEFINITE INNER PRODUCT REPRESENTATIONS

Indefinite inner product representations are a common feature of degenerate systems, starting with the approach of Gupta and Bleuler to electromagnetism [6]. The definitive work was the paper by Strocchi and Wightman and our object is to show the close connection between our approach and theirs in this context. In order to do this we shall distinguish two stages which we call pre-Strocchi-Wightman structures (PSW - structures) and strict Strocchi-Wightman structures (SW - structures). The PSW - structure is defined as follows:

A PSW - structure consists of an indefinite inner product space

(IIP - space) $\{\mathcal{K}, (\ ,\ )\}$ where $(\ ,\ )$ is an indefinite inner product, a unital C* - algebra $\mathcal{F}_{\mathcal{K}} \subset \mathcal{B}(\mathcal{K})$ within which is specified a C* - degenerate system $\mathcal{D}_{\mathcal{K}} \triangleleft \mathcal{O}_{\mathcal{K}} \subset \mathcal{F}_{\mathcal{K}}$ such that there exists a positive subspace $\mathcal{K}' \subset \mathcal{K}$, and a cyclic vector $\Phi_0 \in \mathcal{K}'$ which satisfy:

(i) $\mathcal{O}_{\mathcal{K}} \mathcal{K}' \subset \mathcal{K}'$, (ii) $\mathcal{D}_{\mathcal{K}} \mathcal{K}' \subset \mathcal{K}''$, where $\mathcal{K}''$ is the neutral space of $\mathcal{K}'$ with respect to $(\ ,\ )$. The physical Hilbert space is defined as

$$\mathcal{K}_{phys} = \overline{\mathcal{K}' | \mathcal{K}''},$$

and the physical algebra by:

$$\mathcal{A}_{phys} = \overline{(\mathcal{O}_{\mathcal{K}} | \mathcal{K}') \bmod \mathcal{K}''}.$$

This definition makes sense because of the definitions of $\mathcal{K}''$ and $\mathcal{K}'$. In more descriptive terms, $\mathcal{K}'$ is the set of elements with non-negative norm, $\mathcal{K}''$ those with zero norm, and $\mathcal{K}_{phys}$ with positive definite norm. All the physical descriptions are in $\mathcal{K}_{phys}$.

This PSW - structure leads naturally to the Dirac triplet $\mathcal{D} \subset \mathcal{O} \subset \mathcal{F}$, and conversely. The first implication is in the definition, and the converse is given by the following theorem, where by X is meant a left $\mathcal{F}$ - module which has an IIP and a cyclic element $x_0 \in X$.

Theorem: The collection of objects $(X, (\ ,\ )), \mathcal{F}, \{L_i(\lambda)\}$ defines a PSW structure for each $x \in X$ such that:

(i) $x_0 \in \mathcal{O}x$, (ii) $(Ax, Ax) \geq 0, \forall A \in \mathcal{O}$, (iii) $(Dx, Dx) = 0, \forall D \in \mathcal{D}$.

$\mathcal{A}_{phys}$ as defined by the PSW - structure is isomorphic to $\mathcal{A}$, and so a representation of the latter is induced by this construction.

An SW - structure contains additionally a group structure: $\mathcal{G} \overset{U}{\rightarrow} B(\mathcal{K})$, where $\mathcal{G} \subset \mathrm{Aut}\, \mathcal{F}_{\mathcal{K}}$ and $\alpha_g \subset \Upsilon$, for which:

(i) $U_e = 1, (U_g\Phi, U_g\Psi) = (\Phi, \Psi)$ and $(\Phi, A\Psi) = (U_g\Phi, \alpha_g(A) U_g \Psi)\ \forall A \in \mathcal{F}_{\mathcal{K}}$, $g \in \mathcal{G}, \Phi, \Psi \in \mathcal{K}$ and $(\Psi, U_g \Phi)$ is continuous in g,
(ii) $U_g \mathcal{K}' \subset \mathcal{K}', \forall g \in \mathcal{G}$,
(iii) Φ_0 is the only cyclic vector for which $U_g \Phi_0 = \Phi_0, \forall g \in \mathcal{G}$.

A supply of left $\mathcal{F}$-modules can be obtained from the left ideals of $\mathcal{F}$ with an IIP provided from the hermitian non-positive functionals, $\mathcal{F}_h^*$. If a left ideal is principal it can be generated by a single element $x_0 \in \mathcal{F}$ which does not have a left inverse. Such an element can be taken as a cyclic vector.

Theorem: Let α^P denote the elements of $\mathcal{F}$ which are α- invariant. Then there is a SW - structure for each (f, x_0, F) in $\mathcal{F}_h^* \times \alpha^P \times \mathcal{F}$ which satisfies the following conditions:

(i) $f(x) = f(\alpha_g(x)), \forall x \in \mathcal{F}x_0, \forall g \in \mathcal{G}$, (ii) $x_0 \in \mathcal{O} F x_0$,
(iii) $\mathcal{O}\alpha_g(F) x_0 \subset \mathcal{O} F x_0$, (iv) $f(x_0^* F^* \mathcal{O}_+ F x_0) \geq 0$,
(v) $f(x_0^* F^* \mathcal{D}_+ F x_0) = 0$, (vi) $\alpha^P \cap \mathcal{F}_\ell x_0 = \{x_0\}$ where $\mathcal{F}_\ell = \{x \in \mathcal{F} | x_\ell^{-1} \exists\}$.

Conversely if there exists a SW - structure on the C* -degenerate system above, then there is a (nonpositive) functional θ on $\mathcal{F}$ such that $\theta(x) = \theta(\alpha_g(x)), \forall x \in \mathcal{F}, \forall g \in \mathcal{G}, \theta(\mathcal{O}_+) \geq 0$ and $\theta(\mathcal{D}_+) = 0$.

This functional is called the class functional of the SW - structure. The condition (vi) is imposed in order to ensure a unique vacuum and may be dropped if symmetry breaking is required.

Instead of taking X to be a principal left ideal, it can be taken to be a factor space of left ideals. To do this let $\mathcal{J}$ be an $\mathcal{F}$ - left ideal which contains a sub-left ideal $\mathcal{I} \subset \mathcal{J}$. Then $\mathcal{J}/\mathcal{I}$ is a left $\mathcal{F}$ - module. If $x_0 \in \mathcal{J}$ such that $\mathcal{F}(x_0+\mathcal{I}) = \mathcal{J}$, then it is a cyclic element. In this way we can construct all cyclic $\mathcal{F}$ - modules.

The SW -structures so far described do not assume that X possesses an additional positive definite inner product. If, instead of a general hermitian functional, we choose a positive linear functional ω, this may be used to define a positive definite inner product and a Hilbert space structure using the GNS procedure. If $G \in \mathcal{F}_h$, we can define an IIP by:

$$(A\Omega_\pi, \pi(G)B\Omega_\pi) = \omega(A^*GB).$$

G is called a Gram operator. If further $G^2 \in \{1 + N_\omega\}$, where N_ω is the left kernel of the GNS representation, we have a Krein SW - structure.

If we have two IIP representations, it may be possible to connect them. If this is so, we have a generalized gauge transformation.

A pair of SW - structures are connected by a generalized gauge transformation if there exists a bijection $g : \mathcal{K}_{1phys} \to \mathcal{K}_{2phys}$ such that:

(i) $(\Phi_1, \pi_1(A)\Psi_1)_1 = (\Phi_2, \pi_2(A)\Psi_2)_2$, $\forall A \in \mathcal{O}$, $\forall \Phi_i, \Psi_i \in \mathcal{K}_i'$ with $\eta_{\Psi_2} = g(\eta_{\Psi_1}), \eta_{\Phi_2} = g(\eta_{\Phi_1})$ and η is the canonical map $\mathcal{K}_i' \to \mathcal{K}_i'/\mathcal{K}_i''$

(ii) $\eta_{\Phi_{20}} = g(\eta_{\Phi_{10}})$.

In this case two SW - structures induce the same covariant cyclic representation of $\mathcal{R}$ up to unitary equivalence. Two gauge equivalent SW - structures have class functionals θ_i for which $\theta_1|\mathcal{O} = \theta_2|\mathcal{O}$ and so they give the same results on the observable algebra. In this sense $\mathcal{O}$ is gauge invariant.

A special gauge transformation is one for which $\mathcal{K}_1' = \mathcal{K}_2'$ and then g is the identity. Only the IIP representations π_i are then different.

5. CONCLUSION

The point of trying to find a C^*- algebra framework for quantum fields is that it frees the discussion of such fields from the properties of particular representations. The greater generality so obtained is valuable because interacting fields may only exist in certain representations and not in others. If these representations are unknown it is impossible to describe the behaviour of the fields. In the more abstract formulation however, it may be possible to draw useful conclusions without being committed to a particular representation.

We have shown here how the Dirac quantization method can be cast into such an abstract and mathematically rigorous form, and consequently how representations of a familiar type, such as those with an IIP, may be obtained from the structure of the algebra.

Up to the present, because only linear theories can be quantized in a suitably tractable form, there are few examples which can test the conclusions reached here, and even then, if the constraints are non-hermitian, such as in the Gupta - Bleuler approach, there are problems because one must deal with non-linear functions of the fields. It is clearly necessary, in order to proceed further, to be able to set up the field algebra $\mathcal{F}$ in a rigorous way when the dynamics or the constraints are non-linear. One possible line of development is to relate the C^*- algebra approach to that of path integrals.

REFERENCES

1. Grundling, H. B. G. S., Hurst, C. A. : Algebraic quantisation of systems with a gauge degeneracy. Commun. Math. Phys. 98, 369-390 (1985).
2. Dirac, P. A. M. : Lectures in quantum mechanics. New York: Belfer Graduate School of Science, Yeshiva University, 1964.
3. Segal, I. E. : Mathematical problems of relativistic physics. Providence, R. I. : American Mathematical Society, 1963.
4. Strocchi, F., Wightman, A. S. : Proof of the charge superselection rule in local relativistic quantum field theory. J. Math. Phys. 15, 2198-2224 (1974).
5. Grundling, Hendrik, Hurst, C. A. : Algebraic structures of degenerate systems, physical requirements and the indefinite metric. 1986 (Submitted for publication).
6. Gupta, S. N. : Theory of longitudinal photons in quantum electrodynamics. Proc. Phys. Soc. London A63, 681-691 (1950).
   Bleuler, K. : Eine neue Methode zur Behandlung der longitudinalen und skalaren Photonen. Helv. Phys. Acta 23, 567-586 (1950).

BROKEN SYMMETRY AND THE ACCELERATED OBSERVER

Christopher T. Hill
Fermi National Accelerator Laboratory
P.O. Box 500, Batavia, Illinois, 60510

INTRODUCTION

Spontaneous symmetry breaking affords an interesting probe of the phenomenon of "acceleration radiation" in general relativistic quantum field theory and raises a peculiar paradox. Accelerated observers infer the presence of particles in the vacuum in a thermal distribution with a temperature proportional to the proper acceleration. Nonetheless, by ever increasing the temperature (acceleration) a broken symmetry cannot be restored without violating general covariance. How does the accelerated observer interpret this outcome dynamically?

Coordinate systems possessing horizons lead to certain ambiguities in the definition of a quantum field theory[1,2]. One may consider a collection of observers comoving in such a coordinate system (tacked down to some fixed values of the spatial coordinates; such observers cannot all be freely falling; examples include the comoving observers in Schwarzschild, static deSitter and Rindler coordinates). We attempt to define a Hamiltonian, H_h, which propagates the Schroedinger wavefunctional of the quantum field theory in the accelerated observer's coordinate time (we find this to be the simplest and most conceptual approach, though any conventional formulation of field theory will due; however, there are subtleties with path intgrals since the initial time and final time surfaces in these systems always overlap; the properties of the vacuum state can only be extracted when initial and final surfaces can be separated). However, the singularity in the coordinate system on the horizon translates into ambiguities in the definition of the Hamiltonian density at the horizon and hence the Hamiltonian integral across the horizon. This coordinate system ambiguity represents a *loss of information* which must somehow be reincorporated to maintain consistency with the Minkowski space definition. However, in the literature this subtlety is generally ignored; one simply writes down an expression for the hamiltonian with unspecified constraints upon the field configuration at the horizon. We refer to this "naive" prescription as the "Rindler Hamiltonian". The ground state of the Rindler Hamiltonian is the "Unruh vacuum" and is seen to have *lower* energy than the Minkowski ground state.

These ambiguities are related to the Casimir effect. If one artificially severs continuity normal to some plane in flat space, i.e. neglect the $\nabla\phi \cdot \nabla\phi$ terms in the Hamiltonian on this surface, then the groundstate of the field theory will have different energy than the usual Minkowski vacuum. This owes to the singular field configurations (whose normal derivatives to the plane are nonexistent) which previously had zero amplitude of being found in the vacuum now becoming active and establishing a new groundstate. This is similar to the familiar Casimir effect in which parallel plane conducting surfaces experience a net force due the expulsion of vacuum zero-point fluctuations which are inconsistent with conducting boundary conditions. In fact, this is effectively what happens in the singular coordinate system and leads to the Rindler hamiltonian and Unruh vacuum being physically distinct from the Minkowski case. Indeed, the formal resemblance of the Unruh matrix elements to those of an infinite plane conductor in Minkowski space with *Dirichlet* boundary conditions are striking.

Rindler coordinates are defined in flat space and describe a comoving ensemble of accelerated observers and are given by[3,4]:

$$t = a^{-1}\, e^{a\xi} \sinh(a\eta) \tag{1}$$

$$x = a^{-1}\, e^{a\xi} \cosh(a\eta) \qquad (x > 0) \tag{2}$$

$$x_\perp = x'_\perp \tag{3}$$

where $(-\infty < \eta, \xi < \infty)$. We will presently restrict our attention to the "right hand wedge" corresponding to $x > 0$, though it is straightforward to extend the results to the double wedge case. Eq.(3) describes observers of fixed ξ accelerating with proper acceleration given on the $t = \eta = 0$ time slice by $ae^{-a\xi} = 1/x$, and elapsed proper time $\eta e^{a\xi}$. The metric in Rindler coordinates is given by:

$$ds^2 = e^{2a\xi}(d\eta^2 - d\xi^2) - dx'^2_\perp \tag{4}$$

Presently we will adopt a covariant functional Schroedinger description of the system as developed in ref.(5). We refer the reader to ref.(5) for the formal details. An equivalent approach might be to construct the appropriate Green's functions[6] in the Unruh vacuum and extract local matrix elements from these.

The true physical vacuum is always the usual Minkowski one, and operator matrix elements simply transform covariantly to the accelerating frame. Thus, since $\langle : \phi^2 : \rangle$ is zero (upon renormalization), it will always be measured to be zero by any observer. The Minkowski vacuum in Schroedinger picture is given by a gaussian wave-functional of the form:

$$\Psi_M = \exp\left\{-\tfrac{1}{2}\int dk_x d^{d-1}k_\perp \mid \alpha(k_x, k_\perp) \mid^2 \sqrt{k_x^2 + k_\perp^2 + m^2}\right\} \qquad (5)$$

where we have represented the wave–functional in momentum space. For example, with a plane conducting wall at $x = 0$ and Dirichlet boundary conditions we would have the expansion:

$$\phi(x) = \int_0^\infty dk_x \sqrt{\frac{2}{\pi}} \int \frac{d^{d-1}k_\perp}{(2\pi)^{\frac{d-1}{2}}} e^{ik_\perp \cdot x_\perp}\, \alpha(k_x, k_\perp) \sin k_x x \qquad (6)$$

(recall that in Schroedinger picture the fields are generalized coordinates and carry no time dependence, which is carried by the wavefunctional; we do not indicate the time dependence which is irrelevant presently; we are free to go to the Fourier coefficients as the coordinates of the system). The same field configuration can be represented in the right hand Rindler wedge in terms of massive $d+1$ dimensional Rindler modes as[5]:

$$\phi(x) = \int dk_x \frac{d^{d-1}k_\perp}{(2\pi)^{\frac{d-1}{2}}}\, \beta(k, k_\perp)\, e^{ik_\perp \cdot x_\perp}\, R_{k_x}^{k_\perp}(\varsigma) \qquad (7)$$

where:

$$R_p^{k\perp}(\varsigma) = \frac{1}{\pi}\left(\frac{2p}{a} \sinh \frac{\pi p}{a}\right)^{\frac{1}{2}} K_{\frac{ip}{a}}(a^{-1} e^{a\varsigma} \sqrt{k_\perp^2 + m^2}) \qquad (8)$$

These modes diagonalize the Rindler Hamiltonian. We may then represent the Minkowski wave-functional as a gaussian in the coefficients, $\beta(k)$ (this is equivalent to a Bogoliubov transformation in the usual formalism and is given in ref.(5)) as:

$$\Psi_M = \exp\left\{-\frac{1}{2}\int dk_x d^{d-1}k_\perp \mid \beta(k, k_\perp) \mid^2 k_x \coth\left(\frac{\pi k_x}{2a}\right)\right\} \qquad (9)$$

This is not a groundstate of the Hamiltonian written in Rindler coordinates since the width for each mode has an extra factor of $\coth(\frac{\pi k_x}{2a})$. Indeed, Ψ_M now appears as a state full containing a Bose gas of Rindler particles (these are particles as defined relative to the Rindler Hamiltonian; the full double wedge Minkowski case is similar[5]) with a universal temperature of $T = \frac{a}{2\pi}$.

The groundstate of the Rindler Hamiltonian is the "Unruh" vacuum and is given by:

$$\Psi_U = \exp\left\{-\frac{1}{2}\int dk_x d^{d-1}k_\perp \mid \beta(k, k_\perp) \mid^2 \mid k_x \mid\right\} \qquad (10)$$

Since this state is clearly different than the Minkowski groundstate and since a given Hamiltonian can have only one groundstate, it follows that the Rindler Hamiltonian is different than the Minkowski Hamiltonian. This difference arises due to the singular structure of the Rindler coordinates associated with the horizon.

OPERATOR EXPECTATION VALUES

If we compute local operator matrix elements, such as $\langle\phi^2\rangle$ in the Unruh vacuum, we find that they are not covariant transforms of the same operator evaluated in the Minkowski vacuum, but rather develop generally negative "thermal" corrections[7,8]. For example, we show that $\langle\phi^2\rangle$ becomes $-\frac{T^2}{12}$ in an "high temperature limit" where T is the local Hawking temperature given in terms of the local proper acceleration (at $t = \eta = 0$ we have $T = \frac{1}{2\pi x}$). Thus, *it is the difference between the value of the operator in the Minkowski vacuum and that in the Unruh vacuum which appears as a positive thermal effect.*

Consider now the matrix element in the case of Dirichlet boundary conditions at $x = 0$:

$$\left\langle\phi(x+\frac{\epsilon}{2})\phi(x-\frac{\epsilon}{2})\right\rangle = \int dk_x dp_x d^{d-1}k_\perp d^{d-1}p_\perp\, 2^{2-d}\pi^{-d} \left\{\sin k_x(x+\frac{\epsilon}{2})\sin k_x(x-\frac{\epsilon}{2})\right\} \cdot \langle\alpha(k_x,k_\perp)\alpha(p_x,p_\perp)\rangle\; e^{ik_\perp\cdot x_\perp+ip_\perp\cdot x_\perp} \tag{11}$$

The expectation value is to be taken in the wavefunctional of eq.(5). We have:

$$\begin{aligned}\langle\alpha(k_x,k_\perp)\alpha(p_x,p_\perp)\rangle &= \int D\phi\;\Psi^*_M(\phi)\alpha(k_x,k_\perp)\alpha(p_x,p_\perp)\Psi_M(\phi)\\ &= \frac{\delta(k_x-p_x)\cdot\delta^{d-1}(k_\perp+p_\perp)}{2\left(k_x^2+k_\perp^2+m^2\right)^{\frac{1}{2}}}\end{aligned} \tag{12}$$

Making use of various integral identities, including the d-dimensional solid angle, we arrive at the result[8]:

$$\left\langle\phi(x+\frac{\epsilon}{2})\phi(x-\frac{\epsilon}{2})\right\rangle = 2^{-d}\pi^{\frac{-1-d}{2}} \cdot\left\{\left(\frac{\epsilon}{2m}\right)^{\frac{1-d}{2}} K_{\frac{1-d}{2}}(m\epsilon) - \left(\frac{x}{m}\right)^{\frac{1-d}{2}} K_{\frac{1-d}{2}}(2mx)\right\} \tag{13}$$

We note that the UV singularity of the operator ϕ^2 resides in the Bessel functions with arguments $m\epsilon$. This result is, of course, equivalent to evaluating the Feynman propagator for spacelike interval with Dirichlet boundary conditions. We further note that in the Lorentz invariant vacuum without the presence of the wall at $x = 0$ we obtain the familiar result:

$$\left\langle \phi(x+\frac{\epsilon}{2})\phi(x-\frac{\epsilon}{2})\right\rangle_{Lorentz} = 2^{-d}\pi^{\frac{-1-d}{2}}\left\{\left(\frac{\epsilon}{2m}\right)^{\frac{1-d}{2}} K_{\frac{1-d}{2}}(m\epsilon)\right\} \tag{14}$$

Clearly, the short distance singular part of eq.(13) is not influenced by the boundary conditions, and can be unambiguously subtracted in all coordinate systems (this corresponds to renormalizing the operator matrix element to be zero in the limit of zero acceleration).

The formalism developed in ref.(5) is covariant and we have verified explicitly as a check on the present calculation that if we reexpress the Minkowski vacuum, $\Psi_M(\phi)$ in terms of the Rindler modes and recalculate the $\langle\phi^2\rangle$ we obtain the same result as in eq.(13) Hence, although the Minkowski vacuum appears to be full of particles, the local operator matrix elements are covariant and the acceleration radiation is in a sense fictitious.

It is of interest however to evaluate the matrix element $\langle\phi^2\rangle$ in the physically distinct Unruh vacuum. This is formally equivalent to the preceding analysis, but involves a nontrivial evaluation of a resulting Kontorovich-Lebedev transformation. This is discussed in ref.(8) We obtain the result:

$$\begin{aligned}\left\langle \phi(x+\frac{\epsilon}{2})\phi(x-\frac{\epsilon}{2})\right\rangle_{U} &= 2^{-d}\pi^{\frac{1+d}{2}}\left\{\left(\frac{\epsilon}{2m}\right)^{\frac{1-d}{2}} K_{\frac{1-d}{2}}(m\epsilon)\right.\\ &\quad - 2\int_0^\infty \left\{\frac{\sqrt{x_1^2+x_2^2+2x_1x_2\cosh\omega}}{2m}\right\}^{\frac{1-d}{2}}\\ &\quad \left.\cdot K_{\frac{1-d}{2}}\left(m\sqrt{x_1^2+x_2^2+2x_1x_2\cosh\omega}\right)\frac{d\omega}{\pi^2+\omega^2}\right\}\end{aligned} \tag{15}$$

The second term on the right-hand side is nonsingular in the $\epsilon\to 0$ limit and we thus are led to the result:

$$\begin{aligned}\left\langle \phi(x)^2\right\rangle_{U} &= 2^{-d}\pi^{-\frac{1+d}{2}}\left\{\left(\frac{\epsilon}{2m}\right)^{\frac{1-d}{2}} K_{\frac{1-d}{2}}(m\epsilon)\right.\\ &\quad \left. - 2^{\frac{d+1}{2}}m^{d-1}\int_0^\infty \frac{d\omega}{\pi^2+\omega^2}\, Q^{\frac{1-d}{2}}\, K_{\frac{1-d}{2}}(Q)\right\}.\end{aligned} \tag{16}$$

where $x = x_1 = x_2$:

$$Q = \sqrt{2}mx\,\sqrt{(1+\cosh\omega)}. \tag{17}$$

Thus, the singular structure is identical to that obtained above for the Minkowski, an Dirichlet results. The finite corrections are negative definite and analogous to those obtained for the Dirichlet case. This is not unreasonable mathematically since the Rindler mode functions oscillate infinitely as they approach the horizon, while all normalization integrals have effectively a compact support. As such, we are implicitly forcing the field configuration of eq.(7) to vanish at the horizon by our normalization conventions and this in turn yields the result of eq.(16) not unlike the Dirichlet result.

Eq.(16) yields a more striking result when we consider it in a specific case. Let us specialize to $d = 3$ corresponding to $3 + 1$ dimensional spacetime. We further consider the limit of small x, the "high acceleration" limit. Throwing away the singular ϵ-terms we find the leading behavior:

$$\left\langle : \phi(x)^2 : \right\rangle_U \rightarrow -\frac{1}{4\pi^2 x^2} \int_0^\infty \frac{d\omega}{(1+\cosh\omega)(\pi^2+\omega^2)} = -\frac{T^2}{12} \tag{18}$$

where we define the *local Hawking Temperature* $T(x) = \frac{1}{2\pi x}$ where the local proper acceleration is given by $\frac{1}{x}$ (we have used integrals tabulated in ref.(8) to obtain this latter result as well as the small argument limit of the Bessel function $K_1(x)$).

This result is *minus* the usual thermal correction to the operator ϕ^2 as is easily verified by computing the expectation value with the thermal density matrix. It suggests that locally the Minkowski vacuum expectation value, which is zero upon subtraction, is "hot" by an amount $\frac{T^2}{12}$ when compared to the Unruh result. Nonetheless, there is no conflict with general covariance because the result in Minkowski space is invariant, i.e. zero transforms into zero. *It would be incorrect to conclude that an accelerating observer measuring $\langle\phi^2\rangle$ obtains a thermal result of $\frac{T^2}{12}$.*

THE EFFECTIVE POTENTIAL

The energy operator is the $\{00\}$ component of the stress tensor. We shall presently study the conformal stress-tensor which is traceless in the $\mu \rightarrow 0$ limit and most formally resembles that of radiation[9]:

$$T_{\mu\nu} = \partial_\mu\phi\partial_\nu\phi - \frac{1}{2}g_{\mu\nu}\left(\partial_\rho\phi\partial^\rho\phi - m^2\phi^2\right) - \xi\left((\phi^2)_{;\mu;\nu} - g_{\mu\nu}(\phi^2)_{;\rho;}{}^{\rho}\right) \tag{19}$$

where $\xi = \frac{(1-d)}{4}$ (the conformal tensor which corresponds to a theory which is classically conformally invariant in the limit $\mu^2 \rightarrow 0$).

In Schroedinger picture the time derivative is replaced by the d-dimensional functional derivative:

$$\partial_0\phi(x) \rightarrow -i\frac{\delta}{\delta\phi(x)} \tag{20}$$

The computation of $\langle T_{\mu\nu}\rangle$ in the vacuum states with is straightforward albeit tedious and the details appear in ref.(8). The conformal energy density is in the Unruh vacuum:

$$\langle T^c_{00}\rangle = -2^{-d-1}\,\pi^{\frac{-1-d}{2}}\left(\frac{\epsilon}{2m}\right)^{\frac{-1-d}{2}} K_{\frac{-1-d}{2}}(m\epsilon)$$
$$- 2^{\frac{-d-3}{2}}\,\pi^{\frac{-1-d}{2}}\,m^{1+d}$$
$$\cdot \int_0^\infty \frac{d\omega}{(\pi^2+\omega^2)} \Big\{ (d+1+(1-d)\cosh\omega)\, Q^{\frac{-1-d}{2}} K_{\frac{-1-d}{2}}(Q) \tag{21}$$
$$+\Big(\Big(\frac{2+d}{d}+\frac{2-d}{d}\cosh\omega\Big)Q^2 - 2\,P(\omega)(1+\cosh\omega)\Big)Q^{\frac{-3-d}{2}} K_{\frac{1-d}{2}}(Q)\Big\}$$

where Q is given in eq.(17) and P is defined by:

$$P(\omega) = \frac{12\omega^2 - 4\pi^2}{(\pi^2+\omega^2)^2} \tag{22}$$

Presently we discuss the structure of this result and give a physical interpretation. As we've seen in eq.(18) the leading (high temperature or small x) behavior of $\langle\phi^2\rangle$ is given for $d=3$:

$$\left\langle\phi^2(x)\right\rangle \to -\frac{T^2}{12}; \qquad T = \frac{1}{2\pi x}. \tag{23}$$

What about the analogous results for the energy operator obtained above?

The leading (small–x) behavior of the conformal stress tensor can be evaluated similarly and yields:

$$\langle T^c_{00}\rangle \to -\frac{1}{480\pi^2 x^4} = -aT^4 \tag{24}$$

where $a = \frac{\pi^2}{30}$ is the Stefan-Boltzmann constant. This is consistent with the calculation of ref.(9). Thus the Unruh vacuum produces a singular energy density on the horizon which has the structure of thermal corrections but with the opposite sign. Moreover, we see that the leading behavior of the conformal stress tensor is that of radiation, $T^c_{\mu\nu} \to -aT^4\, diag\,(1, -\frac{1}{3}, -\frac{1}{3}, -\frac{1}{3})$ in the sense of tracelessness.

The effective potential can be understood as a variational calculation in the Schroedinger picture. One constructs a gaussian wavefunctional centered about some "classical" field configuration, ϕ_c (which is considered to be O(1) in an expansion in powers of $\hbar$). The wavefunctional is given an arbitrary mass parameter, μ and one computes the expectation value of the Hamiltonian in this state with some regulator scheme. Then, this regularized expression is varied with respect to the parameter μ to obtain a "mass–gap" equation for μ. The solution to this equation may be substituted back into the regularized expression for the energy. Then the result is renormalized to obtain the effective potential to order $\hbar$. This information is implicit in the stress-tensor expectation values obtained above.

In thermal equilibrium we consider a state centered about the classical minimum of the potential, but described by a thermal density matrix. We can then extract the finite temperature contributions to the energy expectation value in a high temperature expansion. The analyses given by Weinberg, Kirzhnitz and Linde, and Dolan and Jackiw[11] contain essentially this idea, but the technical methods of evaluating effective potentials vary.

We may compare our calculations of the energy density in the Unruh vacuum to these finite temperature analyses. We consider presently a field theory with Hamiltonian density (T_{00}):

$$H = \frac{1}{2}\left\{\pi^2 + (\nabla\phi)^2 - m^2\phi^2 + \frac{\lambda}{12}\phi^4\right\} \tag{25}$$

Choosing to compute in a wavefunctional centered about ϕ_c is equivalent to shifting $\phi \to \phi + \phi_c$ in eq.(25). Furthermore, terms linear in ϕ will produce vanishing contributions to the energy expectation value and may be dropped. The resulting quadratic Hamiltonian to order ϕ^2 in the quantum fluctuation after shifting becomes:

$$H = -\frac{m^2\phi_c^2}{2} + \frac{\lambda\phi_c^4}{24} + \frac{1}{2}\left\{\pi^2 + (\nabla\phi)^2 + (\mu^2)\phi^2\right\} \tag{26}$$

where $\mu^2 = -m^2 + \lambda\phi_c^2/2$ is the mass of the quantum field ϕ.

In the Unruh vacuum the leading behavior of the expectation value of the energy density of eq.(26) follows from the leading plus next to leading terms in the expansion of the stress-tensor matrix elements:

$$\begin{aligned}\langle T_{00}^{c}\rangle &= -\frac{\mu^4}{4\pi^2x^4}\int_0^\infty \frac{d\omega}{(\pi^2+\omega^2)}\left[Q^{-2}K_2(Q)(2-\cosh\omega)\right] \\ &\quad + Q^3K_1(Q)\left[\frac{Q^2}{6}(5-\cosh\omega) - P(\omega)(1+\cosh\omega)\right]\end{aligned} \tag{27}$$

$$\begin{aligned}&\to \text{(leading terms)} \\ &-\frac{\mu^2}{48\pi^2x^2}\int_0^\infty \frac{d\omega}{(\pi^2+\omega^2)(1+\cosh\omega)} \\ &\cdot\left[-1 + 2\cosh\omega - \frac{6(3\omega^2-\pi^2)}{(\omega^2+\pi^2)^2}(1+\cosh\omega)\ln(1+\cosh\omega)\right]\end{aligned} \tag{28}$$

We note that the last term would integrate to zero without the log factor (hence the μ^2T^2 factors in the argument of the log do not contribute; this latter integral is given in ref.(8)). These expressions involve the mass-gap, μ^2 and lead to a different result for the coefficient of the $\lambda\phi_c^2T^2$ term than in ref.(11). This is not surprising because these terms include the mass insertion in the kinetic term loop, which is absent in our naive estimate above in which only the leading behavior of the kinetic terms is kept.

Thus we obtain for the effective potential to this order:

$$\langle H\rangle = -\frac{m^2\phi_c^2}{2} + \frac{\lambda\phi_c^4}{24} - AT^4 - B\lambda\phi_c^2T^4 \tag{29}$$

where A is the Stefan-Boltzman constant in the case of the conformal tensor and we find from eq.(27), $B = \frac{79}{360}$ (while a "naive" estimate gives $B = \frac{1}{24}$). Of course, these terms will vary with the definition of the Hamiltonian as does the Stefan-Boltzmann constant in going from the conventional to the conformal stress-tensor.

This latter result shows that *there is no critical Hawking temperature above which symmetry is restored in the Unruh vacuum.* As one increases T one simply drives the system deeper into a broken symmetry state. Moreover, since the Minkowski vacuum produces the usual $T = 0$ result for the effective potential, we see that it is more in the direction of increasing the symmetry, hence consistent with the interpretation that it is full of a thermal distribution of Rindler particles. In fact, if we always compute the difference between operator matrix elements in the Minkowski and Unruh vacua we will obtain effectively thermal terms as in the difference between matrix elements in the thermal density matrix and the Minkowski vacuum. Thus, we see that symmetries are not restored as seen by accelerating observers (a fundamental consequence of general covariance) and that this is consistent with the dynamics as interpreted by the accelerating observer. Although there is a formal structure resembling radiation, there appears to be no physical manifestation of such. It seems unlikely that the Unruh detector, if carefully defined to avoid singularity difficulties, really responds to any "thermal radiation" since local operator matrix elements do not evidence the presence of such.

We conclude by emphasizing that the Unruh vacuum is a fictitious object (emulating the Boulware vacuum in Schwarzschild geometry). Though matrix elements differ between the Minkowski vacuum and the Unruh vacuum, all physical measurements will produce the usual results given by the Minkowski vacuum suitably transformed to the observers local coordinate system. There is therefore no physical manifestation of the "thermal distribution" of Rindler particles seen by the accelerated observer (e.g. no stress–energy, etc.).

REFERENCES

1. S. Hawking, *Commun. Math. Phys.* **43** (1975) 199
2. W. Unruh, *Phys. Rev.* **D14** (1976) 870
3. W. Rindler, *Am. J. Phys.* **34** (1966) 1174
4. N. Birrell and P. Davies, *(Cambridge University Press, 1982)*
5. K. Freese, C.T. Hill and M. Mueller, *Nucl. Phys.* **B255** (1985) 693

6. W. Unruh, N. Weiss *Phys. Rev.* **D29** (1984) 1656

7. C. T. Hill, *Phys. Letters* **155B** (1985) 343

8. C. T. Hill, FERMILAB–Pub–85/100–T (1986)

9. P. Candelas, D. Deutsch, *Proc. Roy. Soc.* **A354** (1977) 79

10. A. Erdelyi, ed., *Bateman Manuscript Project* **Vol. 2** (1953) 5

11. S. Weinberg, *Phys. Rev.* **D9** (1974) 3357; D. Kirzhnitz, A. Linde, *Phys. Lett.* **42B** (1972) 471; L. Dolan, R. Jackiw, *Phys. Rev.* **D9** (1974) 3320

QUANTUM THEORY OF THE FREE ELECTROMAGNETIC FIELD

C. A. Hurst

University of Adelaide

ABSTRACT

A review of the quantization of the free electromagnetic field from the point of view of C^*-algebras is presented. It is shown how a unified approach to quantization according to the radiation gauge, the Gupta-Bleuler indefinite metric, the Fermi supplementary condition and Dirac's method of constraints can be obtained.

INTRODUCTION

The electromagnetic field is one of the simplest, and certainly one of the most important examples of a gauge field theory, or, to use Dirac's language, of a theory with first class constraints. It therefore provides a very useful way of studying the properties of such theories. Its quantum version, even for a free field, takes many forms, and although these various versions are of course ultimately in some sense equivalent, their relationship is perhaps not as clear as would be desired.

In this review, an approach will be presented which it is hoped will clarify the connection between at least the most commonly used approaches. This approach is algebraic in nature and has the additional virtue that for the free field at least complete mathematical rigor is attainable. It is also more general than a purely Hilbert space approach. So although no new physical consequences are found, the insights that are provided may be enlightening.

Before doing this, it is helpful to have a quick look at the usual heurisitic formulation. The essential starting point in the various approaches that will be studied is to use the four vector potential $A^\mu(x)$, $(x \equiv (\underline{x},t))$ and to postulate the classical action:

$$S = \int d^4x L \equiv -\frac{1}{4}\int d^4x F^{\mu\nu}F_{\mu\nu} = \frac{1}{2}\int d^4x[(\nabla A_0 - \dot{\underline{A}})^2 - (\nabla \times \underline{A})^2], \qquad (1)$$

with $F_{\mu\nu} = \partial_\mu A_\nu - \partial_\nu A_\mu$.

If π_μ are the conjugate field variables, the action S leads to two constraint equations:

$$\pi_0 = 0 = \nabla \cdot \underline{\pi}. \tag{2}$$

These constraint equations are in conflict with the equired canonical commutation relations:

$$[A_\mu(\underline{x},0),\pi_\nu(\underline{y},0)] = i\delta_{\mu\nu}\delta(\underline{x}-\underline{y}), \tag{3}$$

and two ways have been proposed to overcome this. They are:

(a) Modify the Lagrangian density by adding a term $-\frac{1}{2}(\partial_\mu A^\mu)^2$, so that the constraints (2) no longer appear,

(b) Treat the theory by Dirac's method of constraints [1].

For case (a), the additional term must eventually be disposed of, if electromagnetism is to be regained. This is done by imposing a supplementary condition either before or after the theory is quantized. In order for these alternatives to be compatible the following diagram must commute:

Classical theory without constraints	$\rightarrow$	Classical theory with constraints
$\downarrow$		$\downarrow$
Quantum theory without constraints	$\rightarrow$	Quantum theory with constraints

The usual alternatives are the following:

(i) Eliminate unwanted unphysical variables, usually chosen to be A_0 and $\underline{A}^{long}$, at the classical level, and then construct a quantum theory. This is the radiation or Coulomb gauge approach, which has been carried through rigorously by Segal [2].

(ii) Quantize the modified Lagrangian using an indefinite metric and impose the supplementary condition on the physical states:

$$\partial^\mu A_\mu^+ \psi_{physical} = 0.$$

A_μ^+ are the positive frequency or annihilation parts of the quantum field operators $A_\mu(x)$. This is the Gupta-Bleuler [3] approach, and is the most favored.

(iii) Quantize the modified Lagrangian in a definite metric and impose the supplementary condition

$$\partial^\mu A_\mu \psi_{physical} = 0.$$

This is the Fermi [4] approach, and was standard until the appearance of Gupta-Bleuler. It fell into disfavor because it appeared that no normalizable physical states existed.

All these approaches have drawbacks. The radiation gauge is non-local and non-covariant, although it has a positive metric. Gupta-Bleuler has the unfamiliar use of an indefinite metric Hilbert space and, more seriously, implies that the associated representations of the Poincaré group are unbounded [5]. Fermi has the difficulty already mentioned and moreover has a non-invariant vacuum and bounded non-unitary representations of the Poincaré group. However all of them lead to the same self-consistent theory of the physical electromagnetic field.

In terms of annihilation and creation operators in the Fourier transform, the commutation relations (3) are:

$$[a_\mu(\underline{k}),a_\nu^*(\underline{k}')] = -g_{\mu\nu}\delta(\underline{k}-\underline{k}'), \tag{3'}$$

and the Hamiltonian for the modified Lagrangian is:

$$H = \int d\Omega_0(\underline{k})\,|\underline{k}|^2(\underline{a}^*(\underline{k})\cdot\underline{a}(\underline{k}) - a_0(\underline{k})a_0^*(\underline{k})), \tag{4}$$

where $d\Omega_0(\underline{k}) = d^3k/|\underline{k}|$ is the Lorentz invariant measure. The contribution from the scalar photons appears to be negative, and in the Fermi approach this is actually the case. The energy spectrum is then unbounded below and this appears to be another blemish. In Gupta-Bleuler this problem does not arise because of the special properties of the indefinite metric. These three approaches may each be furnished with a Fock representation which has the expected properties. The Dirac approach, whilst free of many of the objections of the other approaches does not have a Fock representation.

All these difficulties are resolved if an algebraic approach is adopted. The remainder of this review will be devoted to showing how this follows for all the cases described. Because of limitations of space, all mathematical proofs are omitted and the interested reader is referred to the original papers [6].

2. ALGEBRAIC APPROACH

A convenient mathematical structure to describe the quantized electromagnetic and other fields is provided by C^*-algebras. Although this is now a very vigorous branch of mathematics with a correspondingly large literature, a short summary of the basic properties will be given for the benefit of those unfamiliar with them. C^*-algebras are well adapted to the needs of quantum mechanics, and were invented by Segal for just this purpose.

A C^*-algebra A is a normed involutive algebra with the following properties:

<u>Involution</u>

(i) There is an <u>involutory operation</u> *: $A \to A$ such that:

(a) $(x^*)^* = x$,
(b) $(x+y)^* = x^* + y^*$,
(c) $(\lambda x)^* = \bar{\lambda}x^*$
(d) $(xy)^* = y^*x^*$, $x,y \in A$.

<u>Norm</u>

(ii) There is a mapping $\|\cdot\|$: $A \to \mathbb{R}_+$:

(a) $\|x\| \geq 0$ and $\|x\| = 0 \iff x = 0$,
(b) $\|\lambda\, x\| = |\lambda|\,\|x\|$,
(c) $\|x+y\| \leq \|x\| + \|y\|$,
(d) $\|xy\| \leq \|x\|\,\|y\|$, $x,y \in A$.

<u>C^*-algebra property</u>

(iii) $\|x^*x\| = \|x\|^2$,

From (iii) and (iid) it follows that $\|x\| = \|x^*\|$. A is closed in the topology induced by the norm, which is usually called the <u>uniform norm</u>. This rather simple set of postulates leads to a very rich structure, which includes the important property that every C^*-algebra may be represented as an algebra of bounded oeprators on a Hilbert space usually in many different ways. This is a converse of the fact that an algebra of bounded operators on a Hilbert space is a C^*-algebra if it is closed with respect to the uniform operator norm, with involution being the taking of the adjoint.

A <u>representation</u> of a C^*-algebra is a mapping $A \to L(\mathfrak{h})$ from A to bounded operators on a Hilbert space $\mathfrak{h}$ with the properties:

(a) $\pi(x+y) = \pi(x) + \pi(y)$, (b) $\pi(\lambda x) = \lambda\pi(x)$,

(c) $\pi(xy) = \pi(x)\pi(y)$, (d) $\pi(x^*) = \pi(x)^*$.

A linear form (functional) is a mapping $f: A \to \mathbb{C}$ which is continuous with respect to the topology on A induced by the norm i.e. $|f(x)| \to 0$ if $\|x\| \to 0$. It satisfies:

(a) $f(x+y) = f(x) + f(y)$, (b) $f(\lambda x) = \lambda f(x)$,

(c) $f^*(x) = \overline{f(x^*)}$. If $f^* = f$, f is hermitian.

(d) If $f(x^*x) \geq 0$, $\forall x \in A$, f is a positive form and is necessarily hermitian.

If $x = y^*y$, x is a positive element, and if $f(x) = 0 \Leftrightarrow x = 0$ for x positive, f is positive definite. A positive definite form for which $f(1) = 1$ is called a state.

From positive definite forms (which always exist) it is possible to construct representations by what is called the GNS (Gel'fand-Naimark-Segal) construction. This construction states that, corresponding to a positive definite form f and a C^*-algebra A there exists a Hilbert space $\mathfrak{h}$, a representation π_f, and a cyclic vector Ω_f such that:

$$\mathfrak{h}_f = \overline{\pi_f(x)\Omega_f}, \quad \forall x \in A,$$

$$(\Omega_f, \pi_f(x)\Omega_f) = f(x).$$

Before we apply these ideas to the electromagnetic field, it is helpful to look at the simpler case of the real scalar field, using Segal's approach. He starts with a real linear vector space M_0 of the complex functions $\phi(k)$ over the positive mass hyperbolid C_+^m: $\{k \in \mathbb{R}^4,\ k_0 > 0,\ k_0^2 = \underline{k}^2 + m^2\}$ with the scalar product

$$(\phi,\psi)_R = \frac{1}{2}\int_{C_+^m} d\Omega_m(\underline{k})\,(\overline{\phi(\underline{k})}\psi(\underline{k}) + \phi(\underline{k})\overline{\psi(\underline{k})}) \tag{5}$$

with $d\Omega_m(\underline{k}) = d^3k/\sqrt{\underline{k}^2 + m^2}$. A real skew-symmetric bilinear form $B(\phi,\psi)$ is defined by:

$$B(\phi,\psi) = -\frac{1}{2}i\int_{C_+^m} d\Omega_m(\underline{k})\,(\overline{\phi(\underline{k})}\psi(\underline{k}) - \phi(\underline{k})\overline{\psi(\underline{k})}). \tag{6}$$

A complex scalar product can be defined if we introduce a real symplectic operator J with the properties:

$$J^2 = -1,\ B(J\phi,J\psi) = B(\phi,\psi),\ B(J\phi,\phi) > 0 \Leftrightarrow \phi \neq 0, \tag{7}$$

and then put:

$$(\phi,\psi) = B(\phi,J\psi) + i\,B(\phi,\psi). \tag{8}$$

If $J = i$, $B(\phi,J\psi) = (\phi,\psi)_R$ and

$$(\phi,\psi) = \int_{C_+} d\Omega_m \overline{\phi(k)}\psi(k). \tag{8'}$$

With the topology induced by (ϕ,ψ), M_0 can be completed to a complex Hilbert space M. From M, a C^*-algebra of the canonical commutation relations can be constructed following a procedure due to Manuceau [7]. Functions $\delta_\phi: M \to \mathbb{C}$ are defined by:

$$\delta_\phi(\phi') = 1 \text{ if } \phi' = \phi, \text{ and zero otherwise,}$$

with the multiplication relations:

$$\delta_\phi \delta_\psi = \delta_{\phi+\psi} \exp -\tfrac{1}{2}i\, B(\phi,\psi),$$

and involution:

$$\delta^*_\phi = \delta_{-\phi}.$$

The closure of the complex linear span of these functions, in a suitable topology originally given by Manuceau, defines the desired C^*-algebra which we shall call $\Delta_c(M)$. A state ω over $\Delta_c(M)$ is uniquely specified by the values of $\omega(\delta_\phi)$ called the generating functional. This generating functional also specifies uniquely (to within unitary equivalence) the corresponding GNS representation. If:

$$\omega(\delta_\phi) = \exp -\tfrac{1}{4}\|\phi\|^2, \tag{9}$$

the GNS representation is the usual Fock representation. A function ϕ on M can then be understood in conventional terms as the state of a single positive energy particle of mass m in this representation. A unitary transformation on M,U: $\phi \to U\phi$ generates the transformation $\delta_\phi \to \delta_{U\phi}$ on $\Delta_c(M)$ and then, because:

$$\omega_{\hat{U}}(\delta_\phi) \equiv \omega(\delta_{U\phi}) = \exp -\tfrac{1}{4}\|U\phi\|^2 = \exp -\tfrac{1}{4}\|\phi\|^2 = \omega(\delta_\phi),$$

ω is invariant under this transformation. This implies that there exists a unitary operator V(U) on $\mathcal{H}_\omega$ such that:

$$\pi_\omega(\delta_{U\phi}) = V(U)\pi_\omega(\delta_\phi)V^{-1}(U), \tag{10}$$

or U is unitarily implementable on the representation π_ω. It is easy to verify that Poincaré transformations on C^m_+ are unitarily implementable on $\mathcal{H}_\omega$.

These considerations can be readily translated to the case of a four-vector massless field although with some important and essential differences. M_0 is the space of $\mathbb{C}^4$ valued functions $\phi_\mu(k)$ over the forward light cone $C^0_+ \equiv C_+$, and the skew form B(,) is:

$$B(\phi,\psi) = \tfrac{1}{2}i \int_{C_+} d\Omega_0(k)(\overline{\phi_\mu(k)}\psi^\mu(k) - \phi_\mu(k)\overline{\psi^\mu(k)}). \tag{6'}$$

If we choose J = i, we have:

$$(\phi,\psi)_R = -\tfrac{1}{2} \int_{C_+} d\Omega_0(k)(\overline{\phi_\mu(k)}\psi^\mu(k) + \phi_\mu(k)\overline{\psi^\mu(k)}), \tag{5'}$$

and

$$(\phi,\psi) = -\int_{C_+} d\Omega_0(k)\overline{\phi_\mu(k)}\psi^\mu(k). \tag{8''}$$

Both these scalar products are no longer positive definite, and the one particle space is that of the Gupta-Bleuler theory.

Instead of introducing a complex structure at this stage, we can follow Segal. He singled out the subspace $N \subset M$ of functions $\phi_\mu(k)$ which satisfy:

$$k^\mu \phi_\mu(k) = 0. \tag{11}$$

On N, $(\phi,\phi) \geq 0$, and the neutral elements define a further subspace $T \subset N$. For $\hat{\phi} \in M' = N/T$, $(\hat{\phi},\hat{\phi}) > 0$, and this factor space may be completed to a Hilbert space in the topology of this scalar product. The Segal-Manuceau procedure then defines a C^*-algebra which is the C^*-algebra of the electromagnetic field. This procedure corresponds to the upper half of the commutative diagram given in the introduction. The Gupta-Bleuler procedure provides one way of going around the bottom half, but it is not well adapted to the C^*-algebra approach because of the indefinite metric. There is however a third procedure which is available, because the complex structure J is not uniquely fixed by B(,), and which naturally defines a C^*-algebra. We choose J_F by the relation:

$$(J_F\phi)_\mu(k) = -i \sum_\nu g_{\mu\nu}\phi_\nu(k), \tag{12}$$

and then:

$$(\phi,\psi)_F = \int_{C_+} d\Omega_0(k)\,(\phi_0(k)\overline{\psi_0(k)} + \overline{\underline{\phi}(k)} \cdot \underline{\psi}(k)). \tag{13}$$

The scalar product is positive definite, and so M_0 may be extended to a complex Hilbert space M and then the corresponding $\Delta_c(M)$ can be constructed.

M contains two closed subspaces T and N with $T \subset N$ which are defined by:

$$\begin{aligned} \phi \in N &\Rightarrow k^\mu \phi_\mu(k) = 0, \\ \phi \in T &\Rightarrow \phi_\mu(k) = k_\mu g(k), \end{aligned} \tag{14}$$

with corresponding C^*-subalgebras $\Delta_c(N)$ and $\Delta_c(T)$. From their definitions it is clear that these subspaces are Poincaré invariant, and this means that $\Delta_c(N)$, and $\Delta_c(T)$ are also invariant. The complement of T in N is denoted by S, and therefore

$$N = S \oplus T. \tag{15}$$

S is defined explicitly by:

$$\phi \in S \Rightarrow \phi_0 = 0 = \underline{k} \cdot \underline{\phi}\,. \tag{16}$$

As S is not invariant, neither is $\Delta_c(S)$. This means that although this algebra appears to be the appropriate candidate for the electromagnetic field, it cannot be so. This means that the "physical photons" do not form a subalgebra of the photon algebra. The physical photons only emerge after the supplementary conditions have been applied according to the following procedure.

From the structure of B(,) it can be easily verified that $\Delta_c(N)$ commutes with $\Delta_c(T)$, and ipso facto so does $\Delta_c(S)$. This means that we can define a continuous homomorphism π_B: $\Delta_c(N) \to \Delta_c(S)$ by a rigorous form of the mapping:

$$\sum \lambda_i \delta_{\phi_i} \to \sum \lambda_i, \quad \phi_i \in T \tag{17}$$

Then the following isomorphisms can be established:

$$\Delta_c(S) \sim \Delta_c(N)/I \sim \Delta_c(N/T). \tag{18}$$

Here $I = \ker \pi_B$, and the third algebra in the chain is Segal's algebra. This process corresponds to the bottom half of the commutative diagram and shows that it commutes. In simple terms, the constraints may be applied either before or after quantization with the same result. Furthermore, in contrast to $\Delta_c(S)$, $\Delta_c(N)/I$ is Poincaré invariant (because both $\Delta_c(N)$ and $\Delta_c(T)$ are) and now physical photons can be correctly described as the particles of Segal's algebra, which is a factor algebra of the unphysical photon algebra $\Delta_c(M)$. This is the proper formulation of the Fermi approach and there is now no problem with non-normalizable states. The use of C^*-algebras is essential for this because they are much better behaved than algebras of bounded operators (von Neumann algebras). Roughly speaking this follows from the observation that if $F(x)$ is a continuous function of $x \in \mathbb{R}$ the mapping $F(x) \to F(x_0)$, x_0 fixed, is well defined. By contrast if $F(x)$ is an L^2-function, the evaluation mapping $F(x) \to F(x_0)$ would be meaningless because $F(x)$ is only defined to within sets of measure zero. It is possible therefore to find a state on a C^*-algebra for which the operator q: $qF(x) \equiv xF(x)$ has eigenvalue x_0. The rigorous proof proceeds essentially along the same lines.

The Poincaré automorphisms of $\Delta_c(N)/I$ are derived from those of $\Delta_c(M)$ and although the would not be expected to be inner, they are isomorphic to a representation by automorphisms inner to $\Delta_c(N)/I$. In particular the Hamiltonian, as the generator of time translations of $\Delta_c(M)$, is given formally by:

$$H_F = \frac{1}{2} \int d\Omega_0(-a_0(k)^* a_0(k) + \underline{a}(k)^* \cdot \underline{a}(k)) \tag{4}$$

and this induces the same time translations on $\Delta_c(N)/I$ as the classical expression:

$$H = \frac{1}{2} \int d^3x(\underline{E}^2 + \underline{B}^2). \tag{19}$$

This is very satisfactory because it demonstrates that the physical algebra $\Delta_c(N)/I$ contains all the relevant information including its space-time transformations.

The Fock representation of $\Delta_c(M)$ is specified by the generating functional:

$$\rho(\delta_\phi) = \exp - \frac{1}{4}\|\phi\|_F^2, \tag{20}$$

and it can be used to induce a representation of the physical algebra, which has all the requisite properties for describing transverse photons -- existence of a unique vacuum, correct Lorentz transformations, etc. However as a representation of non-physical photons, ρ has some peculiarities which are of interest. If (a,Λ) is a Poincaré transformation and $\theta(a,\Lambda)$ is the corresponding automorphism of $\Delta_c(M)$, then:

$$\theta(a,\Lambda)\delta_\phi = \delta_{(a,\Lambda)\phi}. \tag{21}$$

As $\|(a,\Lambda)\phi\|_F \neq \|\phi\|_F$, this Fock representation is not Poincare invariant, although it is invariant under the subgroup of space-time translations and spatial rotations. Lorentz boosts however cannot be unitarily implemented, and so the representation induced from the Fock representation by a Lorentz boost is an inequivalent representation. In order to obtain a completely covariant representation it is necessary to extend the representation space $\mathfrak{h}_\rho$ by constructing the direct integral of a family of Hilbert spaces $\mathfrak{h}_p$ over the homogeneous space:

$$X = \{p \in \mathbb{R}^4 \mid p_0^2 - \underline{p}^2 = 1, \, p_0 > 0\}, \tag{22}$$

with the Lorentz invariant measure $d\mu$ over X. If $\Lambda(p)$ is the Lorentz boost which maps $\hat{p} = (1,0,0,0)$ to p and ρ_p is the generating functional:

$$\rho_p(\delta_\phi) = \exp - \frac{1}{4}\|\Lambda(p)^{-1}\phi\|_F^2, \tag{23}$$

then $\mathfrak{h}_p$ is the GNS Hilbert space corresponding ρ_p. Over the direct integral Hilbert space:

$$\mathfrak{h} = \int_X \oplus \, \mathfrak{h}_p \, d\mu(p), \tag{24}$$

Lorentz boosts act by the mapping $\mathfrak{h}_p \to \mathfrak{h}_{\Lambda p}$ whilst the corresponding representation of $\Delta_c(M)$, π_p, acts in $\mathfrak{h}_p$. In this way we have endowed $\mathfrak{h}$ with the structure of a Hilbert Poincaré bundle in which the Poincaré transformations are unitarily implementable. The vacuum states Ω_p are cyclic under $\pi_p(\Delta_c(M))$ but are not Poincaré invariant. Physically this means that each observer sees a <u>different vacuum state for the non-physical photons</u>, or in other words, there is <u>spontaneous symmetry breaking of the Lorentz group</u>. Despite this the physical photons have a Lorentz invariant vacuum because <u>all the representations of</u> $\Delta_c(N)/I$ <u>induced in each</u> $\mathfrak{h}_p$ are identical.

This representation of the quantum theory of the electromagnetic field is defined by the space M and the skew form B(,). But so far as the physical system is concerned, all that is ultimately required is the physical algebra $\Delta_c(N)/I$, and the isomorphisms (18), together with the values of B(,) over this algebra. So it is possible to have other presentations with other spaces M' and other forms B', for as long as:

$$B(\phi_s,\psi_s) = B'(\phi_s,\psi_s) \tag{25}$$

we shall have the same physical theory. A special case of such a class of theories is that for which B' is given by [9]:

$$B'(\phi,\psi) = B(Z\phi,Z\psi) \tag{26}$$

with Z a real linear densely defined operator on M, and B'(,) also satisfies (25). Such a transformation is called a <u>linear gauge transformation</u> and a variety of them have been investigated by Strocchi and Wightman in the indefinite metric formalism. If we require Z to be translationally invariant, it can be shown that it must be of the form:

$$Z\phi(k) = \zeta(k)\phi(k)$$

with

$$\zeta(k)_{\mu\nu} = g_{\mu\nu} + k_\mu F_\nu(k) + G_\mu(k)k_\nu. \tag{27}$$

Particular cases are:

<u>Radiation gauge</u> $\quad \zeta_{0\nu} = 0, \; \zeta_{ij} = \delta_{ij} - k_i k_j/k^2,$

Kallen-Rollnik-Stech-Nunnemann, $F_\nu = 0,$

$G_\mu = Mk_\mu,$

Evans-Fulton $F_\nu = 0,\ G_\mu = -n_\mu/n\cdot k,$

Valatin $F_\nu = 0,\quad G_\mu = -\frac{n_\mu}{n\cdot k} + \frac{k_\mu}{(n\cdot k)^2},\quad n^2 > 0.$

It is also possible to treat Landau gauge in a similar way, although it is then necessary to enlarge M to include C^∞ functions of fast decrease on momentum space $\mathbb{R}^4$, taking values in $\mathbb{C}^4$ [10]. Z is then defined by:

$$Z = 1 - D_T + P_T g, \tag{28}$$

where

$$(D_T\phi)_\mu(k) = \frac{k_\mu}{2k_0}\frac{\partial}{\partial k_0}(k^\nu\phi_\nu(k)), \tag{29}$$

$$(P_T\phi)_\mu(k) = k_\mu \textstyle\sum_\nu k_\nu\phi_\nu(k)/2k_0^2. \tag{30}$$

Because of the extension of M, the vector potential no longer satisfies $\Box A_\mu = 0$, but rather $\Box^2 A_\mu = 0$. This was also observed by Nakanishi [11].

We have now an algebraic presentation of the quantum theory of the electromagnetic field which is more general than other presentations. It can be further shown [12] that it is possible to construct a perturbation theory which reproduces all the usual results of Gupta-Bleuler and the radiation gauge theory and which validates the heuristic approach of Schwinger [13].

There remains still the Dirac approach which treats the electromagnetic field as an example of a constrained system.

3. C^*-ALGEBRA APPROACH TO DIRAC QUANTIZATION

First of all we shall show how Dirac's approach can be put in rigorous mathematical form [14]. To do this it is necessary to make some initial assumptions which would have to be verified in particular models. The first assumption is:

I (i) $\exists$ C^*-algebra $\mathcal{F}$, the field algebra, with a set of states $\mathcal{I}$,

(ii) $\exists$ a family of one parameter subgroups $\{U_k(\lambda) \mid \lambda \in \mathbb{R},\ k \in I\}$ with I some index set.

Assumption I (ii) expresses the mathematical statement of the heuristic imposition of supplementary conditions:

$$\psi_k\Phi = 0, \tag{31}$$

which can be written as

$$U_k(\lambda)\Phi \equiv e^{i\lambda\psi_k}\Phi = \Phi.$$

The gauge group $\mathcal{G} \subset \mathrm{Aut}\ \mathcal{F}$ contains the elements:

$$\alpha_g[F] = U_k(\lambda)F\,U_k^{-1}(\lambda),\quad \forall\ F \in \mathcal{F}. \tag{32}$$

Dirac states I_D are defined by $\omega \in I_D$:

$$\omega(U_k(\lambda)F) = \omega(F) = \omega(FU_k(\lambda)), \quad \forall\, F \in F.$$

The second assumption is:

II $\quad I_D$ is non-empty.

If we put $L_k(\lambda) = U_k(\lambda)-1$, and $A(L)$ is the norm-completed*-algebra generated by the $L_k(\lambda)$, then it is a C^*-algebra. We define the closed linear space:

$$T = \overline{A(L)\cdot F} + \overline{F \cdot A(L)} \tag{33}$$

It can be shown that

$$\omega \in I_D \iff T \subseteq \ker \omega.$$

The motivation for this definition of T is that:

$$(L_k(\lambda)F) = 0 = \omega(FL_k(\lambda)),$$

and for a cyclic GNS state, this implies the heuristic equation:

$$\pi_\omega(\psi_k)\Omega_\omega = 0, \tag{34}$$

which is reminiscent of Dirac's requirement. However in order to justify this, we need a suitable definition of a two sided ideal. To do this, we define two algebras $S, S^* \subset F$ by: $S \subset F$ is the largest set such that $A(L)S \subset \overline{F\cdot A(L)}$, and $S^* \subset F$ is the largest set such that $S^*A(L) \subset A(L)\cdot F$. They are non-empty norm closed algebras and $S \cap S^* \supset A(L)'$ (the commutator of $A(L)$ in F) is a C^*-algebra.

If we also define the C^*-algebra D by:

$$D = \overline{S^*A(L)S}, \tag{35}$$

then

$$\omega \in I_D \iff \omega(DF) = 0 = \omega(FD), \quad \forall\, D \in D, \quad F \in F,$$

and $T = \overline{D\cdot F} + \overline{F\cdot D}$. This shows that D is doing essentially the same work as $A(L)$, whilst expressing more fully the constraints on the states. The C^*-algebra

$$O = S \cap S^*, \tag{36}$$

contains D as a two sided ideal, and is the largest algebra for which this is so. It is the <u>multiplier algebra</u> of D in F. The factor algebra

$$R = O/D \tag{37}$$

is the algebra of physical observables. The trio of algebras:

$$D \lhd O \subset F,$$

is called a <u>Dirac triplet</u>, and is closely analogous to the triplet of spaces $H'' \subset H' \subset H$ introduced by Strocchi and Wightman. This connection is explored in detail in Ref. [14]. The Dirac states $I_D(O)$ when restricted to O

are identical to the set of states on F which are annihilated by $\mathcal{D}$.

This machinery can be applied to the electromagnetic field, as was done by Dirac. The constraints $\{L_k\}$ are first class constraints for the action (1). The C^*-algebra for the electromagnetic potential can be constructed following Segal and Manuceau, and this is taken to be the field algebra F. The constraint algebra $A(L)$ is generated by the constraints (2). It can then be shown that $\mathcal{O} = S \cap S^* = A(L)'$ and this is just $\Delta_c(N)$ even though we no longer have the Fermi term in the Lagrangian density.

If we attempt to construct a Fock representation from the Hamiltonian defined by the Dirac procedure we find that because of the absence of $\nabla \cdot A$ there is no vacuum state for longitudinal photons.

4. CONCLUSION

In this article we have reviewed various approaches to the quantum theory of the free electromagnetic field from the point of view of C^*-algebras. There are other approaches, such as that of Bongaarts [15], which uses the Borchers algebra, which are not covered, but all the traditional ones are included as special cases. The indefinite metric approach of Gupta and Bleuler, which should be interpreted as a way of representing $\Delta_c(N)$ and $\Delta_c(T)$, is discussed more fully in Ref. 16.

There is another approach, using path integrals, which should also be mentioned. At present that approach lacks a fully rigorous foundation, which is unfortunate because heuristically it is very powerful. As Segal's generating functional bears a close resemblance to a path integral, it suggests that the connection between path integrals and C^*-algebra may be strengthened, at least for free fields.

Of course, the treatment of fields which are non-linear, and so self-interacting, or collections of mutually interacting fields have not been discussed because they have not been capable of rigorous formulation when gauge degeneracies are present. It is believed that if and when such formulations are available they will be describable in this language because of its generality and rigor.

ACKNOWLEDGEMENT

It is a great pleasure to acknowledge the hospitality of Professor Bruno Gruber at Southern Illinois University and the assistance of all my colleagues in Adelaide who have contributed so greatly to this program.

REFERENCES

1. P. A. M. Dirac, Lectures in Quantum Mechanics, Belfer Graduate School of Science, Yeshiva University, New York (1964).
   Canad. J. Math. 2, 129 (1950)
   Canad. J. Math. 3, 1, (1951)
2. I. E. Segal, Mathematical Problems of Relativistic Physics, Providence, R.I.
   American Mathematical Society (1963).
3. S. N. Gupta, Proc. Phys. Soc. London A 63, 681 (1950).
   K. Bleuler, Helv. Phys. Acta 23, 567 (1950).
4. E. Fermi, Atti. Accad. Lincei 9, 881 (1929), ibid. 12, 431 (1930).
   Rev. Mod. Phys. 4, 125 (1932).

5. G. Rideau, Lett. Math. Phys. 2, 529 (1978).
6. A. L. Carey, J. M. Gaffney, C. A. Hurst, J. Math. Phys. 18, 629 (1977).
A. L. Carey, C. A. Hurst, J. Math. Phys. 18, 1553 (1977).
7. J. Manuceau, Ann. Inst. Henri Poincaré 8, 139 (1968).
8. F. Strocchi, A. S. Wightman, J. Math. Phys. 15, 2198 (1974).
9. A. L. Carey, J. M. Gaffney, C. A. Hurst, Reports Math. Phys. 13, 419 (1978).
10. A. L. Carey, C. A. Hurst, Lett. Math. Phys. 2, 227 (1978).
11. N. Nakanishi, Suppl. Prog. Theor. Phys. 51, 1 (1972).
12. J. D. Wright, Aust. J. Phys. 35, 661 (1982).
13. J. Schwinger, Phys. Rev. 75, 651 (1949).
14. H. B. G. S. Grundling, C. A. Hurst, Commun. Math. Phys. 98, 369 (1985).
15. P. J. M. Bongaarts, J. Math. Phys. 23, 1881 (1982).
16. Hendrik Grundling, C. A. Hurst, Algebraic Structures of Degenerate Systems, Physical Requirements and the Indefinite Metric. Adelaide University Preprint. (1985)

DYNAMIC SYMMETRIES IN SCATTERING

F. Iachello

A. W. Wright Nuclear Structure Laboratory, Yale University

New Haven, Connecticut 06511

ABSTRACT

I discuss a recent suggestion that dynamic symmetries and spectrum generating algebras can be used to describe scattering problems.

1. INTRODUCTION

Dynamic symmetries and spectrum generating algebras have been used up to now mostly to describe bound states. Notable examples here are: Gell-Mann-Ne'eman SU(3) which leads to the Gell-Mann-Okubo mass formula[1] and its generalization to SU(6) by Gürsey and Radicati[2] in particle physics; the dynamic symmetries of the interacting boson model[3], U(5), SU(3) and SO(6) in nuclear physics; the symmetry of the Coulomb potential[4], SO(4), and its generalizaton to two-electron atoms[5] in atomic physics; and the symmetries of the vibron model[6], U(3) and SO(4), in molecular physics. Dynamic symmetries as applied to bound state problems have produced classification schemes and mass formulas and have generally led to a new and deeper understanding of the problem at hand. For example, in elementary particle physics, the discovery of SU(3) as a dynamical symmetry of the low-lying hadronic spectra has led to the establishment of the quark model and of quantum chromodynamics. Similarly, the discovery of U(6) as a dynamical symmetry of the low-lying nuclear spectra has led to the establishment of the interacting boson model and of the interpretation of properties of nuclei in terms of correlated pairs.

The structure of all these theories is that the Hamiltonian (or mass operators) is expressed in terms of the generators of some group G, usually of the form

$$H = E_0 + \sum_{\alpha\beta} \varepsilon_{\alpha\beta} G_{\alpha\beta} + \sum_{\alpha\beta\gamma\delta} u_{\alpha\beta\gamma\delta} G_{\alpha\beta} G_{\gamma\delta} + \dots \tag{1.1}$$

where

$$G_{\alpha\beta} \in \mathcal{g} \quad . \tag{1.2}$$

(In some problems 1/H rather than H is expanded in the generators of G).

Dynamic symmetries then correspond to cases in which H is not the most general function of G but it can be expressed as a function of only Casimir invariants of a chain of groups

$$G \supset G' \supset G'' \quad \dots \quad , \tag{1.3}$$

i.e.

$$H = \alpha C(G) + \alpha' C(G') + \alpha'' C(G'') + \dots \quad . \tag{1.4}$$

Taking the expectation value of H in a given representation $\mathcal{R}$ of $G \supset G' \supset G''$ ... leads to mass formulas

$$E = \alpha\langle C(G)\rangle + \alpha'\langle C(G')\rangle + \alpha''\langle C(G'')\rangle + \dots \quad . \tag{1.5}$$

The main property of these mass formulas is that the energy levels are given in terms of the quantum numbers labeling the representations $\mathcal{R}$.

The use of dynamic symmetries in scattering has been so far very limited. The only case discussed in detail has been that of scattering by a Coulomb potential[7] with symmetry group SO(3,1). We have recently begun a series of investigations[8] in an attempt to introduce systematically the use of dynamic symmetries and spectrum generating algebras in scattering problems.

2. DYNAMIC SYMMETRIES IN SCATTERING

I will consider here only non-relativistic problems. The traditional approach has been that of solving the Schrödinger equation

$$[-\frac{\hbar^2}{2\mu}\nabla^2 + V(\vec{r})]\,\psi(\vec{r}) = E\,\psi(\vec{r}) \tag{2.1}$$

with appropriate boundary conditions, for example,

$$\psi(\vec{r}) \xrightarrow[r\to\infty]{} e^{i\vec{k}\cdot\vec{r}} + f(\Omega)\frac{e^{ikr}}{r} \quad . \tag{2.2}$$

This method produces the S-matrix

$$S_\ell(k) = e^{2i\delta_\ell(k)} \quad , \tag{2.3}$$

from which one can compute the scattering amplitude and cross section,

$$f(\theta) = \frac{1}{2ik}\sum_{\ell=0}^{\infty}(2\ell+1)\,[S_\ell(k)-1]\,P_\ell(\cos\theta) \quad . \tag{2.4}$$

It appears that one can obtain S-matrices in a purely algebraic way by defining a dynamic group G which describes the problem in the presence of interactions, and an asymptotic group F which describes the problem in the absence of interactions. The S-matrix appears as coefficients in the expansion of the representations of G into those of F, schematically written as

Representations of G = (S-matrix) × Representations of F	(2.5)

It should be noted that, contrary to the case of bound state problems, where the knowledge of the dynamic group G is sufficient to determine the properties of the system, for scattering problems we need the knowledge of both the dynamic group G and the asymptotic group F. An obvious candidate for the non-relativistic asymptotic group in n-dimensions is the Euclidean group, E(n). I illustrate the result (2.5) with two examples.

Example 1: The Coulomb problem in three dimensions.

The group G here is SO(3,1) with six generators, the angular momentum, $\vec{L}$, and the Runge-Lenz vector, $\vec{A}$. The asymptotic group is E(3) with six generators, the angular momentum, $\vec{L}$, and the momentum, $\vec{P}$.

The representations of G are labelled by

$$G: \quad \left| \begin{array}{ccccc} SO(3,1) & \supset & SO(3) & \supset & SO(2) \\ \downarrow & & \downarrow & & \downarrow \\ (\omega,0) & & \ell & & m \end{array} \right\rangle \quad . \tag{2.6}$$

Those of F by

$$F: \quad \left| \begin{array}{ccccc} E(3) & \supset & SO(3) & \supset & SO(2) \\ \downarrow & & \downarrow & & \downarrow \\ (\pm k,0) & & \ell & & m \end{array} \right\rangle \quad . \tag{2.7}$$

In (2.6) I have used the continuous unitary representations of SO(3,1) with

$$\begin{aligned} \langle C_2 \rangle &= \omega(\omega+2) \quad ; \qquad \omega = -1 + if(k) \\ \langle C'_2 \rangle &= 0 \quad , \end{aligned} \tag{2.8}$$

where $C_2 = \vec{L}^2 - \vec{K}^2$, $C'_2 = \vec{L}\cdot\vec{K}$, $\vec{K} = (\mu\vec{A})/k$. Eq. (2.5) reads in this case

$$|\omega,\ell,m\rangle = A_\ell(k)\ |-k,\ell,m\rangle + B_\ell(k)\ |+k,\ell,m\rangle, \tag{2.9}$$

where $|-k,\ell,m\rangle$ and $|+k,\ell,m\rangle$ are the free incoming and outgoing waves. In order to obtain the S-matrix one writes the generators of G in terms of those of F. This technique, called Euclidean connection, is the crucial point of the whole derivation and is discussed by Y. Alhassid in the accompanying paper[9]. The Euclidean connection yields recursion relations for $A_\ell(k)$, $B_\ell(k)$ and their ratios $R_\ell(k) = B_\ell(k)/A_\ell(k)$,

$$R_{\ell+1}(k) = -\frac{\ell+1+if(k)}{\ell+1-if(k)}\, R_\ell(k) \quad . \tag{2.10}$$

The recursion relations can be solved to yield the S-matrix $S_\ell(k) = e^{i(\ell+1)\pi} R_\ell(k)$, i.e.

$$\boxed{S_\ell(k) = \frac{\Gamma(\ell+1+if(k))}{\Gamma(\ell+1-if(k))}} \tag{2.11}$$

and the cross-section

$$\frac{d\sigma}{d\Omega} = \frac{f^2(k)}{4k^2 \sin^4 \frac{\theta}{2}} \quad . \tag{2.12}$$

It appears that all problems with SO(3,1) dynamic symmetry have S-matrices of this form. Thus the occurrence of a dynamic symmetry in scattering determines the functional form of the S-matrix in terms of the quantum numbers (ℓ in this case), much in the same way in which the presence of a dynamic symmetry for bound states determines the form of the mass formula in terms of the quantum numbers. The unknown function f(k) is determined by the relation between the Hamiltonian and the Casimir invariants of G. For the Coulomb problem,

$$H = -\frac{\mu\beta^2}{2(C_2+1)} \quad , \qquad \beta = Z_1 Z_2 \, e^2 \tag{2.13}$$

gives

$$f(k) = \frac{\mu\beta}{k} \tag{2.14}$$

Example 2: The modified Coulomb problem.

For this problem, we start from the group G ≡ SO(3,2) and consider the asymptotic group F ≡ E(3) ⊗ E(2) composed of a part, E(3), describing the asymptotic <u>space</u> group and a part, E(2), describing the asymptotic <u>internal</u> group. The representations of G are labeled now by

$$G: \left| \begin{array}{ccccccc} SO(3,2) & \supset & SO(3) \otimes SO(2) & \supset & SO(2) & \otimes & SO(2) \\ \downarrow & & \downarrow & & \downarrow & & \downarrow \\ (\omega,0) & & \ell & & m & & \nu \end{array} \right\rangle \quad , \tag{2.15}$$

while those of F are labeled by

$$F: \left| \begin{array}{ccccccc} E(3) \otimes & E(2) \supset & SO(3) \otimes SO(2) & \supset & SO(2) & \otimes & SO(2) \\ \downarrow & \downarrow & \downarrow & & \downarrow & & \downarrow \\ (\pm k,0) & (1) & \ell & & m & & \nu \end{array} \right\rangle \tag{2.16}$$

where again I have used the continuous unitary representations of of SO(3,2) with[8]

$$\langle C_2 \rangle = \omega(\omega+3) \quad ; \quad \omega = -\frac{3}{2} + i f(k) \quad , \tag{2.17}$$

$$\langle C_4 \rangle = 0 \quad .$$

Eq. (2.5) reads in this case

$$|\omega,\ell,m,v\rangle = A_{\ell v}(k)\ |-k,\ell,m,v\rangle + B_{\ell v}(k)\ |+k,\ell,m,v\rangle \qquad . \qquad (2.18)$$

Using the Euclidean connection, one can obtain S-matrices of the form

$$S_\ell(k) = \frac{\Gamma\left(\frac{\ell+v+3/2+if(k)}{2}\right)\Gamma\left(\frac{\ell-v+3/2+if(k)}{2}\right)}{\Gamma\left(\frac{\ell+v+3/2-if(k}{2}\right)\Gamma\left(\frac{\ell-v+3/2-if(k)}{2}\right)} e^{i(2\ell n2)f(k)} \qquad . \qquad (2.19)$$

It is interesting to observe that if one lets v=1/2 in (2.19) one obtains back the Coulomb S-matrix (2.11), as one can see by using the duplication formula

$$\Gamma(z)\ \Gamma(z+\frac{1}{2}) = 2^{1-2z}\ \pi^{1/2}\ \Gamma(2z) \qquad . \qquad (2.20)$$

When v≠1/2, the S-matrix (2.19) appears to describe a modified Coulomb interaction. We have suggested[10] that the S-matrix (2.19) is well suited to describe heavy ion collisions. The interactions between two heavy ions is dominated, at large distances, by their Coulomb repulsion, while, at short distances, it is modified by the short range nuclear interaction. The quantity v can be interpreted as the interaction. The phenomonological analysis of the collisions suggests that v be described by a Woods-Saxon well of the type

$$v(\ell,k) = (v_R - iv_I)\ \frac{1}{1+\exp\left[(\ell-\ell_o)/\Delta\right]} \qquad . \qquad (2.21)$$

Because of the presence of the dynamic symmetry v can only be a function of the quantum numbers ℓ and k. The particular form (2.21) depends explicitly on the physics of the problem at hand. Figure 1 shows the angular distribution obtained from the S-matrix (2.19) with an appropriate choice of the parameters, at a given energy E.

In addition to the S-matrices (2.11) and (2.19) we have constructed[11] S-matrices for all the dynamical groups SO(n,m), where n denotes the number of space dimensions, and m(m-1)/2 the number of potential parameters. Among these, there are the S-matrices describing Coulomb scattering in any number of dimensions.

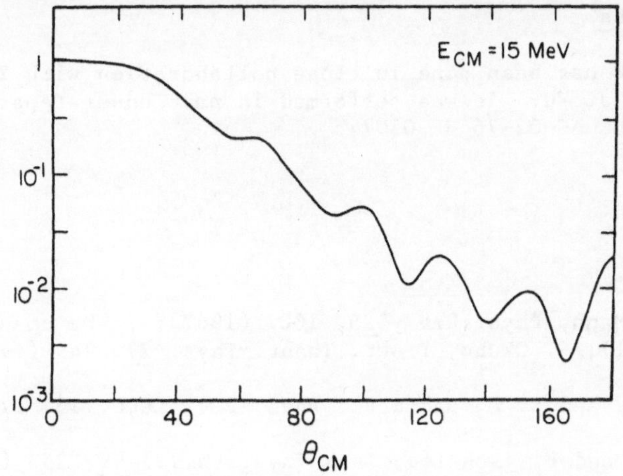

Fig. 1. Angular distribution for the collision of two heavy ions with masses A_1=12, A_2=14 and charges Z_1=6, Z_2=6. The ratio of the differential cross section to the Rutherford cross section is plotted. The parameters used in the calculations are v_R=1.5, v_1=2.5, ℓ_o=15.4, Δ=0.5 and E_{cm}=15 MeV.

3. CONCLUSIONS

It appears that it is possible to extend the notion of dynamic symmetries and spectrum generating algebra to scattering problems in a systematic way, by using the notion of an asymptotic group F and of the Euclidean connection relating the representations of the dynamic group G to those of F. This method provides realistic S-matrices that can be used to analyze the data. The algorithm we have devised, being quite general, can be further generalized to cover relativistic problems. The S-matrices obtained by using the notion of symmetry can play a role similar to that played by the Veneziano model[12] in the late 1960's in the description of collisions between elementary particles. Furthermore, as it has been in the case of bound state problems, it is conceivable that dynamic symmetries will elucidate the physics involved in a particular scattering process. In the case of the Veneziano model, S-matrix theory led to string theory. We have now here a powerful technique to construct, in a systematic way, solvable S-matrices. This may provide further developments.

Acknowledgments

This work has been done in close collaboration with Y. Alhassid, F. Gürsey and J. Wu. It was performed in part under Department of Energy Contract No. DE-AC-02-76 ER 03074.

References

1. M. Gell-Mann, Phys. Rev. 125, 1067 (1962); Y. Ne'eman, Nucl. Phys. 26, 222 (1961); S. Okubo, Progr. Theor. Phys. 27, 949 (1962).

2. F. Gürsey and L. A. Radicati, Phys. Rev. Lett. 13, 173 (1964).

3. A. Arima and F. Iachello, Ann. Phys. (N.Y.) 99, 253 (1976); 111, 201 (1978); 123, 468 (1979).

4. W. Pauli, Z. Physik 36, 336 (1926); V. Fock, Z. Physik 98, 145 (1935); V. Bargmann, Z. Physik 99, 576 (1936).

5. D. R. Herrick and M. E. Kellmann, Phys. Rev. A21, 418 (1980); M. E. Kellmann and D. R. Herrick, J. Phys. B11, L755 (1978).

6. F. Iachello, Chem. Phys. Lett. 78, 581 (1981); F. Iachello and R. D. Levine, J. Chem. Phys. 77, 3046 (1982); O. S. van Roosmalen, F. Iachello, R. D. Levine and A. E. L. Dieperink, J. Chem. Phys. 79, 2515 (1983).

7. D. Zwanziger, J. Math. Phys. 8, 1858 (1957).

8. Y. Alhassid, F. Gürsey and F. Iachello, Ann. Phys. (N.Y.) 148, 356 (1983); 167, (1986); J. Wu, F. Iachello and J. Wu, Ann. Phys. (N.Y.), to be published; A. Frank and K. B. Wolf, J. Math. Phys. 26, 1973 (1985).

9. Y. Alhassid, These Proceedings.

10. Y. Alhassid, F. Iachello and J. Wu, Phys. Rev. Lett. 56, 271 (1986).

11. Y. Alhassid, A. Frank and F. Iachello, Preprint YNT 86-06.

12. G. Veneziano, Nuovo Cimento 57A, 190 (1968).

FACTORIZATION-ALGEBRAIZATION-PATH INTEGRATION AND DYNAMICAL GROUPS

Akira Inomata
Department of Physics
State University of New York
Albany, New York 12222

Raj Wilson
Department of Mathematics
University of Texas
San Antonio, Texas 78285

0. SCHRÖDINGER-INFELD-HULL FACTORIZATIONS

Schrödinger[1] proposed an elegant method of factorization of a quantum-mechanical second-order linear differential equation into a product of two first-order differential operators often referred to as ladder operators. These ladder operators when acting on respective eigenfunctions create new eigenfunctions with a quantum number raised or lowered by one unit. Schrödinger's method was further systematically studied for a class of second-order linear differential equations in particular by Infeld and Hull[2] who have shown that a second-order differential equation which may be brought into the form:

$$\phi_m''(x) + [\rho_m(x) + \lambda]\phi_m(x) = 0;\ m \in \mathbb{N} \tag{0.1}$$

may be factorized into products of two first-order ladder operators $A_m^{\pm}$ such that

$$\begin{aligned} A_m^{+}A_m^{-}\phi_m(x) &= (\lambda - \alpha_m)\phi_m(x)\,, \\ A_m^{-}A_m^{+}\phi_m(x) &= (\lambda - \alpha_{m+1})\phi_m(x)\,. \end{aligned} \tag{0.2}$$

The actions of $A_m^{\pm}$ on $\phi_m(x)$ are given by

$$\begin{aligned} A_m^{+}\phi_m(x) &= -\phi_m'(x) + \kappa_{m+1}(x)\,\phi_m(x) = \left(\lambda - \alpha_{m+1}\right)^{\frac{1}{2}}\phi_{m+1}(x)\,. \\ A_m^{-}\phi_m(x) &= \phi_m'(x) + \kappa_m(x)\,\phi_m(x) = \left(\lambda - \alpha_m\right)^{\frac{1}{2}}\phi_{m-1}(x) \end{aligned} \tag{0.3}$$

where

$$\begin{aligned} 2\left[\rho_{m-1}(x) - \rho_m(x)\right]\kappa_m(x) &= \rho_{m-1}'(x) + \rho_m'(x)\,, \\ \alpha_m &= -\kappa_m^2(x) - \tfrac{1}{2}\left[\rho_{m-1}(x) + \rho_m(x)\right] \end{aligned} \tag{0.4}$$

Clearly, the factorizations (0.2) are possible only when α_m in (0.4) are independent of x. Depending on the form of $\rho_m(x)$ the factorizations are classified as follows[2] (a,b,c,d,p,q are constants):

$$\text{Type A:}\quad \rho_m(x) = -[a^2(m+c)(m+c+1)+d^2]\csc^2[a(x+p)]$$
$$-2ad(m+c+\tfrac{1}{2})\cot[a(x+p)]\csc[a(x+p)]\,.$$

$$\text{Type B:}\quad \rho_m(x) = -d^2 e^{2ax} + 2ad\,(m+c+\tfrac{1}{2})\,e^{ax}$$

$$\text{Type C:}\quad \rho_m(x) = -(m+c)(m+c+1)\frac{1}{x^2} - \frac{b^2}{4}x^2 + b(m-c)$$

$$\text{Type D:}\quad \rho_m(x) = -(bx+d)^2 + b(2m+1)\,.$$

$$\text{Type E:}\quad \rho_m(x) = -a^2 m(m+1)\csc^2[a(x+p)] - 2aq\cot[a(x+p)]$$

$$\text{Type F:}\quad \rho_m(x) = -m(m+1)\frac{1}{x^2} - 2q\frac{1}{x}\,. \tag{0.5}$$

However all these six factorizations are not independent as they are related as follows:

Different factorizations of same equation: Class I { A, E } —→ C, D } degenerate cases

Different factorizations of same equations: Class II { B, F } —→ C, D (0.6)

Besides the six factorizations (0.5) there also exist the so-called artificial factorizations[2] for which (0.4) is not valid.

1. ALGBRAIZATIONS AND DYNAMICAL GROUPS

An elegant group theoretical treatment to the factorizations outlined above has been given very systematically by Miller[3]. Here, the second-order linear differential equation (0.1) appears as the invariant Casimir operator in the Lie algebra of a Lie group, in particular, of rank one, acting on the basis functions belonging to a given unitary irreducible representation of the group and the ladder operators appear as the generators of the concerned group. Certain addition theorems satisfied by these ladder operators and the self-adjointness of these operators have also been studied.[4] Miller[3,5] introduced a four-dimensional Lie algebra $G(a,b)$; $a,b \in \mathbb{R}$, generated by J_i, $i = 1, 2, 3$ and E. The operators J_i realized in terms of differential operators acting on a space of C^∞ functions of one or two variables and E proportional to the identity operators ($E = -iI$; $i = \sqrt{-1}$) satisfy

$$\text{Lie product:}\quad [J^+, j^-] = 2a^2J^3 - 2b\,I\ ;\ J^\pm = \pm J_2 + iJ_1,\ J^3 = iJ_3,$$
$$[J^3, J^\pm] = \pm J^\pm,$$
$$[J^+, I] = [J^3, I] = 0 \tag{1.1}$$

$$\text{Casimir product:}\quad C_{a,b} = J^+J^- + a^2J^3J^3 - (2b + a^2)J^3 \tag{1.2}$$

where $J^\pm$ are the ladder operators. In terms of this Lie algebra $G(a,b)$ and of the algebra of the three dimensional Euclidean Group E[3][5,6] (or E[2,1] as the case may be) the six types of factorization given by Infeld and Hull (0.6) correspond to eight algebraizations as follows:

Class I: $A \Rightarrow G(1,0)$; $E \Rightarrow E(3)$

$C' \Rightarrow G(0,1)$; $C'' \Rightarrow G(0,0)$ } degenerate cases
$D' \Rightarrow G(0,1)$; $D' \Rightarrow G(0,0)$

Class II: $B \Rightarrow G(1,0)$; $F \Rightarrow E(3)$ (1.3)

It is interesting to note that the algebraization splits each degeneracy in (0.6) further into two different cases. In (1.2), $G(1,0) \simeq s\ell(2) \oplus E$, the real form of which is given by $SO(3) \oplus E$ or $SO(2,1) \oplus E$; $G(0,1) \simeq H_4$ the four-dimensional Heisenberg algebra[5,7] (isomorphic to the algebra generated by the field creation operator a^+, the annihilation operator a, the number operator $N = a^+a$ and the identity operator I): and $G(0,0) \simeq E(2) \oplus E$. The degeneracy in factorization corresponds to contraction in algebraization and the contractions[5,8] are given by the diagram:

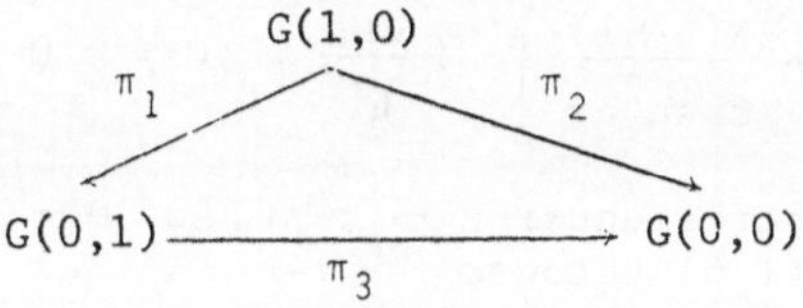

(1.4)

In terms of the solutions to (0,1) for the respective type of factorization, these contractions suggest,

$$\pi_1 \Rightarrow \lim_{\beta \to \infty} P_n^{(\alpha,\beta)}\left(1 - \frac{2x}{\beta}\right) = L_n^{\alpha}(x)$$

$$\pi_2 \Rightarrow \lim_{n \to \infty} \left(\frac{x}{2n}\right)^{\alpha} P_n^{(\alpha,\beta)}\left(\cos\frac{x}{n}\right) = J_{\alpha}(x)$$

$$\pi_3 \Rightarrow \lim_{n \to \infty} \left(\frac{\sqrt{x}}{n}\right)^{\alpha} L_n^{\alpha}\left(\frac{x}{n}\right) = J_{\alpha}(2\sqrt{x}\,). \qquad (1.5)$$

where $P_n^{(\alpha,\beta)}(z)$, $L_n^{\alpha}(z)$ and $J_{\alpha}(z)$ are Jacobi, Laguerre and Bessel functions respectively. In (1.3) the factorizations of type E and F require the faithful irreducible representations of E(3) reduced as $E(3) \supset SO(3) \supset SO(2)$ [or $E(2,1) \supset SO(2,1) \supset SO(2)$].

The group G[a, b], corresponding to the Lie algebra G(a, b) introduced by Miller may be considered as a dynamical group for one-dimensional physical problem. However, for the more realistic three-dimensional problems which admit a Hamiltonian description with, say, spherically symmetric potentials the dynamical group may be taken as $G[a, b] \otimes SO[3]$ or as a larger group which contains the above. The group $SO[4,2] \supset SO[2,1] \otimes SO[3]$ introduced by Barut[9] has been very successful as a dynamical group for composite physical systems in three-dimensions. In a most degenerate unitary irreducible representation the generators of $SO[2,1] \subset SO[4,2]$ may be realized in terms of second-order differential operators (2.1 - 2.3) which contains the Casimir product of the subgroup $SO[3] \subset SO[4,2]$ and thus the physical equations may easily be expressed as linear in $SO[2,1]$ group generators. Similar approach was also made by Casimir[10] who used $SO[4] \sim SO[3] \otimes SO[3]$ as the dynamical group.

We have already pointed out that the ladder operators which are defined for factorization belong to a Lie algebra and are acting on the basis functions of a given unitary irreducible representation. The same factorization can be equivalently achieved[11] in terms of ladder operators which are acting on certain matrix elements of the respective group elements. For example, the Wigner functions[5] $d^{\ell}_{mm^1}(\alpha)$ of $SO[3]$, the Bargmann functions[12] $V^{\ell}_{nn^1}(\alpha)$ of $SO[2,1]$, and their generalization to continuous basis by Barut and Phillips[13] etc. have been considered very extensively. By constructing ladder operators of the two types mentioned above several physical problems have recently been solved[14,15]. Here we very briefly describe the algebraization of second Pöschl-Teller equation.

The second Pöschl-Teller equation is

$$\left[\frac{\partial^2}{\partial r^2} - a^2\left[\frac{\kappa(\kappa-1)}{\sinh^2 ar} - \frac{\lambda(\lambda+1)}{\cosh^2 ar}\right] + \frac{2ME}{\hbar^2}\right]\phi(r) = 0 \qquad r \in [0,\infty) \tag{1.6}$$

The equation is symmetric under $\kappa \to -\kappa+1$ and $\lambda \to -\lambda-1$. With $\kappa = -m-g+\frac{1}{2}$, $\lambda = m-g-\frac{1}{2}$, $\beta = 2ar$ (1.6) becomes

$$\left[\frac{\partial^2}{\partial\beta^2} - \frac{1}{4}\left[\frac{(m+g+\frac{1}{2})(m+g-\frac{1}{2})}{\sinh^2(\beta/2)} - \frac{(m-g-\frac{1}{2})(m-g+\frac{1}{2})}{\cosh^2(\beta/2)}\right] + \Lambda\right]\phi(r) = 0$$

$$\Lambda = \frac{ME}{2a^2\hbar^2} \tag{1.7}$$

which is similar to (0.1). Using factorization of type A we define ladder operators,

$$M^+\phi_{m,g} = e^{i\alpha}\left[-\frac{\partial}{\partial\beta} + \tfrac{1}{2}(m+g+\tfrac{1}{2})\coth(\tfrac{\beta}{2}) + \tfrac{1}{2}(m-g+\tfrac{1}{2})\tanh(\tfrac{\beta}{2})\right]\phi_{m,g}$$

$$= \left[\Lambda + (m+\tfrac{1}{2})^2\right]^{\frac{1}{2}}\phi_{m+1,g}$$

$$M^-\phi_{m,g} = e^{-i\alpha}\left[\frac{\partial}{\partial\beta} + \tfrac{1}{2}(m+g-\tfrac{1}{2})\coth(\tfrac{\beta}{2}) + \tfrac{1}{2}(m-g-\tfrac{1}{2})\tanh(\tfrac{\beta}{2})\right]\phi_{m,g}$$

$$= \left[\Lambda + (m-\tfrac{1}{2})^2\right]^{\frac{1}{2}}\phi_{m-1,g}$$

$$M_3\phi_{m,g} = -i\frac{\partial}{\partial\alpha}\phi_{m,g} = m\,\phi_{m,g}\;; \quad \alpha \in [0,\,2\pi] \tag{1.8}$$

The operators $M^{\pm}$, M_3 close under SU(1,1) and satisfy Lie product,

$$[M^+,\,M^-] = -2M_3,\; [M^{\pm},\,M_3] = \mp M^{\pm}$$

and Casimir product,

$$C_{SU(1,1)}\phi_{m,g} = [-M^+M^- + M_3M_3 - M_3]\phi_{m,g}$$
$$= (-\Lambda - \tfrac{1}{4})\phi_{m,g} = \ell(\ell-1)\phi_{m,g} \qquad (1.9)$$

Thus we obtain the energy spectrum

$$E_n = -\frac{2a^2\hbar^2}{M}(\ell-\tfrac{1}{2})^2 = -\frac{a^2\hbar^2}{2M}(\lambda-\kappa-2n)^2 ;$$
$$n = 0, 1, \ldots < \tfrac{1}{2}(\lambda-\kappa)$$
$$\ell = m - n ;\; m > n \in \mathbb{N}. \qquad (1.10)$$

The equation (1.7) may now be written as

$$\left[C_{SU(1,1)} - \ell(\ell-1)\right]\phi_{m,g} = 0 \qquad (1.11)$$

Since (1.7) remains unchanged under $m \leftrightarrow g$ we obtain additional ladder operators $G^{\pm}$ for the same factorization of type A. $G^{\pm}$ and another operator G_3 close under SU(1,1) and can be easily determined from[16]

$$G_i = \sum_{j=1}^{3} H_{ji}(\alpha,\beta,\gamma)M_j ; \qquad G^{\pm} = G_1 \mp iG_2 ; \quad \gamma \in [0, 2\pi]$$
$$M^{\pm} = M_1 \pm iM_2$$

$$H(\alpha,\beta,\gamma) = \begin{pmatrix} \cos\alpha & -\sin\alpha & 0 \\ \sin\alpha & \cos\alpha & 0 \\ 0 & 0 & 1 \end{pmatrix} \begin{pmatrix} -\cosh\beta & 0 & \sinh\beta \\ 0 & 1 & 0 \\ \sinh\beta & 0 & -\cosh\beta \end{pmatrix} \begin{pmatrix} \cos\gamma & -\sin\gamma & 0 \\ \sin\gamma & \cos\gamma & 0 \\ 0 & 0 & 1 \end{pmatrix} \qquad (1.12)$$

One can easily see that $[M_i, G_j] = 0$. Thus the general solution to (1.6) is given by the eigenfunctions satisfying

$$C_{SU(1,1)}\,\phi_{m,g}(\alpha,\beta,\gamma) = \ell(\ell-1)\,\phi_{m,g}(\alpha,\beta,\gamma)$$
$$M_3\phi_{m,g}(\alpha,\beta,\gamma) = m\phi_{m,g}(\alpha,\beta,\gamma)$$
$$G_3\phi_{m,g}(\alpha,\beta,\gamma) = g\phi_{m,g}(\alpha,\beta,\gamma) \qquad (1.13)$$

From (1.8) and (1.12) we obtain four recurrence relations connecting $\phi_{m,g}(\beta)$, $\phi_{m\pm1,g}(\beta)$, $\phi_{m,g\pm1}(\beta)$ and if we compare these relations with the four recurrence relations[11,12] of Bargmann function for SU[1,1] connecting $V^{\ell}_{n^1 n}(\theta)$, $V^{\ell}_{n^1\pm1,n}(\theta)$, $V^{\ell}_{n^1,n\pm1}(\theta)$ we obtain the normalized solution to (1.6) immediately as

$$\phi(r) = \sqrt{2\ell-1}\,[\sinh(2ar)]^{\frac{1}{2}}\, V^{\ell}_{m,g}(-i\pi-2ar)$$
$$\ell = \tfrac{1}{2}(\lambda-\kappa+1-2n),\; m = \tfrac{1}{2}(\lambda-\kappa+1),\; g = -\tfrac{1}{2}(\kappa+\lambda)$$
$$n = 0, 1, \ldots, < \frac{\lambda-\kappa}{2} \Rightarrow \ell > \tfrac{1}{2}. \qquad (1.14)$$

Here the Bargmann function is expressed in terms of ${}_2F_1$ hypergeometric functions. Besides the discrete spectrum (1.10) and the solution (1.14) the equation (1.6) also has continuous energy spectrum for

$1 + |-\kappa + \frac{1}{2}| - |\lambda + \frac{1}{2}| \geqslant 0$. This can be immediately obtained[15] by using the continuous principal representations of SU[1,1]. It is interesting to note that the Equation (1.6) was also studied by Weyl[17] in 1910 on a different context.

2. PATH INTEGRALS AND DYNAMICAL GROUPS

Recently it has been recognized that the dynamical symmetry consideration is greatly beneficial to path integral calculations.[18] For instance, the Kepler problem in $\mathbb{R}^3$, whose dynamical group is SO[4], has been converted into a harmonic oscillator in $\mathbb{R}^4$ for path integration,[19] and the one-dimensional Pöschl-Teller oscillator, found to carry the SO(3) symmetry, has been path-integrated on $S^3 = SO[4]/SO[3]$.[18] In this section, we shall examine the link between the dynamical group approach and the path integral technique by taking SO[2,1] and SO[3] cases.

Feynman's path integral is Gaussian in nature. Its exact evaluation is limited to quadratic systems.[20] Use of polar coordinates has slightly relaxed the limitation by allowing us to include the inverse-square potential in the exactly path-integrable class.[21] Therefore, path integration can be achieved if the system in question is reducible to the one with a potential of the form $V(r) = ar^2 + br^{-2}$. There is no assurance for other systems to be integrable. This situation may be compared with the limited success of the algebraic method based on the noncompact group SO[2,1]. The spectrum generating algebra has indeed been successful for systems under the same restriction, but not with the restriction.[22]

The generators of SO[2,1] may be realized as[9]

$$\Gamma_0^{(k)} = -(4ak^2)r^{2-k}\nabla^2 + ar^k + br^{-k}, \tag{2.1}$$

$$S^{(k)} = -(4ak^2)r^{2-k}\nabla^2 - ar^k + br^{-k}, \tag{2.2}$$

$$T^{(k)} = \tfrac{1}{2}i[\vec{r}\cdot\vec{\nabla} - \frac{3}{2}]. \tag{2.3}$$

Here k = 1 represents the Kepler problem with an extra inverse-square potential, whereas k = 2 yields a harmonic oscillator with the same extra potential. It is easy to see that the generators of k = 1 transform into those if k = 2 under the replacement $r \to r^2$. Furthermore, the Laplacian multiplied by r changes as

$$r(\frac{d^2}{dr^2} + \frac{1}{r}\frac{d}{dr}) \longrightarrow \tfrac{1}{4}(\frac{d^2}{dr^2} + \frac{1}{r}\frac{d}{dr}), \tag{2.4}$$

$$r(\frac{d^2}{dr^2} + \frac{2}{r}\frac{d}{dr}) \longrightarrow \tfrac{1}{4}(\frac{d}{dr^2} + \frac{3}{r}\frac{d}{dr}). \tag{2.5}$$

This means that the 2-dim. and 3-dim. Kepler problems are algebraically equivalent to the 2-dim. and 4-dim. harmonic oscillators, respectively. In fact, for path integration, the 2-dim. Kepler problem has been reduced by the Levi-Civita transformation into a 2-dim. oscillator,[23] while the 3-dim. Kepler problem has been converted by the Kustaanheimo-Stiefel transformation[24] into a 4-dim. Oscillator.[19] The replacement $r \longrightarrow r^2$ plays an essential role in evaluation of the radial path integral for the Kepler problem.[25] The Morse oscillator which fits into the SO[2,1] scheme can be solved by path integration as well.[26]

The above observations indicate a possible connection between the SO[2,1] scheme and Feynman's path integral. However, they can hardly be an evidence of the link. Let us now pursue a more direct link between the two schemes. In the algebraic method, after an appropriate tilting, the discrete energy eigenvalue problem is written in the form,[9]

$$(\alpha\Gamma_o + \beta)|\psi\rangle = 0. \tag{2.6}$$

The Green function of this equation in coordinate representation may be given by

$$G(\vec{r}'',\vec{r}') = -i\int_0^\infty Q(\vec{r}'',\vec{r}';\sigma)\,d\sigma \tag{2.7}$$

where

$$Q(\vec{r}'',\vec{r}';\sigma) = \langle \vec{r}''|\exp[-i\sigma(\alpha\Gamma_o + \beta)]|\vec{r}'\rangle . \tag{2.8}$$

Dividing the parameter σ into N portions σ_j ($j = 1, 2, \ldots, N$) in such a way that $\max\sigma_j \to 0$ as $N \to \infty$, we can express (2.8) as a path integral,

$$Q(\vec{r}'',\vec{r}';\sigma) = \int \prod_{j=1}^{N} \langle \vec{r}_j|\exp[-i\sigma_j(\alpha\Gamma_o + \beta)]|\vec{r}_{j-1}\rangle \prod_{j=1}^{N-1} d\vec{r}_j \tag{2.9}$$

where $\vec{r}' = \vec{r}_o$ and $\vec{r}'' = \vec{r}_N$. The compact generator Γ_o is related to the Hamiltonian H by $\alpha\Gamma_o + \beta = f(r)(H-E)$ where $f(r)$ is a polynomial of $r = |\vec{r}|$. Replacing the small parameter σ_j by

$$\tau_j = \sigma_j\, f(\sqrt{r_j r_{j-1}}) \tag{2.10}$$

we obtain

$$\langle \vec{r}_j|\exp[-i\sigma_j(\alpha\Gamma_o + \beta)]|\vec{r}_{j-1}\rangle = \langle \vec{r}_j|\exp[-i\tau_j(H-E)]|\vec{r}_{j-1}\rangle . \tag{2.11}$$

Substitution of (2.11) into (2.7) yields

$$G(\vec{r}'',\vec{r}') = \int_0^\infty P(\vec{r}'',\vec{r}';\tau)\,d\tau \tag{2.12}$$

where

$$P(\vec{r}'',\vec{r}';\tau) = Q(\vec{r}'',\vec{r}';\sigma)\,d\sigma/d\tau = \langle \vec{r}''|\exp[-i\tau(H-E)]|\vec{r}'\rangle \tag{2.13}$$

with $\tau = \sum_j \sigma_j$. The entity (2.13) may be identified with what is called the promotor,[27] and the function (2.12) becomes the Green function of the time-independent Schrödinger equation $(H-E)|\psi\rangle = 0$. As is obvious, the promotor (2.13) is expressible in a path integral form.[27] It is this path integral that we calculate for the Kepler problem. Now it is not surprising that the properties of Γ_o directly show up in the results of the path integral calculation. However, we have to realize the fact that the introduction of the time transformation into a path integral has made it possible to establish the link between the two schemes.

There is another class of path integrals that can be evaluated exactly.[28] Any of this class must be reducible to a path integral on

$S^n = SO[n+1]/SO[n]$,

$$P(\hat{q}'',\hat{q}',\tau) = \lim_{N\to\infty} \int \prod_{j=1}^{N} \exp[iW_j] \prod_{j=1}^{N} [2\pi i\tau_j]^{-\frac{1}{2}(n+1)} \prod_{j=1}^{N-1} d^{n+1}\hat{q}_j \tag{2.14}$$

with

$$W_j = (a^2/\tau_j)[1-\cos\Theta_j] + n(n-1)\tau_j/(8a^2) - E\tau_j \tag{2.15}$$

where $\Theta_j = \cos^{-1}(\hat{q}_j \cdot \hat{q}_{j-1})$, $|\hat{q}_j| = a^{-1}$ = constant. This path integral can be easily calculated, the result being[29]

$$P(\hat{q}'',\hat{q}';\tau) = \tfrac{1}{2}\pi^{-\frac{1}{2}(n+1)}\Gamma[\tfrac{1}{2}(n-1)] \sum_{\ell=0}^{\infty} [\ell+\tfrac{1}{2}(n-1)]C_\ell^{\frac{1}{2}(n-1)}(\cos\Theta)\exp[-\frac{i\tau}{2a^2}\ell(\ell+n-1)] \tag{2.16}$$

where $\Theta = \cos^{-1}(\hat{q}'\cdot\hat{q}'')$, $\tau = t''-t'$, and $C_\ell^m(z)$ is the Gegenbauer polynomial.

The algebraic scheme associated with SO[3] whose generators are realized in terms of SU[2] variables corresponds to a class of path integrals on S^3. Naturally the Pöschl-Teller oscillator[18,29] and the Rosen-Morse oscillator,[28,29] both having the SO[3] symmetry, can be treated by the SO[3] algebraic scheme as well as path integration on S^3. This SO[3] or SU[2] scheme can immediately be extended to the SO[2,1] or SU[1,1] case, and has been applied to the Kepler problem in a uniformly curved space. Again, the problem has been solved algebraically[15] and by path integration.[30]

We are thankful to Professors A. O. Barut and Bruno Gruber for very useful discussions and suggestions and for their encouragements.

REFERENCES

1. E. Schrödinger, Proc. Roy. Irish Acad. A46, 9 (1940); 46, 183 (1941); 47, 53 (1941).
2. L. Infeld, Phys. Rev. 59, 737 (1941); T. Inui, Prog. Theor. Phys. 3, 168, 244 (1948); L. Infeld and T. E. Hull, Rev. Mod. Phys. 23, 21 (1951); H. R. Coish, Can. J. Phys. 34, 343 (1956); G. Hadinger, N. Bessis and G. Bessis, J. Math. Phys. 15, 716 (1974).
3. W. Miller, Jr., Mem. Amer. Math. Soc. 50, 5 (1964).
4. B. Kaufman, J. Math. Phys. 7, 447 (1966), A. Joseph, Rev. Mod. Phys. 39, 829 (1967), C. A. Coulson and A. Joseph, Rev. Mod. Phys. 39, 838 (1967).
5. W. Miller, Jr., Lie Theory and Special Functions, (Academic, New York, 1968); Symmetry and Seperation of Variables, (Addison-Wesley, Massachusetts, 1977).
6. W. Miller, Jr., Comm. Pure Appl. Math. 17, 527 (1964), S. Ström, Ark. Fysik, 30 267 (1965), W. Miller, Jr., J. Math. Phys. 9, 1163, 1175, 1434 (1968).
7. W. Miller, Jr. Comm. Pure Appl. Math. 18, 679 (1965); 19, 125 (1966), C. Itzykson, Comm. Math. Phys. 4 92 (1967).
8. E. Celeghini and M. Tarlini, Nuovo Cimento B61, 265 (1982); B65, 172 (1982); B68 133 (1982).

9. A. O. Barut, Dynamical Groups and Generalized Symmetries in Quantum Theory, (University of Canterbury press, Christ-Church, 1972).
10. H.B.G. Casimir, Koninkhijke Academie van Wetenschappen te Amsterdam 34, 844 (1931); Rotation of a Rigid body in Quantum Mechanics, thesis, University of Leiden (1931).
11. C.K.E. Schneider and R. Wilson, J. Maths. Phys. 20, 2380 (1979).
12. V. Bargmann, Ann. Math. 48, 568 (1947).
13. A. O. Barut and E. C. Phillips, Comm. Math. Phys. 8, 52 (1968).
14. N. M. Atakishiev, Theo. and Math. Phys. 56, 735 (1983).
15. A. O. Barut and R. Wilson, Phys. Letts. 110A, 351 (1985), A. O. Barut, A. Inomata and R. Wilson, Algebraization of Pöschl-Teller equation; Algebraization of second Pöschl-Teller equation, Morse-Rosen equation and Eckart-equation, to be published.
16. L. C. Biedenharn and J. D. Louck, The Racah-Wigner Algebra in Quantum Theory (Addison-Wesley, Mass. 1981).
17. H. Weyl, Königlichen Gesellschaft der Wissenschaflin zu Göttingen, Nachrichten page 37 (1909) and page 442 (1910).
18. See, e.g., A. Inomata and R. Wilson, "Path Integral Realization of a Dynamical Group," in Proceedings of the Symposium on Conformal Groups and Structures, eds. A. O. Barut and H. D. Doebner (Springer, Berlin, 1986).
19. I. H. Duru and H. Kleinert, Phys. Lett. 84B, 185 (1979); R. Ho and A. Inomata, Phys. Rev. Lett. 48, 231 (1982).
20. R. P. Feynman and A. R. Hibbs, Path Integrals and Quantum Mechanics (McGraw-Hill, New York, 1965).
21. D. Peak and A. Inomata, J. Math. Phys. 10, 1422 (1969).
22. See, e.g., B. Wybourne, Classical Groups for Physicists (John Wiley, New York, 1974), p. 209.
23. A. Inomata, Phys. Lett. 87A, 387 (1982).
24. P. Kustaanheimo and E. Stiefel, J. Reine Angew. Math. 218, 204 (1965); A. O. Barut, C.K.E. Schneider and R. Wilson, J. Math. Phys. 20, 2244 (1979).
25. A. Inomata, Phys. Lett. 101A, 253 (1984).
26. P. Y. Cai, A. Inomata and R. Wilson, Phys. Lett. 96A, 117 (1983).
27. A. Inomata, in Path Integrals from meV to MeV, eds. M. C. Gutzwiller et al. (World Scientific, Singapore, 1986).
28. G. Junker and A. Inomata, in Path Integrals from meV to MeV, eds. M. C. Gutzwiller et al. (World Scientific, Singapore, 1986).
29. A. Inomata and M. A. Kayed, Phys. Lett. 108A, 9 (1985), J. Phys. A18, 235 (1985).
30. A. O. Barut, A. Inomata and G. Junker, in preparation.

CLASSIFICATION OF OPERATORS IN ATOMIC SPECTROSCOPY BY LIE GROUPS

B. R. Judd

Department of Physics & Astronomy
The Johns Hopkins University
Baltimore, Md. 21218

INTRODUCTION

The use of Lie groups in atomic spectroscopy dates from the 1949 article of Racah.[1] It was shown there that the Coulomb interaction between f electrons could be described as a linear combination of four operators e_i (i = 0, 1, 2, 3) whose descriptions in terms of the irreps W and U of SO(7) and G_2 are (000)(00) for i = 0 and 1, (400)(40) for i = 2, and (220)(22) for i = 3. Highest weights are used to define W and U, and we should also note that Racah used an acute-angled pair of axes for the two numbers (u_1u_2) defining U, rather than the obtuse-angled scheme sometimes preferred today.[2] For the atomic d shell, only three operators of the type e_i are required to represent the Coulomb interaction: two of them correspond to the SO(5) scalar irrep (00), the third to the irrep (22) of that group. At the time of Racah's article on the f shell, all d-electron matrix elements of the Coulomb interaction had already been expressed as linear combinations of the Slater integrals F_k involving the radial parts of the eigenfunctions, so there was little incentive to rework the calculation for the d shell. Shortly after the appearance of Racah's article, the effects of configuration interaction on the configurations d^N and d^Ns began to be studied, but they were analyzed by traditional methods. The effects of two-electron excitations were shown by Trees[3] and Racah[4] to be reproducible by changes in the F_k together with just two new effective operators, L^2 and Q, in the limit where second-order perturbation theory is adequate. The eigenvalues of L^2 are L(L + 1); those of Q are zero for all terms ^{2S+1}L of d^2 except for ^1S. It is not difficult to show that L^2 transforms like a mixture of the irreps (00) and (22) of SO(5), while Q belongs to (00) alone.

By the mid-1950's, studies of the optical spectra of rare-earth ions in crystals were well under way. Elliott et al.[5] showed that the crystal-field splittings of the levels $^{2S+1}L_J$ required the addition to the free-ion Hamiltonian of terms that possessed the same transformation properties in SO(3) as the spherical harmonics Y_{kq} for k = 2, 4 and 6. These 27 objects (allowing for all possible q) transform under the operations of SO(7) and G_2 like the single pair of irreps (200) and (20) respectively. The assignment of (110) and (11) to the spin-orbit interaction was made by McLellan.[6] Since the labels W and U can also be used to define the states of the f shell, we can apply the generalized Wigner-Eckart theorem to SO(7) and G_2 and thereby obtain a wide variety of selection rules and proportionalities between blocks of matrix elements.

In his 1964 lectures at the Collège de France, Racah considered configuration interaction via single-electron excitations. For d^N, he showed that four three-electron effective scalar operators of the type

$$(u_h^{(k)} u_i^{(k')})^{(k")} \cdot u_j^{(k")}$$

were required to second order in perturbation theory, since the triad (kk'k") can assume the four inequivalent forms (222), (224), (244) and (444). His supposition that four new parameters would be required to allow for the interaction was shown by Feneuille[7] to overlook the fact that one linear combination of the four operators transforms like (60) of SO(5) and possesses null matrix elements everywhere, while another transforms like (22) and has matrix elements proportional to a linear combination of the 2 two-electron operators belonging to (22). An analogous reduction occurs in the f shell, where only 6 new three-electron operators are required to represent the effects of single-electron excitations.[8]

All the operators originating in the inter-electronic Coulomb interaction, including the three-electron operators mentioned above, are scalar in the spin and orbital spaces: that is, the associated ranks κ and k are both zero. The non-relativistic limit of the Breit interaction, however, requires operators for which $\kappa = k = 1$ (the so-called spin-other-orbit interaction) and $\kappa = k = 2$ (the spin-spin interaction). For d electrons, seven new operators are introduced in this way, the SO(5) labels being (11) (twice), (21) (twice), (31), (20), and (22).[9] Similar operators (with $\kappa = k = 1$) are required to represent the effect of configuration interaction on the single-electron spin-orbit interaction, a mechanism that is often referred to as EL-SO (electrostatically correlated spin-orbit interaction).

ORTHOGONAL OPERATORS

Operators o_1 and o_2 labelled by different irreps Γ_1 and Γ_2 of a Lie group G possess an interesting property: they are orthogonal in as much as the trace of the matrix of $o_1 o_2$, taken over all states forming a basis for a representation (not necessarily irreducible) of G, vanishes.[10] Put equivalently, the sum

$$\Sigma \langle \psi_1 | o_1 | \psi_2 \rangle \langle \psi_2 | o_2 | \psi_1 \rangle ,$$

taken over all the states ψ_1 and ψ_2 of a configuration C, vanishes provided the states of C form a representation of G. Newman[11] has shown that the parameters p_1 and p_2 that measure the strengths of o_1 and o_2 when a least-squares fit to the energy levels is carried out are minimally correlated: that is, the addition of a new operator o_i minimally affects the strengths of the old ones when o_i is included in the fitting procedure. The mean errors of the parameters p_j are correspondingly reduced when orthogonal operators are used.

Many of the operators introduced above are orthogonal to one another. In cases where two distinct operators correspond to identical sets of irreps, it is often possible to separate them by introducing a new group. An example is provided by e_0 and e_1 of Racah,[1] which both correspond to $\kappa = k = 0$, U = (00), W = (000). The states of f^2 form a basis for the irrep [11] of U(14), corresponding to the antisymmetric Young tableau with two cells in a single column. The matrix elements of e_0 are a constant for all the terms of f^2, and so e_0 can be assigned the scalar [0] of U(14). The 14 components of the creation tensor $a^\dagger$ transform according to the irrep [1] of U(14), while those of the corresponding annihilation tensor transform according to $[00\ldots0\ -1]$, that is, to $[0^{13}-1]$. A two-electron operator is built from products of the type $a_\alpha^\dagger a_\beta^\dagger a_\gamma a_\delta$, and a collection of these forms a basis for $[1^2 0^{10} -1^2]$ of U(14). This is the appropriate label for an operator e_1' orthogonal to e_1, which turns out to be (to within an arbitrary factor) $e_1 - 9e_0/13$.

The symplectic groups can also be brought into play to aid in the separation process. As the classification proceeds, it becomes evident that we may set aside for the time being the physical origin of the operators and concentrate on giving a complete orthogonal set of them. The irreps of the groups U(14), USp(14), SO(7) and G_2 are sufficient for the scalar operators (that is, those for which $\kappa = k = 0$) acting on not more than three f electrons at a time.[12] Recent work on the d shell has revealed that the number of n-electron scalars required in this case is 1, 0, 4, 4, 12 and 0 for n = 0, 1, 2, 3, 4 and 5 respectively.[13] These 21 operators correspond to different sets of irreps of U(10), USp(10) and SO(5). The absence of a five-electron orthogonal operator came as a surprise to us: it does not generalize to the f shell. As far as the fine-structure operators are concerned (that is, those for which $\kappa = k = 1$ or $\kappa = k = 2$), the groups listed above are adequate for two-electron operators in the d and f shells, but multiplicity labels are often required to separate fine-structure operators for which $n \geqslant 3$.

One complication of this kind of work is the coalescence in a single orthogonal operator of parts that derive from different orders of perturbation theory. This is illustrated by e_3 and the generalized Trees operator Ω (in the notation of Racah[1]). Both correspond to the irreps (220) of SO(7) and (22) of G_2, and they are not orthogonal to each other. They can be made so by replacing e_3 by the linear combination $e_3 + \Omega$. The operators $e_3 + \Omega$ and Ω correspond to the irreps $\langle 1^4 0^3 \rangle$ and $\langle 2^2 0^5 \rangle$ of USp(14), and orthogonality is guaranteed since the irreps are different. However, the strengths of both $e_3 + \Omega$ and Ω are expected to be comparable (and of opposite sign) since e_3 appears in first order while Ω is a second-order correction. It would be more appropriate from a physical standpoint to retain e_3, and then combine e_3 and Ω to form its orthogonal companion, which turns out to be $e_3 + 5\Omega$. This operator has no unique USp(14) label, but its strength is a measure of the importance of second-order contributions to the energies of the terms.

We are not interested only in configurations of equivalent electrons. Among the many in which inequivalent electrons appear, attention has so far been directed to $p^N d$ (for all N) and pd^2.[14,15] The obvious way to proceed is to treat the p^N part separately from the d^M part and then couple the two parts to whatever final ranks are required. The group structure descends in a subgroup chain from U(6)×U(10); at each stage a direct product $G_p \times G_d$ is formed, where G_p and G_d are Lie groups whose generators act on the p and d electrons respectively. The irreps of USp(6)×USp(10) are very helpful here. Subject to the overall criteria that all operators be orthogonal to one another and involve at least one electron of each kind, it turns out that for pd^2 there are 6 two-electron and 24 three-electron scalar operators; their 900 matrix elements have been tabulated by Dothe et al.[14] Maintaining the criteria, we find that for $p^N d$ there are 6 two-electron, 11 three-electron, and 9 four-electron scalar operators. Work is just beginning on the fine-structure operators that involve inequivalent electrons; here again, the obvious way to proceed is to break the operators up into parts that separately act on electrons with the same principal and azimuthal quantum numbers, these parts then being coupled to the appropriate final ranks.

COMPARISON WITH EXPERIMENT

In an early test on Pr III $4f^3$, it was established that scalar orthogonal operators lead to reduced errors for the parameters when a least-squares fit to the energies of the levels was carried out.[16] An extension of the method to fine-structure operators was performed last year for several transition-metal ions.[17] It was found that mean errors as small as 1.91 and 3.87 cm^{-1} could be obtained in least-squares fits to the levels of Cr IV $3d^3$ and Ni IV $3d^7$. The configurations themselves extend over some 50000 cm^{-1}. The use of four-electron orthogonal operators in Ne III $2p^3 3d$ was equally striking, a mean error of only 3.4 cm^{-1} being achieved.[15] Several new levels of that configuration were identified as a result of the excellence of the fit. A special study is now being undertaken by Dothe for Fe VI $3d^2 4p$ to test the extent to which the theoretical expressions for the energies of the levels as functions of the parameters (that is, the strengths of the orthogonal operators) depart from linearity, a condition which, if exactly fulfilled, leads to uncorrelated parameters.[18] The mean error of 188 cm^{-1} obtained by Ekberg[19] by traditional methods has already been reduced to 15 cm^{-1} by the introduction of 6 three-electron scalars and 2 new Trees parameters, in spite of significant non-linearity. Fine-structure effects are now being studied.

A more ambitious program involves fitting the $p^N d$ configurations of the third spectra of O, F, Ne, Na and Mg, and the fourth spectra of F, Ne, Na, Mg and Al. Contributions to the parameters from second-order and third-order perturbation theory are in the process of being compared to numerical estimates based on the Hartree-Fock program of Cowan. Preliminary results by Hansen and Lister indicate some remarkable agreements, particularly for the parameters associated with three-electron operators.

ACKNOWLEDGEMENTS

Thanks go to my colleagues and students who have shared the work on orthogonal operators. Their names appear in references 11 through 19. Partial support of the work by the United States National Science Foundation is also acknowledged.

REFERENCES

1. G. Racah, Theory of complex spectra. IV, Phys. Rev., 76:1352 (1949).
2. W. G. McKay and J. Patera, "Tables of Dimensions, Indices, and Branching Rules for Representations of Simple Lie Algebras," Marcel Dekker, New York (1981).
3. R. E. Trees, The L(L + 1) correction to the Slater formulas for the energy levels, Phys. Rev., 85:382 (1952).
4. G. Racah, L(L + 1) correction in the spectra of the iron group, Phys. Rev., 85:381 (1952).
5. J. P. Elliott, B. R. Judd, and W. A. Runciman, Energy levels in rare-earth ions, Proc. Roy. Soc. (London), A240:509 (1957).
6. A. G. McLellan, Selection rules for spin-orbit matrix elements for the configuration f^n, Proc. Phys. Soc. (London), 76:419 (1960).
7. S. Feneuille, Opérateurs à trois particules pour les électrons d équivalents, C. R. Acad. Sc. (Paris), B 262:23 (1966).
8. B. R. Judd, Three-particle operators for equivalent electrons, Phys. Rev., 141:4 (1966).
9. B. R. Judd, Zeeman effect as a prototype for intra-atomic interactions, Physica, 33:174 (1967).

10. B. R. Judd, Operator averages and orthogonalities, in "Group Theoretical Methods in Physics" (Lecture Notes in Physics, vol. 201), G. Denardo, G. Ghirardi, and T. Weber, eds., Springer, Berlin (1984).
11. D. J. Newman, Operator orthogonality and parameter uncertainty, Phys. Lett., 92A:167 (1982).
12. B. R. Judd and M. A. Suskin, Complete set of orthogonal scalar operators for the configuration f^3, J. Opt. Soc. Am., B1:261 (1984).
13. B. R. Judd and R. C. Leavitt, Many-electron orthogonal scalar operators in atomic shell theory, J. Phys. B:At. Mol. Phys., 19:in press, (1986).
14. H. Dothe, J. E. Hansen, B. R. Judd, and G. M. S. Lister, Orthogonal operators for p^Nd and pd^N, J. Phys. B:At. Mol. Phys., 18:1061 (1985).
15. J. E. Hansen, B. R. Judd, G. M. S. Lister, and W. Persson, Observation of four-body effects in atomic spectra, J. Phys. B:At. Mol. Phys., 18:L725 (1985).
16. B. R. Judd and H. Crosswhite, Orthogonalized operators for the f shell, J. Opt. Soc. Am., B1:255 (1984).
17. J. E. Hansen and B. R. Judd, Fine-structure analyses with orthogonal operators, J. Phys. B:At. Mol. Phys., 18:2327 (1985).
18. J. E. Hansen and B. R. Judd, New effective parameters for atomic structure, Comments Atom. Mol. Phys., in press.
19. J. O. Ekberg, Term analysis of Fe VI, Phys. Scr., 11:23 (1975).

AN INTRODUCTION TO SQUEEZED STATES

John R. Klauder

AT&T Bell Laboratories
Murray Hill, NJ 07974

INTRODUCTION

The in- and out-of-phase quadrature components of an oscillator amplitude have, even at zero temperature, an irreducible product of fluctuations as imposed by quantum mechanics and the Heisenberg uncertainty relation. Thus even ideal lasers operating in pure-mode coherent states exhibit amplitude and phase fluctuations, which, as we shall see, have an equal magnitude for any pair of in- and out-of-phase components. Squeezed states, which represent newly observed states of the radiation field,[1] possess fluctuations in either the amplitude or phase (but not both!) that are smaller than that offered by any ideal laser or even by the vacuum itself. This special property of squeezed states should lead to improved channel capacity in communications and to improved phase sensitivity in interferometers, just two of potentially many promising applications.[2]

COHERENT STATES

A proper appreciation of squeezed states begins with a review of coherent states and several of the properties they possess.[3] A coherent state of the radiation field is conveniently defined as

$$|z\rangle = e^{(za^\dagger - z^* a)}\,|0\rangle\ ,$$

where z is an arbitrary complex number, a and $a^\dagger$ are conventional annihilation and creation operators satisfying $[a,a^\dagger] = 1$, and $|0\rangle$ denotes the vacuum, i.e., the normalized ground state of the number operator with $a|0\rangle = 0$.

For an harmonic oscillator Hamiltonian $H = \omega a^\dagger a$ $(\hbar = 1)$, the time dependence of the state $|z\rangle$ is easily seen to be

$$\begin{aligned} &e^{-i\omega a^\dagger a t}\,|z\rangle \\ &\quad = e^{-i\omega a^\dagger a t}\, e^{(za^\dagger - z^* a)}\, e^{i\omega a^\dagger a t}\,|0\rangle \\ &\quad = \exp(z e^{-i\omega t} a^\dagger - z^* e^{i\omega t} a)\,|0\rangle \\ &\quad = |e^{-i\omega t} z\rangle\ . \end{aligned}$$

If z is represented in polar coordinates by

$$z = re^{-i\phi}$$

then it is clear the $\phi = \omega t$ satisfactorily accounts for the time dependence of the coherent state.

Rather than discussing merely the two operators given as the sum and (i times the) difference of a and $a^\dagger$ as is commonly the case, it is especially convenient to consider simultaneously the entire set of possible quadratures as represented for $0 \leqslant \psi < 2\pi$ by

$$X(\psi) \equiv \frac{1}{2} [a\, e^{i\psi} + a^\dagger\, e^{-i\psi}] .$$

The expected value of $X(\psi)$ in the coherent state $|z\rangle$ is readily derived from the eigenproperties

$$a|z\rangle = z|z\rangle, \qquad \langle z|a^\dagger = z^* \langle z| .$$

As a consequence

$$\langle X(\psi)\rangle \equiv \langle z|\, X(\psi)\, |z\rangle$$

$$= \frac{1}{2} (ze^{i\psi} + z^* e^{-i\psi}) ,$$

or expressed in terms of $z = r\, e^{-i\phi}$,

$$\langle X(\psi)\rangle = r \cos(\phi - \psi) .$$

Clearly r represents the amplitude, and $\psi = 0$ and $\psi = \pi/2$, for example, characterize two independent field quadratures. With $\phi = \omega t$ it follows that $\langle X(0)\rangle$ and $\langle X(\pi/2)\rangle$ denote conventional field quadratures.

As a measure of field amplitude fluctuations we shall adopt the variance

$$\Delta X^2(\psi) \equiv \langle X^2(\psi)\rangle - \langle X(\psi)\rangle^2 .$$

In order to evaluate this expression, and future ones like it, it is convenient to decompose X^2 into a normal-ordered component $:X^2:$ and a remainder R,

$$X^2(\psi) = :X^2(\psi): + R(\psi) .$$

A normal-ordered component has all a's to the right and all $a^\dagger$'s to the left. Thus with $X(\psi) = (a\, e^{i\psi} + a^\dagger e^{-i\psi})/2$,

$$:X^2(\psi): = \frac{1}{4} (a^2 e^{2i\psi} + 2a^\dagger a + a^{\dagger 2} e^{-2i\psi}) ,$$

$$R(\psi) = \frac{1}{4} .$$

For any coherent state $|z\rangle$ it follows from the eigenproperties that

$$\langle z|: X^2(\psi): |z\rangle = \langle z|\, X(\psi)\, |z\rangle^2 ,$$

and as a consequence

$$\Delta X^2(\psi) = \frac{1}{4} ,$$

which we observe is *independent* of ψ and z. This independence is not a requirement of principle from quantum mechanics but rather a consequence of the special (coherent) state we have assumed. The only requirements of principle are that

$$\Delta X^2(\psi) \geqslant 0\,,$$

$$\Delta X(\psi)\,\Delta X(\psi+\pi/2) \geqslant \frac{1}{4}\,,$$

the latter being a slightly generalized form of the Heisenberg uncertainty principle. The coherent states are minimum uncertainty states, in that they satisfy the equality in the uncertainty product above, but moreover their uncertainty is constant and uniform for every choice of quadrature. In certain precision measurements, however, the level of quantum noise represented by the variance $\Delta X^2(\psi) = 1/4$ may be unacceptably high.

SINGLE-MODE SQUEEZED STATES

Squeezed states are minimum uncertainty states with unequal variances for different quadratures. In the simplest, and not incorrect, viewpoint such a state arises from an arbitrary dilation of the vacuum. In other words this amounts to taking the ground state of an harmonic oscillator of one frequency and by rescaling the coordinate (dilation) changing it to the ground state of an harmonic oscillator of a different frequency. The standard dilation generator is proportional to $i(a^2-a^{\dagger 2})$, or taking the independence of phase into account, the generator of interest can be taken as $i(a^2e^{2i\phi}-a^{\dagger 2}e^{-2i\phi})$. Thus it is not surprising that a description of single-mode squeezed states[4] begins with the state

$$|r,\phi\rangle \equiv \exp[\frac{1}{2}\,r\,(a^{\dagger 2}e^{-2i\phi}-a^2e^{2i\phi})]\,|0\rangle\,.$$

This state plays the role of an alternative fiducial vector for the purpose of defining canonical coherent states[5] in the sense

$$|z,r,\phi\rangle \equiv \exp(za^\dagger - z^*a)\,|r,\phi\rangle\,.$$

As before we define a family of quadrature operators by the relation

$$X(\psi) = \frac{1}{2}\,(a\,e^{i\psi}+a^\dagger e^{-i\psi})\,.$$

It follows, in the present case, that

$$\langle X(\psi)\rangle \equiv \langle z,r,\phi|\,X(\psi)\,|z,r,\phi\rangle$$

$$= \frac{1}{2}\,(ze^{i\psi}+z^*e^{-i\psi})$$

exactly as in the case of the laser field. To facilitate the calculation of the variance it is convenient to express the coherent and dilation transformations as transformations of the annihilation and creation operators rather than transformations of the ground state $|0\rangle$. To this end we note that

$$D^\dagger(z^*,z)\;a\;D(z^*,z) = a + z\,,$$

$$S^\dagger(r,\phi)\;a\;S(r,\phi) = \cosh r\,a + \sinh r\,a^\dagger e^{-2i\phi}\,,$$

where

$$D(z^*,z) \equiv \exp(za^\dagger - z^*a)\,,$$

$$S(r,\phi) \equiv \exp[\frac{1}{2}\,r\,(a^{\dagger 2}e^{-2i\phi}-a^2e^{2i\phi})]\,.$$

Consequently

$$X'(\psi) \equiv S^\dagger D^\dagger X(\psi) DS$$

$$= \frac{1}{2} (ze^{-i\psi} + z^* e^{i\psi})$$

$$+ \frac{1}{2} [(\cosh r\, e^{-i\psi} + \sinh r\, e^{i\psi - 2i\phi})\, a$$

$$+ (\cosh r\, e^{i\psi} + \sinh r\, e^{-i\psi + 2i\phi})\, a^\dagger] ,$$

from which it follows that

$$X'^2(\psi) = :X'^2(\psi):$$

$$+ \frac{1}{4} |\cosh r + \sinh r\, e^{2i(\psi - \phi)}|^2 .$$

Thus, for single-mode squeezed states we conclude that,

$$\Delta X^2(\psi) = <0| X'^2(\psi) |0> - <0| X'(\psi) |0>^2$$

$$= \frac{1}{4} |\cosh r + \sinh r\, e^{2i(\psi - \phi)}|^2 .$$

For $r = 0$ this expression reduces to the coherent-state variance, but for $r > 0$ it exhibits a nontrivial dependence on ψ (while still being independent of z). For several particular values of $\psi - \phi$ the variance becomes

$$\psi - \phi = \pi/2,\ 3\pi/2 , \qquad \Delta X^2 = \frac{1}{4} e^{-2r} ,$$

$$\psi - \phi = \pi/4,\ 3\pi/4, \ldots , \qquad \Delta X^2 = \frac{1}{4} \cosh(2r) ,$$

$$\psi - \phi = 0, \pi , \qquad \Delta X^2 = \frac{1}{4} e^{2r} .$$

The minimum value of the variance, namely $\Delta X^2 = e^{-2r}/4$, which occurs twice per cycle, can, in principle, be significantly less than the value 1/4 for an ideal laser. As a check on the Heisenberg uncertainty relation we note that

$$\Delta X(\psi)\, \Delta X(\psi + \pi/2) = \frac{1}{4} |1 + (1 - e^{4i(\psi - \phi)}) \sinh^2 r| .$$

Clearly this product is greater than or equal to 1/4, as required, with equality holding only when $\psi - \phi = 0$, $\pi/2$, π, and $3\pi/2$, i.e., four times per cycle.

Thus we see that the single-mode squeezed state offers a reduced variance twice per cycle. However, from an experimental point of view it may be easier to create a slight generalization of the squeezed state as discussed so far.

TWO-MODE SQUEEZED STATES

Two-mode squeezed states involve two independent degrees of freedom, the a and $a^\dagger$ operators introduced previously, and another independent pair, b and $b^\dagger$.[6] The only nonvanishing commutators among these operators are just

$$[a, a^\dagger] = [b, b^\dagger] = 1 .$$

The dilated form of the ground state involves both of these operators and is given by

$$|r,\phi\rangle \equiv T(r,\phi)\,|0\rangle\,,$$

$$T(r,\phi) \equiv \exp[r\,(a^\dagger b^\dagger e^{-2i\phi} - a\,b\,e^{2i\phi})]\,.$$

This operator is similar in construction to the operator S, except for a factor 2 in the term r. The two-mode squeezed state is taken to be

$$|z,r,\phi\rangle \equiv D\,(z^*,z)\;T(r,\phi)\,|0\rangle\,,$$

where in the present case

$$D(z^*,z) = \exp[z\,(a^\dagger + b^\dagger)/\sqrt{2} - z^*(a+b)/\sqrt{2}]\,.$$

Finally, the operator of interest is given by

$$X(\psi) = \frac{1}{2\sqrt{2}}\,[(a+b)\,e^{i\psi} + (a^\dagger + b^\dagger)\,e^{-i\psi}]\,.$$

To facilitate evaluating the required averages we first observe that

$$X'(\psi) = T^\dagger D^\dagger X(\psi)\,DT$$

$$= \frac{1}{2}\,(z\,e^{i\psi} + z^*\,e^{-i\psi})$$

$$+ \frac{1}{2\sqrt{2}}\,[(\cosh r\,e^{-i\psi} + \sinh r\,e^{i\psi-2i\phi})\,(a+b)$$

$$+ (\cosh r\,e^{i\psi} + \sinh r\,e^{-i\psi+2i\phi})\,(a^\dagger + b^\dagger)]\,.$$

As a consequence it follows that

$$\langle X(\psi)\rangle = \langle 0|\,X'(\psi)\,|0\rangle$$

$$= \frac{1}{2}\,(z\,e^{i\psi} + z^* e^{-i\psi})\,.$$

From the relation

$$X'^2(\psi) = :X'^2(\psi): + \frac{1}{4}\,|\cosh r + \sinh r\,e^{2i(\psi-\phi)}|^2$$

we find that

$$\Delta X^2(\psi) = \langle 0|\,X'^2(\psi)\,|0\rangle - \langle 0|\,X'(\psi)\,|0\rangle^2$$

$$= \frac{1}{4}\,|\cosh r + \sinh r\,e^{2i(\psi-\phi)}|^2\,,$$

which gives for the mean and variance exactly the same results as found for the single-mode squeezed states (thanks to a rescaling of r by a factor of 2). Thus all the features discussed for the variance of the single-mode squeezed state apply here without change. In particular, squeezing of a given quadrature takes place twice per cycle.

It is noteworthy that the squeezing variance in all the cases considered is independent of z, or stated otherwise independent of the precise value of $\langle X(\psi)\rangle$. In particular, all the features of the squeezing variance are present when $z = 0$, i.e., $\langle X(\psi)\rangle = 0$, and this is the case we pursue further. In such cases one may legitimately speak of "squeezing the vacuum."

MODELS FOR SQUEEZING STATES

Simple phenomenological models to squeeze the vacuum are readily constructed. For convenience we shall develop a simple model to generate a two-mode squeezed state. As Hamiltonian we adopt

$$H(t) = \omega_a a^\dagger a + \omega_b b^\dagger b - igf(t)\,[ab\,e^{-i(\omega_a+\omega_b)t+2i\phi} - a^\dagger b^\dagger e^{i(\omega_a+\omega_b)t-2i\phi}]$$

where ω_a and ω_b are the frequencies of modes a and b, respectively, g is a coupling constant, and f(t) is a "pulse" function restricted so that $f(t) = 0$ if $t \leqslant 0$ or $t \geqslant T$, and $\int f(t)\,dt \neq 0$. We study Schrödinger's equation

$$i \frac{\partial}{\partial t} |\psi(t)\rangle = H(t)\,|\psi(t)\rangle$$

subject to the initial condition $|\psi(0)\rangle = |0\rangle$. In so doing it is convenient to introduce

$$|\psi(t)\rangle = e^{-i(\omega_a a^\dagger a+\omega_b b^\dagger b)t}\,|\phi(t)\rangle ,$$

where, of course, $|\phi(0)\rangle = |0\rangle$ as well. The Schrödinger equation for $|\phi(t)\rangle$ reads

$$i \frac{\partial}{\partial t} |\phi(t)\rangle$$

$$= -igf(t)\,[a(t)b(t)\,e^{-i(\omega_a+\omega_b)t+2i\phi} - a^\dagger(t)b^\dagger(t)\,e^{i(\omega_a+\omega_b)t-2i\phi}]\,|\phi(t)\rangle ,$$

where

$$a(t) \equiv e^{-i\omega_a a^\dagger a t}\,a\,e^{i\omega_a a^\dagger a t} = e^{i\omega_a t}a ,$$

$$b(t) \equiv e^{-i\omega_b b^\dagger b t}\,b\,e^{i\omega_b b^\dagger b t} = e^{i\omega_b t}b .$$

When these relations are introduced we have

$$i \frac{\partial}{\partial t} |\phi(t)\rangle = igf(t)\,[a^\dagger b^\dagger e^{-2i\phi} - ab\,e^{2i\phi}]\,|\phi(t)\rangle$$

the solution of which is

$$|\phi(t)\rangle = \exp\{g\int_0^t f(s)ds\,[a^\dagger b^\dagger e^{-2i\phi} - ab\,e^{2i\phi}]\}\,|0\rangle .$$

For $t \geqslant T$, and $r \equiv g\int_0^T f(s)ds$, then

$$|\phi(t)\rangle = \exp[r\,(a^\dagger b^\dagger e^{-2i\phi} - ab\,e^{2i\phi})]\,|0\rangle$$

which is recognized as the two-mode squeezed state denoted previously by $|r,\phi\rangle$. Finally, we observe that

$$|\psi(t)\rangle = \exp[-i(\omega_a a^\dagger a+\omega_b b^\dagger b)t]\,|\phi(t)\rangle$$

$$= \exp[r\,(a^\dagger b^\dagger\,e^{-i(\omega_a+\omega_b)t-2i\phi} - ab\,e^{i(\omega_a+\omega_b)t+2i\phi})]\,|0\rangle$$

$$\equiv |r,\ \phi + \frac{1}{2}(\omega_a+\omega_b)t\rangle .$$

Hamiltonians of the indicated type arise, in a phenomenological way, in at least two fashions. A parametric amplifier mixes three field components and contains interaction terms proportional to

$$a^\dagger b^\dagger c + c^\dagger ab$$

which may be singled out by wave vector and frequency conservation. If the c field is strong, serving as the pump field, then it may be treated as a classical (c-number) field giving rise to the adopted form of the Hamiltonian. The pulse f(t) can arise as a descriptor of how long the signal stays inside the amplifier medium.

In a four-wave mixer four fields are coupled, and a term of the form

$$a^\dagger b^\dagger cd + c^\dagger d^\dagger ab$$

can be isolated by wave vector and frequency conditions. If the fields c and d are both strong, and can be treated classically, then we again arrive at the phenomenological Hamiltonian.

Fully Quantum Mechanical Model

For an improved treatment of squeezed state generation we need to introduce a more realistic atom-field interaction in place of the phenomenological field-field interaction. One such model that has been studied[7] is based on the Hamiltonian for a four-wave mixer given by

$$H = \int H_r \, d^3r/\delta V ,$$

$$H_r = \sum_{j=1}^{4} \omega_j a_j^\dagger a_j + \sum_{k=1}^{N_0} \frac{1}{2} \omega_0 \sigma_{zk}$$

$$+ ig \sum_{k=1}^{N_0} (\sigma_k^\dagger a_r - a_r^\dagger \sigma_k) + \sum_{k=1}^{N_0} (\Gamma^\dagger \sigma_k + \sigma_k^\dagger \Gamma) ,$$

$$a_r = \sum_{j=1}^{4} a_j e^{ik_j \cdot r} .$$

Here the model is specialized to two-level atoms with energy separation ω_0, and $\sigma_k (\sigma_k^\dagger)$ lowers (raises) the atomic level. The operators Γ and $\Gamma^\dagger$ refer to the reservoir, here regarded as uninteresting modes of the radiation field, and account for loss and degradation mechanisms in the model. The operator H_r refers to an elementary volume δV in which there are N_0 atoms, and the ultimate integration over space will insure the wave-vector condition $k_1 + k_2 = k_3 + k_4$, where k_1 and k_2 refer to pump modes and k_3 and k_4 refer to signal and idler modes, respectively.

A full account of this model has been presented elsewhere[7] and need not be reproduced here. We only observe that the theoretical predictions of this fully quantum mechanical model agree with present squeezed-state experiments[1] in the regime in which they have been performed.

SQUEEZED STATE APPLICATIONS

Finally, we wish to make a few comments on the utility of squeezed states of the radiation field. In one application they can be used to increase markedly the phase sensitivity of suitable interferometers. An interferometer with N photons entering may exhibit a phase sensitivity $\Delta\phi \approx 1/\sqrt{N}$, e.g., with an ideal laser input. However, by using a squeezed laser input and by introducing a squeezed state instead of the vacuum into an unused port of the interferometer the phase sensitivity can, in principle, be increased to $\Delta\phi \approx 1/N$, approaching the limit imposed by the Heisenberg uncertainty relation.[8] If $N \approx 10^{12}$, as may be obtained for lasers, this change in sensitivity would be sizeable indeed.

Squeezed states have a role to play in communications as well. On the one hand, taking advantage of a reduced phase uncertainty would make more channels available leading to an approximate doubling of the channel capacity when other parameters are held fixed. On the other hand, taps along an optical fiber that make use of squeezed state inputs will have the tendency, in a given quadrature, to reduce the noise as well as the signal in the main fiber thus exhibiting a uniform signal-to-noise ratio for each of the many successive taps.

It would certainly appear that generation and application of squeezed states is indeed a promising frontier.

ACKNOWLEDGEMENTS

It is a pleasure to thank S. L. McCall, R. E. Slusher, and B. Yurke for numerous discussions pertaining to squeezed states.

REFERENCES

1. R. E. Slusher, L. W. Hollberg, B. Yurke, J. C. Mertz, and J. F. Valley, Phys. Rev. Lett. *55*, 2409 (1985).
2. See, e.g., D. F. Walls, Nature *306*, 141 (1983).
3. See, e.g., J. R. Klauder and E. C. G. Sudarshan, "Fundamentals of Quantum Optics," (W. A. Benjamin, New York, 1968).
4. D. Stoler, Phys. Rev. D*1*, 3217 (1970).
5. See, e.g., J. R. Klauder and B.-S. Skagerstam, "Coherent States," (World Scientific, Singapore, 1985).
6. C. M. Caves, Phys. Rev. D*26*, 1817 (1982).
7. J. R. Klauder, S. L. McCall, and B. Yurke, "Squeezed states from Nondegenerate Four-wave Mixers," Phys. Rev. A (in press).
8. See, e.g., B. Yurke, S. L. McCall, and J. R. Klauder, "SU(2) and SU(1,1) Interferometers," Phys. Rev. A (in press).

ON GEOMETRY OF PHENOMENOLOGICAL THERMODYNAMICS

Jerzy Kocik

Department of Mathematics, Southern Illinois University
Carbondale, Illinois 62901

1. MOTIVATIONS

Thanks to works of Caratheodory [4] and Gibbs [5] phenomenological thermodynamics of equilibrium (PTE) has become a standard axiomatic theory. Formulated in the general way [11,3], it reveals the structure which is universal in the sense that the later statistical and quantum statistical mechanics have not replaced it but rather serve as models of the general scheme.[1]

Phenomenological thermodynamics is claimed to be a typical theory, the formalism of which could not be understood without the language of differential geometry. Nonetheless, in the traditional approach the geometric situation is often unclear, "differential calculus" used there is highly formal, without specified underlying manifolds often arbitrary and confusing.

Recently, thanks to some attempts [6,9,10] PTE occurred in the mainstream of the geometrization of physics. This presentation is intended to be such an attempt suggesting however alternate recognition of geometrical objects underlying the theory.

It can be assumed that in classical textbooks the basic manifold of PTE is that of n+1 extensive variables (like S, V, N, and U).[2] New attempts extend the above configuration space of PTE by including n intensive coordinates (like T, p, μ). Thus one obtains a universal space E (the phase space of PTE - according to [6,10]), where admissible states of a given theory form some submanifold $\Lambda \subset E$. The phase space E is provided with the distinguished Gibbs differential one-form θ, and an admissible submanifold Λ has to satisfy the Gibbs relation $\theta|_\Lambda = 0$. In [6,9,10] the space is *odd*-dimensional and θ constitutes a *contact* structure, and thus a *pre*-symplectic structure on E determined by the bi-form $d\theta$.

---

[1] for models other than PT see e.g. [2] (cosmology of black holes).

[2] Carathéodory assumes even that each coordinate description of a given system demands independent treatment (space).

In this paper we would like to justify that the accurate geometrical description of PTE should be based on the assumption that the phase space of PTE is an *even*-dimensional manifold. Moreover, the internal energy U, no longer a basic coordinate, forms, together with other thermodynamical potentials, an algebraic lattice (Sec. 3). Thus,

> *PTE is a superposition of symplectic structure of the phase space and of lattice structure of thermodynamical potentials.*

Classical Mechanics vs Phenomenological Thermodynamics

Some authors (see e.g. [10]) stress the similarity between structure of PTE and that of classical mechanics (CM), whereas they should be rather contrasted.

Although both PTE and CM rely essentially on a manifold with a closed bi-form ω [(pre-)symplectic structure (Poisson bracket)], yet the one-form θ, $d\theta = \omega$, plays different roles in these theories:

- in CM θ is defined up to an external derivative of any function on the manifold of the phase space, i.e. the gauge $\theta \to \theta' = \theta + df$ does not change the classical equation of motion $X \lrcorner d\theta = dH$, since $d\theta = d\theta'$ (nor even quantum measurable effects [7]: the term df only shifts the global phase factor of the wave function);
- whereas in PTE things seem to be quite opposite: one of the one-forms (the Duhem-Gibbs form) generating the symplectic structure, say θ_0, is distinguished, and this fact is crucial for the theory. The equation for the submanifold Λ of admissible states, $\theta_0|_\Lambda = 0$, is not invariant under the gauge $\theta_0 \to \theta_0 + df$ (however the resulting Maxwell identity $d\theta_0|_\Lambda = 0$ is invariant).

  And this is the source of importance of thermodynamical potentials in PTE.

Odd- vs Even-Dimensional Phase Space

Earlier approaches [6,9,10] assume for the phase space of PTE the odd-dimensional manifold routinely constructed from <u>all</u> variables appearing in the Gibbs relation $dU - TdS + pdV - \mu dN = 0$.

Unfortunately such an approach overlooks some important subtlety of the structure of PTE. If we take into account that the symplectic form $d\theta$ conjugates thermodynamical variables into pairs (T-S, P-V, μ-N,...) leaving internal energy U out, and the fact that U is actually of a different genre than above variables, having common features with other thermodynamical potentials rather, it should be obvious that the natural basic PHASE SPACE of PTE is an even-dimensional symplectic manifold. Then U becomes a distinguished function on the space (i.e. of 2n variables (e.g. T, S, P, V, μ, N)), as well as other potentials.

In fact it will appear, that the function U is natural in terms of the geometry of phase space, i.e. can be constructed in the coordinate-free manner (see Sec. 2).

Legendre vs Gauge Transformations

The appearance of different potentials in PTE is often referred to the Legendre transformation (justifying parallelity of structures of CM and PT). Legendre transformation in CM is essentially the map between tangent and cotangent bundles over configuration space $TQ \leftrightarrow T^*Q$ so that the induced transformation sends the symplectic form $d(\partial_v L) \wedge dx$ defined by the Lagrangian L on TQ to the canonical form $dp \wedge dx$ on T^*M (Darboux variables) [1]. Whereas this context does not appear in PT, where the (pre-)symplectic form $d\theta = dT \wedge dS + dV \wedge dp + d\mu \wedge dN$ has already the canonical form. The potentials appear when one would like to express the same θ in another basis of differentials on M (like e.g. $TdS = -SdT + d(TS)$).

In order to geometrize the notion of potentials one has to extend the manifold of phase space M^{2n} to a real linear fiber bundle E over M and reinterpret the Gibbs form in terms of the connection structure on E, so that the potentials, interpreted as sections of the fiber bundle, refer to different "gauges" of the connection, as shown in the Section 4.

On the other hand, if the manifold M^{2n} has a distinguished splitting into (Lagrangian) subspaces of intensive and extensive variables, the potentials, as functions on M, constitute the lattice structure shown in Section 3.

2. PHASE SPACE OF PTE

Let us state the general definitions constituting PTE.

- By THERMODYNAMICAL PHASE SPACE we will understand the pair (M,α), where M is a 2n-dimensional manifold and $\alpha \in \Lambda^1 M$ is a distinguished differential 1-form of the maximal rank. The form α will be called the Gibbs form.

It follows from the definition that the phase space is a symplectic manifold, since the bi-form $\omega =: d\alpha$ is nondegenerate.

- A STATE is a point in M.
- A PROCESS is a curve $c: \mathbb{R} \to M$ (not all are admissible).

The one-form α has a physical sense of the work W[c] one has to put into the system to carry out the process c:

$$W[c] = \int_c \alpha \tag{1}$$

A THEORY (or a SYSTEM) is a submanifold Λ of the phase space with an embedding

$$\iota : \Lambda \longrightarrow M \tag{2}$$

such that

$$\iota^*\alpha = 0. \tag{3}$$

Eq. (2) is called the EQUATION OF STATE.

Eq. (3) is called the GIBBS-DUHEM RELATION and means that the directions tangent to Λ lie in the kernel of α, i.e. that α restricted to Λ vanishes, $\alpha|_\Lambda = 0$.

Cor. Since the external derivative commutes with ι^*, it immediately follows from (3) that

$$\iota^* d\alpha = 0, \tag{4}$$

i.e. that Λ is a Lagrangian submanifold of M (of course, eq. (3) are more restrictive conditions for Λ).

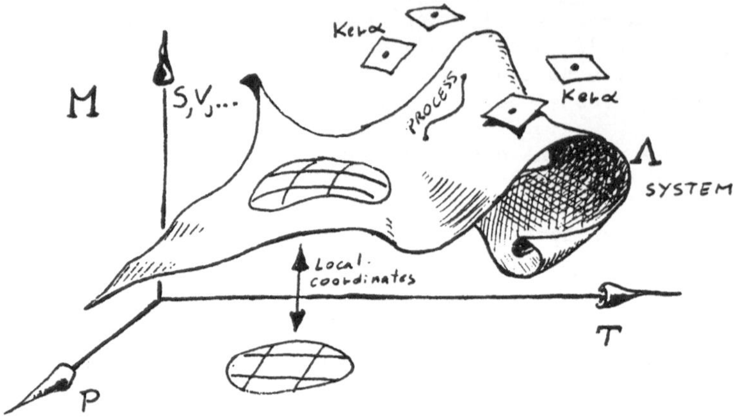

FIG. 1

Cor. The dimension of Λ is at most $\frac{1}{2}\dim M$.

Eq. (4) is called a MAXWELL IDENTITY and is, in fact, the integrability condition on Λ (it is an analog of the Hamilton-Jacobi equation in CM).

- By the SPECIAL PHASE SPACE we will understand the cotangent bundle over the linear space Q

$$\begin{array}{ccc} M = T^*Q & \cong & \mathbb{R}^{2n} \\ \downarrow \pi & & \\ Q & \cong & \mathbb{R}^n \end{array} \tag{5}$$

The form $-\alpha$ is here defined as the natural Liouville form of the cotangent bundle. Q will be named INTENSIVE SPACE (analogously to the intensive parameters of PTE).

If $\{x'^i\}$ are any linear coordinates on Q, and $\{p_i\}$ induced coordinates on T*Q, called EXTENSIVE PARAMETERS, then

$$\alpha = -p_i \, dx^i$$
$$\text{and} \quad \omega = d\alpha = -dp_i \wedge dx^i. \qquad (x^i =: x'^i \circ \pi) \tag{6}$$

The above geometry (5) of the special phase space distinguishes the natural function defined in coordinate-independent way:

- The INTERNAL ENERGY is the function $U \in FM$

$$T^*Q \ni \quad p \longmapsto U(p) =: p(V_{\pi(p)}) \equiv p(\kappa_{\pi(p)} \circ \pi \, p) \tag{7}$$

where $V \in \mathfrak{X}Q$ is the natural Liouville vector field on the intensive space, κ is the natural canonical isomorphism on linear spaces $\kappa_q: Q \longmapsto T_qQ \quad \forall q \in Q$, and p as a point of T*Q is a differential form on Q.

In the coordinates (6) $\qquad U = p_i x^i$,

and, since $dU = p_i dx^i + x^i dp_i$, the Gibbs form may be written in the more familiar form:

$$\alpha = x^i dp_i - dU \tag{8}$$

so that the Gibbs relation (3) is

$$\iota^*(x^i dp_i - dU) = 0 \tag{9}$$

i.e. $(x^i \circ \iota) d(p_i \circ \iota) - d(U \circ \iota) = 0.$

Since we have the distinguished function U on M, another possibility of description of the system Λ in M appears. Let ρ be a projection on some manifold D of $\dim D \leq \frac{1}{2} \dim M$. We can set the problem: for a given function $u \in FD$ find the submanifold $\iota: \Lambda \to M$ such that

$$\iota^* U = (\rho \circ \iota)^* u \tag{10}$$

Eq. (10) is called the FUNDAMENTAL RELATION. Often ρ and D are π and Q of (5). In this case (10) is equivalent to (2).

Let us call u a DEFINING FUNCTION of internal energy. It ought to be clearly distinguished from the function of internal energy $U \in FM$. The function $u \circ \iota$ is the INTERNAL ENERGY OF THE SYSTEM ι.

Summarizing this section:

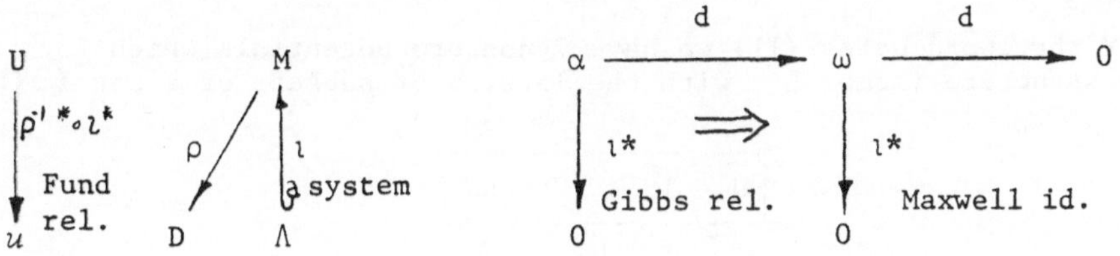

The standard model is 6-dimensional:

$$\begin{aligned} x^1 &= T \text{ -temperature} & p_1 &= S \text{ -entropy} \\ x^2 &= -P \text{ -pressure} & p_2 &= V \text{ -volume} \\ x^3 &= \mu \text{ -chemical potential} & p_3 &= N \text{ -number of molecules} \end{aligned} \tag{11}$$

The Gibbs form is

$$\begin{aligned} \alpha &= -SdT + VdP - Nd\mu \\ &= TdS - PdV + \mu dN - dU \end{aligned} \tag{12}$$

where

$$U = ST - PV + \mu N. \tag{13}$$

The embedding ι of a system (2) may be set by "constraints" of the type pV^{α} = const. T, etc., or - equivalently - by the fundamental relation. Let e.g. $u \in FQ$, i.e. $u = u(T,P,\mu)$. Thus eq. (10) is an equation for such an embedding $S \circ \iota = S(T,P,\mu)$, $V \circ \iota = V(T,P,\mu)$ and $N \circ \iota = N(T,P,\mu)$ that

$$u(T,P,\mu) = S(T,P,\mu)\cdot T - V(T,P,\mu)\cdot P + N(T,P,\mu)\cdot\mu. \tag{14}$$

3. SEVEN (OR ∞) POTENTIALS OF PTE

The Gibbs form α of the phase space can always be expressed in the canonical coordinates as $\alpha = p_i dx^i$ (Darboux theorem). If for any reason we would like to have a term $x^{\dot{k}} dp_{\dot{k}}$ (the dot designs no summation over k) instead of $p_{\dot{k}} dx^{\dot{k}}$ for some k, then the external derivative of $p_{\dot{k}} x^{\dot{k}}$ must appear:

$$\alpha = p_i dx^i = \sum_{i \neq k} p_i dx^i - x^{\dot{k}} dp_{\dot{k}} + d(p_{\dot{k}} x^{\dot{k}}) \tag{15}$$

Such a "coordinate flip" can be done for any such pair of conjugated canonical variables. Although the above coordinate trick in the general case of M has an underlying geometry in the case of the special phase space, let us simplify it here to the statement, that for distinguished 2n coordinates on M each subset J of I =: {1,2,...,n} is related to one of n! distinguished functions:

$$f_J =: \sum_{k\in J} p_k x^k, \qquad f_\emptyset =: 0 \tag{16}$$

called the Jth POTENTIAL, so that the Gibbs form can be expressed

$$\alpha = \sum_{i\in I\setminus J} p_i dx^i - \sum_{k\in J} x^k dp_k + df_J. \tag{17}$$

For the coordinates (11) we have 7 nonzero potentials which form a lattice structure isomorphic with the lattice of subsets of a set I={1,2,3}

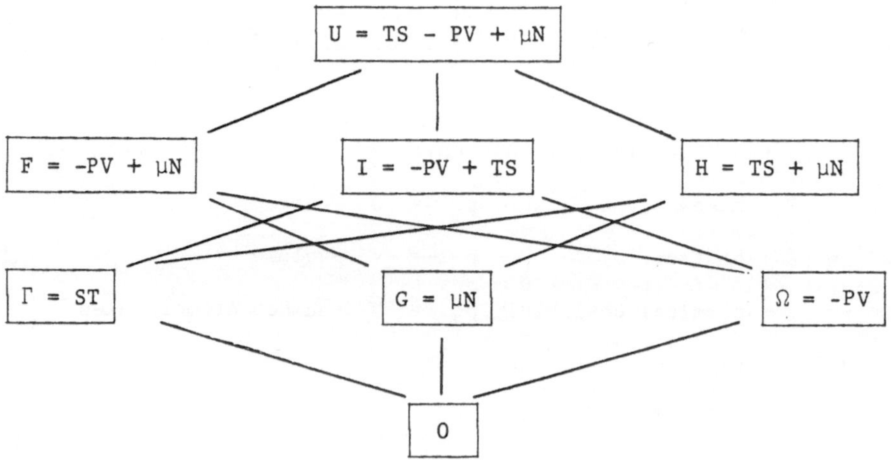

FIG. 2

where the functions are called:

U -	internal energy	Γ -	nameless
F -	Helmhotz free energy	G -	Gibbs potential
I -	nameless	Ω -	grand potential
H -	enthalpy	0 -	just zero! (unrecognized trivial potential)

In fact, the geometry of the special phase space allows us to define thermodynamical potentials in coordinate-independent way:

Each linear decomposition $\mathcal{A}$ of Q into two subspaces

$$Q = A \oplus B, \qquad \pi_A : Q \to A \tag{18}$$

determines the function $f_{\mathcal{A}} \in \mathcal{F}M$ (cf. (7))

$$f_{\mathcal{A}}(p) =: p(\kappa_{\pi p} \circ \pi_A \circ \pi p) \tag{19}$$

called the THERMODYNAMICAL POTENTIAL for the splitting $\mathcal{A}$.

Thus we have the infinite lattice L of thermodynamical potentials with the structure induced from the natural lattice of subspaces A, B of the linear space Q. The maximal element in the lattice, U, corresponds to the trivial decomposition $Q = Q \oplus \emptyset$. The minimal element of the decomposition $Q = \emptyset \oplus Q$ is the constant null function.
The lattice in Fig. 2 is a finite sublattice of L.

NOTE 1: If the euclidean structure is introduced to Q, then π_A is determined by only one subspace and the orthogonal projection on it. Thus the lattice L is isomorphic to the Grassman lattice of all subspaces of Q.

NOTE 2: Decomposition (18) induces a decomposition of M into the symplectic spaces $M = T^*A \oplus T^*B$. This shows the way to generalize Def. (19) to the case of the general phase space.

Generally for a potential $f \in \mathcal{F}M$ we will write

$$\alpha \equiv \alpha_f - df. \tag{20}$$

In fact, for the special phase space (5) the statement (1) concerns the integral of α_U rather.

Introduced potentials enables to express the Gibbs relation (3) in a way described here in coordinates (11). If a neighborhood U of Λ is projectible on some subspace of M, then coordinates of the subspace can be used (by pull-back, $\iota^* p_i =: p_i \circ \iota$, $\iota^* x_i$) to parametrize $U \subset \Lambda$ (see Fig. 1). Each choice of those coordinates brings through (3) the "definitions" of the rest. E.g. if the variables T,V,N on Λ are chosen, then the convenient expression for the form α on M is

$$\alpha = -SdT + PdV + \mu dN - dF \tag{21}$$

and thus

$$\iota^*\alpha = -S(T,V,N)dT - P(T,V,N)dV + \mu(T,V,N)dN - dF(T,V,N) \tag{22}$$

so $\iota^*\alpha = 0$ means:

$$\begin{aligned} \iota^*S = S(T,V,N) &= -\left.\frac{\partial F \circ \iota}{\partial T}\right|_{V,N} \\ P(T,V,N) &= -\left.\frac{\partial F \circ \iota}{\partial V}\right|_{T,N} \\ \mu(T,V,N) &= \left.\frac{\partial F \circ \iota}{\partial N}\right|_{P,T} \end{aligned} \tag{23}$$

For other choices of coordinates e.g. entropy can be expressed similarly:

$$\iota^* S = -\left.\frac{\partial G}{\partial T}\right|_{P,N} = -\left.\frac{\partial \Omega}{\partial T}\right|_{N,\mu} = \left.\frac{\partial \Gamma}{\partial T}\right|_{\mu,P} \tag{24}$$

By analogy, all other of 3·7 such relations can be expressed.

4. GAUGE INTERPRETATION OF PTE

Let $\{E,\pi,M\}$ be a real linear fiber bundle over thermodynamical phase space M, with a projection

$$\pi: E \to M, \qquad \pi^{-1}(m) \cong \mathbb{R} \quad \forall m \in M \tag{25}$$

Def. The Gibbs connection on E is the connection ∇ defined by the vector-valued one-form

$$\boldsymbol{\alpha} =: \partial_u \otimes \alpha \tag{26}$$

The curvature of the connection is

$$\mathrm{Curv}(\nabla) \equiv \boldsymbol{\omega} =: \partial_u \otimes d\alpha \tag{27}$$

Thus a system is defined now as a submanifold Λ of M such that curv ∇ on the subbundle $(\pi^{-1}(\Lambda),\pi,\Lambda)$ vanishes.

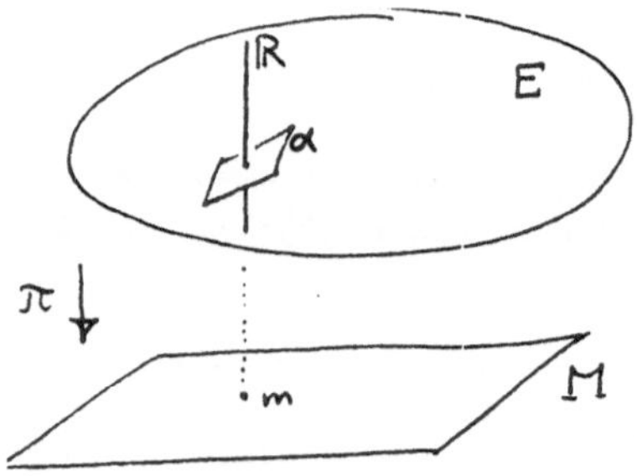

FIG. 3

Now a potential is a section of the fiber bundle

$$\Gamma(E,\pi,M) \ni \quad f: M \to E, \qquad \pi \circ f = id_M \tag{28}$$

Each such section induces a differential one-form α_f on M (see (20)) by the formula:

$$\Gamma(E,\pi,M) \ni \quad f \longmapsto \alpha_f =: \nabla f \tag{29}$$

$$\text{i.e.} \quad \alpha_f(v) = \nabla_v f \qquad \forall v \in TM.$$

Thus the known potentials correspond to convenient "choices of gauge" of the Gibbs connection $\boldsymbol{\alpha}$.

REMARK: If the connection (26) is built rather as $\boldsymbol{\alpha} =: \partial_v \otimes \alpha_U$, then it has a clear physical sense: if $\gamma: I \to M$ is a process in the phase space, then the curve $\tilde{\gamma}$ lifted through the connection rises by the energy put into the system.

CONCLUSIONS

The geometrization of PTE in terms of even-dimensional symplectic space seems to be more profound than the usually suggested odd-dimensional contact manifold approach. [An interesting inverse shift may be noted: CM is usually presented in terms of symplectic geometry, whereas here the odd-dimensional contact formulation seems to be more profound [8] as reflecting the particle-wave duality in Lagrangian description.]

Such an approach is supported by the fact that now thermodynamical potentials can be defined in a rigorous geometrical way and they reveal the structure of a lattice.

The phase space may be extended by one more dimension by the use of the fiber bundle technique. Then, PTE may be reinterpreted in terms of "gauge theory".

REFERENCES

[1] R. Abraham, J. E. Marsden, Foundation of Mechanics, Benjamin, New York, 1967.

[2] J. W. Barden, B. Carter, and S. W. Hawking, Commun. Math. Phys. 31, (1973), 161-170.

[3] H. Callen, Thermodynamics, Wiley, New York, 1960.

[4] C. Caratheodory, Undersuchungen uber die Grunlagen der Thermodynamik, Math. Ann. 67, (1909), 355.

[5] J. W. Gibbs, Collected Works, Vol. I, Thermodynamics, Longmans, New York, 1928.

[6] R. Hermann, "Geometry, Physics, and Systems", Marcel Dekker, New York, 1973.

[7] P. A. Horvathy, Classical Action, the Wu-Yang Phase Factor and Prequantization, Marseille Prep. CPT 80, (1978), 1182.

[8] J. Kocik, Lagrangian vs Newtonian Mechanics, Wroclaw Univ. Prep. No. 538, (1981).

[9] L. Mistura, Nuovo Cimento, 51B, (1979), 125.

[10] R. Mrugała, Rep. Math. Phys. 14, (1978), 419.
Acta Phys. Pol., Vol. A58, No. 1, (1980), 19.

[11] L. Tisza, Generalized Thermodynamics, The MIT Press, 1966.
Ann. Phys., 13, (1961), 1.

ON PROJECTIONS OF SPINOR SPACES ONTO MINKOWSKI SPACE

J. Kocik* and J. Rzewuski**

*Department of Mathematics, Southern Illinois University at Carbondale, Carbondale, IL 62901

**Department of Physics and Astronomy, Southern Illinois University at Carbondale, Carbondale, IL 62901
On leave of absence from Institute of Theoretical Physics, University of Wroclaw, Wroclaw, Poland

1. INTRODUCTION

The problem of projecting spinor spaces on Minkowski space is rather old. The first formula of this type is

$$x_\mu = \xi^*_a \sigma^{\dot{a}b}_\mu \xi_b, \quad a,b = 1,2, \quad \mu = 0,1,2,3 \tag{1.1}$$

projecting the two-dimensional complex spinor space $\mathbb{C}^2$ onto the light cone in M_4. One can consider (1.1) also as a projection on E_3. These projections are consistent with the group in the sense that SU(2) transformations of the ξ_a induce SO(3) transformations of x_1, x_2, x_3 and leave x_0 invariant. Transformations of $SL(2,\mathbb{C})$ induce SO(3.1) transformations of x_μ.

The spinor space $\mathbb{C}^2$ with group $SL(2,\mathbb{C})$ or SU(2) has a great advantage over the Euclidean space E_3 with the group SO(3) because the representations of SU(2) in the Hilbert space of functions $\mathbb{C}^2 \cong \mathbb{R}^4 \to \mathbb{C}$ are complete whereas the representations in the Hilbert space of functions $E_3 \to \mathbb{C}$ contain only integer spin. This is best seen when one compares the two Laplace-Beltrami operators

$$\frac{\partial^2}{\partial\xi^*_a \partial\xi_a} = r\left\{\frac{\partial^2}{\partial x_i \partial x_i} + \frac{1}{r^2 \sin^2\theta}\left[2\cos\theta \frac{\partial^2}{\partial\phi\partial\psi} + \frac{\partial^2}{\partial\psi^2}\right]\right\} \tag{1.2}$$

in $\mathbb{C}^2 \cong \mathbb{R}^4$ and in E_3. The fourth variable ψ is just the Eulerian angle (besides θ and ϕ) defined in terms of the ξ_a as

$$e^{i\psi} = \frac{\xi_1 \xi_2}{|\xi_1||\xi_2|} . \tag{1.3}$$

The angular part of the operator in (1.2) is the well known operator of generalized (three index) spherical harmonics and its eigenvalues are $j(j+1)$ with $j = 0, 1/2, 1, 3/2, \ldots$ (cf. e.g., [1]).

The spinor space $\mathbb{C}^2$ has also disadvantages from the point of view of the projection itself. Firstly, it projects $\mathbb{C}^2$ on a three-dimensional space (E_3 or the light-cone). Secondly it is inconsistent with the group of translations.

To remove the first difficulty projections were introduced [2] from more complex dimensions. In the case of two complex spinors ξ_a and η_a one has the three projections

$$x_\mu = \pm\frac{1}{2}(\xi^*_a\sigma_\mu^{\dot{a}b}\xi_b + \eta^*_a\sigma_\mu^{\dot{a}b}\eta_b),$$
$$y_\mu = \frac{1}{2}(\xi^*_a\sigma_\mu^{\dot{a}b}\xi_b - \eta^*_a\sigma_\mu^{\dot{a}b}\eta_b), \tag{1.4}$$

the first two being on the inside of the future and past light cones, the second on the outside of the light cone. Light cone itself being obtained by putting $\xi = \eta$ in the formula for x_μ.

The four additional real variables (besides x_μ or y_μ) were considered to correspond to some internal degrees of freedom and invariant differential equations were studied, analogous to (1.2) [2].

These projections still have the disadvantage of not incorporating translations of the Minkowski variables. The next step was taken by again doubling the complex dimension [3] and considering two bispinors rather than two spinors. The projection can in this case be written in the form

$$z_\mu = x_\mu + iy_\mu = \lambda\frac{v_\mu}{s + p}, \tag{1.5}$$

where

$$s = \xi_a\varepsilon^{ab}\eta_b,\ p = \xi_a\gamma_5^{ab}\eta_b,\ v_\mu = \xi_a\gamma_\mu^{ab}\eta_b \qquad a,b = 1,\dots,4 \tag{1.6}$$

are six antisymmetric bilinear forms and λ is an arbitrary constant with the dimension of length.

This projection satisfies already more than all demands. It is on the whole of a complex Minkowski space and consistent not only with the Poincaré but with the whole conformal (in fact with the whole $GL(4,\mathbb{C})$, see Section 2) group in the sense that simultaneous $SU(2,2)$ transformations of ξ and η induce conformal transformations of the complex variables z_μ.

It is easily seen that projection (1.5) is invariant with respect to another group of transformations mixing the variables ξ and η:

$$\xi'_a = a\xi_a + b\eta_a \qquad \eta'_a = c\xi_a + d\eta_a \qquad \begin{vmatrix} a & b \\ c & d \end{vmatrix} \neq 0. \tag{1.7}$$

Indeed, the antisymmetric forms change under (1.7) by the factor $\begin{vmatrix} a & b \\ c & d \end{vmatrix}$ which cancels in the ratio (1.5). It means that, in fact, a point in Minkowski space corresponds to a 2-dimensional complex subspace in $\mathbb{C}^4$. Thus we have the situation that transformations of the direct product $GL(4,\mathbb{C}) \times GL(2,\mathbb{C})$ can be defined on the matrix ξ^A_a,

$$\xi^A_a \to \xi'^A_a = h^A_B\xi^B_b g^b_a \tag{1.8}$$

where $\xi_a^1 := \xi_a$, $\xi_a^2 := \eta_a$, $g \in GL(4,\mathbb{C})$, $h \in GL(2,\mathbb{C})$, such that the transformations of the second factor do not affect the projection, whereas the transformations of the first factor induce non-linear transformations of the z_μ among themselves.

$$z_\mu \to z'_\mu = z'_\mu(z_0,z_1,z_2,z_3). \tag{1.9}$$

The transformations of $GL(2,\mathbb{C})$ can be considered, therefore to contain internal symmetries.

Since $GL(4,\mathbb{C}) \times GL(2,\mathbb{C})$ contains the physical symmetries $SU(2,2) \times SU(2)$ or $P \times SU(2)$, with $SU(2)$ as internal symmetry only, the projection (1.5) calls for generalization to $GL(m,\mathbb{C})$ as the second factor, which would contain $SU(m)$ as internal symmetry (with arbitrary $m \geq 2$). Here however we are stopped by a difficulty [4]. If we consider more than two bispinors ξ_a^A ($a = 1,\dots,4$, $A = 1,\dots,m$) we can construct $\frac{1}{2}m(m-1)$ projections $z_\mu^{(A,B)}$ of the type (1.5), each corresponding to a pair of indices (A,B) out of $1,\dots,m$. So the projection would not be unique. Moreover it would not be consistent with the group since the antisymmetric forms (1.6) corresponding to different pairs of second indices would transform into each other linearly according to a matrix representation of the second factor $GL(m,\mathbb{C})$ thus introducing objects of new type.

In this paper we want to show how this difficulty can be overcome by restricting the projection to a submanifold $O_2^{(4,m)}$ of the space $\mathbb{C}^{4m}$ of the variables ξ_a^A. On this occasion we are going to study also other projections which appear automatically when considering orbits of the group $GL(n,\mathbb{C}) \times GL(m,\mathbb{C})$ in $\mathbb{C}^{nm}$ (Section 2).

We found it interesting that there exists, for arbitrary m, only one submanifold $O_2^{(4,m)}$ of $\mathbb{C}^{4m}$ which admits a projection onto complex Minkowski space. A projection which is unique and consistent with the group $GL(4,\mathbb{C}) \times GL(m,\mathbb{C})$ (Section 2). When the symmetry is restricted to the physical case $SU(2,2) \times SU(m)$ or $P \times SU(m)$ this submanifold decomposes further into a 2 or 3 parametric set of constituents each admitting the projection. Section 3 contains a detailed study of these constituents. Finally, Section 4 is devoted to representations of generators of the group in the Hilbert space of functions $\mathbb{C}^{4m} \cong \mathbb{R}^{8m} \to \mathbb{C}$ and their reduction to $O_2^{(4,m)}$. In analogy with the two-dimensional complex example mentioned at the beginning one obtains representations containing all quantum number of external as well as internal symmetries.

Our aim is to obtain an understanding of all aspects of the geometrical objects which appear in the procedure described above. We shall use therefore geometrical as well as group-theoretical methods and investigate global as well as local properties of the manifolds in question.

2. TISSUE SUBMANIFOLDS OF $\mathbb{C}^n \otimes \mathbb{C}^m$.

A. Motivations: $\mathbb{R}$-Physics vs $\mathbb{C}$-Physics

There is a gap between descriptions of the two worlds, the particles live in. On one hand the "visual" space-time is expressible in terms of real numbers. On the other we have the world of internal degrees of freedom $\mathbb{C}^m$ where the complex structure appears in a natural way. We believe that there exists a unified picture of both worlds, "internal" and "external". Thus the group $SU(2,2)$ covering the conformal group $C_+^\uparrow(3,1)$

seems to be more fundamental symmetry structure of "external" degrees of freedom, and the space $C^{2,2}$, i.e., C^4 with a hermitian scalar product of signature (+ + - -), should be considered as primary object possessing both attributes: that of space-time structure and that of complex structure.

Thus $U \cong C^{2,2} \otimes C^m$ will be assumed as the universal unifying space describing internal and external degrees of freedom.

The goal of this and the next chapter is to specify how this space is related to Minkowski space M. The relation will appear to be a projection of some submanifold $O_2 \subset U$ onto M. This submanifold appears in a natural decomposition of U generated by $\mathbb{C}$-linearity only, so in this chapter we will study the general case of the tensor product $\mathbb{C}^n \otimes \mathbb{C}^m$ (without any additional hermitian structure).

If the basis in $\mathbb{C}^n$ and $\mathbb{C}^m$ is chosen, the element ξ of the universal space $U =: \mathbb{C}^n \otimes \mathbb{C}^m \cong \mathbb{C}^{n \cdot m}$ can be represented as an $n \times m$ matrix and one can roughly think of it as consisting of m n-spinors (or n m-spinors). We would like to interpret U as the space of all homomorphisms $\xi: \mathbb{C}^n \to \mathbb{C}^m$.

B. General Geometry of $U = \mathbb{C}^n \otimes \mathbb{C}^m$

The universal space

$$U =: \mathrm{Hom}_{\mathbb{C}}\ (\mathbb{C}^n, \mathbb{C}^m) \cong \mathbb{C}^n \otimes \mathbb{C}^m \cong \mathbb{C}^{nm} \tag{2.1}$$

splits in submanifolds of maps of fixed rank. Let us recall that if $\xi \in U$ and rank $\xi = k$, then $\dim_{\mathbb{C}} \mathrm{Im}\ \xi = k$ and $\dim_{\mathbb{C}} \ker \xi = n-k$. According to the first isomorphism theorem one can decompose ξ into three steps:

$$\mathbb{C}^n \xrightarrow{\ \pi\ } \mathbb{C}^n/\ker \xi \xrightarrow{\ \tilde{\xi}\ } \mathrm{Im}\ \xi \xrightarrow{\ \iota\ } \mathbb{C}^m \tag{2.2}$$

where π-canonical projection on coset space, ι - an embedding of the image of the map ξ, and $\tilde{\xi}$ is an isomorphism, i.e., $[\tilde{\xi}] \in GL(k,\mathbb{C})$.

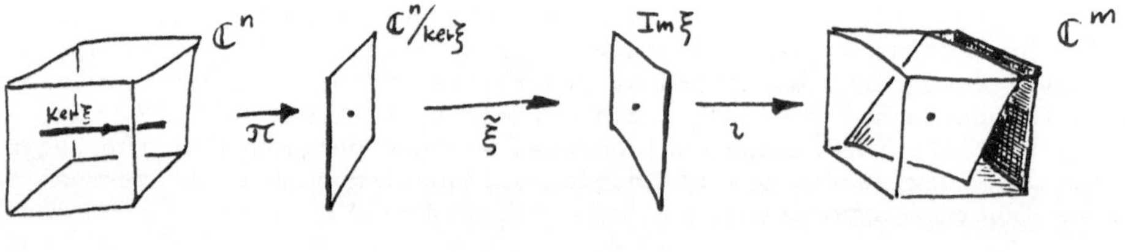

FIGURE 1.

Obviously $k = 0,1,\ldots,\min(n,m)$. Define a kth TISSUE in $U = \mathrm{Hom}(\mathbb{C}^n,\mathbb{C}^m)$ as the manifold

$$\mathcal{O}_k^{(n,m)} =: \{\xi \in U\colon \mathrm{rank}\ \xi = k\} \subset U. \tag{2.3}$$

Of course

$$U = \bigcup_{i=0}^{\min(n,m)} \mathcal{O}_i^{(n,m)} \tag{2.4}$$

and

$$\mathcal{O}_k^{(n,m)} \cap \mathcal{O}_i^{(n,m)} = \emptyset \text{ for } k \neq i. \tag{2.5}$$

C. Group Interpretation

Since only linearity of U is used in the splitting (2.4), a dual description can be realized through the most general group acting on $U = \mathrm{Hom}\,(\mathbb{C}^n,\mathbb{C}^m)$, namely $G \cong GL(n,\mathbb{C}) \times GL(m,\mathbb{C})$, so that the following diagram commutes

$$\begin{array}{ccc} \mathbb{C}^n & \xrightarrow{\ \xi\ } & \mathbb{C}^m \\ \big\downarrow g \in GL(n,\mathbb{C}) & & \big\downarrow h \in GL(m,\mathbb{C}) \\ \mathbb{C}^n & \xrightarrow{\ \xi'\ } & \mathbb{C}^m \end{array} \qquad (2.6)$$

i.e., $GL(n,\mathbb{C}) \times GL(m,\mathbb{C}) \ni (g,h)\colon \xi \rightsquigarrow \xi' = h \circ \xi \circ g^{-1} \in U$. Now the splitting (2.4) appears as a decomposition of U into orbits of the group. (Each tissue is an orbit of G.)

D. Histology of $\mathcal{O}_k^{(m,n)}$

The decomposition (2.2) of ξ is crucial to see the geometry of the kth tissue $\mathcal{O}_k^{(n,m)}$ (the upper indices will sometimes be omitted). Roughly speaking ξ is totally characterized by three objects: the (n-k) dimensional subspace $\ker\xi < \mathbb{C}^n$, the k-dimensional subspace $\mathrm{Im}\,\xi < \mathbb{C}^m$, and by isomorphism $\tilde{\xi} \in GL(k,\mathbb{C})$. Nevertheless, $\mathcal{O}_k$ is not just a cartesian product of (complex) Grassman manifolds G^n_{n-k}, G^m_k *) and $GL(k,C)$. It is a fiber bundle with $G^n_{n-k} \times G^m_k$ as a basis (compact), a fiber homomorphic to $GL(k,\mathbb{C})$, and projection:

$$\pi\colon \mathcal{O}_k \mapsto G^n_{n-k} \times G^m_k\colon \qquad\qquad (2.7)$$
$$\xi \mapsto (\ker\xi,\ \mathrm{Im}\,\xi).$$

The fiber bundle $(\mathcal{O}_k, G^n_{n-k} \times G^m_k, \pi)$ is nontrivial. It is also seen that more detailed fibering appears:

$$\begin{array}{ccccc} & \swarrow^{\pi_1} & \mathcal{O}_k & {}^{\pi_2}\searrow & \\ & & \big\downarrow \pi & & \\ G^n_{n-k} & \xleftarrow{(id,0)} & G^n_{n-k} \times G^m_k & \xrightarrow{(0,id)} & G^m_k \end{array} \qquad (2.8)$$

The construction we will carry out for the case $n = 4$, $k = 2$ is based on the fibering (projection) $\pi_1 = (id,0) \circ \pi$. The manifold $\pi_1(\mathcal{O}_k^{(4,m)}) \cong G^4_2$ is homeomorphic to the compactified complex Minkowski manifold $M^{\mathbb{C}}$, the real part of it is compactified Minkowski manifold $M \cong S^3 \times S^1$ (space with a point in infinity and closed time) [6].

---

*) G^m_k is the set of all k-dimensional complex linear subspaces in $\mathbb{C}^m$. Moreover, $G^n_{n-k} = G^n_k$.

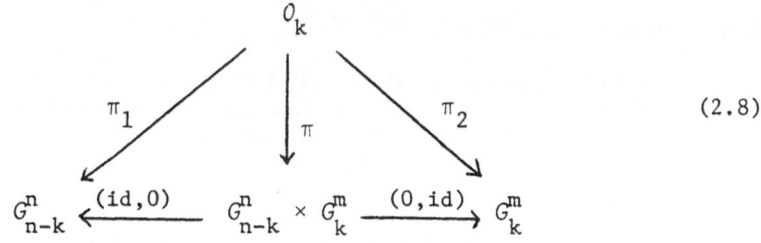

E. On Dimension of Tissues $\mathcal{O}_k$

Since the complex dimension of G_k^n is $(n-k)k$,

$$\dim_{\mathbb{C}} \mathcal{O}_k^{(n,m)} = (n + m - k)k. \tag{2.9}$$

The quantity $(a-b)b$ appears in our considerations fairly often, so let us draw the analogue of the Pascal triangle for $[{a \atop b}] = (a-b)b$.

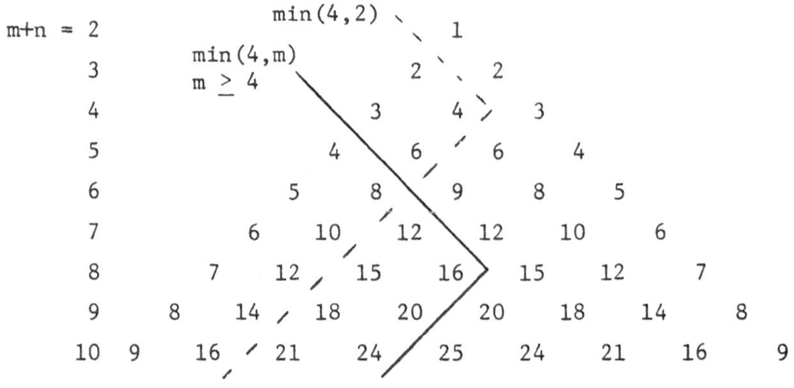

FIGURE 2

Dimensions of $\mathcal{O}_k^{(n,m)}$ appear here as $[{m+n \atop k}]$. The right side of the triangle is cut out by the requirement $k \leq \min(n,m)$. In the case $n = 4$, $m \geq 4$, (see Fig. 2) we have the decomposition (2.4)

$$U = \mathcal{O}_0 \cup \mathcal{O}_1 \cup \mathcal{O}_2 \cup \mathcal{O}_3 \cup \mathcal{O}_4$$

with corresponding complex dimensions 0, 7, 12, 15, 16. Admissible projections π_1 (2.8) are then on Grassmanians with complex dimensions 0, 3, 4, 3, 0, respectively. The case $n = 4$, $m = 2$ is also drawn in Fig. 2 since it corresponds to the Penrose model [3] (see Chapter 3).

F. Group Picture of a Tissue $\mathcal{O}_k^{(n,m)}$

Any (complex) Grassman manifold G_q^p being a set of q-dimensionsl subspaces of p-dimensional space can be described as a coset of $GL(p,\mathbb{C})$ with isotropy group $GL^q(q,\mathbb{C})$ being a set of automorphism leaving some q-dimensional subspace invariant. In properly chosen coordinates $GL^q(q,\mathbb{C}) = \{g = ({ab \atop 0d})$: $a \in GL(q,\mathbb{C})$, $d \in GL(p-q,\mathbb{C})$, $b \in L(\mathbb{C}^{p-q},\mathbb{C}^q) \cong \mathbb{C}^{q(p-q)}$. Now taking into account the decomposition (2.2) it becomes obvious that the orbit $\mathcal{O}_k^{(n,m)}$ can be described also as the coset space

$$\mathcal{O}_k^{(n,m)} \cong GL(n,C) \times \mathbb{C}L(m,\mathbb{C})/H_k^{(n,m)} \tag{2.10}$$

the isotropy group being

$$H_k^{(n,m)} \cong \left\{ \begin{pmatrix} a & b_1 \\ 0 & d_1 \end{pmatrix} \times \begin{pmatrix} (a^T)^{-1} & b_2 \\ \mathcal{O} & d_2 \end{pmatrix} \right\}$$

with $a_1 \in GL(k,\mathbb{C})$, $d_1 \in GL(n-k,\mathbb{C})$, $b_1 \in \mathbb{C}^{k(n-k)}$, $d_2 \in GL(m-k,\mathbb{C})$, $b_2 \in \mathbb{C}^{k(m-k)}$ (the isotropy group of the point $\left(\begin{array}{c|c} \mathbf{1} & 0 \\ \hline 0 & 0 \end{array}\right)$ in U, $\mathbf{1}$ being a $k \times k$ unit matrix) [7]. Moreover the two projections π_1 and π_2 (2.8) are on the coset spaces of $GL(n,\mathbb{C})$ and $GL(m,\mathbb{C})$ defined by the isotropy groups $GL^k(n,\mathbb{C})$ and $GL^k(m,\mathbb{C})$, respectively.

G. Morphology of the Tissue Decomposition $\bigcup_k \mathcal{O}_k$

All submanifolds $\mathcal{O}_k$ have elements arbitrarily close to the $0 \in \mathbb{C}^{nm}$ (in the sense of the natural metric induced from $\mathbb{C}^{nm}$), however they have different dimensions. In fact, they form a flag of manifolds (in the analogy to the notion of the flag of subspaces) in the sense that

$$\mathcal{O}_k \subset \overline{\mathcal{O}}_{k+1} \tag{2.11}$$

for any admissible $k = 1,\ldots,\min(n,m)-1$. (The bar means closure.)

All closed orbits $\mathcal{O}_k$ meet at $0 \in \mathbb{C}^{nm}$, and their tangent spaces form the flag in the usual sense:

$$0 = T_0\overline{\mathcal{O}}_0 < T_0\overline{\mathcal{O}}_1 < \ldots < T_0\overline{\mathcal{O}}_k < \ldots < T_0\overline{\mathcal{O}}_{\min(n,m)} \cong \mathbb{C}^{nm} \tag{2.12}$$

being an element of the flag manifold $F_{\left[{m+n \atop 0}\right],\left[{m+n \atop 1}\right],\ldots,\left[{m+n \atop \min(m,n)}\right]}$.

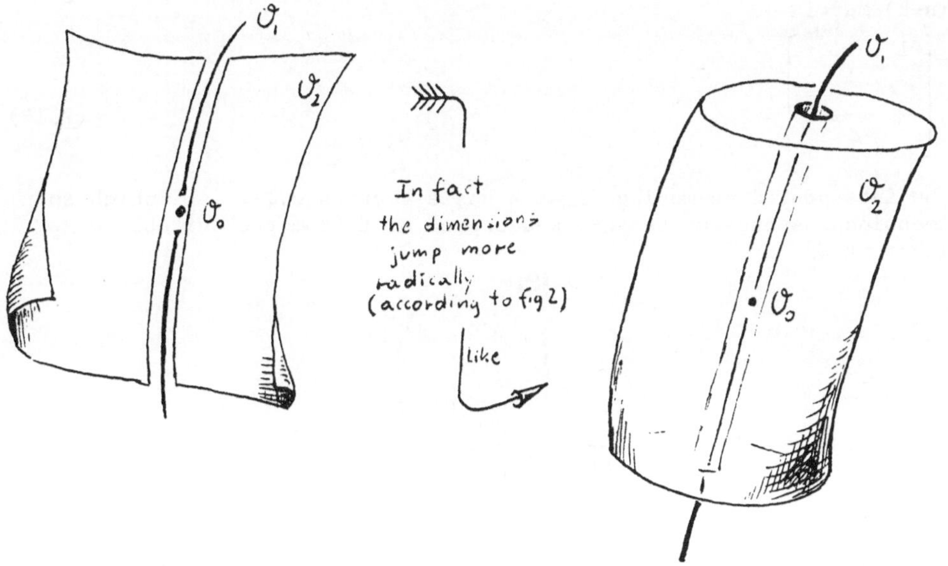

Of course the manifolds $\mathcal{O}_k$ possess all topological nontrivialities of their components (Grassmanians).

H. Local Coordinates on $\mathcal{O}_k$

a) Grassman Manifold. Before we introduce a convenient coordinate system on $\mathcal{O}_k$ let us first consider the case of a Grassman manifold G^n_k of k-dimensional subspaces of n-dimensional space E over any field $\mathbb{R}$, $\mathbb{C}$, $\mathbb{H}$.

Recall the simplest case of the projective space $G_1^n = P^{n-1}$. Let $\{e_1,\dots,e_n\}$ be the basis in E, $\{\varepsilon^1,\dots,\varepsilon^n\}$ - the dual basis. Each vector $v = \begin{bmatrix} a_1 \\ \vdots \\ a_n \end{bmatrix} \in E$ is related to some $[v'] = \begin{bmatrix} 1 \\ a_2/a_1 \\ \vdots \\ a_n/a_1 \end{bmatrix}$ in the hyperplane $E_1 = \{v \in E\colon \varepsilon^1(v) = 1\}$ which can be obtained by applying the one-element matrix $[a_1^{-1}]$ on the right side: $[v^1] = [v] \cdot [a^{-1}]$.

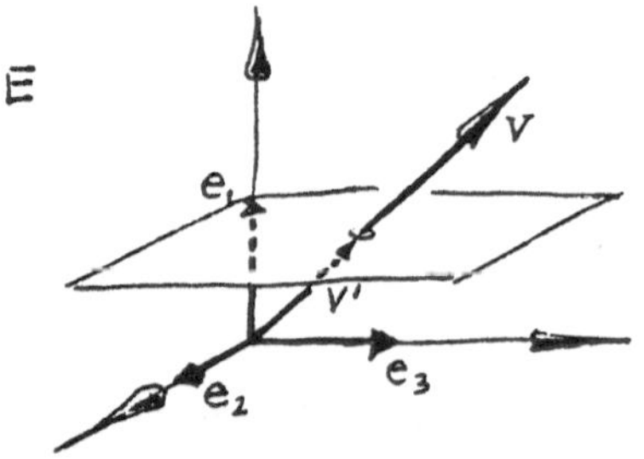

The set of (n-1) coefficients a_i/A_1 may be considered as a map on the open neighborhood E_1 of P^{n-1} and have clear geometrical interpretation of a vector from subspace spanned by $\{e_2,\dots,e_n\}$ one has to add to e_1 to move to considered point in E_1 (and then -- in P^{n-1}).

So one can think of $[v^1]$ as a $n \times 1$ matrix which, acting on the vector e_1 ranges over whole $E_1 \subset P^{n-1}$. In the similar way in the case of G_k^n the $n \times (n-k)$ matrix

$$\left[\begin{array}{c} 1 \quad \\ \quad 1 \\ \quad\quad 1 \\ \quad\quad\quad \ddots \\ \hline A \end{array}\right] = Z' \tag{2.13}$$

can be interpreted as moving first k basis vectors $e_1,\dots,e_k$, which span k-dimensional subspace, to some k-subspace of E from the suitable open neighbor of G_k^n.

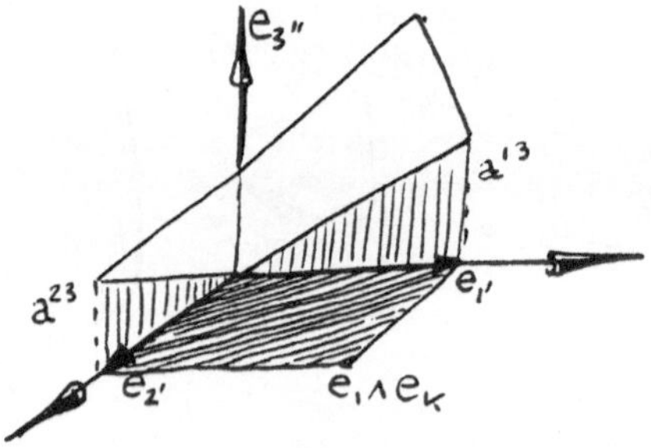

As previously, any other $n \times (n-k)$ matrix is related to some matrix of the shape (2.13):

$$Z = \left[\frac{K}{A}\right] \to \left[\frac{K}{A}\right] \cdot [K^{-1}] = \left[\begin{array}{c} 1 \quad\quad \\ \quad 1 \quad \\ \quad\quad 1 \\ \quad\quad\quad \ddots \\ \quad\quad\quad\quad 1 \\ \hline A\, K^{-1} \end{array}\right] = Z' \tag{2.14}$$

The k(n-k) coefficients of the submatrix $A \cdot K^{-1}$ are then the natural coordinates on an open neighborhood of G_k^n with the clear geometrical meaning.

It will appear important for the analysis of the model that each element of the matrix $A \cdot K^{-1}$ can be expressed in terms of the subdeterminants of Z. Let i',j' range from 1 to k, i",j" = k+1,...,n.

Let $\Omega_k = e_1 \wedge e_2 \wedge \dots \wedge e_k$ be a k-vector built from the first k basis vectors. Any simple k-vector determines k-dimensional plane in E, and thus - an element of G^n_k. A plane "moved" from position $_k$ can be described by some $\Omega = \Omega_k + \Omega'$ where $\Omega' = a^{i'j''} e_1 \wedge e_{i'-1} \wedge e_{j''} \wedge e_{i'+1} \wedge \dots \wedge e_k$ with some coefficients $a^{i'j''}$ being, in fact, elements of $A \circ K^{-1}$. The k-vector Ω is simple and thus spans some k-plane in E. To get the coefficient $a^{i'j''}$ from Ω one can just apply a k-form $\varepsilon^{i'j''} =: \varepsilon^1 \wedge \dots \varepsilon^{i'-1} \wedge \varepsilon^{j''} \wedge \varepsilon^{i'+1} \wedge \dots \wedge \varepsilon^k$. It holds $\varepsilon^{i'j''}(\Omega) = \varepsilon^{i'j''}(\Omega') = a^{i'j''}$. But the value of the polivector in the external form becomes in coordinates a determinant. This is why

$$a^{i'j''} = Z' \begin{pmatrix} 1,2,\dots & ,k \\ 1,\dots,i'-1,j'',i'+1,\dots,k \end{pmatrix}, \tag{2.15}$$

where the right side designs the subdeterminant made up from the elements in cross of the columns 1,...,k and the rows 1,...,i'-1,j",i'+1,...,k.

The matrix Z has the meaning of the same k-vector Ω (and thus of the same element of G^n_k) but with basic k-vector Ω_k expressed by $e'_1,\dots,e'_k$ "turned" in the plane of Ω_k by matrix K. This is generally why

$$a^{i'j''} = \frac{Z \begin{pmatrix} 1,2,\dots & ,k \\ 1,\dots,i'-1,j'',i'+1,\dots,k \end{pmatrix}}{\det k} . \tag{2.16}$$

b) <u>Tissue $\mathcal{O}_k$</u>. The same picture works in the case of $\mathcal{O}_k$, containing two Grassmanians as factors of the projection (2.8).

Let $\{e_\alpha\} \subset \mathbb{C}^n$ and $\{f_A\} \subset \mathbb{C}^m$ be the pair of bases, and $\{\varepsilon^\alpha\}$ and $\{\phi^A\}$- dual bases. Then $\{f_A\} \times \varepsilon^\alpha\}$ with $\alpha = 1,\dots,n$, and $A = 1,\dots,m$ is an induced basis in $U = \mathrm{Hom}(\mathbb{C}^n,\mathbb{C}^m)$, and any $\xi \in U$ "becomes" a matrix:

$$\xi = \xi^A_\alpha f_A \times \varepsilon^\alpha \tag{2.17}$$

Let $\xi \in \mathcal{O}_k^{(m,n)}$ such that $\xi_*(e_1 \wedge \dots \wedge e_k) \neq 0$. (2.18)

(ξ_* is the induced map from k-vectors in $\mathbb{C}^n$ to k-vectors in $\mathbb{C}^m$). Then ξ as a matrix can be described according to the decomposition (2.2) and to (2.13), as follows

$$\left[\begin{array}{c|c} \begin{matrix}1 & & \\ & 1 & \\ & & \ddots \; 1\end{matrix} & 0 \\ \hline A & 0 \end{array}\right] \cdot \left[\begin{array}{c|c} k & 0 \\ \hline 0 & 0 \end{array}\right] \cdot \left[\begin{array}{c|c} \begin{matrix}1 & & \\ & 1 & \\ & & \ddots \; 1\end{matrix} & B \\ \hline 0 & 0 \end{array}\right] = \left[\begin{array}{c|c} K & KB \\ \hline AK & AKB \end{array}\right] \tag{2.19}$$

So, now starting with a general matrix,

$$[\xi] = \left[\begin{array}{c|c} K & B \\ \hline A & Y \end{array}\right] \tag{2.20}$$

(where K is a $k \times k$ matrix, i.e., $K \in M_{k,k}$, $A \in M_{n-k,k}$, $B \in M_{k,m-k}$, $Y \in M_{n-k,m-k}$), it belongs to the considered neighborhood (2.18) of $\mathcal{O}_k$ if $\det K \neq 0$ and $Y = AK^{-1}B$.

Define $\mathcal{A} = AK^{-1}$ and $\mathcal{B} = K^{-1}B$. By analogy to (2.15) one can introduce an equivalence relation

$$\left[\begin{array}{c|c} K & B \\ \hline A & Y \end{array}\right] \qquad \left[\begin{array}{c|c} K & B \\ \hline A & Y \end{array}\right] \cdot \left[\begin{array}{c|c} K^{-1} & 0 \\ \hline 0 & 0 \end{array}\right] = \left[\begin{array}{c|c} \begin{matrix} 1 & & \\ & \ddots & \\ & & 1 \end{matrix} & 0 \\ \hline AK^{-1} & 0 \end{array}\right] \tag{2.21}$$

and interpret this as a projection π_1 (2.8)

$$\pi_1 : \mathcal{O}_k^{(n,m)} \mapsto G_{n-k}^n$$

expressed in the considered coordinates. Similarly the "left" equivalence relation

$$\left[\begin{array}{c|c} K & B \\ \hline A & Y \end{array}\right] \qquad \left[\begin{array}{c|c} K^{-1} & 0 \\ \hline 0 & 0 \end{array}\right] \cdot \left[\begin{array}{c|c} K & B \\ \hline A & Y \end{array}\right] = \left[\begin{array}{c|c} \begin{matrix} 1 & & \\ & \ddots & \\ & & 1 \end{matrix} & K^{-1}B \\ \hline 0 & 0 \end{array}\right]$$

corresponds to the second projection, π_2

$$\pi_2 : \mathcal{O}_k^{(n,m)} \to G_k^m .$$

Repeating the considerations (2.14) the coordinates of the image $\pi_1(\xi)$, i.e. the elements of the matrix AK^{-1}, can be expressed in terms of proper determinants:

$$\mathcal{A} =: AK^{-1}, \; \mathcal{A}^{a'}_{a''} = \xi^{A'}_{a''}(K^{-1})^{a'}_{A'} = (\det K)^{-1} \cdot \xi \begin{pmatrix} 1, \ldots & , k \\ 1, \ldots, a'-1, a'', a'+1, \ldots, k \end{pmatrix} \tag{2.22}$$

Similarly for the projection π_2 on the second Grassmanian:

$$\mathcal{B} := K^{-1}B, \; \mathcal{B}^{A''}_{A'} = (K^{-1})^{a'}_{A'} \xi^{A''}_{a'} = (\det K)^{-1} \cdot \xi \begin{pmatrix} 1, \ldots, A'-1, A'', A'+1, \ldots, k \\ 1, \ldots, & \ldots, k \end{pmatrix} \tag{2.23}$$

As before we assume $A', a' = 1, \ldots, k$, $a'' = k+1, \ldots, n$, $A'' = k+1, \ldots, m$. So that $K = \{\xi^{A'}_{a'}\}$, $A = \{\xi^{A'}_{a''}\}$, $B = \{\xi^{A''}_{a'}\}$, $Y = \{\xi^{A''}_{a''}\}$. The projections expressed in the above coordinates are restricted to the neighborhood (2.17). However there are $\binom{n}{k}\binom{m}{k}$ maps consisting of neighborhoods

$$\{\xi \in \mathcal{O}_k : \; \xi \begin{pmatrix} A_1, \ldots, A_k \\ a_1, \ldots, a_k \end{pmatrix} \} \tag{2.24}$$

for each choice of $a_1, \ldots, a_k$ and $A_1, \ldots, A_k$, and of the analogous to (2.18) coordinates, constituting an atlas for $\mathcal{O}_k^{(n,m)}$.

We would like to remind that the projections (2.7) and (2.8) are independent of the coordinates defined by (2.16). Moreover obviously coordinates of A are invariant with respect to the second factor $GL(m,\mathbb{C})$ of the group (2.6) and transform among themselves with respect to the first, $GL(n,\mathbb{C})$.

In the particular case of $4 \times m$ matrices of $O_2^{4,m}$ seem to coincide with the "centre of mass" problem described in [4].

3. REDUCTION OF THE SYMMETRY

Now let us get into further surgery of the tissues O_k. Till now they have only been fibered over $G^n_{n-k} \times G^m_k$ (2.8). Reduction of symmetry means that some additional structure has been introduced, which may further decompose the space U. The dual thinking of this is in terms of reduction of the group of symmetry acting on U (2.6) to some subgroup $H \times G$ of $GL(n,\mathbb{C}) \times GL(m,\mathbb{C})$. The resulting decomposition is connected with an appearance of invariants of $H \times G$.

The slightest reduction $GL(n,\mathbb{C}) \times GL(m,\mathbb{C}) \to SL(n,\mathbb{C}) \times SL(m,\mathbb{C})$ does not produce new invariants unless $k = n = m$. In this case the orbit O_k splits into a set of manifolds each of the dimensions n^2-1.

A. <u>The $SU(n-p,p) \times SU(m-q,q)$ Case</u>. Physical motivations described in Section 1 lead to the concept of universal space U built up from the pair of spaces $\{(\mathbb{C}^{2,2},\mathbf{f}), (\mathbb{C}^m,\mathbf{h})\}$ where $\mathbf{f}$ is a hermitian structure in $\mathbb{C}^4$ with the signature (++--), and $\mathbf{h}$ - hermitian product in $\mathbb{C}^m$. Keeping this in mind let us first consider a general case

$$U = \mathrm{Hom}\,(\mathbb{C}^{n-p,p},\ \mathbb{C}^{m-q,q}) \tag{3.1}$$

where $\mathbb{C}^{n-p,p} \cong \{\mathbb{C}^n,\mathbf{f}\}$,

$\mathbf{f}$ - hermitian product of sgn $(\underbrace{++\ldots+}_{n-p},\ \underbrace{-\ \ldots\ -}_{p})$

and $\mathbb{C}^{m-q,q} \cong \{\mathbb{C}^m,\mathbf{h}\}$ with hermitian structure $\mathbf{h}$ of sgn $(\underbrace{+\ldots+}_{n-q},\ \underbrace{-\ldots-}_{q})$.

In the group theory language (2.10) this corresponds to the reduction of the group (2.6) to

$$SU(n,n-p) \times SU(m,m-p) \tag{3.2}$$

The space U has an induced hermitian scalar product equal to the tensor product $\mathbf{g} \otimes \mathbf{h}^*$. Its value on $\xi \in U$ has in coordinates the form

$$J_1(\xi) = \bar{\xi}^A_a\, \mathbf{f}^{\dot{a}b}\, \mathbf{h}_{AB}\, \xi^B_b \tag{3.3}$$

and is the first of a family of invariants of the group(3.2):

$$J_i(\xi) = \xi^* \begin{pmatrix} A_1,\ldots,A_i \\ a_1,\ldots,a_i \end{pmatrix} \mathbf{f}^{\dot{a}_1 b_1} \ldots \mathbf{f}^{\dot{a}_i b_i}\, \mathbf{h}_{\dot{A}_1 B_1} \ldots \mathbf{h}_{\dot{A}_i B_i}\, \xi \begin{pmatrix} B_1,\ldots,B_i \\ b_1,\ldots,b_i \end{pmatrix} \tag{3.4}$$

The geometrical meaning of them is as follows. A map ξ induces the unique "pull back" ξ^* on dual spaces $\xi^*: (\mathbb{C}^m)^* \to (\mathbb{C}^n)^*$. Since the products $\mathbf{f}$ and

$\mathbf{h}$ can be considered as maps $\mathbf{f}\colon (\mathbb{C}^n) \to (\mathbb{C}^n)^*$ and $\mathbf{h}\colon \mathbb{C}^m \to (\mathbb{C}^m)^*$ so for a given ξ we can construct a not commuting diagram of maps:

$$\begin{array}{ccc} (\mathbb{C}^n)^* & \xleftarrow{\ \xi^*\ } & (\mathbb{C}^m)^* \\ \uparrow \mathbf{f} & & \uparrow \mathbf{h} \\ \mathbb{C}^n & \xrightarrow{\ \xi\ } & \mathbb{C}^m \end{array} \tag{3.5}$$

The one turn around, $\mathbf{f}^{-1} \circ \xi^* \circ \mathbf{h} \circ \xi$ is an automorphism of $\mathbb{C}^n$ and its determinant corresponds to $J_1(\xi)$. Extending ξ to the map from k-vectors in $\mathbb{C}^n$ to k-vectors in $\mathbb{C}^m$ the not commuting diagram

$$\begin{array}{ccc} \Lambda^k(\mathbb{C}^n)^* & \xleftarrow{\ \xi^*\ } & \Lambda^k(\mathbb{C}^m)^* \\ \uparrow \mathbf{f} & & \uparrow \mathbf{h} \\ \Lambda^k\mathbb{C}^n & \xrightarrow{\ \xi\ } & \Lambda^k\mathbb{C}^n \end{array}$$

gives invariants $J_i(\xi)$ of the group (3.2). The invariant $J_i(\xi)$ says how much the map ξ "extends" the elements of k-volume. So $J_i(\xi)$ considered on $\mathcal{O}_k^{(n,m)}$ is different from zero only for first k this invariants. Subsets of $\mathcal{O}_k$ given by the k equations

$$J_i(\xi) = c_i, \qquad c_i \in \mathbb{R}, \qquad i = 1,\dots,k, \tag{3.6}$$

form a k-parametric family of submanifolds (of rather complex topological properties).

The tissue decomposition $U = \bigcup_k \mathcal{O}_k$ enriched by (3.6) for each $\mathcal{O}_k$, provides the decomposition of U into orbits of the group (3.2).

There is another equivalent way of constructing of the invariants of (3.2). According to the diagram (3.5) of maps we have

$$\begin{aligned} &\mathbf{f}^{-1} \circ \xi^* \circ \mathbf{h} \circ \xi \in \operatorname{Aut} \mathbb{C}^n \\ \text{and} \\ &\xi \circ \mathbf{f}^{-1} \circ \xi^* \circ \mathbf{h} \in \operatorname{Aut} \mathbb{C}^m \end{aligned} \tag{3.7}$$

As automorphisms they have their eigenvectors, and their eigenvalues are solutions to the algebraic equations

$$\operatorname{Det} \{r_{\dot\alpha\beta} - \lambda \mathbf{f}_{\dot\alpha\beta}\} = 0, \quad \operatorname{Det} \{s^{AB} - \lambda h^{AB}\} = 0, \tag{3.8}$$

where

$$r =: \xi^* \circ \mathbf{h} \circ \xi \qquad \text{i.e.} \qquad r_{\dot a b} = \bar\xi^A_a\, \mathbf{h}_{AB}\, \xi^B_b$$

$$s =: \xi \circ \mathbf{f}^{-1} \circ \xi^* \qquad \text{i.e.} \qquad s^{AB} = \xi^A_a\, \mathbf{f}^{ab}\, \bar\xi^B_b\,.$$

Each set of λ's (for r, and for s) provides an equivalent set of invariants.

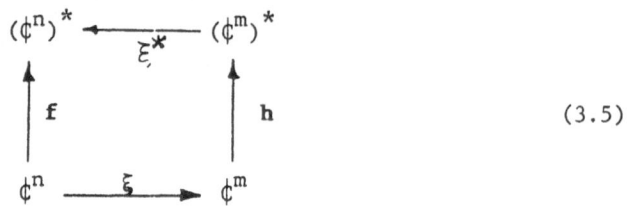

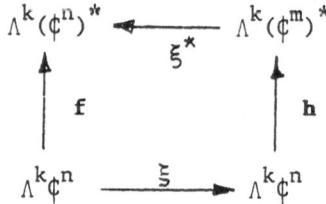

It can be shown that they are proportional to each other, and moreover, that the invariants (3.4) are homogeneous functions of the solutions to (3.8).

B. Geometry of Subspaces of $(\mathbb{C}^n, \mathbf{f})$. Before we get into the case $SU(2,2) \times SU(m)$ let us first study a little more about the general case using now the structure approach rather than the group description. Since now $\mathbb{C}^n$ and $\mathbb{C}^m$ have hermitian structures, the subspaces $\operatorname{Ker}\xi < \mathbb{C}^n$ and $\operatorname{Im}\xi < \mathbb{C}^m$ of the decomposition (2.2) carry the induced structures $\mathbf{f}|_{\operatorname{Ker}\xi}$ and $\mathbf{h}|_{\operatorname{Im}\xi}$. The signature of them is invariant under the action of the group (3.2).

So the first approach to visualize the geometry of the new decomposition of $_k$ leads through the decomposition of the basis of the fibering (2.10) of $_k$.

To do that let us consider the general case: we shall show how to find admissible signatures of k-dimensional subspace of m-dimensional space with given hermitian (or pseudo-euclidean) non-degenerate structure $\mathbf{f}$ with a signature $(m-q,q) = (\underbrace{++\ldots}_{m-q}, \underbrace{--\ldots}_{q})$. It resolves itself into the following algorithm: *)

1. Construct the infinite lattice L of formal words of signs:

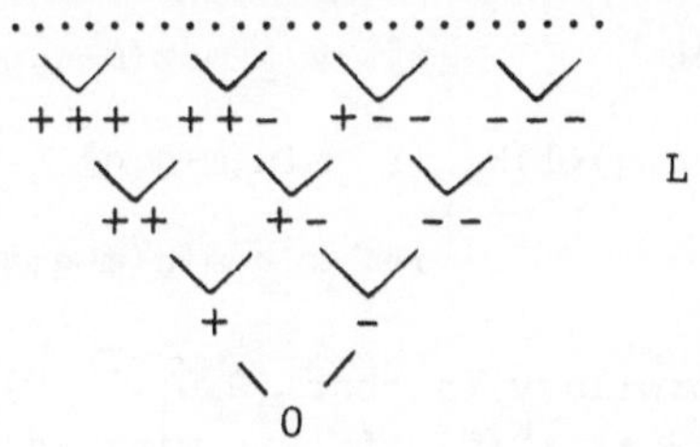

where two elements are joined when one is a reduced form of the other (by cutting off one sign). The lattice is the ordered set with $\prec$.

2. Find the signature of $\mathbf{f}$ of the considered space and reduce L to the finite sublattice $L(\mathbf{f}) =: \{x \in L : x \prec \mathbf{f}\}$. (See Fig. 6).

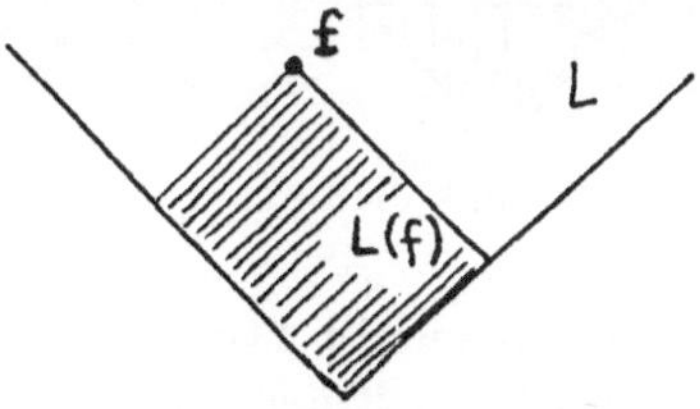

FIGURE 6

3. Find in the lattice $L(\mathbf{f})$ the kth row R^k (consisting of k-sign signatures), and build a sublattice (see Fig. 7)

$$L^k(\mathbf{f}) = \{x \in L(\mathbf{f}) \mid x \prec y \text{ and } y \text{ is k-signature} \Rightarrow y \in R^k\}.$$

*) To be proved elsewhere.

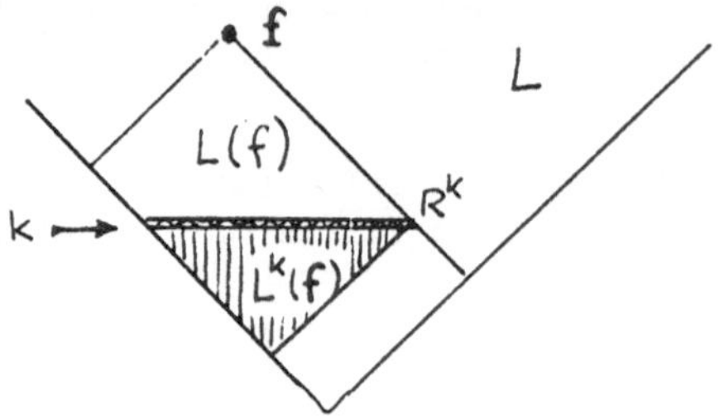

FIGURE 7

4. All elements of $L^k(\mathbf{f})$ filled out with zeros up to k signs form the set of all admissible signatures of the product $\mathbf{f}$ restricted to k-plane. Call this lattice $\mathcal{L}^k(\mathbf{f})$.

The Grassmanian G_k^m splits into some regions of different signatures (and different dimension) and if $\mathbf{f}_1 \prec \mathbf{f}_2$ in the lattice $\mathcal{L}^k(\mathbf{f})$, then the region of k-planes with induced signature $\mathbf{f}_1$ lies in the border of the region with the induced signature $\mathbf{f}_2$.

The number $N\binom{m\ ,k}{m-q,q}$ of pieces of G_k^m of splitting determined by the $\mathbf{f}$ of signature (m-q,q) is

$$N\binom{m\ ,k}{m-q,q} = \begin{cases} \frac{1}{2}(m-k+2)k & \text{if} \quad k \geq \max(m-q,q) \\ \frac{1}{2}(\min(m-q,q)+1)k & \text{if} \quad \min(m-q,q) \leq k \leq \max(m-q,q) \\ \frac{1}{2}(k+1)k & \text{if} \quad k \leq \min(m-q,q). \end{cases} \quad (3.9)$$

The simple corollary is that (0,0,...,0) (corresponding to null k-plane) belongs to $\mathcal{L}^k(\mathbf{f})$ iff $k \leq \min(m-q,q)$ (see Fig. 8).

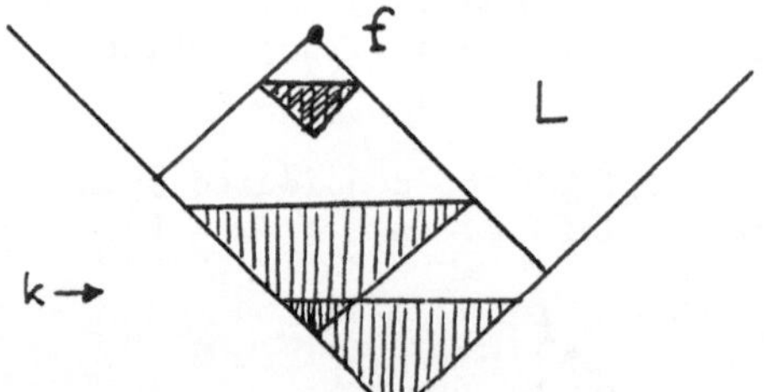

FIGURE 8

The dimension (over the assumed field) of the region $R_{(a,b)}^{(m,-q,q)}$ of of signature (a,b) ["a" pluses, "b" minuses and "k-a-b" zeros] in the Grassmanian G_k^{m-k} is $\dim R_{(a,b)}^{(m-q,q)} = \dim G_k^m - (k-a-b)$.

Let us go back to the tissue $\mathcal{O}_k^{(n,m)}$. According to (2.8) $\mathcal{O}_k$ is fibered over $G_{n-k}^n \times G_k^m$. Each Grassmanian is now spitted into $N_1 = N\binom{n,\ k}{n-q,q}$ and $N_2 = N\binom{m,\ k}{m-p,p}$ pieces respectively, so $G_1 = A_1 \cup \ldots \cup A_{N_1}$ and $G_2 = B_1 \cup \ldots \cup B_{N_2}$, and then

$$G^n_{n-k} \times G^m_k = A_1 \times B \cup A_1 \times B_2 \cup \dots \cup A_i \times B_j \cup \dots \cup A_{N_1} \times B_{N_2}.$$

The inverse of the projection, π^{-1}, decomposes $\mathcal{O}_k$ into chimneys $\pi^{-1}(A_i \times B_j)$ over the basis. Moreover the fibers of $\mathcal{O}_k$ being isomorphic to $GL(k,\mathbb{C})$ are sliced into pieces of the constant determinant $\det[\tilde{\xi}]$ (2.2) (well-defined since $\mathbf{f}$ and $\mathbf{h}$ (3.1) determine the suitable forms of volume). The group $SU(m-q,q) \times SU(m-p,p)$ does not lead points out of the described parts of $\mathcal{O}_k$.

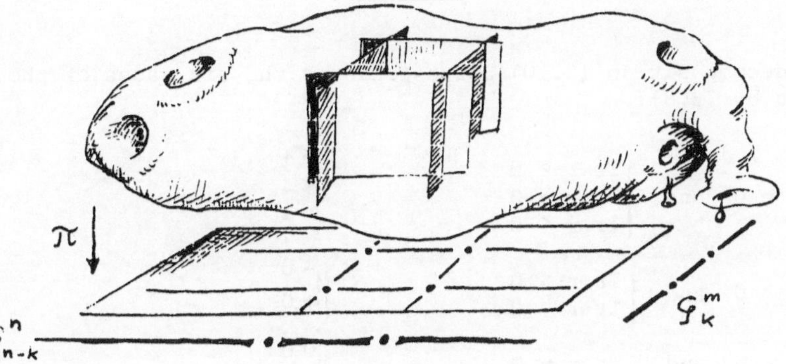

FIGURE 9

Summarizing, the decomposition of the space $U = \mathbb{C}^{nm}$ into orbits of the group (3.2) described already in terms of invariants (3.6) or (3.8) correspond to the superposition of decompositions: into the tissues $\mathcal{O}_k^{(n,m)}$ and into the chimneys $\pi^{-1}(A_i \times B_j)$ sliced by the value of the determinant $\det[\tilde{\xi}]$.

<u>Application to $\mathbb{C}^{2,2}$</u>. In the key case of $\mathbb{C}^{2,2} \cong \{\mathbb{C}^4, \mathbf{f}(++--)\}$ the complex Grassman manifold G^4_2 is homeomorphic to compactified complex Minkowski space $M_{\mathbb{C}}$ [6] (see the end of Sec. 2.D).

According to the lattice $L^2(++--)$:

```
++     +-     --
  \   /  \   /
   +0     -0
     \   /
      00
```

we get the decomposition of $M_{\mathbb{C}} = G^4_2$ into six pieces

$$M_{\mathbb{C}} = M^{++} \cup M^{+-} \cup M^{--} \cup M^{+0} \cup M^{-0} \cup M^{00} \tag{3.10}$$

where complex dimensions are:

$\dim_{\mathbb{C}} M^{++} = \dim M^{+-} = \dim M^{--} = 4$

$\dim M^{-0} = \dim M^{+0} = 3$

$\dim M^{00} = 2.$

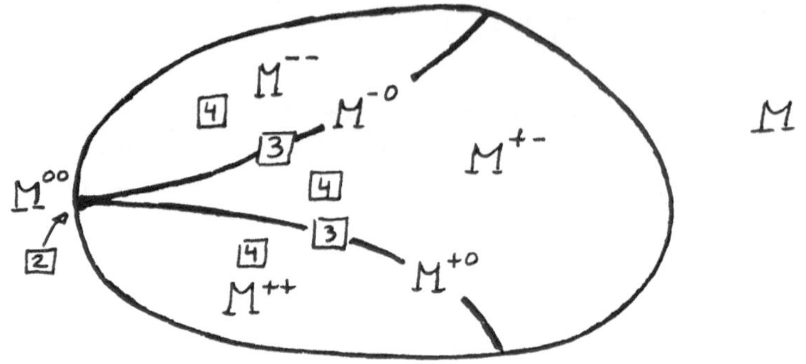

The above decomposition (3.10) corresponds to the six cases of the invariant $r_{\dot{a}b}$ defined by (3.8):

$$\det r = 0 \text{ and } \begin{cases} \operatorname{Tr} r > 0 & \ldots M^{++} \\ \operatorname{Tr} r = 0 & \ldots M^{+-} \\ \operatorname{Tr} r < 0 & \ldots M^{--} \end{cases}$$

$$\det r > 0 \text{ and } \begin{cases} \operatorname{Tr} r < 0 & \ldots M^{-0} \\ \operatorname{Tr} r > 0 & \ldots M^{+0} \end{cases}$$

$$\det r = 0 \text{ and } \operatorname{Tr} r = 0 \qquad \ldots M^{00}$$

Note on the Penrose Model. The Penrose model [3] can be interpreted as a special case of the model described in this paper for $m = 2$.

In such a case elements $[\xi] \in \mathcal{O}_2^{(4,2)}$ are 4×2 complex matrices (two 4-spinors (twistors)). The first projection $\pi_1: \mathcal{O}_2^{(4,2)} \to G_2^4$ onto complex Minkowski space has typical fiber GL(2,$\mathbb{C}$). The second one is trivial $\pi_2: \mathcal{O}_2^{(4,2)} \to G_2^2 \cong \{\mathbb{C}^2\}$.

Penrose's subsets M^0, M^-, M^+ correspond to the above decomposition (3.10)

$$M^0 = M^{00}, \quad M^- = M^{0-} \cup M^{--}, \quad M^+ = M^{++}.$$

(The part M^{+-} is missing.) The null part $M^0 = M^{00}$ has real dimension 4 and corresponds to compactified real Minkowski space $M_{\mathbb{R}} \cong S^3 \times S^1$ ([6], [3]).

C. The SU(2,2) × SU(m) Case. Let us carry out the above classifications with details for the key case of the group SU(2,2) × SU(m) (see 1.11).

There are five orbits of GL(4,$\mathbb{C}$) × GL(m,$\mathbb{C}$) i.e. tissues $\mathcal{O}_k^{(4,m)}$, $k = 0,1,2,3,4$, the first of which is trivial ($\xi = 0$) and the last with no fibering. We are left with the three cases $k = 1,2,3$. In $\mathcal{O}_1$ and $\mathcal{O}_3$ the projections are on three-dimensional complex spaces and cannot contain the complex Minkowski space. The dimensions of manifolds

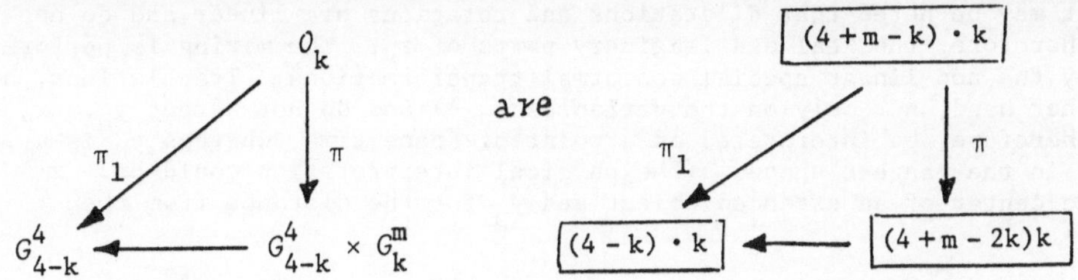

and the dimensions of G^4_{4-k} are 0,3,4,3,0 for $k = 0,\ldots,4$. Thus only the case of $O_2^{(4,m)}$ leaves the hope for constructing the required projection.

In the case of $O_2^{(4,m)}$ the variables $A^{a'}_{a''}$ form a 2×2 complex matrix. It can be shown that the related coordinates

$$z_\mu = x_\mu + iy_\mu = \frac{i\lambda}{2}(\sigma_\mu)^{a''}_{a'} A^{a'}_{a''} \tag{3.11}$$

as functions of the variables $\xi^{A'}_\alpha$, and $\xi^{A''}_\alpha$, have all the necessary properties of a projection onto Minkowski space M for all $m \geq 2$.

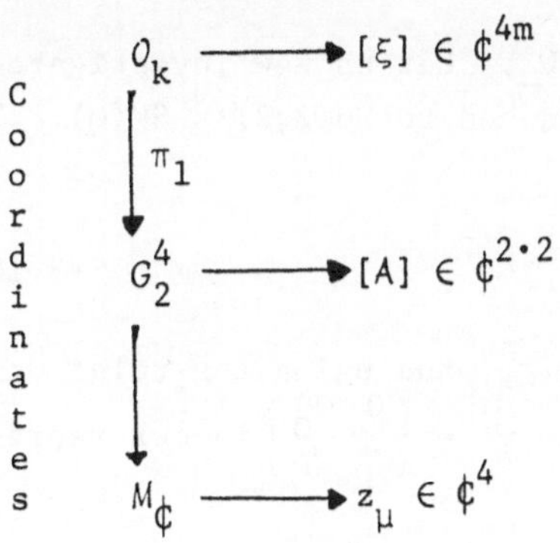

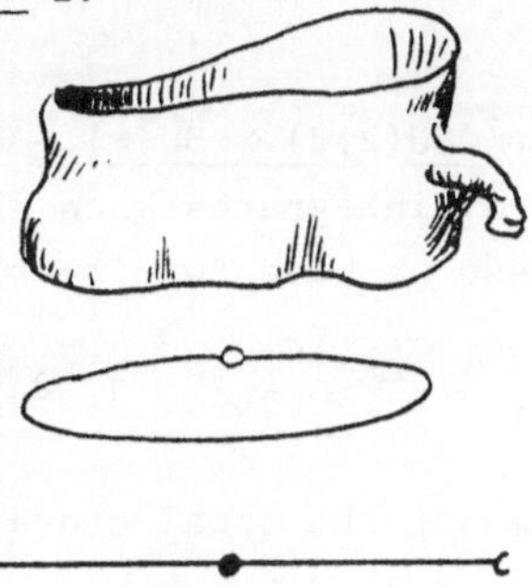

FIGURE 11

Indeed, infinitesimal $SU(2,2) \times \mathbf{1}$ transformations of the ξ given by the generators

$$\begin{aligned} d &= -\frac{i}{2}\gamma_5, \quad p_\mu = i\lambda^{-1}\gamma_-\gamma_\mu, \quad k_\mu = -i\lambda\gamma_+\gamma_\mu \\ m_{\mu\nu} &= \frac{i}{4}[\gamma_\mu,\gamma_\nu], \quad (\gamma_\pm = \frac{1}{2}(1 \pm \gamma_5)) \end{aligned} \tag{3.12}$$

acting on the upper index (rows) of the matrix $\xi = \{\xi^A_\alpha\}$ induce conformal transformations of the z_μ:

$$\begin{aligned} dz_\mu &= -iz_\mu, \quad k_\lambda z_\mu = -ig_{\lambda\mu}z^2 - 2iz_\lambda z_\mu, \\ p_\lambda z_\mu &= -ig_{\lambda\mu}, \quad m_{\lambda\rho}z_\mu = -ig_{\lambda\mu}z_\rho + ig_{\rho\mu}z_\lambda. \end{aligned} \tag{3.13}$$

In (3.11) and (3.12) λ is an arbitrary constant with dimension of length, γ_μ are Dirac's 4×4 matrices, d, p_μ, k_μ, $m_{\mu\nu}$, are the generators of dilatations, translations, special conformal transformations and rotations respectively.

According to the general results of Section 2 the z_μ are invariant with respect to $\mathbf{1} \times GL(m,¢)$ and domain independent.

It may be noted that dilatations and rotations are linear and do not mix, therefore, the real and imaginary parts of z_μ. The mixing is performed only by the non-linear special conformal transformations. Translations, on the other hand, set only on the variables x_μ *) and do not affect y_μ. x_μ can, therefore, be interpreted as a point of space time, whereas y_μ is a vector in the tangent space. (The physical interpretation could be: x_μ for the center of an extended object and y_μ for the distance from the center).

The second rank determinants appearing in the numerator $A^{A''}_{B'}$ in (3.11) (cf. also (2.72) specialized for $\mathcal{O}_2^{(4,m)}$) after multiplication with $(\sigma_\mu)^{A''}_{A'}$ become just proportional to the $\mathcal{O}_\mu$ of (1.6) and η_a any pair of ξ^A_a, ξ^B_b from the set $\{\xi^A_\alpha\}_{\alpha=1,\dots,m}$ (domain independent). Formulae (1.5) and (3.11) are then, of course, valid on the corresponding neighborhood $\xi\binom{A,B}{1,2} = \frac{1}{2}(s+p)$. Generally formulae (1.5) and (3.11) are valid on the cross section of all the $\binom{m}{2}$ neighborhoods $\xi\binom{\alpha,\beta}{1,2} \neq 0$. The representation of the γ_μ matrices is determined by the representation of the σ_μ matrices in (3.11)

$$\gamma_\mu = i\begin{pmatrix} 0 & \sigma_\mu \\ -\tilde{\sigma}_\mu & 0 \end{pmatrix}, \quad \gamma^{ab}_\mu = \varepsilon^{ac}(\gamma_\mu)^b_c, \quad \varepsilon = \begin{pmatrix} \varepsilon & 0 \\ 0 & \varepsilon \end{pmatrix} \tag{3.14}$$

D. <u>SU(2,2) × SU(m) - Decomposition of $\mathcal{O}^k$</u>. Let us now investigate how $\mathcal{O}_2^{(4,m)}$ disintegrates when the symmetry is reduced to SU(2,2) × SU(m). In this case we have two invariants

$$J_1 = r^{\dot{A}A} \quad \text{and} \quad J_2 = 2(J_1^2 - r^{\dot{A}B}\, r_{\dot{B}A}). \tag{3.15}$$

In terms of the local coordinates x_μ, y_μ and ξ^A_a, and using the relation inverse to (3.11) we obtain successively ($f = \gamma_\mu = \begin{pmatrix} 0 & \sigma_0 \\ \sigma_0 & 0 \end{pmatrix}$ in our representation (3.14))

$$r^{\dot{A}B} = \xi^{*A}_a\, f^{\dot{a}b}\, \xi^B_b = A^{*a'}_{\ a}\, \xi^{*A}_{a'}\, f^{\dot{a}b}\, \xi^B_{b'} = 2\lambda^{-1}\, y^\mu\, r^{\dot{A}B}_\mu \tag{3.16}$$

where

$$r^{\dot{A}B}_\mu = -\xi^{*A}_{a'}(\tilde{\sigma}_\mu)^{\dot{a}'b'}\, \xi^B_{b'} \tag{3.17}$$

$(\tilde{\sigma}_0 = -\sigma_0,\ \tilde{\sigma}_i = \sigma_i)$. It follows that

$$r^{\dot{A}A} = 2\lambda^{-1}\, y^\mu\, r_\mu, \quad r^{\dot{A}B}\, r_{\dot{B}A} = 4\lambda^{-1} y^\mu y^\nu r_{\mu\nu} \tag{3.18}$$

where

$$r_\mu = r^{\dot{A}A}_\mu, \quad r_{\mu\nu} = r^{\dot{A}B}_\mu\, r_{\nu\dot{B}A} = -\frac{1}{2}g_{\mu\nu} r_\lambda r^\lambda + r_\mu r_\nu. \tag{3.19}$$

Introducing the second equation (3.14) into the second relation (3.13) one obtains

---

*)From hermiticity of the generator (4.1) it follows that $X_k(\xi_a\gamma^{ab}\xi_b)^* = -(X_k\, \xi_a\, \gamma^{ab}\xi_b)^*$ and consequently, $P_\lambda x_\mu = -i\, g_{\lambda\mu}$, $P_\lambda y_\mu = 0$.

$$r^{\dot{A}B}\, r_{\dot{B}A} = 4\lambda^{-2}\{(r_\mu y^\mu)^2 - \tfrac{1}{2} r_\lambda r^\lambda y_\mu y^\mu\} \tag{3.20}$$

We can use, therefore, instead of (3.10), the two invariants $r_\mu y^\mu$ and $r_\mu r^\mu y_\nu y^\nu$ and describe the submanifolds of $\mathcal{O}_2^{(4,m)}$ by the two equations

$$r_\mu y^\mu = -c_{12} \qquad r_\mu r^\mu y_\nu y^\nu = c \tag{3.21}$$

where c_{12} and c are arbitrary real constants. The coordinates x_μ do not appear in the invariants (3.16). This is, of course, a consequence of the fact that they are not translationally invariant.

Let us note that the form of conditions (3.21) is independent on m. The dimension of the interval group enters only in the defining equation for r_μ (cf. (3.19), (3.17)). Another important fact is that r_μ is necessarily time like. This is a consequence of the SU(m) symmetry which does not admit negative terms in the sum $r_\mu = r_\mu^{\dot{A}A}$ (cf. equations (1.4)).

The projection $r_\mu^{\dot{A}A} = -\xi^{*A}_{a'}(\sigma_\mu)^{\dot{a}'b'}\xi^B_{b'}$ (cf. (3.17), (3.18)) is for the case m = 2 (A = 1,2) the same as the first projection on the future light cone in (1.4) studied in [2]. It is seen that the vector r_μ plays here a different role to that assumed in (1.4). It reminds (after normalization) of the time-like unit vector n_μ used frequently to bring nonrelativistic equations (like the equation of a sphere) into covariant form (cf. the first two equations (3.21)). E.g. the conditions (3.16) in a coordinate system in which $r_i = 0$ are just the conditions for the constancy of the product of two spheres $r_0^2 \vec{y}^2 = c - c_{12}^2$, r_0 being the radius of S^{4m-1} in the space $\mathbb{C}^{2m} = \mathbb{R}^{4m}$ of the variables $\xi^A_{a'}$, $\xi^{*A}_{a'}$ (cf. definition of r_0 in (3.13) and (3.12)).

Equations (3.21) describe two surfaces in the eight-dimensional real space of the variables r_μ, y_μ. Introducing the variables

$$\begin{gathered} y = \sqrt{y_1^2 + y_2^2 + y_3^2}, \qquad r = \sqrt{r_1^2 + r_2^2 + r_3^2} \\ \cos\theta = (yr)^{-1}(y_1 r_1 + y_2 r_2 + y_3 r_3) \\ c_{22} = -r_\mu r^\mu \geq 0 \end{gathered} \tag{3.22}$$

we can reduce the problem to the two equations

$$yr\cos\theta - y_0 r_0 = -c_{12}, \qquad (y^2 - y_0^2)(r^2 - r_0^2) = c \tag{3.23}$$

for the five variables y, r, y_0, r_0, θ, and express any two of them by the remaining three. Introducing the notation

$$y^2 - y_0^2 = -c_{11} \qquad r^2 - r_0^2 = -c_{22} \qquad (c_{22} \geq 0) \tag{3.24}$$

we have e.g.

$$y = \frac{c_{12} r\cos\theta \pm \sqrt{(r^2 + c_{22})[c_{12}^2 - c_{11}(c_{22} + r^2\sin^2\theta)]}}{c_{22} + r^2\sin^2\theta} \tag{3.25}$$

or

$$\cos\theta = \frac{-c_{12} \pm y_0 r_0}{yr} \tag{3.26}$$

where in (3.20) the quantities c_{11} and c_{22} are expressed in terms of r and r_0: $c_{11} = -\frac{c}{r^2 - r_0^2}$, $c_{22} = r_0^2 - r^2$ so that $y = y(r, r_0, \theta)$ and in (3.21) any of the four variables y, r, y_0, r_0 can be expressed in terms of the other three by means of equation (3.23).

E. $P \times SU(m)$ - Decomposition of $\mathcal{O}_2^{(4,m)}$. If we further reduce the symmetry to $P \times SU(m)$, another invariant appears, namely $r_\mu r^\mu$. In this case both quantities c_{11} and c_{22} become constant and we have three $P \times SU(m)$-invariant equations for the five variables y, r, y_0, r_0,

$$\begin{aligned} y_\mu r^\mu &= yr\cos\theta - y_0 r_0 = -c_{12}, \quad y_\mu y^\mu = y^2 - y_0^2 = -c_{11} \\ r_\mu r^\mu &= r^2 - r_0^2 = -c_{22} \quad (c_{22} \geq 0). \end{aligned} \tag{3.27}$$

Solution (3.25) depends now on 2 variables r and θ and solution (3.26) on say, y and r.

It is seen from (3.25) that the condition for the two surfaces (3.23) to intersect is

$$c_{12}^2 - c \geq c\,\frac{r^2 \sin^2\theta}{c_{22}} \tag{3.28}$$

which is at the same time a condition for the positive quantity $c_{22}^{-1} r^2 \sin^2\theta$ and for the constants c_{12} and c

$$c_{12}^2 - c \geq 0 \tag{3.29}$$

If $c < 0$, both conditions are automatically satisfied. For the case of Poincaré invariance condition (3.28) takes the form

$$\det c \leq 0 \tag{3.30}$$

and condition (3.27) remains valid with a constant c_{22} $(c = c_{11}c_{22})$.

In the space of the variables y_μ the surface can be described as the intersection of a space-like plane, perpendicular to the time-like vector r_μ, and a hyperboloid. These surfaces always intersect if the hyperboloid is time-like (outside of the light cone, $c < 0$). If the hyperboloid is space-like (inside the light cone, $c > 0$) condition (3.27) must be satisfied for the two equations (3.23) to have common solutions.

The situation which presents itself is now the following: For the $P \times SU(m)$ symmetry the orbit $\mathcal{O}_2^{(4,m)}$ decomposes into Minkowski space with the coordinates x_μ and the manifolds with the coordinates y_μ and ξ_a^A, restricted by the three conditions (3.26). For the larger symmetry $SU(2,2) \times SU(m)$ Minkowski space does not split off because the x_μ and y_μ transform into each other with respect to special conformal transformations according to (3.13).

4. REPRESENTATIONS IN HILBERT SPACE

At the beginning of this paper we have emphasized that the representations of the rotation group in the Hilbert space of functions over S^2 are not complete because there appear only even values of angular momentum. On the other hand, the representations in the Hilbert space of functions of S^3 in the spinor space are complete. A similar situation presents itself now. If we would consider the representations in the Hilbert space of functions over Minkowski space we would obviously obtain integral spin only and no interval quantum numbers at all. Considering, however, that x is a projection from a larger space we are led in a natural way to representations in the Hilbert space of functions over this space.

Let us first consider representations of $GL(n,\mathbb{C}) \times GL(m,\mathbb{C})$ in the Hilbert space of functions over the whole $\mathbb{C}^{nm}$:

$$X_k^{(1)} = -\frac{\partial}{\partial\xi_a^A}(X_k^{(1)})_a^b\,\xi_b^A + \text{h.c.}\quad X_k^{(2)} = -\frac{\partial}{\partial\xi_a^A}(X_k^{(2)})_B^A\,\xi_a^B + \text{h.c.} \tag{4.1}$$

where $X_k^{(i)}$, i = 1,2 are generators of the first or second factor of $GL(n,\mathbb{C}) \times GL(m,\mathbb{C})$ resp. and $X_k^{(i)}$, i = 1,2, their Hilbert space counterparts, chosen to be hermitian.

To reduce these representations to $O_2^{(4,m)}$ we use in the second case (of internal group) the coordinate system z_μ, $\xi^A_{a'}$, $\mu = 0,1,2,3$, $a' = 1,2$, $A = 1,\dots,m$, and obtain

$$X_k^{(2)} = -\frac{\partial}{\partial\xi_{a'}^A}(X_k^{(2)})_B^A\,\xi_{a'}^B + \text{h.c.} \tag{4.2}$$

These generators do not contain the variables z_μ because of their invariance with respect to the second factor. It is seen that all representations are obtained in this way.

For the representation of the external group it is convenient to use another coordinate system. Equation (3.11) is invertible for any pair of indices A, B out of 1,...,m. We can, therefore, replace y_μ by r^{AB} which is conformally invariant (cf. (3.10) and (3.11)) and the corresponding derivatives will not appear in the representation of the conformal group. We shall write down here only the generators of the Poincaré group (for a complete treatment cf. [5] and the literature quoted there).

$$P_\mu = -i\frac{\partial}{\partial x^\mu}$$

$$M_{\mu\nu} = i\left(x_\mu\frac{\partial}{\partial x^\nu} - x_\nu\frac{\partial}{\partial x^\mu}\right) - \frac{i}{4}\xi^{A^*}_{a'}[\gamma_\nu^+,\gamma^+]^{a'}_{b'}\frac{\partial}{\partial\xi^{*A*}_{b'}} - \frac{i}{4}\frac{\partial}{\partial\xi^A_{a'}}[\gamma_\mu,\gamma_\nu]^{b'}_{a'}\,\xi^A_{b'}\,. \tag{4.3}$$

For the physical interpretation the Casimirs $P_\mu P^\mu$ and $W_\mu W^\mu$ of the Poincaré group are of importance because their eigenvalues represent the masses and spins occurring in the theory.

It is seen from (4.3) that the generator of translations P_μ contains derivatives with respect to x_μ only and has the same form as in the conventional theory in Minkowski space. This is due to the fact that the variables y_μ and $\xi^A_{a'}$ are translationally invariant and the corresponding generator has, therefore, no components in these directions.

The square of the Pauli-Lubanski vector

$$W_\mu = \frac{1}{2}\,\varepsilon_{\mu\nu\rho\lambda} P^\nu M^{\rho\lambda} \tag{4.4}$$

is

$$W_\mu W^\mu = \frac{1}{2} M_{\nu\lambda} M^{\mu\lambda} \{\delta^\nu_\mu \Box_x - 2\,\frac{\partial^2}{\partial x^\mu \partial x_\nu}\} \tag{4.5}$$

The angular momentum in (4.5) consists of two parts,

$$M_{\mu\nu} = M^{(1)}_{\mu\nu} + M^{(2)}_{\mu\nu} \tag{4.6}$$

where

$$M^{(1)}_{\mu\nu} = i\,(x_\mu \frac{\partial}{\partial x^\nu} - x_\nu \frac{\partial}{\partial x^\mu}) \tag{4.7}$$

One easily checks that the quantity

$$\overset{(1)}{M}_{\nu\lambda} (\delta^\nu_\mu \Box_x - 2\,\frac{\partial^2}{\partial x^\mu \partial x^\nu}) \tag{4.8}$$

is symmetric in λ and μ. The terms $\overset{(1)}{M}_{\nu\lambda} \overset{(2)}{M}{}^{\mu\lambda}$, $\overset{(2)}{M}_{\nu\lambda} \overset{(1)}{M}{}^{\mu\lambda}$ and $\overset{(1)}{M}_{\nu\lambda} \overset{(2)}{M}{}^{\mu\lambda}$ in $M_{\nu\mu} M^{\mu\lambda}$ do not contribute to (4.5). It remains

$$W_\mu W^\mu = \frac{1}{2}\, \overset{(2)}{M}_{\nu\lambda} \overset{(2)}{M}{}^{\mu\lambda} \{\delta^\nu_\mu \Box_x - 2\,\frac{\partial^2}{\partial x^\mu \partial x_\nu} \;. \tag{4.9}$$

In a representation in which $\frac{1}{i}\frac{\partial}{\partial x^\mu} = k_\mu$, $k^2 = -m^2$ and in a coordinate system in which $k_i = 0$ $(i = 1,2,3)$, $k_0 = m$ we obtain finally

$$W_\mu W^\mu = \frac{1}{2}\, m^2\, \overset{(2)}{M}_{ik} \overset{(2)}{M}_{ik} = m^2\, \overset{(2)}{M}_i \overset{(2)}{M}_i , \tag{4.10}$$

with

$$\overset{(2)}{M}_i = \frac{1}{2}\,\varepsilon_{ijk} \overset{(2)}{M}_{jk} = \frac{1}{2}\,\frac{\partial}{\partial \xi^A_{a'}}\,(\sigma_i)^{b'}_{a'}\,\xi^A_{b'} + \text{h.c.} \tag{4.11}$$

$\overset{(2)}{M}_i$ consists of m identical parts $\overset{(2)}{M}{}^A_i$ $(A = 1,\ldots,m)$. The square of each of these parts is just the angular part of the differential operator (1.2)

$$\frac{1}{\sin\theta}\frac{\partial}{\partial\theta}(\sin\theta\,\frac{\partial}{\partial\theta}) + \frac{1}{\sin^2\theta}\frac{\partial^2}{\partial\theta^2} + \frac{\partial^2}{\partial\psi^2} + 2\cos\theta\,\frac{\partial^2}{\partial\phi\partial\psi} \;. \tag{4.12}$$

The eigenfunctions are generalized spherical harmonics [1] and the eigenvalues are $j(j+1)$ with $j = 0,\frac{1}{2},1,\frac{3}{2},\ldots$. It is seen that all spins are represented in the theory.

Consider now, finally, the projection of the representation on the submanifolds of $O_2^{(4.m)}$ in the case of $P \times SU(m)$ invariance (they are described by equations (3.21)). This projection does not affect the variables x_μ and, therefore, also the form of the generator of translations P_μ and the part of the generator of rotations (4.3) which contains derivatives with respect to x_μ. The part of $M_{\mu\nu}$ containing the derivatives with respect to ξ_a^A, and the generators of internal symmetries (4.2) are subject to two conditions: $r_\mu y^\mu = -c_{12}$, $r_\mu r^\mu = -c_{22}$. Since these conditions are invariant with respect to $P \times SU(m)$ they cannot affect the generators of this group.

For the physical interpretation of the internal degrees of freedom it is essential to consider the full Laplace-Beltrami operator on $O_2^{(4,m)}$ and its submanifolds. Attempts in this direction were described in [5] (cf. also the literature quoted in [5]). We believe that this operator, corresponding to the full physical symmetry group, will describe the internal degrees of freedom of elementary particles in an analogous way as the operator (1.2) in the case of SU(2) describes the full spin content, the half integer spin content being connected with the third Eulerian angle ψ.

ACKNOWLEDGEMENTS

One of the authors (J.R.) would like to thank Professor S. Bose and the authorities of Southern Illinois University for the invitation. Special thanks are due to Professor and Mrs. Gruber for their warm hospitality and care extended to me during my stay at Carbondale.

It is a pleasure for the other author (J.K.) to thank Dr. Zbigniew Oziewicz for much encouragement and many fruitful discussions during the joyful years they have worked together.

REFERENCES

[1] N. Ja. Vilenkin, Special functions and the theory of group representtations, Translation of mathematical monographs Vol. 22, American Mathematical Society, 1968.
[2] J. Rzewuski, Acta Phys. Polon. 18:549 (1959);
J. Mozrzymas, J. Rzewuski, Bull. Acad. Polon. Sci. Cl. III. 9:225 (1961).
[3] R. Penrose, Ann. Phys. 10:171 (1960).
[4] L. P. Hughstrone, Twistors and particles in Lecture Notes in Physics, Vol. 97, Springer, 1979.
[5] J. Rzewuski, Reports on Math. Phys.
[6] N. H. Kuiper, Ann. Math. 50:916 (1949).
[7] A. Crumeyrolle, Reports on Math. Phys.

s.d.g. IBM AND THE EMBEDDING METHOD

Yinsheng Ling, Shenxin Dong, and Dehuang Ji

Suzhou University
People's Republic of China

THE SUBALGEBRA CHAINS OF s.d.g. IBM

According to the microscopic calculation, there are some influences of g boson for the deformed nuclei [1]. The dynamic group of s.d.g. IBM is su(15)(A_{14}). The physical subalgebra chains in Lie algebra A_{14} should end with some so(3)(A_1). There are a lot of subalgebras A_1 in A_{14}. In most of the cases, these subalgebras can be distinguished by the embedding index. The embedding index β of A_1 in the simple algebra A_{n-1} can be expressed by [2]

$$\beta = \sum_i \frac{1}{6} K_i (K_i+1)(K_i+2). \tag{1}$$

Here $K_i = 2\ell_i$ (i = 1-s), ℓ_i (i = 1-s) are the angular momentum contents of the defining space of A_{n-1}. In the s.d.g. IBM, the angular momentum contents of the defining space of A_{14} are ℓ = 0,2,4. So that the embedding index of the physical A_1 in A_{14} for the s.d.g. IBM is β = 140.

Using Dynkin-Gruber's embedding method, [3],[4],[5], we can find the physical subalgebra chains in the s.d.g. IBM as follows (Figure 1).

THE GENERATORS OF THE SUBALGEBRA CHAIN $A_{14} \supset A_2^{35}$

For the s-subalgebras of A_{14}, we are only interested in the subalgebra chain $A_{14} \supset A_2^{35}$ (su(15) $\supset$ su(3)). The defining matrix of A_2^{35} in A_{14} is

$$f_{A_2^{35} \subset A_{14}} = \frac{1}{3}\begin{pmatrix} 8 & 5 & 5 & 2 & 2 & 2 & -1 & -1 & -1 & -1 & -4 & -4 & -4 & -4 & -4 \\ -4 & -1 & -4 & 2 & -1 & -4 & 5 & 2 & -1 & -4 & 8 & 5 & 2 & -1 & -4 \\ -4 & -4 & -1 & -4 & -1 & 2 & -4 & -1 & 2 & 5 & -4 & -1 & 2 & 5 & 8 \end{pmatrix}. \tag{2}$$

It is easy to see that the columns of the defining matrix $f_{A_2^{35} \subset A_{14}}$ are the weights of the representation (4,0) of A_2^{35}. The fundamental representation [1] of A_{14} is reduced to the irreducible representation (4,0) of A_2^{35}. Suppose the weight states of the representation (4,0) of A_2^{35} are $a_i^+|0\rangle$ (i = 1-15). The generators of the Cartan space of A_2^{35} are

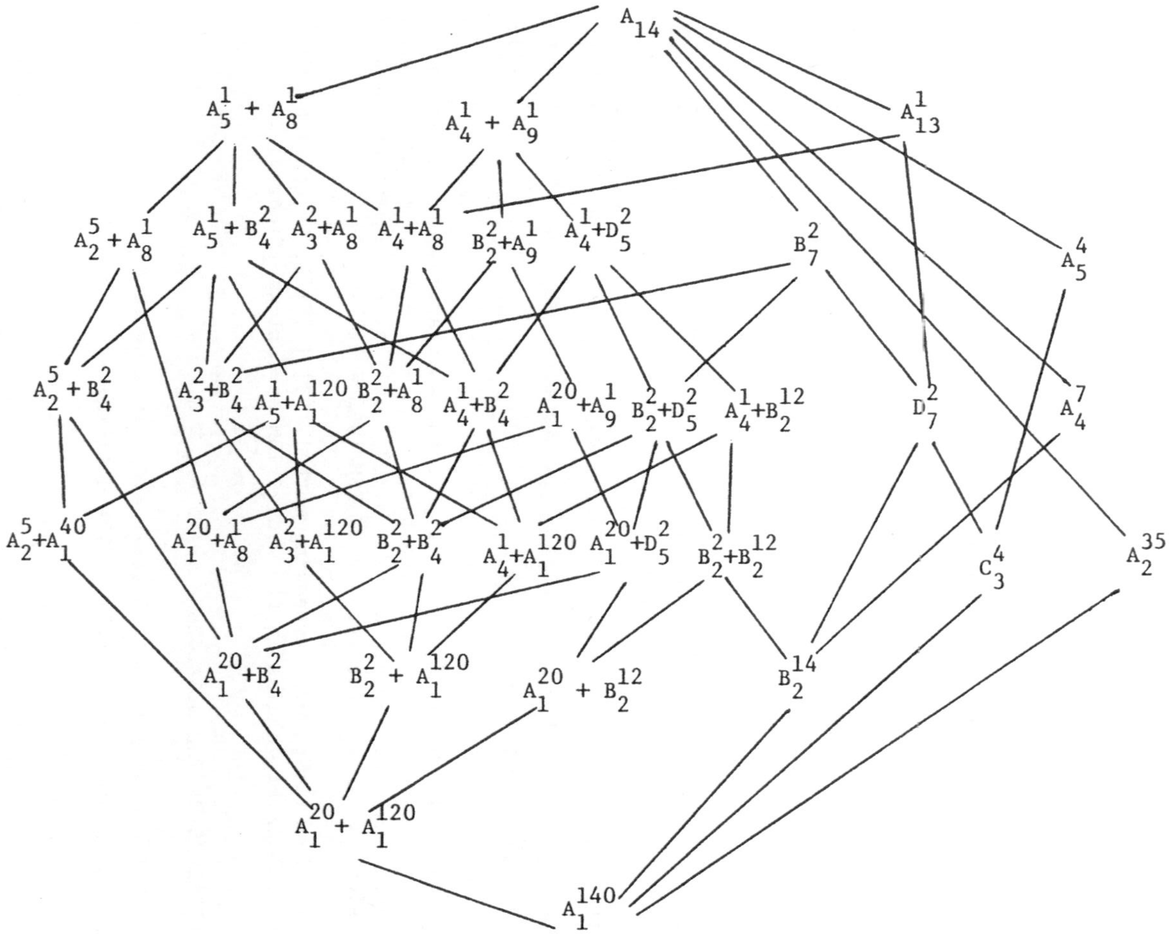

Fig. 1. The physical chains of the s.d.g. IBM.

$$f(H_1') = \frac{1}{3}(8a_1^+a_1+5a_2^+a_2+5a_3^+a_3+2a_4^+a_4+2a_5^+a_5+2a_6^+a_6-a_7^+a_7-a_8^+a_8-a_9^+a_9$$
$$-a_{10}^+a_{10}-4a_{11}^+a_{11}-4a_{12}^+a_{12}-4a_{13}^+a_{13}-4a_{14}^+a_{14}-4a_{15}^+a_{15}$$
$$f(H_2') = \frac{1}{3}(-4a_1^+a_1-a_2^+a_2-4a_3^+a_3+2a_4^+a_4-a_5^+a_5-4a_6^+a_6+5a_7^+a_7+2a_8^+a_8-a_9^+a_9 \qquad (3)$$
$$-4a_{10}^+a_{10}+8a_{11}^+a_{11}+5a_{12}^+a_{12}+2a_{13}^+a_{13}-a_{14}^+a_{14}-4a_{15}^+a_{15}),$$
$$f(H_3') = \frac{1}{3}(-4a_1^+a_1-4a_2^+a_2-a_3^+a_3-4a_4^+a_4-a_5^+a_5+2a_6^+a_6-4a_7^+a_7-a_8^+a_8+2a_9^+a_9$$
$$+5a_{10}^+a_{10}-4a_{11}^+a_{11}-a_{12}^+a_{12}+2a_{13}^+a_{13}+5a_{14}^+a_{14}+8a_{15}^+a_{15}).$$

The shift operators of A_2^{35} are

$$f(E_{12}') = 2a_1^+a_2+\sqrt{6}a_2^+a_4+\sqrt{3}a_3^+a_5+\sqrt{6}a_4^+a_7+2a_5^+a_8+\sqrt{2}a_6^+a_9+2a_7^+a_{11}$$
$$+\sqrt{3}a_8^+a_{12}+\sqrt{2}a_9^+a_{13}+a_{10}^+a_{14},$$
$$F(E_{21}') = 2a_2^+a_1+\sqrt{6}a_4^+a_2+\sqrt{3}a_5^+a_3+\sqrt{6}a_7^+a_4+2a_8^+a_5+\sqrt{2}a_9^+a_6+2a_{11}^+a_7$$
$$+\sqrt{3}a_{12}^+a_8+\sqrt{2}a_{13}^+a_9+a_{14}^+a_{10},$$
$$f(E_{23}') = a_2^+a_3+\sqrt{2}a_4^+a_5+\sqrt{2}a_5^+a_6+\sqrt{3}a_7^+a_8+2a_8^+a_9+\sqrt{3}a_9^+a_{10}+2a_{11}^+a_{12}$$
$$+\sqrt{6}a_{12}^+a_{13}+\sqrt{6}a_{13}^+a_{14}+2a_{14}^+a_{15}, \qquad (4)$$
$$f(E_{32}') = a_3^+a_2+\sqrt{2}a_5^+a_4+\sqrt{2}a_6^+a_5+\sqrt{3}a_8^+a_7+2a_9^+a_8+\sqrt{3}a_{10}^+a_9+2a_{12}^+a_{11}$$
$$+\sqrt{6}a_{13}^+a_{12}+\sqrt{6}a_{14}^+a_{13}+2a_{15}^+a_{14},$$
$$f(E_{13}') = 2a_1^+a_3+\sqrt{3}a_2^+a_5+\sqrt{6}a_3^+a_6+\sqrt{2}a_4^+a_8+2a_5^+a_9+\sqrt{6}a_6^+a_{10}+a_7^+a_{12}$$
$$+\sqrt{2}a_8^+a_{13}+\sqrt{3}a_9^+a_{14}+2a_{10}^+a_{15},$$
$$f(E_{31}') = 2a_3^+a_1+\sqrt{3}a_5^+a_2+\sqrt{6}a_6^+a_3+\sqrt{2}a_8^+a_4+2a_9^+a_5+\sqrt{6}a_{10}^+a_6+a_{12}^+a_7$$
$$+\sqrt{2}a_{13}^+a_8+\sqrt{3}a_{14}^+a_9+2a_{15}^+a_{10}.$$

Let

$$f(H_1') = \frac{1}{2}L_0 + \frac{1}{2\sqrt{3}}Q_0,$$
$$f(H_2') = -\frac{1}{\sqrt{3}}Q_0, \qquad (5)$$
$$f(H_3') = -\frac{1}{2}L_0 + \frac{1}{2\sqrt{3}}Q_0.$$

In this case, $a_i^+|0\rangle$ (i = 1-15) are the common eigenstates of L_0 and Q_0. The operators a_i^+ (i = 1-15) are called the cylindrical bosons [6]

$$D_2^+ = a_1^+, \quad D_1^+ = a_3^+, \quad D_0^+ = a_6^+, \quad D_{-1}^+ = a_{10}^+, \quad D_{-2}^+ = a_{15}^+; \qquad (6)$$
$$\Delta_{3/2}^+ = a_2^+, \quad \Delta_{1/2}^+ = a_5^+, \quad \Delta_{-1/2}^+ = a_9^+, \quad \Delta_{-3/2}^+ = a_{14}^+;$$

$$P_1^+ = a_4^+, \quad P_0^+ = a_8^+, \quad P_{-1}^+ = a_{13}^+;$$

$$\Lambda_{1/2}^+ = a_7^+, \quad \Lambda_{-1/2}^+ = a_{12}^+;$$

$$\bar{s}^+ = a_{11}^+.$$

The s^+, d_m^+, g_m^+ are called the spherical bosons. The relationships between the cylindrical bosons and the spherical bosons are

$$\begin{aligned}
g_4^+ &= D_2^+, & & \\
g_3^+ &= \Delta_{3/2}^+, & & \\
g_2^+ &= \sqrt{1/7}D_1^+ + \sqrt{6/7}P_1^+, & d_2^+ &= -\sqrt{6/7}D_1^+ + \sqrt{1/7}P_1^+, \\
g_1^+ &= \sqrt{3/7}\Delta_{1/2}^+ + \sqrt{4/7}\Lambda_{1/2}^+, & d_1^+ &= -\sqrt{4/7}\Delta_{1/2}^+ + \sqrt{3/7}\Lambda_{1/2}^+ \\
g_0^+ &= \sqrt{3/35}D_0^+ + \sqrt{24/35}P_0^+ + \sqrt{8/35}\,\bar{s}^+, & d_0^+ &= -\sqrt{8/21}D_0^+ - \sqrt{1/21}P_0^+ + \sqrt{4/7}\,\bar{s}^+, \\
g_{-1}^+ &= \sqrt{3/7}\Delta_{-1/2}^+ + \sqrt{4/7}\Lambda_{-1/2}^+, & d_{-1}^+ &= -\sqrt{4/7}\Delta_{-1/2}^+ + \sqrt{3/7}\Lambda_{-1/2}^+ \\
g_{-2}^+ &= \sqrt{1/7}D_{-1}^+ + \sqrt{6/7}P_{-1}^+, & d_{-2}^+ &= -\sqrt{6/7}D_{-1}^+ + \sqrt{1/7}P_{-1}^+, \\
g_{-3}^+ &= \Delta_{-3/2}^+ & s^+ &= -\sqrt{8/15}D_0^+ + \sqrt{4/15}P_0^+ - \sqrt{1/5}\,\bar{s}^+. \\
g_{-4}^+ &= D_{-2}^+. & &
\end{aligned} \tag{7}$$

From the commutation relations

$$\begin{aligned}
[L_q, L_{q'}] &= (-1)^{q+q'+1}\sqrt{6}\begin{pmatrix} 1 & 1 & 1 \\ q & q' & -q-q' \end{pmatrix} L_{q+q'}, \\
[L_q, Q_{q'}] &= (-1)^{q+q'}\sqrt{30}\begin{pmatrix} 1 & 1 & 1 \\ q & q' & -q-q' \end{pmatrix} Q_{q+q'}, \\
[Q_q, Q_{q'}] &= (-1)^{q+q'}\sqrt{30}\begin{pmatrix} 2 & 2 & 2 \\ q & q' & -q-q' \end{pmatrix} L_{q+q'},
\end{aligned} \tag{8}$$

we can prove

$$\begin{aligned}
L_1 &= -f(E'_{12}) - f(E'_{23}), \\
L_{-1} &= f(E'_{21}) + f(E'_{32}), \\
Q_2 &= \sqrt{2}f(E'_{13}), \\
Q_1 &= f(E'_{23}) - f(E'_{12}), \\
Q_{-1} &= f(E'_{21}) - f(E'_{32}) \\
Q_{-2} &= \sqrt{2}f(E'_{31}).
\end{aligned} \tag{9}$$

Using (7), we can get

$$L_m = \sqrt{10}(d^+\tilde{d})^{(1)}_m + \sqrt{60}(g^+\tilde{g})^{(1)}_m ,$$

$$Q_m = 4\sqrt{7/15}[(s^+\tilde{d})^{(2)}_m + (d^+\tilde{s})^{(2)}_m] + 11\sqrt{2/21}(d^+\tilde{d})^{(2)}_m - 36\cdot\sqrt{1/105}[(d^+\tilde{g})^{(2)}_m + (g^+\tilde{d})^{(2)}_m] + 2\cdot\sqrt{33/7}(g^+\tilde{g})^{(2)}_m . \tag{10}$$

THE ADAPTED STATES AND THEIR APPLICATIONS

Recently a new magnetic dipole mode was observed for the deformed nuclei in the 3-4 Mev. energy region [7],[8]. In s.d.g IBM, the M1 operator can be written as

$$\begin{aligned} T(M1) &= \sqrt{3/4\pi}\{g_2\sqrt{10}(d^+\tilde{d})^{(1)} + g_4\sqrt{60}(g^+\tilde{g})^{(1)}\} \\ &= \sqrt{3/4\pi}\{(g_2-g_4)\sqrt{10}(d^+\tilde{d})^{(1)} + g_4 L\}. \end{aligned} \tag{11}$$

In the $A_{14} \supset A_2^{35}$ limit, the M1 strength

$$\begin{aligned} B(M1;\ 0_1^+ \to 1_1^+) &= \langle [N](4N,0)0\|T(M1)\|[N](4N-6,3)1\rangle^2 \\ &= (3/2\pi)(g_2-g_4)^2 \langle [N](4N,0)0\|d^+\|[N-1](4N-4,0)2\rangle^2 \\ &\quad \cdot\langle [N](4N-6,3)1\|d^+\|[N-1](4N-4,0)2\rangle^2 . \end{aligned} \tag{12}$$

Using Gruber's method [9], the relationships between the adapted states of A_{14}

$$\begin{aligned} &|[N]\tfrac{1}{15}(14n_1-n_2-\ldots-n_{15}, -n_1+14n_2-n_3-\ldots-n_{15}, \ldots, \\ &\qquad -n_1-n_2-\ldots-n_{14}+14n_{15})\rangle \\ &= [n_1,n_2,\ldots,n_{15}] \\ &= \frac{1}{\sqrt{n_1!n_2!\ldots n_{15}!}} (a_1^+)^{n_1}(a_2^+)^{n_2} \ldots (a_{14}^+)^{n_{15}}|0\rangle , \end{aligned} \tag{13}$$

the adapted states of $A_{14} \supset A_2^{15}$

$$|[N](\lambda,\mu)(w_1,w_2,w_3)\rangle \tag{14}$$

and the adapted states of $A_{14} \supset A_2^{35} \supset A_1^{140}$ is

$$|[N](\lambda,\mu)LM\rangle \tag{15}$$

are shown in Table 1.

From Table 1 we can get

$$|[N-1](4N-4,0)4N-4,4N-4\rangle = [N-1],$$

TABLE 1(a)

		[N]	[N-1,1]	[N-1,0,1]	[N-2,2]	$[N-1,0^2,1]$	[N-2,1,1]	$[N-1,0^3,1]$	[N-3,3]	[N-2,1,0,1]	$[N-1,0^5,1]$
4N,0	$\frac{1}{3}(8N-4N,-4N)$	1									
	$\frac{1}{3}(8N-3,-4N+3,-4N)$		1								
	$\frac{1}{3}(8N-3,-4N,-4N+3)$			1							
	$\frac{1}{3}(8N-6,-4N+6,-4N)$				$\sqrt{\frac{4(N-1)}{4N-1}}$	$\sqrt{\frac{3}{4N-1}}$					
	$\frac{1}{3}(8N-6,-4N+3,-4N+3)$						$\sqrt{\frac{4(N-1)}{4N-1}}$	$\sqrt{\frac{3}{4N-1}}$			
	$\frac{1}{3}(8N-9,-4N+9,-4N)$								$\sqrt{\frac{8(N-1)(N-2)}{(2N-1)(4N-1)}}$	$\sqrt{\frac{18(N-1)}{(2N-1)(4N-1)}}$	$\sqrt{\frac{3}{(2N-1)(4N-1)}}$
4N-4,2	$\frac{1}{3}(8N-6,-4N+6,-4N)$				$\sqrt{\frac{3}{4N-1}}$	$-\sqrt{\frac{4(N-1)}{4N-1}}$					
	$\frac{1}{3}(8N-6,-4N+3,-4N+3)$						$\sqrt{\frac{3}{4N-1}}$	$-\sqrt{\frac{4(N-1)}{4N-1}}$			
	$\frac{1}{3}(8N-9,-4N+9,-4N)$								$\sqrt{\frac{9(N-2)}{(N-1)(4N-1)}}$	$\frac{5-2N}{\sqrt{(N-1)(4N-1)}}$	$-\sqrt{\frac{6}{4N-1}}$
4N-6,3	$\frac{1}{3}(8N-9,-4N+9,-4N)$								$\sqrt{\frac{3}{(N-1)(2N-1)}}$	$-\sqrt{\frac{3(N-2)}{(N-1)(2N-1)}}$	$\sqrt{\frac{2(N-2)}{2N-1}}$

$$|[N](4N,0)4N-2,4N-2\rangle = -\sqrt{\frac{8N-2}{8N-1}}[N-1,0,1] + \sqrt{\frac{4N-4}{(4N-1)(8N-1)}}[N-2,2]$$

$$+\sqrt{\frac{3}{(4N-1)(8N-1)}}[N-1,0^2,1], \qquad (16)$$

$$|[N](4N-6,3)4N-3,4N-3\rangle = \sqrt{\frac{3}{(N-1)(2N-1)}}[N-3,3] - \sqrt{\frac{3(N-2)}{(N-1)(2N-1)}}[N-2,1,0,1]$$

$$+\sqrt{\frac{2(N-2)}{2N-1}}[N-1,0^5,1].$$

So that the matrix elements are

$$\langle[N](4N,0)4N-2,4N-2|d_2^+|[N-1](4N-4,0)4N-4,4N-4\rangle$$

$$= \langle[N](4N,0)4N-2,4N-2\left|-\sqrt{\frac{6}{7}}a_3^+ + \sqrt{\frac{1}{7}}a_4^+\right|[N-1](4N-4,0)4N-4,4N-4\rangle$$

$$= \sqrt{\frac{3(8N-1)}{7(4N-1)}}, \qquad (17)$$

$$\langle[N](4N-6,3)4N-3,4N-3|d_1^+|[N-1](4N-4,0)4N-4,4N-4\rangle$$

$$= \langle[N](4N-6,3)4N-3,4N-3\left|\sqrt{\frac{4}{7}}a_5^+ + \sqrt{\frac{3}{7}}a_7^+\right|[N-1](4N-4,0)4N-4,4N-4\rangle$$

$$= \sqrt{\frac{6(N-2)}{7(2N-1)}}. \qquad (18)$$

The reduce matrix elements are

$$\langle[N](4N,0)4N-2||d^+||[N-1](4N-4,0)4N-4\rangle = \sqrt{\frac{3(8N-1)(8N-3)}{7(4N-1)}}, \qquad (19)$$

$$\langle[N](4N-6,3)4N-3||d^+||[N-1](4N-4,0)4N-4\rangle = \sqrt{\frac{3(N-2)(8N-5)}{7(N-1)}}, \qquad (20)$$

$$\langle[N](4N,0)0||d^+||[N-1](4N-4,0)2\rangle$$

$$= \sqrt{\frac{1}{8N-3}}\cdot\frac{\langle(4N-4,0)2(4,0)2||(4N,0)0\rangle}{\langle(4N-4,0)4N-4(4,0)2||(4N,0)4N-2\rangle}$$

$$\cdot\langle[N](4N,0)4N-2||d^+||[N-1](4N-4,0)4N-4\rangle$$

$$= \sqrt{\frac{16N(N-1)(4N+1)}{7(4N-1)(4N-3)}}, \qquad (21)$$

$$\langle[N](4N-6,3)1||d^+||[N-1](4N-4,0)2\rangle$$

$$= \sqrt{\frac{3}{8N-5}}\cdot\frac{\langle(4N-4,0)2(4,0)2||(4N-6,3)1\rangle}{\langle(4N-4,0)4N-4(4,0)2||(4N-6,3)4N-3\rangle}$$

$$\cdot\langle[N](4N-6,3)4N-3||d^+||[N-1](4N-4,0)4N-4\rangle$$

$$= \sqrt{\frac{48(N-2)}{7(4N-5)}}. \qquad (22)$$

TABLE 1(b)

	(4N,0)					(4N-4,2)			(4N-6,3)
	4N 4N	4N-2	4N-3	4N-2 4N-2	4N-3	4N-2 4N-2	4N-3	4N-3 4N-3	4N-3 4N-3
(4N,0)									
$\frac{1}{3}(8N,-4N,-4N)$	1								
$\frac{1}{3}(8N-3,-4N+3,-4N)$	1								
$\frac{1}{3}(8N-3,-4N,-4N+3)$		$\sqrt{\frac{1}{8N-1}}$		$-\sqrt{\frac{2(4N-1)}{8N-1}}$					
$\frac{1}{3}(8N-6,-4N+6,-4N)$		$\sqrt{\frac{2(4N-1)}{8N-1}}$		$\sqrt{\frac{1}{8N-1}}$					
$\frac{1}{3}(8N-6,-4N+3,-4N+3)$			$\sqrt{\frac{3}{8N-1}}$		$-\sqrt{\frac{4(2N-1)}{8N-1}}$				
$\frac{1}{3}(8N-9,-4N+9,-4N)$			$\sqrt{\frac{4(2N-1)}{8N-1}}$		$\sqrt{\frac{3}{8N-1}}$				
(4N-4,2)									
$\frac{1}{3}(8N-6,-4N+6,-4N)$						1			
$\frac{1}{3}(8N-6,-4N+3,-4N+3)$							$\sqrt{\frac{1}{2N-1}}$	$-\sqrt{\frac{2(N-1)}{2N-1}}$	
$\frac{1}{3}(8N-9,-4N+9,-4N)$							$\sqrt{\frac{2(N-1)}{2N-1}}$	$\sqrt{\frac{1}{2N-1}}$	
(4N-6,3)									
$\frac{1}{3}(8N-9,-4N+9,-4N)$									1

The M1 strength is

$$B(M1;\ 0_1^+ \rightarrow 1_1^+) = \frac{2^7 \cdot 3^2}{7^2}(g_2 - g_4)^2 \frac{N(N-1)(N-2)(4N+1)}{(4N-1)(4N-3)(4M-5)} \binom{2}{N}. \qquad (23)$$

REFERENCES

1. N. Yoshinaga, et al., Phys. Lett. 143B:5 (1984).
2. B. Gruber and M.S. Thomas, Kinam 2:133 (1980).
   Kinam 2:381 (1980).
   Kinam 4:139 (1982).
   Kinam 5:173 (1983).
3. E. B. Dynkin, Am. Math. Soc. Transl. (2) 6:111 (1965).
   Am. Math. Soc. Transl. (2) 6:245 (1965).
4. B. Gruber and M. T. Samuel, in Group Theory and Its Applications Vol. III, e.d. E. M. Loebl, Academic Press, New York (1975).
5. M. Lorente and B. Gruber, J. Math. Phys. 13:1639 (1972).
6. H. T. Cheng and A. Arima, Phys. Rev. Lett. 51 (1983).
7. A. Richter, Invited talk presented at the Niels Bohr Centennial Symposium on Nuclear Structure, May 20-24 (1985).
8. D. Bohle, et al., Distribution of Orbital Magnetic Dipole Strength in ^{156}Gd (preprint).
9. H. P. Fritzer and B. Gruber, Symmetry Chains and Adaptation Coefficients J. Math. Phys. 26:1128 (1985).

ARE THE DIRAC STRINGS HARMLESS?

Harry J. Lipkin[a] and Murray Peshkin[b]

Argonne National Laboratory, Argonne, IL 60439-4843 USA
and
Weizmann Institute of Science, Rehovot, Israel

The material presented at the Symposium "Symmetries in Science II" is included in a paper that has since been submitted for publication. Therefore, the present report is limited to a summary of the main points, with emphasis on speculations about what we learn from the analysis of a simple model that appears to embody some essential features of the interaction of magnetic charges with electric currents.

Magnetic monopole charges are usually discussed in the context of their interactions with electric charges. In classical theory, the introduction of magnetic charges and currents encounters no fundamental difficulty. The Maxwell equations and the Lorentz force law are easily generalized to include the magnetic sources.

$$\begin{aligned} &\nabla\cdot E = 4\pi\rho_e \\ &\nabla\cdot B = 4\pi\rho_m \\ &\nabla\times B - (1/c)(\partial E/\partial t) = 4\pi j_e \\ &\nabla\times E + (1/c)(\partial B/\partial t) = 4\pi j_m \\ &F = q_e[E+(v/c)\times B] + q_m[B-(v/c)\times E] \end{aligned} \tag{1}$$

However, there is no vector potential and consequently no Lagrangian or Hamiltonian formulation of the theory, and therefore the usual way of quantizing the theory is not available.

The impossibility of defining a Hamiltonian is not surprising when one considers the interaction of a magnetic charge with an external magnetic field B_1 whose source is an electric current. Let B_2 be the magnetic field whose source is the magnetic charge. The magnetic field term in the Hamiltonian is usually taken to be the field energy, $(1/8\pi)\int(B_1+B_2)^2d^3x$. Then the forces on the sources of B_1 and B_2, which are usually currents, are given correctly by the gradients of the interaction energy

$$E_{int} = (1/4\pi)\int B_1\cdot B_2 d^3x\ . \tag{2}$$

However, in our case B_1, being due to an electric current, is the curl of a vector potential A_1, while B_2, being a monopole field, is minus the gradient of a scalar potential M_2.

$$E_{int} = -(1/4\pi) \int (\nabla\times A_1) \cdot \nabla M_2 d^3x = + (1/4\pi) \int M_2 \nabla\cdot\nabla\times A_1 d^3x = 0 . \qquad (3)$$

Then the force on the monopole should be the negative gradient of E_{int} with respect to the position of the monopole, but that vanishes, so the use of the conventional Hamiltonian is certainly wrong. Another aspect of the same paradox appears when one considers the energy of a monopole in the magnetic field due to an externally fixed current in a wire. Carrying the monopole around a loop encircling the wire results in a net gain or loss of energy, so there can be no Hamiltonian which depends only upon the variables of the monopole and the local fields or vector potentials.

Dirac[1] first addressed them problem by introducing the vector potential

$$A_\psi = g(1-\cos\theta)/r\sin\theta$$

$$A_r = A_\theta = 0 \qquad (4)$$

to represent the magnetic field due to a monopole of strength g at the origin. The curl of Dirac's vector actually equals the monopole field

$$B = g\hat{r}/r^2 \qquad (5)$$

plus a singular field along the negative z axis which carries an amount of flux equal to $4\pi g$ from minus infinity to the origin. Dirac showed that when the electric and magnetic charges of the particles obey the quantization condition

$$eg/c = n\hbar/2 , \qquad (6)$$

with integer n, a gauge transformation that commutes with the observables can move the singular string from the negative z axis to any other line connecting infinity with the origin. In that sense, the singular string is unobservable, and Dirac's vector potential represents a physical monopole. The energy paradox is avoided in Dirac's case because the B_2 is the curl of a vector potential A_2 instead of the gradient of a scalar potential M_2. Physically, that is because the monopole has really been replaced by a semi-infinite solenoid of zero thickness with its end at the origin, so the magnetic field B_2 is actually due to an electric current rather than a magnetic charge.

Other authors have avoided the singular strings by using noncanonical variables[2] or new topologies.[3] However, these methods have been applied only to the simplest systems, typically one electric and one magnetic charge, where it is unnecesary to describe the motion of the strings. Unsolved problems remain in their application to more complicated systems where the positions of several electric and magnetic charges are quantum mechanical variables described by wave functions and probability amplitudes. The positions of singular strings or the boundaries between different topological sections must also be functions of time, described by wave functions and probability amplitudes. Furthermore, the many-particle Hilbert space is not simply the product of one-particle spaces because the points where electric and magnetic charges coincide have to be excluded.[2]

Dirac's formulation, with its implication that the strings themselves are unobservable when the charge g is quantized in units of $\hbar c/2e$, seems at first to resolve these difficulties. However, there are some indications that the strings can give rise to spurious observable effects and that the Dirac formulation does not contain the same physics as monopoles without strings. If a monopole carrying a Dirac string is carried around an electric current, the string must be prevented from passing through the current in order to avoid unphysical effects. The string must therefore wind around the current, producing a topological memory of the difference between the initial and final configurations. This possible role of the string configuration is neglected in Dirac's theory, and the issue has remained confusing and obscure.

Also, the hyperfine interaction between an atomic electron and the magnetic dipole moment of a nucleus has been shown[4] to depend upon whether the dipole moment is produced by magnetic charges or by a current loop. Since a Dirac string connecting a monopole and an antimonopole effectively converts a dipole produced by magnetic charges to a dipole produced by currents, the addition of the string seems to introduce observable effects in this case. This suggests that the theory does not really describe local interactions between the particles and the electromagnetic fields or potentials at the positions of the particles.

In this paper, we explore the physics involved in the interaction of a magnetic charge with an external magnetic field due to electric currents by means of a simple soluble model which includes the variables of the source of the magnetic field. We find that when those variables are included in the dynamical theory, there is no difficulty. The Hamiltonian exists, the energy is conserved, and the equations of motion are correct. However, in the limit where the current is externally fixed, paradoxes arise which appear to have serious implications for local theories based on the Maxwell fields or the vector potentials. Those paradoxes can be avoided by introducing Dirac strings, but then the strings appear not to be physically harmless. They imply the existence of a topological winding number variable which would not be present for amonopole without a Dirac string.

Consider a magnetic charge which moves freely on a circle of radius r in the xy plane. A long resistanceless wire along the z axis connects two capacitor plates far from the circle where the monopole moves. The dynamical variables are the charge q on the capacitor, the angle coordinate θ of the monopole on its circle, and their time derivatives. The current $\dot{q}$ in the wire is the source of a magnetic field in the θ direction

$$B_\theta(t) = 2\dot{q}(t)/rc \tag{7}$$

and the moving magnetic charge is in turn the source of an electric force on the current. These interactions are described by the Lagrangian

$$\mathcal{L} = (1/2)I\dot{\theta}^2 + (1/2)L\dot{q}^2 - (1/2C)q^2 - (2g/c)q\dot{\theta} \ , \tag{8}$$

where L is the self inductance of the wire, C is the capacitance of the two plates, and $I=mr^2$. The corresponding Hamiltonian is given by

$$H = (\frac{1}{2L})\ p_q^2 + \frac{1}{2}\ L\omega^2 q^2 + \frac{1}{2I}\ p_\theta^2 + \frac{4g}{Ic}\ qp_\theta \ , \tag{9}$$

where
$$\omega^2 = \frac{1}{L}\ (\frac{1}{C} + \frac{4g^2}{I}) \ . \tag{10}$$

The Hamiltonian (10) equals the conserved total energy,

$$E = \frac{1}{2}\ L\dot{q}^2 + \frac{1}{2C}\ q^2 + \frac{1}{2}\ I\dot{\theta}^2 \ . \tag{11}$$

Since p_θ is a constant of the motion, this Hamiltonian describes a harmonic oscillator of frequency ω in the variable q, with the equilibrium value displaced from the origin by the constant

$$q_0 = \frac{4g}{ILc}\ \omega^2 p_\theta \ . \tag{12}$$

The physics of this model is transparent. The magnetic field due to the current exerts a force on the monopole which changes the angular velocity of the monopole around the current. The general solutions of the equations of motion are oscillations around the equilibrium solutions, given by

$$q(t) = q_0 + \frac{i_0}{\omega} \sin\omega(t-t_0)$$

$$\dot{\theta}(t) = - \frac{c}{2gC} q_0 + \frac{2g}{\omega Ic} i_0 \sin\omega(t-t_0) , \tag{13}$$

where i_0 is a constant.

The case of an externally fixed magnetic field produced by a stationary current is represented in our model by the limit $L\to\infty$, $\omega\to 0$, while $L\omega^2$ remains finite as do q_0 and i_0. The capacitance C may be finite or infinite. For any finite time t, the current and magnetic field approach the constant values

$$\lim\{\dot{q}(t)\} = i_0$$

$$\lim\{B_\theta(t)\} = 2i_0/rc , \tag{14}$$

but the energy of the electrical system

$$E_q = \frac{1}{2} L \dot{q}^2 + \frac{1}{2C} q^2 \tag{15}$$

approaches

$$\lim\{E_q\} = \frac{1}{2} L i_0^2 + \frac{1}{2C} [q_0+i_0(t-t_0)]^2 - \frac{1}{2} (L\omega^2) i_0^2(t-t_0)^2 \tag{16}$$

for finite values of $(t-t_0)$.

The first term on the rhs of eq. (16) is an infinite constant amount of energy. The last two terms represent a finite amount of energy which is exchanged between the magnetic charge and the fixed external magnetic field in which the magnetic charge moves. This energy compensates the changing kinetic energy of the magnetic charge to conserve the total energy of the interacting system.

Because of the last two terms in eq. (16), it is impossible to obtain a Hamiltonian depending only on the monopole variables θ and p_θ in the external field limit by subtracting (16) from the full Hamiltonian. When the variables of the electrical system are included in the dynamical description, the Hamiltonian theory works correctly. However, the Hamiltonian does not separate into an electrical part and a magnetic charge part, even in the limit where the inertia of the electrical system becomes infinite and the magnetic field becomes fixed, because the energy of the electrical system is then infinite and a finite amount of energy passes between the magnetic charge and the current as the magnetic charge moves.

All these results for the external field case are physically correct. However, they cannot be expressed in terms of a Hamiltonian with the interactions of the monopole described by a local interaction involving only the dynamical variables of the monopole and the electromagnetic fields or potentials at the position of the monopole. Newton's laws using only the field and monopole variables describe the motion correctly, but there is no Hamiltonian description without including the interaction with the current source, and that interaction must be nonlocal. Therefore the ordinary ways of doing quantum mechanics do not work. The interaction terms in the Lagrangian (8) and the Hamiltonian (9) between the monopole and the electrical system involve the product of the monopole velocity or canonical momentum and the charge q. The charge q is a variable which is far away from the monopole, and the interaction is manifestly nonlocal. Since the magnetic field at the position of the monopole is proportional to the current and therefore to the variable q, the variable q can be expressed as a time integral of the magnetic field or vector potential at the position of the monopole. This gives an interaction which is local in space but nonlocal in time, and it is equally unsatisfactory.

In principle, all these troubles may be avoidable in a complete quantum field theory, where the sources of the fields are always part of the dynamical system, and where we know the charges will have to be quantized so that the limiting case of a continuous current does not necessarily exist. However, such a theory may not be easily be achieved, and it may require some very new idea. What our simple model appears to be telling us is that such a theory cannot have the usual simple limits and therefore it must be constructed quite differently from the usual quantum field theories. For instance, the free field term in the Hamiltonian desnity, $(E^2+B^2)/8\pi$, certainly cannot work in the same way. Perhaps that is an important part of the reason why we have no satisfactory many-body theory or quantum field theory at present.

To this point, the model speaks only to the problem of a monopole without the Dirac string. The string can be added by making the point $\theta=0$ inequivalent to the point $\theta=2\pi$. That is what happens if the Lagrangian $\mathcal{L}$ of eq. (8) is replaced by

$$\mathcal{L}' = \mathcal{L} + \frac{d}{dt}(q\theta) = \frac{1}{2} I\dot{\theta}^2 + \frac{1}{2} L\dot{q}^2 - \frac{1}{2C} q^2 + \frac{2q}{c}\dot{q}\theta \quad (17)$$

The equations of motion are unchanged and the interaction term now has the form of the product of the monopole co-ordinate θ and the current variable which is proportional to the field at the position of the monopole. In the external field limit, the interaction term in the Lagrangian (17) reduces to $-(2i_o/rc)\Theta$, which reminds one of the interaction term in the case of an electric charge in an external electric field. However, whether or not one considers the external field limit of the interacting system, the variable θ is not a well-defined operator because it is not single valued. It carries additional information beyond the actual position of the monopole, a kind of winding number describing the number of times the monopole has moved around the current. That represents the motion of a monopole carrying a string, with the string winding and unwinding around the current as the monopole moves back and forth.

This suggests to us that the introduction of a singular string may not be harmless, in spite of Dirac's quantization condition which requires that the external electric current must be granular and not strictly stationary. As in the discussion of a monopole with no string, a theory which relies upon granularity to avoid unphysical effects cannot have the usual simple limiting cases with steady currents, and it remains to be seen whether one can construct a successful quantum field theory without those limits.

This work supported in part by the U. S. Department of Energy, Nuclear Physics Division, under contract W-31-109-ENG-38, and in part by the Minerva Foundation, Munich, Germany.

References

[a]Permanent address: Weizmann Institute of Science, Rehovot, Israel
[b]Permanent address: Argonne National Laboratory, Argonne, IL 60439-4843 USA

1. P. A. M. Dirac, Proc. Roy. Soc. (London) A133 (1931) 60.
2. H. J. Lipkin, W. I. Weisberger and M. Peshkin, Ann. Phys. 53 (1969) 203.
3. T. T. Wu and C. N. Yang, Phys. Rev. D12 (1975) 3845; and Nucl. Phys. B107 (1976) 365.
4. J. D. Jackson, "The Nature of Intrinsic Dipole Moments", CERN Report 77-17, Theory Division (1977); F. Bloch, Phys. Rev. 50 (1936) 259; and J. S. Schwinger, Phys. Rev. 51 (1937) 544.

5. E. N. Parker, Astrophysical Journal 160 (1970) 383; and E. M. Purcell, in "Magnetic Monopoles", eds. R. A. Carrigan, Jr. and W. P. Trower, Plenum Press, New York and London (1982) p. 141.

PHYSICAL MODELS ON DISCRETE SPACE AND TIME

Miguel Lorente

Facultad de Ciencias Físicas

Universidad Complutense de Madrid

INTRODUCTION

The idea of space and time quantum operators with a discrete spectrum has been proposed frequently spacially after the discovery that some physical quantities exhibit measured values that are multiple of a fundamental unit. In 1935 Heisenberg proposed an elementary length in analogy with the observed unit of charge and action. It would be impossible enumerate all the papers written in the same direction. We want to concentrate ourselves in several papers published recently, where other references can be found.

The physical motivations in these papers are multifold, some of which we will mention:
(i) New method to solve field operator equations based in the finite difference technique (Bender et al., Moncrief).
(ii) The physical assumption that the particle has some discrete internal motion in a finite space with the order of magnitud of its Compton wave length. (Santhanam et al., Tolar).
(iii) The assumption that the space and time are composed of some elementary domains in order to avoid the divergence problem in quantum field theory (Yukawa, Yamamoto).
(iv) The assumption that the space-time variables are nothing more than some discrete parameters labelling the quantum field operators that appears in the Hamiltonian of interacting particles (Kaplunovsky, Weinstein).
(v) The fundamental theory that unifies Quantum theory and Relativity in which the structure of space and time is a physical consequence of the elementary processes (Weizsäcker, Finkelstein, Penrose, Lorente).

The mathematical consequences of these assumptions can be divided in two classes:
(i) One uses difference equations to describe physical laws on some space-time lattice and then takes the limit to obtain the correspondence with the continuous case.
(ii) One uses difference equations without taking the continuous limit.

1. BENDER ET AL.: THE METHOD OF FINITE ELEMENTS

In recent papers[1] these authors have proposed a solution of operator field equations on a Minkowski lattice that is equivalent to a numerical evaluation of a functional integral. The method consists of three steps: (i) Decompose the domain of solutions into a set of contiguous non-overlappling patches called finite elements. (ii) On each finite element approximate the solution to the differential equation by a low-order polynomial such that are continuous across adjacent patches. (iii) Impose differential equations at one point on every patch, and boundary conditions on the extreme patches.

To be more specific, consider the Heisenberg equations of motion for a one-dimensional quantum system. The Hamiltonian is

$$H = \frac{1}{2} p^2 + V(q)$$

and the Heisenberg equations are

$$\frac{dq(t)}{dt} = p(t) \quad , \quad \frac{dp(t)}{dt} = -V'(q)$$

The quantum mechanical problem consists of solving these equations for the operators p(t) and q(t) that satisfy the equal time commutation relations $[q(t),p(t)] = i$.

The solve this problem, Bender approximates the solution on the first patch by linear functions of t:

$$q(t) = (1-t/h)q_o + (t/h)q_1 \qquad 0 \leq t \leq h$$
$$p(t) = (1-t/h)p_o + t/h\ p_1$$

Then applying the differential equations on the point $t_o = h/2$, he obtains:

$$\frac{(q_1-q_o)}{h} = \frac{1}{2}(p_o+p_1) \quad , \quad \frac{p_1-p_o}{h} = -V'(\tfrac{1}{2}(q_o+q_1))$$

that preserves the equal time commutation relations at the end point of the patch, namely:

$$[q_1,p_1] = [q_o,p_o] = i$$

The same result can be obtain at the end points of different patches.

Bender has applied the method of finite elements to solve the operator field equations for the scalar and fermion field, with the following results: the operator difference equations (i) are consistent with the equal time commutation relations, (ii) are consistent with unitarity, (iii) preserve chiral symmetry in the massless case and (iv) avoid the problem of fermion doubling.

In the limit, as the spacing h approaches zero, the solution of the theory approaches that of the continuous theory. But in this model for every value of h we have a fully consistent quantum theory in which time is a continuous parameter.

Moncrief[2] has pointed out that Bender's method gives only implicit solutions for non linear systems and that the proof of unitarity becomes awkards in the case of many degrees of freedom. He proposes an alternative scheme, related to the "leap frog" method for difference equations, with the following properties: (i) gives explicit evolution equations, (ii) is unitary and preserves equal time commutation relations, (iii) is applicable to non-linear sistems, (iv) is local, (v) convergent, for systems with many degrees of freedom and a self-adjoint Hamiltonian, to the true solution of the Heisenberg equations of motion in the limit in which the time step size goes to zero.

For the case of Heisenberg equations of motion of a system of n particles of equal mass we have

$$\frac{dq^i}{dt} = \frac{p_i}{m} \quad , \quad \frac{dp_i}{dt} = F_i(q^k) - \frac{\partial V}{\partial q^i}$$

The discrete approximation to these equations is carried out by defining the operators $q^i(t)$, $p_i(t)$ at discrete time steps $t = jh$ $(i = 1,2,\ldots)$ and writing the Heisenberg equations in the form

$$\frac{q^i(j+1)-q^i(j)}{h} = \frac{1}{m} \pi_i(j+\tfrac{1}{2})$$

$$\frac{\pi_i(j+\frac{1}{2})-\pi_i(j-\frac{1}{2})}{h} = F_i\,(q^k(j))$$

where the π_i are auxiliary momenta related to the true momenta by

$$p_i(j) = \frac{1}{2}\left[\pi_i(j+\tfrac{1}{2}) + \pi_i(j-\tfrac{1}{2})\right]$$

The auxiliary momenta can be readily eliminate to yield the explicit evolution equation

$$q^i(j+1) = q^i(j) + \frac{h}{m}\left[p_i(j) + \frac{1}{2}\, h\, F_i\,(q^k(j)\right]$$

$$p_i(j+1) = p_i(j) + \frac{1}{2}\, h\, F_i(q^k(j) + \frac{1}{2}\, h\, F_i(q^k(j+1))$$

These equations can be written in the form:

$$q^i(j+1) = u^+(j,h)\; q^i(j)\; u(j,h)$$
$$p_i(j+1) = u^+(j,h)\; p_i(j)\; u(j,h)$$

where

$$u(j,h) = \exp\,(-\frac{i}{2}\, h\, V(q^k(j))\; \exp\,(-\frac{ih}{2m}\sum_{\ell} p_{\ell}^2(j))\; \exp\,(-\frac{i}{2}\, h\, V(q^m(j))$$

therefore it follows inmediately that the model is unitary and preserves equal time commutation relations.

Following Yakawa's suggestion that space and time are divided in elementary domains, in order to avoid the divergence difficulties encountered in field theory, Yamamoto[3] develops a physical model with the following assumptions:
(i) there exsists an absolute minimum distance of time and space (τ, λ)
(ii) all differential equations with respect to space-time should be replace by difference equations; the difference equations should tend to the original differential equations if λ and τ go to zero.
(iii) the difference equations are not Lorentz invariant, because the discretizacion of space and time is done refering to a particular coordinate system. Lorentz invariant is recovered in the limit to the differential equations.

Using the symmetric difference

$$\Delta_o \psi(x) = \frac{1}{a} \left[\psi\left(x + \frac{1}{2} a \hat{0}\right) - \psi\left(x - \frac{1}{2} a \hat{0}\right)\right]$$

$$\Delta_i \psi(x) = \psi\left(x + \frac{1}{2} \hat{i}\right) - \psi\left(x - \frac{1}{2} \hat{i}\right)$$

where a is the relation between λ and τ, and $\hat{0}$, $\hat{1}$, $\hat{2}$, $\hat{3}$ are unit vector in the directions of x_o, x_1, x_2, x_3 respectively, Yamamoto writes down the action for the scalar field as:

$$S_\phi = \sum_x \frac{1}{2} \{[\Delta_\mu \phi(x)] \; [\Delta^\mu \phi(x)] - m^2 \phi^2 (x)\}$$

from which he obtains the field equation

$$(\Delta_\mu \Delta^\mu + m^2) \; \phi(x) = 0$$

The solution to this equation is

$$\phi(x) = \int_R d^3\theta \; [A(\vec{\theta}) e^{-i(\omega t/a - \vec{\theta}.\vec{r})} + A^*(\vec{\theta}) e^{i(\omega t/a - \vec{\theta}.\vec{r})}]$$

where $A(\vec{\theta})$ is an arbitrary function and ω is given by

$$\sin^2 \frac{\omega}{2} = a^2 \left[\sin^2 \frac{\theta_1}{2} + \sin^2 \frac{\theta_2}{2} + \sin^2 \frac{\theta_3}{2} + \frac{m^2}{4}\right] ,$$

and the domain of integration R is $-\pi \leq \theta_i \leq \pi$ $(i = 1,2,3)$. If $a \leq (3+m^2/4)^{-1}$ then ω is real for any θ.

Yamamoto applies the canonical quantization to the scalar field, obtaining

$$\phi(x) = N \int_R \frac{d^3}{(\sin \omega)^{1/2}} \; [a(\vec{\theta}) e^{-i(\omega t/a - \vec{\theta}.\vec{r})} + a^+(\vec{\theta}) e^{i(\omega t/a - \vec{\theta}.\vec{r})}]$$

where $a(\theta)$ and $a^+(\theta)$ satisfy the standard commutation relations

$$[a(\theta), a^+(\theta')] = \delta^3(\theta-\theta')$$

$$[a(\theta), a(\theta')] = [a^+(\theta), a^+(\theta')] = 0$$

Using the vacuum defined by $a(\theta)|0\rangle = 0$ he calculates the propagator and investigates the divergence in the case $x = 0$, finding

$$D_F(0;m^2) = \frac{1}{4(2\pi)^3} \int_R \frac{d^3}{[\sin^2 \frac{\vec{\theta}}{2} + \frac{m^2}{4}]^{1/2} \, [1-a^2(\sin^2 \frac{\vec{\theta}}{2} + \frac{m^2}{4})]^{1/2}}$$

which is finite regardless of $m = 0$ or $m \neq 0$ because $a^2 \leq (3+m^2/4)^{-1/2}$.

3. SANTHANAM ET AL.: FINITE DIMENSIONAL QUANTUM MECHANICS

This model[4] gives a description of the physics of a particle totally confined to move within a finite region of space. The configuration space has a structure dependent on its mass and the possible extent of its motion in the space and time is considered to be a continuous parameter as in the quantum theory. This model is based on the following assumptions:

i) The position eigenvalues of the particle form a discrete and finite set $\{q_n\}$, given by

$$q_n = n\varepsilon \quad , \quad n = -J, \ldots J-1, J .$$

where ε is a positive small number.

The quantum mechanical system space of the particle is a $(2J+1)$-dimensional vector space, and the position operator Q is given by

$$\langle n|Q|n'\rangle = n \,\epsilon\, \delta_{nn'}$$

Physically $2J\varepsilon$ gives the dimension of the region of confinement of the particle.

ii) The momentum eigenvalues of the particle form a discrete and finite set $\{p_n\}$

$$p_n = n\eta \quad , \quad n = -J, \ldots J-1, J$$

where η is a positive real number. The momentum operator P conjugate to Q, is given by

$$\langle n|P|n'\rangle = \begin{cases} 0 \quad , & \text{if} \quad n = n' \\ \frac{i\eta}{2} \csc\left(\frac{2\pi J(n-n')}{2J+1}\right) \quad , & \text{if} \quad n \neq n' \end{cases}$$

iii) If an observable K of the particle is represented by the operator $K(q,p)$ in the quantum mechanical formalism, then it will be represented by the matrix $K(Q,P)$ obtained by the replacement

$$q \to Q \qquad , \qquad p \to P$$

and the eigenvalues of the matrix K(Q,P) are the values that the observable K can take. For example

$$H = \frac{P^2}{2m} + \frac{1}{2} m\omega^2 Q^2$$

will represent the hamiltonian operator of the harmonic oscillator of frequency ω.

iv) Time is regarded as an independent continuous parameter, and the temporal development of the state vector is governed by the Schrödinger equation

$$i\hbar \frac{\partial}{\partial t} \psi(t)> = H|\psi(t)>$$

with H the Hamiltonian matrix.

v) The fundamental quantities, quantum of position ε and quantum of momentum η, are related to the fundamental quantities rest mass m, Planck's constant $\hbar$, the speed of light c, and Compton wave length $\lambda\!\!\!^{-}$, by

$$\varepsilon = (\frac{2\pi}{2J+1})^{1/2} \cdot \lambda\!\!\!^{-} \quad , \quad \eta = (\frac{2\pi}{2J+1})^{1/2} \cdot mc$$

therefore

$$\frac{\varepsilon\eta}{\hbar} = \frac{2\pi}{2J+1}$$

and consequently Q and P are canonically conjugate in the sense that

$$\exp(\frac{i\,\eta\,\varepsilon P}{\hbar}) \exp(\frac{im\eta Q}{\hbar}) = \exp\frac{imn\varepsilon\eta}{\hbar} \exp(\frac{im\eta Q}{\hbar}) \exp(\frac{in\varepsilon P}{\hbar})$$

vi) When $J \to \infty$, $\varepsilon \to 0$, $\eta \to 0$, we recover the continuous spectra of the position and momentum operators.

vii) This model has been proved[5] to be invariant under a (1+1)-dimensional Galilei group.

4. KAPLUNOVSKY ET AL.: SPACE-TIME AS LATTICES OF N-SIMPLICES

In dynamical theories space-time is treated, according to these authors, as an arena, even when one takes into account the Einstein approach to gravity as geometry. In quantum theory the role of the x_μ's are nothing but labels for independent operators which define the Hamiltonian of the system. They construct[6] and Hamiltonian with the help of the boson field $\phi(\ell)$ and fermion field $\psi(\ell)$ operators satisfying the canonical commutation relations, where the quantum degrees of freedom ℓ are to be considered as playing the role of site variables in a lattice model.

In addition to these site fields they introduce link fields, $P_{<\ell m>}$ and $X_{<\ell m>}$ connecting the fields $\psi(\ell)$ and $\psi(m)$ in the site ℓ, and m respectively, that allow to write a kinematic term without the introduction of differentiation. The Hamiltonian reads as follows:

$$H = \sum_{\ell m} \Big[\phi^2(\ell) + X^2_{\ell m} \; \phi(\ell)\,\phi(m) + V(\phi) - iX_{\ell m} \{\psi^+(\ell)\,\psi(m) - \psi^+(m)\psi(\ell)\} + \frac{1}{2} P^2_{\ell m} + V(X_{\ell m}) \Big]$$

The idea is now to see whether some $X_{\ell m}$ adquire non-vanishing expectation value for the ground state. In this case the fields $\phi(\ell)$ and $\psi(m)$ yield a solvable zeroth order Hamiltonian for the matter fields. In the case of boson fields their contribution to the vacuum energy becomes zero when the expectation values is minimized, and so the boson fields are uninteresting to the model.

When the Hamiltonian has no boson fields and $V(X_{\ell m}) = \frac{1}{2}\mu^2 X^2_{\ell m}$, then the vacuum expectation values for the link fields becomes non-zero, but the fermions themselves tend to push the system to higher dimensions. If $V(X_{\ell m}) = \frac{1}{2}\mu^2 X^{2r}_{\ell m}$ with $r > 1$, the analysis leads to the result that the configuration which minimizes the vacuum expectation values corresponds to the all links fields $X_{\ell m}$, in other words, a configuration similar to the so called an N-simplex, i.e. an object for which every point ℓ is a nearest neighbour of every other point m. This system generates a theory whose low energy degrees of freedom tend in the continuous limit to a 1+1-dimensional field theory.

In order to construct higher dimension systems Kaplunovsky and Weinstein introduce a potential

$$V = \sum (X^A_{\ell m})^{2r} + \sum f_{ABCD}(X^A_{jk}\ X^B_{k\ell}\ X^C_{\ell m}\ X^D_{mj})^2$$

where f_{ABCD} are completely symmetric coefficients. After calculating the vacuum expectation values for the link variables they obtain the result of some configuration in which the lattice breaks into n subsets of points, which are connected by x-links and form an x-simplex. Different x-simplices are connected to each other by y-links. When only y-links are considered, the lattice breaks into y-simplices, that are connected to each other by x-links. This configuration is called a SLAC lattice, and leads to a (2+1)-dimensional theory when the continuous limit is taken. Higher dimensions can be obtained if a generalized SLAC lattice is used.

5. WEIZSAECKER ET AL.: THEORY OF ELEMENTARY PROCESSES

Weizsäcker[7] does not pressuposes the existence of space and time but a finite number of simple alternatives, called "urs". The Hilbert space of the urs consists on the complexed vectors with two components (yes-not decision). Its symmetry group is U(2) and systems of urs are described by spinor functions of this group. This group is homomorphic to the group of rotations in the real three dimensional space, and this is the reason why ordinary space is three dimensional.

Weizsäcker has carried out a systematic program for the urs. All the objects can be reduced to simple alternatives of binary objects. The actual number of the simple alternatives in the universe is finite, but the number of possible alternatives in the future is infinite and they are governed by laws of probability.

In order to have a relativistic picture of the model Castell[7] introduces a second 2-dimensional Hilbert space the elements of which an the antiurs. The urs plus the antiurs are represented by complex 4-vectors satisfying the equation of motion

$$i\gamma_o \frac{\partial}{\partial\alpha}\begin{pmatrix}\psi\\ \phi\end{pmatrix} = \omega\begin{pmatrix}\psi\\ \phi\end{pmatrix}$$

where ψ and ϕ represent the ur and antiur respectively. The group of symmetry is now SU(2,2), which is locally isomorphic to the conformal group, and it contains as a subgroup the Poincaré group.

According to Finkelstein[8] the world is a network of elementary quantum processes, of creation and destruction, which are ensembled as a network of mutual relations. This connection between succesive processes is graphically represented by a checkerboard. This image leads to a discrete manifolds of space and time.

6. OUR MODEL: (n+1)-DIMENSIONAL SPACE-TIME LATTICE

We have proposed[9] an interpretation of the structure of space and time, by which there are fundamental entities interacting among themselves 1 to 2n, in order to bild up a n-dimensional cubic lattice, as a ground field where the physical interactions take place. The matter fields (fermion and boson) operators act with the ground field and among themselves. The space-time coordinates are nothing but the labelling of the ground field, and they take only discrete values. The physical laws must be described by difference equations and the eigenvalues take only integer values.

One can relax this assumption by taking the continuous limit, because the lattice structure is not observed, then the difference equations should go to the differential equations, with the condition that the convergence, stability and symmetries of the difference equations are preserved in the limit.

6.1 Rational trigonometric and hiperbolic functions

We have define[9] the rational functions

$$\cos k\alpha = \frac{1}{2}\left(\frac{z^{2k}}{|z|^{2k}} + \frac{z^{-2k}}{|z|^{2k}}\right), \quad \sin k\alpha = \frac{1}{2i}\left(\frac{z^{2k}}{|z|^{2k}} - \frac{z^{-2k}}{|z|^{2k}}\right)$$

where $z = m + ni$, m,n,k integers, and α represents the length of the chord between two consecutive unit vectors P_k and P_{k+1}, namely,

$$\alpha^2 = \left|\frac{z^{2k+2}}{|z|^{2k+2}} - \frac{z^{2k}}{|z|^{2k}}\right|^2 = \frac{4n^2}{m^2+n^2}$$

These rational trigonometric functions satisfy the addition formulas and the following difference equations

$$\Delta \sin k\alpha = \alpha \cos k\alpha \quad , \quad \Delta \cos k\alpha = -\alpha \sin k\alpha$$

where Δ is the central operator

$$\Delta f(k\alpha) = f((k+\tfrac{1}{2})\alpha) - f((k-\tfrac{1}{2})\alpha)$$

We can define also the rational exponential function

$$\exp ik\alpha = \cos k\alpha + i \sin k\alpha = \frac{z^{2k}}{|z|^{2k}}$$

satisfying $\Delta \exp k\alpha = i\alpha \exp k\alpha$.

Making k=m, we can recover the continuous functions and the differential equations when m goes to infinity.

$$\exp m\,\alpha = \frac{(1+\frac{in}{m})^m}{(1-\frac{in}{m})^m} \xrightarrow[m\to\infty]{} e^{i2n}$$

Also $m\,\alpha \equiv \xrightarrow[m\to\infty]{} 2n$, and $\alpha \xrightarrow[m\to\infty]{} 0$, therefore, we can divide by α the difference equation and we get the limit

$$\frac{\Delta \exp m\,\alpha}{\alpha} = \exp m\,\alpha \to \frac{d \exp i(2n)}{d(2n)} = \exp\ i(2n)$$

and similar expression for the trigonometric functions.

We have defined also the rational hyperbolic functions

$$\mathrm{ch}\ k\,\beta = \frac{1}{2}\left(\frac{u^{2k}}{|u|^{2k}} + \frac{u^{-2k}}{|u|^{2k}}\right) \quad , \quad \mathrm{sh}\ k\,\beta = \frac{1}{2\sigma}\left(\frac{u^{2k}}{|u|^{2k}} - \frac{u^{-2k}}{|u|^{2k}}\right)$$

where $u = m + n\,\sigma$, m,n,k integers and σ an hypercomplex number satisfying $\sigma^2 = 1$. The argument β represents the length of the chord between two consecutive unit vectors, namely

$$-\beta^2 = \left| \frac{u^{2k+2}}{|u|^{2k+2}} - \frac{u^{2k}}{|u|^{2k}} \right|^2 = - \frac{4n^2}{m^2-n^2}$$

These functions satisfy the usual addition formulas and the following difference equations

$$\Delta\ \mathrm{ch}\ k\,\beta = \beta\ \mathrm{sh}\ k\,\beta \quad , \quad \Delta\ \mathrm{sh}\ k\,\beta = \beta\ \mathrm{ch}\ k\,\beta$$

We also define the exponential function

$$\exp k\,\beta = \mathrm{ch}\ k\,\beta + \mathrm{sh}\ k\,\beta = \left(\frac{m+n}{m-n}\right)^k$$

satisfying

$$\Delta \exp k\,\beta = \beta \exp k\,\beta$$

This equation becomes the differential equation when we make $k = m$ go to infinity

$$\frac{\Delta \exp k\,\beta}{\beta} = \exp k\beta \xrightarrow[m\to\infty]{} \frac{d \exp (2n)}{d(2n)} = \exp\ (2n)$$

6.2 Rational representations of the classical groups

Using the Cayley parametrization of the classical groups we have described[9] a method to find all elements of the fundamental representations with rational matrix elements. The matrix S of the fundamental representation is decomposed as

$$S = \frac{1-H}{1+H} = \frac{(1-H)^2}{1-H^2}$$

where H satisfies $H^T+H = 0$, for the orthogonal groups, $H^+ + H = 0$ for the unitary groups and $H^TJ+JH = 0$ for the symplectic groups. If we impose on H to take only integer values S will be rational.

In the case of pseudoorthogonal groups of physical interest of low dimension, the calculation can be simplified by the use of the isomorphism between real forms. It is well known the isomorphism between SL(2,C) and SO(3,1). From the isomorphism between Sp(4,R) and SO(3,2) we define a (4x4) real matrix, satisfying $A^TJ=JA$ and TrA=0.

The bijection of an element (x_1,x_2,x_3,x_4,x_5) of R^5 into A is the following

$$\begin{bmatrix} x_1 & x_2+x_3 & 0 & x_4+x_5 \\ x_2-x_3 & -x_1 & -x_4-x_5 & 0 \\ 0 & x_4-x_5 & x_1 & x_2-x_3 \\ -x_4+x_5 & 0 & x_2+x_3 & -x_1 \end{bmatrix} = A$$

The transformation $A' = SAS^{-1}$ with $S \in Sp(4,R)$ maps A into it self, namely, $A'^TJ = JA'$, $TrA' = 0$. Since

$$\det A = (x_1^2 + x_2^2 - x_3^2 - x_4^2 + x_5^2)^2 = \det A'$$

this transformation induces the desired automorphism. If S is given by a rational matrix the induced transformation of SO(3,2) will also be a rational matrix. Similar constructions can be carried out for the groups SO(4,1), SO(4,2), SO(5,1).

For one parameter subgroups we define the rational exponential function

$$S(k\alpha) = \left(\frac{1-\frac{n}{m}H}{1+\frac{n}{m}H}\right)^k \quad , \qquad H = \begin{matrix} 0 & 1 & 0 \\ -1 & 0 & 0 \\ 0 & 0 & 0 \end{matrix}$$

corresponding to the rotation around the x_3-axis. Here m,n,k are integers and α represents the "chord" between two consecutive vectors, as in the case of trigonometric functions. Therefore

$$\Delta\, S(k\,\alpha) = -\alpha\, H\, S\,(k\,\alpha) \quad , \qquad \alpha = \frac{2n}{\sqrt{m^2+n^2}}$$

This difference equation becomes the differential equation when we take the continuous limit:

$$\frac{\Delta\, S(m\,\alpha)}{\alpha} = -H\, S(m\,\alpha) \xrightarrow[m\to\infty]{} \frac{d\,\exp\,(-2n\,H)}{d(2n)} = -H\,\exp\,(-2n\,H)$$

6.3 Integral real forms

In the assumption that the structure of the space-time is a (n+1)-dimensional cubic lattice, and the continuous limit is not taken, we have to use the real forms with integer matrix elements that keep the lattice invariant. In the case of compact groups this condition is fulfilled by a finite number of group elements. For instance, for the orthogonal groups we have matrix elements with ± 1 only, and the number of integral forms acting in an N-dimensional space is $N!2^N$, which must be divided by 2, if one takes the proper orthogonal group.

For non-compact groups the number of integral forms is infinite, but this condition imposes strong restrictions on the symmetries of some physical systems, as we will see latter on. A very easy way to get integral forms is to take the Cayley decomposition and make the generator H nihilpotent. Then

$$S = 1-2H \quad , \quad S^k = 1-2kH \quad , \quad H^2 = 0 \ .$$

For the pseudo-orthogonal groups, this condition is applied to the isomorphism between real forms in order to simplify the calculations. For instance, if we make in the SL(2,¢) group

$$S = 1-2H = \begin{bmatrix} 1 & -2p+2qi \\ 0 & 1 \end{bmatrix}$$

we obtain the integral Lorentz transformation:

$$\Lambda = \begin{bmatrix} 1 & 0 & 2p & -2p \\ 0 & 1 & -2q & 2q \\ -2p & 2q & 1-2p^2-2q^2 & 2p^2+2q^2 \\ -2p & 2q & -2p^2-2q^2 & 1+2p^2+2q^2 \end{bmatrix} \quad , \ p,q \in Z$$

6.4 Integral trigonometric and hyperbolic functions

Given any integral form A of the proper orthogonal group acting in N-dimensional space we can construct the integral trigonometric functions. Consider the matricial function

$$A(x) = A^x \quad , \quad x \in Z \quad , \quad A^N = 1 \ ,$$

satisfying the periodicity condition

$$A(x+N) = A(x)$$

Using the matrix elements of this function we define

$$a_{ij}(x) = [A(x)]_{ij} \quad , \quad i,j = 1,2,\ldots,N$$

From the definition of A(x) we derive inmediately the following properties:

i) $a_{ij}(x+y) = \sum_{m=1}^{N} a_{im}(x)\, a_{mj}(y)$, $x,y \in Z$

ii) $\sum_{j=1}^{N} a_{ij}^2(x) = 1$, $i = 1,2,\ldots,N$

iii) $a_{ij}(x+N) = a_{ij}(x)$

iv) $E^k a_{ij}(x) = a_{i,j+k \ (\mathrm{mod}\ N)}(x)$, $E^N a_{ij}(x) = a_{ij}(x)$

where E is the translation operator.

We have defined[9] the integral hyperbolic function with the help of the vector

$$v = v_o+v_1e_1+v_2e_2+v_3e_3$$

where e_i are unit vectors satisfying $e_ie_j + e_je_i = 2\delta_{ij}$, and $v_1^2+v_2^2+v_3^2-v_o^2 = 1$. Hence, for $(\beta = 2v_o)$, we have defined

$\cosh k\beta = v^{2k} + \bar{v}^{-2k}$, $\sinh_i k\beta = v^{2k} e_i - e_i \bar{v}^{-2k}$, $i = 1,2,3$

satisfying $ch^2 k\beta - sh_1^2 k\beta - sh_2^2 k\beta - sh_3^2 k\beta = 1$

and $\Delta^2 ch\, k\beta = \beta^2 ch\, k\beta$, $\Delta^2 sh_i k\beta = \beta^2\, sh_i k\beta$, $i = 1,2,3$

6.5 Physical applications

In the assumption of a discrete space-time, the physical laws must be described by difference equations, which become differential equations in the continuous limit.

Let $x_\mu = (ct, x_1, x_2, x_3)$ be a 4-vector defined on the lattice. The interval between two events is written Δx_μ, which transforms under the Lorentz group as

$$\Delta x'_\mu = \Lambda_\mu^\nu \Delta x_\nu$$

The Lorentz invariant proper time in the discrete version reads

$$\Delta\tau = \frac{1}{c} (\Delta x_\mu \Delta x^\mu)^{1/2}$$

The 4-velocity U_μ and 4-acceleration A_μ are also covariant expressions

$$U_\mu = \frac{\Delta x_\mu}{\Delta\tau}, \quad U^2 = c^2$$

$$A_\mu = \frac{\Delta U_\mu}{\Delta\tau}, \quad U.A = 0$$

The 4-momentum is a covariant expression in the discrete form

$$p_\mu = m_o U_\mu = (\frac{E}{c}, \vec{p}) \quad , \quad p^2 = m_o^2 c^2$$

from which the relativistic mechanics fundamental law is derived

$$\vec{F} = \frac{\Delta\vec{p}}{\Delta t} \quad , \quad \vec{F}.\vec{u} = \frac{\Delta E}{\Delta t}$$

or in covariant form

$$F_\mu = \frac{\Delta p_\mu}{\Delta\tau}, \quad \text{where } F_\mu = (\frac{\vec{F}.\Delta\vec{x}}{c\ \Delta\tau}, \vec{F}\frac{\Delta t}{\Delta\tau})$$

is the Minkowski covariant force.

We can use also partial difference equations for the electromagnetic field, but in this case these equations are not Lorentz invariant, which is restored when the continuous limit is taken. If this limit is not taken we have to use the integral Lorentz transformations, but in this case it can be proved that the allowed velocity between two inertial systems must be equal or greater than $(\sqrt{3}/2)c$. This strong restriction would be eliminated when one uses inertial frame in de Sitter Universe of constant curvature.

For the wave equation we can use the two aforementioned assumptions: i) The space-time intervals are finite and the wave function is a rational function that satisfies a partial difference equation; this approach ensures the stability and the exactness of the solution and thus it is very suitable for numerical calculations. ii) The space-time coordinates take only integer values and the wave function is a rational or integral function; in the last case the solution is periodical and invariant under integral Lorentz transformations.

For the first assumption we can use the rational function

$$\psi = \left[\frac{z_o^2}{|z_o|^2}\right]^{k_o\ell_o} \cdot \left[\frac{\bar{z}_1^2}{|z_1|^2}\right]^{k_1\ell_1} \cdot \left[\frac{\bar{z}_2^2}{|z_2|^2}\right]^{k_2\ell_2} \cdot \left[\frac{\bar{z}_3^2}{|z_3|^2}\right]^{k_3\ell_3}$$

with $z_\mu = m_\mu + in_\mu$, m,n,k,ℓ integers. Taking the partial symmetric difference operator Δ_μ with respect to ℓ_μ (μ = 0,1,2,3) the wave function satisfies

$$\left[\frac{\Delta_o^2}{\Delta\alpha_o^2} - \frac{\Delta_1^2}{\Delta\alpha_1^2} - \frac{\Delta_2^2}{\Delta\alpha_2^2} - \frac{\Delta_3^3}{\Delta\alpha_3^3}\right] \psi = 0$$

where $\alpha_o^2 = 4n_o^2 (m_o^2 + n_o^2)^{-1}$ and so on. The solvability condition reads as follows

$$\frac{1-\cos k_o\alpha_o}{\alpha_o^2} - \frac{1-\cos k_1\alpha_1}{\alpha_1^2} - \frac{1-\cos k_2\alpha_2}{\alpha_2^2} - \frac{1-\cos k_3\alpha_3}{\alpha_3^2} = 0$$

It can be proved, from the definition of the α's that the numerators of the last expressions are always multiple of α^2, namely, $2(1-\cos \alpha) = \alpha^2$, $2(1-\cos 2\alpha) = \alpha^2(4-\alpha^2)$, $2(1-\cos 3\alpha) = \alpha^2(9-6\alpha^2+\alpha^4)$ and so on.

In the continuous limit, namely when $m_o = \ell_o \to \infty$, and so on the wave function goes to the exponential function

$$\psi = e^{i2(n_ok_o-n_1k_1-n_2k_2-n_3k_3)}$$

as required, and the wave equation becomes the differential equation, while the solvability condition goes to the Lorentz invariant expression

$$k_o^2 - k_1^2 - k_2^2 - k_3^2 = 0$$

We can also introduce the first assumption in the quantum mechanics formalism. The Heisenberg equation of motion will be described by difference equations and the solution by rational functions. Let us write down the Hamiltonian for the harmonic oscillator

$$H = \frac{p^2}{2m} + \frac{1}{2} m q^2$$

where p an q are the momentum and position operator respectively satisfying the canonical commutation relations. The Heisenberg equations of motion read:

$$\frac{\Delta q}{dt} = \frac{1}{m} p(t) \quad , \quad \frac{\Delta p}{\Delta t} = -m q(t)$$

therefore $\dfrac{\Delta^2 p}{\Delta t^2} = -p(t)$

whose solutions are expressed in rational form:

$$p(t) = p_o \cos t - m q_o \operatorname{sen} t$$

$$q(t) = \frac{1}{m} p_o \operatorname{sen} t + q_o \cos t$$

p_o and q_o being at t = 0. Obviously, if $[p_o,q_o] = i$ it follows that $[p(t),q(t)] = i$ hence, the equal time commutation relations are preserved. In the continuous limit we obtain the standard differential equation and the stability of the solution is garanteed.

Combining the solution for the wave equation as above, and the canonical commutation relation we can also solve the field operator equations for the scalar and fermion free fields.

Some philosophical foundation for our model can be found elsewhere[10]. A more complete classification of the foundational theories for the structure of space and time has been published by García-Sucre[11].

REFERENCES

1. C.M. Bender, D.H. Sharp, Phys. Rev. Lett. 50, 1535 (1983)
   C.M. Bender, K.A. Milton, D.H. Sharp, Phys. Rev. Lett. 51, 1815 (1983)
   C.M. Bender, K.A. Milton, D.H. Sharp, Phys. Rev. D31, 383 (1985)
   C.M. Bender, K.A. Milton, D.H. Sharp, L.M. Simmons, Jr., R. Stong, Phys. Rev. D32, 1476 (1985)
   C.M. Bender, F. Cooper, V.P. Gutschick, M. Martin Nieto, Phys. Rev. D32, 1486 (1985)
   C.M. Bender, K.A. Milton, S.S. Pinsky, L.M. Simmons, Phys. Rev. D33, 1692 (1986)
   C.M. Bender, L.M. Simmons, Jr., R. Stong, Phys. Rev. D33, 2362 (1986)

2. V. Moncrief, Phys. Rev. D28, 2485 (1983). See also
   L. Vâzquez, Phys. Rev. D32, 2066 (1985)

3. H. Yamamoto, Phys. Rev. D30, 1727 (1984), 32, 2659 (1985)

4. R. Jagannathan, T.S. Santhanam, R. Vasudevan, Inter. J. Theor. Phys. 20, 755 (1981)
   R. Jagannathan, T.S. Santhanam, Inter. J. Theor. Phys. 21, 351 (1982)
   R. Jagannathan, Inter. J. Theor. Phys. 22, 1105 (1983)

5. P. Stovicek, J. Tolar, Rep. Math. Phys. 20, 157 (1984)

6. V. Kaplunovsky, M. Weinstein, Phys. Rev. D31, 1879 (1985)

7. L. Castell, C.F. von Weizsäcker, Quantum Theory and the Structure of Space and Time, V. 5, C. Hanser Verlag, Munich 1983

8. D. Finkelstein, Phys. Rev. D9, 2219 (1974), Inter. J. Theor. Phys. 21, 489 (1982)

9. M. Lorente, Inter. J. Theor. Phys. 11, 213 (1974), 12, 927 (1976), 25, 55 (1986)

10. M. Lorente, "A Causal Interpretation of the Structure of Space and Time", in Foundations of Physics (P. Weingarther, G. Dorn ed.) Hölder-Pichler-Tempsky, Vienna 1986

11. M. García-Sucre, Inter. J. Theor. Phys. 24, 441 (1985)

THE MANY SYMMETRIES OF HUBBARD ALTERNANT POLYENES

F. A. Matsen

Departments of Chemistry and Physics
The University of Texas
Austin, Texas 78712

ABSTRACT

In the freeon unitary-group-formulation of quantum chemistry the relevant group is U(n) where n is the number of freeon orbitals. The Hamiltonian is a second degree polynomial in the U(n) generators so the Hilbert space of the Hamiltonian is the direct sum of the U(n) irreducible representation spaces (IRS). The Pauli principle is imposed by restricting the physically significant IRS to those labeled by the partitions $[\lambda] = [2^{(N/2)-S}, 1^{2S}]$ where N is the number of electrons and S is the spin. The IRS have the following properties: i) For each IRS labeled by $[\lambda]$ there exists a conjugate IRS labeled by $[\underline{\lambda}] = [2^{(\underline{N}/2)-S}, 1^{2S}]$ where $\underline{N} = 2n-N$ is the number of holes in $[\lambda]$. ii) The dimension of the $[\underline{\lambda}]$th IRS equals the dimension of the $[\lambda]$th IRS. iii) The symmetry-adaptation of the $[\underline{\lambda}]$th IRS with respect to any group yields the same decomposition as does the symmetry-adaptation of the $[\lambda]$th IRS. iv) There is defined a selfconjugate Hamiltonian such that the $[\lambda]$th and the $[\underline{\lambda}]$th spectra differ by only a constant energy shift, $\Delta E = \Delta E^{\circ}(n-N)$. v) For n = N the conjugate group, G_k, is a group of the Hamiltonian and supplies the conjugation quantum number.

The Hubbard alternant polyene is defined by a selfconjugate Hamiltonian and exhibits the properties described above. The exact group chain of the neutral (n = N) Hubbard alternant polyene is $U(n) \supset R^Q(n) \supset (G, G_k)$ where $R^Q(n)$, G and G_k are the quasispin, point and conjugate groups, respectively. We take the n = 3 site (the allyl system) as an example. The freeon ℓ-shell atom ($n = 2\ell + 1$), the isofreeon j-shell nucleus ($n = 2j + 1$) and the baryon octet (n = 3) exhibit equivalent symmetries.

GROUP THEORETICAL INTRODUCTION

In the freeon unitary group formulation of quantum chemistry [1],[2] the Hamiltonian is realized by a second degree polynomial in the generators $\{E_{rs}; r,s = 1 \text{ to } n\}$ of U(n) where n is the number of freeon orbitals (one-particle states). It follows that the Hilbert space of a U(n) Hamiltonian is the direct sum of the irreducible representation spaces (IRS), denoted $V_n[\lambda]$ are labeled by partitions $[\lambda]$ of the electron number, N. The physically significant freeon IRS are labeled by

$$[\lambda] = [2^{(N/2)-S}, 1^{2S}] \tag{1}$$

where S is the spin. Each IRS is graphically labeled by a Young diagram with no more than two columns of lengths $n_1 \geq n_2$ where $N = n_1 + n_2$ and where

$$S = (n_1 - n_2)/2 = N/2 - n_2 \tag{2}$$

The associated spin space is labeled by the partition

$$[\lambda] = [N - S, S] \tag{3}$$

The freeon IRS are spanned by Gel'fand states, $\{|G\rangle\}$ [3],[4],[5] which are graphically labeled by Gel'fand arrays, A_G and Gel'fand-Weyl tableaux, T_G. For the $[\lambda]$th (primary) IRS there exists a conjugate IRS [5],[6] of the same dimension and spin labeled by the partition

$$[\underline{\lambda}] = [2^{(\underline{N}-S)}, 1^{2S}] \tag{4}$$

with the particle number

$$\underline{N} = 2n - N \tag{5}$$

Since a filled shell contains 2n particles, $\underline{N}$ is the number of holes in the primary IRS. See Figure 1a.

The Gel'fand states are constructed by the Moshinsky-Nagel construction [4] denoted

$$|G\rangle = G|0\rangle \tag{6}$$

where G is a polynomial in the generators and $|0\rangle$ is the highest-weight state in $V_n[\lambda]$. See Figure 1b. This construction permits the direct Lie-algebraic evaluation of matrix elements. For each $|G\rangle$ in the primary space $V_n[\lambda]$ there exists in the conjugate space $V_n[\underline{\lambda}]$ a conjugate Gel'fand state

$$|\underline{G}\rangle = K|G\rangle \tag{7}$$

where the conjugation operator, K is defined by $K = K^+$, $KK = 1$ and

$$KE_{rs}K = 2\delta(r,s) - (-1)^{r+s}E_{sr} \tag{8}$$

We relate the spectra of the primary and conjugate spaces by means of the shift Hamiltonian $\Delta H = H - \underline{H}$, where $\underline{H} = KHK$. Then

$$\begin{aligned}\langle G \mid \Delta H \mid G'\rangle &= \langle G \mid H \mid G'\rangle - \langle G \mid \underline{H} \mid G'\rangle \\ &= \langle G \mid H \mid G'\rangle - \langle \underline{G} \mid H \mid \underline{G}'\rangle\end{aligned} \tag{9}$$

Finally, we define a selfconjugate Hamiltonian as a Hamiltonian such that

$$\Delta H = \Delta E^0(n - N) \tag{10}$$

For the special case of $n = N$, $[H,K] = 0$, the conjugation group

$$G_K = \{1,K\} \tag{11}$$

is a group of the Hamiltonian which supplies the conjugation quantum numbers $\{+ \ -\}$.

$N = 2$

$V_3[1^2]$:

$N = 4$

$V_3[2,1^2]$:

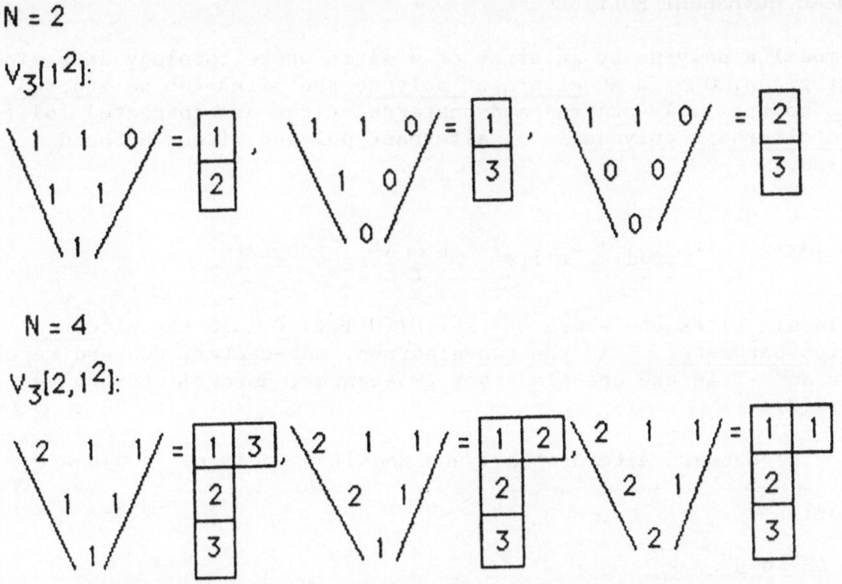

Fig. 1a. Triplet (S = 1) Freeon Spaces. Note that orbitals which are unoccupied in the primary space are doubly occupied in the conjugate space.

$$|0\rangle = \boxed{\begin{matrix}1\\2\end{matrix}} = \diagup\!\!\!\diagdown \begin{matrix}1 & 1 & 0\\ & 1 & 1\\ & & 1\end{matrix}\!\!\diagup = \text{the highest-weight state}$$

$$\boxed{\begin{matrix}1\\3\end{matrix}} = \diagup\!\!\!\diagdown \begin{matrix}1 & 1 & 1\\ & 1 & 0\\ & & 1\end{matrix}\!\!\diagup = NL_{32}|0\rangle = E_{32}|0\rangle$$

$$\boxed{\begin{matrix}2\\3\end{matrix}} = \diagup\!\!\!\diagdown \begin{matrix}1 & 1 & 0\\ & 1 & 0\\ & & 0\end{matrix}\!\!\diagup = NL_{21}L_{32}|0\rangle = NE_{21}E_{32}|0\rangle$$

$$= N(E_{32}E_{21} - E_{31})|0\rangle = -E_{31}|0\rangle$$

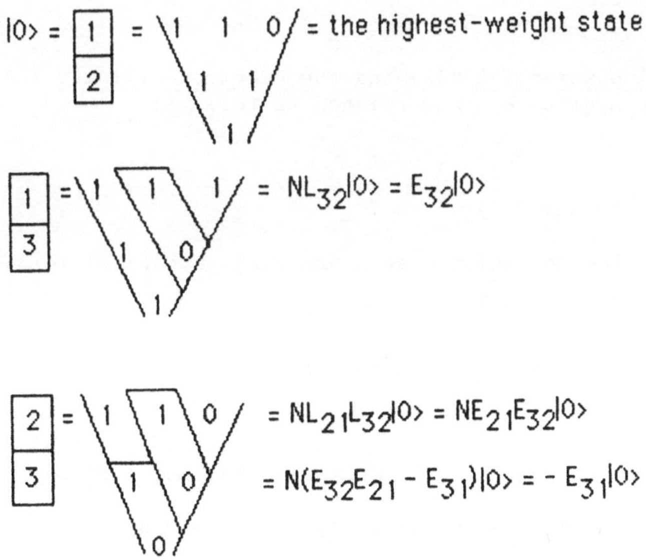

Fig. 1b. The Moshinsky-Nagel construction of the N = 2, Triplet space $G = N (L_{ij})^{a_{ij}}$ where L_{ij} is the Moshinsky operator and $a_{ij} = h_{ji} - h_{j(i-1)}$.

THE HUBBARD ALTERNANT POLYENE

We model a polyene by an array of n sites whose topology determines its point group, G. In an alternant polyene the sites can be starred and unstarred so that only starred and unstarred sites are connected [6],[8]. A Hubbard alternant polyene is an alternant polyene with a Hubbard Hamiltonian,

$$H = (x-1)\sum_{r}^{n-1}(E_{r,r+1} + E_{r+1,r}) + x\sum_{r}(E_{rr}^2 - E_{rr})/2 \tag{12}$$

where r labels sites and where $0 \leq x = U/(U + T) \leq 1$ is the electron correlation parameter, U is the two-electron, one-center, Hubbard repulsive parameter and -T is the one-electron, two-center, Hueckel attractive parameter (β).

THEOREM. The Hubbard alternant polyene Hamiltonian is selfconjugate.

Proof:

i) $\Delta H^0 = H^0 - \underline{H}^0$

$$= (x-1)\sum_{r}^{n-1}(E_{r,r+1} + E_{r+1,r}) - (E_{r+1,r} + E_{r,r+1}) = 0$$

ii) $\Delta H = \Delta V = V - \underline{V} = x\sum_{r}(E_{rr}^2 - E_{rr}) - ((2 - E_{rr})^2 - (2 - E_{rr}))/2$

$$= x\sum_{r}(E_{rr} - 1)$$

so

$$\Delta E = x(N - n) \qquad \text{QED}$$

The ionization potential (IP) and the electron affinity (EA) of a neutral alternant hydrocarbon are related as follows:

$$IP(\pi^n) = E(\pi^{n-1}) - E(\pi^n)$$

$$EA(\pi^n) = E(\pi^n) - E(\pi^{n+1})$$

Then since the cation and anion states are conjugate particle-hole states

$$IP(\pi^n) + EA(\pi^n) = E(\pi^{n-1}) - E(\pi^{n+1}) = \Delta E$$

$$= x(N - n) = x((n - 1) - n)$$

$$= -x \tag{13}$$

Equivalent results were obtained by McLachlan [9] using Slater determinants.

For our example we take the allyl system: (n = 3)

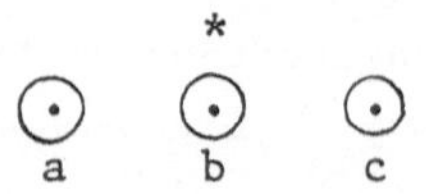

The orbital space is

V_3: $\{|a>,|b>,|c>\}$

where the orbitals are site (Wannier) orbitals localized on the allylic sites. The exact group chain is $U(3) \supset C_2$ where N = 2, the point group is realized in terms of U(3) generators by

$$C_2 = \{I = (E_{aa} + E_{bb} + E_{cc})/2, \quad C_2 = \frac{1}{2}(E_{aa} + E_{bb} + E_{cc})^2 - 2)\} \quad (14)$$

The symmetry-adapted states for the allyl cation (N = 2) and the allyl anion (N = 4) are shown in Figure 2 and their Hubbard spectra in Figure 3.

The neutral alternant polyene (N = n) has additional symmetry, the conjugation group (see Section 1) and the quasispin group $R^Q(n)$ [10],[11], a Lie group whose algebra is

$$LAR^Q(n): \{\Omega_{rs} = (\frac{1}{2})(E_{rs} - (-1)^{r-s}E_{sr})\} \quad (15)$$

and whose quadratic invariant is

$$I_2^Q = \sum_r \sum_s \Omega_{rs}\Omega_{sr}$$

with eigenvalues given by

$$I_2(N,S,Q) = (\frac{1}{2})N(N + 2) - Q(Q + 1) - S(S + 1)$$

We construct a new (and simpler) invariant, called the quasispin operator, by combining it with the U(n) invariant, I_2^U. Thus

$$Q^2 = (\frac{1}{2})((I_2^U - 2I_2^Q) = \sum_{rs} (-1)^{r+s}E_{rs}E_{rs} - N$$

with eigenvalues $I_2(Q) = Q(Q + 1)$ where Q is the quasispin.

The many symmetries of the neutral Hubbard alternant polyene are summarized in the exact group chain: $U(n) \supset R^Q \supset \{G, G_k\}$. The quasispin and the conjugation quantum numbers are related by $K = (-1)^Q$. The allyl radical spectrum is plotted in Figure 4.

ATOMS, NUCLEI AND BARYONS

The freeon unitary group formulation of Hubbard alternant polyene exhibits strong similarities to the unitary group formulations of atoms, nuclei and baryons:

a) The p-shell atom: $U(3) \supset R(3)$.

V_3: $\{|1\bar{1}> = |\bar{1}>, |10> = |0>, |11> = |1>\}$

$LAR(3)$: $\{L_z = E_{11} - E_{\bar{1}\bar{1}}, L^+ = \sqrt{2}(E_{10} + E_{0\bar{1}}), L^- = \sqrt{2}(E_{01} + E_{\bar{1}0})\}$

Then symmetry-adapted states for N = 2 and 4 are shown in Figure 5. The p-shell Hamiltonian is selfconjugate with $\Delta E = (5A - 10B)(N-3)$ where A and B are the Condon parameters. Then

$$E(p^2;{}^1S) - E(p^2;{}^1D)/\{(p^2;{}^1D) - E(p^2;{}^1D) - E(p^2;{}^3P)\}$$

$$= E\{(p^4;{}^1S) - E(p^4;{}^1D)\}/\{(p^4;{}^1D) - E(p^4;{}^1D) - E(p^4;{}^3P)\} = 1.5.$$

SPIN SINGLET STATES

FREEON SINGLET STATES

d = 0
$|\pi^2;1^1A\rangle = \boxed{a\ c}$

$|\pi^2;2^1A\rangle = \boxed{a\ b} + \boxed{b\ c}$

$|\pi^2;1^1B\rangle = \boxed{a\ b} - \boxed{b\ c}$

d = 1
$|\pi^2;3^1A\rangle = \boxed{b\ b}$

$|\pi^2;4^1A\rangle = \boxed{a\ a} + \boxed{c\ c}$

$|\pi^2;2^1B\rangle = \boxed{a\ a} - \boxed{c\ c}$

d = 1
$|\pi^4;1^1A\rangle = \begin{array}{|c|c|}\hline a & b\\\hline b & c\\\hline\end{array}$

$|\pi^4;2^1A\rangle = \begin{array}{|c|c|}\hline a & b\\\hline c & c\\\hline\end{array} + \begin{array}{|c|c|}\hline a & a\\\hline b & c\\\hline\end{array}$

$|\pi^4;1^1B\rangle = \begin{array}{|c|c|}\hline a & b\\\hline c & c\\\hline\end{array} - \begin{array}{|c|c|}\hline a & a\\\hline b & c\\\hline\end{array}$

d = 2
$|\pi^4;3^1A\rangle = \begin{array}{|c|c|}\hline a & a\\\hline c & c\\\hline\end{array}$

$|\pi^4;4^1A\rangle = \begin{array}{|c|c|}\hline b & b\\\hline c & c\\\hline\end{array} + \begin{array}{|c|c|}\hline a & a\\\hline b & b\\\hline\end{array}$

$|\pi^4;2^1B\rangle = \begin{array}{|c|c|}\hline b & b\\\hline c & c\\\hline\end{array} - \begin{array}{|c|c|}\hline a & a\\\hline b & b\\\hline\end{array}$

SPIN TRIPLET STATES

FREEON TRIPLET STATES

d = 0
$|\pi^2;^3A\rangle = \begin{array}{|c|}\hline a\\\hline b\\\hline\end{array} - \begin{array}{|c|}\hline b\\\hline c\\\hline\end{array}$

$|\pi^2;1^3B\rangle = \begin{array}{|c|}\hline a\\\hline b\\\hline\end{array} + \begin{array}{|c|}\hline b\\\hline c\\\hline\end{array}$

$|\pi^2;2^3B\rangle = \begin{array}{|c|}\hline a\\\hline c\\\hline\end{array}$

d = 1
$|\pi^4;^3A\rangle = \begin{array}{|c|c|}\hline a & c\\\hline b\\\cline{1-1} c\\\cline{1-1}\end{array} - \begin{array}{|c|c|}\hline a & a\\\hline b\\\cline{1-1} c\\\cline{1-1}\end{array}$

$|\pi^4;1^3B\rangle = \begin{array}{|c|c|}\hline a & c\\\hline b\\\cline{1-1} c\\\cline{1-1}\end{array} + \begin{array}{|c|c|}\hline a & a\\\hline b\\\cline{1-1} c\\\cline{1-1}\end{array}$

$|\pi^4;2^3B\rangle = \begin{array}{|c|c|}\hline a & b\\\hline b\\\cline{1-1} c\\\cline{1-1}\end{array}$

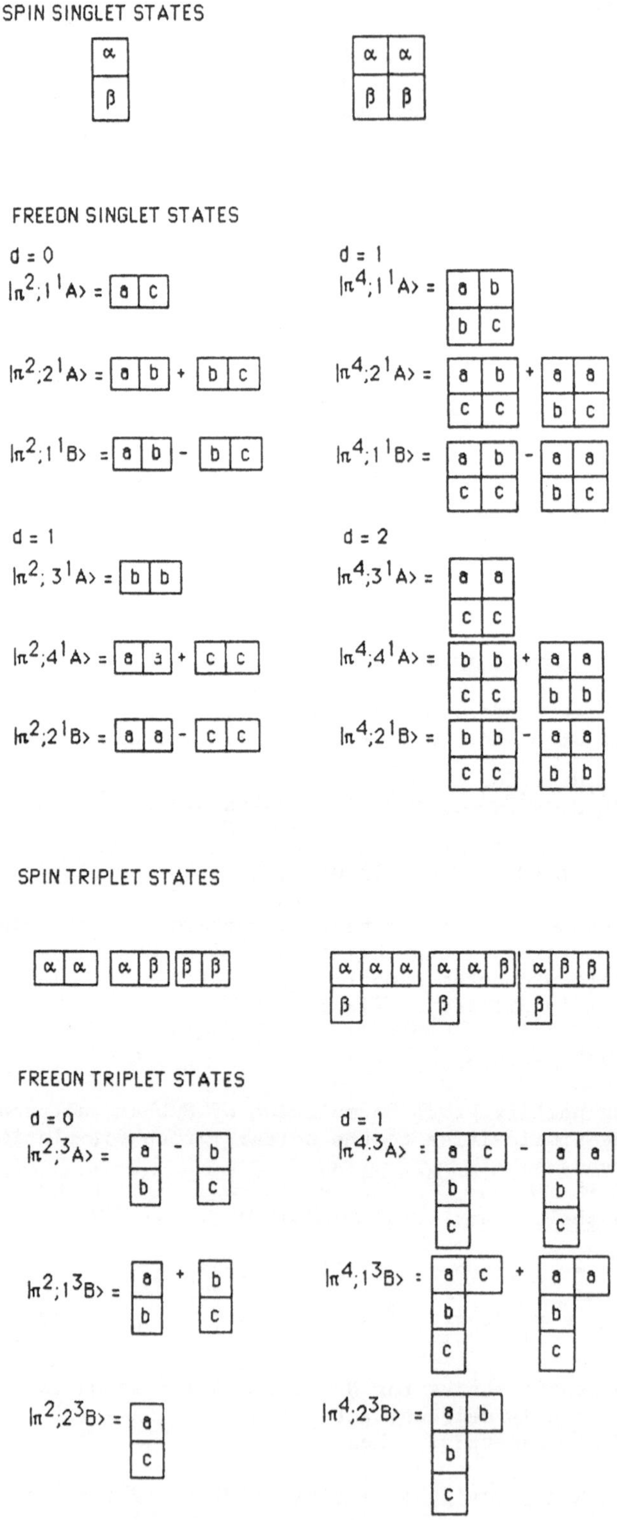

Fig. 2. Symmetry-adaptation of the allylic cation and anion states with respect to $U(3) \supset C_2$.

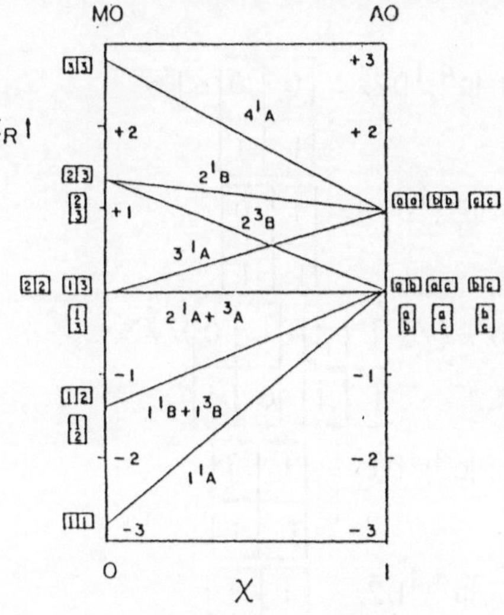

Fig. 3. Electronic structure diagram for allyl monocation.

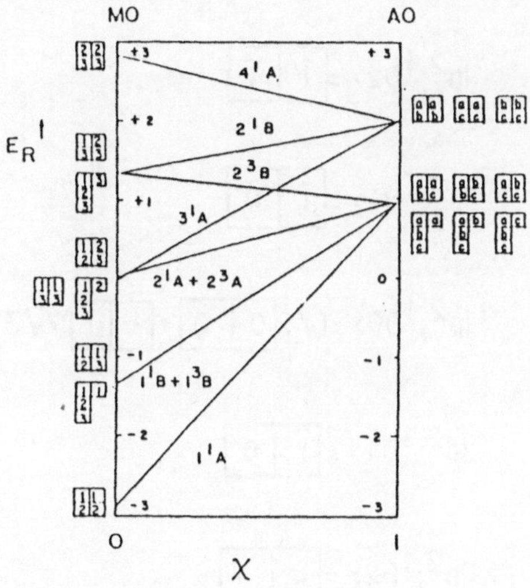

Electronic structure diagram for allyl monoanion.

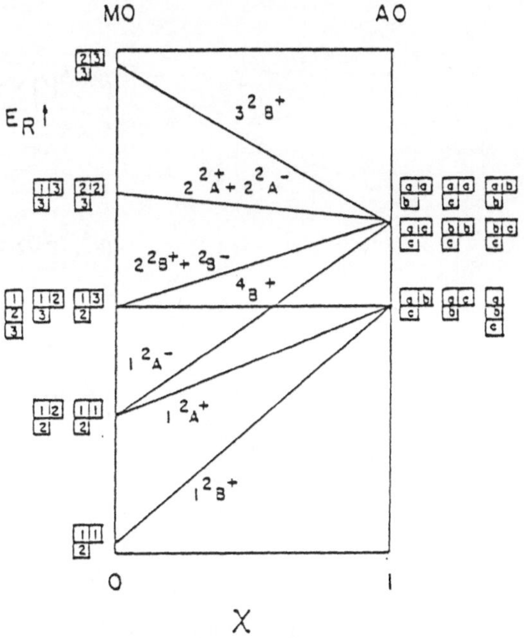

Fig. 4. Electronic structure diagram for allyl radical.

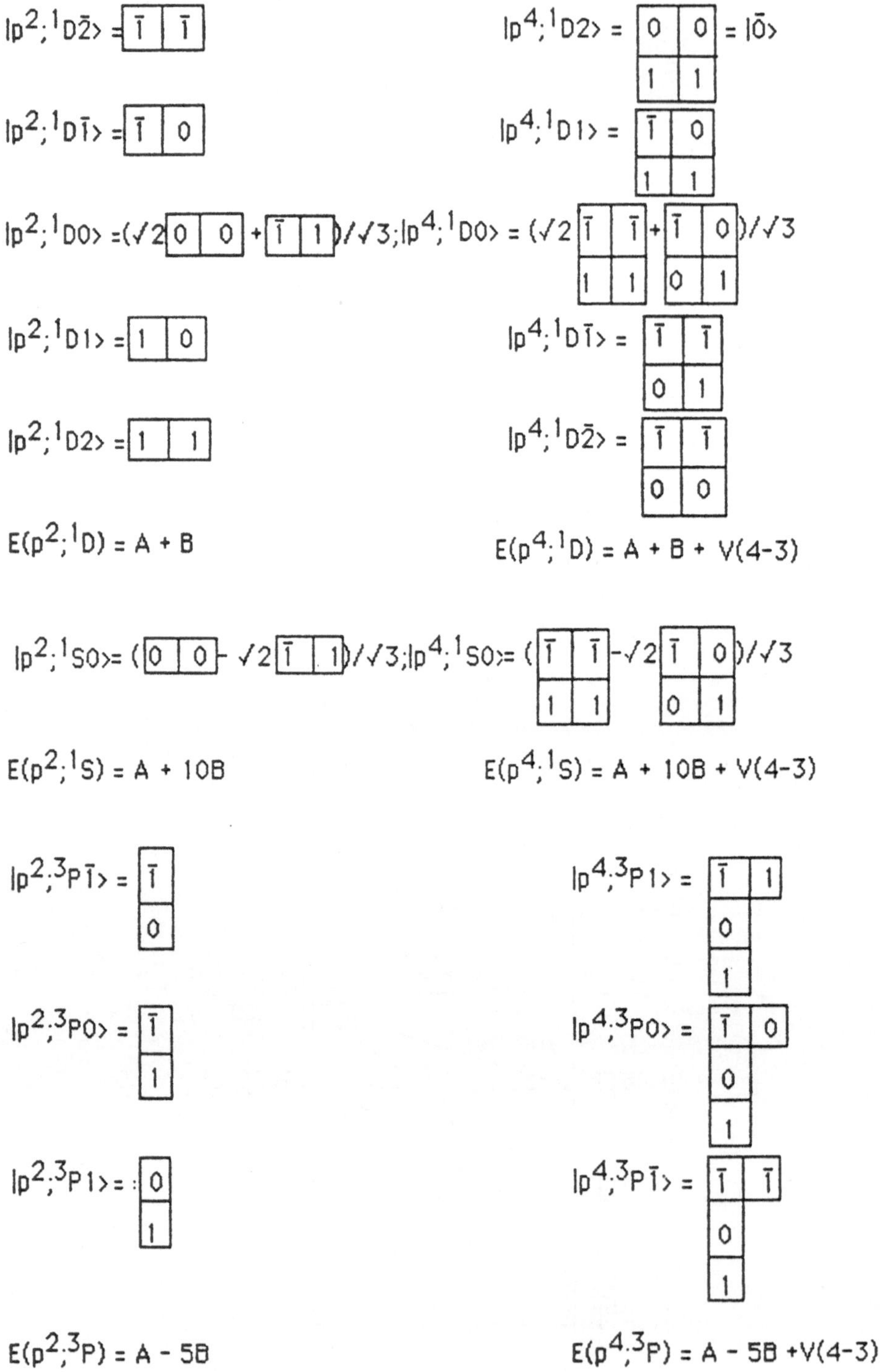

Fig. 5. The Atomic p-shell. $(U(3) \supset R(3))$ $\Delta E = V = 5A - 10B$

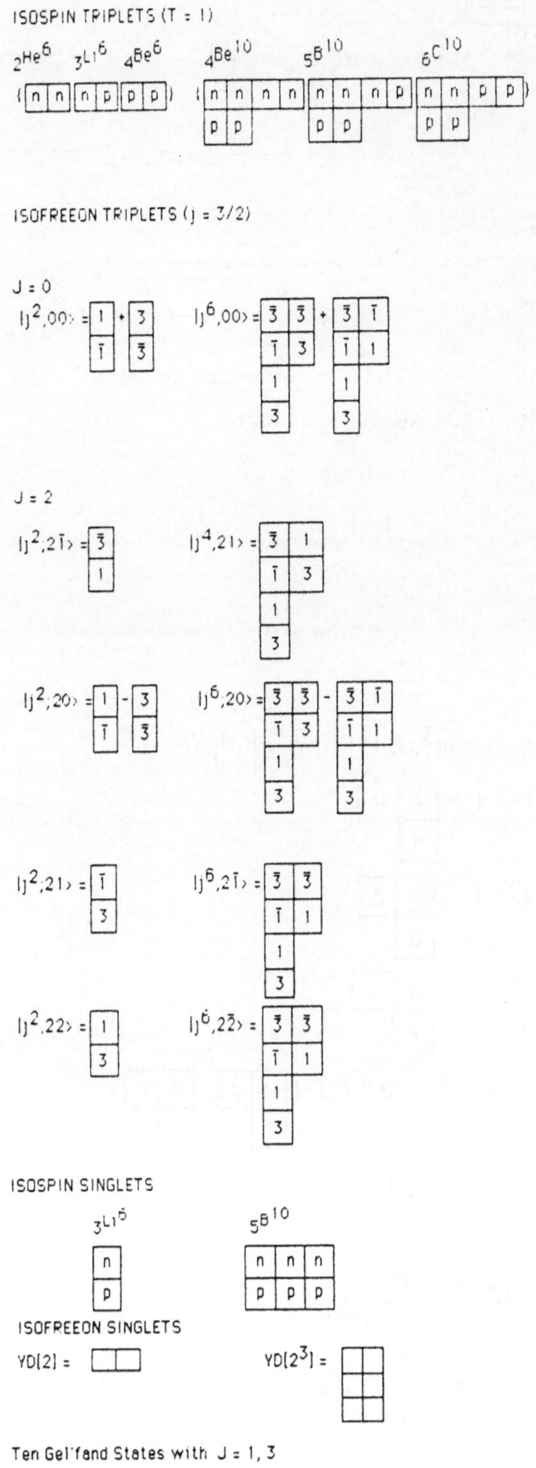

Fig. 6. Isofreeon states A(4) ⊃ SU(2).

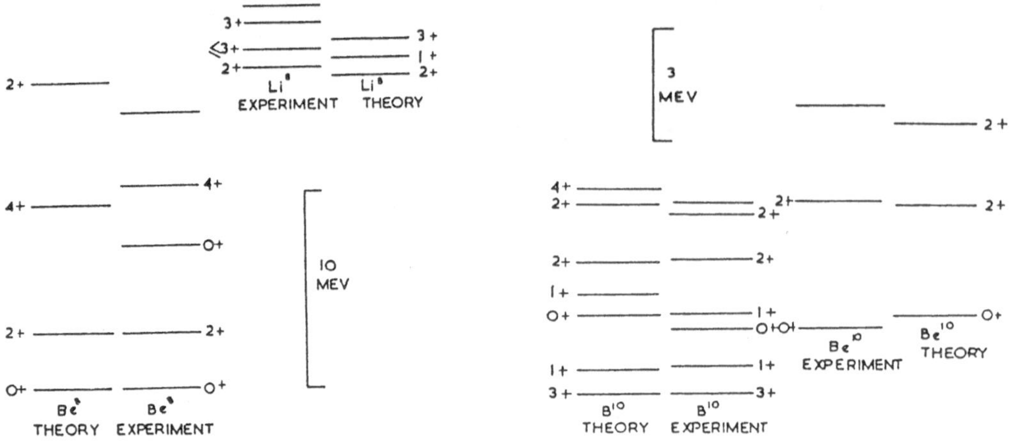

Fig. 7. Nuclear Spectra

V_3: { $|u\rangle$, $|d\rangle$, $|s\rangle$ } $|q^3;[2,1]$ $I,M_I,Y\rangle = |I,M_I,Y\rangle$

$|p\rangle = |1/2,1/2,1\rangle = $
$$\begin{array}{|c|c|} \hline u & u \\ \hline d \\ \cline{1-1} \end{array}$$

$|n\rangle = |1/2,\overline{1/2},1\rangle = $
$$\begin{array}{|c|c|} \hline u & d \\ \hline d \\ \cline{1-1} \end{array}$$

$|\Sigma^+\rangle = |1,1,0\rangle = $
$$\begin{array}{|c|c|} \hline u & u \\ \hline s \\ \cline{1-1} \end{array}$$

{ $|\Sigma^0\rangle = |1,0,0\rangle$, $|\Lambda^0\rangle = |0,0,0\rangle$ } = {
$$\begin{array}{|c|c|} \hline u & d \\ \hline s \\ \cline{1-1} \end{array}$$
,
$$\begin{array}{|c|c|} \hline u & s \\ \hline d \\ \cline{1-1} \end{array}$$
}

$|\Sigma^-\rangle = |1,\overline{1},0\rangle = $
$$\begin{array}{|c|c|} \hline d & d \\ \hline s \\ \cline{1-1} \end{array}$$

$|\Xi^0\rangle = |1/2,1/2,\overline{1}\rangle = $
$$\begin{array}{|c|c|} \hline u & s \\ \hline s \\ \cline{1-1} \end{array}$$

$|\Xi^-\rangle = |1/2.\overline{1/2},\overline{1}\rangle = $
$$\begin{array}{|c|c|} \hline d & s \\ \hline s \\ \cline{1-1} \end{array}$$

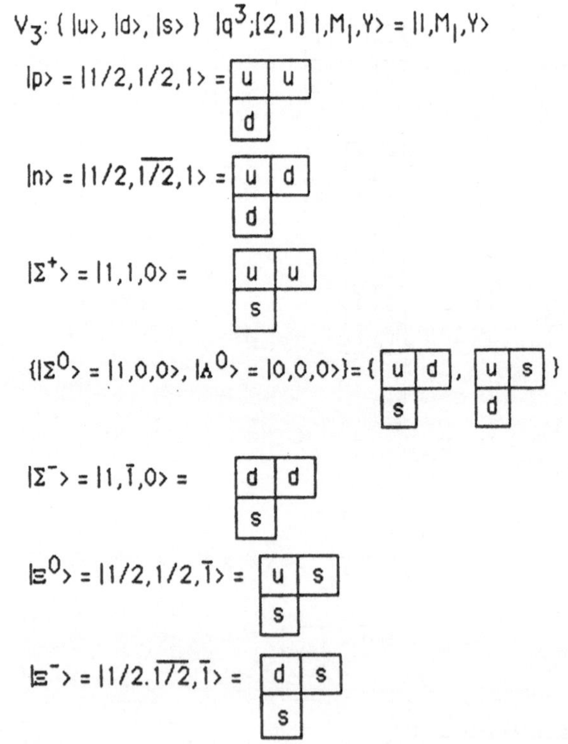

Fig. 8. Baryon octet (n = N). Symmetry-adaptation to
U(3)(SU(3)) ⊃ SU(2).

These results were obtained by Shortley and Freed [12] using Slater determinants.

b) Isofreeon theory. The Pauli-allowed isofreeon IRS are labeled by $[\lambda] = [2^{(A/2-T)}, 1^{2T}]$, T = isospin. The associated isospin IRS is labeled by $[\Lambda] = [N-T, T]$. For the j = 3/2 shell the isofreeon chain is U(4) ⊃ SU(2) and the isofreeon orbital space is denoted

$$V_4: \quad \{|3/2,\overline{3/2}\rangle = |\bar{3}\rangle, |3/2,\overline{1/2}\rangle = |1\rangle, |3/2,\overline{1/2}\rangle = |1\rangle, |3/2,3/2\rangle = |3\rangle\}$$

The SU(2) algebra in terms of the U(4) generators is then

$$\text{LAS U(2): } \{J_z = (\tfrac{1}{2})(3E_{33} + E_{\bar{1}1} - E_{\bar{1}\bar{1}} + 3E_{\bar{3}\bar{3}}),$$

$$J^+ = \sqrt{3}(E_{31} + E_{\bar{1}\bar{3}}) + 2E_{11},\ J^- = \sqrt{3}(E_{13} + E_{\bar{3}\bar{1}}) + 2E_{\bar{1}1}\}$$

The symmetry adapted states for A = (4) + 2 and A = (4) + 6 are shown in Figure 6 and their experimental and theoretical spectra 13 in Figure 7. Note that this nuclear pair does not exhibit particle-hole symmetry since the lowest pair of levels are reversed. Agreement with experiment is obtained by increasing the spin-orbit interaction in the A = 10 Hamiltonian thereby destroying the selfconjugacy.

c) The baryon octet. SU(3) ⊃ SU(2)

$$V_3: \quad |u\rangle, |d\rangle, |s\rangle\}$$

$$\text{LAS U(2): } \{T_z = (E_{uu} - E_{dd})/2,\ T^+ = E_{ud},\ T^- = E_{du}\}$$

(the isospin group)

The symmetry-adapted states are shown in Figure 8. The (Gell-Mann-Okubo) Hamiltonian is

$$H = a + bY + c(T^2 - Y^2/4)(Y = (E_{uu} + E_{dd} - 2E_{ss})/3)$$

is selfconjugate.

REFERENCES

1. F. A. Matsen, Int. J. Quantum Chem. 8S:379 (1974); Adv. Quantum Chem. (1978); F. A. Matsen and R. Pauncz, "The Unitary Group in Quantum Chemistry," Elsevier, Amsterdam (in press).
2. J. Paldus, J. Chem. Phys. 61:5321 (1974); "Theoretical Chemistry, Advances and Perspectives," 2:131, Academic Press (1976).
3. I. M. Gel'fand and M. I. Graev, Am. Math. Soc. Translation 64:116 (1964).
4. M. Moshinsky, J. Math. Phys. 4:1128 (1963); "Group Theory and the Many-Body Problem," Gordon and Breach (1968).
5. G. Baird and L. C. Biedenharn, J. Math. Phys. 4:463 (1963).
6. F. A. Matsen and T. L. Welsher, Int. J. Quantum Chem. 12:985, 1001 (1977).
7. J. Nagel and M. Moshinsky, J. Math. Phys. 6:683 (1965); Rev. Mexicana de Fis. 14:29 (1965).
8. J. Koutecky, J. Paldus and J. Cizek, J. Chem. Phys. 83:1722 (1985).
9. A. D. McLachlan, Mol. Phys. 2:271 (1959), 4:49 (1961).
10. D. R. Herrick, Adv. Chem. Phys. 52:1 (1983).
11. L. Cizek, R. Pauncz and E. R. Vrscay, J. Chem. Phys. 78:2486 (1983).
12. G. H. Shorley and B. Fried, Phys. Rev. 54:739 (1938).
13. D. Kurath, Phys. Rev. 101:216 (1956).

SYMMETRIES AND POLARIZATION

Michael J. Moravcsik

Department of Physics
Institute of Theoretical Science
University of Oregon
Eugene, Oregon 97403
USA

Abstract

In unpolarized cross sections constraints imposed by symmetries produce only quantitative changes which, in the absence of the precise knowledge of dynamics, cannot be used to test the validity of those symmetries. In polarization observables, in sharp contrast, imposition of symmetries produces qualitative changes, such as the vanishing of some observables or linear relationships among observables, which can be used to check the validity of symmetries without a detailed knowledge of dynamics. Such polarization observables can also separate the different constraints caused by different symmetries imposed simultaneously. This is illustrated for the two cases when Lorentz invariance and parity conservation, and Lorentz invariance and time reversal invariance, respectively, hold. It is also shown that it is impossible to construct, in any reaction in atomic, nuclear, or particle physics, a null experiment that would unambiguously test the validity of time-reversal invariance independently of dynamical assumptions. Finally, for a general quantum mechanical system undergoing a process, it is shown that one can tell from measurements on this system whether or not the system is characterized by quantum numbers the existence of which is unknown to the observer, even though the detection equipment used by the observer is unable to distinguish among the various possible values of the "secret" quantum number and hence always averages over them. This allows us to say whether the spin of a particle in a reaction is zero or not even if we can measure nothing about that particle's polarization.

The aim of my brief account is to emphasize the pivotal role of polarization experiments in testing the validity of various symmetries in quantum mechanical particle reactions, that is, in molecular, atomic, nuclear, and high energy physics. I will do this both through a general discussion of the features of polarization which bring about its ability to test conservation laws, and through the mentioning of specific results which are in this category, and which are already available in the published literature.

The emphasis will be on presenting an overview, and hence I will omit details available from the references. One remark, however, on formalisms may nevertheless be in order.

The derivation of many of the results regarding polarization quantities in recent years has been done using the so-called optimal set of polarization formalisms[1]. These optimal frames are characterized by the requirement that in them the relationship of the bilinear products of reaction amplitudes ("bicoms") to the experimental observables is made as simple as possible. It can be shown that a complete diagonalization of the matrix giving the relationship between bicoms and observables is not possible, since it would violate the reality of the observables (or Hermiticity). The best one can do is to choose a basis set of amplitudes and a corresponding set of observables in such a way that the matrix connecting the bicoms with the observables has only small submatrices along its main diagonal and zeroes elsewhere. These small submatrices, for any four-particle reaction regardless of the values of the spins of the four particles, are always only 1-by-1, or 2-by-2, or 4-by-4, or 8-by-8. This is so when only Lorentz invariance is imposed on the reaction. When additional symmetries also hold, the submatrices get even smaller.

The requirement of optimality in the above sense does not determine the representation completely. On the contrary, even after imposing optimality we are left, in most cases, with an infinite set of choices. Optimality fixes the structure of the matrices describing spins and density matrices in a representation in which the rows and columns of these matrices are labeled by the spin projections of the various particles, but it leaves us free to choose the directions in which we choose to quantize the particles. When only Lorentz invariance holds and no other symmetry, we can choose the quantization direction of each of the four particles separately and arbitrarily. When additional symmetries are also imposed, the quantization directions of different particles may need to be coordinated, but even in the most constrained situation when parity conservation, time reversal invariance, and identical particle constraints all hold, we still have an infinite set of optimal formalisms. This freedom is extremely important in the utilization of the optimal formalism for various purposes, since it allows us to choose a formalism best suited for the particular purpose we face.

At the same time when we stress the advantages of the optimal formalisms, we must also emphasize that the general properties of the polarization observables and their superiority, in particular, in testing symmetries in no way depends on the kind of formalism we use. A good formalism simply allows us to discover these properties more easily.

So let me return now to these general properties. Let me first discuss the unpolarized (differential) cross section for an arbitrary reaction and explain why it is unsuitable for testing symmetries.

A reaction is in general characterized by a set of complex reaction amplitudes, each labeled by some specification of the spin projection of each particle in the reaction. There are $2s+1$ such projections for a particle with spin s. Denoting $2s+1$ by x, we see that the total number of amplitudes is the product of the four x's. This is true when only Lorentz invariance holds. When additional symmetries are also imposed, some of the former amplitudes either vanish or become dependent on each other. Thus the number of independent amplitudes is decreased by the imposition of such additional symmetries.

The unpolarized differential cross section, in any formalism, contains the sum of the magnitude-squares of all amplitudes. In particularly clumsy formalisms (in particular, in formalisms in which the basis vectors are not chosen to be orthogonal to each other), the unpolarized differential cross section may contain, in addition to the magnitude-squares, also some other bilinear products of amplitudes, but in any case, the complete set of magnitude-squares are always present. If then the additional imposition of symmetries reduces the number of independent amplitudes, all that happens to the differential cross section is that it assumes a different numerical value, being now the sum of a different set of magnitude-squares. But the numerical value of the differential cross section, in any case, is something we can predict only if we have complete knowledge of the dynamics, which determines the values of the amplitudes and hence of the magnitude-squares. Thus, in the absence of a complete knowledge of dynamics, measuring unpolarized cross sections alone will not help us to test symmetries.

In contrast, polarization observables contain either sums and differences of magnitude-squares or bilinear products of different amplitudes. In such a situation the additional imposition of symmetries and the resulting new constraints on the reaction amplitudes can make a polarization observable zero when it was not before, or can establish relationships among polarization quantities which hold independently of dynamics and at any energy and angle. These vanishings or relationships, which did not exist before the imposition of the additional symmetries, can therefore serve as tests of such symmetries.

Let me now turn to the interesting question of disentangling tests of various symmetries in a situation when two or more of such symmetries are imposed simultaneously. As I just explained, the violation of a conservation law will manifest itself in a polarization quantity ceasing to vanish, or a linear relationship between polarization quantities ceasing to hold. In most cases, however, if we focus only on one such vanishing observable or one such linear relationship, it can undergo such an alteration in several ways and for several reasons.

To illustrate this on a hypothetical and artificially simple example, let us consider a reaction with four complex amplitudes which we will call a,b,d, and d. All four amplitudes are in general non-zero if only Lorentz invariance holds. Let us now consider two symmetries, designated as R and Q, and let us assume that if R is imposed on the reaction in addition to Lorentz invariance, the two amplitudes a and c vanish, while if Q is imposed in addition to Lorentz invariance, a and d vanish. If we now consider the polarization observable which is given by the bilinear combination of amplitudes Re $(ab^* + cd^*)$, we see that this observable will vanish either when R is additionally imposed or when Q is additionally imposed. Thus this particular observable cannot be used as an unambiguous test of the validity of either of the two above symmetries. We can, however, come to a firm conclusion by considering also another polarization observable which is given in terms of the amplitudes by $\mathrm{Re}(ac^* + bd^*)$, the vanishing of which can tell the two possibilities apart.

This is the essence of the effect that was used in two papers to discuss tests of parity conservation[2] and of time reversal invariance,[3] respectively, from the point of view of whether they might also indicate apparent violation of these two symmetries when actually these symmetries are unviolated but Lorentz invariance is instead violated. In both cases the answer is that indeed, the commonly used tests for parity conservation and time reversal invariance, respectively, do have this ambiguity. In both cases the ambiguity can be resolved by additional experiments. Since in these cases one of the two symmetries is Lorentz invariance itself, the violation of which would bring about a gigantic change in the structure of the reaction matrix, the additional experiments needed to separate it from parity conservation or time reversal invariance are subtle and in fact need to be based on some assumption about the way Lorentz invariance is broken. The particular way illustrated in the two papers was in terms of a macroscopic unisotropy of space, in other words, a macroscopic direction (a cosmological one) which is differentiated from other directions.

Let me now turn to a different result which pertains to time reversal invariance alone. We saw that the constraints of an additionally imposed symmetry can cause either a polarization observable to vanish or a linear relationship among polarization observables to develop. From an experimental point of view, the former possibility is much to be preferred, because it allows a "null experiment", that is, an experiment in which deviations from zero are to be measured. If the breaking of the symmetry is slight and hence the deviation of the observable from zero is small, such null-experiments can be performed to very high degrees of accuracy.

In contrast, the experimental checking of linear relationships among observables can be carried out to a much lesser accuracy only, since in effect we have to make absolute measurements of several different observables.

This can be illustrated well in the case of testing parity conservation in strong interaction reactions, where a tiny violation is expected from the simultaneous presence of weak interactions. In that case, choosing a polarization observable which vanishes in the case when parity conservation is unbroken (such as a component of the vector polarization of one particle in the reaction plane with the other three particles unpolarized) a measurement to very high degree of accuracy (one part in 10 million) can be made to measure the admixture of weak interaction.

The question then arises whether such a null-experiment is also available to test time reversal invariance. The commonly suggested dynamics-independent tests of time reversal invariance involve the comparison of the unpolarized cross sections of a reaction and its time-reversed reaction, but such a comparison can be performed only to a very limited degree of accuracy.

Unfortunately, the answer to the above question is in the negative. It can be shown in complete generality[4] (using the optimal formalism) that it is impossible to construct, in any reaction in atomic, nuclear, or particle physics, a null experiment that would unambiguously test the validity of time-reversal invariance independently of dynamical assumptions.

The proof is quite simple. With the optimal formalism it is easy to write down an expression for the completely general case of any four-particle reaction, containing particles with arbitrary values of spins, which gives the relationship between the various polarization quantities and the bilinear products of amplitudes. In that formula one can then change the various arguments to their time-reversed values, and see whether these two expressions can possibly be the negatives of each other, in which case the observable would be a vanishing one under time reversal invariance. Inspection of the expression reveals that the task is impossible.

The significance of this result is that we cannot count, in the foreseeable future, on testing time reversal invariance to a high degree of accuracy in a dynamics-independent way in any reaction, even though doing so is a task of very high priority, especially in elementary particle physics.

As a final example for the way polarization experiments provide important results regarding symmetries, let me turn to an even more general situation, of which polarization considerations form a special case.

Let us consider a general quantum mechanical system undergoing some process of transformation. This system is characterized in the initial state by a set of quantum numbers (each having a set of possible values), and similarly in the final state by another (or the same) set of quantum numbers. Let us also assume that we are aware of the existence of all these quantum numbers except for one, and that we have therefore measuring equipment which is sensitive to the various possible values of each of the quantum numbers except for the one that we are unaware of. It can then be shown[5] that if we measure the various transition amplitudes from the initial states to the final states (but of course always averaging over the values of the one quantum number the existence of which we are unaware of), then simply from the lack of consistency of the measurements performed we can conclude that there is a "secret" quantum number that we missed.

The proof of this theorem is based on the observation that while the entire assembly of quantum numbers with all their values form a regular complex vector space for the transition amplitudes pertaining to these quantum numbers, the incomplete set of all quantum numbers minus one does not form such a vector space, and hence certain general and basically geometric relationships among the magnitudes and phases of the amplitudes do not hold. This failure for these relationship to hold, which can be experimentally established through polarization experiments, indicates that we missed a quantum number.

Applying this theorem to polarization phenomena, we can, for example, conclude that in "spinclusive" reactions (in which nothing is ever measured about the polarization of one particular particle), we can nevertheless conclude whether that particle has zero spin or not. There are also applications to various types of other inclusive reactions, and to applications of polarization measurements to the exploration of dynamics.

This concludes this short presentation, the aim of which was to illustrate the important role polarization experiments play in the exploration of symmetries, a role that cannot be assumed by unpolarized cross section measurements.

References

1 The original paper formulating the optimal formalisms was G.R. Goldstein and M.J. Moravcsik, Ann. Phys. (N.Y.) 98, 128 (1976). For a very recent overview of this approach with a complete list of references, see M.J. Moravcsik, "Polarization as a Probe of High Energy Physics", in the Proceedings of the Tenth Hawaii Topical Conference on High Energy Physics, (to be published) (1986).

2 G.R. Goldstein and M.J. Moravcsik, Phys. Rev. D25, 2934 (1982).

3 M.J. Moravcsik, Phys. Rev. Lett. 48, 718 (1982).

4 F. Arash, M.J. Moravcsik, and G.R. Goldstein, Phys. Rev. Lett. 54, 2649 (1985).

5 M.J. Moravcsik, "Detecting 'Secret' Quantum Numbers", Phys. Rev. Lett. 6908 (1986).

BOSON REALIZATION OF SYMPLECTIC ALGEBRAS AND

THE MATRIX REPRESENTATION OF THEIR GENERATORS

Marcos Moshinsky*

Departamento de Física Teórica
Instituto de Física, UNAM
Apdo. Postal 20-364, 01000 Mexico, D.F., Mexico

ABSTRACT

The purpose of this note is to show that a procedure for the matrix representation of Lie algebras introduced by Gruber and Klimyk can be applied to symplectic ones and, when combined with a Dyson type boson realization of these algebras, it provides a convenient technique for the matrix representation of their generators. We illustrate the procedure for the case of sp(4,R) and indicate how its extension to sp(6,R) can be useful for applications to the symplectic model of the nucleus.

One of the important microscopic models for collective phenomena in nuclei is based on the symplectic Lie algebra sp(6,R) also known as sp(3,R). In this model the Hamiltonian is in the enveloping algebra of sp(6,R), where the generators of the latter can be built in terms of the coordinates and momenta or, equivalently, the creation and annihilation operators, associated with the A-nucleon problem. To obtain the eigenvalues of this Hamiltonian as well as the matrix elements of transition operators, one requires the matrix representation of the generators of symplectic algebras with respect to states in a basis associated with an irrep in the positive discrete series of the latter.

The purpose of this note is to indicate that a procedure originally introduced by Gruber and Klimyk[1] for some Lie algebras, can be extended to the symplectic ones and, when combined with the Dyson boson realization of these algebras[2,3], provide a convenient technique for the matrix representation of their generators.

We shall first illustrate our procedure in the simple

*Member of El Colegio Nacional

case of su(2) and then apply it in sp(4,R), leaving for the last paragraph how our results can be extended to symplectic algebras of arbitrary dimension.

We designate the generators of su(2) in spherical components as

$$S_1=-(1/\sqrt{2})(S_x+iS_y),\quad S_0=S_z,\quad S_{-1}=(1/\sqrt{2})(S_x-iS_y), \tag{1}$$

with the commutation rules

$$[S_0,\ S_{\pm 1}]=\pm S_{\pm 1}\ ,\quad [S_{-1},\ S_1]=S_0\ . \tag{2}$$

In the usual basis matrix representation of the generators is

$$(s\sigma'|S_{\pm 1}|s\sigma)=\mp(1/\sqrt{2})\left[(s\mp\sigma)(s\pm\sigma+1)\right]^{1/2}\delta_{\sigma',\sigma\pm 1}\ , \tag{3a}$$

$$(s\sigma'|S_0|s\sigma)=\sigma\delta_{\sigma',\sigma}\ . \tag{3b}$$

An alternative, though non-normalized, basis can be introduced by the definition

$$|s\sigma>\equiv S_1^{s+\sigma}|s,-s> \tag{4}$$

where $|s,-s>$ is the lowest weight state satisfying

$$S_{-1}|s,-s>=0\ ,\quad S_0|s,-s>=-s|s,-s>\ . \tag{5}$$

Applying then $S_{\pm 1}$, S_0 to the state (4) and using the commutation relations (2) we obtain

$$S_q|s\sigma>=\sum_{\sigma'}|s\sigma'>M^q_{\sigma'\sigma}\ ;q=1,0,-1\ , \tag{6}$$

where

$$M^1_{\sigma'\sigma}=\delta_{\sigma'\sigma+1}\ ,\quad M^0_{\sigma'\sigma}=\sigma\delta_{\sigma'\sigma}\ ,\quad M^{-1}_{\sigma'\sigma}=-(1/2)(s+\sigma)(s-\sigma+1)\delta_{\sigma'\sigma-1} \tag{7}$$

gives us a matrix representation of the su(2) Lie algebra which is different from the standard one indicated in (3).

A way to obtain the matrix representation (7) of the generators of su(2) by simple differentiation instead of using the commutation relations, is through the boson realization of this Lie algebra. We start with the boson creation α and annihilation $\bar{\alpha}$ operators satisfying the commutation rule

$$[\bar{\alpha},\alpha]=1 \tag{8}$$

From coherent state considerations[4] we find out that we can express the generators S_q, $q=1,0,-1$ of su(2) in terms of $\alpha,\bar{\alpha}$ as

$$S_1=-\alpha(\alpha\bar{\alpha}-2s)\ , \tag{9a}$$

$$S_0=\alpha\bar{\alpha}-s\ , \tag{9b}$$

$$S_{-1}=-(1/2)\bar{\alpha}\ . \tag{9c}$$

From (8) we can immediately check that the S_q satisfy the commutation rules (2) of an su(2) Lie algebra. Furthermore, again from (8), we can interpret

$$\bar{\alpha}=\partial/\partial\alpha \tag{10}$$

when applied to any polynomial expression in the creation operators.

From the differential form (9,10) of the S_q we see that the effect of these generators on the state

$$|s\sigma\}=[(s-\sigma)!]^{-1}\alpha^{s+\sigma}|0\} , \tag{11a}$$

where

$$\bar{\alpha}|0\}=0 , \tag{11b}$$

is the same as that of S_q on the states $|s\sigma>$ of (4). Thus the matrix representation of the generators of su(2) is again given by (7) and it can be obtained by simple differentiation.

Furthermore with the help of (7) we can get a recursion relation for the overlap of the states (11) i.e.

$$\{s\sigma|s\sigma\}=-\{s\sigma-1|S_{-1}|s\sigma\}=-(1/2)(s+\sigma)(s-\sigma+1)\{s\sigma-1|s\sigma-1\} , \tag{12}$$

and once this overlap is known we can determine the normalized states $|s\sigma)$ associated with the $|s\sigma\}$.

We proceed now to extend the above analysis from su(2) to sp(4,R) as an illustration of the procedure one can follow for all symplectic algebras. In reference 3 we gave the generators of sp(4,R) in vector form with cartesian components. In this paper we prefer to express the vectors in spherical components q=1,0,-1 as in (1). We thus have the ten generators as

$$N,\ B^+_q,\ J_q,\ B_q\ ;\ q=1,0,-1 , \tag{13}$$

with well known[3,5] commutation relations.

The set of generators (13) can be divided into three subsets of raising, weight and lowering type, which are separated by semicolors[3,5]

$$B^+_q, J_1;\ N, J_o;\ B_q, J_{-1} \tag{14}$$

The lowest weight state, which we designate by $|ws>$ can now be characterized by

$$B_q|ws>=0\ ;\ q=1,0,-1\ ;\ J_{-1}|ws>=0\ ; \tag{15a,b}$$

$$N|ws>=w|ws>\ ;\ J_o|ws>=-s|ws> , \tag{15c,d}$$

where w,s are integers or semi-integers[3] that characterize the

the irreps of sp(4,R) in the positive discrete series.

The analysis of Gruber and Klimyk[1] as well as a previous discussion[3,5] indicates that the full basis for the irrep (ws) of sp(4,R) is given by applying powers of the raising generators B_1^+, B_o^+, B_{-1}^+, J_1 i.e. elements of the enveloping algebra, to the lowest weight state $|ws\rangle$. Thus we can characterize this basis by

$$|(ws)NM\mu\sigma\rangle=$$

$$(B_1^+)^{(1/2)(N+M-\mu-\sigma)}(B_o^+)^{\mu}(B_{-1}^+)^{(1/2)(N-M-\mu+\sigma)}(J_1)^{s+\sigma}|ws\rangle \tag{16}$$

where the choice of the exponents guarantees[5] that the ket (16) is an eigenstate of the weight generators $\mathcal{N}$, J_o i.e.

$$\mathcal{N}\,|(ws)NM\mu\sigma\rangle=(N+w)\,|(ws)NM\mu\sigma\rangle\ , \tag{17a}$$

$$J_o|(ws)NM\mu\sigma\rangle=M|(ws)NM\mu\sigma\rangle\ . \tag{17b}$$

By applying the generators (13) of sp(4,R) to the states (16) we can obtain, with the help of the commutation relations as we did in (6), (7) for su(2), the matrix representation of these generators for the irrep (ws) of sp(4,R). This procedure is greatly simplified by the use of the Dyson boson realization[2,5] of sp(4,R), as we proceed to show.

In previous references[3,5] we have indicated how the generators of sp(4,R) can be expressed in terms of those of the direct sum of $\omega(3)$ and su(2), where the former is a Weyl Lie algebra in three dimensions whose generators are the creation operators β_q and annihilation operators $\bar{\beta}^q=(-1)^q\bar{\beta}_{-q}$, q=1,0,-1, satisfying

$$[\beta_q,\beta_{q'}]=[\bar{\beta}^q,\bar{\beta}^{q'}]=0;\ [\bar{\beta}^q,\beta_{q'}]=\delta^q_{q'}\ , \tag{18a,b}$$

while the latter has as generators the spin operators satisfying

$$[S_o,S_{\pm1}]=S_{\pm1},\ [S_{-1},S_1]=S_o\ . \tag{18c,d}$$

Furthermore $\omega(3)$ and su(2) are independent so that

$$[\beta_q,S_{q'}]=[\bar{\beta}_q,S_{q'}]=0\ . \tag{18e}$$

In vector notation the realization of the ten generators (13) in terms $\underline{\beta}$, $\underline{\bar{\beta}}$, $\underline{S}$ and the eigenvalue w in (15c) is given by[3,5]

$$\underline{B}^+=\underline{\beta};\ \underline{J}=\underline{L}+\underline{S};\ \mathcal{N}=n+w\ , \tag{19a,b,c}$$

$$\underline{B}=-\underline{\beta}(\underline{\bar{\beta}}\cdot\underline{\bar{\beta}})+(2n+2w)\underline{\bar{\beta}}-2i(\underline{\bar{\beta}}x\underline{S})\ . \tag{19d}$$

where

$$\underline{L}=-i(\underline{\beta}\times\underline{\bar{\beta}}) \ , \quad n=\underline{\beta}\cdot\underline{\bar{\beta}} \ . \tag{20a,b}$$

We can immediately check from (18) that N, B_q^+, J_q, B_q of (19) satisfy the commutation rules for the generators of sp(4,R)[3,5]. Note though that from the hermitian properties of these generators we see that

$$\beta_q^+ \neq \beta^q \tag{21}$$

and thus we have what is known as a Dyson type boson realization[2,3] and not a Holstein-Primakoff one.

We can furthermore express S_q, $q=1,0,-1$, in the boson realization (9) which is again of the Dyson type, and therefore we can write N, B_q^+, J_q, B_q; $q=1,0,-1$, in terms of the creation operators β_q, α and the annihilation ones $\bar{\beta}^q$, $\bar{\alpha}$ which besides the commutation rules (18) satisfy

$$[\beta_q,\alpha]=[\bar{\beta}_q,\alpha]=[\beta_q,\bar{\alpha}]=[\bar{\beta}_q,\bar{\alpha}]=0 \ ; \ [\bar{\alpha},\alpha]=1 \ . \tag{22a,b}$$

We now define the boson states

$$|(ws)NM\mu\sigma\}= (\beta_1)^{(1/2)(N+M-\mu-\sigma)}(\beta_0)^{\mu}(\beta_{-1})^{(1/2)(N-M-\mu+\sigma)}[(s-\sigma)!]^{-1} \times \alpha^{s+\sigma}|0\} \tag{23}$$

where the boson vacuum has the property

$$\bar{\beta}_q|0\}=0; q=1,0,-1; \bar{\alpha}|0\}=0 \tag{24}$$

We immediately conclude that applying the operators (19) in which S_q is replaced by (9), to the states (23), gives the same result as applying the generators (13) to the state (16). Thus the matrix representation of the generators in the basis (16) can be determined applying the operators (19) to (23), where these operators are of a differential type as from the commutation rules (18b), (22b) we can make the replacements

$$\bar{\beta}^q=\partial/\partial\beta_q, q=1,0,-1, \bar{\alpha}=\partial/\partial\alpha \tag{25}$$

If we express generically by X one of the generators of sp(4,R) we will have

$$X|(ws)NM\mu\sigma> = \sum_{N'M'\mu'\sigma'} |(ws)N'M'\mu'\sigma') \ M^X_{N'M'\mu'\sigma',NM\mu\sigma} \tag{26}$$

where

$$\underset{\sim}{X} \equiv \| M^{X}_{N'M'\mu'\sigma',NM\mu\sigma} \| \tag{27}$$

is the matrix representation of X in the enveloping algebra basis (16). The explicit expressions for the elements of these matrix representations are given in a paper by Castaños and Moshinsky.[5]

Unfortunately the basis $|(ws)NM\mu\sigma\rangle$ is non-orthonormal and thus we have to determine the overlaps

$$\langle (ws)NM\mu'\sigma' | (ws)NM\mu\sigma \rangle \tag{28}$$

where w,s,N,M, are the same in bra and ket as they are eigenvalues of Hermitian operators in (15c,d), (17a,b). From the definition (16) of the states it is clear that the matrix representation (26) provides a recursion relation for the determination of the overlap (28) for the states associated with irreps in the positive discrete series of sp(4,R), in a similar way that (7) does it for su(2) as indicated in (12).

With the help of $\underset{\sim}{X}$ of (27) we can then carry out calculations for eigenvalues of the Hamiltonian in the enveloping algebra of sp(4,R), as well as matrix elements of transition operators. The extension to the Lie algebras sp(2d,R) where d is any integer is immediate as again we can divide the set of d(2d+1) generators in raising, weight and lowering ones.[3] We thus can define a lowest weight state in a way similar to (15) and construct our full basis by applying raising generators as in (16). The Dyson realization of sp(2d,R) is also available[6] and thus a representation of the type (26) can again be determined by differentiation. Thus our procedure is applicable for all symplectic algebras and, in particular, we plan to implement it for sp(6,R) in view of its relevance for collective phenomena in nuclei.

REFERENCES

1. B. Gruber and A.U. Klimyk, J. Math. Phys. 25; 755 (1984)
2. F.J. Dyson, Phys. Rev. 102; 1217 (1956)
3. O. Castaños, E. Chacón, M. Moshinsky and C. Quesne, J. Math. Phys. 26; 2107 (1985);
   O. Castaños, P. Kramer, M. Moshinsky, J. Math. Phys. 27; 924 (1985)
4. P. Kramer and M. Saraceno, Geometry of the Time Dependent Variational Principle, Springer N.Y. (1981) p. 50
5. O. Castaños and M. Moshinsky, J. Phys. A. Math. Gen. (submitted for publication)
6. M. Moshinsky, Nucl. Phys. A421; 81 (1984).

INTEGRAL TRANSFORMS ON HOMOGENEOUS SPACES OF THE DE SITTER AND CONFORMAL GROUPS

P. Moylan

Department of Science and Mathematics
Saint Louis University
Cahokia, ILL. 62206 USA

INTRODUCTION

Solutions to many problems in physics often invoke the use of integral transforms which are associated with certain invariance or transformation groups of the particular physical system under consideration. The most well-known example is the use of the Fourier transform in problems of non-relativistic and relativistic physics. There are other integral transforms which are also very useful. Fock's solution to the bound state hydrogen atom problem[1] uses an integral transform which is related, thru stereographic projection from S^3 onto R^3, to a Knapp-Stein intertwining operator[2] of the associated $SO_o(1,4)$ transformation group of the problem.[3] The Wightman function for a masseless scalar field is a kernel of an intertwining operator between two elementary representations (ERs) of the conformal group.[4] We also wish to note the use of the Radon transform in problems of radiography.[5]

In this paper we shall consider integral transforms which are generalizations of usual Fourier transforms associated with Euclidean or pseudo-Euclidean motion groups to transforms on Riemannian or pseudo-Riemannian symmetric spaces of semisimple Lie groups, which are related to the corresponding motion groups via group contractions.[6] Specifically we construct transforms on de Sitter space, $SO_o(1,4)/SO_o(1,3)$ and on a certain subset of the Einstein universe or "universal cosmos" of Segal's theory.[7,8] These transforms play a useful role in solutions of problems in conformal quantum field theory and in quantum field theory on de Sitter space-time.[9,10] Since the constructions are very similar we shall only treat the de Sitter case in detail. First, however, it is necessary to introduce some facts about wave equations in Robertson-Walker space-times which shall be needed in the analysis of the integral transforms.

WAVE EQUATIONS IN ROBERTSON-WALKER SPACE-TIMES

There is one class of space-times on which, in special cases, the quantum field theory is highly developed and reasonably well understood. These are the so-called Robertson-Walker space-times. De Sitter and anti-de Sitter space-times as well as Minkowski space-time and the Einstein universe provide interesting examples of Robertson-Walker cosmologies. A Robertson-Walker space-time, $\mathcal{M}$, is characterized by a metric

$$ds^2 = g_{\mu\nu}dx^\mu dx^\nu = dt^2 - S^2(t)\,d\sigma^2 . \tag{1}$$

$d\sigma^2$ is the metric of a three space of constant curvature scalar, κ , which for suitable choice of units can be chosen to have the value +1, 0 or -1. For us it is sufficient to assume that $\mathcal{M}$ is diffeomorphic to $\mathcal{S} \times R$ where $\mathcal{S}$ is an orientable three manifold. We denote the component connected to the identity of the isometry group of $\mathcal{M}$ by $I(\mathcal{M})$. Robertson-Walker space-times are at least locally conformally flat.[11] To see this introduce the conformal time parameter

$$\eta = \int^t S^{-1}(t')\,dt' . \tag{2}$$

Then locally we have

$$ds^2 = C(\eta)\,(d\eta^2 - d\sigma^2) . \tag{3}$$

For $\kappa = +1$

$$d\sigma^2 = d\chi^2 + \sin^2\chi\,(d\theta^2 + \sin^2\theta\,d\phi^2). \tag{4}$$

Introduce Minkowskian coordinates x^0 and r thru[12]

$$\begin{aligned} 2x^0 &= \tan\left(\tfrac{1}{2}[\eta+\chi]\right) + \tan\left(\tfrac{1}{2}[\eta-\chi]\right) \\ 2r &= \tan\left(\tfrac{1}{2}[\eta+\chi]\right) - \tan\left(\tfrac{1}{2}[\eta-\chi]\right) \end{aligned} \quad (0 \le \eta < \chi < 2\pi) . \tag{5}$$

A simple calculation gives

$$dx^{0\,2} - dx^{i\,2} = \tfrac{1}{4}\sec^2\left[\tfrac{1}{2}(\eta-\chi)\right]\sec^2\left[\tfrac{1}{2}(\eta+\chi)\right](d\eta^2 - d\sigma^2). \tag{6}$$

Combining (3) and (6) we obtain the desired conclusion in the case $\kappa = +1$. Similar arguments establish the result for $\kappa = -1$.

The massive scalar wave equation on $\mathcal{M}$ is[13]

$$(\Box - \tfrac{1}{6}K + \mu^2)\,\phi = 0 . \tag{7}$$

K is the Ricci-scalar and $\Box$ is the Laplace-Beltrami operator on $\mathcal{M}$. According to Penrose it is conformally invariant for $\mu = 0$.[13] A generalization of the usual Minkowskian norm is

$$\|\phi\|^2 = -i \int_\Sigma \{\phi^*(x) \overleftrightarrow{\partial_\mu} \phi(x)\} [-g(x)]^{\frac{1}{2}} \eta^\mu d\Sigma \quad (g(x) = \det(g^{\mu\nu}(x))) \tag{8}$$

where Σ is a space-like hypersurface and $d\Sigma$ is the volume element on Σ. η^μ is a future time-like unit normal vector. Gauss' theorem implies $\|\phi\|$ is independent of Σ. For the Dirac equation on $\mathcal{M}$ we take the equation[14]

$$\gamma^\mu [\partial_\mu + i\Gamma_\mu(x)] \Psi(x) = -im \Psi(x) \tag{9}$$

where $\Psi \in C^\infty(\mathcal{M}, \mathbb{C}^2 \oplus \mathbb{C}^2)$, $\{\gamma^\mu, \gamma^\nu\} = 2g^{\mu\nu}$, $\Gamma_\mu = g_{\mu\alpha} \{\lambda^\alpha_\beta \partial_\rho \lambda^\beta_\gamma - \Gamma^\alpha_{\nu\rho}\} S^{\rho\nu}$, $S^{\mu\nu} = \frac{i}{4}[\gamma^\mu, \gamma^\nu]$ and λ^α_β are the vierbein fields. ($\Gamma^\alpha_{\nu\rho}$ are the Christoffel symbols.) The generalization of the Minkowskian norm is

$$\|\Psi\|^2 = \int_\Sigma \langle \Psi, \gamma_\mu \Psi \rangle \eta^\mu (-g)^{\frac{1}{2}} d\Sigma \tag{10}$$

$\langle\ ,\ \rangle$ is the $SL(2,\mathbb{C})$ invariant hermitian form on $\mathbb{C}^2 \oplus \mathbb{C}^2$.

THE DE SITTER CASE

A. Generalities

The metric of de Sitter space-time is obtained from equation (1) by choosing $\varkappa = +1$ and $S(t) = ch(t)$. The manifold, which we shall denote by V_4, can be imbedded into $\mathbb{R}^5$ as the hypersurface

$$V_4 = \{x^\mu \in \mathbb{R}^5 \mid x_0^2 - x_1^2 - x_2^2 - x_3^2 - x_4^2 = -R^2\} \quad (R = 1) \tag{11}$$

and is diffeomorphic to $S^3 \times \mathbb{R}$. $I(V_4) = SO_0(1,4) = G$. Let

$$H = \left\{ \begin{bmatrix} 1 & 0 \\ 0 & A \end{bmatrix} \middle| A \in SO_0(1,3) \right\} \text{ and } H' = \left\{ \begin{bmatrix} \pm 1 & 0 \\ 0 & * \end{bmatrix} \in G \right\}. \tag{12}$$

Then $V_4 \cong G/H$ and let $V_4' \cong G/H'$. Any $f \in C^\infty(G/H')$ can be identified with a C^∞ function on V_4 such that $f(x) = f(-x)\ (x \in V_4)$.

The Lie algebra $\mathfrak{g}$ of G is defined as

$$\mathfrak{g} = \{ a_{\mu\nu} \mid 0 \le \mu \le \nu \le 4;\ a_{00} = a_{ii} = 0;\ a_{ij} = -a_{ji}\ (1 \le i < j \le 4);\ a_{0j} = a_{j0}\ (1 \le j \le 4) \}. \tag{13}$$

Let $(E_{\mu\nu})^\rho_\sigma = \delta^\rho_\mu \delta_{\nu\sigma}$ and set $X_{ij} = E_{ij} - E_{ji}$, $Z_i = E_{i,0} + E_{0,i}$. Let $\mathfrak{k}$, $\mathfrak{a}$ and $\mathfrak{n}$ be subalgebras generated by X_{ij}, Z_4 and $X_{4,i} + Z_i\ (1 \le i \le 3)$, respectively. Let $\mathfrak{m}$ be the centralizer of $\mathfrak{a}$ in $\mathfrak{k}$. Denote the corresponding analytic subgroups of G by K, M, A and N. We have that P = MAN is a minimal parabolic subgroup of G.

There are several decompositions of G which play an important role in the representation theory. The most informative mathematical results are obtained using the so-called "noncompact picture", which is associated with the (Gelfand-Naimark) Bruhat decomposition.[15] The decompositions which we shall need are the Hannabuss decomposition[16]

$$g = h\,a\,n \quad h \in H',\ a \in A,\ n \in N \quad \text{for almost all } g \in G, \tag{14}$$

and the decomposition of G into the cartesian product $H \times V_4$.[17] These decompositions are associated with actions of G on the momentum hyperboloid, T^3 and on de Sitter space, V_4.

B. Representations

The elementary representations (ERs) or non-unitary principal series[18] representations of G are characterized by two numbers s and ν with $s = 0, 1/2, 1, 3/2 \ldots$ and $\nu \in \mathbb{C}$. Various realizations of the ERs of G are explicitly given in [9]. We shall use two of them. For simplicity we consider only $s = 1/2$ and $\nu = 3/2 + i\rho$, $\rho \in \mathbb{R}^+$. One realization is on the space of solutions of the usual Dirac equation in momentum space. The Dirac equation reads

$$\begin{gathered} \bar{\gamma}^{\mu} p_{\mu} \Psi(p) = m \Psi(p) \quad (p \in T^3 \text{ i.e. } p_\mu p^\mu = m^2) \\ \{\bar{\gamma}^{\mu}, \bar{\gamma}^{\nu}\} = 2\eta^{\mu\nu}, \quad \eta^{\mu\nu} = \mathrm{diag}(1, -\mathbb{1}_3) \quad (m = \tfrac{\rho}{R}). \end{gathered} \tag{15}$$

For the norm we take the usual one of the Dirac theory. Denote the space of normalizable solutions of the Dirac equation by $\mathcal{R}^{\nu,s}$ ($s = 1/2$). The representation of G on $\mathcal{R}^{\nu,s}$ is given by

$$(U(g)\Psi)(p) = |\mu(\bar{g}^{-1}, \bar{\gamma}(p))|^{-\nu+s}\, D(\bar{g})\, \Psi(m\,\bar{g}^{-1}\tfrac{p}{m}) \tag{16}$$

where $\bar{\gamma}(p) \in SO_0(1,3)$ is the matrix representative of the rotation free Lorentz boost corresponding to momentum p, and

$$|\mu(\bar{g}^{-1}, \bar{\gamma}(p))| = |a(\bar{\gamma}^{-1}(p)\, g\, \bar{\gamma}(m\bar{g}^{-1}\tfrac{p}{m}))| \quad (|a(x)| = e^{t(x)},\ a(x) = e^{I_{04} t(x)}). \tag{17}$$

$\bar{\gamma}^{-1}(p)\, g\, \bar{\gamma}(m\bar{g}^{-1}\tfrac{p}{m}) = h\, a(\bar{\gamma}^{-1}(p)\, g\, \bar{\gamma}(m\bar{g}^{-1}\tfrac{p}{m}))\, n$ by equation (14) and

$$D(\bar{g}) = \exp(i \tfrac{\omega^{ab}}{2} \bar{S}_{ab}), \quad \bar{S}_{\mu\nu} = -\tfrac{i}{4}[\bar{\gamma}_\mu, \bar{\gamma}_\nu], \quad \bar{S}_{4\rho} = -\tfrac{i}{2}\bar{\gamma}_\rho, \tag{18}$$

and $p \to m\,\bar{g}^{-1}\frac{p}{m}$ is the action of $g \in G$ on T^3 associated with the Hannabuss decomposition. (ω^{ab} are the infinitesimal rotation parameters of $g \in G$.)

The other realization is the realization on the space of solution of the Dirac equation in de Sitter space. The equation which Dirac proposed in 1935 is [19]

$$S^{ab}M_{ab}F(\xi) = \bar{\sigma}F(\xi) \quad (\bar{\sigma} = -\sigma - 3\ ,\ \sigma = -\tfrac{3}{2} + s - \nu)\ (s = \tfrac{1}{2}) \tag{19}$$

with, in our case, $F(\xi) \in V_4'$, and

$$M_{ab} = i(\xi_a \tfrac{\partial}{\partial \xi^b} - \xi_b \tfrac{\partial}{\partial \xi^a}) \quad \text{and} \quad S_{ab} = \tfrac{i}{4}[\gamma_a, \gamma_b],\ \{\gamma_a, \gamma_b\} = 2\eta_{ab},$$
$$\eta_{ab} = \mathrm{diag}(1, -1_3, -1).$$

The norm is defined by equation (10). Denote the Hilbert space of normalizable solutions of equation (19) by $\mathcal{H}^{\nu,s}$. The representation on $\mathcal{H}^{\nu,s}$ is, with $\xi \to g^{-1}\xi$ being the action of G on V_4,

$$\pi(g)F(\xi) = D(g)F(g^{-1}\xi) \quad (D(g) = \exp\{\tfrac{i}{2}\omega^{ab}S_{ab}\}). \tag{20}$$

The norm is K-invariant. We can and do arrange things so that

$$S^{\mu\nu} = \bar{S}^{\mu\nu} \quad \text{and} \quad \bar{\gamma}^{\mu} = -\gamma^4\gamma^{\mu}. \tag{21}$$

We note that equation (19) is equivalent to the usual Dirac equation on $\mathcal{M} = V_4'$ which we presented in the first section.[20] The equivalence between $(U, \mathcal{R}^{\nu,s})$ and $(\pi, \mathcal{H}^{\nu,s})$ $(s = \frac{1}{2},\ \nu = \frac{3}{2} + i\rho)$ is accomplished by the following integral transform:

$$F(\xi) = (\Pi^{\varkappa}\phi)(\xi) = \varkappa_0^R(s,\rho)\int_{T^3} d\Omega_{T^3}\, \phi(\tfrac{p}{m})\, |\xi^\mu \tfrac{p_\mu}{m} - \xi^4|^{-\bar{\nu}-s}. \tag{22}$$

It is argued in [9] for any s that this is an intertwining operator between $(U, \mathcal{R}^{\nu,s})$ and $(\pi, \mathcal{H}^{\nu,s})$. There appears to be several possible determinations of the factor $|\varkappa_0^R(s,\rho)|^2$. One answer is obtained by requiring $\Pi^{\varkappa}$ to be a unitary equivalence. At least in the spin zero case ($s = 0$), another possibility is to relate $|\varkappa_0^R(s,\rho)|^2$ to the Plancherel factor in the decomposition of $\mathcal{L}^2(V_4')$ into eigenspaces of the Laplace-Beltrami operator on V_4'. This is accomplished by first using stereographic projection from T^3 into S^3 and then performing an analysis similar to that of Strichartz [21] and Rossmann [22]. We shall apply this method to the determination of the factor, corresponding to $|\varkappa_0^R(0,\rho)|^2$, for the conformal case in the next section.

The integral formula (22) for s = 0 expressed as an integral over S^3 is related to a special case of a general formula for the integral transform of K finite eigenfunctions of the $SO_0(p,q)$ invariant Laplace-Beltrami operator on $\mathcal{L}^2(SO_0(p,q)\ /\ (SO_0(p,q-1))$, which is presented in [21] and [22]. Our arbitrary spin result for the de Sitter case suggests that the methods employed in these references are capable of generalization to

Hilbert spaces of vector-valued functions on hyperboloids.

THE CONFORMAL CASE

Most of the above results generalize to the conformal group, $SO_0(2,4)$. A classification of the unitary irreducible representations (UIRs) of the four-fold covering group, SU(2,2), of $SO_0(2,4)$ is given in [23]. Since $SO_0(2,4)$ is a quotient of SU(2,2), their classification also serves as a classification of the UIRs of $SO_0(2,4)$. According to the program announced in [24], concrete realizations of the ERs and explicit constructions of the intertwining operators between the various ERs of SU(2,2) are given in [25].

In [21] and [22] the decomposition of $\mathcal{L}^2(SO_0(2,4)/SO_0(2,3))$ into UIRs of $SO_0(2,4)$ is described. In the notation of [24] and [25], the relevant ERs are characterized by (we need only the continuous part of the spectral decomposition)

$$\chi_2 = [1,1,\varepsilon, i\rho] \quad (\varepsilon = 0,1;\ \rho \in \mathbb{R}) \tag{23}$$

The UIRs are realized on the following representation spaces (see [21] and [22]):

$$\mathcal{L}^2_\varepsilon(\bar{M}) = \{ f : \bar{M} \to \mathbb{C} \mid f(-v,-u) = (-1)^\varepsilon f(v,u),\ \varepsilon = 0,1, \int_{\bar{M}} |f(y)|^2 dy < \infty \} \tag{24}$$

where y = $(v,u) \in \bar{M} \cong \tilde{M}/\mathbb{Z} \cong S^1 \times S^3$ and $v = (v_1, v_2) \in S^1$, $u = (u_1,u_2,u_3,u_4) \in S^3$ $(|v|^2 = |u|^2 = 1)$. ($\tilde{M} = \mathbb{R} \times S^3$ is the "universal cosmos".) dy is the curved measure on $\bar{M}$.[8] Let $\bar{\bar{M}} = \{(v,u) \in \bar{M} \mid v^1 + v^4 > 0\}$. If $f \in \mathcal{L}^2_{\varepsilon=0}(\bar{M})$ then f can be identified with a function in $\mathcal{L}^2(\bar{\bar{M}})$ where

$$\mathcal{L}^2(\bar{\bar{M}}) = \{ f : \bar{\bar{M}} \to \mathbb{C} \mid \int_{\bar{\bar{M}}} |f(y)|^2 dy < \infty \} \tag{25}$$

Introduce coordinates on $\bar{\bar{M}}$ as follows

$$M_0 \cong \mathbb{R}^1 \times \mathbb{R}^3 \longrightarrow \bar{\bar{M}},\ (x^0, \vec{x}) \to y(x^0,\vec{x}),$$
$$y(x^0,\vec{x}) = p(x^0,x)(1-x^2,\ 2x^0,\ 2\vec{x},\ 1+x^2) \tag{26}$$

with

$$p(x^0,x) = [(1-x^2)^2 + (2x^0)^2]^{-\frac{1}{2}},\quad x^2 = x^{0^2} - (\vec{x})^2$$

Note: Equations (5) can be obtained from these by the introduction of angular coordinates on $\bar{M}$ (see[8]). The relation between the standard measure dx on Minkowski space, M_0, and dy is:[8]

$$16\, dx = p^{-4} dy \tag{27}$$

This enables us to define an isometric isomorphism between $\mathcal{L}^2_{\varepsilon=0}(\bar{M})$ and $\mathcal{L}^2(M_0)$ as follows:

$$f(y) = 2^{-2-\frac{i}{2}}[\rho(x^0,x)]^{-2+i\rho}\tilde{f}(x^0,\vec{x}) \tag{28}$$

This isomorphism establishes unitary equivalences between the UIRs of $SO_0(2,4)$ on $\mathcal{L}^2_{\varepsilon=0}(\bar{M})$ and those on $\mathcal{L}^2(M_0)$ (see [8] and [21] for definitions).

The integral transform given in [21] and [22] is:

$$F(\xi) = c(\varepsilon,\rho)\int_{\bar{M}} |(\xi,y)|^{-2-i\rho}\,\mathrm{sign}^{\varepsilon}(\xi,y)\,\varphi(y)\,dy \tag{29}$$

One can show that $F(\xi)$ is an eigenfunction of the Laplace-Beltrami operator, Δ, on $X = SO_0(2,4)/SO_0(2,3)$ with eigenvalue $-4-\rho^2$ and that it is a G-map (Lemma 5 of [22]). We require that $|c(\varepsilon,\rho)|^{-2}d\rho$ be the Plancherel measure in the decomposition of $\mathcal{L}^2(X)$ into eigenspaces of Δ. For $\varepsilon=0$, we determine the measure as follows: From (b) of Lemma 8 in [22] we have

$$|c(\varepsilon,\rho)|^2\varphi(y) = \left[C_-(\varepsilon,i\rho)\left(C_-(\varepsilon,-i\rho)\varphi\right)\right](y) \quad \left(\varphi\in\mathcal{L}^2_{\varepsilon=0}(\bar{M})\right) \tag{30}$$

where (Lemma 7 of [22]):

$$\left(C_-(\varepsilon,i\rho)\varphi\right)(y) = \int_{\bar{M}'} |(y,y')|^{-2-i\rho}\,\mathrm{sign}^{\varepsilon}(y,y')\,\varphi(y')\,dy' \tag{31}$$

Using (26), (27) and (28) with $\varepsilon=0$ we transform this expression into an integral transform on M_0

$$\widetilde{\left(C_-(\varepsilon=0,i\rho)\varphi\right)}(x^0,\vec{x}) = 2^3\int_{M^{0\prime}} |(x-x')^2|^{-2-i\rho}\,\tilde{\varphi}(x^{0\prime},\vec{x}')\,dx' \tag{32}$$

The RHS of the equation is $2^3/M_2(\chi_2)$ times the expression, given in equation (4.51b) of [25], for the Knapp-Stein intertwining operator associated with one of the ERs of equation (23), provided we let $\tilde{\varphi}$ be the function in (4.51b) of [25] which has constant z dependence. Therefore specializing (4.59b) of [25] to the case at hand we obtain

$$|c(\varepsilon=0,\rho)|^2 = \frac{2^6}{M_2(s\chi_2)M_2(\chi_2)} = \frac{2^{10}\tan^2\left(\frac{\pi i\rho}{2}\right)}{\rho^2(\rho^2+1)} \tag{33}$$

This result is the same as that obtained in [22]. The case $\varepsilon=-1$ can be handled similarly. The methods employed here and in [21] and [22], hold quite generally. There are formulae analogous to (29) for spherical functions on Riemannian symmetric spaces and on semisimple Lie groups, and there are relations similar to (33) for the Plancherel measure in these cases.[26, 27]

ACKNOWLEDGMENT

The author thanks John Gilbert for useful discussions.

REFERENCES

1. V. Fock, Z.Physik 98 (1935), 145; J. Schwinger, J. Math.Phy., 5, (1964), 1606.
2. P. Moylan, Fortsch d. Phys., 28, (1980), 269-284.
3. M. Bander, C. Itzykson, Rev. Mod. Phys., 38, (1966), 330.
4. G. Mack, Comm. Math. Phys., 55, (1977), 1-28.
5. S. Helgason, Groups and Geometric Analysis..., (Academic Press, New York, 1984), 130.
6. G. W. Mackey, in Lie Groups and their Representations, Summer School of the Bolyai Janos Mathematical Society, Proceedings, Budapest, 1971, ed. I. M. Gel'fand, (Wiley, London, 1975), 361.
7. I. E. Segal, Proc. Math. Acad. Sci. USA, 79, (1982), 7961.
8. S. M. Paneitz, I. E. Segal, Journ. Funct. Anal., 47, (1982), 78-142 and 49, (1982), 335-414.
9. P. Moylan, Fortsch. d. Phys., 11, (1986) (in press).
10. G. Mack, I. T. Todorov, Phys. Rev. D., 8 # 6, (1972), 1764-1787.
11. Gregg Zuckermann, in Lecture Note in Mathematics 1077, Lie Group Representations III, Proceedings, University of Maryland (1982-1983, Eds. R. Herb. et.al., (Springer-Verlag, New York, 1984), p. 437.
12. S. W. Hawking, G.F.R. Ellis, The Large Scale Structure of Space-Time, (Cambridge, New York, 1973).
13. R. Penrose, Proc. Roy. Soc. London, A284, (1965), 163.
14. D. R. Brill, J. A. Wheeler, Rev. Mod. Phys. 29, #3, (1957), 467.
15. G. Warner, Harmonic Analysis on Semisimple Lie Groups I, Springer-Verlag, (New York, 1972).
16. K. C. Hannabuss, Proc. Camb. Phil. Soc., 70 (1971), 283.
17. J. Hebda, P. Moylan, Homogeneous Spaces and Associated Group Decompositions, St. Louis University, (1986).
18. J. E. Gilbert, R. A. Kunze, P. A. Tomas, Intertwining Kernels and Invariant Operators in Analysis, Univ. of Texas at Austin, (1984).
19. P.A.M. Dirac, Ann. Math, 36, (1935), 657.
20. P. Moylan, Dissertation, U.T. Austin (1982).
21. R. Strichartz, Jour. Funct. Anal., 12, (1973), 341-383.
22. W. Rossmann, Jour. Funct. Anal., 30, (1978), 448-447.
23. A. W. Knapp, B. Speh, Jour. Funct. Anal., 45, (1982), 41.
24. V. K. Dobrev, Jour. Math. Phys., 26 (2), (1985), 235-251.
25. V. K. Dobrev, P. Moylan, MPI Preprint, MPI-PAE/PTh 49/85, (to appear in Jour. Math. Phys.).
26. A. W. Knapp, E. M. Stein, Ann. Math., 93, (1971), 489.
27. S. Helgason, Differential Geometry and Symmetric Spaces, Adacemic Press, (New York, 1962).

VECTOR-SCALAR VERSION OF THE DIRAC LAGRANGIAN

Patrick L. Nash

Division of Earth and Physical Sciences
The University of Texas at San Antonio
San Antonio, Texas 78285-0663

ABSTRACT

Using the recent formulation of the exceptional equivalence of a pair of real Dirac spinors and a real spacetime vector plus four real scalars, the vector-scalar equivalent of the Dirac Lagrangian is derived. The field equations associated with this new Lagrangian describe the dynamical evolution of a spin-1/2 fermion whose corresponding wave function is not a complex Dirac spinor, but is instead a real spacetime vector plus four real scalars.

1. Introduction

As everyone knows, a complex Dirac spinor may be utilized as a spin-1/2 fermion wavefunction. Less well known is the fact that this choice is not unique. In actuality one may employ either a complex Dirac spinor or a complex spacetime vector as a spin-1/2 wavefunction. Moreover, since a complex vector is equivalent to a pair of real vectors, and a vector is in turn equivalent to a set of four scalars (in virtue of the standard tetrad formalism), a real spacetime vector plus four real scalars may also be used as a spin-1/2 wavefunction. Let us see how this somewhat counter-intuitive result arises.

Clearly the standard fermion wave function, a complex Dirac spinor, may be associated with a pair of real Dirac spinors. It is convenient to formulate this association as follows: let D_4 denote real four dimensional Dirac space, D_4^* its dual; let $\lambda \in D_4$ and $\xi \in D_4^*$ be real Dirac spinors of type (1,0) and (0,1), respectively. ξ transforms inversely to λ under $\overline{SO(3,1)}$. We denote the symplectic form on D_4 by ε, and use a standard notation for Dirac's gamma matrices. Our convention for a fermion wave function ϕ is

$$\sqrt{2}\,\phi = \lambda - i\varepsilon^{-1}\tilde{\xi}, \tag{1.1}$$

where the tilde denotes transpose.

Next, let us construct a real eight component spinor ψ according to

$$\psi = \begin{pmatrix} \lambda \\ \tilde{\xi} \end{pmatrix} . \tag{1.2}$$

ψ may be considered as an $\overline{SO(3,4)}$ spinor, where, however, we restrict the action of $\overline{SO(3,4)}$ to a $\overline{SO(3,1)}$ subgroup. More generally, such a real eight component spinor may be regarded as an element of a real eight dimensional vector space $V_{4,4}$, which carries a real irreducible representation of $\overline{SO(3,4)}$ (and also $\overline{SO(4,4)}$, whence the notation). Now, it is known that every element of $V_{4,4}$ is equivalent to a SO(3,4) vector plus a SO(3,4) scalar [see Eq. (2.13) below][1]. This equivalence is nothing more than a realization of E. Cartan's principle of triality[2,3]. Under the restriction of SO(3,4) to SO(3,1), this relationship is manifested in the equivalence of a pair of real Dirac spinors (that is, ψ) and a real SO(3,1) vector plus four real SO(3,1) scalars. [By "equivalence" we mean that there exists a linear one-to-one invertible map from a pair of real Dirac spinors to a real spacetime vector plus four real scalars. This map is explicitly constructed in Ref. (1).]

In light of this relationship it is clear that the Dirac equation is equivalent to a first order wave equation in which the wave function that describes a fermion is given by a real spacetime vector plus four real scalars. This paper is devoted to the construction of the Lagrangian whose associated Euler-Lagrange equations are the vector-scalar version of the Dirac equation.

A summary of the remainder of this paper is as follows. In the next section the exceptional equivalence of a pair of real Dirac spinors and a real SO(3,1) vector plus four real scalars is reviewed. Here we recall the tau matrices from Ref. (1), which are the generators of two inequivalent real 8 × 8 irreducible representations of $\overline{SO(4,4)}$. The exceptional equivalence may be concisely formulated using one of the properties of these remarkable matrices. The Dirac Lagrangian is then given in terms of the real eight component spinor ψ and the tau matrices. We next substitute for ψ using Eq. (2.13) in this eight component version of the Dirac Lagrangian. This yields a fermion Lagrangian whose wave function is a set of spacetime tensors (a real spacetime vector plus four real scalars).

2. The tau matrices and equivalence

The main body of this section begins with the definition of a set of real 8 × 8 matrices that generate two inequivalent irreducible representations of $\overline{SO(4,4)}$ [and by restriction, two equivalent irreducible representations of $\overline{SO(3,4)}$]. Collectively, we call these matrices the tau matrices. Next we state a basic identity satisfied by the tau matrices. This identity is then utilized in formulating the equivalence of a $V_{4,4}$ spinor ψ and a SO(3,4) vector-scalar pair.

Let us first endow $V_{4,4}$ with a $\overline{SO(4,4)}$-invariant pseudo-Riemannian metric $\sigma \longleftrightarrow \sigma_{ab}$, where a,b,... = 1,...,8 are $\overline{SO(4,4)}$ spinor indices. In an admissible frame σ has components

$$\sigma = \begin{pmatrix} 0 & 1 \\ 1 & 0 \end{pmatrix}. \tag{2.1}$$

[A $V_{4,4}$ frame is said to be admissible if it is the image of the canonical frame, to be introduced in Section 3, under the action of $\overline{SO(4,4)}$.] As a quick check on consistency we can verify that $V_{4,4}$ endowed with σ admits so(4,4) as the Lie algebra of its maximal symmetry group: since $\sigma^2 = 1$, σ has eigenvalues ± 1; since $\mathrm{tr}(\sigma) = 0$, these eigenvalues occur with equal multiplicity. Hence σ can be brought to the form diag(1,-1) by a real orthogonal similarity transformation. The set of derivations of the bilinear form σ is just so(4,4).

We now define the generators $(\tau^{A'}, \bar{\tau}^{A'})$, A', B',... = 1,...,8, of two inequivalent real 8 × 8 irreducible representations of $\overline{SO(4,4)}$ by demanding that the tau matrices verify [1]

$$\tilde{\tau}^{A'}\sigma = \sigma\bar{\tau}^{A'} \quad \text{and} \tag{2.2}$$

$$\begin{aligned} \tau^{A'}\bar{\tau}^{B'} + \tau^{B'}\bar{\tau}^{A'} &= 2IG^{A'B'} \\ &= \bar{\tau}^{A'}\tau^{B'} + \bar{\tau}^{B'}\tau^{A'}, \text{ where} \end{aligned} \tag{2.3}$$

$$G_{A'B'} \longleftrightarrow G = \begin{pmatrix} g & 0 \\ 0 & -g \end{pmatrix}, \tag{2.4}$$

$$g_{\alpha\beta} \longleftrightarrow g = \mathrm{diag}(1,1,1,-1) \tag{2.5}$$

is the metric tensor on M_4 (Greek indices run from 1 to 4), and I denotes the 8 × 8 unit matrix. We associate spinor indices according to $\psi \longleftrightarrow \psi^a$ and $\tau^{A'} \longleftrightarrow \tau^{A'a}{}_b$, and use σ (resp., σ^{-1}) to lower (resp.,raise) spinor indices. [We shall spell out the irreducible representation of $\overline{SO(4,4)}$ under which ψ transforms in a moment.] With these conventions Eq. (2.2) may be rewritten as

$$\tau^{A'}{}_{ba} = \bar{\tau}^{A'}{}_{ab}. \tag{2.6}$$

As shown in Ref. (1), the generators of the two inequivalent real 8 × 8 irreducible representations $D_1^{A'B'}$ and $D_2^{A'B'}$ of $\overline{SO(4,4)}$ are given by

$$-4D_1{}^{A'B'} = \tau^{A'}\bar{\tau}^{B'} - \tau^{B'}\bar{\tau}^{A'} \quad \text{and} \tag{2.7}$$

$$4\tilde{D}_2{}^{A'B'} = \bar{\tau}^{A'}\tau^{B'} - \bar{\tau}^{B'}\tau^{A'}. \tag{2.8}$$

Under the action of $\overline{SO(4,4)}$, ψ transforms under the irreducible representation D_1; we say that ψ is a real contravariant $\overline{SO(4,4)}$ spinor. A covariant spinor χ is defined to transform under the action of $\overline{SO(4,4)}$ according to $\chi \longrightarrow \chi D_1(g^{-1})$, $g \in \overline{SO(4,4)}$. We shall usually restrict our attention to the $\overline{SO(3,4)}$ subgroup of $\overline{SO(4,4)}$ for which $\tilde{D}_2(g) = D_1(g^{-1})$, or to an appropriate $\overline{SO(3,1)}$ subgroup.

Let us now turn briefly to the statement of a simple but useful identity satisfied by the tau matrices[1]: if T is an arbitrary 8×8 matrix that transforms according to $T \rightarrow D_1 T D_1^{-1}$ under $\overline{SO(4,4)}$ such that σT is symmetric, i.e., $\widetilde{\sigma T} = \sigma T$, then

$$\bar{\tau}_{A'} T \tau^{A'} = I \,\text{trace}(T), \tag{2.9}$$

where $\bar{\tau}_{A'} = G_{A'B'}\bar{\tau}^{B'}$. With the aid of this identity we may easily formulate the equivalence of a $\overline{SO(3,4)}$ spinor ψ and a SO(3,4) vector-scalar pair.

We chose a real constant spinor $J \in V_{4,4}$ normalized such that

$$\tilde{J}\sigma J = 1, \tag{2.10}$$

but being otherwise arbitrary. The condition that J be constant is not necessary and is imposed solely for the sake of simplicity. We can obtain a resolution of the identity on $V_{4,4}$ utilizing J as follows. We define a matrix T according to

$$T = J\tilde{J}\sigma. \tag{2.11}$$

We observe that T is symmetric and satisfies the conditions upon which Eq. (2.9) is contingent. Moreover $tr(T) = 1$, since $tr(T) = tr(J\tilde{J}\sigma) = \tilde{J}\sigma J = 1$ by Eq. (2.10). Upon substituting $T = J\tilde{J}\sigma$ and $tr(T) = 1$ into Eq. (2.9) we find that

$$\begin{aligned} I &= \bar{\tau}_{A'} T \tau^{A'} \\ &= \bar{\tau}_{A'} J\tilde{J}\sigma\tau^{A'}. \end{aligned} \tag{2.12}$$

Let us now write ψ as $\psi = I\psi = (\bar{\tau}_{A'} J\tilde{J}\sigma\tau^{A'})\psi = \bar{\tau}_{A'} J(\tilde{J}\sigma\tau^{A'}\psi)$, that is

$$\psi = \bar{\tau}_{A'} J\theta^{A'}, \quad \text{where} \tag{2.13}$$

$$\theta^{A'} = \tilde{J}\sigma\tau^{A'}\psi. \tag{2.14}$$

For $A,B,\ldots = 1,\ldots,7$ we have shown that the θ^A transform as a SO(3,4) vector, while θ^8 is a SO(3,4) scalar, when one employs a representation of the tau matrices in which $\tau^8 = \bar{\tau}^8 = I$. We shall adopt this convention here. Eqs. (2.13) and (2.14) express the equivalence of the $\overline{SO(3,4)}$ spinor ψ and the SO(3,4) vector-scalar pair $\theta^{A'}$. In fact, $\{\bar{\tau}_{A'}, J\}$ $A' = 1,\ldots,8$ is a $V_{4,4}$ orthonormal frame, and the $\theta^{A'}$ are the components of the spinor ψ with respect to this frame. Accordingly we put

$$E^a_{A'} = \bar{\tau}^a_{A'b} J^b, \text{ and} \tag{2.15}$$

$$E^{A'}_a = G^{A'B'} \sigma_{ab} E^b_{B'} \longleftrightarrow \tilde{J}\sigma\tau^{A'}. \tag{2.16}$$

The $E^a_{A'}$ verify

$$E^a_{A'} E^{A'}_b = \delta^a_b \text{ and} \tag{2.17}$$

$$E^{A'}_a E^a_{B'} = \delta^{A'}_{B'}. \tag{2.18}$$

Using this new notation Eqs. (2.13) and (2.14) read $\psi^a = E^a_{A'}\theta^{A'}$ and $\theta^{A'} = E^{A'}_a \psi^a$ $(=\psi^{A'})$.

3. The vector-scalar version of the Dirac Lagrangian

In terms of the real eight component ψ the Dirac Lagrangian is

$$2L = \tilde{\psi}\sigma\tau^6[\tau^\alpha(\partial_\alpha - eA_\alpha\tau^5\tau^7) + m\tau^6]\psi. \tag{3.1}$$

Henceforth for simplicity we consider the case $eA_\alpha = 0$. To obtain the vector-scalar version of the Dirac Lagrangian we substitute $\psi = \bar{\tau}_{A'}J\theta^{A'}$ in Eq. (3.1) and simplify. In this process we encounter the SO(3,4) tensor $(A,B,\ldots = 1,\ldots,7)$

$$w^{ABC} = \tilde{J}\sigma\tau^A\tau^B\tau^C J \tag{3.2}$$

and its dual

$$w_{ABCD} = (1/3!)\varepsilon_{ABCDRST}w^{RST} \tag{3.3}$$

We note without proof that

$$w^{ABC}w_{CRS} = \delta^{AB}_{RS} - w^{AB}{}_{RS}. \tag{3.4}$$

Using Eqs. (3.2) - (3.4) and (2.13) in Eq. (3.1) we find that the vector-scalar version of the Dirac Lagrangian is given by

$$2L = \theta^\alpha \overleftrightarrow{\partial}_\alpha \theta^6 + w^{\alpha A6}\theta_A \overleftrightarrow{\partial}_\alpha \theta^8$$
$$-w^{\alpha AB6}\theta_B\theta_{A,\alpha} + m\theta_{A'}\theta^{A'}. \tag{3.5}$$

Let us now evaluate the w tensors. We first define the canonical $V_{4,4}$ frame to be the frame $\bar{\tau}_{A'}, J$ in which J has the special form

$$\sqrt{2}\, J = (0,0,0,1,0,0,0,1). \tag{3.6}$$

By direct evaluation we find that in the canonical frame the components of w^{ABC} are given by (6 = 1', 5 = 2', and 4 = 3'; we use the concrete representation of the tau matrices defined in Ref. [1])

$$w_{jhk} = \varepsilon_{jhk} , \tag{3.7}$$

$$w_{jh'7} = \delta_{jh} , \tag{3.8}$$

$$w_{jh'k'} = \varepsilon_{jhk} , \text{ and} \tag{3.9}$$

$0 = w_{j'h'k'} = w_{jhk'} = w_{jh7} = w_{j'h'7}$. In addition we find that the non-vanishing components of w_{ABCD} are

$$w_{j'h'k'7} = -\varepsilon_{jhk} \tag{3.10}$$

$$w_{jhk'7} = -\varepsilon_{jhk} , \text{ and} \tag{3.11}$$

$$w^{jh}{}_{i'k'} = -\delta^{jh}_{ik}. \tag{3.12}$$

In conclusion we remark without proof that the $w^{AB}{}_{C} = E^{A}_{a} \bar{\tau}^{Ba}{}_{b} E^{b}_{C}$ are a subset of the multiplication constants for a certain real form of the complexified octonion algebra.

Acknowledgement

I wish to thank Ms. Becky Jimenez for typing this manuscript.

References

1. P.L. Nash, J. Math. Phys 27, 1185 (1986).
2. E. Cartan, Lecons sur la theorie des spineurs (Hermann and Cie, Paris, 1938), Vols. I and II.
3. C. C. Chevalley, The Algebraic Theory of Spinors (Columbia U.P., New York, 1954).

DOUBLE COVERING OF DIFFEOMORPHISMS FOR SUPERSTRINGS IN GENERIC CURVED SPACE[1]

Yuval Ne'eman* and Djordje Šijački**

*Sackler Faculty of Exact Sciences, Tel Aviv University
Tel Aviv, Israel
Wolfson Chair Extraordinary in Theoretical Physics
Also on leave from the University of Texas, Austin

**Institute of Physics, P.O. Box 57, Belgrade
Yugoslavia

[1]Supported in part by the US DOE Grant DE-FG05-85ER40200
the US-Israel Binational Science Foundation and by
RZNS (Belgrade)

ABSTRACT

The embedding of the superstring in a generic curved space involves the use of world-spinors behaving according to the (infinite) unitary representations of $\overline{SL}(10,\mathbb{R})$, the double-covering of the linear group on R^{10}.

A supersymmetric extension is provided by the embedding of $\overline{GL}(10,\mathbb{R})$ in the supergroup $\overline{GQ}(10,\mathbb{R})$ whose flat limit reproduces Poincaré supersymmetry.

INTRODUCTION

Closed, unoriented, type I superstrings [1] may provide a finite theory [2] of quantum gravity. The hope that this is indeed the quantum version of the gravitational field is based upon the positive fit at the level of on-mass-shell amplitudes [3], but the picture is otherwise still somewhat obscure. Indeed, in the conventional Lagrangian formulation [1,4,5] for strings or superstrings, the world-sheet R^2 (locally-reparametrizable, or "generally-covariant" as a curved 2-space with coordinates ζ^μ, $\mu = 0, 1$) is embedded in a flat D-dimensional Minkowski manifold $M^{1,D-1}$. On the other hand, macroscopic gravity is described classically by Einstein's theory, corresponding to a curved Riemannian R^4 manifold. However, the attempts to embed the string in a curved manifold have encountered three fundamental difficulties: (a) The fermionic $\theta(\zeta)$ frame-fields required by supersymmetry and constructed at any point ζ^μ of the world-sheet as spinors in $M^{1,D-1}$, cannot be embedded (ref. [6] notwithstanding) in a curved generic Riemannian R^D, since there exist no finite-dimensional spinorial representations of $GL(D,\mathbb{R})$ on the one hand, and on the other hand one cannot apply the usual tetrad formulation, as we shall explain. The current discussion of the more restricted program of Kaluza-Klein spontaneous compactification of 6 dimensions in $M^{1,9}$ hopefully leading to the observed spinor fields in flat physical $M^{1,3}$ avoids this issue by concentrating on Ricci-flat vacuum solutions. This would be consistent if spinor fields were made to appear only in the solutions, but not in the equations, which have to be general-covariant. (b) The "critical" dimensionality [4] of the embedding space $D = 26$ (for strings) or $D = 10$ (for superstrings) is modified [6,7] in the presence of generic curvature, thus destroying the essential conditions for a ghost-free finite-theory. (c) Without spinors we also lose supersymmetry, essential as a constraint for the removal of a tachyon [4] whose unwanted presence now makes the theory unphysical.

We suggest solutions [8,9] for all three problems, based upon the (anholonomic) application of the doubly-covered groups of Diffeomorphisms and Superdiffeomorphisms in the "tangent" at ζ^μ. We use the (infinite) spinorial representations of the double-covering $\overline{GL}(10,\mathbb{R})$, whose existence was pointed out some years ago [10] and which we have recently constructed [11] for $D = 4$ and utilized [12] for the construction of world-spinors in R^4, thus answering the quest in (a) while realizing the Principle of General Covariance. Further, we resolve problems (b) and (c) by embedding $\overline{GL}(10,\mathbb{R})$, in the doubly-covered real-form $\overline{GQ}(10,\mathbb{R})$, a supergroup generated by the superalgebra q(10) [13]. Under these conditions, the curving of $M^{1,9} \to R^{10}$ preserves supersymmetry (i.e. no tachyons) and the resulting critical dimension (i.e. no ghosts) both on and off mass shell.

Note that the possible emergence of finite quantum gravity (and even supergravity) in one sector of a large string-unified theory is especially interesting after the recent demonstration [4] of the existence of residual infinities in Einstein gravity. The latter theory is thus apparently just the low-energy limit of a more elaborate structure, fitting the quantum regime. The obvious alternative of quadratic Lagrangians provides a renormalizable theory, but it appears to suffer from ghosts. It is therefore important to thoroughly explore the possibility of deriving quantum gravity from superstrings.

FRAME-FIELDS IN THE GREEN-SCHWARZ MODEL

In the Polyakov [15] formulation of the Green-Schwarz [1,5] quantized superstring, every point ζ^μ of the evolving-string world-sheet (greek indices) carries a global D-dimensional (latin lower case indices) "Poincaré" supersymmetric frame $(X^m(\zeta),\ \theta^{\alpha a}(\zeta))$, $m = 0,1,\dots,D-1$ representing the components of a bosonic D-vector frame and $a = 1,\dots,2^{(D-2)/2}$ standing for the fermionic components of a real (Majorana) chiral (Weyl) spinor frame. Note that the "Majorana plus Weyl" condition for a Minkowski metric exists only for $D = 2 \bmod 8$. The index $\alpha = 1,2$ stands for the (somewhat trivial) spinor index in the sheet dimensionality. Note that the expressions appearing in the Lagrangian

$$e^m_\mu(\zeta') := \frac{\partial}{\partial\zeta^\mu} X^m \Big|_{\zeta=\zeta'} = \partial_\mu X^m(\zeta'), \tag{1}$$

$$\psi^a_\mu(\zeta') := \frac{\partial}{\partial\zeta^\mu} \theta^a \Big|_{\zeta=\zeta'} = \partial_\mu \theta^a(\zeta'), \tag{2}$$

are generalized (rectangular) "tetrads" with the latin indices supporting the corresponding action of the 10-dim. local Lorentz group $\overline{SO}(1,9)$ or its subgroups. In the analogy to General Relativity with tetrads, the 2-sheet plays the role of curved space-time (holonomic, "greek" coordinates) and the embedding 10-dimensional Lorentz boson or fermion indices fulfill the role of the "latin" (anholonomic) tetrad-indices.

CURVATURE AND SPINORS

In the attempts [6,7] to introduce curvature in the embedding, going from $M^{1,9}$ to R^{10}, one replaces $\eta_{mn} \to G_{\tilde{m}\tilde{n}}(\zeta)$, a curved metric. Ref. [6] provides a full series of additional replacements, necessary for an "effective" field theory action in curved space, and the equations of motion indeed reproduce the Einstein equations in the low-energy limit. However, the method fails for the spinors, a result that leads to some confusion in ref. [6]. In the usual technique in General Relativity a spinor would be defined in the local tangent (= flat embedding $M^{1,9}$, coordinate x^m) space at $x^{\tilde{m}}$, where a local Lorentz group can act on it (this is known as "anholonomic" action). This is done through the introduction of tetrad frames

$$e^m_{\tilde{m}}(\tilde{x}') := \frac{\partial}{\partial x^{\tilde{m}}} x^m \Big|_{x^{\tilde{m}}=x^{\tilde{m}'}} \tag{3}$$

We would have in addition as in Supergravity spinorial frames,

$$\psi^a_{\tilde{m}}(\tilde{x}') := \frac{\partial}{\partial x^{\tilde{m}}} \theta^a \Big|_{x^{\tilde{m}} = x^{\tilde{m}'}} \tag{4}$$

However, in the string formalism, this method is used up prior to curving of the embedding space, since the original spinor field is indeed already defined (equ. 1,2) as a local (fiber) frame-field on the tangent to the curved string (the base manifold) at the string coordinate ζ^μ. Indeed, the coordinates x^m (flat) or $x^{\tilde{m}}$ (curved) do not appear in the formalism, having been replaced for the flat case by the action of their isotropy group on that vector frame-field $X^m(\zeta)$ as a fiber on the local tangent to the curved string at ζ^μ in an Associated Bundle. As a result, the usual transition for the gamma matrices $\gamma^n \to \gamma^{\tilde{n}}$ cannot be performed by an ordinary tetrad-like matrix, since the two sets of "coordinates" x^m or $x^{\tilde{m}}$ are replaced in the string formation by fields X^m or $X^{\tilde{m}}$ over ζ^μ. The usual definition of a tetrad in equ. (3), (4), would now read

$$\text{"}e^{\tilde{m}}_m\text{"} = \frac{\partial}{\partial X^m(\zeta)} X^{\tilde{m}}(\zeta),$$

a meaningless expression for (quantized) fields. Our $X^m(\zeta)$ are in the flat tangent at ζ^μ, and there are no "tangents to the tangent", frames over frames. Infinite frames supporting the action of $\overline{\Delta}$. Were it not for the spinor, generic curving could have been achieved (as in ref. [6]) by replacing $X^m(\zeta)$ by $X^{\tilde{m}}(\zeta)$, a world-vector (= "holonomic") carrying finite linear representations of GL(10,$\mathbb{R}$) and non-linear representations of the General Covariance Group Δ (the analytical diffeomorphisms). In other words - changing the structure group of the bundle from $\overline{SO}(1,9)$ to Δ, as if we were "gauging" the entire diffeomorphisms. We therefore indeed replace the action of $\overline{SO}(1,9)$ by that of $\overline{GL}(10,\mathbb{R})$ and $\overline{\Delta}$, the double-covering of the General Linear and General Covariance groups, as the new bundle structure groups. For spinors, where the double-covering is required, these can, however, only be infinite-dimensional. We thus replace the $\theta^a(\zeta)$ by $\psi^A(\zeta)$ infinite frames transforming according to the unitary infinite dimensional representations of the linear subgroup $\overline{SL}(10,\mathbb{R})$. In the Equivalence Principle (flat) limit, necessary for the definition of particle states and for covariant quantization, these frames reduce to infinite reducible non-unitary representations $\psi^A(\zeta)$ of the $\overline{SO}(1,9)$ subgroup, an infinite direct sum of finite conventional spinor frames, the lowest level of which is identical to the $\theta^A(\zeta)$.

With this method, we have found a way in which the information on curvature and gravity enters the formalism through the action of $\overline{\Delta}$, i.e. the coordinates $\tilde{x}^m$ now appear as parameters of the local gauge group on R^2 at ζ^μ. Moreover, even the "active" action of $\overline{\Delta}$ necessary for the definition of boundary conditions (e.g. a specific vacuum configuration) can be implemented along these lines.

In the following pages we provide the group-theory basis for the construction of infinite spinorial and tensorial frames, to support that $\overline{\Delta}$ action.

PRESERVING SUPERSYMMETRY IN THE CURVED VERSION

There are, however, two further known difficulties to overcome, for the superstring to be embeddable in a generic curved space. Beyond the fitting in of the spinors, we have to preserve the supersymmetry they support, otherwise we cannot get rid of an unphysical tachyon state [4]. Moreover, we also have to preserve the critical role played by the dimension $D = 10$ in removing the conformal anomaly and ensuring the segregation of ghosts (in analogy to the role of the BRS equations in gauge theories). In Lovelace's calculation [7] for an S^{N-1} compactification with SO(N) symmetry of the original flat "internalizable" dimensions for the bosonic string, the sum of the free field conformal anomaly in the residual [D-(N-1)] flat dimensions, plus that contributed by the string within the compactified subspace (equivalent to an O(N) sigma model in R^2, or N-2 effective "flat" dimensions) results in a critical dimensionality $[D-(N-1)] + [N-2] = D-1$, so that the Louiville scalar field kinetic term has a coefficient $\gamma = -[26-(D-1)]$ or $(27-D)$. The effect is a nonconservation of the Virasoro charges and the ghost state does not decouple. In a similar calculation [6] with generic curvature but no spinors the coefficient $\gamma = -[26 - (D - \frac{3}{2}\alpha' R)]$, where R is the embedding curvature. For the Neveu-Schwarz-Ramond model [4], ref. [6] got $\gamma = -[10-(D-\alpha' R)]$. These examples display the importance of preserving the constraints of supersymmetry in the generic curved case.

GL(10,ℝ) GROUP STRUCTURE

The $\overline{GL}(10,\mathbb{R})$ group is the double-covering of the non-compact group $GL(10,\mathbb{R})$ of the $M^{1,9}$ space linear transformations. The maximal compact subgroup of this group is $\overline{SO}(10) \simeq \text{Spin}(10)$, the double-covering group of the SO(10) group. There is a four-element center of $\overline{SO}(10)$ which is isomorphic to Z_4. The factor group of $\overline{GL}(10,\mathbb{R})$ or $\overline{SO}(10)$ w.r.t. a two-element subgroup Z_2 of Z_4 is isomorphic to $GL(10,\mathbb{R})$ or SO(10) respectively. Let H_{mn}, m, n = 0,1,...,9 be the $\overline{GL}(10,\mathbb{R})$ generators. The $\overline{GL}(10,\mathbb{R})$ commutation relations read

$$[H_{mn}, H_{k\ell}] = ig_{nk} H_{m\ell} - ig_{m\ell} H_{kn}, \tag{5}$$

where for the structure constants g_{mn} one can take the invariant metric tensor: either $\delta_{mn} = (+1,+1,\ldots,+1)$ w.r.t. the $\overline{SO}(10)$ subgroup or $\eta_{mn} = (+1,-1,\ldots,-1)$ w.r.t. the 10 dimensional Lorentz subgroup $\overline{SO}(1,9)$ of the $\overline{GL}(10,\mathbb{R})$ group. The metric tensor is $\overline{GL}(10,\mathbb{R})$ covariant. The antisymmetric operators (when $g_{mn} = \eta_{mn}$) $L_{mn} = H_{[mn]}$ generate the metric-preserving Lorentz subgroup; the traceless symmetric operators $T_{mn} = H_{\{mn\}} - \frac{1}{10} \cdot \eta_{mn} H^k{}_k$ (10 shear) generate the (non-trivial) 10-volume-preserving transformations, and the trace $D = H^k{}_k$ generates the dilation subgroup. The traceless part H^o_{mn} of H_{mn}, i.e. L_{mn} and T_{mn}, generates the $\overline{SL}(10,\mathbb{R})$ group with the commutation relations (5), where H^{o}_{mn} is substituted for H_{mn}. In terms of L and T, the $\overline{SL}(10,\mathbb{R})$ commutation relations are

$$[L,L] \subset L, \quad [L,T] \subset T, \quad [T,T] \subset L. \tag{6}$$

$\overline{SL}(10,\mathbb{R})$ is the double-covering group of the $SL(10,\mathbb{R})$ group; $\overline{SL}(10,\mathbb{R})/Z_2 \simeq SL(10,\mathbb{R})$. In the (1+9) notation the $\overline{SL}(10,\mathbb{R})$ generators are the compact J_{ij} (angular momentum), and $N_i = T_{oi}$, and the noncompact $K_i = L_{oi}$ (boost), T_{ij} (9-shear), and T_{oo}, i,j,...,9. The relevant $s\ell(10,\mathbb{R})$ subalgebras are: the maximal compact subalgebra so(10: J_{ij} and N_i, the Lorentz so(1,9): J_{ij} and K_i, and $s\ell(9,\mathbb{R})$: J_{ij} and T_{ij}. The commutation relations (5) are invariant under the "deunitarizing automorphism"

$$\begin{aligned} A : J_{ij} &\to J_{ij}, \quad T_{ij} \to T_{ij}, \quad T_{oo} \to T_{oo}, \\ D &\to D, \quad N_k \to iK_k, \quad K_k \to iN_k. \end{aligned} \tag{7}$$

This automorphism allows us to identify the finite (unitary) representations of the abstract $\overline{SO}(10)$ compact subgroup (J_{ij}, N_i) with non-unitary representations of the physical Lorentz group (J_{ij}, K_i), while the infinite (unitary) representations of the abstract Lorentz group of (J_{ij}, K_i) now represent (non-unitarily) the compact (J_{ij}, N_i). The $m \neq 0$ stability subgroup non-Abelian part $\overline{SL}(9,\mathbb{R})$ of the inhomogeneous $\overline{GL}(10,\mathbb{R})$ group $\overline{GA}(10,\mathbb{R})$ (containing the 10-dimensional Poincaré group) is unaffected by A, and we use its unitary representations to characterize the particle states. The unitary irreducible representations of the abstract $\overline{GL}(10,\mathbb{R})$ are thus used as non-unitary representations of the physical $\overline{GL}(10,\mathbb{R})$ and thereby avoid a disease common to infinite component equations (for 4-dimensional case cf. [11,12]). The $\overline{SL}(10,\mathbb{R})$ group can be contracted (à la Wigner-Inönü) w.r.t. its $\overline{SO}(10)$ subgroup to yield the semidirect-product group $T_{54} \circledS \overline{SO}(10)$. The T_{54} is an Abelian group generated by 54 operators U_{mn}, which form a second rank symmetric operator (□□) w.r.t. $\overline{SO}(10)$. The commutation relations are

$$[J,J] \subset J, \quad [J,U] \subset U, \quad [U,U] = 0 \tag{8}$$

$\overline{SL}(10,\mathbb{R})$ UNITARY REPRESENTATIONS

Owing to the fact that dilations commute with the $\overline{SL}(10,\mathbb{R})$ subgroup of the $\overline{GL}(10,\mathbb{R})$ group, the essential part of the $\overline{GL}(10,\mathbb{R})$ unitary representations is given by the $\overline{SL}(10,\mathbb{R})$ unitary (finite-dimensional) spinorial and tensorial representations. An efficient way of constructing explicitly the $\overline{SL}(10,\mathbb{R})$ unitary representations is based on the decontraction formula [16], which is an inverse of the Wigner-Inönü contraction (cf. [17,18] for $\overline{SL}(n,\mathbb{R})$, n = 3,4). According to the decontraction formula, the following operators

$$T_{mn} = p\, U_{mn} + \frac{i}{2}(U\cdot U)^{-1/2}[C_2(\overline{SO}(10)),\, U_{mn}], \tag{9}$$

together with J_{mn} satisfy the $\overline{SL}(10,\mathbb{R})$ commutation relations. The parameter p is an arbitrary real number, and C_2 is the $\overline{SO}(10)$ second-rank Casimir operator.

For the representation Hilbert space we take the homogeneous space of L^2 functions of the maximal compact subgroup $\overline{SO}(10)$ parameters. The $\overline{SO}(10)$ unirrep labels are given either by the Dynkin labels $(\lambda_1,\lambda_2,...,\lambda_5)$ or by the highest weight vector which we denote by $[M_1,M_2,...,M_5]$. The $\overline{SL}(10,\mathbb{R})$ commutation relations (6) are invariant w.r.t. an automorphism defined by: s(L) = +L, s(T) = -T, which for real matrices becomes transposition symmetry (cf. [19] for SU(3)). This enables us to define an "s-parity" to each $\overline{SO}(10)$ unirrep of an $\overline{SL}(10,\mathbb{R})$ representation. In terms of Dynkin labels (generalizing the $\overline{SL}(n,\mathbb{R})$, n = 3,4 [20,21]) we find

$$s = (-)^{\lambda_1+\lambda_2+\lambda_3+\frac{1}{2}(\lambda_5-\lambda_4-\varepsilon)} \tag{10}$$

where $\varepsilon = 0(+1)$ if $\lambda_5-\lambda_4$ is even (odd). The 54 representations of $\overline{SO}(10)$, i.e. (20000) = □□, has

$$s(20000) = +1 \tag{11}$$

A basis of an $\overline{SO}(10)$ unirrep is provided by the Gel'fand-Zetlin pattern characterized by the maximal weight vectors of the group chain $\overline{SO}(10) \supset \overline{SO}(9) \supset \ldots \supset \overline{SO}(2)$. We write the basic vectors as

$$\left| \begin{matrix} [M] \\ (m) \end{matrix} \right\rangle ,$$

where (m) corresponds to $\overline{SO}(9) \supset \ldots \supset \overline{SO}(2)$ subgroup chain weight vectors. When necessary, we generalize the basis to

$$\left| \begin{matrix} (k) \\ [M] \\ (m) \end{matrix} \right\rangle$$

to accomodate both the group action to the right and left, and work in the homogeneous space over $\overline{SO}(10) \otimes \overline{SO}(10)$. The (k) and (m) labels correspond to the two subgroup chains.

The 54 Abelian generators $\{U_{mn}\} = \{U[\square\square]\}$ of $T_{54} \circledS \overline{SO}(10)$ commute, mutually, and thus in the most general case we find

$$U^{[\square\square]}_{(\mu)} = \Sigma_{i=o}^{4} \tilde{p}^{(i)} D^{(k_i)}_{(\mu)}[\square\square] (\{\alpha\}), \tag{12}$$

where

$$D^{(k)}_{(\mu)}[\square\square] (\{\alpha\}) = \left\langle \begin{matrix} [\square\square] \\ (k) \end{matrix} \right| g(\{\alpha\}) \left| \begin{matrix} [\square\square] \\ (\mu) \end{matrix} \right\rangle ,$$

are the $\overline{SO}(10)$ - Wigner functions, $g(\{\alpha\}) \in \overline{SO}(10)$, and (k_i) are all "sub-labels" for which s-parity is +, as required by (11). For the 10-shear we obtain from (9), (12) and the orthogonality properties of the D-functions:

$$\left\langle \begin{matrix} (k') \\ [M'] \\ (m') \end{matrix} \right| T^{[\square\square]}_{(\mu)} \left| \begin{matrix} (k) \\ [M] \\ (m) \end{matrix} \right\rangle = \begin{pmatrix} [M'] & [\square\square] & [M] \\ (m') & (\mu) & (m) \end{pmatrix} \left\langle \begin{matrix} [M'] \\ (k') \end{matrix} \right\| T \left\| \begin{matrix} [M] \\ (k) \end{matrix} \right\rangle , \tag{13}$$

where the reduced matrix elements are

$$\left\langle \begin{matrix}[M']\\(k')\end{matrix} \middle\| \; T \; \middle\| \begin{matrix}[M]\\(k)\end{matrix} \right\rangle = i(N([M'])N([M]))^{1/2} \Big\{ [p^{(o)}$$

$$+ \tfrac{1}{2}(C_2[M'] - C_2([M]))] \begin{pmatrix}[M'] & [\square\square] & [M]\\(k') & (o) & (k)\end{pmatrix}$$

$$+ \Sigma_{i=1}^{4} p^{(i)} \begin{pmatrix}[M'] & [\square\square] & [M]\\(k') & (k_i) & (k)\end{pmatrix} \Big\} ,$$

N([M]) is the representation dimension, and $(\begin{smallmatrix}\cdot & \cdot & \cdot\\ \cdot & \cdot & \cdot\end{smallmatrix})$ are the $\overline{SO}(10)$ "3-j" symbols. For the representation parameters we could have a priori $p^{(o)}, p^{(1)}, \ldots, p^{(5)} \in ¢$. Imposing the hermiticity condition (unitary representations) we obtain several series of (infinite) unitary spinorial representations; e.g. Principal series when all $p^{(1)}$ are pure imaginary, and all [M'] and [M] half integer.

Let us concentrate on the simplest - multiplicity free (each $\overline{SO}(10)$ irrep appears at most once) spinorial representations [22]. These are obtained from (13) by setting $(k_i') = (k_i) = (0)$, $i = 0, \ldots, 4$ and $p^{(1)}, \ldots, p^{(4)} = 0$, and by an analytical continuation of [M] labels to the half-integer values. By making use of the $\overline{SO}(10)$ Clebsch-Gordan series and the s-parity we find the following simplest unitary spinorial systems (belonging to discrete series) characterized by their $\overline{SO}(10)$ irrep dimensionality content:

$$\begin{aligned} \mathcal{D}^{disc}(16) &= \{16, 144, 720, 2640, 7920, \ldots\}, \\ \mathcal{D}^{disc}(560) &= \{560, 3696'; 8800, 15120, \ldots\}, \\ \mathcal{D}^{disc}(672) &= \{672, 1440, 11088, \ldots\}, \end{aligned} \tag{14}$$

each of them supplemented by a system of conjugated states, e.g.

$$\mathcal{D}^{disc}(\overline{16}) = \overline{16}, \overline{144}, \overline{720}, \overline{2640}, \overline{7920}, \tag{15}$$

The "ladder" tensorial unirreps can be most easily obtained as the limiting cases of the totally symmetrized 2n-boxes ($n \to \infty$) products of the SU(10) unirreps applied to the "vacuum" consisting of either singlet or 1-box state: $\cdot$, □□, □□□□,...; □, □□□, □□□□□, ... Thus we find

$$\begin{aligned} \mathcal{D}^{ladd}(1) &= \{1, 54, 660, 4290, 19305, \ldots\}, \\ \mathcal{D}^{ladd}(10) &= \{10, 210', 1782', 9438, 37180, \ldots\}. \end{aligned} \tag{16}$$

FIELD EQUATIONS

The above constructed unitary $\overline{SL}(10,\mathbb{R})$ representations define the corresponding $\overline{SL}(10,\mathbb{R})$ covariant infinite-component fields - "manifields". There are two crucial physical requirements to be fulfilled at this point. First: the intrinsic Lorentz generators represented on fields must be non-unitarily represented. Otherwise, the Lorentz boosts would excite a given state to other spins and masses contrary to experience. We resolve this by applying the deunitarizing automorphism $\mathcal{A}(5)$ to the $\overline{SL}(10,\mathbb{R})$ unitary representations. The non-unitarity in the intrinsic boost parts now cancels their physical action precisely as in finite tensors or spinors, the boost thus acting kinetically only. Second: in the absence of gravity, by the Principle of Euivalence, only the Lorentz group $\overline{SO}(1,3) \subset \overline{SL}(4,\mathbb{R})$ survives (or, in fact, the double-covering of the Poincaré group when we adjoin the translations as well). We generalize this requirement, namely, that for $D=4$ our $\overline{SL}(D,\mathbb{R})$ manifields satisfy only $\overline{SO}(1,3)$ covariant equations [23], to $D=10$ and $\overline{SO}(1,9)$ covariance.

For the tensorial representations the simplest choices are either $\mathcal{D}^{ladd}(1)$ or $\mathcal{D}^{ladd}(10)$ with a Klein-Gordon-like (infinite-component) equation for the corresponding manifield $X(x) = \{X_M(x) \mid M \leftrightarrow \mathcal{D}^{ladd}(1) \text{ or } \mathcal{D}^{ladd}(10)\}$,

$$(\partial_m \partial^m + \mu^2) X(x) = 0. \tag{17}$$

Spinor manifields obey a first-order equation (cf. [23] for $D=4$) with infinite Γ_m matrices generalizing Dirac's. The Γ_m behave as $\overline{SO}(1,9)$ 10 vectors, and we are forced in the simplest case to use the reducible pair of $\overline{SL}(10,\mathbb{R})$ representations $\mathcal{D}^{disc}(16) \oplus \mathcal{D}^{disc}(\overline{16})$ for the manifield $\psi(x) = \{\psi_A(x) \mid A \leftrightarrow \mathcal{D}^{disc}(16) \oplus \mathcal{D}^{disc}(\overline{16})\}$,

$$(i\,\Gamma_m \partial^m - \quad)\psi(x) = 0. \tag{18}$$

The $\Gamma_m = \{(\Gamma_m)^A_B\}$ operators connect various $\overline{SO}(10)$ spinorial states of an $\overline{SL}(10,\mathbb{R})$ representation; e.g. $\Gamma \otimes 16 \supset \overline{16} \oplus \overline{144}$, $\Gamma \otimes 144 \supset \overline{16} \oplus \overline{144} \oplus \overline{720}$, $\Gamma \otimes 720 \supset \overline{144} \oplus \overline{720} \oplus \overline{2640}$, $\Gamma \otimes 2640 \supset \overline{720} \oplus \overline{2640} \oplus \overline{7920}$, ...

$\overline{GQ}(10,\mathbb{R})$ SUPERGROUP

The $Q(10,\mathbb{R})$ supergroup is generated by the operators of the simple $q(10)$ superalgebra. This supergroup can be enlarged to the non-simple group of general linear supermatrices $GQ(10,\mathbb{R})$ on a $M^{1,9/10}$ superspace with real $M^{1,9}$ bosonic subspace. The odd (symmetric) brackets for the $GQ(10,\mathbb{R})$ algebra are given (in contradistinction to the $Q(10,\mathbb{R})$ case) by the anti-commutators. The even (bosonic) operators of the $GQ(10,\mathbb{R})$ algebra generate the $GL(10,\mathbb{R})$ group. We take the double-covering $\overline{GQ}(10,\mathbb{R})$ of the $GQ(10,\mathbb{R})$ supergroup, with the corresponding $\overline{GL}(10,\mathbb{R})$ even subgroup. The $\overline{Q}(10,\mathbb{R})$ generators are $\sigma_\alpha \otimes \lambda_a$, where σ_α, $\alpha = 0,1$ are the relevant Pauli matrices and the λ_a, $a = 1,2,\ldots,99$ are the traceless generators of $SL(10,\mathbb{R})$. For $GQ(10,\mathbb{R})$ we relax the traceless condition for the even subalgebra, as we shall do here for the odd (fermionic) generators as well for convenience only. For the even and odd generators we take respectively

$$H_{mn} = \sigma_o \otimes \lambda_{mn}, \quad F_{mn} = \sigma_1 \otimes \lambda_{mn}, \tag{19}$$

where $(\lambda_{mn})_{pq} = i g_{mp} g_{nq}$, $m,n,p,q = 0,1,\ldots,9$, and for g_{mn} one can take the invariant metric tensor: either $\delta_{mn} = (+1,+1,\ldots,+1)$ w.r.t. the $\overline{SO}(10)$ subgroup or $\eta_{mn} = (+1,-1,\ldots,-1)$ w.r.t. the 10-dimensional Lorentz subgroup $\overline{SO}(1,9)$ of the $\overline{GL}(10,\mathbb{R})$ group. The $\overline{GQ}(10,\mathbb{R})$ brackets are

$$\begin{aligned}
[H_{mn}, H_{k\ell}] &= i(g_{nk}\, H_{m\ell} - g_{m\ell}\, H_{kn}), \\
[H_{mn}, F_{k\ell}] &= i(g_{nk}\, F_{m\ell} - g_{m\ell}\, F_{kn}), \\
\{F_{mn}, F_{k\ell}\} &= i(g_{nk}\, H_{m\ell} + g_{m\ell}\, H_{kn}).
\end{aligned} \tag{20}$$

There are two relevant subgroup chains of the $\overline{GQ}(10,\mathbb{R})$ even part

$$\overline{GL}(10,\mathbb{R}) \supset \overline{SL}(10,\mathbb{R}) \supset \overline{SO}(10) \supset \overline{SO}(2) \otimes \overline{SO}(8)$$
and
$$\overline{GL}(10,\mathbb{R}) \supset \overline{SL}(10,\mathbb{R}) \supset \overline{SO}(10) \supset \overline{SO}(9). \tag{21}$$

The traceless symmetric components of both H_{mn} (10-shear) and F_{mn} (10-super-shear), i.e. $H_{\{mn\}}$ and $F_{\{mn\}}$, transform w.r.t. $\overline{SO}(10) \supset \overline{SO}(9) \supset \overline{SO}(8)$ and $\overline{SO}(10) \supset \overline{SO}(2) \otimes \overline{SO}(8)$ respectively as

$$54 \supset 44 \oplus 9 \oplus 1 \supset 35_v \oplus 8_v \oplus 1 \oplus 8_v \oplus 1 \oplus 1,$$
$$54 \supset (1,35_v) \oplus (2,8_v) \oplus (2,1) \oplus (1,1). \tag{22}$$

The antisymmetric components of both H_{mn} (10-Lorentz) and F_{mn} (10-super Lorentz), i.e. $H_{[mn]}$ and $F_{[mn]}$, transform w.r.t. $\overline{SO}(10) \supset \overline{SO}(9) \supset \overline{SO}(8)$ and $\overline{SO}(10) \supset \overline{SO}(2) \otimes \overline{SO}(8)$ respectively as

$$45 \supset 36 \oplus 9 \supset 28 \oplus 8_v \oplus 8_v \oplus 1,$$
$$45 \supset (1,28) \oplus (2,8_v) \oplus (1,1). \tag{23}$$

GRADED ALGEBRA $\overline{GQ}(10,\mathbb{R})$ AND THE SPIN-STATISTICS THEOREM

The $\overline{GQ}(10,\mathbb{R})$ algebra is a graded algebra, i.e. it is defined by the brackets with appropriate antisymmetry or symmetry properties and by the graded Jacobi identity requirements. The graded brackets for $\overline{GQ}(10,\mathbb{R})$ are given by commutators and anticommutators. However, even though the odd generators are a subject of anticommutation relations they do not transform w.r.t. either the even subalgebra $\overline{GL}(10,\mathbb{R})$ or the maximal even orthogonal subalgebra $\overline{SO}(1,9)$ as spinors. Indeed, (22) and (23) reveal their tensorial character in the defining $(1,9/10) \times (1,9/10)$ representation. This fact is due to the non-existence of finite $\overline{SL}(10,\mathbb{R})$ spinors, and the way in which $\overline{SO}(10)$ is embedded in $\overline{SL}(10,\mathbb{R})$. It is only for the infinite-dimensional representations of $\overline{GQ}(10,\mathbb{R})$, defined in the superspace of $\overline{SL}(10,\mathbb{R})$ infinite tensors and spinors that the odd operators connect bosons (fermions) to fermions (bosons), i.e. behave as physical spinorial operators, and thus the spin-statistics theorem is recovered.

CONTRACTION, QUOTIENT SPACE AND THE STABILITY GROUP

Let us split the even and odd generators of $\overline{GQ}(10,\mathbb{R})$ according to 2+8 notation, i.e. w.r.t. the "$\overline{GQ}(2,\mathbb{R})$" $\times \overline{GQ}(8,\mathbb{R})$ subgroup: $H_{mn} = \{H_{\alpha\beta}, H_{ij}, H_{\alpha i}, H_{i\alpha}\}$, $F_{mn} = \{F_{\alpha\beta}, F_{ij}, F_{\alpha i}, F_{i\alpha}\}$, $\alpha,\beta = 0,9$ and $i,j = 1,2,\ldots,8$. If we rescale the generators in the $\overline{GQ}(10,\mathbb{R})$ algebra, $H_{\alpha\beta} \to \varepsilon^2 H_{\alpha\beta} = H'_{\alpha\beta}$, $H_{oi} \to \varepsilon H_{oi} = H'_{oi}$, $H_{io} \to \varepsilon H_{io} = H'_{io}$, $H_{ij} \to H_{ij}$, $F_{\alpha\beta} \to \varepsilon F_{\alpha\beta} = F'_{\alpha\beta}$, $F_{\alpha i} \to \varepsilon F_{\alpha i} = F_{\alpha i}$, $F_{i\alpha} \to \varepsilon F_{i\alpha} = F'_{i\alpha}$, $F_{ij} \to F_{ij}$, and take $\varepsilon \to 0$ (Wigner-Inönü contraction), we obtain the $\overline{GQ}(10,\mathbb{R})_{con}$ superalgebra. It is straightforward to check that this superalgebra contains the flat-space limit stability super subalgebra generated by the even operators: $H_{[ij]}$, $H_{[9i]}$, H'_{oo}, H'_{99}, and the odd operators F'_{oi}, F'_{io}, F'_{9i}, F'_{i9}. Here the $H_{[ij]}$ generate the $\overline{SO}(8)$ subgroup, while $H_{[ij]}$ and $H_{[9i]}$ generate the $\overline{SO}(9)$ subgroup. Let us define

$$S_{\alpha i} = \frac{1}{2} F'_{\alpha i}, \quad \bar{S}_{\alpha i} = \frac{1}{2} F'_{i\alpha}, \quad \alpha = 0,9, \quad i = 1,2,\ldots,8. \tag{24}$$

The anticommutators read

$$\{S_{\alpha i}, S_{\beta j}\} = \{\bar{S}_{\alpha i}, \bar{S}_{\beta j}\} = 0 \tag{25}$$

$$\{S_{\alpha i}, \bar{S}_{\beta j}\} = 2(\sigma_o h_o + \sigma_3 h_9)_{\alpha\beta}, \tag{26}$$

where σ_o and σ_3 are the Pauli matrices, and where h_o and h_9 are the eigenvalues of the operators $\frac{1}{2}(H'_{oo} + H'_{99})$ and $\frac{1}{2}(H'_{99} - H'_{oo})$ respectively. For a unitary (infinite-dimensional) $\overline{SL}(10,\mathbb{R})$ representation, h_o and h_9 are real. The super subalgebras containing the $\overline{SO}(8)$ or $\overline{SO}(9)$ subalgebras correspond to the m = o or m ≠ o supersymmetry stability algebras respectively, and we denote them by ssa(8) and ssa(9). These supersymmetry stability algebras generate respectively the stability groups $\overline{SSG}(8)$ and $\overline{SSG}(9)$ of the homogeneous superspaces

$$\overline{GQ}(10,\mathbb{R})/\overline{SSG}(8) \quad \text{and} \quad \overline{GQ}(10,\mathbb{R})/\overline{SSG}(9).$$

THE FLAT-SPACE LIMIT SPECTRUM

Owing to the fact that the operators H'_{oo} and H'_{99} commute with all generators of the $\overline{GQ}(10,\mathbb{R})_{con}$ contracted group, in the $\overline{GQ}(10,\mathbb{R})$ unitary representation space, we can renormalize the relevant odd operators and have $h_o, h_9 = 0, \pm 1$ in (26). For the choice $h_o = 1$, $h_9 = 0$, we find the m ≠ o supersymmetry Clifford algebra with the brackets given by (25) and by

$$\{S_{oi}, \bar{S}_{oj}\} = \{S_{9i}, \bar{S}_{9j}\} = 2\delta_{ij}. \tag{27}$$

For the choice $h_o = 1$, $h_9 = \pm 1$, we find the m = o supersymmetry Clifford algebras with the brackets given by (25) and respectively by

$$\{S_{oi}, \bar{S}_{oj}\} = 4\delta_{ij}, \ \{\bar{S}_{9i}, S_{9j}\} = 0. \tag{28}$$

or

$$\{S_{oi}, \bar{S}_{oj}\} = 0, \quad \{S_{9i}, \bar{S}_{9j}\} = 4\delta_{ij}. \tag{29}$$

The correspondence to the superstring spectrum of states [4] is now evident. Note that the 16-dimensional representation of the $\overline{SSG}(8)$ supergroup corresponds to the m = o superstring Fock space ground state, while our operators $H_{[9i]}$ and S_{oi} yield, in the representation space, results corresponding to the action of the superstring spectrum-generating operators α^I_{-1} and $\mathcal{S}^a_{-1}$ respectively.

$\overline{SO}(10)$ TENSOR PRODUCTS: 10-VECTOR, 54-(10-SHEAR) × SPINORIAL IRREPS

$16 \otimes 10 = \overline{16} \oplus \overline{144}$
$144 \otimes 10 = \overline{16} \oplus \overline{144} \oplus \overline{560} \oplus \overline{720}$
$560 \otimes 10 = \overline{144} \oplus \overline{560} \oplus \overline{1200} \oplus \overline{3696'}$
$672 \otimes 10 = \overline{1440} \oplus \overline{5280}$
$720 \otimes 10 = \overline{144} \oplus \overline{720} \oplus \overline{2640} \oplus \overline{3696'}$
$1200 \otimes 10 = \overline{560} \oplus \overline{1200} \oplus \overline{1440} \oplus \overline{8800}$
$1440 \otimes 10 = \overline{672} \oplus \overline{1200} \oplus \overline{1440} \oplus \overline{11088}$
$2640 \otimes 10 = \overline{720} \oplus \overline{2640} \oplus \overline{7920} \oplus \overline{15120}$
$3696' \otimes 10 = \overline{560} \oplus \overline{720} \oplus \overline{3696'} \oplus \overline{8064} \oplus \overline{8800} \oplus \overline{15120}$
$5280 \otimes 10 = \overline{672} \oplus \overline{11088} \oplus \overline{17280} \oplus \overline{23760}$
$7920 \otimes 10 = \overline{2640} \oplus \overline{7920} \oplus \overline{20592} \oplus \overline{48048}$
$8064 \otimes 10 = \overline{3696'} \oplus \overline{8064} \oplus \overline{25200} \oplus \overline{43680}$

$16 \otimes 54 = 144 \oplus 720$
$144 \otimes 54 = 16 \oplus 144 \oplus 560 \oplus 720 \oplus 2640 \oplus 3696'$
$560 \otimes 54 = 144 \oplus 560 \oplus 720 \oplus 1200 \oplus 3696' \oplus 8800 \oplus 15120$

$672 \otimes 54 = 1440 \oplus 11088 \oplus 23760$

$720 \otimes 54 = 16 \oplus 144 \oplus 560 \oplus 720 \oplus 2640 \oplus 3696' \oplus 7920 \oplus 8064 \oplus 15120$

$1200 \otimes 54 = 560 \oplus 1200 \oplus 1440 \oplus 3696' \oplus 8800 \oplus 11088 \oplus 38016$

REFERENCES

1. M. B. Green and J. H. Schwarz, Phys. Lett. 109B:448 (1982) and 136B:367 (1984).
2. M. B. Green and J. H. Schwarz, Phys. Lett. 149B:117 (1984) and 151B:21 (1985); J. Thierry-Mieg, Phys. Lett. 156B:199 (1985).
3. J. Scherk and J. H. Schwarz, Nucl. Phys. B81:118 (1974).
4. M. Jacob, ed., "Dual Theory", Physics Reports reprint book (North-Holland, Amsterdam/Oxford, American Elsevier, New York) (1974).
5. J. H. Schwarz, ed., "Superstrings" (World-Scientific, Singapore) (1985).
6. E. S. Fradkin and A. A. Tseytlin, Phys. Lett. 158B:316 (1985).
7. C. Lovelace, Phys. Lett. 135B:75 (1984).
8. Y. Ne'eman and Dj. Šijački, "Spinors for Superstrings in a Generic Curved Space", TAUP N174/86.
9. Y. Ne'eman and Dj. Šijački, "Superstrings in a Generic Supersymmetric Curved Space", TAUP N175/86.
10. Y. Ne'eman, Ann. Inst. Henri Poincare A28:369 (1978).
11. Dj. Šijački and Y. Ne'eman, J. Math. Phys. 26:2457 (1985).
12. Y. Ne'eman and Dj. Šijački, Phys. Lett. 157B:267 and 275 (1985).
13. Y. Ne'eman, "Cosmology and Gravitation", P. G. Bergmann and V. de Sabbata, eds., Nato Advanced Study Inst. Series, (Plenum Press, New York) B58:pp. 177-226 (1980). See in part. page 192.
14. M. H. Goroff and A. Sagnotti, Phys. Lett. 160B:81 (1985).
15. A. M. Polyakov, Phys. Lett. 103B:207 and 211 (1981).
16. Y. Dothan and Y. Ne'eman, "Symmetry Groups in Nuclear and Particle Physics", F. J. Dyson, ed., (Benjamin, New York) (1966).
17. Y. Ne'eman and Dj. Šijački, J. Math. Phys. 21:1312 (1980).
18. Y. Ne'eman and Dj. Šijački, Ann. Phys. (N.Y.), 120:292 (1979).
19. L. C. Biedenharn, Phys. Lett. 28B:537 (1969).
20. Dj. Šijački, J. Math. Phys. 16:289 (1975).
21. Dj. Šijački, Ann. Isr. Phys. Soc., 3:35 (1980).
22. Dj. Šijački, Lect. Notes in Phys. 201:88 (1984).
23. A. Cant and Y. Ne'eman, J. Math. Phys. 26:3180 (1985).

SYSTEMS, SUBSYSTEMS AND INDUCED REPRESENTATIONS

D.J. Newman

Department of Physics
University of Hong Kong
HONG KONG

In conventional (Racah) state labelling schemes a sequence of groups $G \supset \ldots \supset H$ is employed, where G is a covering group in which the physical bases are uniquely labelled by irreducible representations, or irreps, and H is the symmetry group of the system. Intermediate groups in the sequence are chosen to ensure that the irreps of H spanned by the physical bases are uniquely specified. This makes it possible to use standard group theoretical techniques to determine matrix elements for complex systems. In some cases of interest, however, sequences which satisfy these criteria do not exist (e.g. relating R_3 to the octahedral group). We are then forced to look at induction techniques. Even when we are not forced it may, nevertheless, be preferable to use induced representations for labelling because of the better insight they provide into the structure of the physical eigenstates.

In some cases, such as in Altmann's treatment of the irreps of space group[1], the full power of the theory of induced representations has been employed to good effect. However, in other, more experimentally related problems, its relevance has often been completely unrecognised. This is a pity, as induced representation theory can provide a rather direct insight into the underlying physics. Even when the relevance of induced representation theory has been understood, only the relatively simple theorems of Frobenius have been exploited, so that much work remains to be done, for example, in the development of applications of Mackey's theorems[1,2,3].

The aim of this article is to give a brief survey of several topics in which induced representations of discrete groups play an important role. In particular we investigate the physical consequences in the case of systems for which the states can be viewed as induced representations of the states of subsystems, and show that such consequences occur in widely diverse contexts.

1. BASIC IDEAS

Following the approach of Newman[4] and Ceulemans[5] we shall focus attention primarily on systems (with symmetry group G) which can be viewed as constructions from sets of identical subsystems (with symmetry group $H \subset G$), distinguished by a label i. The basis vectors of the complete

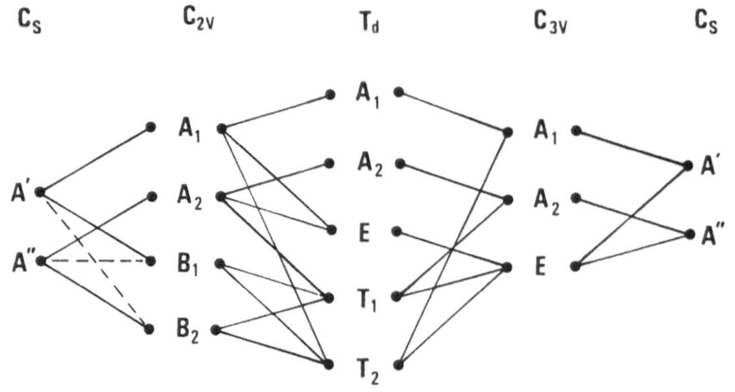

Fig. 1. Correlation diagram for $T_d \supset C_{2v} \supset C_s$ and $T_d \supset C_{3v} \supset C_s$.

system may then be generated from the subsystem bases using the same set of operators g_i which provides the decomposition of G into its left cosets g_iH. This process generates induced representations of H in G. The subsystem groups H_i form conjugate subgroups in G: $H_i = g_iHg_i^{-1}$.

Applications of induced representation theory involve the use of Frobenius' reciprocity theorem[1,2,3]. This has also been called the "correlation theorem[6,7]. Its substance is that the correlation diagram may be used both ways. As an example we take $T_d \equiv G$ and $C_{3v} \equiv H$. The $T_d \supset C_{3v}$ correlation diagram shown in figure 1 shows that we may subduce

$$T_1(T_d) \downarrow C_{3v} = A_2 + E$$

or induce

$$A_2(C_{3v}) \uparrow T_d = A_2 + T_1.$$

Both operations are transitive, e.g. (from figure 1)

$$A'(C_s) \uparrow C_{2v} \uparrow T_d = A'(C_s) \uparrow C_{3v} \uparrow T_d = A'(C_s) \uparrow T_d.$$

Notice that in subduction the dimension of the representation remains unchanged, while in induction from H to G it is multiplied by $|G|/|H|$, which in this example is $|T_d|/|C_{3v}| = 4$.

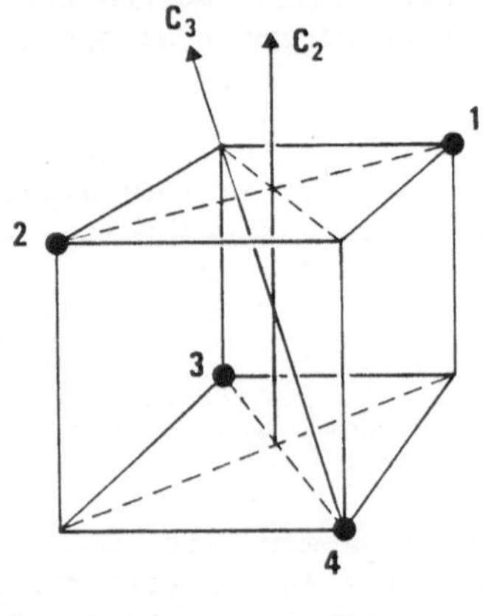

Fig. 2. C_3 and C_2 axes of the cube, showing two reflection planes.

To go a little further we consider the tetrahedral structure of four atoms shown in figure 2. Each of the four C_{3v} subsystem bases will be assumed to correspond to the displacements of a single atom $\Gamma_{\underset{\sim}{R}}(C_{3v}) \equiv A_1 + E$. The T_d representations corresponding to the displacements of all four atoms are then given by

$$A_1(C_{3v}) \uparrow T_d = A_1 + T_2$$

$$E(C_{3v}) \uparrow T_d = E + T_1 + T_2$$

We note that the two occurrences of the T_2 irrep of T_d are thus distinguished by the subgroup (C_{3v}) labels A_1 and E. Hence, not only can the T_d labels be used to distinguish the occurrences of C_{3v} irrep labels (as is well known), but the C_{3v} labels, using the same correlation table, can be used to distinguish occurrences of T_d irrep labels.

Induction thus has some useful formal properties:

(a) It can provide a unique group theoretical labelling in situations where suitable higher order groups do not exist (as in the case $R_3 \supset O$).

(b) The symmetry adapted bases of induced representations can easily be determined in terms of subsystem symmetry adapted bases (e.g. see ref. 8).

(c) Because the labelling is unique, it should not be necessary to determine the symmetry adapted bases in order to evaluate matrix elements, if suitable general formulae can be obtained. It should be possible to apply the product theorem (ref. 1 p.155) to this problem.

Eigenstates corresponding to induced representations may also show some interesting physical properties:

(i) Approximate degeneracies will occur between levels corresponding to different irreps, but induced from the same subgroup irrep, if the coupling between subsystems is weak (see Section 3).

(ii) Degeneracies in the subsystem energies will be reflected as degeneracies in the averaged energies of the states induced by a single irrep of the subsystem. This leads to "sum rules" for the eigenvalues of the complete system (see Section 4).

It can happen that the most appropriate group (H, say) for defining irrep labels of the physical basis is not a subgroup of the symmetry group K (e.g. see Section 5). In this case we seek a covering group G which contains both H and K as subgroups. The physically significant irreps of K are then labelled using irreps of H induced into G and subduced into K. In such circumstances, it may be useful to apply Mackey's theorem[1], viz.

$$\Gamma(H) \uparrow G \downarrow K = \sum_{\delta}(\Gamma^{\delta}(H) \downarrow L^{\delta}) \uparrow K \qquad (1)$$

where $\Gamma^{\delta}(s) = \Gamma(\delta s \delta^{-1})$ and the sum is carried over a set of subgroups $L^{\delta} = (\delta^{-1}H\delta) \cap K$ where the elements $\delta \in G$ are chosen to span the double coset decomposition of $G = \sum_{\delta} H\delta K$. Thus Mackey's theorem relates the representations obtained in a labelling via G to an alternative labelling via the subgroups L^{δ} of K.

An example of Mackey's theorem may be illustrated by reference to figures 1 and 2 taking $H \equiv C_{2v}$, $K \equiv C_{3v}$. The subgroup structure is identified in figure 1. Two distinct $L^{\delta} = C_s^x$ $(x = 1,2)$ subgroups are required in the double coset decomposition of T_d, corresponding to different reflection planes, related by $(13)(24)C_s^1(13)(24) = C_s^2$ where the labels are defined through figure 2 and the alternative $C_s^x \subset C_{2v}$ correlations are shown in figure 1 where dashed lines are used for $C_s^2 \subset C_{2v}$. We can now check, for example, that

$$B_1(C_{2v}) \uparrow T_d \downarrow C_{3v} = (T_1)A_2 + (T_1)E + (T_2)A_1 + (T_2)E$$

gives the same result as combining the two terms on the right hand side of equation 1 with $L^{\delta} \equiv C_s^1$ and C_s^2:

$$B_1(C_{2v}) \downarrow C_s^1 \uparrow C_{3v} = A_1 + E$$

$$B_1(C_{2v}) \downarrow C_s^2 \uparrow C_{3v} = A_2 + E$$

2. INDUCED REPRESENTATIONS AND CRYSTAL CLUSTERS

Many of the properties of paramagnetic ions in crystals can be related to the coupling between the open-shell electrons (which give rise to its magnetic properties) and the displacements of the surrounding cluster of atoms. These clusters are conveniently viewed as consisting of successive "shells" of atoms surrounding the paramagnetic ion of interest such that all the atoms in a given shell can be exchanged by the operations of the point symmetry group centred on the paramagnetic ion (i.e. they form an "orbit").

We are interested in the general problem of finding unique group theoretical labels for the symmetry coordinates of shell displacements. As an example we consider the case of shells in clusters of octahedral (O_h) symmetry. It is convenient to factorize the operators g of O_h into direct products of the operators g_P of O_h^P which permute the equilibrium atomic positions[5,10,11] without changing the direction of their displacements from equilibrium, and operators g_R of O_h^R which correspondingly rotate the directions of the atomic displacements: $g = g_P \otimes g_R$ (see figure 3). The group O_h^R then operates on the representation T_{1u}, corresponding to a vector displacement, and O_h^P can act, at most, on a 48-atom shell corresponding to the regular representation.

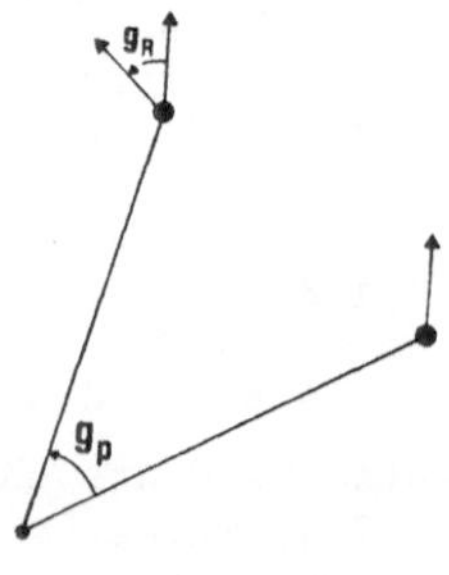

Fig. 3. Factorization of a rigid rotation into a permutation (g_P) of atomic positions and a rotation (g_R) of atomic displacements.

The most appropriate subsystem group to use in this case is C'_{2v} which, as was shown in ref. 10, gives a unique labelling of the regular representation:

$$A_1 \uparrow O_h = A_{1g} + E_g + T_{2g} + T_{1u} + T_{2u}$$

$$A_2 \uparrow O_h = A_{1u} + E_u + T_{1g} + T_{2g} + T_{2u}$$

$$B_1 \uparrow O_h = A_{2g} + E_g + T_{1g} + T_{1u} + T_{2u}$$

$$B_2 \uparrow O_h = A_{2u} + E_u + T_{1g} + T_{2g} + T_{1u}$$

It then follows that, because O_h is simply reducible, the atomic displacements can also be expressed in terms of uniquely labelled symmetry coordinates. Ref. 8 shows how the corresponding induced symmetry coordinates can be obtained using very simple algebraic manipulations, at least compared with the more conventional use of projection operators.

Although the approach described above provides an adequate formalism for most applications, it has the formal defect that displacements in different shells are not distinguished with group theoretical labels. Ref. 13 suggests a possible way of extending the formalism to provide such a labelling. The concept of a "supershell" within a cyclic region is introduced. In a supershell group the atoms are not only related by point symmetry operations, but also by cyclic translations. Nevertheless, the supershell remains finite because of the cyclic nature of the translations.

An example of such a supershell[13], is shown in figure 4. This

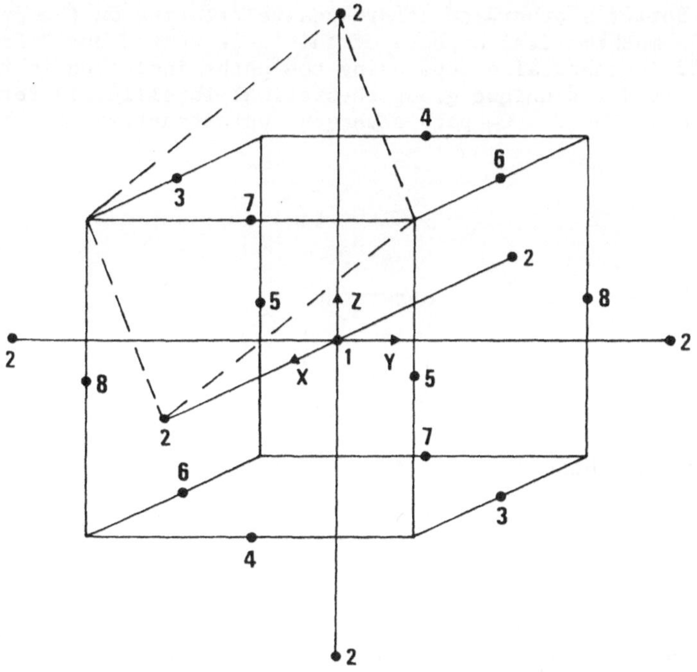

Fig. 4. Eight distinct atoms in a supershell defined in a face-centred cubic cyclic region. A segment of the surface of the cyclic region is defined by the dashed lines.

supershell contains eight distinct nodes, and equivalences between nodes due to the cyclic boundary conditions are indicated by identical numbering. A segment of the surface of the region which defines the cyclic boundary conditions (i.e. the cyclic region) is also shown in figure 4. In ref. 13 the character table of the group describing this supershell is given. It consists of the octahedral group supplemented by eight cyclic translation operators with 192 elements overall. It is a factor group of the full space group with respect to a Bravais lattice group with primitive translations twice those of the space group. Apart from labelling the symmetry coordinates of atomic clusters this group has been found useful in labelling many-electron states (see Section 5). In relation to our discussion of induced representations, it should be noted that each node in the cyclic region represents a sublattice of the crystal. Crystal sublattices as subsystems will be discussed in Section 4.

3. EIGENVALUE CLUSTERING

Harter and Patterson[14] have discussed the clustering of the rotational energy levels of spherical top molecules, such as SF_6, and pointed out that the observed clusters are related to irreps of C_n groups induced into the molecular symmetry group. Their work provides many details and comparisons; in particular it makes a link with the "tight binding" model in solid state physics. This clustering phenomenon was described by B.R. Judd[15] in the first Symmetries in Science symposium, where the link between clustering of molecular rotation levels and crystal field levels was stressed (see also ref. 16). We shall therefore only make a passing reference to it here.

My interest in this subject arose from the possibility of using induced representation labels to distinguish repeated occurrences of octahedral irreps in the reduction of R_3 irreps[17]. Arbitrary labels had been used in Butler's otherwise comprehensive treatise on the point groups[18]. The mathematical aspects of this were worked out[19] for octahedral and icosahedral groups using the paths indicated in figure 5. This indeed provides a unique group theoretical labelling in terms of the irreps defined by the C_∞, C_n path although, unfortunately, the basis

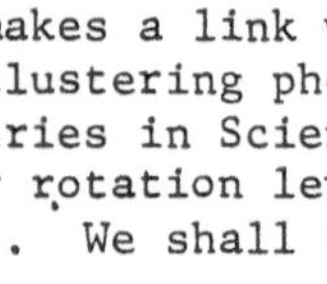

Fig. 5. Group chain used to provide a unique labelling for $R_3 \downarrow O$.

functions defined by these labels are non-orthogonal. In practice we can use Schmidt orthogonalisation to overcome this problem, but such a technique has evident disadvantages.

A more practical consequence of the clustering phenomenon is that the induced symmetry adapted states provide good approximations to eigenstates[16]. This is because, if we ignore the weak coupling to states outside the cluster, the cluster states can be determined from symmetry considerations alone.

4. SHORT RANGE INTERACTIONS AND SUM RULES

Can the range of interactions in molecules and crystals be expressed as a symmetry property? The quantum chemists have studied this problem, especially in relation to the Hückel[20] and Pariser-Parr-Pople models of π-electron bonding in <u>alternant</u> planar hydrocarbons[21]. Alternant hydrocarbons have the property that the carbon atoms can be divided into two sets, usually denoted "starred" and "unstarred", such that starred carbons only bond with unstarred carbons, no two members of the same set being bonded together. The example of naphthalene is shown in figure 6. It has been shown that an <u>alternancy</u> symmetry operator can be found for model Hamiltonians of the above type which both changes the sign of the starred orbitals and produces particle-hole conjugation[22].

The main consequences of this new symmetry operation are to relate properties of the positive and negative ions and to provide a two-valued quantum number for the states of neutral molecules which restricts possible transitions to those between states of opposite type.

Alternant molecules with equal numbers of starred and unstarred atoms have additional properties which cannot be explained in terms of alternancy symmetry, however. In particular their one-electron energies are <u>paired</u> so that $E_i = A \pm B_i$, where A is the same for all pairs and can be identified with the orbital energy for a free carbon atom. Hence we have a "sum rule" for the state pairs. This property can be understood in terms of induced representations. In such molecules the starred and unstarred atoms form equivalent subsystems. The orbital energies in these systems are not perturbed by interatomic interactions, so that all the subsystem π-orbitals are degenerate. The induced states form pairs with the energies E_i given above, according to whether the subsystem states are bonding or antibonding.

A related application of induced representation theory in solid state physics is in the theory of "special points" in the Brillouin zone[23-27]. This subject has recently been reviewed[28], and might not therefore be worth discussing here, were it not for the fact that the review does not mention the concept of induced representations. We shall, however, concentrate on explaining the physical ideas and the relevance of induced representation theory following the approach in an earlier exposition[29].

Consider a square planar Bravais lattice with (s-state) wavefunctions

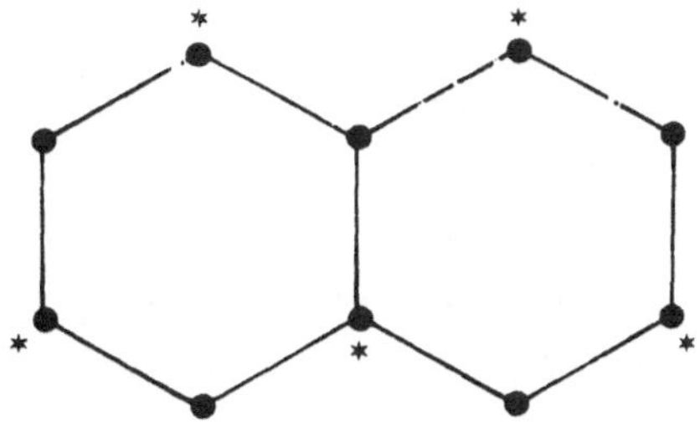

Fig. 6. Naphthalene structure.

at the nodes as shown below

ϕ_{00}		ϕ_{10}
+ + + +		+ - + -
+ + + +		+ - + -
+ + + +		+ - + -
+ + + +		+ - + -

ϕ_{01}		ϕ_{11}
+ + + +		+ - + -
- - - -		- + - +
+ + + +		+ - + -
- - - -		- + - +

The above diagram shows four distinct sets of phases at the nodes, corresponding to different $\underset{\sim}{k}$ values of the Bloch function $\phi_{\underset{\sim}{k}}$. The construction below indicates how these may be combined to produce a "subsystem" in which the amplitudes vanish on three out of every four nodes:

2 0 2 0 $\quad \tfrac{1}{2}(\phi_{00} + \phi_{01} + \phi_{10} + \phi_{11}) = \phi^{(00)}$
0 0 0 0
2 0 2 0
0 0 0 0

This subsystem corresponds to a sublattice of the original lattice, the primitive unit vectors being multiplied by 2. We see therefore that the irreducible representations of the sublattice translation group each generate a four-dimensional induced representation on the original lattice.

We now generalize these considerations a little and relate them to the two-dimensional Brillouin zone. Let the four possible positions of the sublattice be denoted by $\underset{\sim}{R} = (R_xR_y) = (00),(10),(01)$ and (11), then the four possible subsystem states can be written

$$\phi^{(\underset{\sim}{R})} = \tfrac{1}{2}(\phi_{00} + e^{i\pi R_x}\phi_{10} + e^{i\pi R_y}\phi_{10} + e^{i\pi(R_x+R_y)}\phi_{11}), \qquad (2)$$

where the suffices correspond to $\underset{\sim}{k}$ vectors at the centre and boundary points of the Brillouin zone. These transformations (for all four $\underset{\sim}{R}$ values) have been called "decoupling transformations"[29]. A set of interaction energies between the sublattice, or quasi-localized, states can now be written using a similar decoupling transformation:

$$I^{(\underset{\sim}{R})} = \langle\phi^{(00)}|H|\phi^{(\underset{\sim}{R})}\rangle$$

$$= \tfrac{1}{4}(E_{00} + e^{i\pi R_x}\,E_{10} + e^{i\pi R_y}\,E_{01} + e^{i\pi(R_x+R_y)}\,E_{11}) \qquad (3)$$

where $E_{\underset{\sim}{k}}$ are the energies at $\underset{\sim}{k}$, and $\underset{\sim}{k} = (00)$ corresponds to the point Γ in the Brillouin zone, $\underset{\sim}{k} = (10)$ and (01) to X and $\underset{\sim}{k} = (11)$ to M (see figure 7). Inverting the equations (3), we obtain

$$E_\Gamma = I^{(00)} + 2I^{(10)} + I^{(11)},$$

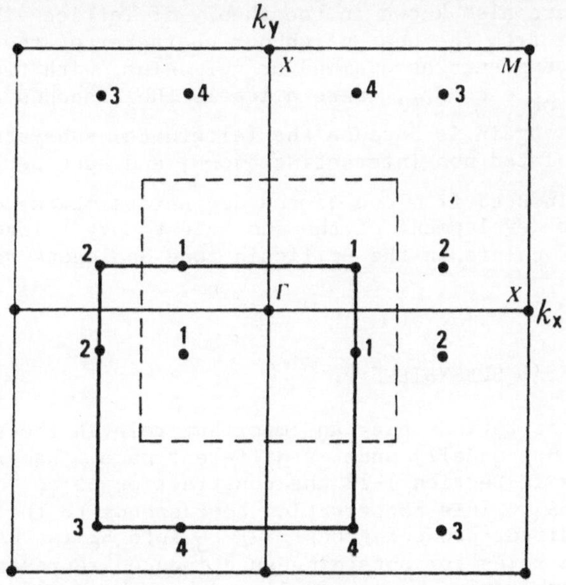

Fig. 7. Brillouin zone and decoupling point equivalences for a square planar lattice. The reduced Brillouin zone, corresponding to the sublattice, is outlined by the dashed square.

$$E_X = I^{(00)} - I^{(11)}$$

$$E_M = I^{(00)} - 2I^{(10)} + I^{(11)}.$$

The mean energy is just $I^{(00)} = \frac{1}{4}(E_\Gamma + 2E_X + E_M)$, which corresponds to energy of the sub-lattice states. We now note that this result is quite independent of the four points in the reduced Brillouin zone used to generate the transformation. The four points corresponding to an induced representation form the corners of a square in the Brillouin zone (see figure 7). In particular, if this square is centred at $\underset{\sim}{k} = 0$ (or Γ), the four points ($\pm\frac{1}{2}$, $\pm\frac{1}{2}$) all correspond to the same irrep, with energy $I^{(00)}$. For this reason the point ($\frac{1}{2}$,$\frac{1}{2}$) is known as a "special point" in the Brillouin zone[26].

The general form of decoupling transformations, relating symmetry adapted states on the lattice with those on the sublattices, has been described by Newman[30]. Its applications to the problem of relating the energies and interaction energies of localised states to band energies has been discussed in a number of papers[28]. Most applications of the decoupling idea have, however, been centred on the question of obtaining good approximations to integrals over the Brillouin zone, but we shall not pursue the details here (see ref. 24).

Another aspect of this formalism is the current interest in the use of cyclic regions of "large unit cells" in calculations which involve localised perturbations of the crystal[24]. Such perturbations can be viewed (from a group theoretical point of view) as splitting the degeneracies predicted by the cyclic region groups.

Because of the group theoretical character of the decoupling transformation it can be applied to any problems in which sublattice states interact. For example, it is possible to similarly analyse the lattice vibration dispersion curves for covalent and metallic systems[31,32].

"Sum rules" are also known in the theory of lattice vibrations[33,34,35]. These again derive from the near-neighbour character of the interactions. As applied to systems such as diamond or germanium, with two atoms per unit cell, the sum $\sum_n \omega^2_{n\underset{\sim}{k}}$ = const., where n labels the branches and $\underset{\sim}{k}$ is the wave-vector. This again is because the (alternate) subsystems consist of sublattices of isolated non-interacting atoms, and each set of $\omega^2_{n\underset{\sim}{k}}$ values for a given n is induced from one of the degenerate states of this sub-lattice. A recent development of the sum rule theory[36] involves the values of $\omega^2_{n\underset{\sim}{k}}$ at sets of points in the Brillouin zone analogous to those described above.

5. CONFIGURATIONS AS SUBSYSTEMS

Induced representations play an important role in the theory of symmetric groups, but usually under a different name. Hamermesh[37] (for example) describes in Section 7-12 the construction of an "outer" direct representation of S_6. This construction corresponds to the induction of the irrep of the direct product group $S_3 \otimes S_3$ into S_6 (as has been remarked by Coleman[3]). The rules for obtaining such induced representations are well known and many results have been listed by Wybourne[38] (Table B-1) who refers to them as outer products of S-functions. According to Frobenius' theorem, these rules can be obtained directly from the correlation table for the subduction $S_3 \otimes S_3 \subset S_6$. This is noted in the literature, e.g. see ref. 39, but omitting the link with Frobenius' theorem.

Hamermesh[37] also provides a physical interpretation of the outer direct product (see ps. 249-50) in terms of building up the states of a system of six particles from those of two subsystems of three particles. In order to avoid confusion in nomenclature we shall use the term "configuration" to describe the compound states corresponding to irreps of direct product groups, such as $S_3 \otimes S_3$. Configurations can then be regarded as subsystems of an S_6 system, in the sense used elsewhere in this article.

This approach to build up systems from configurations has been used by Chan and Newman[40,41]. The induction of representations into S_n from the possible configurations of m electrons in s-states on n sites, allows a unique labelling of the S_n states using chains of the type

$$S_n \supset S_{n-r} \otimes S_r \supset S_{m-2r} \otimes S_{n-m+r} \otimes S_r$$

for configurations with n+r-m empty sites. Subduction, then provides a unique labelling of the states for various geometric arrangements of the n sites. The chain of groups used to solve the 8-site case is

$$\begin{aligned} S_8 \supset A_8 \supset [8] + [4^2] + [2^4] + [1^8] \supset \\ [8] + [62] + 2[4^2] + [42^2] + [3^2 1^2] + 2[2^4] + [2^2 1^4] + [1^8] \\ \supset K(\text{spatial symmetry group}). \qquad (4) \end{aligned}$$

Here we have used the notation for subgroups given by Littlewood[42]: the representations are those induced into S_8 from the invariant representation of the subgroup. Note that the 192 element group in the second line of equation (4) is isomorphic to the cyclic region group described by figure 4.

This work was later developed[41] along alternative lines using

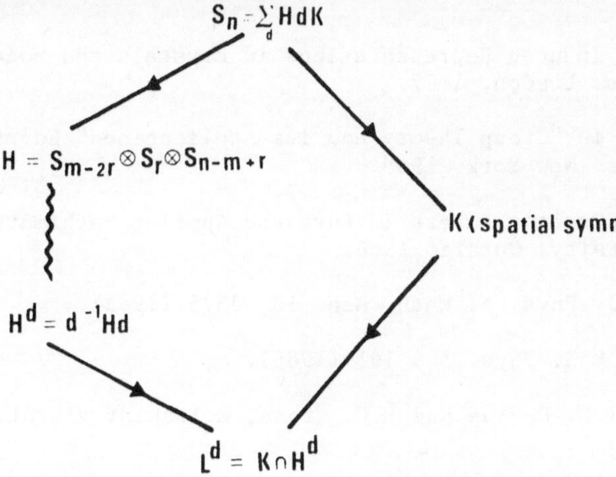

Fig. 8. Group structure used in the application of Mackey's theorem to many-electron systems.

Mackey's theorem[1]. In order to apply this theorem it is necessary to obtain the set of intersection groups L^d between the configuration groups H^d and spatial symmetry groups K. The relationship between these various groups is shown in Figure 8. This method was found to be more involved, but nevertheless gave greater insight, especially as the relationship between the configuration groups H and spatial symmetry is explored in detail. It also seems preferable as a starting point for a formalism to calculate matrix elements, in view of the theorems (based on Mackey's theorem) which relate the inner direct products of induced representations to intersection group representations[1].

The general question of obtaining matrix element formulae for induced representations has been attached by Klein and Seligman[43,44]. Their results are of considerable interest although it is not yet clear to the author excatly how they can be applied to multicentre systems.

It is also worth remarking that the standard combinatorial theorems, such as those due to Pólya and de Bruijn, can be viewed as relatively simple consequences of symmetric group theory[40]. This, in conjunction with the fact that standard combinatorics deals only with classical objects (i.e. not fermions), makes it inappropriate to use combinatorial theorems for most physical problems.

6. SUMMARY

The relevance of induced representation theory to a wide range of physical problems has been described. In particular we have discussed the consequences of interactions having a restricted range in some detail, partly because the literature is very diverse and disconnected, and also because in this area there is no general understanding of the role of induced representation theory. It will clearly take some time before the use of induced representation theory becomes commonplace, and even longer before double coset decompositions and Mackey's theorems are routinely applied in the evaluation of matrix elements. My aim has been to bring such times a little closer.

REFERENCES

1. S.L. Altmann, Induced Representations in Crystals and Molecules, Academic Press, London, 1977.

2. A.J. Coleman, in "Group Theory and its Applications" Ed. E.M. Loebl, Academic Press, New York, 1968.

3. A.J. Coleman, Queen's Papers in Pure and Applied Mathematics No.4, Queen's University, Ontario 1966.

4. D.J. Newman, J. Phys. A: Math. Gen. 16, 2375 (1983).

5. A. Ceulemans, Mol. Phys. 54, 161 (1985).

6. E.B. Wilson, J.C. Decius and P.C. Cross, Molecular Vibrations, Dover, New York, 1980.

7. R.L. Flurry, Jr., Theor. Chim. Acta 31, 221 (1973).

8. S.C. Chen and D.J. Newman, J. Phys. A: Math. Gen. 15, 331 (1982).

9. G. Fieck, Theor. Chim. Acta 44, 279 (1977).

10. D.J. Newman, J. Phys. A: Math. Gen. 14, 3143 (1981).

11. T. Lulek, Acta Phys. Polon., A57, 407 (1980).

12. M. Kuzma, I. Kuplowski and T. Lulek, Acta Phys. Polon. A57, 415 (1980).

13. K.S. Chan and D.J. Newman, J. Phys. A: Math. Gen. 15, 3383 (1982).

14. W.G. Harter and C.W. Patterson, J. Chem. Phys. 66, 4872 (1977).

15. B.R. Judd, in Symmetries in Science, eds. B. Gruber and R.S. Millman, Plenum Press, New York, 1980.

16. R.J. Elliott, R.T. Harley, W. Hayes and S.K.P. Smith, Proc. Roy. Soc. A328, 217 (1972).

17. D.J. Newman, Phys. Letters 97A, 153 (1983).

18. P.H. Butler, Point Group Symmetry Applications, Plenum Press, New York, 1981.

19. Y.Y. Yeung and D.J. Newman, J. Chem. Phys. 83, 4691 (1985).

20. K. Yates, Hückel Molecular Orbital Theory, Academic Press, New York, 1978.

21. A.D. McLachlan, Mol. Phys. 2, 271 (1959).

22. J. Koutecky, J. Paldus and J. Cizek, J. Chem. Phys. 83, 1722 (1985).

23. R.N. Euwema, D.J. Stukel, T.C. Collins, J.S. DeWitt and D.G. Shankland, Phys. Rev. 178, 1419 (1969).

24. R.A. Evarestov and V.A. Lovchikov, Phys. Stat. Sol.(b) 93, 469 (1979).

25. D.J. Chadi and M.L. Cohen, Phys. Rev. B8, 5747 (1973).

26. D.J. Chadi, Phys. Rev. B16, 1746 (1977).

27. S.L. Cunningham, Phys. Rev. B10, 4988 (1974).

28. R.A. Evarestov and V.P. Smirnov, Phys. Stat. Sol.(b) 119, 9 (1983).

29. D.J. Newman, J. Phys. C: Solid State Physics, 6, 458 (1973).

30. D.J. Newman, J. Phys. Chem. Solids 35, 1187 (1974).

31. D.J. Newman, J. Phys. C: Solid State Phys.

32. Y.Y. Yeung, J. Chem. Phys. (to be published 1986).

33. R. Brout, Phys. Rev. 113, 43 (1959).

34. H.B. Rosenstock, Phys. Rev. 129, 1959 (1963).

35. H.B. Rosenstock and G. Blanken, Phys. Rev. 145, 546 (1966).

36. V. Frei and P. Deus, Phys. Stat. Sol.(b)125, 121 (1984).

37. M. Hamermesh, Group Theory and its Applications to Physical Problems, Addison-Wesley, Reading, Mass. 1962.

38. B.G. Wybourne, Symmetry Principles and Atomic Spectroscopy, Wiley-Interscience, New York 1970.

39. J.P. Elliott and P.G. Dawber, Symmetry in Physics, Vol. II, MacMillan, London 1979.

40. K.S. Chan and D.J. Newman, J. Phys. A: Math. Gen. 16, 2389 (1983).

41. K.S. Chan and D.J. Newman, J. Phys. A: Math. Gen. 17, 253 (1984).

42. D.E. Littlewood, The Theory of Group Characters, Clarendon Press Oxford, 1950.

43. T.H. Seligman, in Permutation Groups in Chemistry, J. Hinze ed., Springer-Verlag, Berlin 1979.

44. D.J. Klein and T.H. Seligman, Kinam 4, 349 (1982).

SYMMETRY BETWEEN BOSONS AND FERMIONS*

Yoshio Ohnuki

Department of Physics
Nagoya University
Nagoya 464, Japan

Susumu Kamefuchi

Institute of Physics
University of Tsukuba
Ibaraki 305, Japan

INTRODUCTION

By definition Bosons and Fermions behave quite differently as regards statistics. It is equally true, however, that in some other respects they do behave similarly or even symmetrically. In the present paper we would like to show that such similarity or symmetry can be exhibited most fully when the theory is formulated in a specific manner, i.e. in terms of annihilation and creation operators a_j and $a_j^\dagger$ or what will be termed g-numbers.

The difference between Bosons and Fermions can, of course, be traced back to the difference in the signatures $(jj) = +,-$ attached to the brackets in the basic commutation relations: $[a_j, a_j^\dagger]_{-(jj)} = 1$, $[a_j, a_j]_{-(jj)} = 0$. However, the substantial part of the theory can in fact be formulated without specifying the individual signatures (jj). And this is the reason why it is possible to treat Bosons and Fermions in a unified manner, and to thereby consider, among the two, super- or more general, g-symmetry transformations.

MATHEMATICAL PRELIMINARIES: G-NUMBERS

Let us consider a set of generalized or *g-numbers* $x_j (j=1,2,\cdots,n)$ such that

$$[x_i, x_j]_{-(ij)} = 0 \quad , \quad x_j^N \neq 0 \quad \text{for } (jj) = + \text{ and any } N = 1,2,\cdots , \tag{2.1}$$

where $(ij) = (ji) = +$ or $-$ will be called *relative signatures.*[1] The suffices $i,j,\cdots$ may be so arranged that $(jj) = +$ for $j = 1,2,\cdots,b$ and $(jj) = -$ for $j = b+1, b+2, \cdots, b+f \equiv n$. The former (latter) set of x_j's are called generalized Bose or *gB-* (generalized Fermi or *gF-*) *numbers.* In the special case when all $(ij) = +(-)$ for $i \neq j$ the gB- (gF-) numbers turn out to be complex (Grassmann) numbers. Sometimes we also consider the *-conjugates x_j^* of x_j, assuming that x_j and x_j^* have the same relative signatures. More generally, the commutation property between any two g-numbers is assumed to be determiend solely by (ij).

---

*delivered by S. Kamefuchi.

Algebraic Properties

We call $x=(x_1,x_2,\cdots,x_n)$ a *g-vector*, and $T_{j_1j_2\cdots j_\ell}$ a *g-tensor* if it behaves in the same way as $x_{j_1}x_{j_2}\cdots x_{j_\ell}$ so far as the commutation properties are concerned. Given a g-tensor M_{ij} of second rank we construct a *g-matrix* $M=\|M_{ij}\|$, and define the following operations:

$$\text{transpose} \qquad (\tilde{M})_{ij} \equiv (ii)(ij)M_{ji} \quad , \quad (\bar{M})_{ij} \equiv (ij)(jj)M_{ji} \quad , \tag{2.2}$$

$$\text{trace} \qquad \mathrm{Tr}\, M \equiv \sum_{j=1}^{n} (jj)M_{jj} \quad , \tag{2.3}$$

$$\text{g-determinant} \qquad \mathrm{Det}\, M \equiv \exp[\mathrm{Tr}\log M] \quad . \tag{2.4}$$

We then have

$$\begin{aligned} &\widetilde{MM'} = \tilde{M}'\tilde{M} \quad , \quad \overline{MM'} = \bar{M}'\bar{M} \quad , \quad \bar{\tilde{M}} = \tilde{\bar{M}} = M \quad , \\ &\tilde{\tilde{M}} = \bar{\bar{M}} \quad , \quad \tilde{\tilde{\tilde{M}}} = \bar{M} \quad , \quad \bar{\bar{\bar{M}}} = \tilde{M} \quad , \\ &\mathrm{Tr}\,\tilde{M} = \mathrm{Tr}\,\bar{M} = \mathrm{Tr}\, M \quad , \quad \mathrm{Tr}(MM') = \mathrm{Tr}(M'M) \quad , \\ &\mathrm{Det}\,\tilde{M} = \mathrm{Det}\,\bar{M} = \mathrm{Det}\, M \quad , \quad \mathrm{Det}(MM') = \mathrm{Det}\, M\cdot\mathrm{Det}\, M' \quad , \\ &[\mathrm{Det}\, M, \hat{x}_j] = 0 \quad , \end{aligned} \tag{2.5}$$

where $\hat{x}_j$ stands for both x_j and x_j^*, and the last relation implies that Det M is a g-scalar. Klein transformations with which to change (ij)'s can be formulated through the intermediary of other gB-numbers.

Analytical Operations

For functions $f(x,x^*)$ of $\hat{x}_j$'s , expressed e.g. as formal series in $\hat{x}_j$'s or *g-functions*, we can define various analytical operations. Left- and right-differentiations can be defined in an obvious manner, and in either case we have $[\partial/\partial\hat{x}_i, \partial/\partial\hat{x}_j]_{-(ij)} = 0$.

On the other hand, integration can be defined for double integrals for a pair of x and x*, or more generally as

$$\begin{aligned} &\int x_1^{r_1} x_1^{*s_1} x_2^{r_2} x_2^{*s_2} \cdots x_n^{r_n} x_n^{*s_n} \exp[-x^*\cdot x](dxdx^*) \\ &\qquad = \prod_{j=1}^{n} (\delta_{r_js_j} r_j!) \quad , \end{aligned} \tag{2.6}$$

where $r_j,s_j=0,1,2,\cdots (r_j,s_j=0,1)$ for $j=1,2,\cdots,b(j=b+1,b+2,\cdots,b+f)$, $x^*\cdot x \equiv \Sigma_{j=1}^n x_j^* x_j$ and $(dxdx^*) \equiv dx_1dx_1^*dx_2dx_2^*\cdots dx_ndx_n^*$. The formula remains invariant under *-conjugation. For the case of a pair of gB-numbers x and x* (i.e. b=1, f=0) (2.6) implies

$$\int x^r x^{*s} \exp[-x^*x]dxdx^* \equiv \iint_{-\infty}^{\infty} (u+iv)^r(u-iv)^s \exp[-(u^2+v^2)]\frac{dudv}{\pi} = \delta_{rs} r! \quad . \tag{2.6'}$$

For the case of a pair of gF-numbers x and x* (i.e. b=0, f=1) (2.6) is consistent with the well-known formulae for Grassmann integrals: $\int dx = \int dx^* = 0$, $\int x dx = \int x^* dx^* = i$.[2]

In connection with the integration as specified by (2.6) we can introduce operations such as partial integration, ((ij)-conserving) change of variables and its Jacobian, Fourier transformaiton, etc. In particular, the function defined by

$$\delta_n(x^*-x'^*, x-x') \equiv \lim_{\varepsilon\to+0} \int \exp[(x^*-x'^*)\cdot\bar{x} - \bar{x}^*\cdot(x-x') - \varepsilon\bar{x}^*\cdot\bar{x}](d\bar{x}d\bar{x}^*)$$

serves as the delta function:

$$\begin{aligned} &\hat{x}_j\delta_n(x^*-x'^*, x-x') = \hat{x}_j'\delta_n(x^*-x'^*, x-x') \quad , \\ &\int f(x,x^*)\delta_n(x^*-x'^*, x-x')(dxdx^*) = f(x',x'^*) \quad . \end{aligned} \tag{2.7}$$

Further, in the functional space spanned by g-functions $f(x,x^*)$, $g(x,x^*)$, $\cdots$ it is convenient to define inner products (g,f) by

$$(g,f) \equiv \int g^*(x,x^*)f(x,x^*)\exp[-x^*\cdot x](dxdx^*) \quad . \tag{2.8}$$

<u>Group-theoretical Properties</u>

Our g-numbers enable us to generalize supersymmetry transformations of the usual kind. Let G be a group whose elements are specified by g-numbers $x_j (j=1,2,\cdots,n)$. A formal extension of the usual Lie-group arguments then leads to the following relations for Lie generators L_j $(j=1,2,\cdots,n)$:[3]

$$[L_i, L_j]_{-(ij)} = (ij)C^k_{ij}L_k \quad , \tag{2.9}$$

$$\begin{aligned} (ki)[L_i,[L_j,L_k]_{-(jk)}]_{-(ki)(ij)} + (ij)[L_j,[L_k,L_i]_{-(ki)}]_{-(ij)(jk)} \\ +(jk)[L_k,[L_i,L_j]_{-(ij)}]_{-(jk)(ki)} = 0 \quad , \end{aligned} \tag{2.10}$$

where the structure constants C^k_{ij} are g-tensors subject to

$$\begin{aligned} &C^k_{ij} = -(ij)C^k_{ji} \quad , \\ &(\ell k)C^m_{jk}C^i_{\ell m} + (\ell j)C^m_{k\ell}C^i_{jm} + (jk)C^m_{\ell j}C^i_{km} = 0 \quad . \end{aligned} \tag{2.11}$$

The transformations specified by G will hereafter be called *g-symmetry* transformations.

SYSTEM OF BOSE AND FERMI OSCILLATORS

Let us consider a system consisting of (non-relativistic) Bose fields $\phi_j(x) (j=1,2,\cdots,b)$ and Fermi fields $\phi_j(x) (j= b+1,b+2,\cdots,b+f \equiv n)$. The most general form of equal-time commutation relations is given as

$$[\phi_i(x) \ , \ \phi_j^\dagger(x')]_{-(ij)} = \delta_{ij}\delta(x-x') \quad , \quad [\phi_i(x) \ , \ \phi_j(x')]_{-(ij)} = 0 \quad , \tag{3.1}$$

where $(jj) = +(-)$ for $j=1,2,\cdots,b$ $(j = b+1,b+2,\cdots,b+f)$. As is well known, (ij)'s with $i \neq j$ should be so fixed as to be consistent with the locality condition, and possible choices of (ij)'s are usually classified into the normal and anomalous cases. In the normal case $(ij) = -$ for $i,j = b+1,b+2,$

$\cdots$,b+f and (ij) = + otherwise, whereas in any anomalous case at least one of (ij)'s with $i \neq j$ should be different from the corresponding (ij) of the normal case. Furthermore, Lüders' theorem states that the so-called Klein transformations enable us to go from one case to another, thereby establishing physical equivalence of the respective field theories.

For simplicity let us now replace each field $\phi_j(x)$ by a single, harmonic oscillator, describing the latter by the annihilation and creation operators a_j and $a_j^\dagger$ ($j=1,2,\cdots,n$). Corresponding to (3.1) we then have the following commutation relations:

$$[a_i, a_j^\dagger]_{-(ij)} = \delta_{ij} \quad , \quad [a_i, a_j]_{-(ij)} = 0 \quad , \tag{3.2}$$

with the same (ij)'s as given in (3.1). For the time being let us restrict ourselves to such a general system of n oscillators.

In order to describe the above system we make use of the g-numbers $\hat{x}_i$ with the same (ij)'s as given in (3.2), and supplement (2.1) and (3.2) by

$$[\hat{a}_i, \hat{x}_j]_{-(ij)} = 0 \quad , \quad \hat{x}_j|0\rangle = |0\rangle\hat{x}_j \quad , \quad \hat{x}_j\langle 0| = \langle 0|\hat{x}_j \quad , \tag{3.3}$$

where $|0\rangle$ denotes the ground state of the system, satisfying $a_j|0\rangle = 0$, and $\hat{a}_j$ stands for both a_j and $a_j^\dagger$. It should be understood that under hermitian conjugation the order of factors in each term is reversed, and operators, c-numbers and g-numbers contained therein undergo hermitian-, complex- and *-conjugations, respectively.

In terms of g-numbers we can introduce a special representation, to be referred to hereafter as the *g-representation*. To do this let us first define coherent states (unnormalized) by

$$|x\rangle \equiv \exp[a^\dagger \cdot x]|0\rangle \quad , \quad \langle x^*| \equiv \langle 0|\exp[x^* \cdot a] \quad , \tag{3.4}$$

where $a^\dagger \cdot x \equiv \Sigma_{j=1}^n a_j^\dagger x_j$. These are simultaneous eigenstates of $\hat{a}_j$'s:

$$a_j|x\rangle = x_j|x\rangle \quad , \quad \langle x^*|a_j^\dagger = \langle x^*|x_j^* \quad , \tag{3.5}$$

and their inner products and the completeness relation are given respectively by

$$\langle x^*|x\rangle = \exp[x^* \cdot x] \quad , \tag{3.6}$$

and

$$\int |x\rangle\langle x^*|\exp[-x^* \cdot x](dxdx^*) = 1 \quad . \tag{3.7}$$

Next, for a given state vector $|\rangle$ we define the *g-wavefunction* $\psi(x^*)$ and its dual $\bar{\psi}(x)$ by

$$\psi(x^*) \equiv \langle x^*|\rangle \quad , \quad \bar{\psi}(x) \equiv \langle|x\rangle \quad , \tag{3.8}$$

the normalization condition being

$$\int \bar{\psi}(x)\psi(x^*)\exp[-x^* \cdot x](dxdx^*) = 1 \quad . \tag{3.9}$$

For observables $F(a^\dagger, a)$, given as normal-ordered polynomials in $\hat{a}_j$'s, we have

$$\langle x^*|F(a^\dagger, a)|x\rangle = F(x^*, x)\exp[x^* \cdot x] \quad ,$$

$$<2|F(a^\dagger,a)|1> = \int \bar{\psi}_2(x')F(x'^*,x)\psi_1(x^*) \times \exp[x'^*\cdot x - x^*\cdot x - x'^*\cdot x'](dxdx^*)(dx'dx'^*) \quad . \tag{3.10}$$

When the Hamiltonian of the system is also given in the normal-ordered form $H(a^\dagger,a)$, the Schrödinger equation reads

$$i\frac{\partial\psi(x^*,t)}{\partial t} = \int H(x^*,x')\psi(x'^*,t)\exp[x^*\cdot x' - x'^*\cdot x'](dx'dx'^*) \quad . \tag{3.11}$$

Our result implies that the transformation theory of quantum mechanics can be extended further to the g-representation.

UNIFIED DESCRIPTION

Path Integral Formulae

Let us first consider the transition amplitude $<F|U(t_F,t_I)|I>$ for a transition: $|I>$ at time $t_I \rightarrow |F>$ at time t_F, where $U(t_F,t_I)=\exp[-iH(t_F-t_I)]$. As usual we divide the time interval (t_F-t_I) into N equal segments: $(t_F-t_I)/N = \Delta t = t_k-t_{k-1}$, where $t_I = t_0$, $t_F = t_N$ and $k = 1,2,\cdots,N$, and write

$$U(t_F,t_I) = U(t_N,t_{N-1})\cdots U(t_2,t_1)U(t_1,t_0) \quad . \tag{4.1}$$

Now, for N sufficiently large we have, by use of (3.10),

$$\begin{aligned}<x^*_{(k)}|U(t_k,t_{k-1})|x_{(k-1)}> &= <x^*_{(k)}|[1-i\Delta tH(a^\dagger,a)]|x_{(k-1)}> \\ &= \exp[x^*_{(k)}\cdot x_{(k-1)} - i\Delta tH(x^*_{(k)},x_{(k-1)})] \quad .\end{aligned} \tag{4.2}$$

Thus, by inserting LHS of (3.7) in between all neighbouring $U(t_k,t_{k-1})$'s in (4.1) and then using (4.2) we find

$$\begin{aligned}&<F|U(t_F,t_I)|I> \\ &= \lim_{N\to\infty} \int\cdots\int \bar{\psi}_F(x_{(N)})\exp[\sum_{k=0}^{N} i\Delta tL_k]\psi_I(x^*_{(0)}) \prod_{k=0}^{N} (dx_{(k)}dx^*_{(k)}) \quad ,\end{aligned} \tag{4.3}$$

where

$$L_k \equiv i(x^*_{(k)}\cdot \frac{x_{(k)} - x_{(k-1)}}{\Delta t} - H(x^*_{(k)},x_{(k-1)}) \quad , \tag{4.4}$$

with $x_{(-1)} = H(x^*_{(0)},x_{(-1)}) = 0$ and $L_0 = ix^*_{(0)}\cdot x_{(0)}/\Delta t$.

As a second example, let us consider $\mathrm{tr}\exp[-\beta H(a^\dagger,a)]$. Again we divide β into N equal segments $\beta/N = \Delta\beta = \beta_k-\beta_{k-1}$ $(k=1,2,\cdots,N)$ and write the exponential function as a product of N operators $\exp[-\Delta\beta H(a^\dagger,a)]$. For N sufficiently large we can deal with the resulting expression in a similar manner, to obtain

$$\begin{aligned}&\mathrm{tr}\exp[-\beta H(a^\dagger,a)] \\ &= \lim_{N\to\infty} \int\cdots\int \exp[-\sum_{k=0}^{N} \{x^*_{(k)}\cdot(x_{(k)}-x_{(k-1)}) + \Delta\beta H(x^*_{(k)},x_{(k-1)})\} \\ &\qquad + x^*_{(0)}\cdot \tilde{x}_{(N)}] \prod_{k=0}^{N} (dx_{(k)}dx^*_{(k)}) \quad ,\end{aligned} \tag{4.5}$$

where $\tilde{x}_{(N)j} \equiv (jj)x_{(N)j}$ and $x_{(-1)} = H(x^*_{(0)}, x_{(-1)}) = 0$. We note that in the pure gF-case (i.e. $b=0$, $f=n \geq 1$) the integration over $x^*_{(0)}$ and $x_{(N)}$ can be easily carried out, and the relation $\tilde{x}_{(N)i} = (ii)x_{(N)i}$ gives rise to the so-called antiperiodic boundary condition $x_{(N)} = -x_{(0)}$.

Functional Space

G-functions $f(x,x^*)$ with finite norms in the sense of (2.8) form a linear functional space, and as its basis functions we can employ

$$\prod_{j=1}^{n} \chi_{\ell_j \nu_j}(x_j, x_j^*) \quad . \tag{4.6}$$

Here in the product $\chi_{\ell\nu}$'s are placed, for example, in the increasing order of j,

$$\chi_{\ell\nu}(x,x^*) \equiv \left(\frac{(\ell-|\nu|)}{\ell!}\right)^{\frac{1}{2}} L_{\ell-|\nu|}^{|\nu|}(xx^*)\,\tilde{x}^{|\nu|} \quad , \tag{4.7}$$

$$\tilde{x}^{|\nu|} \equiv \begin{cases} x^{\nu} & \text{for } \nu \geq 0 \ , \\ x^{*|\nu|} & \text{for } \nu < 0 \ , \end{cases} \tag{4.7'}$$

$\ell = 0,1,2,\cdots$; ν = integers such that $|\nu| \leq \ell$ and $L_k^{\alpha}(z)$'s are Laguerre polynomials. For x, x^* = gB-numbers $(\chi_{\ell\nu}, \chi_{\ell'\nu'}) = \delta_{\ell\ell'}\delta_{\nu\nu'}$, whereas for x,x^* = gF-numbers only $\chi_{00}, \chi_{11}, \chi_{10}$ and $\chi_{1,-1}$ are nonvanishing and $(\chi_{\ell\nu}, \chi_{\ell'\nu'}) = \varepsilon_{\ell\nu}\delta_{\ell\ell'}\delta_{\nu\nu'}$ with $\varepsilon_{00} = \varepsilon_{1,-1} = 1$, $\varepsilon_{11} = \varepsilon_{10} = -1$. Hence our functional space is of indefinite metric.

Weyl Operators

The Weyl operator $W(p,q)$ is usually defined for a Bose degree of freedom as $W(p,q) \equiv \exp[i(pQ-qP)] = \exp[a^{\dagger}x - x^*a]$, where $Q \equiv (a+a^{\dagger})/\sqrt{2}$, $P \equiv (a-a^{\dagger})/\sqrt{2}i$, $q \equiv (x+x^*)/\sqrt{2}$ and $p \equiv (x-x^*)/\sqrt{2}i$. Let us thus define the Weyl operator for our system of n oscillators by

$$W(x,x^*) \equiv \exp[a^{\dagger}\cdot x - x^*\cdot a] \quad . \tag{4.8}$$

Then, as in the usual Bose case we have $W(x,x^*)|0\rangle = \exp[-x^*\cdot x/2]|x\rangle$, that is, normalized coherent states.

It is also an easy matter to derive

$$\begin{aligned} W^{\dagger}(x,x^*) &= W^{-1}(x,x^*) = W(-x,-x^*) \quad , \\ W^{\dagger}(x,x^*)\hat{a}_j W(x,x^*) &= \hat{a}_j + \hat{x}_j \quad , \\ W(x,x^*)\, W(x',x'^*) &= \exp[-(x^*\cdot x' - x'^*\cdot x)/2]W(x+x', x^*+x'^*) \quad . \end{aligned} \tag{4.9}$$

These relations imply that W provides a unitary ray representation of the group of displacement $\hat{a}_j \rightarrow \hat{a}_j + \hat{x}_j$, where $(dxdx^*)$ is an invariant measure in the group manifold with the coordinates x_j, x_j^* $(j=1,2,\cdots,n)$. The ortho-completeness relation of group elements is given as

$$\mathrm{Tr}[W(x,x^*)W^{\dagger}(x',x'^*)] = \delta_n(x^*-x'^*, x-x') \quad , \tag{4.10}$$

where $\mathrm{Tr}\, F \equiv \Sigma_{\ell}\langle\ell|(-1)^{N_f}F|\ell\rangle$ with N_f being the number operator of Fermi quanta and with $|\ell\rangle$'s forming an ortho-normal basis. As a consequence of (4.10) we can expand any $f(x,x^*)$ in the form:

$$f(x,x^*) = \sum_{\ell,\ell'} \langle\ell|W(x,x^*)|\ell'\rangle c_{\ell'\ell} \, ,$$
$$c_{\ell'\ell} = \int \langle\ell'|W^\dagger(x,x^*)(-1)^{N_f}|\ell\rangle f(x,x^*)(dxdx^*) \, . \tag{4.11}$$

Further, for any two states such that $|\alpha\rangle \equiv \alpha[a^\dagger]|0\rangle$ and $|\beta\rangle \equiv \beta[a^\dagger]|0\rangle$ there holds the relation

$$\int W(x,x^*)|\alpha\rangle(\alpha\beta)\langle\beta|W^\dagger(x,x^*)(dxdx^*) = \langle\beta|\alpha\rangle \, , \tag{4.12}$$

where $(\alpha\beta)$ is the relative signature of the operators $\alpha[a^\dagger]$ and $\beta[a^\dagger]$.

From the above results we are thus led to conclude the following: when described in terms of g-numbers, Bose and Fermi oscillators behave very similarly, so much so that no separate treatments are needed for them.

PARA-QUANTIZATION: GROUP-THEORETICAL ASPECTS

The relation between Bosons and Fermions can be exhibited more clearly and systematically when viewed within a wider framework, i.e. the field theory based on *para-quantization.*[4] As is well known, when a field is quantized in this manner, the resulting field quanta turn out to be *para-Bosons* or *para-Fermions*.

Let a_k and $a_k^\dagger$ $(k=1,2,\cdots)$ be the annihilation and creation operators of a given field $\phi(x)$. For the time being the number of possible modes k is limited to n, so that $k=1,2,\cdots,n$. The basic *para-commutation relations* for $\hat{a}_k$'s are given as follows:

$$[a_k,N_{\ell m}] = \delta_{k\ell}a_m \, , \quad [a_k,L_{\ell m}] = \delta_{k\ell}a_m^\dagger \pm \delta_{km}a_\ell^\dagger \, , \quad [a_k,M_{\ell m}] = 0 \, ; \tag{5.1}$$

$$[a_k^\dagger,a_\ell]_\pm = 2N_{k\ell} \, , \quad [a_k^\dagger,a_\ell^\dagger]_\pm = 2L_{k\ell} \, , \quad [a_k,a_\ell]_\pm = 2M_{k\ell} \, ; \tag{5.2}$$

$$[N_{k\ell},N_{mn}] = \delta_{\ell m}N_{kn} - \delta_{kn}N_{m\ell} \, , \quad [L_{k\ell},L_{mn}] = [M_{k\ell},M_{mn}] = 0 \, ,$$

$$[L_{k\ell},N_{mn}] = -\delta_{kn}L_{m\ell} \mp \delta_{\ell n}L_{mk} \, , \quad [M_{k\ell},N_{mn}] = \delta_{km}M_{n\ell} \pm \delta_{\ell m}M_{nk} \, , \tag{5.3}$$

$$[L_{k\ell},M_{mn}] = -\delta_{km}N_{\ell n} \mp \delta_{kn}N_{\ell m} - \delta_{\ell n}N_{km} \mp \delta_{\ell m}N_{kn} \, ;$$

where by definition

$$N_{\ell m}^\dagger = N_{m\ell} \, , \quad L_{\ell m}^\dagger = M_{m\ell} \, , \quad L_{\ell m} \mp L_{m\ell} = M_{\ell m} \mp M_{m\ell} = 0 \, . \tag{5.4}$$

Here and in the following the upper (lower) signs correspond to para-Bose (para-Fermi) quantization.

In order to derive the above commutation relations group-theoretically we consider "infinitesimal" (super-) transformations such as[3]

$$a_k \to a_k' = a_k - i\sum_{m=1}^{n}(\nu_{km}a_m + \lambda_{km}a_m^\dagger \mp \theta_m^* M_{km} \pm N_{mk}\theta_m) \, ,$$
$$a_k^\dagger \to a_k'^\dagger = a_k^\dagger + i\sum_{m=1}^{n}(\nu_{mk}a_m^\dagger + \mu_{mk}a_m \mp L_{mk}\theta_m \pm \theta_m^* N_{km}) \, ; \tag{5.5}$$

$$N_{k\ell}\to N'_{k\ell} = N_{k\ell} + i\sum_{m=1}^{n}(N_{m\ell}\nu_{mk} - \nu_{\ell m}N_{km} - \lambda_{\ell m}L_{km} + M_{\ell m}\mu_{km}) + \frac{i}{2}(\theta^*_k a_\ell - a^\dagger_k\theta_\ell) \quad ,$$

$$L_{k\ell}\to L'_{k\ell} = L_{k\ell} + i\sum_{m=1}^{n}(L_{m\ell}\nu_{mk} + \nu_{m\ell}L_{km} + \mu_{km}N_{\ell m} + N_{km}\mu_{m\ell}) - \frac{i}{2}(\theta^*_k a^\dagger_\ell - a^\dagger_k\theta^*_\ell) \quad , \tag{5.6}$$

$$M_{k\ell}\to M'_{k\ell} = M_{k\ell} - i\sum_{m=1}^{n}(M_{m\ell}\nu_{km} + \nu_{\ell m}M_{km} + N_{m\ell}\lambda_{km} + \lambda_{m\ell}N_{mk}) - \frac{i}{2}(a_k\theta_\ell - \theta_k a_\ell) \quad ;$$

where infinitesimal complex parameters $\nu_{km},\lambda_{km},\mu_{km}$ are subject to

$$\nu^*_{km} = \nu_{mk} \quad , \quad \lambda^*_{km} = \mu_{mk} \quad , \quad \lambda_{km} \mp \lambda_{mk} = 0 \quad , \tag{5.7}$$

and Grassmann or complex parameters $\hat{\theta}_k$ are subject to

$$[\hat{\theta}_k,\hat{\theta}_\ell]_\pm = [\hat{\theta}_k,\hat{a}_\ell]_\pm = [\hat{\theta}_k,N_{\ell m}] = [\hat{\theta}_k,L_{\ell m}] = [\hat{\theta}_k,M_{\ell m}] = 0 \quad . \tag{5.8}$$

Further, we assume that the above transformations are generated by a "unitary" operator U in such a way that (primed operator) = U^{-1}(unprimed operator)U with U given by

$$U = 1 - i\sum_{\ell,m}(\nu_{\ell m}N_{\ell m} + \frac{1}{2}\lambda_{\ell m}L_{\ell m} + \frac{1}{2}\mu_{\ell m}M_{\ell m}) - \frac{i}{2}\sum_\ell(\theta^*_\ell a_\ell + a^\dagger_\ell\theta_\ell) \quad . \tag{5.9}$$

Now, in the case of upper (lower) signs the transformations of (5.5) and (5.6) form Z_2-graded Sp(2n,R)(SO(2n+1)), and the relations (5.1),(5.2) and (5.3) all follow as the Lie commutation relations of the respective groups. Oddly enough it is in the para-Bose case that Grassmann numbers have to be introduced. On the other hand, if we set all $\hat{\theta}_k = 0$ and consider only the transformation (5.5), then the group shrinks to Sp(2n,R) (SO(2n)). In this case (5.1) and (5.3) follow from group theory, whereas (5.2) should be regarded as the equations to define $N_{k\ell}$, $L_{k\ell}$ and $M_{k\ell}$ in terms of $\hat{a}_k$ and $\hat{a}_\ell$.

We can argue in a similar manner even when a number of ordinary and para-fields coexist. Any permissible set of commutation relations, bilinear or trilinear in the field operators, can then be derived as the Lie commutation relations of a suitable g-symmetry group, whose group parameters are given, in general, by g-numbers.

STATE VECTORS AND THE SYMMETRIC GROUP

The irreducible representations of the operators $\hat{a}_k$ satisfying (5.1)-(5.3) which allow of the Fock vacuum $|0\rangle$ with $a_k|0\rangle = 0$ are uniquely fixed by a single parameter, called the *order* of para-quantization, p=1,2,··· (p=0: trivial) through

$$a_k a^\dagger_\ell|0\rangle = p\delta_{k\ell}|0\rangle \quad . \tag{6.1}$$

The operator defined by

$$N_k \equiv N_{kk} \mp \frac{p}{2} \tag{6.2}$$

plays the role of the number operator for the mode k. By use of (5.1), (5.2) and (6.1) we can easily see that any state vector belonging to our Fock space V can be constructed by applying to $|0>$ a suitable polynomial only in $a_k^\dagger$'s.

It has been known also that the representation specified by p can be expressed as (Green decomposition[5])

$$\hat{a}_k = \sum_{\alpha=1}^{p} \hat{a}_k^{(\alpha)} , \tag{6.3}$$

with

$$[a_k^{(\alpha)}, a_\ell^{(\alpha)\dagger}]_\mp = \delta_{k\ell} , \quad [a_k^{(\alpha)}, a_\ell^{(\alpha)}]_\mp = 0 , \tag{6.3'}$$
$$[\hat{a}_k^{(\alpha)}, \hat{a}_\ell^{(\beta)}]_\pm = 0 \quad (\alpha \neq \beta) .$$

Thus p=1 corresponds to the ordinary Bose or Fermi case. In addition to $[\hat{a}_k, \hat{a}_\ell]_\pm$ let us introduce two kinds of (anti-) symmetrized brackets, undotted and dotted:

$$[\hat{a}_1, \hat{a}_2, \cdots, \hat{a}_m]_\mp \equiv \sum_{\sigma \in S_m} \binom{\mathrm{sign}\,\sigma}{1} \hat{a}_{\sigma 1} \hat{a}_{\sigma 2} \cdots \hat{a}_{\sigma m} , \tag{6.4}$$
$$\vdots [\hat{a}_1, \hat{a}_2, \cdots, \hat{a}_m]_\mp \vdots \equiv n! \sum_{\alpha_1, \alpha_2, \cdots, \alpha_m}' \hat{a}_1^{(\alpha_1)} \hat{a}_2^{(\alpha_2)} \cdots \hat{a}_m^{(\alpha_m)} ,$$

where S_m denotes the symmetric group of order m! and Σ' means the summation over all different values of $\alpha_1, \alpha_2, \cdots, \alpha_m$.

Our basic theorem, which can be proved by repeated use of (5.1) and (5.2), is that any monomial $\hat{a}_1 \hat{a}_2 \cdots \hat{a}_m$ can be expressed as a sum of terms such as

$$P_{st}(\hat{a}) \equiv [\hat{a}_{i_1}, \hat{a}_{j_1}]_\pm [\hat{a}_{i_2}, \hat{a}_{j_2}]_\pm \cdots [\hat{a}_{i_s}, \hat{a}_{j_s}]_\pm [\hat{a}_{k_1}, \hat{a}_{k_2}, \cdots, \hat{a}_{k_t}]' , \tag{6.5}$$

where $2s + t \leqslant m$ and $[\ ,\ ,\cdots]'$ stands for both undotted and dotted brackets.[4] A corollary of this theorem is that our Fock space V consists of state vectors of the form $P_{st}(a^\dagger)|0>$.

Let us now consider an m-particle subspace $V^{(m)}$ spanned by $a_{\sigma 1}^\dagger a_{\sigma 2}^\dagger \cdots \times a_{\sigma m}^\dagger |0>$ with $\sigma \in S_m$, and decompose it into irreducible subspaces with respect to S_m. On the basis of the above result we can then prove that $V^{(m)}$ for the para-Bose (para-Fermi) case contains one and only one irreducible subspace corresponding to each Young diagram whose first column (row) has no more than p squares.[4] Thus, if a certain Young diagram is allowed in the para-Bose case, then its conjugate Young diagram is allowed in the para-Fermi case, and vice versa. The order p in the para-Bose (para-Fermi) case gives the maximum number of those particles that can form antisymmetric (symmetric) states with nonvanishing norms.

The structures of the state-vector space V for the para-Bose and para-Fermi cases are symmetrical in the above sense.

Let F(V) be an observable defined in a spatial domain V. For the reason mentioned above it suffices to consider observables of the form $F(V) = \int_V dx_1 \int_V dx_2 \cdots P_{st}(\hat{\phi}(x_j))$, where $\hat{\phi}(x)$ stands for both $\phi(x)$ and $\phi^{\dagger}(x)$. All such F(V)'s are required, however, to satisfy the locality condition of either of the following forms:

$$\text{strong locality} \qquad [\hat{\phi}(x), F(V)] = 0 \qquad \text{for } x \sim V \quad , \tag{7.1}$$

$$\text{weak locality} \qquad [F(V), F'(V')] = 0 \qquad \text{for } V \sim V' \quad , \tag{7.2}$$

where $A \sim B$ means that two spatial domains A and B are mutually space-like.

Under the para-commutation relations the above condition severely restricts possible forms of F(V)'s in the following way: any F(V) should be a functional of brackets of the kinds as given in Table 1 (in each term dotted brackets appear at most linearly).[4] From this table we see that observables in the para-Bose and para-Fermi cases have the symmetrical structure for p=even, but dissymmetry appears for p=odd. Note that for p=1 the dotted brackets consist of a single $\hat{\phi}(x)$.

Corresponding to the Green decomposition (6.3) we can write $\hat{\phi}(x)$ as

$$\hat{\phi}(x) = \sum_{\alpha=1}^{p} \hat{\phi}^{(\alpha)}(x) \quad , \tag{7.3}$$

where $\hat{\phi}^{(\alpha)}(x)$'s have the same relative signatures $(\alpha\beta)$ as those of $\hat{a}^{(\alpha)}$'s. By applying Klein transformations to $\hat{\phi}^{(\alpha)}(x)$'s we can go over to a new set of operators $\hat{\psi}^{(\alpha)}(x)$: $\hat{\phi}^{(\alpha)}(x) \equiv (i)^{1-\rho(\alpha)} K_{\alpha+\rho(\alpha)} \hat{\psi}^{(\alpha)}(x)$ where $\rho(\alpha) \equiv 0(1)$ for α = even (odd), such that the relative signatures $(\alpha\beta)'$ of $\hat{\psi}^{(\alpha)}(x)$'s are all +(-) for the para-Bose (para-Fermi) case; that is to say, $\hat{\psi}^{(\alpha)}(x)$'s are Bose (Fermi) fields of the normal case. Moreover, the two kinds of brackets which constitute all F(V)'s can be rewritten in terms of $\hat{\psi}^{(\alpha)}(x)$'s as follows:

$$[\hat{\phi}(x), \hat{\phi}(x')]_{\pm} = \sum_{\alpha=1}^{p} [\hat{\psi}^{(\alpha)}(x), \hat{\psi}^{(\alpha)}(x')]_{\pm} \quad , \tag{7.4}$$

$$\vdots [\hat{\phi}(x_1), \hat{\phi}(x_2), \cdots, \hat{\phi}(x_p)]_{\mp} \vdots$$

Table 1. The Structure of Observables and Gauge Groups: $F(V) = F(V;A,B)$, $A \equiv [\hat{\phi}(x), \hat{\phi}(x')]_{\pm}$, $B \equiv \vdots [\hat{\phi}(x_1), \hat{\phi}(x_2), \cdots, \hat{\phi}(x_p)]_{\mp} \vdots$

locality	order p	para-Bose	para-Fermi
strong	even	A O(p)	A O(p)
	odd	A,B SO(p)	
weak	even	A,B SO(p)	A,B SO(p)
	odd		A O(p)

$$= p!(-1)^{[p/2]} \sum_{\alpha_1,\alpha_2,\ldots,\alpha_p} \varepsilon_{\alpha_1\alpha_2\cdots\alpha_p} \hat{\psi}^{(\alpha_1)}(x_1)\hat{\psi}^{(\alpha_2)}(x_2)\cdots\hat{\psi}^{(\alpha_p)}(x_p) \quad , \tag{7.5}$$

where $\varepsilon_{\alpha_1\alpha_2\cdots\alpha_p}$ is antisymmetric with respect to the p indices and $\varepsilon_{12\cdots p} = 1$.

We are now in a position to define a *gauge group* G as the one whose transformations $g \in G$

$$\psi^{(\alpha)}(x) \rightarrow \sum_{\beta=1}^{p} g_{\alpha\beta}\psi^{(\beta)}(x) \tag{7.6}$$

leave all F(V)'s invariant. We then find immediately that the brackets (7.4) and (7.5) are precisely invariants of G = O(p) and G = SO(p), respectively. Thus, for each of the cases considered in Table 1 we have the G of the type as indicated in the lower right-hand corner thereof.[6,4] If further conditions are to be imposed on F(V)'s, G will obviously be enlarged to $G' \supset G$. Suppose, for instance, that we require as well the conservation of particle numbers. In such a case G of e.g. a para-Fermi field of p=3 undergoes $G = O(3) \rightarrow G' = U(3)$.

In view of the above results we can say that the symmetry between para-Bosons and para-Fermions of p=odd, including ordinary Bosons and Fermions as a special case p=1, is less than the symmetry between para-Bosons and para-Fermions of p=even. The case in which more than one field coexists can be argued similarly, but the gauge structure becomes more complex.[4]

REFERENCES

1. Y. Ohnuki and S. Kamefuchi, Fermi-Bose Similarity, Supersymmetry and Generalized Numbers, Nuovo Cimento 70A:435 (1982), 73A:328, 77A:99 (1983), 83A:275 (1984).
2. Y. Ohnuki and T. Kashiwa, Coherent States of Fermi Operators and the Path Integral, Prog. Theor. Phys. (Kyoto) 60:548 (1978).
3. M. Omote, Y. Ohnuki and S. Kamefuchi, Fermi-Bose Similarity, Prog. Theor. Phys. (Kyoto) 56:1948 (1976).
4. Y. Ohnuki and S. Kamefuchi, "Quantum Field Theory and Parastatistics", Univ. of Tokyo Press, Tokyo/Springer Verlag, Berlin-Heidelberg-New York (1982).
5. H. S. Green, A Generalized Method of Field Quantization, Phys. Rev. 90:270 (1953).
6. K. Drühl, R. Haag and J. E. Roberts, On Parastatistics, Commun. Math. Phys. 18:204 (1970).

GENERAL DYNKIN INDICES AND THEIR APPLICATIONS

Susumu Okubo

Department of Physics and Astronomy
University of Rochester
Rochester, NY 14627

INTRODUCTION

The notion of general Dynkin indices for representations of simple Lie algebras has been defined, and their various applications have been discussed.

Let L be a Lie algebra over a field F which is assumed here to be the complex number field for simplicity. Let $t_1, t_2, .., t_d$ be a basis of L with multiplication table

$$[t_\mu, t_\nu] = C^\lambda_{\mu\nu}\, t_\lambda \tag{1}$$

where $C^\lambda_{\mu\nu} \epsilon F$ are structure constants and the standard summation convention on repeated indices is understood hereafter. Let $g^{\mu_1\mu_2..\mu_p} \epsilon F$ be totally symmetric in p indices $\mu_1, \mu_2, .., \mu_p$. If

$$J_p = g^{\mu_1\mu_2..\mu_p}\, t_{\mu_1} t_{\mu_2} \ldots.\, t_{\mu_p} \tag{2}$$

commutes with all elements $t_\lambda \epsilon L$, i.e., if $[J_p, t_\lambda] = 0$, then J_p is called a p-th order Casimir invariant of L. It is known[1)] that a simple Lie algebra of rank n possesses precisely n algebraically independent Casimir invariants which we call fundamental. Their orders are listed below for convenience:

$A_n (n\geq 1)$	: 2, 3, ..., n+1
B_n and $C_n (n\geq 2)$	: 2, 4, 6, ..., 2n
$D_n (n\geq 4)$	: 2, 4, 6, ..., 2(n-1), and n
G_2	: 2, 6

F_4 : 2, 6, 8, 12

E_6 : 2, 5, 6, 8, 9, 12

E_7 : 2, 6, 8, 10, 12, 14, 18

E_8 : 2, 8, 12, 14, 18, 20, 24, 30 (I)

First, we note that the 2nd Casimir invariant is always unique. As usual, we set

$$g_{\mu\nu} = \mathrm{Tr}(\mathrm{ad}t_\mu \mathrm{ad}t_\nu) \tag{3}$$

and define $g^{\mu\nu}$ as its inverse matrix. We will raise and lower indices by means of $g^{\mu\nu}$ and $g_{\mu\nu}$ in the standard way. Then, the 2nd order Casimir invariant J_2 may be defined by

$$J_2 = g^{\mu\nu}\, t_\mu\, t_\nu \quad . \tag{4}$$

However, higher order Casimir invariants are not in general unique. Consider the case of the 4-th order. If J_4 is a 4-th order invariant, then so is

$$J_4' = J_4 + C(J_2)^2 + C'J_2$$

for arbitrary numbers C and C'. This corresponds to a change

$$g^{\mu\nu\alpha\beta} \to g'^{\mu\nu\alpha\beta} = g^{\mu\nu\alpha\beta} + \frac{1}{3}C\{g^{\mu\nu}g^{\alpha\beta} + g^{\mu\alpha}g^{\nu\beta} + g^{\mu\beta}g^{\nu\alpha}\} \quad .$$

We can eliminate this ambiguity by imposing orthogonality condition

$$g^{\mu\nu\alpha\beta}\, g_{\mu\nu}\, g_{\alpha\beta} = 0 \quad . \tag{5}$$

Then, J_4 is uniquely determined except for its over-all constant multiplicative factor. Its explicit expression as well as their eigen-values for any simple Lie algebra is given in ref. 2. The situation will become more involved for higher order invariants. For example, J_6' given by

$$J_6' = J_6 + C_1 J_2 J_4 + C_2(J_2)^3 + C_3(J_3)^2$$

is also a six-order invariant for arbitrary constants C_1, C_2, and C_3. However, these ambiguities can be eliminated similarly by imposing orthogonality condition

$$g^{\mu\nu\lambda\alpha\beta\gamma}g_{\mu\nu}g_{\lambda\alpha\beta\gamma} = g^{\mu\nu\lambda\alpha\beta\gamma}g_{\mu\nu}g_{\lambda\alpha}g_{\beta\gamma} = g^{\mu\nu\lambda\alpha\beta\gamma}g_{\mu\nu\lambda}g_{\alpha\beta\gamma} = 0 \quad , \tag{6}$$

which determines J_6 uniquely except for over-all constant factor. In this way, we can always determine the p-th order Casimir invariant J_p uniquely for all simple Lie algebras except for the case of $L = D_n$ (n=even), where we have two independent n-th order invariants J_n and $\hat{J}_n$. However, for this case, a canonical construction of J_n and $\hat{J}_n$ is well-known[2),3)]. In this way, we construct fundamental Casimir invariants J_p's for any simple Lie algebra. Note that they are non-zero only for values of p listed in (I). For other values of p not in the Table (I), we have identically $J_p = 0$ as has been emphasized in references 2 and 3.

Let ω be a generic finite-dimensional representation of a simple Lie algebra L, and set

$$X_\mu = \omega(t_\mu) \quad . \tag{7}$$

Then, any generic element $t \in L$ may be expressed as

$$t = \xi^\mu t_\mu \tag{8}$$

where $\xi^\mu \in F$ are constants. Correspondingly, we set

$$X = \xi^\mu X_\mu = \omega(t) \tag{9}$$

to be a generic representation matrix in ω of a generic element $t \in L$. Then, the p-th order Dynkin index $D_p(\omega)$ is now defined by[3)]

$$D_p(\omega) = \mathrm{Tr}\, J_p = \mathrm{Tr}\left\{g^{\mu_1\mu_2\ldots\mu_p} X_{\mu_1} X_{\mu_2} \ldots X_{\mu_p}\right\} \tag{10}$$

where the trace is over the representation space ω. Note that $D_p(\omega)$ is identically zero for values of p not listed in (I). Since the absolute normalization of $D_p(\omega)$ is not fixed, it is often convenient to define

$$Q_p(\omega) = \frac{D_p(\omega)}{D_p(\lambda)} \tag{11}$$

where λ is an arbitrary but fixed representation satisfying $D_p(\lambda) \neq 0$. In general, we can choose λ to be the lowest dimensional irreducible representation of L except for the case of $\hat{J}_n$ for the Lie algebra D_n. In that case, we have to choose λ to be the fundamental spinor representation of D_n. The explicit formulas of $Q_p(\omega)$ for all classical Lie algebras A_n, B_n, C_n, and D_n have been given in ref. 4, while those for some low-dimensional irreducible representations ω for exceptional Lie algebras have been computed in ref. 5.

(I) Self-Conjugate Representations

Let ω^* be the conjugate representation of ω so that

$$\omega^*(t) = -X^T \tag{12}$$

where X^T is the transpose matrix of X. Then, we have

$$D_p(\omega^*) = (-1)^p\, D_p(\omega) \quad . \tag{13}$$

A representation ω is self-conjugate, if ω^* is equivalent to ω, i.e., if there exists a fixed non-singular matrix S satisfying

$$S^{-1} X S = -X^T \quad . \tag{14}$$

Then, we have $D_p(\omega) = 0$ for any self-conjugate representation ω whenever p is an odd integer. From the Table (I), only Lie algebras $A_n (n \geq 2)$, D_{2n+1}, $(n \geq 2)$, and E_6 possess an odd-order fundamental Casimir invariant J_p (p=odd). For any other simple Lie algebra, they have no odd-order Casimir invariant. This implies that any representation ω of these Lie algebras is always self-conjugate. Let ω be irreducible. Then, from Eq. (14) together with Schur's lemma, we can easily prove

$$S^T = \pm S \quad . \tag{15}$$

Any self-conjugate representation ω obeying $S^T = S$ or $S^T = -S$ is called real (orthogonal) or pseudo-real (symplectic). The question whether a given self-conjugate representation is real or pseudo-real has been studied by many authors[6)].

(II) Clebsch-Gordan Decomposition

Let ω_A and ω_B be two irreducible representations of L. Then, its Kronecker product $\omega_A \otimes \omega_B$ is in general reducible as a sum of irreducible components ω_j:

$$\omega_A \otimes \omega_B = \sum_j \oplus \omega_j \quad . \tag{16}$$

We can now prove[3)] the validity of sum rules

$$d(\omega_A)\, D_p(\omega_B) + d(\omega_B)\, D_p(\omega_A) = \sum_j D_p(\omega_j) \tag{17}$$

where $d(\omega)$ is the dimension of the representation ω. The special case of p = 2 has been already noted by Dynkin[7)], while the case of p = 3 and 4 has been proved in reference 2. We have many other index sum rules involving polynomials of $D_p(\omega)$, although we will not go into detail[3)]. Study of these sum rules is in general sufficient to uniquely determine the Clebsch-Gordon decomposition Eq. (16) at least for many low dimensional representations without considering details of representation theory. Actually, the method has been originated by Patera, Sharp and Winternitz[8)], where they use another indices $\ell_2(\omega)$ and $\ell_4(\omega)$ which are related to $D_2(\omega)$ and $D_4(\omega)$. Since an extensive table for $d(\omega)$, $\ell_2(\omega)$ and $\ell_4(\omega)$ is available[9)], the uses of $\ell_2(\omega)$ and $\ell_4(\omega)$ are in practice more convenient. For details, see references 3)-5).

(III) Branching Sum Rules

Let L_o be a sub-Lie algebra of L. Then, an irreducible representation ω of L is now decomposed as a direct sum of irreducible representations ω_j of L_o:

$$\omega \downarrow \sum_j \oplus \omega_j \quad . \tag{18}$$

Assuming L_o also to be simple, then we can prove[3)] a sum rule

$$\xi_p \, D_p(\omega) = \sum_j D_p^{(o)}(\omega_j) \tag{19}$$

where $D_p(\omega)$ and $D_p^{(o)}(\omega_j)$ are p-th order indices of L and L_o, respectively, and where ξ_p is a constant which may depend upon p, L, and L_o, but not upon specific irreducible representation ω. Hence, once we determine ξ_p from a simplest known branching rule of $L \downarrow L_o$, then we can use Eq. (19) to determine the decomposition Eq. (18) for general ω. The special case of p = 2 for Eq. (19) has been previously noted by Dynkin[7)]. When L is one of exceptional Lie algebras, we can find many other mixed branching sum rules. Also, if L_o is not simple but a product of two simple Lie algebra $L_o = L_A \otimes L_B$, we can find analogous sum rules[5)]. All these are useful[5)] for practical determination of the branching rule, although we will not go into detail.

(VI) Decomposition of Tensor-Power Representation

Let □ be a representation of L. Because of the Pauli principle, it is often important to consider tensor-power representations of □ with a definite permutation symmetry Γ. This fact can be conveniently specified by use of Young tableau[10),11)]. Let Γ be one such Young tableau. In general, Γ is reducible as

$$\Gamma = \sum_j \oplus \omega_j \quad . \tag{20}$$

We then have an index sum rule

$$D_p(\Gamma) = \sum_j D_p(\omega_j) \quad . \tag{21}$$

We can compute $D_p(\Gamma)$ from a formula given in ref. 4 and we can utilize[4] this relation to determine the decomposition Eq. (20). An extensive study of such a decomposition by use of index sum rules has been made by Schellenken, Koh and Kang[12)]. Also, such a decomposition has been applied to study of preon models in particle physics by some authors[13)]. More recently, the same technique has been used[14)] to systematically derive various non-associative algebras such as Octonion, Jordan, and Freudenthal's triple system from representation theories of G_2, F_4, and E_7.

(V) Trace Identities

Let X_μ, and X be defined by Eqs. (7)-(9). For simplicity, we define a p-th order monomial $f_p(\xi)$ of variables ξ_μ's by

$$f_p(\xi) = g^{\mu_1\mu_2\cdots\mu_p}\xi_{\mu_1}\xi_{\mu_2}\cdots\xi_{\mu_p} \quad , \tag{22}$$

which does not depend upon ω. Moreover, we designate the adjoint representation of L by ρ_o hereafter, i.e., $\rho_o = \omega_{\text{adjoint}}$.

First, we note

$$\mathrm{Tr}(X_\mu X_\nu) = g_{\mu\nu}\, D_2(\omega)/d(\rho_o) \tag{23}$$

so that we find

$$\mathrm{Tr}\, X^2 = f_2(\xi)\, D_2(\omega)/d(\rho_o) \quad . \tag{24}$$

Similarly, we can prove[15] the validity of

$$\mathrm{Tr}\, X^3 = C_3\, f_3(\xi)\, D_3(\omega) \tag{25}$$

where $C_3^{-1} = g^{\mu\nu\lambda}g_{\mu\nu\lambda}$. We note that this relation has been earlier observed by Biedenharn[16] for the case of L = A_2(=su(3)). Since any simple Lie algebra other than $A_n(n\geq 2)$(=su(n+1)) possesses no third order Casimir invariant (see Table (I)), we have $D_3(\omega) = 0$ and hence[15]

$$\mathrm{Tr}\, X^3 = 0 \tag{26}$$

for any simple Lie algebras other than L = $A_n(n\geq 2)$. The condition Eq. (26) is crucial for renormalizability of chiral gauge field theories, since it implies[17] the absence of the triangle anomaly. For L = su(n+1), the condition Eq. (26) requires[15),18] $D_3(\omega) = 0$ which imposes a severe constraint to choice of representation ω. In this connection, we remark that these facts are also important for a classification[19] of a class of non-associative algebra called flexible Lie-admissible algebras.

For $p \geq 4$, the situation is more involved. Hereafter we assume ω to be irreducible. First, we can show[2),3] that we can write

$$\frac{1}{4!}\sum_P \mathrm{Tr}(X_\mu X_\nu X_\alpha X_\beta) = F_1(\omega)\, g_{\mu\nu\alpha\beta} + F_2(\omega)\{g_{\mu\nu}g_{\alpha\beta} + g_{\mu\alpha}g_{\nu\beta} + g_{\mu\beta}g_{\nu\alpha}\}$$

where the summation is over 4! permutations P of μ, ν, α, and β. The unknown constants $F_1(\omega)$ and $F_2(\omega)$ are determined by multiplying $g^{\mu\nu\alpha\beta}$ and $g^{\mu\nu}g^{\alpha\beta}$ to both sides and noting the orthogonality condition Eq. (5). In this way, we finally find[2] the 4-th order trace identity

$$\mathrm{Tr}\, X^4 = C_4\, f_4(\xi)\, D_4(\omega) + K(\omega)(\mathrm{Tr}X^2)^2 \quad , \tag{27a}$$

$$K(\omega) = \frac{1}{2[2+d(\rho_o)]}\left\{6\,\frac{d(\rho_o)}{d(\omega)} - \frac{D_2(\rho_o)}{D_2(\omega)}\right\} \quad , \tag{27b}$$

where C_4 is a constant which is independent of ξ_μ and ω. We remark that these trace identities together with no triangle anomaly condition Eq. (26) have been applied[20),21),22)] to determine all possible gauge groups for unified weak-electro theory as well as GUT theory without any exotic particles.

We can proceed to derive higher order identities. We consider here only the case of p = 6. Although I have derived the most general 6-th order trace identity, it is quite complicated (especially for $L = A_n$), so that I will here restrict to a consideration of L being one of exceptional Lie algebras G_2, F_4, E_6, E_7, and E_8 as well as A_1. Since these Lie algebra possess no 3rd and 4-th order fundamental Casimir invariants, we have $D_3(\omega) = D_4(\omega) = 0$ which simplify the problem considerably. The result is

$$\mathrm{Tr}\, X^6 = C_6\, f_6(\xi)\, D_6(\omega) + A(\omega)(\mathrm{Tr}X^2)^3 \quad , \tag{28a}$$

$$A(\omega) = \frac{15}{[2+d(\rho_o)][4+d(\rho_o)]}\left\{\left[\frac{d(\rho_o)}{d(\omega)}\right]^2 - \frac{1}{2}\,\frac{d(\rho_o)}{d(\omega)}\,\frac{D_2(\rho_o)}{D_2(\omega)} + \frac{1}{12}\left[\frac{D_2(\rho_o)}{D_2(\omega)}\right]^2\right\}. \tag{28b}$$

Especially, the cases of $L = A_1$ or E_8 are simple, since $D_6(\omega) = 0$ identically for these algebras. In that case, Eq. (28a) gives[21)]

$$\mathrm{Tr}\, X^6 = A(\omega)(\mathrm{Tr}X^2)^3 \quad . \tag{29}$$

Actually, the validity of $D_6(\rho_o) = 0$, i.e., absence of the 6-th order index for the adjoint representation ρ_o is very important for the super-string theory since the theory can be only consistent for that case alone. As has been noted by Green and Schwarz[23)], the condition is obeyed only for $L = E_8 \times E_8$ and SU(32), when we will also take into account the absence of the so-called gravitational anomaly. A general group-theoretical consideration for the condition $D_6(\omega) = 0$ has been made in ref. 24. Note that $D_6(\omega)$ is the anologue of the triangle anomaly coefficient $D_3(\omega)$, since we are now working in 10 dimensional space-time[23)] instead of the customary 4-dimension.

Finally, as other applications of trace identities, we first note that all exceptional Lie algebras as well as A_1 and A_2 have <u>no</u> 4-th order Casimir invariant and hence $D_4(\omega) = 0$. In that case, Eq. (27) is reduced to[25)]

$$\mathrm{Tr}\, X^4 = K(\omega)(\mathrm{Tr}X^2)^2 \quad . \tag{30}$$

If we choose further $L = A_1$ with $X = J_3$ being the 3rd component of the angular momentum, then Eqs. (29) and (30) give[4),25)]

$$\sum_{m=-j}^{j} m^4 = \frac{1}{15} j(j+1)(2j+1)\{3j(j+1)-1\} \quad , \tag{31a}$$

$$\sum_{m=-j}^{j} m^6 = \frac{1}{21} j(j+1)(2j+1)\{3[j(j+1)]^2-3j(j+1)+1\} \tag{31b}$$

for any integral or half integral values of j, i.e., for j = 0, 1/2, 1, 3/2, 2, ...

Next, we identify L to be A_2 = su(3) with ω being 3-dimensional representation. Then, Eq. (30) is rewritten as

$$\mathrm{Tr}\, X^4 = \frac{1}{2} (\mathrm{Tr}X^2)^2 \tag{32}$$

for any traceless 3×3 matrix X. This relation is equivalent to validity of an identity

$$a^4 + b^4 + c^4 = \frac{1}{2} (a^2+b^2+c^2)^2 \tag{33}$$

whenever we have a + b + c = 0. Let X and Y be two traceless 3×3 matrices. Replacing X by X + zY for complex variable z, Eq. (32) gives

$$2\mathrm{Tr}(X^2Y^2) + \mathrm{Tr}(XY)^2 = \frac{1}{2} \mathrm{Tr}\, X^2\, \mathrm{Tr}\, Y^2 + (\mathrm{Tr}XY)^2 \quad . \tag{34}$$

We now define a non-associative product $X * Y$ by

$$X * Y = \mu XY + \nu YX - \frac{1}{3} (\mathrm{Tr}XY)\, E \tag{35}$$

where E is the unit 3×3 matrix and where μ and ν are constants satisfying

$$\mu + \nu = 3\mu\nu = 1 \quad . \tag{36}$$

Further, we define a non-degenerate symmetric bi-linear form by

$$(X,Y) = \frac{1}{6} \mathrm{Tr}(XY) \quad . \tag{37}$$

Then, Eqs. (34)-(37) enable us to deduce

$$(X * Y, X * Y) = (X,X)(Y,Y) \quad , \tag{38}$$

$$(X * Y, Z) = (X, Y * Z) \quad . \tag{39}$$

Especially, Eq. (38) implies that our algebra is a eight-dimensional composition algebra. Nevertheless, this algebra is not equivalent to the Octonion algebra[26], since ours can be shown[27] to possess no unit element. Because of this, it has been named[27] as pseudo-octonion algebra. It is a flexible Lie-admissible algebra with automorphism group SU(3). Moreover, if we restrict ourselves to uses of only hermitian matrices for X's, then we can show[28] that it defines a new class of eight-dimensional real division algebra.

In ending this section, we briefly note that there exists a connection between Casimir invariants and vector (or adjoint) operators as well as existence of polynomial identities among generators of simple Lie algebras. This topic as well as a general method of computing eigen-values of Casimir invariants is studied in detail in ref. 29.

Acknowledgement

This paper is in part supported by the U.S. Department of Energy under Contract No. DE-AC02-76ER13065.

References

1. J. Dixmier; "Enveloping Algebras" (North-Holland, Amsterdam, 1977).
2. S. Okubo; Jour. Math. Phys. 23, 8 (1982).
3. S. Okubo and J. Patera; Jour. Math. Phys. 25, 219 (1984).
4. S. Okubo and J. Patera; Jour. Math. Phys. 24, 2722 (1983).
5. S. Okubo; Jour. Math. Phys. 26, 2127 (1985).
6. M. L. Mehta; Jour. Math. Phys. 7, 231 (1966).
   M. L. Mehta and P. K. Srivastava; ibid 7, 1833 (1966).
   A. K. Bose and J. Patera; ibid 11, 2231 (1970).
   J. Tits; in Lecture Notes in Mathematics No. 40 (Springer, New York, 1967).
7. E. B. Dynkin; Math. Sb. SSSR 30 349 (1952) [Am. Math. Soc. Trans. Ser. 2 6, 111 (1957)].
8. J. Patera, R. T. Sharp and P. Winternitz; Jour. Math. Phys. 17, 1972 (1976), Erratum 18 1519 (1977).
   J. Patera and R. T. Sharp; ibid 22, 2352 (1981).
9. W. G. McKay and J. Patera; "Tables of Dimensions, Indices and Branching Rules for Representations of Simple Lie Algebras" (Dekker, NY, 1981). We have, however, changed notation $I^{(2p)}$ there to $\ell_{2p}(\omega)$ here.
10. M. Hammermesh; "Group Theory and its Application to Physical Problems" (Addison-Wesley, Reading, MA, 1962).
11. H. Weyl; "Classical Groups" (Princeton U.P., Princeton, NJ, 1939).
12. A. N. Schellenkens, I.-G. Koh and K. Kang; Jour. Math. Phys. 23, 2244 (1982).
13. A. N. Schellenkens, K. Kang and I.-G. Koh; Phys. Rev. D26, 658 (1982).
    Y. Tosa and R. E. Marshak; ibid D27, 616 (1983).
14. S. Okubo; "Construction of Non-Associative Algebras from Representation Modules of Simple Lie Algebras", University of Rochester Report UR-920 (1985).
15. S. Okubo; Phys. Rev. D16, 3528 (1977).
16. L. C. Biedenharn; "Group Theory and the Classification of Elementary Particles", CERN Report 64-41 (1965).

17. D. Gross and R. Jackiw; Phys. Rev. D6, 477 (1972).
C. Bouchiat, J. Iliopoulos and Ph. Meyer; Phys. Lett. 35B, 519 (1972).
H. Georgi and S. L. Glashow; Phys. Rev. D6, 429 (1972).

18. J. Banks and H. Georgi; Phys. Rev. D14, 1159 (1976).

19. S. Okubo and H. C. Myung; Trans. Amer. Math. Soc. 264, 459 (1981).

20. S. Okubo; Phys. Rev. D18, 3792 (1978).
Y. Tosa and S. Okubo; Phys. Rev. D23, 2486 and 3058 (1981).
Y. Tosa, ibid D25, 1714 (1982).
S. Okubo; ibid D26, 2893 (1982) and Had. Jour. 5, 7 (1981).

21. Y. Tosa, R. E. Marshak and S. Okubo; Phys. Rev. D27, 444 (1983).

22. Y. Tosa; "How Many U(1) Symmetries are Allowed for Quarks and Leptons", University of Colorado Reprint (1986).

23. M. B. Green and J. H. Schwarz; Phys. Lett. 149B, 117 (1984), Nucl. Phys. B255, 93 (1985).

24. S. Okubo and J. Patera; Phys. Rev. D31, 2669 (1985).

25. S. Okubo; Jour. Math. Phys. 20, 586 (1979).

26. R. D. Schafer; "An Introduction to Non-Associative Algebras" (Academic Press, New York, 1966).

27. S. Okubo; Hadronic Jour. 1, 1250 (1978).

28. S. Okubo and H. C. Myung; Jour. Algebra 67, 479 (1980).
G. M. Benkart and J. M. Osborn; Pacific Jour. of Math. 96, 265 (1981).

29. S. Okubo; Jour. Math. Phys. 18, 2382 (1977).

GROUP THEORETICAL APPROACHES TO MANY-ELECTRON CORRELATION PROBLEM

Josef Paldus

Department of Applied Mathematics and (GWC)[2]
University of Waterloo
Waterloo, Ontario, Canada N2L 3G1

The unitary group approach (UGA) to many-fermion correlation problem may be regarded as a direct outgrowth of the original ideas as laid down by Hermann Weyl[1] in his "Gruppentheorie und Quantenmechanik". Although these advances have in the past been overshadowed by the simplicity of Slater determinant based formalisms, and have even been referred to[1] at one time as a "group pest", today we find Weyl's original ideas very much alive and well, particularly in the many-electron correlation problem. For instance, the Gelfand-Tsetlin (GT) representation theory[2] of the unitary group is a natural extension of Weyl's branching rule presented in the very last Section of his book.[1]

The attempts to employ the unitary group structure can be traced back to the pioneer days of Quantum Mechanics.[3] However, it was not until 1968 when the fundamental ideas of exploiting UGA in the (nuclear) many-body problem were set forth by Moshinsky.[4] Later, this approach proved to be particularly suitable and simple for the many-electron correlation problem.[5] These early ideas,[4,5] were quickly developed into a very flexible and versatile algorithm[6-10] which, during the past decade, witnessed numerous, often very ingenious and highly original practical exploitations[11-19] in molecular electronic structure calculations. This rapid development has been amply documented in numerous reviews[20-24] and even monographs.[25,26] Consequently, in this overview, we shall only briefly outline the most basic principles and aims of UGA while concentrating on the most recent developments which have yet to be exploited in actual computations (Sec. V).

We should also stress that the unitary group is not the invariance or symmetry group of the molecular electronic Hamiltonian employed but, in a certain sense, may be regarded as a dynamical group. Following the developments in nuclear and elementary particle physics, the exploitation of dynamical groups for one-electron systems[27] was initiated by Barut.[28] Recently, very remarkable advances in the molecular vibronic problem have been made by Iachello and Levin.[29] The most recent extension to resonance widths[30] and scattering problems[31,32] also holds great promise.

I. MANY-ELECTRON CORRELATION PROBLEM AND UGA

The foundation of our understanding and of most computational approaches to the electronic structure of atoms, molecules and solids is undoubtedly the independent particle model (IPM), which approximates the exact non-relativistic wavefunction by a single antisymmetrized product (Slater determinant) of (atomic, molecular or crystal) orbitals. In the molecular structure context these molecular orbitals (MO's) are usually determined variationally as finite linear combinations of a chosen atomic orbital (AO) basis (LCAO approximation) using the Hartree-Fock-Roothaan self-consistent field (SCF) procedure. Although in this approximation [approaching the true Hartree-Fock (HF) limit for sufficiently large dimension of the space spanned by the AO's] the long-range inter-electronic Coulomb forces are only accounted for in an averaged way through a separable one-electron HF potential, the resulting total energies are very close to the exact ones. Indeed, the difference between the exact (non-relativistic) and HF energies, called the correlation energy,[33] seldom exceeds 1% of the total energy and for larger systems is invariably considerably less than 0.5% of the total energy. Nonetheless, in view of very small energy differences which are decisive for various molecular properties and govern most chemical phenomena, the HF description is not only quantitatively inadequate but often qualitatively incorrect: the dissociation energies of stable molecules (such as F_2) may become negative (this is always the case for the so-called van der Waals molecules), the dipole moments have wrong sign, the order of excited states is permuted, etc. Consequently, a considerable effort has been made to go beyond the IPM or HF approximations using both perturbative and variational approaches.

The most often used method, perhaps due to its simplicity and universality, is the variational configuration interaction (CI) or shell-model approach.[26,34] With the exception of one- and two-electron systems, this approach provides presently the most accurate results for both small and medium size molecules. The principal disadvantage of this approach, at least as applied with the existing MO bases, is its slow convergence requiring the diagonalization (factual or effective) of very large (even though very sparse) matrices.

The dimension of the CI problem can be considerably reduced by exploiting various symmetry or invariance properties of molecular electronic Hamiltonians. The most significant reduction arises from the spin-independence of these Hamiltonians. It is precisely the spin-adaptation aspect of the CI expansion where UGA provides a particularly efficient tool. Needless to say that the spin-adaptation problem resurfaces in perturbation-type approaches as well, and may become rather demanding once degenerate states with a non-vanishing spin are considered. We just note here that UGA has also been exploited in perturbative[11] and Green function[35] calculations, and may prove to be very useful in coupled cluster approaches to degenerate systems.[36] We also recall that a similar, though more demanding (involving both spin and isospin), symmetry-adaptation problem occurs in nuclear physics, where UGA originated,[4] as well as for nuclear spin NMR problems.[37]

Although most UGA developments, which took place during the past decade, are limited to spin-½ fermions, these techniques can be extended to larger spin (or spin and isospin) cases,[38] even though at the expense of their simplicity and efficiency. However, the most recently proposed development[39,40] exploiting the spinorial Clifford algebra basis extends easily to an arbitrary spin.

II. PROBLEM FORMULATION

We shall consider an N-electron system which can be described by a spin-independent Hamiltonian $\hat{H}$ involving at most two-body forces. We introduce a one-electron Hilbert space H_1 with an orthonormal orbital basis

$\{|a\rangle\}$ and a corresponding spin-orbital space $H_1' = H_1 \otimes \Sigma$, where the two-dimensional spin-space Σ is spanned by spin-states $|\sigma\rangle$, $\sigma = \pm\frac{1}{2}$, and we write a general spin-orbital $|A\rangle$ (a monomial in H_1') as $|A\rangle = |a\rangle|\sigma\rangle$. We also introduce the creation (annihilation) operators $X_A^\dagger$ (X_A) of the second quantization formalism which are defined on the spin-orbital basis $\{|A\rangle\}$ and satisfy the usual anticommutation relations.

The molecular electronic Hamiltonian $\hat{H}$ in the Born-Oppenheimer approximation then takes the form

$$\hat{H} = \sum_{a,b}\sum_{\sigma}\langle a|\hat{z}|b\rangle X_{a\sigma}^\dagger X_{b\sigma} + \tfrac{1}{2}\sum_{a,b,c,d}\sum_{\sigma,\tau}\langle ab|\hat{v}|cd\rangle X_{a\sigma}^\dagger X_{b\tau}^\dagger X_{d\tau}X_{c\sigma}\ , \qquad (1)$$

where $\hat{z}$ represents the kinetic energy and external potential due to the clamped nuclei and $\hat{v}$ the interelectronic Coulomb potential. Various quantum chemical models are then defined by selecting an appropriate finite-dimensional subspace $\mathbf{V}$ of H_1 (or $\mathbf{V}'$ of H_1'). We distinguish

(a) semi-empirical models, for which $n{:=}\dim \mathbf{V} = N$, and the one- and two-electron integrals $\langle a|\hat{z}|b\rangle$ and $\langle ab|\hat{v}|cd\rangle$, respectively, are determined by fitting selected experimental data (for a few model compounds) after their number has been drastically reduced using various simplifying assumptions, and

(b) ab-initio models,[41] which are defined by selecting an appropriate spanning set for $\mathbf{V}$ consisting of atomic-like orbitals localized on various nuclei of the molecular system (off-nuclear position localized orbitals are seldom used). Generally $n{:=}\dim \mathbf{V} = km + \ell$, where m is the number of types of occupied atomic orbitals, resulting in a minimal (k=1), double-zeta (k=2) or triple-zeta (k=3) basis with ($\ell \neq 0$) or without (ℓ=0) polarization (higher angular momentum than the ground state IPM orbital) bases. The one- and two-electron integrals are then evaluated numerically or analytically and subsequently transformed into an orthonormal molecular orbital (MO) basis [usually the HF(SCF) or MC-SCF basis in the same $\mathbf{V}$].

The problem is then to find eigenvalues and eigenstates of $\hat{H}$, defined on the N-particle component of an antisymmetric Fock space $W = {}^{A}\mathbf{V}_{(N)} \equiv \mathbf{V}^{\wedge N}$. This can be simply achieved by constructing a matrix representative of $\hat{H}$ in W [using, for example, a simple anti-symmetrized product (Slater determinant) basis for W] and by diagonalizing it (full CI). However, the dimension of W quickly becomes unwieldy so that further simplifications and approximations must be made. A very substantial reduction of the dimension of W, without invoking any approximation, is achieved by exploiting a symmetry adapted basis for W implied by the spin-independence of $\hat{H}$, so that $[\hat{H},\hat{S}^2] = [\hat{H},\hat{S}_z] = 0$, where $\hat{S}^2$ and $\hat{S}_z$ are the total spin operators. Thus

$$W = \bigoplus W_{S,S_z}\ , \qquad (2)$$

where W_{S,S_z} is a simultaneous eigenspace of $\hat{S}^2$ and $\hat{S}_z$ characterized by spin-quantum numbers S, S_z.

Even this reduction does not suffice in general (say, for a double-zeta model of H_2O when N=10, n=20 we have $\dim W_{0,0} \approx 5\times10^6$) and further truncation of each W_{S,S_z} is necessary (limited CI).

It is the spin-adaptation of the required tensorial bases for W (or its subspaces) and an efficient evaluation of the matrix representative of $\hat{H}$ which we address in this review. Although we mention explicitly only the CI problem, needless to say that similar spin-adaptation problem arises in other approximate approaches to the solution of the non-relativistic

Schrödinger equation. We only mention in passing that the UGA discussed here is also being exploited for this purpose in perturbation theory,[11] MC-SCF,[42,43] Green's function[35] and coupled-cluster formalisms.[36,44]

III. SPIN-ADAPTATION PROBLEM

The spin-adaptation of N-electron basis states (called configurations) represents a special case of a general symmetry-adaptation procedure when $\hat{H}$ possesses other integrals of motion than spin. It arises in applications to atomic, molecular, nuclear and solid-state problems and has been addressed by numerous authors. For molecular applications an excellent and comprehensive treatment may be found in the monograph by Pauncz.[25] All pre-UGA approaches were based either on the symmetric group S_N representation theory[45] or on the angular momentum Racah formalism based on the SU(2) group. The many-electron UGA is based on the group chain [46]

$$U(2n) \supset U(n) \otimes U(2) , \tag{3}$$

relating the spin-orbital group U(2n) with the relevant subgroup given by the product of the orbital U(n) and spin U(2) groups. Their respective generators e_{AB}, E_{ab} and $E_{\sigma\tau}$ can be represented as follows

$$e_{AB} = X_A^{\dagger}X_B , \quad E_{ab} = \sum_{\sigma} X_{a\sigma}^{\dagger}X_{b\sigma} , \quad E_{\sigma\tau} = \sum_{a} X_{a\sigma}^{\dagger}X_{a\tau} , \tag{4}$$

with $[E_{ab}, E_{\sigma\tau}] = 0$, so that the spin-independent Hamiltonian $\hat{H}$, Eq. (1), can be expressed as a two-form in terms of the orbital generators E_{ab} only,

$$\hat{H} = \sum_{a,b} \langle a|\hat{z}|b\rangle E_{ab} + \tfrac{1}{2} \sum_{a,b,c,d} \langle ab|\hat{v}|cd\rangle \, (E_{ac}E_{bd} - \delta_{bc}E_{ad}) . \tag{5}$$

Since only totally antisymmetric irreps of U(2n) are admissible (Pauli principle) it follows that U(n) and U(2) irreps must be mutually conjugate, so that in fact we can restrict our attention to the two-column irreps of the orbital group U(n), thus achieving a spin-free formulation of our problem (see Matsen[21,45,47]). Since the U(n) representation theory is closely related with that of S_N, we can easily understand the relationship between different spin-adaptation approaches based on S_N, U(n) or U(2) [or SU(2)] groups, as schematically indicated below[48]

$$\underbrace{S_N \leftrightarrow U(n) \leftrightarrow U(2)}_{\uparrow\qquad\qquad\uparrow} \supset SU(2) . \tag{6}$$

The usefulness of this inter-relationship can hardly be overemphasized.[48] Although either of these groups is sufficient to achieve our goal, it is the relationship between their distinct group structures which provides us with new and useful insights. To illustrate this fact, let us mention that while the symmetric group has been employed for a very long time to construct the appropriate many-electron spin functions (see, e.g., Refs. 45,49,50) of various types, it was only very recently[51] that the computational effectiveness of this approach was significantly advanced by exploiting the so-called distinct row table (DRT) structure,[9] which naturally appears in UGA but is not so easily discernible in the S_N based approaches. We shall see that likewise the graphical methods of spin algebras,[52-56] based on the SU(2) group, can be most helpful in devising of efficient algorithms for the evaluation of U(n) generator matrix elements.[57,58]

IV. UNITARY GROUP APPROACH TO CI

We shall now outline the basic features of UGA as employed in many-electron correlation problem. We consider an N-electron model problem characterized by the Hamiltonian (1) defined by n=dim **V** orthonormal orbitals $|i\rangle$, i=1, ..., n. Then for the states with multiplicity (2S+1) we have to consider the U(n) irrep $\langle 2^a\ 1^b\ 0^c\rangle$ with

$$a = \tfrac{1}{2}N-S, \quad b = 2S \quad \text{and} \quad c = n-a-b = n-\tfrac{1}{2}N-S , \tag{7}$$

since all components of a given multiplet are exactly degenerate (so that it is irrelevant which S_z component we take) and at most two-column irreps of U(n) can occur in view of the spin-independence of $\hat{H}$. Consequently, any basis of the carrier space for the U(n) irrep $\langle 2^a\ 1^b\ 0^c\rangle$ will be automatically spin adapted. In UGA approach one employs the canonical GT basis,[2] which is associated with the subduction chain

$$U(n) \supset U(n-1) \supset \cdots \supset U(1) , \tag{8}$$

and whose vectors are uniquely labeled by triangular Gelfand tableaux $[m_{ij}]$, whose rows give the highest weights for the chain (8).

(i) Configuration representation

Since the Gelfand tableaux $[m_{ij}]$ contain much superfluous information in our special case, we use instead more compact 3-column ABC (nowadays referred to as[9,23-26] Paldus) tableaux $[a_i,b_i,c_i]$ with each row defining the corresponding U(i) irrep label

$$\langle 2^{a_i}\ 1^{b_i}\ 0^{c_i}\rangle$$

of the Gelfand tableau, so that

$$a_i + b_i + c_i = i . \tag{9}$$

It is also convenient to define a differential tableau[6] ΔABC with entries $\Delta x_i = x_i - x_{i-1}$ (x=a,b,c; i=1, ..., n; x_0=0), in which case

$$\Delta a_i + \Delta b_i + \Delta c_i = 1 . \tag{10}$$

It is easily seen that the only possible values of Δa_i and Δc_i are 0 or 1, so that in view of Eq. (10) we can uniquely label each electronic GT state by ternary arrays (d_i), i=1, ..., n with entries (called <u>step numbers</u>) $d_i = 2\Delta a_i + \Delta\bar{c}_i = 1 + 2\Delta a_i - \Delta c_i$ or, equivalently, by the corresponding ΔAC or $\Delta A\bar{C}$ tableaux $[\Delta\bar{x}_i := 1-\Delta x_i]$. The corresponding orbital <u>occupation numbers</u> n_i

$$n_i = 2\Delta a_i + \Delta b_i = 1 + \Delta a_i - \Delta c_i = \Delta a_i + \Delta\bar{c}_i , \tag{11}$$

are then given by the digital sum of the $\Delta A\bar{C}$ tableau entries at each level (row) [see Fig. 1]. The four possible values of step numbers d_i are listed in Table I below.

Table I. Permissible values of step numbers d_i, corresponding occupation numbers n_i, intermediate spin quantum numbers $\Delta S_i = \tfrac{1}{2}\Delta b_i$ and Δx_i (x=a,b,c) values.

d_i	Δa_i	$\Delta\bar{c}_i$	Δc_i	Δb_i	ΔS_i	n_i
0	0	0	1	0	0	0
1	0	1	0	1	$\tfrac{1}{2}$	1
2	1	0	1	-1	$-\tfrac{1}{2}$	1
3	1	1	0	0	0	2

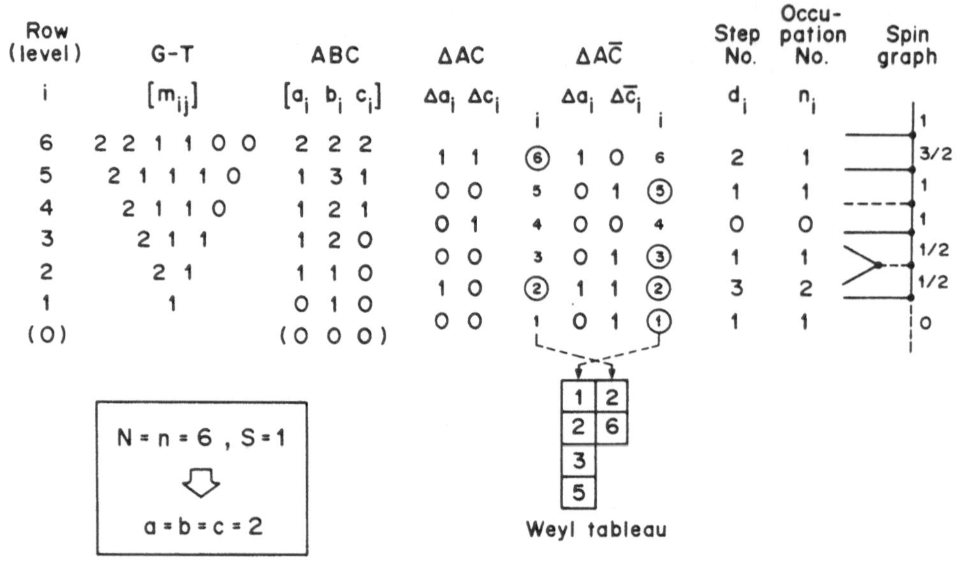

Fig. 1. An example of different representations of the same 6-electron triplet configuration (or canonical GT state) of U(6) using minimum basis set [i.e., n=N=6, S=1]. A very close connection[8] between the $\Delta A\bar{C}$ and Weyl tableaux is also pointed out. In the diagrammatic representation[10] based on graphical methods of spin algebras[52-56] the lines are labeled by intermediate spin quantum numbers. The unlabeled (full) lines carry the spin $\frac{1}{2}$ and the dashed lines spin 0.

It is now easy to construct the N-electron basis labeled by the ABC tableaux, starting with the highest weight [a,b,c] as given by Eqs. (7), and subtracting at each level the four possible ΔABC rows ([0,0,1], [0,1,0], [1,-1,1] and [1,0,0], cf. Table I), while keeping only lexical rows (all entries non-negative). The $\Delta A\bar{C}$ or equivalent step number labeling is also closely related with the corresponding Weyl tableau labeling,[8] whose first column contains indices i for which $\Delta\bar{c}_i = 1$ and the second column those for which $\Delta a_i = 1$. The relationship of various configuration labeling schemes is illustrated in Fig. 1, where we also provide the graphical SU(2) representation.[10]

As Table I shows, each level i (associated with the i-th orbital) is either unoccupied (d_i=0), singly (d_i=1,2) or doubly (d_i=3) occupied. For unoccupied and doubly occupied orbitals, no spin change occurs ($\Delta b_i = \Delta S_i = 0$) indicating that the spin of both electrons in the doubly occupied case must be pre-coupled to a singlet. For singly occupied orbitals, the total spin can either increase (d_i=1) or decrease (d_i=2) the total spin value. We thus see that, up to a phase, the GT electronic states are identical with Yamanouchi-Kotani (YK) states.[50] We can represent each such state by indicating the sequential spin couplings by representing the corresponding Clebsch-Gordan (CG) coefficients (or Wigner's 3j-m symbols) by three lines issuing from the same vertex and carrying the spin angular momentum labels (cf. Fig. 1).[52-56]

The dimension of the U(n) irrep [a,b,c] is given by the following simple expression[5]

$$\dim[a,b,c] = \frac{b+1}{n+1}\binom{n+1}{a}\binom{n+1}{c}, \tag{12}$$

where $\binom{m}{n} = m!/(m-n)!n!$ is the usual binomial coefficient.

(ii) Global representation

In order to better visualize the structure of the GT basis, Shavitt[9] introduced its graphical representation in which each distinct i-th row $[a_i,b_i,c_i]$ of the ABC tableaux is represented by a vertex at the i-th level of the diagram, while the allowed step numbers are represented by edges connecting corresponding vertices. An equidistant horizontal arrangement of vertices (according to decreasing a_i and b_i values with maximal values at left) is such that the (negative) slope of the edge increases with the increasing step number (Fig. 2a). In this two-rooted graph, with the top vertex (root) representing the highest weight [a,b,c], which defines the U(n) irrep used, and the bottom vertex (root) associated with the weight [0,0,0] introduced for convenience, each path between the two roots represents one spin-adapted configuration (a GT basis vector).

The structure of the basis is thus described by distinct rows, together with at most four chaining (or incidence) indices for each distinct row (distinct row table or DRT[9]). On the basis of this information and the dimensions of irreps associated with distinct rows [cf. Eq. (12)] one can easily calculate[9] the lexical index of a given configuration or, conversely, find its ABC tableau. Thus, the structure of the GT basis leads naturally to an efficient DRT representation (e.g., for the triplet states of the double-zeta model of H_2O [N=10, n=20, S=1] when a=4, b=2, c=14 so that dim[4,2,14] = 99 419 400 $\approx 10^8$, there are only 355 distinct rows in the DRT). We also note that the DRT concept can be exploited in the S_N approach.[51]

The Shavitt graph can also be regarded as an expanded form of the Yamanouchi-Kotani branching diagram by rearranging the vertices according to their intermediate spin values as indicated in Fig. 2.

(iii) Matrix element evaluation

Once we know the explicit matrix representatives for the orbital U(n) generators and their products it is straightforward to obtain the matrix representative of the Hamiltonian, Eq. (1), as follows from Eq. (5). The actual computational procedure which accomplishes this step will be very much different for different implementations of UGA and we shall briefly address this point in the next subsection. Regardless of the procedure used, however, it is essential to have as simple an algorithm as possible to achieve an efficient computation of Hamiltonian matrix elements, particularly when large scale CI computations are intended. Calculation of explicit matrix representatives of U(n) generators thus received considerable attention and we now briefly summarize the most effective options which are now available. We restrict our attention to raising generators E_{ij} (i<j) since they are simply related with the lowering generator matrix elements (i.e., $\langle[m']|E_{ij}|[m]\rangle = \langle[m]|E_{ji}|[m']\rangle$) and since it is straightforward to calculate the weight generator E_{ii} matrix elements (i.e., the occupation numbers n_i).

(A) Elementary generators $E_{i,i+1}$ are easily evaluated as 0, 1 or the squareroot of a ratio of two consecutive integers [cf. Eqs. (31,32) of Ref. 5 or Eq. (5.7) and Fig. 13 of Ref. 6 or Eq. (37) of Ref. 20]. Any row (column) of the matrix representative of any elementary generator has at most two nonvanishing entries.

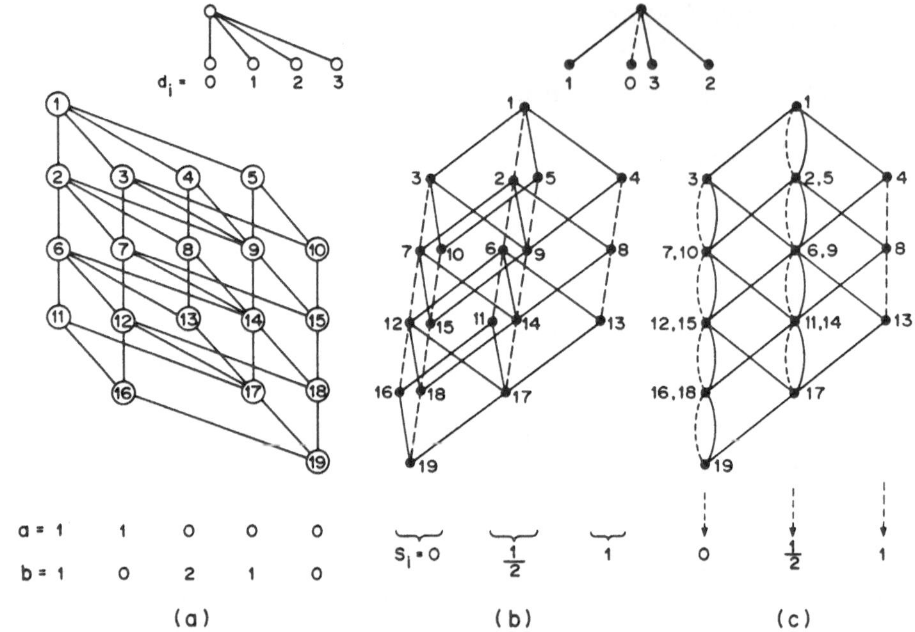

Fig. 2. Graphical representation of the canonical GT basis for the irrep [1,1,3] ≡ <2,1,0> of U(5) yielding the spin-adapted basis for N=3, n=5 and S=½. Diagram (a) is a standard Shavitt graph,[9] while diagram (c) may be regarded as a subgraph of a standard Yamanouchi-Kotani branching diagram.[50] An intermediate representation (b), which still uniquely represents the individual configurations by distinct paths as the diagram (a), shows the transition from (a) to (c) by ordering the vertices according to their intermediate spins. The slope of the edges determines the pertinent step number in each case and is indicated on the top.

(B) Non-elementary generators

(α) Iterative procedures. Once the elementary generator matrix elements are known, we can compute the remaining generator matrix elements by exploiting the commutation relation

$$E_{i,j+1} = [E_{ij}, E_{j,j+1}] , \quad (j=i+1, \ldots, n-1) . \tag{13}$$

When non-truncated basis (full CI) is used and all the generator matrix elements are computed, this is perhaps the fastest procedure. However, in limited CI, the intermediate states which are needed in evaluating the right-hand-side commutator in Eq. (13) may be missing. One can carry out the basis truncation in such a way that Eq. (13) can be used at any stage of truncation using the harmonic level excitation diagram (HLED).[20] However, this may be inconvenient for certain purposes. A similar problem is then encountered in calculation of matrix elements of products of generators. In any case it is desirable to have an algorithm for an independent evaluation of generator matrix elements.

(β) Segmentation formulas. Shavitt[9,59] was first to notice that the elementary generator formulas of Paldus[5,6] can be expressed as a product of two factors, each one associated with the orbital level involved (i.e., i and i+1 for $E_{i,i+1}$). He also showed[9] by induction how this factorization can be extended to non-elementary generators. However, this factorization follows most naturally from the SU(2) representation of GT states, particularly when using the graphical methods of spin algebras.[52-56] In this representation[10] an E_{ij} matrix element is given by the spin graph formed from two diagrams of the type shown in Fig. 1 by interconnecting all the spin-½ lines at the same orbital level except at levels i and j, which are interconnected to represent the generator E_{ij}. Since such graph has at most three angular momentum lines at each level, it can be immediately separated into trivial "oyster"-type[56] diagrams (yielding at most a phase factor) and a product of simple 6j diagrams providing the required segment values. We can thus generally write[10]

$$<[m']|E_{ij}|[m]> = \prod_{r\in\Omega} W_r \;, \tag{14}$$

where the operator range Ω is the set

$$\Omega \equiv \Omega(E_{ij}) = \{k|\min(i,j) \leq k \leq \max(i,j)\} \;. \tag{15}$$

The required segment values have been evaluated for various phase conventions.[10,60] We also note that very much the same technique can be employed for the S_N based approach and was first exploited by Gouyet et al.[61] and Drake and Schlesinger.[62]

(γ) Generator products. For efficient evaluation of two-electron contributions (which represents the major hurdle in all correlation problems since the number of two-electron integrals is $\sim n^4/8$, while the one-electron part involves only $\sim n^2/2$ terms) the segment formulas are particularly convenient. In this case we have[10]

$$<[\tilde{m}]|E_{ij}\;E_{i'j'}|[m]> = \prod_{k\in\Theta} W_k \sum_{X=0}^{1} \prod_{k'\in\Theta'} W_{k'}(X) \;, \tag{16}$$

where

$$\Theta = \Omega\Delta\Omega' = (\Omega\cup\Omega')\backslash(\Omega\cap\Omega') \text{ and } \Theta' = \Omega\cap\Omega' \tag{17}$$

designate the non-overlap and overlap generator product ranges and $\Omega\equiv\Omega(E_{ij})$, $\Omega'\equiv\Omega(E_{i'j'})$ are defined by Eq. (15). All the required segment values have been tabulated in Ref. 10.

(iv) Computer implementations

The efficiency, simplicity and versatility which is inherent in UGA enabled numerous computer implementations of this formalism, which in turn made feasible large scale CI computations (with over 10^6 configurations[18]). These various implementations can be roughly divided into the two categories:

(a) conventional (formula tape, matrix element driven) approaches which first compute the expressions for the Hamiltonian matrix elements in terms of one and two-electron integrals and subsequently evaluate them after one or several passes through the integral list, and

(b) direct (integral driven) approaches[63] in which the actual Hamiltonian matrix elements are never stored or even evaluated but the integrals (in an arbitrary order) are accessed sequentially, their contribution to possible Hamiltonian matrix elements is evaluated and immediately employed in the diagonalization algorithm. Clearly, the direct approach is only feasible when the Hamiltonian matrix elements, or their components associated with a given integral, can be evaluted so fast that it is more economical to recompute them when needed rather than to store them and search for them. This approach was first employed by Roos and Siegbahn[63] for closed shell singlet ground states with CI limited to singly and doubly excited configurations. The UGA developments enabled its extension to a general multi-reference case.[12,15]

We conclude this section by giving (a possibly incomplete) list of various computational implementations of UGA (in an approximate chronological order):

(a) Matrix element driven algorithm[11] based on HLED.[20]
(b) Loop driven algorithm.[13,14]
(c) Integral driven algorithm.[12,16]
(d) Loop-recursion algorithm.[64]
(e) Shape driven algorithm.[18]
(f) Internal interaction block driven algorithm.[17]
(g) Vectorized recursion algorithm.[19]

V. RECENT DEVELOPMENTS

We now briefly review some of the most recent developments in UGA which not only provide a new insight but will hopefully enable both more extensive and efficient implementation of UGA.

(i) Matrix element calculation using Green-Gould (GG) method[65,66]

Gould and Chandler[67] recently published a new derivation of UGA matrix elements in segmented form [cf. IV. (iii)(b)(β,γ)] following Gould's representation theory[66] for unitary and orthogonal groups based on Green's polynomial identities[65] for generators of semi-simple Lie groups. This procedure provides an alternative derivation of factorized expressions for arbitrary generator or generator product matrix elements which is solely based on U(n) formalism. More importantly, however, this technique also easily extends to other problems, such as determination of spin-orbital U(2n) generator matrix elements in U(n) GT bases,[68] which is needed when considering the spin-orbital interactions, or to general "group function" approaches* when the generator matrix elements must be evaluated in the basis that is adapted to the chain[69,70]

$$U(n) \supset U(n_1) \otimes U(n_2) \otimes \cdots \otimes U(n_k); \quad \sum_{i=1}^{k} n_i = n \,. \tag{18}$$

Note that a similar problem (with k=2) also occurs when employing the particle-hole formalism.[71]

To briefly outline the basic ideas of this approach, we introduce the n×n matrix $\mathbf{E}$ whose entries are U(n) generators E_{ij},

$$\mathbf{E}^{(n)} \equiv \mathbf{E} := [E_{ij}]_{(n\times n)} \,, \qquad (i,j=1, \ldots, n) \tag{19}$$

and note that its powers are uniquely defined since

$$(\mathbf{E}^{m})_{ij} = E_{ik} (\mathbf{E}^{m-1})_{kj} = (\mathbf{E}^{m-1})_{ik} E_{kj} \,, \tag{20}$$

(summation over repeated indices is now implied). It can be shown[65,66] that for any irrep $\langle\boldsymbol{\lambda}\rangle$ of U(n) the following polynomial identities hold on the irrep carrier space $\mathbf{V}\langle\boldsymbol{\lambda}\rangle$

$$\prod_{r=1}^{n} (\mathbf{E}-\varepsilon_r) = 0 \quad \text{and} \quad \prod_{r=1}^{n} (\bar{\mathbf{E}}-\bar{\varepsilon}_r) = 0 \,, \tag{21}$$

---

*Here a "group" means a chemical functional group and not a group in a mathematical sense.

where $\varepsilon_r = \lambda_r+n-r$, $\bar{\varepsilon}_r = r-1-\lambda_r = n-1-\varepsilon_r$ and for a conjugate (contragredient) representation we have that $(\bar{\mathbf{E}})_{ij} = \bar{\mathbf{E}}_{ij} = -E_{ji}$. For the two-column irrep [a,b,c] of U(n), which is required in UGA, these identities are of the third order (n=3) with $\varepsilon_1 = 1+c$, $\varepsilon_2 = n+2-a$, $\varepsilon_3 = 0$ and $\bar{\varepsilon}_1 = n-c$, $\bar{\varepsilon}_2 = a-1$ and $\bar{\varepsilon}_3 = -2$.

With these identities we can associate mutually orthogonal projectors P[r]

$$P[r] = \prod_{\substack{\ell=1 \\ (\ell\neq r)}}^{n} \frac{\mathbf{E}-\varepsilon_\ell}{\varepsilon_r-\varepsilon_\ell} \; , \quad \sum_{r=1}^{n} P[r] = \mathbf{1} \; , \tag{22}$$

and similarly $\bar{P}[r]$ for a conjugate representation.

An arbitrary U(n) vector operator $\boldsymbol{\Psi}$ with components Ψ_i (i=1, ... n) may then be resolved into a sum of shift components when acting on $\mathbf{V}\langle\boldsymbol{\lambda}\rangle$,

$$\Psi_i = \sum_r \boldsymbol{\Psi}[r]_i \; , \tag{23}$$

such that $\Psi[r]_i|v\rangle \in \mathbf{V}\langle\boldsymbol{\sigma}_r\rangle$ when $|v\rangle \in \mathbf{V}\langle\boldsymbol{\lambda}\rangle$ and $\boldsymbol{\sigma}_r = \boldsymbol{\lambda}+\boldsymbol{\Delta}_r$, with $\boldsymbol{\Delta}_r$ having all components equal to zero except the i-th one which equals 1, i.e.

$$\boldsymbol{\Delta}_r = (0, \ldots, 0,1,0, \ldots, 0) \; . \tag{24}$$

In particular, for the irrep [a,b,c] of U(n) we have that

$$\boldsymbol{\Psi}[r] : \mathbf{V}\langle\boldsymbol{\lambda}\rangle \rightarrow \mathbf{V}\langle\boldsymbol{\lambda}+\boldsymbol{\mu}_r\rangle \; , \tag{25}$$

where $\boldsymbol{\mu}_1 = \boldsymbol{\Delta}_{a+b-1}$, $\boldsymbol{\mu}_2 = \boldsymbol{\Delta}_{a+1}$ and $\boldsymbol{\mu}_3 = \boldsymbol{\Delta}_1$, and similarly for a contragredient vector operator $\boldsymbol{\Psi}^\dagger$,

$$\boldsymbol{\Psi}^\dagger[r] : \mathbf{V}\langle\boldsymbol{\lambda}\rangle \rightarrow \mathbf{V}\langle\boldsymbol{\lambda}-\boldsymbol{\nu}_r\rangle \; , \tag{26}$$

with $\boldsymbol{\nu}_1 = \boldsymbol{\Delta}_{a+b}$, $\boldsymbol{\nu}_2 = \boldsymbol{\Delta}_a$ and $\boldsymbol{\nu}_3 = \boldsymbol{\Delta}_n$. However, since we restrict ourselves in this case to at most 2-column irreps, we must disregard $\boldsymbol{\Psi}[3]$ and $\boldsymbol{\Psi}^\dagger[3]$ components, which yield 3-column irreps. Thus, for the remaining two comonents we have

$$\begin{aligned} \boldsymbol{\Psi}[r]_i &: [a_i,b_i,c_i] \rightarrow [a_i,b_i,c_i] + \boldsymbol{\delta}_r \\ \boldsymbol{\Psi}^\dagger[r]_i &: [a_i,b_i,c_i] \rightarrow [a_i,b_i,c_i] - \boldsymbol{\delta}_r, \qquad (r=1,2), \end{aligned} \tag{27}$$

where

$$\boldsymbol{\delta}_1 = [0,1,-1] \text{ and } \boldsymbol{\delta}_2 = [1,-1,0] \; . \tag{28}$$

Considering now any two adjacent subgroups from the GT canonical chain (8), i.e., $U(m+1) \supset U(m)$, we see easily that the generators $E_{i,m+1} =: \Theta_i$ constitute a vector operator $\boldsymbol{\Theta}$ of U(m) since

$$[E_{ij}, \Theta_k] = \delta_{jk}\, \Theta_i \; . \tag{29}$$

Resolving Θ into the shift components we get

$$\boldsymbol{\Theta}[r]_i : [a_i,b_i,c_i] \rightarrow [a_i,b_i,c_i] + \boldsymbol{\delta}_r \qquad (r=1,2) \tag{30}$$

and likewise for $\Theta^{\dagger}[r]_i$. We can thus write

$$E_{m,m+1}|\{\mathbf{P}\}> = \sum_{r=1}^{2} \Theta\,[r]_m|\{\mathbf{P}\}> = \sum_{r=1}^{n} N_m^r|\{\mathbf{P}+\delta_r^m\}> \;, \tag{31}$$

where

$$|\{\mathbf{P}\}> = \left|\begin{matrix}\mathbf{P}_{m+1}\\ \mathbf{P}_m\\ (\mathbf{P})\end{matrix}\right\rangle \qquad , \; |\{\mathbf{P}+\delta_r^m\}> = \left|\begin{matrix}\mathbf{P}_{m+1}\\ \mathbf{P}_{m}+\delta_r\\ (\mathbf{P})\end{matrix}\right\rangle \quad , \tag{32}$$

so that N_m^r gives the desired matrix element [note that we label the basis vectors by Paldus' ABC tableaux, so that $\mathbf{P}_i = [a_i,b_i,c_i]$ with $(\mathbf{P})$ designating the remainder of the tableau]. Choosing the phase to be real we thus can write

$$N_m^r = <\Theta^{\dagger}[r]_m\,\Theta[r]_m>^{\frac{1}{2}} \tag{33}$$

with the mean value evaluated for the state $|\{\mathbf{P}\}>$, Eq. (32).

Exploiting the Wigner-Eckart theorem for U(n) we find that[72]

$$\Theta^{\dagger}[r]_m\,\Theta[r]_m = R_m^r\,\overline{C}_m^r \;, \tag{34}$$

where R_m^r is a U(m) invariant (reduced matrix element) and $\overline{C}_m^r = \overline{P}[r]_{mm}$ is a U(m-1) invariant operator which is diagonal in the GT basis. These invariants can then be evaluated in terms of the U(m) and U(m+1) dimension formulas, Eq. (12), and the first two Casimir invariants $I_k = tr(\mathbf{E}^k)$, k=1,2. Designating the constant value which I_k takes on $V\langle\lambda\rangle$ by $\chi_{\lambda}(I_k)$, we find that

$$\begin{aligned}\chi_{\lambda}(I_1) &= N \;,\\ \chi_{\lambda}(I_2) &= a(n+3-a) + (n-c)(1+c) \;,\end{aligned} \tag{35}$$

where $\langle\lambda\rangle \equiv [a,b,c]$. This gives finally

$$<\{\mathbf{P}+\delta_r^m\}|E_{m,m+1}|\{\mathbf{P}\}> = \left[\frac{(1+b_m)(b_m+4-2r)}{(1+b_{m-1})(1+b_{m+1})}\right]^{\frac{1}{2}} \;, \quad (r=1,2) \tag{36}$$

which is identical with Eq. (31,32) of Ref. 5.

As we already indicated, a great advantage of this approach lies in its generality, since it can be extended to handle similarly not only the non-elementary generators and generator products,[67] but also general "group function" bases[69,70] associated with the group chains of the type (18).

(ii) Clifford algebra unitary group approach (CAUGA) [40]

This approach employs an imbedding of U(n) into a much larger group $U(2^n)$ exploiting the chain*

$$U(2^n) \supset Spin(m) \supset SO(m) \supset U(n) \;, \qquad m = 2n \text{ or } 2n+1 \;. \tag{37}$$

The usefulness of this imbedding stems from the fact[39] that any p-column irrep of U(n) is contained in the totally symmetric p-box irrep of $U(2^n)$.

---

*Note that $U(n^2) \supset U(n)$ imbedding was used by Biedenharn et al.[72c] to resolve the U(n) multiplicity problem.

Moreover, Nikam and Sarma[73] showed how the U(n) generators can be conveniently represented in terms of the $U(2^n)$ generators when we exploit the spinorial Clifford algebra type basis in $\mathbf{V}_{2n}$ by generalizing Ichimura's work[74] for SO(5). One can in fact give a simple and explicit expression for this relationship.[39,40]

Avoiding any derivations and justifications, we shall illustrate the basic features of this approach on a simple example and refer the reader to Ref. 40 for more detail. Consider, thus, U(3) with the imbedding (37), i.e., $U(8) \equiv U(2^3) \supset U(3)$, and the spinorial basis $|m_{1k}m_{2k}m_{3k}\rangle$ with $|m_{ik}| = \frac{1}{2}$. Using a simplified notation $\pm\frac{1}{2} \rightarrow (\pm)$ and a binary ordering, we can write

$$\begin{aligned} |1\rangle &= |(+++)\rangle\,, & |5\rangle &= |(-++)\rangle\,, \\ |2\rangle &= |(++-)\rangle\,, & |6\rangle &= |(-+-)\rangle\,, \\ |3\rangle &= |(+-+)\rangle\,, & |7\rangle &= |(--+)\rangle\,, \\ |4\rangle &= |(+--)\rangle\,, & |8\rangle &= |(---)\rangle\,. \end{aligned} \tag{38}$$

We designate the U(8) generators by E_{ij} (i,j = 1, ..., 8) as usually, defining $E_{ij} = |i\rangle\langle j|$, while the U(3) generators we now label by Λ_{ij} (i,j = 1,2,3). We find immediately the following relationship between the U(3) and U(8) generators

$$\begin{aligned} \Lambda_{12} &= E_{35} + E_{46} \\ \Lambda_{23} &= E_{23} + E_{67} \\ \Lambda_{13} &= -E_{25} + E_{47}\,. \end{aligned} \tag{39}$$

Indeed, for a given Λ_{ij} $(i \neq j)$ this relationship is easily discerned when we find pairs of vectors (38) in which the (-+) labels in positions i and j change into (+-) while all the remaining labels do not change [the sign is $(-1)^k$ where k is the number of +'s which appear between the i'th and j'th positions]. The reader can easily verify that Λ_{ij}, Eq. (39), satisfy standard U(3) commutation relations.

The general relationship of the type (39) for an arbitrary n and its structure may be found in Ref. 40. We only note that while generally each Λ_{ij} $(i \neq j)$ of U(n) is expressed through 2^{n-2} generators E_{ij} of $U(2^n)$, we only need

$$\binom{n-2}{a-1} + \binom{n-2}{a+b-1}(1-\delta_{b0})$$

terms for each specific [a,b,c] irrep of U(n).

The GT basis for any k-column irrep of U(n) may then be characterized by the k-box totally symmetric Weyl tableaux or their linear combinations. Specifically, for UGA two-column irreps [a,b,c] we only need two-box tableaux. For notational convenience we define

$$[m|n] := |\boxed{m|n}\rangle = |\boxed{n|m}\rangle\,. \tag{40}$$

The lexically highest (HWS) and the lowest (LWS) weight states of the [a,b,c] irrep of U(n) are thus represented as $[2^c|2^{b+c}]$ and $[1+2^n-2^{a+b}|1+2^n-2^a]$, respectively. Generally, any GT state can be expressed as a linear combination of two-box tableaux

$$|[m]\rangle = \sum_k c_k^{[m]}\,[i_k|j_k]\,. \tag{41}$$

One can give rules for the determination of coefficients $c_k^{[m]}$ or, preferably, for the basis generation in the HLED form by starting with the

HWS and applying elementary lowering generators.[40] In fact, the HLED procedure leads directly to the canonical valence-bond (VB) type basis. Schmidt orghogonalization at each level gives then the GT basis. There is also a possibility[36] to exploit directly the 2-box basis as such, which is orthonormal, but which involves in addition to a given irrep also the states of other irreps with higher multiplicity. The latter are of course automatically eliminated once the Hamiltonian is diagonalized. Thus, it might be advantageous to exploit the simplest partially adapted 2-box basis at the expense of a slight increase in the dimension of the CI problem. This possibility will be discussed in greater detail elsewhere.[36]

Irrespective of the basis actually employed, all states are represented as (linear combinations of) 2-box tableau(x) (40) so that it is straightforward to calculate matrix elements of any generator or of their products using the representation of the type (39) and the following simple rules

$$
\begin{aligned}
E_{ij}\,[j|r] &= [\underline{i}|r]\,, && (r\neq i,j)\\
E_{ij}\,[j|j] &= \sqrt{2}\,[i|j]\,, &&\\
E_{ij}\,[i|j] &= \sqrt{2}\,[i|i]\,, &&\\
E_{ij}\,[r|s] &= 0\,, && (j\neq r,s)\,.
\end{aligned}
$$

To conclude this brief outline we list the principal features and advantages of CAUGA:

- CAUGA effectively reduces an N-electron problem to a number of two-boson problems,
- CAUGA can be applied to particle-number non-conserving operators,
- CAUGA enables an exploitation of any coupling scheme, and is particularly suited for VB-type basis, unlike UGA which is restricted to a rather unphysical GT basis,
- CAUGA can exploit many UGA features, such as the DRT concept, so that it can be regarded as an extension rather than replacement of UGA,
- CAUGA drastically simplifies the algorithm for matrix element evaluation for an arbitrary generator or a product of such generators,
- CAUGA can be easily extended to systems with arbitrary spin-s fermions.

(iii) Parafermi algebras and correlation problem[75]

A canonical quantization of a classical system described by the Hamiltonian H (q_i, p_i), where q_i and p_i represent canonical coordinates and momenta, respectively, consists in the reinterpretation of q_i and p_i as quantum mechanical operators satisfying canonical commutation relations. However, these commutation relations are only sufficient but not necessary conditions for the validity of the Heisenberg's equation of motion and their consistency with the classical Euler-Lagrange equations of motion.[76] Indeed, it has been shown that a more general quantization scheme is possible, which is usually referred to as a paraquantization,[77] and is characterized by the following tri-linear commutation relations[77,78]

$$
\begin{aligned}
[X_k, [X_\ell^\dagger, X_m]] &= 2\,\delta_{k\ell}\,X_m\\
[X_k, [X_\ell^\dagger, X_m^\dagger]] &= 2\,\delta_{k\ell}\,X_m^\dagger - 2\,\delta_{km}\,X_\ell^\dagger\\
[X_k, [X_\ell, X_m]] &= 0
\end{aligned}
$$

and the vacuum property

$$X_k\,X_\ell^\dagger\,|0\rangle = p\,\delta_{k\ell}\,|0\rangle\,,$$

where p is the order of parastatistics (note that parafermions of order 1 correspond to normal fermions). Even though so far there is no experimental evidence of the existence of para-particles which would necessitate such a quantization,[78] there exists an extensive literature on parafield theory, that is a generalized method of field quantization, originally introduced by Green.[79] A detailed and extensive account of these developments may be found in the monograph by Ohnuki and Kamefuchi.[77]

Very recently we have shown[75] that second order (p=2) parafermi creation and annihilation operators which, respectively, create and annihilate spin-averaged para-particles, occur naturally in the spin-independent many-electron problem and the Hamiltonian (1) (or in fact its particle-nonconserving analogue) can be directly expressed in terms of these parafermion operators. In fact, it has been shown by Bracken and Green[80] that the parafermion Fock space carries the irrep (p/2,p/2, ..., p/2) of O(2n+1) and that it decomposes into a direct sum of U(n) irreps with at most p columns, and that all such irreps occur exactly once. A more detailed relationship between parastatistics and CAUGA (and the many-electron correlation problem in general), together with an explicit matrix representation for the parafermi algebra of order 2, obtained with GG approach, for a canonical U(n) basis, will be appearing shortly.[81]

ACKNOWLEDGEMENTS

The author wishes to express his gratefulness to Prof. Bruno Gruber for giving him the opportunity to take part in this very fascinating Symposium on Symmetries in Science II. A continued support of author's research by the Natural Sciences and Engineering Research Council of Canada is also greatly appreciated.

REFERENCES

1. H. Weyl, "Gruppentheorie und Quantenmechanik", Hirzel, Leipzig; Germany, (1928); English translation: "The Theory of Groups and Quantum Mechanics", Dover, New York (1964).
2. I. M. Gelfand and M. L. Tsetlin, Dokl. Akad. Nauk SSSR 71, 825, 1070 (1950).
3. P. Jordan, Z. Phys. 94, 531 (1935).
4. M. Moshinsky, in "Many-Body Problems and Other Selected Topics in Theoretical Physics", M. Moshinsky, T. A. Brody and G. Jacob, eds., p. 289, Gordon and Breach, New York (1966); also published separately as "Group Theory and the Many-Body Problem", Gordon and Breach, New York (1968).
5. J. Paldus, J. Chem. Phys. 61, 5321 (1974).
6. J. Paldus, in "Theoretical Chemistry: Advances and Perspectives", Vol. 2, H. Eyring and D. Henderson, eds., p. 131, Academic Press, New York (1976).
7. J. Paldus, Int. J. Quantum Chem., Symp. 9, 165 (1975).
8. J. Paldus, Phys. Rev. A14, 1620 (1976).
9. I. Shavitt, Int. J. Quantum Chem., Symp. 11, 131 (1977), Symp. 12, 5 (1978).
10. J. Paldus and M. J. Boyle, Phys. Scripta 21, 295 (1980).
11. M. Downward and M. A. Robb, Theor. Chim. Acta 46, 129 (1977); D. Hegarty and M. A. Robb, Mol. Phys. 38, 1795 (1979).
12. P. E. M. Siegbahn, J. Chem. Phys. 70, 5391 (1979); 72, 1647 (1980).
13. B. R. Brooks and H. F. Schaefer, III, J. Chem. Phys. 70, 5092 (1979).
14. B. R. Brooks, W. D. Laidig, P. Saxe, N. C. Handy and H. F. Schaefer, III. Phys. Scripta 21, 312 (1980); B. R. Brooks, W. D. Laidig, P. Saxe, J. D. Goddard, Y. Yamaguchi and H. F. Schaefer, III, J. Chem. Phys. 72, 4652 (1980).

15. P. E. M. Siegbahn, J. Chem. Phys. 75, 2314 (1981); Chem. Phys. 66, 443 (1982).
16. H. Lischka, R. Shepard, F. D. Brown and I. Shavitt, Int. J. Quantum Chem., Symp. 15, 91 (1981).
17. V. R. Saunders and J. H. van Lenthe, Mol. Phys. 48, 923 (1983).
18. P. Saxe, D. J. Fox, H. F. Schaefer, III, and N. C. Handy, J. Chem. Phys. 77, 5584 (1982).
19. P. E. M. Siegbahn, Chem. Phys. Lett. 109, 417 (1984).
20. J. Paldus, in "Electrons in Finite and Infinite Structures", P. Phariseau and L. Scheire, eds., p. 411, Plenum, New York (1977).
21. F. A. Matsen, Advan. Quantum Chem. 11, 223 (1978).
22. J. Paldus, in "Group Theoretical Methods in Physics", Lecture Notes in Physics, Vol. 94, W. Beiglböck, A. Böhm and E. Takasugi, eds., p. 51, Springer, New York (1979).
23. J. Hinze, ed., "The Unitary Group for the Evaluation of Electronic Energy Matrix Elements", Lectures Notes in Chemistry, Vol. 22, Springer, Berlin (1981).
24. M. A. Robb and U. Niazi, Comp. Phys. Rep. 1, 127 (1984).
25. R. Pauncz, "Spin Eigenfunctions: Construction and Use", Plenum, New York (1979).
26. S. Wilson, "Electron Correlation in Molecules", Clarendon, Oxford (1984).
27. J. Čížek and J. Paldus, Int. J. Quantum Chem. 12, 875 (1977); Phys. Scripta 21, 364 (1980); J. Čížek, M. Clay and J. Paldus, Phys. Rev. A22, 793 (1980); B. G. Adams, J. Čížek and J. Paldus, Int. J. Quantum. Chem. 21, 153 (1982); Advan. Quantum. Chem. 18, in press and references therein.
28. A. O. Barut, "Dynamical Groups and Generalized Symmetries in Quantum Theory", University of Canterbury, Christchurch, New Zealand (1971).
29. F. Iachello and R. D. Levine, J. Chem. Phys. 77, 3046 (1982); O. S. van Roosmalen, F. Iachello, R. D. Levine, and A. E. L. Dieperink, J. Chem. Phys. 79, 2515 (1983); O. S. van Roosmalen, I. Benjamin and R. D. Levine, J. Chem. Phys. 81, 5986 (1984); R. D. Levine, J. Phys. Chem. 89, 2122 (1985) and references therein.
30. Y. Alhassid, F. Iachello and R. D. Levine, Phys. Rev. Lett. 54, 1746 (1985).
31. F. Iachello, this volume.
32. P. C. Ojha, SO(2,1) Lie Algebra and the Jacobi-Matrix Method for Scattering, preprint.
33. P.-O. Löwdin, Advan. Chem. Phys. 2, 207 (1959).
34. See, e.g., I. Shavitt, in "Methods of Electronic Structure Theory", H. F. Schaefer, III, ed., p. 189, Plenum, New York (1977).
35. G. Born and I. Shavitt, J. Chem. Phys. 76, 558 (1982); G. Born, Int. J. Quantum Chem., Symp. 16, 633 (1982); 28, 335 (1985).
36. B. Jeziorski and J. Paldus, unpublished results.
37. R. D. Kent and M. Schlesinger, Phys. Rev. B27, 46 (1983); P. S. Ponnapalli, M. Schlesinger and R. D. Kent, Phys. Rev. B31, 1258 (1985); R. D. Kent, M. Schlesinger and P. S. Ponnapalli, Phys. Rev. B31, 1264 (1985).
38. S. Rettrup and C. R. Sarma, Phys. Lett. A75, 181 (1980); J. Phys.: Math. Gen. A13, 2267 (1980).
39. C. R. Sarma and J. Paldus, J. Math. Phys. 26, 1140 (1985).
40. J. Paldus and C. R. Sarma, J. Chem. Phys. 83, 5135 (1985).
41. Cf., e.g., P. Čárský and M. Urban, "Ab Initio Calculations. Methods and Applications in Chemistry", Lecture Notes in Chemistry, Vol. 16, Springer, Berlin (1980).
42. B. R. Brooks, W. D. Laidig, P. Saxe, J. D. Goddard and H. F. Schaefer, III, in Ref. 23, p. 158.
43. P. E. M. Siegbahn, A. Heiberg, B. O. Roos and B. Levy, Phys. Scripta 21, 323 (1980).
44. A. Banerjee and J. Simons, Int. J. Quantum Chem. 19, 207 (1981); J.

Chem. Phys. 76, 4548 (1982).
45. F. A. Matsen, Advan. Quantum Chem. 1, 60 (1964).
46. M. Moshinsky and T. H. Seligman, Ann. Phys. (N.Y.) 66, 311 (1971).
47. F. A. Matsen, this volume.
48. J. Paldus, Ref. 23, p. 1.
49. R. Serber, Phys. Rev. 45, 461 (1934); J. Chem. Phys. 2, 697 (1934).
50. M. Kotani, Proc. Phys. Math. Soc. Japan 19, 460 (1937); M. Kotani, A. Amemiya, E. Ishiguro and T. Kimura, "Tables of Molecular Integrals", Maruzen, Tokyo (1955).
51. W. Duch and J. Karwowski, Int. J. Quantum Chem. 22, 783 (1982); Comp. Phys. Rep. 2, 93 (1985).
52. A. P. Jucys, I. B. Levinson, and V. V. Vanagas, "Mathematical Apparatus of the Theory of Angular Momenta", Israel Program for Scientific Translations, Jerusalem (1962) and Gordon and Breach, New York (1964).
53. D. M. Brink and G. R. Satchler, "Angular Momentum", 2nd ed., Chap. VII, Clarendon, Oxford (1968).
54. E. ElBaz and B. Castel, "Graphical Methods of Spin Algebras", Dekker, New York (1972).
55. For a brief outline of this technique and of the required rules, see Appendix I of Ref. 56.
56. J. Paldus, B. G. Adams and J. Čížek, Int. J. Quantum Chem. 11, 813 (1977).
57. P. E. S. Wormer and J. Paldus, Int. J. Quantum Chem. 16, 1307 (1979); 18, 841 (1980).
58. J. Paldus and P. E. S. Wormer, Int. J. Quantum Chem. 16, 1321 (1979).
59. I. Shavitt, Ref. 23, p. 50.
60. P. W. Payne, Int. J. Quantum Chem. 22, 1085 (1982).
61. J.-F. Gouyet, R. Schranner and T. H. Seligman, J. Phys. A8, 285 (1975).
62. G. W. F. Drake and M. Schlesinger, Phys. Rev. A15, 1990 (1977).
63. B. O. Roos, Chem. Phys. Lett. 15, 153 (1972); B. O. Roos and P. E. M. Siegbahn, in "Methods of Electronic Structure Theory", H. F. Schaefer, III, ed., p. 277, Plenum, New York (1977).
64. M. Bénard, unpublished.
65. H. S. Green, J. Math. Phys. 12, 2106 (1971); A. J. Bracken and H. S. Green, J. Math. Phys. 12, 2009 (1971).
66. M. D. Gould, J. Math. Phys. 21, 444 (1980); 22, 15, 2376 (1981).
67. M. D. Gould and G. S. Chandler, Int. J. Quantum Chem. 25, 553, 603 (1984); 27, (E)787 (1985).
68. M. D. Gould and G. S. Chandler, Int. J. Quantum Chem. 25, 1089 (1984); 26, 441 (1984); 27, (E)787 (1985).
69. P. E. S. Wormer and A. van der Avoird, J. Chem. Phys. 57, 2498 (1972); P. E. S. Wormer, in "Electron Correlation: Proceedings of the Daresbury Study Weekend" (17-18 November, 1979), M. F. Guest and S. Wilson, eds., p. 49, Science Research Council, Daresbury Laboratory, U.K. (1980).
70. M. D. Gould and J. Paldus, Int. J. Quantum Chem., in press.
71. J. Paldus and M. J. Boyle, Phys. Rev. A22, 2299 (1980); M. J. Boyle and J. Paldus, Phys. Rev. A22, 2316 (1980).
72. L. C. Biedenharn and J. D. Louck, (a) "Angular Momentum and Quantum Mechanics: Theory and Application", and (b) "The Racah-Wigner Algebra in Quantum Theory", Encyclopedia of Mathematics and Its Applications, Vols. 8 and 9, respectively, Addison-Wesley, Reading, Mass. (1981); (c) L. C. Biedenharn, A. Giovannini and J. D. Louck, J. Math. Phys. 8, 691 (1967).
73. R. S. Nikam and C. R. Sarma, J. Math. Phys. 25, 1199 (1984); R. S. Nikam, G. G. Sahasrabudhe, and C. R. Sarma, J. Phys. A, in press.
74. M. Ichimura, Progr. Theor. Phys. 33, 215 (1965).
75. M. D. Gould and J. Paldus, Phys. Rev. A, in press.
76. E. P. Wigner, Phys. Rev. 77, 711 (1950).
77. Y. Ohnuki and S. Kamefuchi, "Quantum Field Theory and Parastatistics", Springer, New York (1982).

78. O. W. Greenberg and A. M. L. Messiah, Phys. Rev. B138, 1155 (1965).
79. H. S. Green, Phys. Rev. 90, 270 (1953).
80. A. J. Bracken and H. S. Green, Nuovo Cimento A9, 349 (1972).
81. M. D. Gould and J. Paldus, to be published.

ON THE REPRESENTATIONS OF THE BASIC LIE SUPERALGEBRAS:

GEL'FAND ZETLIN BASIS FOR sl(1,n)

Tchavdar D. Palev

Institute of Nuclear Research and Nuclear Energy

1184 Sofia, Bulgaria

Introduction

In the present note we touch shortly certain points from the representation theory of the basic Lie superalgebras (LS's). We consider in more details the special linear Lie superalgebra(LS) sl(1,n) and indicate how one can introduce a concept of a Gel'fand-Zetlin basis in the finite-dimensional irreducible modules (fidirmods)[1] of sl(1,n).

The place of the basic LS's among all LS's is illustrated on Fig. 1 (see Ref. 2).

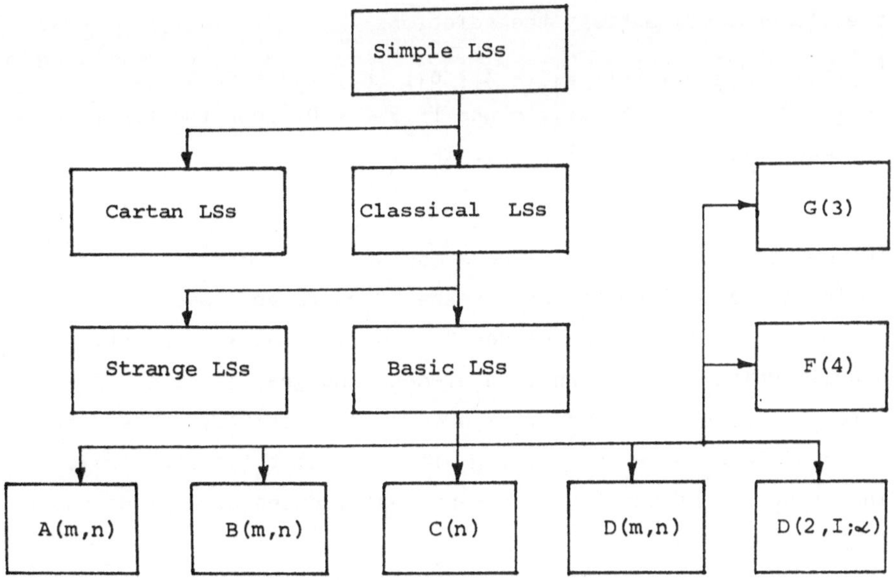

Fig. I

We recall that the simple LS A is said to be classical (resp. Cartan) LS if its even (= bosonic) subalgebra is reductive (resp. is not reductive). The classical LS A is basic (resp. strange) if its Killing form is nondegenerate (resp. degenerate).

The basic LS's which are not LA resolve into four countable series (m,n = 1,2,...), one continuous class of 17-dimensional algebras $D(2,1;\alpha)$ and two exceptional LS's G(3) and F(4).

The structure of the basic LS's resembles in many respects the structure of the simple Lie algebras (LA's). Every such algebra A can be represented as a direct-space sum $A = N^- \oplus H \oplus N^+$ of its Cartan subalgebra H, which is the Cartan subalgebra of the even part, and the subalgebras N^- and N^+ spanned on the negative and the positive root vectors, respectively. The root vectors l_α are in one-to-one correspondence with their roots α, which are elements from the dual to H space $\overset{*}{H}$: $\alpha \in \overset{*}{H}$. The correspondence $l_\alpha \leftrightarrow \alpha$ is determined from the relation

$$[h,l_\alpha] = \alpha(h) l_\alpha, \ \forall\ h \in H, \tag{1}$$

where $[,]$ denotes the product in A. One can always choose a canonical system of 3r elements (r = dim H)

$$e_i,\ h_i,\ f_i \qquad i = 1, 2, \ldots, r, \tag{2}$$

which generate A and have the following properties:

(a) $h_1, h_2, \ldots, h_r$ constitute a basis in H;

(b) $e_i \in N^+$ and $f_i \in N^-$ are positive and negative root vectors;

(c) the elements (2) satisfy the relations

$$[e_i, f_j] = \delta_{ij}\, h_i,\ [h_i,\ e_j] = \alpha_{ij}\, e_j,\ [h_i,\ f_j] = -\alpha_{ij}\, f_j, \tag{3}$$

where $\alpha_{ii} = 0$ or 2, i = 1, ..., r and if $\alpha_{ii} = 0$, then the first nonzero element among $\alpha_{i,i+k}$, k = 1,2, ... is 1.

The matrix $\alpha = (\alpha_{ij})$ is called a Cartan matrix of A. The Kac-Dynkin diagram, which determines the LS up to an isomorphism, consists of r white, grey and black nodes denoted as O, ⊗ and ●, respectively. The i-th node is white if e_i is an even element and gray or black if e_i is an odd (= fermionic) element and α_{ii} = 0 or 2, respectively. The i-th and the j-th nodes are joined by $|\alpha_{ij}\alpha_{ji}|$ lines (except for $D(2,1;\alpha)$).

The structure of the finite-dimensional modules of a given basic LS

A(i.e. of the finite-dimensional representation spaces of A) is illustrated graphically on Fig. 2 (Ref. 3)

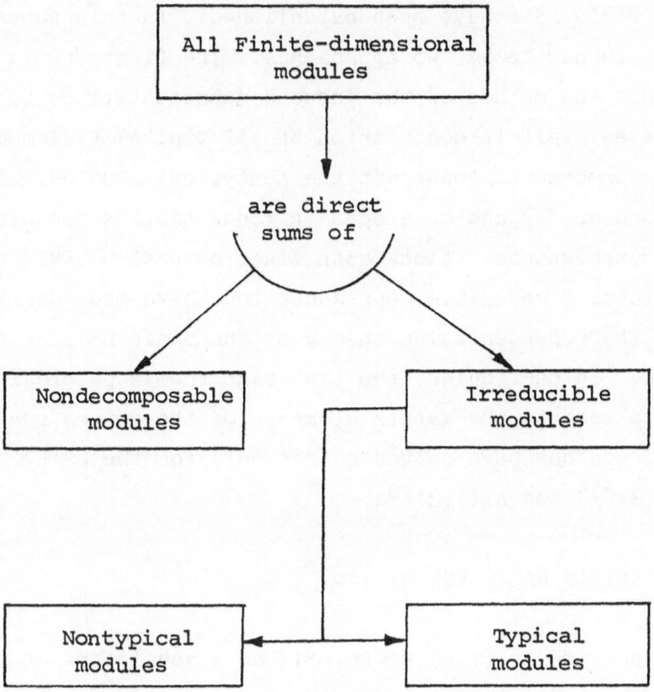

Fig.2

Apart from the algebras B(o,n), every basic LS has indecomposable (i.e., non fully reducible)finite-dimensional modules. Several examples of such modules are avilable. However, at present there exists no theory to tell us how to construct all indecomposable representations. In contrast to this, all finite-dimensional irreducible modules are fully classified[3]. Every fidermod V is either typical or nontypical. The A-module V is called typical, if whenever V is submodule of a larger A-module W, then there always exists a compliment to V subspace V', which is also an A-module. If this is not always the case, V is said to be nontypical. Therefore, the nontypical irreducible modules are building blocks for the nondecomposable modules.

As in the Lie algebra representation theory, one defines for the LS's such concepts as weight vectors and weights, highest weight, etc. In particular, all irreducible modules are with a highest weight. Each such A-module V is characterized by the coordinates $\alpha_i = \Lambda(h_i)$, $i = 1,\ldots,r$ of its highest weight Λ in the dual to $h_1,\ldots,h_r$ basis $h^1,\ldots, h^r$, or graphically, by a Kac-Dynkin diagram (⊖ = 0, ⊗ or ●)

$$\overset{\alpha_1}{0}----\overset{\alpha_2}{0}---- \ldots ---\overset{\alpha_{r-1}}{0}-----\overset{\alpha_r}{0} \, . \qquad (4)$$

In the last years several important results in the representation theory of the basic LS's have been established. In this respect of particular use turned to be two approaches. The first one is due to Kac[3], who extends the method of the induced representations to the LS's and in this way gives explicit description of all typical fidirmods and shows how one has to proceed to construct the nontypical modules. The second approach introduces [4,5] and developes[6] a Young tableux technique to the case of the superalgebras. Along both lines several useful results of the finite-dimensional irreducible representations have been derived.[7] Nevertheless, the representation theory of the basic LS's is far from being complete. In particular, the important from a physical point of view problem to compute the matrix elements of the generators within an arbitrary fidirmod has been solved so far only for the LS's osp(2,1), sl(1,2)[8,9] and sl(1,3)[10].

II. GEL'FAND ZETLIN BASIS FOR sl(1,n)

We shall consider sl(1,n) ≡ A(o,n-1) as a subalgebra of the general linear LS gl(1,n). The latter consists of all (n+1) x (n+1) matrices. We label the rows and the columns of these matrices with indices A,B,C,D,... = 0,1,2,..., n and assign to each A a degree (A), which is zero for A = 0 and 1 for A = 1,2 ..., n. Let e_{AB} gl(1,n) be a matrix with 1 on the Ath row and the Bth column and zero elsewhere. The even (resp. odd) part of gl(1,n) is defined as the linear envelope of all e_{AB}, for which (A) + (B) is an even (resp. an odd) number. The multiplication [,] on gl(1,n) is given with the linear extension of the relations

$$[e_{AB} \quad e_{CD}] = \delta_{BC}\, e_{AD} - (-1)^{[(A)+(B)]\ [(C)+(D)]}\delta_{AD}\, e_{CB}. \qquad (5)$$

The basic LS sl(1,n) is a subalgebra of gl(1,n) consisting of all matrices a gl(1,n) which supertrace (≡ str) is zero, i.e.

$$sl(1,n) = \{ a \quad a \quad gl(1,n),\ \mathrm{str}a \quad \sum_{A=0}^{n} (-1)^{(A)} a_{AA} = 0 \}. \qquad (6)$$

Its even subalgebra

$$sl_o(1,n) = \mathrm{lin.env.} \{ E_{ij}\ E_{ij} = e_{ij} + \delta_{ij}\, e_{oo},\ i,j = 1,2,\ldots,n \} \qquad (7)$$

is isomorphic to the Lie algebra gl(n). The Cartan subalgebra H of sl(1,n) consists of all diagonal matrices which supertrace is zero and it is a Cartan subalgebra also of gl(n). The set of all e_{AB}, $A < B = 0,1,\ldots n$, (resp. $A > B = 0,1,\ldots, n$) gives the positive (resp. the negative) root vectors of sl(1,n). The canonical system (2) of generators is

$$\begin{array}{lll} h_1 = E_{11} & e_1 = e_{01} & f_1 = e_{10} \\ h_2 = E_{11} - E_{22} & e_2 = e_{12} & f_2 = e_{21} \\ \cdots\cdots & \cdots\cdots & \cdots\cdots \\ h_n = E_{n-1,n-1} - E_{nn} & e_n = e_{n-1,n} & f_n = e_{n,n-1} \end{array} \tag{8}$$

with e_1 and f_1 being the only odd generators in it. Since, moreover, $[h_1, e_1]_+ = 0$, i.e. $\alpha_{11} = 0$, the Kac-Dynkin diagram contains one gray and n-1 white nodes:

```
⊗----0----0---- ... ----0----0.      (9)
1    2    3             n-1  n
```

Every fidirmod of sl(1,n) is characterized by a Kac-Dynkin diagram[3]

```
α1    α2    α3            αn-1  αn
⊗-----0-----0--- ... -----0-----0      (10)
```

where $\alpha_1 \in \mathbb{C}$ is an arbitrary complex number, and $\alpha_2, \ldots, \alpha_n \in \mathbb{Z}_+$ are arbitrary nonnegative integers. If x_Λ is the highest weight vector, corresponding to the highest weight Λ, then

$$h_i x_\Lambda = \alpha_i x_\Lambda, \quad i = 1, \ldots, n. \tag{11}$$

The fidirmods of the LA gl(n) will be essentially used in the remaining part of the paper. Throughout we use the Gel'fand and Zetlin notation (see Ref. 11) accepting also some abbreviation from Ref. 12. Every fidirmod of gl(n) is labeled by a lexical n-triple

$$[m]_n \equiv [m_{1n}, m_{2n}, \ldots, m_{nn}], \tag{12}$$

i.e. by n in general complex numbers $m_{1n}, \ldots, m_{nn}$ such that $m_{in} - m_{i+1,n}$ are nonnegative integers

$$m_{in} - m_{i+1,n} \in \mathbb{Z}_+, i = 1, 2, \ldots, n-1. \tag{13}$$

Let $V_0([m_{1n}, m_{2n}, \ldots, m_{nn}]) \equiv V_0([m]_n)$ be an gl(n)- module carrying the representation $[m]_n$. As a basis in $V_0([m]_n)$ we choose the Gel'fand-Zetlin basis (GZ-basis)

$$\left| \begin{matrix} m_{1n}, m_{2n}, \ldots, m_{nn} \\ \cdot\ \cdot\ \cdot\ \cdot\ \cdot\ \cdot \\ m_{1i}, m_{2i}, \ldots, m_{ii} \\ \cdot\ \cdot\ \cdot\ \cdot \\ m_{11} \end{matrix} \right\rangle \equiv \left| \begin{matrix} [m]_n \\ \\ [m]_i \\ \\ m_{11} \end{matrix} \right\rangle \equiv \left| \begin{matrix} [m]_n \\ (m) \end{matrix} \right\rangle, \qquad (13)$$

where the numbers $m_{1n}, m_{2n}, \ldots, m_{nn}$ are fixed and label $V_o([m]_n)$. The other numbers m_{ij} distinguish the basis vectors and take all possible values consistent with the "betweenness" condition

$$m_{i,j+1} - m_{ij} \in \mathbb{Z}_+ , \quad m_{ij} - m_{i+1,j} \in \mathbb{Z}_+. \qquad (14)$$

The highest weight vector x_Λ, $\Lambda = [m]_n$ in $V_o([m]_n)$ is the one for which $m_{ii} = m_{i,i+1} = \ldots = m_{i,n}$, $i = 1, \ldots, n$. In this case

$$E_{ii}\, x_\Lambda = m_{in}\, x_\Lambda \qquad (15)$$

and, therefore, $m_{1n}, \ldots, m_{nn}$ are the coodinates of the gl(n)- highest weight Λ in the dual to

$E_{11}, E_{22}, \ldots, E_{nn}$ basis $E^1, E^2, \ldots, E^n$, i.e.

$$\Lambda = \sum_{i=1}^{n} m_{in}\, E^i. \qquad (16)$$

We now proceed to introduce, following Kac[3], the sl(1,n) module $\overline{W}([m]_n)$, induced from the gl(n)- fidirmod $V_o([m]_n)$. To this end denote by P_+ the linear envelope of all odd positive root vectors of sl(1,n),

$$P_+ = \text{lin.env.} \{ e_{oi} | i = 1,2,\ldots, n \} \qquad (17)$$

and let $P = sl_o(1,n) \oplus P_+ \equiv gl(n) \oplus P_+$. Extend $V_o([m]_n)$ to a P- module, assuming

$$P_+ V_o([m]_n) = 0. \qquad (18)$$

The sl(1,n)- module $\overline{W}([m]_n$, induced from the gl(n)- fidirmod $V_o([m]_n)$, is the factor space

$$\overline{W}([m]_n) = U \otimes V_o([m]_n)/I([m]_n) \qquad (19)$$

of the tensor product of the sl(1,n)- universlal enveloping algebra U with $V_o([m]_n)$ factorized by the subspace

$$I([m]_n) = \text{lin.env.}\ \{ up\otimes v - u\otimes pv \mid u \in U,\ p \in P \subset U,\ v \in V_o \}. \quad (20)$$

$W([m]_n)$ is equipped with a structure of an $sl(1,n)$- module in a natural way.

$$g(u\otimes v) = gu\otimes v,\ g\in sl(1,n),\ u\otimes v \in \overline{W}[m]_n), \quad (21)$$

Similarly as for $sl(1,3)$ (see Ref. 10) one shows that

$$\overline{W}([m]_n) = T \otimes V_o([m]_n), \quad (22)$$

Where

$$T = \text{lin.env.}\ \{(e_{10})^{\theta_1}(e_{20})^{\theta_2}\ldots (e_{no})^{\theta_n} \mid \theta_i = 0,1; i=1,\ldots,n\} \subset U. \quad (23)$$

Since $[gl(n),T] \subset T$, T can be viewed as a $gl(n)$ module. Moreover, for any $a \in gl(n)$ and $t\otimes v \in T \otimes V_o([m]_n)$

$$a(t\otimes v) = (ada)t\otimes v + t\otimes av,\ (ada)t = [a,t] \quad (24)$$

and, therefore, $\overline{W}([m]_n)$ (see (22)) is a tensor product of the $gl(n)$ modules T and $V_o([m]_n)$. We now proceed to decompose $\overline{W}([m]_n)$ into a direct sum of irreducible $gl(n)$ modules.

Proposition 1 For any integer $0 \leq N \leq n$ the subspace

$$T_N = \text{lin.env.}\ \{(e_{10})^{\theta_1}\ldots(e_{no})^{\theta_n} \mid \sum_{i=1}^{n} \theta_i = N,\ \theta_i = 0,1\} \subset T \quad (25)$$

is an irreducible $gl(n)$ module with coordinates of the highest weight $m_{in} = 1-N$, $m_{jn} = -N$, $i = 1,2,\ldots, N$, $j = N+1,\ldots, n$. Therefore, in the Gel'fand-Zetlin notation

$$T_N = V([\overset{1,}{1-N},\ldots,\overset{N,}{1-N},\overset{N+1,}{-N},\ldots,\overset{n}{-N}]). \quad (26)$$

The proof is straightforward. The subspace T_o is one dimensional and it is spannned in the unity $\mathbb{1}$ of U. The subspace T_1 = lin.env. $\{e_{io} \mid i=1,\ldots, n\}$ is n-dimensional and the correspondence root vector ↔ root is

$$e_{io} \leftrightarrow (-1+\delta_{1i},-1+\delta_{2i},\ldots,-1+\delta_{ni}). \quad (27)$$

Hence, e_{10} is the highest weight vector in T_1. The subspace T_N is the antisymmetrized N-times tensor product $T_1 \wedge T_1 \wedge \ldots \wedge T_1$, i.e. using the language of the Young diagrams and denoting

$$T_1 = \square \ , \text{ then } T_N = \begin{array}{|c|}\hline \ \\ \hline \ \\ \hline \vdots \\ \hline \ \\ \hline \end{array}\begin{array}{l} 1 \\ 2 \\ \\ N \end{array} . \tag{28}$$

Since $T = T_o \oplus T_1 \oplus \cdots \oplus T_n$, in the notation (26) and (28) one has

$$T = \sum_{N=0}^{n} \oplus \ V([1-N,\ldots, 1-N,-N,\ldots,-N) = \sum_{N=0}^{n} \oplus \ \begin{array}{|c|}\hline 1 \\ \hline 2 \\ \hline \vdots \\ \hline N \\ \hline \end{array} . \tag{29}$$

This result together with (22) gives

$$\overline{W}([m]_n) = \sum_{N=0}^{n} \oplus \quad (\ \begin{array}{|c|}\hline 1 \\ \hline 2 \\ \hline \vdots \\ \hline N \\ \hline \end{array} \otimes V_o(m]_n)). \tag{30}$$

The decomposition of $T_N \otimes V_o([m]_n)$ can be easily carried out ($\theta_1,\ldots,$ $\theta_n,=0,1$):

$$\begin{array}{|c|}\hline 1 \\ \hline 2 \\ \hline \vdots \\ \hline N \\ \hline \end{array} \otimes V_o([m]_n) = \sum_{\theta_1+..+\theta_n=N} \oplus \qquad V([m_{1n}+\theta_1-N,\ldots,m_{nn}+\theta_n-N]). \tag{31}$$

In the right-hand side of (31) one has to delete all terms $V([m_{1n},m_{2n},\ldots m_{nn}])$ which are nonlexical, i.e. for which the condition $m_{1n}\geq m_{2n}\geq\ldots\geq m_{nn}$ is not fulfilled. Combining (30) with (31) we have

Proposition 2 The induced sl(1,n) module $\overline{W}([m]_n)$ decomposes into a direct sum of gl(n) fidirmods as follows:

$$\overline{W}([m]_n) = \sum_{\theta_1,\ldots,\theta_n=0,1} \oplus \qquad V([m_{1n}+\theta_1-\sum_{i=1}^{n}\theta_i,\ldots, m_{nn}+\theta_n-\sum_{i=1}^{n}\theta_i]). \tag{32}$$

Observe that each gl(n) fidirmod appears in the sum (32) only once, i.e. the decomposition of $W([m]_n)$ into gl(n) fidirmods is simple. This suggests immediately that one can introduce a basis in $\overline{W}([m]_n)$ taking as basis vecors all those vectors, which belong to one and only one gl(n) fidirmod, which appear when $\overline{W}([m]_n)$ is decomposed along the chain

$$sl(1,n)\supset gl(n)\supset gl(n-1)\supset\ldots\supset gl(2)\supset gl(1). \tag{33}$$

This is a natural generalization of the way Gel'fand and Zetlin have introduced an orthonormal basis in the fidirmods of the classical LA's. At this place it is convenient to change slightly the notations, replacing everywhere m_{in} by $m_{i,n+1}, i=1,\ldots, n$, i.e.

$$[m]_n = [m_{1n},m_{2n},\ldots,m_{nn}] \to [m_{1,n+1},m_{2,n+2},\ldots,m_{n,n+1}] \equiv [m]_{n+1}. \tag{34}$$

Then (32) reads

$$\overline{W}([m]_{n+1}) = \sum_{\theta_1,\ldots,\theta_n=0,1} \oplus V([m_{1,n+1}+\theta_1-\sum_{i=1}^{n}\theta_i,\ldots,m_{n,n+1}+\theta_n-\sum_{i=1}^{n}\theta_i]). \quad (35)$$

Using the abbreviation (34) and $[m_{1i},\ldots,m_{ii}] = [m]_i$, $i = 1,2,\ldots,n$, we denote the basis in $W([m]_{n+1})$ as

$$\left|\begin{matrix}[m]_{n+1}\\ [m]_n\\ \vdots\\ [m]_2\\ m_{11}\end{matrix}\right\rangle \equiv \left|\begin{matrix}[m]_{n+1}\\ (m)\end{matrix}\right\rangle, \quad (36)$$

where

$$m_{in} = m_{i,n+1} + \theta_i - \sum_{k=1}^{n} \theta_k, \; i = 1,\ldots, n \quad (37)$$

and $\theta_1,\ldots,\theta_n$ take all possible values 0,1 consistent with the conditions

$$m_{i,n+1} \le m_{in} \le m_{i+1,n+1}, \; i = 1,2,\ldots, n. \quad (38)$$

By definition

$$\left|\begin{matrix}[m]_{n+1}\\ (m)\end{matrix}\right\rangle \in V(m_{11}) \subset V([m]_2) \subset \ldots \subset V([m]_n) \subset \overline{W}([m]_{n+1}). \quad (39)$$

In order to define a basis in the sl(1,n) fidirmods we use

<u>Proposition 3</u>(Refs. 2,13) The sl(1,n) module $\overline{W}([m]_{n+1})$ is typical (and, hence, irreducible) iff

$$m_{k,n+1} \ne k - 1, \; \forall\, k = 1,2,\ldots, n. \quad (40)$$

In this case we write $\overline{W}([m]_{n+1}) = W([m]_{n+1})$. If for certain $k = 1,\ldots, n$ $m_{k,n+1} = k-1$, then $\overline{W}([m]_{n+1})$ is nondecomposible and contains a nontrivial maximal invariant subspace $\overline{I}([m]_{n+1})$. The factor module $W([m]_{n+1}) = \overline{W}([m]_{n+1})/\overline{I}([m]_{n+1}$ is nontypical. Every sl(1,n) fidirmod is one of $W([m]_{n+1})$.

<u>Proposition 4.</u> If $m_{k,n+1} = k-1$, then

$$\overline{I}\,([m]_{n+1}) = \sum \oplus V([m_{1,n+1}+\theta_1-\sum_{i=1}^{n}\theta_i,\ldots,m_{n,n+1}+\theta_n-\sum_{i=1}^{n}\theta_i]), \quad (41)$$

where $\theta_k = 1$ and the sum is over all others $\theta_i = 0,1$, $i \ne k$, which are consistent with (38). Therefore, the factor-space $W([m]_{n+1})$ has a basis consisting of all those vectors (36), for which $\theta_K = 0$.

We scip the proof. The highest weight vector x_Λ in $W([m]_{n+1})$ is the one from (36), for which $m_{ii} = m_{i,i+1} = \dots = m_{i,n+1}$, $i = 1,\dots, n$. In this case (15) yields $E_{ii}x_\Lambda = m_{i,n+1}x_\Lambda$, $i=1,\dots, n$. Therefore, (see(8) and (11)), the Kac indices corresponding to the fidirmod $W([m]_{n+1}$ are

$$\alpha_1 = m_{1,n+1}, \quad \alpha_i = m_{i-1,n+1} - m_{i,n+1}, \quad i = 2,3,\dots, n. \tag{42}$$

The results obtained so far can be summarized also in the following way.

Corollary. In the finite dimensional irreducible sl(1,n) module $W([m]_{n+1})$ one can choose as a basis the weight vectors

$$\left| \begin{matrix} m_{1,n+1}, & m_{2,n+2}, & \dots, & m_{n,n+1} \\ m_{1n}, & m_{2n}, & \dots, & m_{nn} \\ & \dots & \dots & \\ & m_{1i} & \dots\ m_{ii} & \\ & & \dots & \\ & & m_{11} & \end{matrix} \right\rangle \equiv \left| \begin{matrix} [m]_{n+1} \\ [m]_n \\ \vdots \\ [m]_i \\ \vdots \\ m_{11} \end{matrix} \right\rangle \tag{43}$$

where $m_{1,n+1},\dots,m_{n,n+1}$ are fixed complex numbers, whereas m_{ij}, $i\le j = 1,\dots, n$ label the basis vectors and take all possible values consistent with the conditions($\mathbb{Z}_+$ = all nonnegative integers):

1) $m_{i,n+1} - m_{i+1,n+1} \in \mathbb{Z}_+$;

2) $m_{i,n+1} - m_{i,n} = 0, 1, \dots, n-1$;

3) $|(m_{i,n+1} - m_{i,n}) - (m_{j,n+1} - m_{j,n})| \le 1$;

4) $\frac{1}{n-1} \sum_{i=1}^{n} (m_{i,n+1} - m_{i,n}) \in \mathbb{Z}_+$;

5) $m_{i,j+1} - m_{ij} \in \mathbb{Z}_+$, $m_{i,j} - m_{i+1,j} \in \mathbb{Z}_+$;

6) If $m_{k,n+1} = k-1$, $\sum_{i=1}^{n} (m_{i,n+1} - m_{i,n}) = (n-1)(m_{k,n+1} - m_{k,n})$

It is natural to call the above basis (43) a Gel'fand-Zetlin basis (GZ basis). The transformation of this basis under the action of the even subalgebra gl(n) is known (Ref. 11). It remains to determine how the GZ basis transforms under the action of the odd generators. For the LS's sl(1,2) and sl(1,3) this problem has been solved in Refs. 9 and 10.

References

1. Throughout the paper we use the following abbreviations and notations:
   fidermod(s) - finite-dimensional irreducible modules(s)
   LS,LSs - Lie superalgebra, Lie superalgebras
   LA,LAs-Lie algebra, Lie algebras
   [,] - product in the LS.
   $[m]_i = [m_{1i}, m_{2i} \ldots, m_{ii}]$, $i = 1,2,\ldots, n$
   $[m]_{n+1} = [m_{1,n+1}, m_{2,n+1}, \ldots, m_{n,n+1}]$
2. V. G. Kac, Adv. Math. 26, 8 (1977).
3. V. G. Kac, Lecture Notes in Math. 626, 597 (1978).
4. P. H. Dondi and P. D. Jarvis, Z. Phys. C4, 201 (1980); J. Phys. A14, 547 (1981).
5. A. B. Balantekin and I. Bars, J. Math. Phys. 22, 1149 (1981); 22, 1810 (1981).
6. A. B. Balantekin, and I. Bars, J. Math. Phys. 23, 1239 (1982); A. B. Balantekin, J. Math. Phys. 23, 486 (1982).
7. I. Bars, B. Morel and H. Ruegg, J. Math. Phys. 24, 2253 (1983); I. Bars and M. Günaydin, Comm. Math. Phys. 91, 31 (1983); F. Delduc and M. Gourdin, J. Math. Phys. 25, 1651 (1984); 26, 1865 (1985); I. Bars, Physica 15D, 42 (1985).
8. A. Pais and V. Rittenberg, J. Math. Phys. 16, 2062 (1975); M. Scheunert, W. Nahm and V. Rittenberg, J. Math. Phys. 18,155 (1977); M. Marcu, J. Math. Phys. 21, 1277 (1980); B. Gruber, T. S. Santhanam and R. Wilson, J. Math. Phys. 25, 1253 (1984).
9. A. H. Kamupingene and T. D. Palev, Trieste prepreint IC/85/146 (1985).
10. T. D. Palev, J. Math. Phys. 26, 1640 (1985); Triestse preprint IC/85/130 (1985).
11. I. M. Gel'fand and M. L. Zetlin, Doklady Akad. Nauk SSSR 71, 825 (1950); see also G. E. Baird and L. C. Biedenharn, J. Math. Phys. 4, 1449 (1963).
12. J. D. Louck, Am. J. Phys. 38, 18 (1970).
13. T. D. Palev and O. Ts. Stoytchev, C. R. Acad. Bulg. Sci. 35, 733 (1982).

THE USE OF SYMMETRY IN MOLECULAR AB-INITIO SCF CALCULATIONS

Michael Ramek

Institut für Physikalische und Theoretische Chemie
Technische Universität Graz, A-8010 Graz, Austria

Todays molecular ab-initio SCF calculations approximate Hartree-Fock solutions of the electronic Schrödinger equation by a linear combination ansatz[1-3] and the well known iterative process (which is given here for the closed shell case):

1. Define nuclear geometry, total charge, and multiplicity
2. Define basis set {f}
3. Calculate the one electron integrals:
   $S_{ij} = \int f_i f_j dV$, $H_{ij} = \int f_i (-\frac{1}{2}\nabla^2 - \sum_{k \in nuc} Z_k/r_k) f_j dV$
4. Calculate the two electron integrals:
   $(ij,kl) = \iint f_i(1) f_j(1) \frac{1}{r_{12}} f_k(2) f_l(2) dV_1 dV_2$
5. Assume an initial charge and bond order matrix P
6. Calculate transformation matrix to orthogonalize the basis set
7. Calculate Fock matrix according to
   $F_{ij} = H_{ij} + \sum_k \sum_l P_{kl} \left[(ij,kl) - \frac{1}{2}(ik,jl)\right]$
8. Transform F into orthogonal basis
9. Calculate eigenvalues E and eigenvectors U of F
10. Transform U back into original basis
11. Calculate the new charge and bond order matrix:
    $P_{ij} = 2 \sum_{k \in occ} U_{ki} U_{kj}$
12. If this new matrix P differs from the previous one more than a certain threshold, return to step 7
13. Calculate expectation value of electronic energy, population analysis, dipole moment, etc.
14. Eventually define a new geometry and return to step 3.

Two types of basis functions are currently used: exponential type functions (ETF) of the form

$$f = N\, r^{n} \exp(-br)\, Y_{lm}(\theta,\phi)$$

and Gauß-type functions (GTF) of the form

$$f = N\, x^{a} y^{b} z^{c} \exp(-dr^{2}).$$

Each of these two types has different advantages and disadvantages: ETF on the one hand are qualitatively "good" basis functions due to their similarity with the exact solutions for the hydrogen atom, but formulae for the necessary integrals are still in development;[4] GTF on the other hand are qualitatively "poor" basis functions but sophisticated formulae and algorithms for all occuring integrals are available.[5,6] Standard computer programs for ab-initio molecular orbital calculations[7-9] are therefore based on GTF, and a whole library of GTF basis sets has been developed.[10]

The minor quality of GTF, however, demands rather large basis sets, which lead to a very large number of two electron integrals. This large number of integrals and the fact, that their computation still is a time consuming process, limits routinely performed ab-initio calculations to a maximum number of basis functions between 80 and 130. (Programs capable of much larger basis sets[11,12] are exotic exceptions far away from routine.)

As a consequence, the most important use of molecular symmetry in ab-initio calculations is the reduction of the large number of two electron integrals. This is achieved by a strategy of calculating, storing, and processing only symmetry-unique and non-zero integrals,[13-18] which reduces the number of two electron integrals (and therefore also the computer time) often more than 50%. Three specific examples for the effect of this method, which is widely used today, are given below.

A related approach aiming at a use of local symmetries in over all unsymmetric or low symmetric molecules[19] has, however, not achieved general application so far.

A second use of molecular symmetry, which is independent of the use of symmetry in the integral calculation, is the symmetry adaptation of the basis set.[20-26] This adaptation can be combined with the transformation to the orthogonal basis to one single transformation, which also factorizes the Fock matrices. Computational advantage then arises from the fact, that factorized matrices can be diagonalized blockwize, and that this blockwize diagonalization can be performed quicker than the diagonalization of the whole matrix. The typical rate of computer time reduction by symmetry adapted basis sets in ab-initio calculations is 20% in the iterative part. In addition to this saving of computer time, numerical problems with almost degenerate molecular orbitals from different irreducible representations, which easily occur for example when the Givens-Householder matrix diagonalization routine[27] is used, can be avoided completely by symmetry adaptation of the basis. Sample calculations using this technique are also given below.

As examples "single-point" calculations were carried out with the Univac version of GAUSSIAN 80[9,28] for $H_3BOH_2BH_3$, $H_3BNH_2CH_3$, and FN_3 (at their 4-31G optimized geometries with point groups C_{2v}, C_s, and C_s, resp., given in tables 1 - 3 and shown in fig. 1), each of them with no use of symmetry at all, use of symmetry only in the integral calculation, and full use of symmetry. In all calculations the 4-31G* basis set[29-31] and atomic initial densities[32] were used; convergence criterion was a rms difference of the charge and bond order matrix less than 10^{-5} in two successive SCF

Table 1. Cartesian coordinates (Å) of the 4-31G equilibrium geometry of $H_3BOH_2BH_3$

Nucleus	x	y	z
O	0.000000	0.000000	0.000000
B	0.000000	0.000000	2.046738
B	1.880595	0.000000	-0.807773
H	-0.444029	0.792868	-0.292533
H	-0.444029	-0.792868	-0.292533
H	-1.187402	0.000000	2.150017
H	1.506866	0.000000	-1.939549
H	0.579405	1.023552	2.192112
H	0.579405	-1.023552	2.192112
H	2.242838	-1.023552	-0.332775
H	2.242838	1.023552	-0.332775

iterations. Dynamical damping[33] was used in the FN_3 calculation. The computer times necessary for these calculations are given in tables 4 - 6 and are also shown in fig. 2.

Table 2. Cartesian coordinates (Å) of the 4-31G equilibrium geometry of $H_3BNH_2CH_3$

Nucleus	x	y	z
N	0.000000	0.000000	0.000000
B	0.000000	0.000000	1.695661
C	1.364032	0.000000	-0.576975
H	-0.513813	0.811856	-0.286843
H	-0.513813	-0.811856	-0.286843
H	-1.166906	0.000000	2.000139
H	1.336790	0.000000	-1.657666
H	0.582717	1.010070	2.003999
H	0.582717	-1.010070	2.003999
H	1.882997	-0.874811	-0.222917
H	1.882997	0.874811	-0.222917

Table 3. Cartesian coordinates (Å) of the 4-31G equilibrium geometry of FN_3

Nucleus	x	y	z
N	0.000000	0.000000	0.000000
N	0.000000	0.000000	1.286510
N	-0.123828	0.000000	2.385091
F	1.418597	0.000000	-0.324225

Table 4. Computer times (CPU) and number of two electron integrals for a 4-31G* calculation of $H_3BOH_2BH_3$ with

a) no use of symmetry at all,
b) use of symmetry only in the integral calculation,
c) full use of symmetry.

	a)	b)	c)
Program unit:		Computer time:	
1,2	1.4 sec	1.4 sec	1.4 sec
3	11.6 sec	11.6 sec	11.6 sec
4	1459.0 sec	563.2 sec	563.2 sec
5	12.2 sec	12.2 sec	12.2 sec
6-13	1113.5 sec	725.6 sec	361.6 sec
Integrals:	1 053 880	343 181	343 181

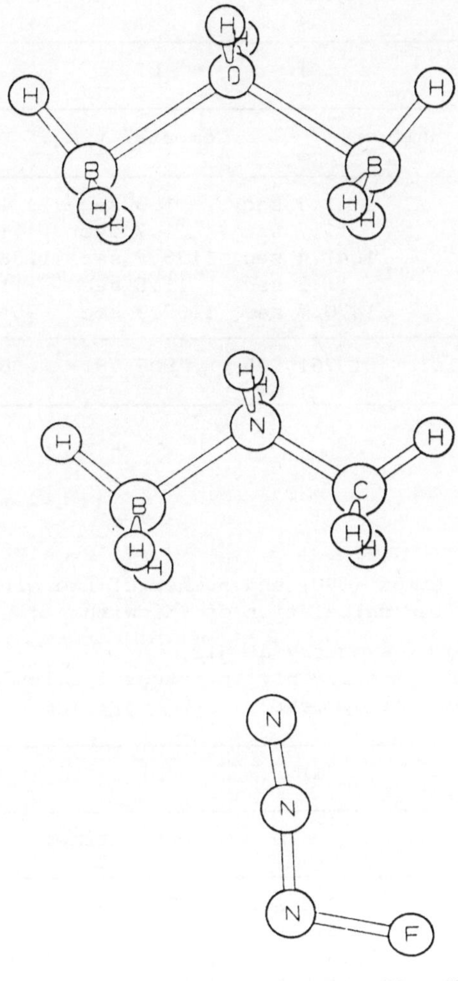

Fig. 1. 4-31G optimized geometries of $H_3BOH_2BH_3$, $H_3BNH_2CH_3$, and FN_3.

Table 5. Computer times (CPU) and number of two electron integrals for a 4-31G* calculation of $H_3BNH_2CH_3$ with

a) no use of symmetry at all,
b) use of symmetry only in the integral calculation,
c) full use of symmetry.

	a)	b)	c)
Program unit:		Computer time:	
1,2	1.3 sec	1.3 sec	1.3 sec
3	11.7 sec	11.7 sec	11.7 sec
4	1641.4 sec	1136.7 sec	1136.7 sec
5	11.5 sec	11.5 sec	11.5 sec
6-13	1470.7 sec	1167.9 sec	954.6 sec
Integrals:	1 261 295	806 781	806 781

Table 6. Computer times (CPU) and number of two electron integrals for a 4-31G* calculation of FN_3 with

a) no use of symmetry at all,
b) use of symmetry only in integral calculation,
c) full use of symmetry.

	a)	b)	c)
Program unit:		Computer time:	
1,2	0.5 sec	0.5 sec	0.5 sec
3	3.4 sec	3.4 sec	3.4 sec
4	1196.7 sec	1196.7 sec	1196.7 sec
5	11.6 sec	11.6 sec	11.6 sec
6-13	1547.6 sec	1547.6 sec	1283.4 sec
Integrals:	647 702	647 702	647 702

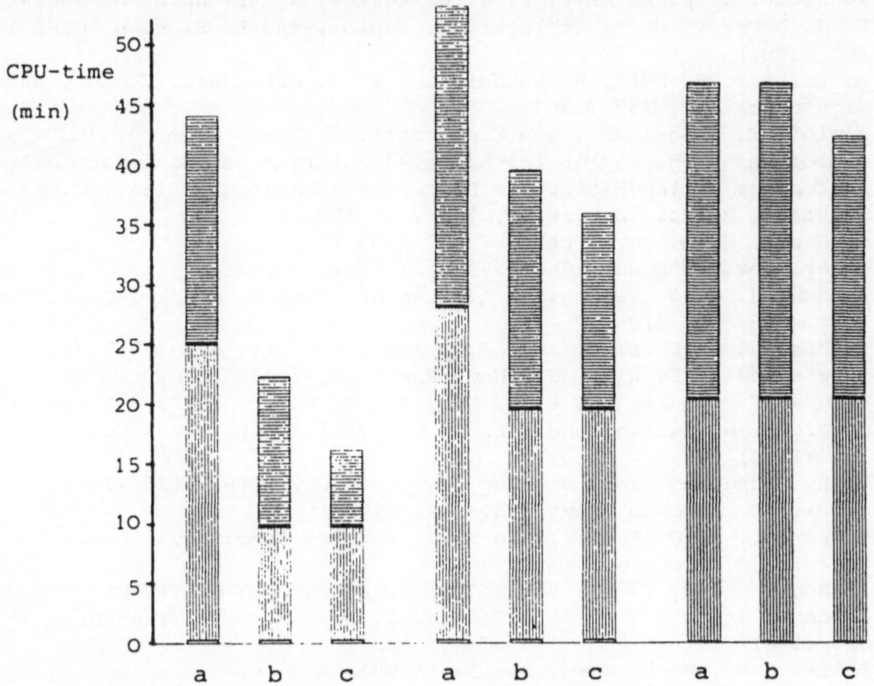

Fig. 2. Comparison of computer times (CPU) for 4-31G single point calculations of $H_3BOH_2BH_3$ (left), $H_3BNH_2CH_3$ (middle), and FN_3 (right) with

a) no use of symmetry at all,
b) use of symmetry only in integral calculation, and
c) full use of symmetry.

geometry and basis set definition, calculation of one electron integrals;

calculation of two electron integrals;

definition of initial charge and bond order matrix;

SCF iteration.

References

1. C. C. J. Roothaan, Rev. Mod. Phys. 23, 69 (1951)
2. J. A. Pople and R. K. Nesbet, J. Chem. Phys. 22, 571 (1954)
3. C. C. J. Roothaan, Rev. Mod. Phys. 32, 179 (1960)
4. E. O. Steinborn in: Methods in Computational Molecular Physics (eds.: G. H. F. Diercksen and S. Wilson), Reidel, Dordrecht, 1983, p. 37 ff
5. V. R. Saunders in: Methods in Computational Molecular Physics (eds.: G. H. F. Diercksen and S. Wilson), Reidel, Dordrecht, 1983, p. 1 ff
6. V. R. Saunders in: Computational Techniques in Quantum Chemistry and Molecular Physics (eds.: G. H. F. Diercksen, B. T. Sutcliffe, and A. Veillard), Reidel, Dordrecht, 1975, p. 347 ff
7. D. Goutier and R. Macauly, QCPE 11, 241 (1974)
8. M. Dupuis, J. Rys, and H. F. King, QCPE 12, 336 (1977)

9. J. A. Pople, J. S. Binkley, R. A. Whiteside, R. Krishnan, R. Seeger, D. J. DeFrees, H. B. Schlegel, S. Topiol, and L. R. Kahn, QCPE 13, 406 (1981)
10. R. A. Poirier, R. Kari, R. Daudel, and I. G. Csizmadia, J. Mol. Struct. (Theochem) 122, 259 (1985)
11. J. Almlöf, K. Faegri Jr., and K. Korsell, J. Comp. Chem. 3, 385 (1982)
12. J. Almlöf and P. R. Taylor in: Advanced Theories and Computational Approaches to the Electronic Structure of Molecules (ed.: C. E. Dykstra), Reidel, Dordrecht, 1984, p. 107 ff
13. P. D. Dacre, Chem. Phys. Lett. 7, 47 (1970)
14. M. Elder, Int. J. Quant. Chem. 7, 75 (1973)
15. L. J. Schaad, D. J. Wilson, B. A. Hess Jr., and P. Čársky, Chem. Phys. Lett. 105, 433 (1984)
16. M. Dupuis and H. F. King, Int. J. Quant. Chem. 11, 613 (1977)
17. M. Dupuis and H. F. King, J. Chem. Phys. 68, 3998 (1978)
18. T. Takada, M. Dupuis, and H. F. King, J. Chem. Phys. 75, 332 (1981)
19. E. Ortoleva, G. Castiglione, and E. Clementi, Comp. Phys. Comm. 19, 337 (1980)
20. A. L. H. Chung and G. L. Goodman, J. Chem. Phys. 56, 4125 (1972)
21. R. M. Pitzer, J. Chem. Phys. 58, 3111 (1973)
22. N. W. Winter, W. C. Ermler, and R. M. Pitzer, Chem. Phys. Lett. 19, 179 (1973)
23. P. S. Bagus and U. I. Wahlgren, Comput. & Chem. 1, 95 (1976)
24. B. Jordanov and W. J. Orville-Thomas, J. Mol. Struct. (Theochem) 76, 323 (1981)
25. L. Skâla, Comp. Phys. Comm. 24, 135 (1981)
26. M. Ramek and H. P. Fritzer, Comput. & Chem. 5, 79 (1981)
27. J. A. Pople and D. L. Beverdige, Approximate Molecular Orbital Theory, McGraw Hill, New York, 1970, p. 190 ff
28. M. Ramek, QCPE-bulletin 2, 77 (1982)
29. R. Ditchfield, W. J. Hehre, and J. A. Pople, J. Chem. Phys. 54, 724 (1971)
30. W. J. Hehre and J. A. Pople, J. Chem. Phys. 56, 4233 (1972)
31. P. C. Hariharan and J. A. Pople, Theoret. Chim. Acta 28, 213 (1973)
32. M. Ramek, QCPE 13, 503 (1985)
33. M. Ramek and H. P. Fritzer, Comput. & Chem. 6, 165 (1982)

"CLIFFORD ALGEBRAIC SYMMETRIES IN PHYSICS"

Nikos Salingaros

University of Texas at San Antonio

San Antonio, Texas 78285

ABSTRACT

This talk reviews some of the many appearances of Clifford algebras in Theoretical Physics. The full extent of the role of Clifford algebras is not easy to appreciate, since the various applications are disguised by an entirely distinct notation in each case. We propose that based on the almost universal application of the Clifford algebras, this is a mathematical scheme which is somehow intrinsic to the physical world.

INTRODUCTION

Much of the physical world appears to have an intrinsic algebraic structure, as is evident from the success of symmetry models in describing natural phenomena. We will present evidence in this talk that a large part of this intrinsic symmetry is in fact a Clifford algebraic symmetry. As there exist an infinite number of Clifford algebras, the possible applicability of the Clifford algebras is theoretically limitless. This last point contrasts with the attempts to employ division algebras, of which there are only four (if one requires alternativity and a unit) $\mathbb{R}$, ¢, $\mathbb{H}$ and ϕ. The series of division algebras ends with the octonions ϕ. On the other hand, the Clifford algebras contain the first three division algebras: the real numbers $\mathbb{R}$, the complex numbers ¢, and the quaternions $\mathbb{H}$, as well as many other useful algebras such as the Pauli algebra $\$_1$, the Majorana algebra $\mathbb{N}_3$, and the Dirac algebra $\$_2$. This includes also the larger algebras used in strings and supersymmetry.

Our claim here concerns the logical structure of theoretical physics, i.e., that the Clifford algebras are more than a tool; they are somehow intrinsic. This proposition cannot be rigorously proven, but may be addressed in the spirit of experimental mathematics. The claim is supported by the extraordinarily wide number of instances where a Clifford-algebraic description is the only possible description of a physical system or, in the presence of an alternate mathematical description, the

Clifford-algebraic one is at once more powerful, concise, and elegant. The evidence for this is contained in the various appearances of Clifford algebras in theoretical physics.

A historical justification for our claim is that many of the Clifford algebras were discovered by physicists. After $\mathbb{R}$ and $\mathbb{C}$, the quaternions $\mathbb{H} = \mathbb{N}_2$ were discovered by Hamilton (who was Professor of Astronomy in Dublin), and the dihedral Clifford algebra $\mathbb{N}_1$ by Hermite. Clifford discovered the biquaternions Ω_2, while Hamilton and later Pauli found the first spinor algebra $\$_1$ (isomorphic to the complex quaternions). Eddington, and independently Majorana, discovered the Clifford algebra $\mathbb{N}_3$, and Dirac the second spinor algebra $\$_2$. Soon after that time, the mathematicians Brauer and Weyl, simultaneously with the physicist Majorana, gave a general construction of all the Clifford algebras. This construction was later put in a more practical form by the author. For some historical remarks, see [1].

The classic applications of Clifford algebras in theoretical physics include the following: Explicit references, which are very numerous, are given in our review papers [2, 3].

1. Statistical Mechanics

The solution of the two-dimensional Ising model was intrinsically Clifford algebraic (Onsager). The algebraic method of solution was not well understood at the time and, as a result, many alternate methods have been given since (Lieb, Green). The papers by Kaufmann contain the most lucid presentation of the role of the Clifford algebras. In recent years, the underlying algebraic structure has given rise to new incentives for the study of more general Ising systems, and a possible key to the solution of the three-dimensional Ising model. (Miwa, Jimbo, Rasetti, Sato, Samuel). It should also be mentioned that a separate branch of mathematical research has grown out of this particular application, especially by the Kyoto school. (see Tracy and Palmer).

2. General Relativity

Gödel gave his general set of solutions to the Einstein equations which depended upon the Clifford algebra $\mathbb{N}_1$. This algebra is often confused with the ordinary quaternions, even though it is a distinct Clifford algebra which is not a division algebra. A recent discussion is given by Jantzen. See also the more mathematical work by Klainerman and Christodoulou on general Minkowskian metric spaces.

3. Quantum Electrodynamics

The Dirac equation is intrinsically Clifford-algebraic, as Dirac had to create a new Clifford algebra in order to factor the relativistic Hamiltonian. Most of the usual results on spin are representation-independent, and follow strictly from the algebraic structure. This is the thrust of the work of Sauter, Kähler, and Greider. (see [2]). There is an immediate connection through the Clifford algebra to gauge theories on a lattice (see below).

Perturbation theory in quantum electrodynamics is just the calculation of the invariants of the Clifford algebra. This point was emphasized by Caianiello and his school, who obtained a very elegant renormalization method. As a much simpler alternative to the usual gamma-matrix manipulations, the author, in parallel with Caianiello and Fubini, obtained sets of identities for the trace of products of Dirac spinors. (see also Chisholm)

4. Internal Symmetries

Schemes of classification of elementary particles which are more general than the usual su(3) to su(6) Lie algebraic symmetries have been studied for some time by Biedenharn, Barut, Basri, and Horwitz. At this time, the advantage of a Clifford algebraic particle classification scheme over the traditional ones based on Lie algebras is not overwhelming. Nevertheless, there is a high degree of optimism that the Clifford algebra can provide a much broader and more intrinsic model than the ones which have been traditionally used in particle physics. This point has been emphasized in particular by Wilczek, Zee, Mohapatra, Sakita, Kugo and Townsend. The classification symmetries tie in a remarkable way to the appearance of larger Clifford algebras in supersymmetry, supergravity, and the current theory of strings. For traditional supersymmetry using Clifford algebras, see Salam, Wess, Zumino, and Coquereaux. The supersymmetry implicit in the Dirac equation was shown by Iachello, and later by Gürsey. Slightly different work which is more explicitly algebraic includes the spinor theories of Budinich and his school, and related work by Horwitz and Biedenharn. Lepton-quark symmetries were formulated by Gatto and his school based upon Clifford algebras. Finally, the Clifford algebra foundation of modern superstring theories is clearly discussed in the work of Schwarz and Green. All of the above have an implicit, if not always evident, connection.

5. The vee Product

A breakthrough in the applications of Clifford algebras occurred with the creation of the Clifford algebra of differential forms by the mathematician Kähler. His work was unfortunately ignored and was later rederived by the author. Both independently introduced the vee product $\vee$ of differential forms. The importance of this representation, which was in line with Grassmann's original ideas, was in providing a geometrical basis for the Clifford algebras. As all physical fields have well-defined transformation properties and therefore tensor or spinor rank, there is a decided advantage in having an algebraic formalism which not only realizes the spinor structure, but also maintains the tensor structure as well. It is then possible to combine all the tools of differential forms developed by E. Cartan with the purely algebraic formalism of Clifford algebras.

Kähler himself formulated the Dirac equation and its solutions directly in terms of differential forms. This parallels early work done by Sauter, which was later extended by Greider. There exists an intrinsic connection with the theories of lattice gauge fields, and Kähler's work was appreciated quite recently by establishing a link from that direction (Becher, Joos, Dothan, Rabin, Göckeler, Mitra, Gliozzi). In this context, the author was able to settle a long-standing controversy over the inequivalence of the real forms of the spacetime metrics with signature (+++−) and (−−−+). It is impossible to have real spin one-half massive fermions in the second metric, thus contradicting an earlier claim by Hestenes.

6. Classical Electrodynamics

In addition to providing a tool for extending field theory in new directions, the Clifford algebra formalism has also been very fruitful in re-formulating old theories. This is particularly true with classical electromagnetic fields. Maxwell's equations assume a particularly simple form in the Clifford algebra, as was noticed by M. Riesz (see Teitler, Hestenes, and Greider). Going one step further, the author established the link between the Cauchy-Riemann and the Maxwell equations [4]. The mathematical problem of generalizing holomorphic fields from two to three and four dimensions was already solved by Maxwell. It is remarkable that a solution to such a fundamental mathematical problem should be provided by experimental physics.

It should be emphasized that the above is not a unique generalization (Stein and Weiss). An entirely different direction is pursued in the presently developing field of "Clifford Analysis" (Delanghe, Brackx, Lounesto, Ryan).

7. Motion of Charged Particles

Utilizing the intrinsic description of electromagnetic fields in terms of Clifford algebras, the author (following Hestenes and Taub) solved for the motion of a charged particle in an external electromagnetic field. The two published solutions for the relativistic particle in a constant homogeneous field were found to contain errors. Also, the accepted expression for the vectorial relativistic gyrofrequency was shown to be unphysical, and the correct one was derived. (see [3]).

8. The Lorentz Group

Even at this late stage, it is possible to give a new result in special relativity. The combination of two successive (nonparallel) Lorentz boosts is equivalent to a net boost along with a spatial rotation. The exact finite result, derived by the author [5], is apparently new. The finite rotation angle part was derived earlier by Hestenes and Ben-Menahem, but is not widely known. The total combination of net boost with rotation angle reduces to the Thomas precession in the infinitesimal case.

9. Conclusion

In the final analysis, what is the reason for the power of the Clifford-algebraic description? We believe it to reside in the ability of performing representation-free calculations which are generalizations of the traditional vector algebra. This considerable computational asset, in combination with the intrinsic symmetry, provides a practical framework for much of theoretical physics.

REFERENCES

1. N. Salingaros, "Some Remarks on the Algebra of Eddington's E Numbers" Found. Phys. 15 (1985), 683.

2. N. A. Salingaros, G. P. Wene, "The Clifford Algebra of Differential Forms," Acta Applicandae Mathematicae 4 (1985), 291.

3. N. Salingaros, "Clifford Algebras and the Vee Product," in "Proceedings of the XIV th International Colloquium on Group-Theoretical Methods in Physics," Seoul, Korea, 1985. Edited by B. H. Cho., World Scientific, Singapore, 1986.

4. N. Salingaros, "Electromagnetism and the Holomorphic Properties of Spacetime," J. Math. Phys. 22 (1981), 1919.

5. N. Salingaros, "The Lorentz Group and the Thomas Precession II. Exact Results for the Product of Two Boosts," J. Math. Phys. 27 (1986), 157.

SQUEEZED STATES AND QUADRATIC HAMILTONIANS:

A WIGNER DISTRIBUTION APPROACH

R. Simon

The Institute of Mathematical Sciences

Madras - 600 113, India

INTRODUCTION

For some time now there has been much interest in a special class of quantum states of the radiation field known as squeezed coherent states or simply squeezed states[1-5]. In fact the first experimental observation of squeezing has been reported very recently[6]. Squeezed states form a generalization of the coherent states in some sense, the latter being states which behave in the most possible classical way[7]. Whereas coherent states are minimum uncertainty states having equal fluctuations (zero-point fluctuations) in the two quadratures of the annihilation operator of the field mode, squeezed states have less fluctuation in one at the expense of increased fluctuation in the other quadrature in a manner consistent with the uncertainty principle. From an operator point of view squeezed states are produced from the coherent states by what is equivalent to the Bogolubov transformation. It is interesting to note that different authors discovered or rediscovered the squeezed states under different names - two photon coherent states[8], new coherent states[9], correlated coherent states[10], pulsating states[11] and twisted states[12] - emphasising different view points of what is essentially one and the same thing.

Squeezed states have many potential applications; we mention just two of them. In optical systems which use coherent laser beam (and such a beam is in a coherent state in the quantum mechanical sense) propagating in optical fibers, the ultimate limit to the noise level is set by the quantum noise represented by the zero-point fluctuation; in many practical systems this is already the dominant source of noise. Clearly, if the beam was in a squeezed state instead, and if information was transmitted in the phase quadrature with reduced fluctuation, then the effective quantum noise limit can be pushed substantially below the standard zero-point fluctuation[13]. Another application of the concept of squeezed states is in the detection of gravitational waves using the standard Weber bar detector[14]. Here too, the bar treated as a harmonic oscillator has zero-point fluctuation in its position whose magnitude is of the same order as the expected displacement of the bar by the gravitational radiation.

In all these applications the underlying object is a harmonic oscillator prepared in a state with minimum uncertainty, which then evolves under the action of a quadratic Hamiltonian. The basic features of all these cases become more transparent in the position representation of the oscillator system. In this representation the most general state with minimum fluctuation has a Gaussian wave function. So the present analysis is addressed to the unitary evolution of general Gaussian states (squeezed, and displaced in phase space) under the action of quadratic Hamiltonians[15,16]. The technique we use was developed in the context of problems involving a class of partially coherent light beams known as the Gaussian Schell-model beams acted on by first order optical systems[17]. The basic idea is to focus attention on the Wigner distribution of the state rather than the state function itself. This shift of focus reduces this problem to one of 2 x 2 real matrices and their transformations. Further, it leads to an elegant geometric picture where the Gaussian states are represented by vectors and the unitary evolution by Lorentz transformation in a (2+1) dimentional Minkowski space[17].

GAUSSIAN STATES AND THEIR WIGNER DISTRIBUTION

We consider a harmonic oscillator described by the Hamiltonian

$$\hat{H} = \omega\left(\frac{\hat{p}^2}{2m\omega} + \frac{1}{2}m\omega\hat{q}^2\right) \tag{1}$$

with position coordinate $\hat{q}$ and conjugate momentum $\hat{p}$:

$$[\hat{q}, \hat{p}] = i\hbar. \tag{2}$$

From the commutation relation (2) it follows that in every state of the oscillator the fluctuations in $\hat{q}$ and $\hat{p}$ are obliged to respect the inequality

$$(\Delta q)^2(\Delta p)^2 \geqslant \hbar^2/4 . \tag{3}$$

The states which minimize this inequality are given, in the position representation, by

$$\psi(q) = \frac{a^{1/2}}{\pi^{1/4}} \exp\left[-a^2(q-q_0)^2 + i\, p_0 q/\hbar\right]. \tag{4}$$

If $a = (m\omega/\hbar)^{1/2}$, we have the usual coherent states[7]. In all other cases we have squeezed coherent states - squeezed in position or in momentum depending on whether a is greater or less than $(m\omega/\hbar)^{1/2}$.

It will be shown in the sequel that the evolution of the centre of gravity in phase space, namely (q_o, p_o), can be handled independently of the structure around it; hence it suffices to consider the case with $q_o = p_o = 0$ as far as the behaviour of the fluctuations is concerned. Further when such a state evolves under a quadratic Hamiltonian it generally picks up a phase factor quadratic in q (phase curvature), and hence we include such a phase curvature described through a real parameter R and define as Gaussian states those states of the oscillator with wave functions of the following form:

$$\psi(q) = \frac{a^{1/2}}{\pi^{1/4}} \exp[-a^2q^2/2] \exp[iq^2/2R],$$

$$0 < a^2 < \infty, \qquad 0 < |R| \leq \infty . \tag{5}$$

Next we compute the Wigner distribution corresponding to this two-parameter family of Gaussian states. The Wigner distribution function of any state is related to its density operator $\hat{\rho}$ through[18]

$$W(q,p) = \frac{1}{(\pi\hbar)} \int dy \, \langle q - y | \hat{\rho} | q + y \rangle \exp[2ipy/\hbar] . \tag{6}$$

The Wigner distribution for our Gaussian states (5) can be easily calculated and can be written in a compact form using a two-element column vector

$$Q = \begin{pmatrix} q \\ p \end{pmatrix} . \tag{7}$$

We have,

$$W(q,p) = W(Q) = \frac{1}{\pi\hbar} \exp[-\frac{1}{\hbar} Q^T G Q] ,$$

$$G = \begin{pmatrix} \hbar a^2 + \dfrac{\hbar}{R^2 a^2} & -\dfrac{1}{Ra^2} \\ -\dfrac{1}{Ra^2} & \dfrac{1}{\hbar a^2} \end{pmatrix} . \tag{8}$$

We note the following properties of G:

$$G^T = G ,$$

$$G > 0 ,$$

$$\det G = 1 . \tag{9}$$

Thus, we see that there is a one-to-one correspondence between the two-parameter family of Gaussian states and the 2 x 2 real symmetric positive definite unimodular matrices.

It is useful to relate the fluctuations in $\hat{q}$ and $\hat{p}$ directly to the G-matrix elements. Since $\langle\hat{q}\rangle = \langle\hat{p}\rangle = 0$ for our Gaussian states (5), we have

$$(\Delta q)^2 = \int dq\, dp\; q^2\, W(q,p) = (\hbar/2)\, G_{22}\,,$$

$$(\Delta p)^2 = \int dq\, dp\; p^2\, W(q,p) = (\hbar/2)\, G_{11}\,. \tag{10}$$

Now the unimodularity of G implies that

$$(\Delta q)(\Delta p) = \hbar/2, \text{ if and only if } R=0. \tag{11}$$

That is, all those Gaussian states whose phase curvature is zero are minimum uncertainty states for the pair of variables $\hat{q}$, $\hat{p}$. This is to be expected for, while a phase curvature does not affect the width of the wave function in the q domain it definitely increases the width in the Fourier transform p domain.

UNITARY EVOLUTION AND SL(2,R) REPRESENTATION

As already noted we are interested in the unitary evolution of Gaussian states under quadratic Hamiltonians. Denoting such a Hamiltonian by $\hat{\Omega}$ and arranging $\hat{q}$, $\hat{p}$ as a two-element column, we have

$$\hat{\Omega} \text{(quadratic in } \hat{q}, \hat{p}\text{)}\,;$$

$$\hat{\Omega}^{\dagger} = \hat{\Omega}\,;$$

$$\hat{Q} = \begin{pmatrix} \hat{q} \\ \hat{p} \end{pmatrix}\,;$$

$$\exp(i\hat{\Omega}t/\hbar)\, Q\, \exp(-i\hat{\Omega}t/\hbar) = \exp(Jt)\hat{Q}$$
$$= S\hat{Q}\,;$$

$$S = \exp(Jt) \in Sp(2,R) \sim SL(2,R)\,;$$

$$J : \frac{-i}{\hbar}\,[\,Q\,,\hat{\Omega}] = J\hat{Q}. \tag{12}$$

We see that under the unitary evolution induced by $\hat{\Omega}$ $\hat{q}$, $\hat{p}$ undergo a linear canonical transformation given by the 2 x 2 matrix S whose generator is J. It is useful to have some examples (we choose scales such that $m\omega = 1$ numerically):

$$\hat{\Omega}_0 = \frac{\hat{p}^2}{2m}\,, \quad J_o = \begin{pmatrix} 0 & 1/m \\ 0 & 0 \end{pmatrix}\,, \quad S_o = \begin{pmatrix} 1 & t/m \\ 0 & 1 \end{pmatrix}\,;$$

$$\hat{\Omega}_1 = \omega\left(\frac{\hat{p}^2}{2} + \frac{\hat{q}^2}{2}\right)\,, \quad J_1 = \omega\begin{pmatrix} 0 & 1 \\ -1 & 0 \end{pmatrix}\,, \quad S_1 = \begin{pmatrix} \cos\omega t & \sin\omega t \\ -\sin\omega t & \cos\omega t \end{pmatrix}\,;$$

$$\hat{\Omega}_2 = \alpha\left(\frac{\hat{p}^2}{2} - \frac{\hat{q}^2}{2}\right)\,, \quad J_2 = \alpha\begin{pmatrix} 0 & 1 \\ 1 & 0 \end{pmatrix}\,, \quad S_2 = \begin{pmatrix} \cosh\alpha t & \sinh\alpha t \\ \sinh\alpha t & \cosh\alpha t \end{pmatrix}\,;$$

$$\hat{\Omega}_3 = \beta\left(\frac{\hat{q}\hat{p} + \hat{p}\hat{q}}{2}\right) \ , \ J_3 = \beta\begin{pmatrix} 1 & 0 \\ 0 & -1 \end{pmatrix}, \ S_3 = \begin{pmatrix} e^{\beta t} & 0 \\ 0 & e^{-\beta t} \end{pmatrix}. \quad (13)$$

We find that while the harmonic oscillator Hamiltonian $\hat{\Omega}_1$, produces just a rotation in the $\hat{q}, \hat{p}$ plane the maps produced by the anti-oscillator $\hat{\Omega}_2$ and the magnifier $\hat{\Omega}_3$ are non-periodic. We will have occation to come back to this result later. But now we examine the effect of the unitary evolution on the Wigner distribution.

From the definition of the Wigner distribution it can be shown that under the unitary evolution which induces a linear canonical transformation S on $\hat{Q}$ the phase space function $W(q,p) = W(Q)$ transforms in the following way[17]:

$$W(Q) \longrightarrow W'(Q) = W(S^{-1}Q) \ . \quad (14)$$

For our Gaussian states (8) this transformation takes an especially simple form:

$$W(Q) = \frac{1}{\pi\hbar} \exp\left[-\frac{1}{\hbar} Q^T G Q\right] \rightarrow \frac{1}{\pi\hbar} \exp\left[-\frac{1}{\hbar} Q^T G' Q\right],$$

$$G' = (S^{-1})^T G \, S^{-1}. \quad (15)$$

The properties (9) for G imply

$$G'^T = G' \ ,$$
$$G' > 0 \ ,$$
$$\det G = 1 \ ; \quad (16)$$

and hence it follows that the canonical transformation maps the set of Gaussian states onto itself in a one-to-one manner.

Now it is easy to show what was promised following (4). A Gaussian state whose centre of gravity is at the phase space point $(q^o, p^o) = Q^o$ can be represented by the pair (G, Q^o). The transformation of this pair readily follows from (14) and (15). We have

$$(G, Q^o) \longrightarrow (G', Q^{o\prime}) = ((S^{-1})^T G \, S^{-1} \ , \ SQ^o) \ , \quad (17)$$

showing that the centre of gravity follows the classical trajectory in phase space while the structure about the centre of gravity itself transforms in the same way as that of an undisplaced Gaussian state.

Equiped with these results we now focus attention on the Hamiltonian

$$\hat{H} = \omega\left(\frac{\hat{p}^2}{2} + \frac{\hat{q}^2}{2}\right) + \alpha\left(\frac{\hat{p}^2}{2} - \frac{\hat{q}^2}{2}\right) + \beta\,\frac{(pq + qp)}{2} \ . \quad (18)$$

This is the most general form homogeneous and quadratic in $\hat{q}, \hat{p}$. We will later relate this Hamiltonian to degenerate parametric amplifiers. The generator of the SL(2,R) transformation produced by this Hamiltonian is

$$J = \begin{pmatrix} \beta & \alpha+\omega \\ \alpha-\omega & -\beta \end{pmatrix} . \tag{19}$$

We combine ω, α, β into a new parameter Λ:

$$\Lambda = +(\omega^2 - \alpha^2 - \beta^2)^{1/2}. \tag{20}$$

Exponentiating J in (19) we get

$$S = \exp(Jt) = \begin{pmatrix} \cos\Lambda t + \frac{\beta}{\Lambda}\sin\Lambda t & \frac{\alpha+\omega}{\Lambda}\sin\Lambda t \\ \frac{\alpha-\omega}{\Lambda}\sin\Lambda t & \cos\Lambda t - \frac{\beta}{\Lambda}\sin\Lambda t \end{pmatrix} . \tag{21}$$

Now it is clear that if $(\alpha^2+\beta^2) < \omega^2$, then S is periodic with period $2\pi/\Lambda$; Q^o oscillates with period $2\pi/\Lambda$ whereas G whose transformation (17) is quadratic in S oscillates with period π/Λ . All the parameters of the state including the parameter of squeezing are periodic. On the other hand if the perturbation is strong enough that $(\alpha^2+\beta^2) > \omega^2$, then Λ becomes imaginary making the trignometric functions in (21) hyperbolic. Hence in this case the evolution of the state is aperiodic and in particular the squeezing increases as a monotone function of time. Using the relation

$$\operatorname*{Lt}_{\Lambda\to\infty} \frac{\sin\Lambda t}{\Lambda} = t , \tag{22}$$

it can be seen that the last statement is true also in the case $\alpha^2+\beta^2 = \omega^2$.

The detailed behaviour of the state parameters can be worked out using (21) in (17). But it turns out that this can be done in a more transparent way through a geometric picture which we develop presently.

SO(2,1) REPRESENTATION AND ANALYSIS

It is possible to picture the Gaussian states and their transformation in a (2+1) - dimensional space. To this end we define the generators

$$J_o = \frac{i\sigma_2}{2} ,$$
$$J_1 = \frac{\sigma_3}{2} ,$$
$$J_2 = \frac{\sigma_1}{2} . \tag{23}$$

They obey the commutation relations

$$[J_1, J_2] = J_o ,$$
$$[J_2, J_o] = -J_1,$$
$$[J_o, J_1] = -J_2 . \tag{24}$$

While the J's form a complete set for the SO(2,1) algebra, we have also

$$i\sigma_2(J_0, J_1, J_2) = \frac{1}{2}(\mathbb{1}, \sigma_1, -\sigma_3)$$
$$= \text{complete set of } 2 \times 2 \text{ real symmetric matrices.} \qquad (25)$$

Hence we can express our G matrix in the following way:

$$G = x^0\mathbb{1} + x^1\sigma_1 - x^2\sigma_3$$
$$= \begin{pmatrix} x^0 - x^2 & x^1 \\ x^1 & x^0 + x^2 \end{pmatrix} . \qquad (26)$$

The unimodularity of G reads

$$x_0^2 - x_1^2 - x_2^2 = 1 , \qquad (27)$$

while its positive definiteness reads

$$x_0 \geqslant 1 . \qquad (28)$$

Thus the Gaussian states are in one-to-one correspondence with points on the positive branch of the hyperboloid (27); and the action of the Hamiltonian (18) becomes SO(2,1) Lorentz rotation about the axis $(x^0, x^1, x^2) = (\omega, \beta, \alpha)$. That is, writing the expansion coefficients of G in (26) as a column

$$x = \begin{pmatrix} x^0 \\ x^1 \\ x^2 \end{pmatrix} , \qquad (29)$$

we have

$$G \rightarrow (S^{-1})^T G S^{-1} \Rightarrow x \rightarrow x' = A(S)x ;$$

$$A(S)^T g A(S) = g , \quad g = \begin{pmatrix} 1 & 0 & 0 \\ 0 & -1 & 0 \\ 0 & 0 & -1 \end{pmatrix} . \qquad (30)$$

The mapping $S \in SL(2,R) \longrightarrow A(S) \in SO(2,1)$ is two-to-one. It is easy to work out the elements of A(S) in terms of the elements of S[19,17].

Since Lorentz rotation maps the hyperboloid (27) onto itself in an invertible way, it becomes clear once again that under general quadratic Hamiltonians (no restriction on the relative magnitudes of ω, α, β) the two-parameter family of Gaussian states is mapped onto itself in an one-to-one way.

Further insight can be gained by analysing the orbits of points on the hyperboloid (the Gaussian states) under the action of a given quadratic Hamiltonian. For instance, if the Hamiltonian is that of a free harmonic oscillator, $\alpha = \beta = 0$ and the Lorentz rotation in this case becomes "spatial" rotation in the x^1x^2 plane; the orbits are parallel circles normal to the x^0 axis formed by intersections of planes orthogonal to the x^0 axis with the hyperboloid. The vector x which remains invariant under this rotation is

$$x = \begin{pmatrix} 1 \\ 0 \\ 0 \end{pmatrix} , \tag{31}$$

which corresponds to

$$G = \begin{pmatrix} 1 & 0 \\ 0 & 1 \end{pmatrix} . \tag{32}$$

Comparing (32) with (8) we see that this eigen x corresponds to $\hbar a^2 = 1$; reinstating m this becomes

$$a = (m\omega/\hbar)^{1/2} , \tag{33}$$

which is the familiar result for the width of the ground state eigen function of a harmonic oscillator.

On the other hand if the perturbation terms α, β were present we have to take the intersection of the hyperboloid with planes Lorentz-orthogonal to the direction (ω, β, α). Now two cases are to be distinguished:

Case 1:

The vector (ω, β, α) is time-like $\omega^2 - \beta^2 - \alpha^2 > 0$.

In this case the planes orthogonal to (ω, β, α) definitely cut the hyperboloids along closed curves. These planes are parallel to the plane which is tangential to the hyperboloid at the point where the vector (ω, β, α) when extended cuts the hyperboloid. Every Gaussian state belongs to one unique member of this family of closed curves, and evolves along this curve with a frequency Λ/π, where Λ is the "norm" of the vector (ω, β, α). Clearly, the eigen state of this Hamiltonian corresponds to the vector $x = (\omega, \beta, \alpha)^T$ or, equivalently, the G matrix

$$G = \frac{1}{(\omega^2 - \beta^2 - \alpha^2)^{1/2}} \begin{pmatrix} \omega - \beta & \alpha \\ \alpha & \omega + \beta \end{pmatrix} . \tag{34}$$

We find that for Hamiltonians with $\alpha \neq 0$, the eigenstate is not a minimum uncertainty state for the pair $\hat{q}, \hat{p}$.

Case 2:

The vector (ω, β, α) is space-like $\omega^2 - \beta^2 - \alpha^2 < 0$.

In this case the vector (ω, β, α) when extended never meets the hyperboloid. The planes Lorentz-orthogonal to this vector cut the hyperboloid along open curves. As a result every Gaussian state undergoes a non-periodic evolution under this Hamiltonian; there is no Gaussian state which is an eigenstate of this Hamiltonian. In fact it is possible to make a stronger statement. From (17) it follows that no coherent state comes back to itself on evolution under this Hamiltonian. Hence by virtue of the completeness (overcompleteness) of the coherent states we conclude not only that this Hamiltonian has no eigen states but also that no state is periodic under this Hamiltonian.

EXAMPLE: DEGENERATE PARAMETRIC AMPLIFIER

As a simple illustrative example of the above formalism we show how detuning affects squeezing in an ideal degenerate parametric amplifier[20,21]. This system can be modelled by the Hamiltonian

$$\hat{H} = \hbar \omega_0 \hat{a}^\dagger \hat{a} + \hbar[g\hat{a}^2 \exp(i\omega' t) + g^* \hat{a}^{\dagger 2} \exp(-i\omega' t)], \tag{35}$$

where ω_0 is the frequency of the degenerate signal/idler mode, ω' is the frequency of the classical pump field, g is the product of the amplitude of the pump field and the coupling constant between the pump and the signal modes, and $\hat{a}$ is the annihilation operator for the signal mode. Introduce a new operator $\hat{A}$ through

$$\hat{A} = \exp(i\omega' t/2)\hat{a}. \tag{36}$$

This amounts to going over to a coordinate system rotating about the x^0 axis with frequency $\omega'/2$. The Hamiltonian for this new operator is easily seen to be

$$\hat{H}' = \hbar(\omega_0 - \omega'/2)\hat{A}^\dagger \hat{A} + \hbar(g\hat{A}^2 + g^* \hat{A}^{\dagger 2}). \tag{37}$$

We can introduce two hermitian operators $\hat{q}, \hat{p}$ and write the non-hermitian operator $\hat{a}$ as

$$\hat{a} = \frac{1}{(2\hbar)^{1/2}} (\hat{q} + i\hat{p}). \tag{38}$$

Now, Hamiltonian (37) can be rewritten in terms of the hermitian operators $\hat{q}, \hat{p}$ and real parameters g_1, g_2 where $g = g_1 + ig_2$:

$$H' = (\omega_0 - \omega'/2)\left(\frac{\hat{p}^2}{2} + \frac{\hat{q}^2}{2}\right) - 2g_1\left(\frac{\hat{p}^2}{2} - \frac{\hat{q}^2}{2}\right) - 2g_2\left(\frac{\hat{p}\hat{q} + \hat{q}\hat{p}}{2}\right). \quad (39)$$

The Hamiltonian (39) is of the same form as (18) with

$$\omega = \omega_0 - \omega'/2,$$
$$\alpha = -2g_1,$$
$$\beta = -2g_2. \quad (40)$$

From (20) we see that if the detuning $|\omega_0 - \omega'/2|$ is small enough so that

$$|\omega_0 - \omega'/2| \leq 2|g|, \quad (41)$$

then the squeezing increases as a monotone function of time, whereas if the detuning is large compared to the strength of the interaction such that

$$|\omega_0 - \omega'/2| > 2|g|, \quad (42)$$

then the evolution is periodic, consistent with the results of Ref.[20].

CONCLUDING REMARKS

We have analysed the evolution of Gaussian states of a harmonic oscillator under the action of general quadratic Hamiltonians. The method of Wigner distribution functions leads to a simple geometric way of picturing this evolution. Since rotation about the x^0 axis simply amounts to including a c-number phase factor in the annihilation operator, all Gaussian states related to one another through such a rotation have the same amount of squeezing. Thus, for Gaussian states squeezing is a montone increasing function of x^0.

This formalism can be easily applied to thermal states. The Wigner distribution function of a thermal state has a form similar to (8):

$$W(Q) = \frac{[\det G]^{1/2}}{\pi\hbar} \exp\left[-\frac{1}{\hbar} Q^T G Q\right]. \quad (43)$$

G is again symmetric and positive definite, but its determinant obeys

$$0 < \det G \leq 1. \quad (44)$$

At zero temperature the thermal state degenerates into the ground state with det G=1. As temperature increases the value of det G decreases and asymptotically approaches zero as the temperature becomes arbitrarily large.

This analysis for a single mode system can be extended to a two-mode system. In this case G becomes a 4 x 4 matrix, and the quadratic Hamiltonians are now represented by Sp(4,R) matrices. Again it is possible to develop a geometric picture with the Gaussian states represented as second rank anti-symmetric tensors in a (3+2) - dimensional space where the unitary evolution becomes SO(3,2) de Sitter transformations[22].

It is a pleasure to thank Profs.N.Mukunda and E.C.G.Sudarshan for illuminating discussions.

REFERENCES

1. D.F.Walls, Squeezed states of light, Nature 306: 141 (1983).

2. M.M.Nieto, What are squeezed states like? (Los Alamos Preprint).

3. H.Takahasi, Information theory of quantum-mechanical channels, Ad.Commun. Syst. 1: 227 (1965).

4. D.Stoler, Equivalence classes of minimum uncertainty packets, Phys.Rev.D 1: 3217 (1970).

5. J.N.Hollenhorst, Quantum limits on resonant-mass gravitational-radiation detectors, Phys.Rev.D 19: 1669 (1979).
T.S.Santhanam, Generalized coherent states, in "Symmetries in Science", B.Gruber and R.S.Millman, eds., Plenum: New York (1980).
M.Venkata Satyanarayana, Generalized coherent states and generalized squeezed coherent states, Phys.Rev.D 132: 400 (1985).
J.R.Klauder, These proceedings.

6. R.E.Slusher, L.W.Hollberg, B.Yurke, J.C.Mertz, and J.F.Valley, Observation of squeezed states generated by four-wave mixing in an optical cavity, Phys.Rev.Lett. 55: 2409 (1985).

7. J.R.Klauder and E.C.G.Sudarshan, "Fundamentals of Quantum Optics", Benjamin, New York (1968).

8. H.P.Yuen, Two photon coherent states of the radiation field, Phys.Rev.A 13: 2226 (1976).

9. A.K.Rajagopal and J.T.Marshal, New coherent states with applications to time-dependent systems, Phys.Rev.A. 26: 2977 (1982).

10. V.V.Dodonov, E.V.Kurmyshev, and V.I.Man'ko, Generalized uncertainty relations and correlated coherent states, Phys.Lett. 79A: 150 (1980).

11. I.Fugiwara and K.Miyoshi, Pulsating states for quantal harmonic oscillator, Prog.Theor.Phys. 64: 715 (1980).

12. H.P.Yuen, Contractive states and the standard quantum limits for monitoring free-mass positions, Phys.Rev.Lett. 51: 719 (1983).

13. H.P.Yuen and J.H.Shapro, Optical communication with two photon coherent states - Part I: Quantum state propagation and quantum noise reduction, IEEE Trans. Inform. Theory IT-24: 657 (1978).
J.H.Shapiro, H.P.Yuen, and Jesus A.Machado Mata, Optical Communication with two-photon coherent states-Part II: Photoemissive detection and structured receiver performance, IEEE Trans. Inform. Theory IT-25:179(1979).

14. C.M.Caves, K.S.Thorne, R.W.P.Drever, V.D.Sandberg, and M.Zimmermann, On the measurement of weak classical force coupled to a quantum-mechanical oscillator I.Issues of principle, Rev.Mod.Phys. 52: 341 (1980).

15. P.Broadbridge and C.A.Hurst, Canonical forms for quadratic Hamiltonians, Physica 108A: 39 (1981).
P.Broadbridge and C.A.Hurst, Existence of complex structure for quadratic Hamiltonians, Ann.Phys. 131: 104 (1981).

16. M.Moshinsky and T.H.Seligman, Canonical transformations to action and angle variables and their representations in quantum mechanics, Ann.Phys. 114: 243 (1978).

17. R.Simon, E.C.G.Sudarshan, and N.Mukunda, Generalized rays in first-order optics: Transformation properties of Gaussian Schell-model fields, Phys.Rev.A 29: 3273 (1984).

18. M.Hillery, R.F.O'Connell, M.O.Scully, and E.P.Wigner, Distribution functions in Physics: Fundamentals, Phys.Rep. 106: 121 (1984).

19. N.Mukunda, A.P.Balachandran, Jan S.Nilsson, E.C.G.Sudarshan, and F.Zaccaria, Evolution, symmetry and canonical structure in dynamics, Phys.Rev.D 23: 2189 (1981).

20. H.J.Carmichael, G.J.Milburn and D.F.Walls, Squeezing in a detuned parametric amplifier, J.Phys.A 17: 469 (1984).

21. M.T.Raiford, Degenerate parametric amplification with time-dependent pump amplitude and phase Phys.Rev.A 9: 2060 (1974).

22. R.Simon, E.C.G.Sudarshan, and N.Mukunda, Anisotropic Gaussian Scheel-model beams: Passage through first order optical systems and associated invariants, Phys.Rev.A 31: 2419 (1985).
G.J.Milburn, Multimode minimum uncertainty squeezed states, J.Phys.A 17: 737 (1984).

LIGHT AND GROUP THEORY*

E. C. G. Sudarshan

Center for Particle Theory
University of Texas
Austin, Texas 78712

1. RADIATIVE TRANSFER AND ELECTRODYNAMICS

Light was recognized as the subtlest kind of matter quite a long time ago. Light has dominated human experience both in the natural form as well as when artifically produced. From the study of the propogation of light it was learned that light propogates rectilinearly.

When light is received from a great distance we find that it comes in the form of a beam but for artificially produced light we always see it as a pencil, originating at the source considered as a point: more generally, for an extended source we have a collection of pencils with vertices at the source. In both these cases it is as if the pencil (or beam) is composed of a multitude of individual rays each propogating rectilinearly. In practice, however, we never see individual rays; rather, we can think of arbitrarily narrow pencils which subtend smaller and smaller solid angles at the vertex.

The transmission of light energy by such complexes of pencils has been the subject of study in radiative transfer. In such studies we attempt to treat light as a collection of pencils, whose rays propogate rectilinearly. Since we can consider arbitrarily narrow pencils the radiant energy transfer can be reduced to integrodifferential equations.

We must expect substantial modification of the whole framework of radiative transfer when quantum ideas are invoked. Light has the attribute of polarization, which cannot be satisfactorily taken account of in conventional radiative transfer. Further since quantum theory of light demands that the wave aspect is essential, arbitrarily narrow pencils cannot be entertained; diffraction effects would then prevent rectilinear propogation.

In the context of quantum electrodynamics we have still further elaborations to be made. The field strengths now become quantum operators. However, that the first fundamental theorem of quantum optics asserts that every quantum illumination can be expressed as an ensemble of coherent wave fields with a possibly singular ensemble density. Each coherent field can

*Talk presented at the Symmetries in Science II Conference held at Southern Illinois University, Carbondale, March 23-26, 1986

for many purposes be thought of as a classical wavefield. So at least as far as lossless transmission in passive media is considered, the full quantum electrodynamic behaviour can be described as if we have an ensemble of classical wave fields. We shall rely on this fundamental theorem and treat optics as if it is classical wave optics with ensembles of wavefields as representative illuminations.

Light waves are transverse and therefore a full definition of the illumination involves the specification of polarization.

2. POLARIZED LIGHT AND TRANSFORMATIONS

When a parallel thin beam of light or a narrow paraxial pencil of light of a definite color is considered it has the additional quality of polarization. The polarization is transverse to the axis of propogation. We recognize two linearly independent states of polarization which may be chosen to be two linear orthogonal polarizations. The general state of polarization of a single wave would be a complex linear combination of these two. If the intensity of light is normalized the polarization state in general is a normalized complex combination; it therefore requires two complex numbers the sum of the absolute value being unity:

$$\underline{e} = \alpha_1 \underline{e}_1 + \alpha_2 \underline{e}_2 \; ; \qquad |\alpha_1|^2 + |\alpha_2|^2 = 1.$$

Since measurable quantities are bilinear in the field, the overall phase of $\underline{e}$ is irrelevant (provided we have only one beam!) we may recognize that three real numbers the sum of whose squares is unity define the polarization state. For a pure state the polarization is thus specified not by a unimodular vector in a complex two-dimensional space but by a ray which is the equivalence class of all complex 2-vectors which differ from each other by a phase.

The special cases $\alpha_2/\alpha_1 = \pm i$ are circular polarizations. If α_2/α_1 is real we get a linear polarization. Both these are special cases of the general elliptic polarization.

In a general statistical light beam the polarization may not be complete and has to be represented by a polarization matrix ρ, which in turn is completely defined by the Stokes' parameters:

$$\underline{\Sigma} = \mathrm{tr}(\rho \underline{\sigma})$$

where σ_1, σ_2, σ_3 are 2x2 Pauli matrices. From the nonnegativity of ρ and its normalization

$$\mathrm{tr}(\rho) = 1 \; ; \qquad \Sigma^2 \leq 1$$

with the equality being realized only for pure states with idempotent density matrices.

Consider the action of phase plates and optically active media on a beam of light and the manner in which the density matrix is transformed. To the extent that the propogation is lossless the set of transformations constitute the unitary unimodular group SU(2). Any such action can always be synthesized in terms of three preassigned kinds of objects using the familiar decomposition of a general rotation in 3-dimensions to be products of rotations respectively along z-axis, y-axis and z-axis. We may choose to synthesize any combination of elements affecting polarizations by three elements, respectively a phase plate, an optically active medium and a phase plate:

$$A = \begin{pmatrix} e^{i\theta/2} & 0 \\ 0 & e^{-i\theta/2} \end{pmatrix} \begin{pmatrix} \cos\phi & -\sin\phi \\ \sin\phi & \cos\phi \end{pmatrix} \begin{pmatrix} e^{i\psi/2} & 0 \\ 0 & e^{-i\psi/2} \end{pmatrix}$$

In writing down this we have used a basis of orthogonal linear polarizations and absorbed overall phases in the representative of phase plates. The associated transformation of the density matrix is:

$$\rho \to A\rho A^{\dagger} = \rho' \quad ; \quad tr(\rho') = tr(\rho).$$

When absorption is involved like in a polarizer the matrix A is no longer unimodular:

$$\rho \to A\rho A^{\dagger} = \rho' \quad ; \quad tr(\rho') < tr(\rho).$$

More generally any linear map of $\rho \to \rho'$ which preserves the positivity is an acceptable map. If this is accomplished by beam splitting and recombination the map would belong to the subclass of "completely positive maps" with a standard form:

$$\rho \to \rho' = \sum_n A_n \rho A_n^{\dagger} \quad ; \quad \sum_n A_n^{\dagger} A_n - 1 \leq 0.$$

For the lossless case

$$\sum_n A_n^{\dagger} A_n = 1.$$

Given any two maps

$$\rho \to \sum_n A_n \rho A_n^{\dagger} \quad , \quad \rho \to \sum_m B_m \rho B_m^{\dagger}$$

we could form the convex combination

$$\rho \to \cos^2 \alpha \Sigma A_n \rho A_n^{\dagger} + \sin^2 \alpha \Sigma B_m \rho B_m^{\dagger}$$

and this could be rewritten in the form

$$\rho \to \sum_{\ell} C_{\ell} \rho C_{\ell}^{\dagger}$$

A map which cannot be expressed in this form is called extremal. The unitary maps

$$\rho \to A\rho A^{\dagger} \quad , \quad A^{\dagger}A = 1$$

are extremal. But we can also construct an extremal family of the form

$$\rho \to A_1 C\rho C A_1^{\dagger} + A_2 S\rho S A_2^{\dagger}$$

where A_1, A_2 are unitary and C, S are real matrices of the form

$$C = \begin{pmatrix} \cos\theta & 0 \\ 0 & 1 \end{pmatrix} \quad ; \quad S = \begin{pmatrix} \sin\theta & 0 \\ 0 & 0 \end{pmatrix} \quad .$$

These two families exhaust the set of extremal completely positive maps. The antiunitary map

$$\rho \to A\rho^T A^\dagger \quad , \quad A^\dagger A = 1$$

is extremal but not completely positive and cannot be realized in terms of beam splitting and recombinations. The set of extremal not completely positive maps have been determined elsewhere.

3. RELATIVISTIC PARAXIAL PROPOGATION AND GROUPS

When we take account of relativity the light quantum appears as a realization of the Poincaré group with lightlike momenta. The irreducible realizations correspond to light quanta with positive or negative helicity. Thus circular polarizations are more natural in the context of relativistic transformations. Under any Lorentz transformation a left circularly polarized photon of a definite momentum goes into a photon of the same helicity and a new momentum with a phase factor which depends on the particular Lorentz transformation. A right circularly polarized photon would do the same but with the opposite phase.

If we temporarily neglect the polarization property and deal with scalar optics we could transform the scalar wave equation

$$\left(\frac{\partial^2}{\partial t^2} - \frac{\partial^2}{\partial x^2} - \frac{\partial^2}{\partial y^2} - \frac{\partial^2}{\partial z^2} \right) \phi\ (xyz) = 0$$

into the form

$$\left(\frac{\partial^2}{\partial\tau\partial\sigma} - \frac{\partial^2}{\partial x^2} - \frac{\partial^2}{\partial y^2} \right) \phi = 0$$

by introducing the light cone coordinates

$$\sigma = \frac{1}{2}\ (t-z) \quad , \quad \tau = \frac{1}{2}\ (t+z).$$

The elementary wave solution propogating along the z-axis has the alternate expressions

$$\exp\ \{i(k_3 z - \omega t)\} = \exp\ \{-i(\omega + k_3)\sigma - i(\omega - k_3)\tau\}$$

In the paraxial approximation $k_3 \approx \omega$ and for a pencil with small line width

we may replace $\omega + k$ by a quantity 2M; when $\omega + k_3$ is constant we refer to it as being henochromatic, an idealization that we shall systematically use in paraxial optics. The coefficient of $-i\tau$ is

$$\omega - k_3 \approx \frac{k_1^2 + k_2^2}{2M} \quad ; \quad k_1^2 + k_2^2 \ < < M^2\ .$$

Thus the differential equation for scalar wave propogation in the paraxial henochromatic idealization becomes

$$i\frac{\partial}{\partial\tau}\psi(x,y,\tau) + -\frac{1}{2M}\ (\frac{\partial}{\partial x^2} + \frac{\partial}{\partial y^2})\ \psi(x,y,\tau)$$

reminiscent of the nonrelativistic Schrodinger equation with the henochromatic frequency M taking the place of the nonrelativistic mass.

The dynamical variables acting on the scalar wave amplitude are composed of $-i\frac{\partial}{\partial x}$, $-i\frac{\partial}{\partial y}$, x and y. The propogation "Hamiltonian" is quadratic in the momenta

$$\mathcal{H} = \frac{1}{2m}(p_1^2 + p_2^2);$$

$$p_1 = -i\frac{\partial}{\partial x} \quad , \quad p_2 = -i\frac{\partial}{\partial y} \quad .$$

We may consider the three bilinear quantities

$$\mathcal{H} = \frac{1}{2}(p_1^2 + p_2^2) \ ,$$

$$\mathcal{M} = (p_1 q_1 + q_2 p_2)$$

$$\mathcal{L} = \frac{1}{2}(q_1^2 + q_2^2)$$

which form the O(2,1) Lie algebra and the ten bilinear quantities

$$\mathcal{H} = \frac{1}{2}(p_1^2 + p_2^2), \ \mathcal{K}_1 = \frac{1}{2}(p_1^2 - p_2^2), \ K_2 = p_1 p_2$$

$$\mathcal{M} = (q_1 p_1 + p_2 q_2), \ \mathcal{N} = (q_1 p_1 - q_2 p_2)$$

$$\mathcal{J} = q_1 p_2 - q_2 p_1 \qquad \mathcal{N}_2 = (q_1 p_2 + p_2 q_1)$$

$$\mathcal{L} = \frac{1}{2}(q_1^2 + q_2^2) \qquad \mathcal{R}_1 = \frac{1}{2}(q^2 - q_2^2), \ \mathcal{R}_2 = q_1 q_2$$

constitute the O(3,2) Lie algebra. We shall see that the corresponding groups do play an important role in paraxial wave propogation in axial optical systems. In passing we also note that if the quantities q_1, q_2, p_1, p_2, 1 may be added to the Lie algebras to close them into their inhomegeneous semidirect product extensions. We shall see their relevance in the context of paraxial wave propogation in axial optical systems.

The inclusion of polarizations involve a nontrivial modification. If the scalar wave amplitude ϕ is replaced by a vector wave amplitude $\underline{V}$ (which may be taken to be the transverse divergence-free part of the vector potential) it is clear that multiplication by the transverse coordinates would take it outside the class of transverse vectors. Such an operator is therefore not defined but we must have a replacement for it so that we can deal with lens systems (see below). For this purpose one notes that the stabilizer group, that is the subgroup of the Lorentz group which leaves the light-like vector (1,0,0,1) is the Euclidean group E(2) made up of

$$G_1 = K_1 - J_2$$

$$G_2 = K_2 + J_1$$

$$J_3 \ .$$

In the scalar case

$$G_1 = (\omega + k_3)\, x_1 = 2M\, x_1$$

$$G_2 = (\omega + k_3)\, x_2 = 2M\, x_2$$

$$J_3 = 0.$$

Therefore

$$x_1 = \frac{1}{2M} G_1 \quad ; \quad x_2 = \frac{1}{2M} G_2 .$$

But both in the spinless and in the spinning case G_1 and G_2 commute; we may therefore define as transverse coordinates

$$q_1 = \frac{1}{2M} (K_1 - J_2) = x_1 + \frac{1}{M} g_1 \ ;$$

$$q_2 = \frac{1}{2M} (K_2 + J_1) = x_2 + \frac{1}{M} g_2 \ .$$

These augmentations of x_1 and x_2 preserve the transverse nature of the vector field $\underline{V}$ and contribute precisely the small longitudinal components:

$$g_1 = \begin{pmatrix} 0 & 0 & 0 \\ 0 & 0 & 0 \\ 1 & 0 & 0 \end{pmatrix} \quad ; \quad g_2 = \begin{pmatrix} 0 & 0 & 0 \\ 1 & 0 & 0 \\ 0 & 1 & 0 \end{pmatrix}$$

As the frequency becomes higher the importance of these augmentations becomes less.

In either case the transverse coordinates are obtained from the generators of the stabilizer group.

4. GROUP THEORY OF AXIAL OPTICAL SYSTEMS

The passage of light through a system of optical elements like lenses, optically active media, or even free space displays realizations of the groups SO(2,1) and SO(3,2) and their inhomogeneous extensions. On encountering a thin spherical lens the geometrical thickness at any point (x_1, x_2) is of the form

$$\delta - \left(\frac{1}{2R_1} + \frac{1}{2R_2}\right) (x_1^2 + x_2^2)$$

and hence the wavefront suffers a phase shift

$$(n - 1)\delta - \frac{1}{2f} (x_1^2 + x_2^2)$$

where n is the refractive index of the lens medium and

$$f = (n - 1) \left(\frac{1}{R_1} + \frac{1}{R_2}\right)$$

is the focal length of the lens. The constant term $(n - 1)\delta$ may be omitted without any loss of generality; for simplicity we have assumed that the lens is placed normal to the axis and centered. On the other hand the free particle propogation through a distance D is given by

$$D \quad (p_1^2 + p_2^2) \; / \; 2M.$$

These may therefore be identified with elements on the one-parameter subgroups generated by $\mathcal{L}$ and by $\mathcal{H}$. Since $\mathcal{L}$, $\mathcal{H}$ and $\mathcal{M}$ Lie algebra of SO(2.1) it follows that an exial system made up of a sucession of any number of spherical lenses separated b y arbitrary intervals can be identified with an element of the Lorentz group SO(2.1).

A more interesting question connected with the group properties of the axially symmetric system is that of synthesis: Given an arbitrary element of SO(2,1) how many lenses would be necessary and in what configuration to realize it? A more basic question is whether all such elements can be realized. These questions are nontrivial since while we can use both positive (convex) and negative (concave) lenses, the propogation distances D have to be chosen positive. By a straightforward but elaborate analysis we can show that all such SO(2,1) elements can be realized by a combination of at most three lenses with three separate propogation distances. In particular we could synthesize "propogation through a negative distance".

Noncentered axially symmetric lenses and lenses placed not precisely normal to the axis lead to terms linear in coordinates and momenta and therefore to the inhomogeneous extensions of the SO(2,1) group (which is better seen as the symplectic group Sp(2,R)).

Note that the linear terms in momenta and in coordinates are appropriate for a prism and for a phase plate.

If the lenses are not axially symmetric the optical phase shift on encountering a lens is a general quadratic form spanned by $\mathcal{L}$, $\mathcal{K}_1$ and $\mathcal{K}_2$. If the lenses are properly centered and oriented these suffice. Propogation introduces phases proportional to $\mathcal{H}$ and this together with $\mathcal{L}$, $\mathcal{K}_1$, $\mathcal{K}_2$ close on the Lie algebra SO(3,2). Any centered oriented system of lenses and free propogation realizes an element of SO(3,2). The synthesis question can be raised here also and can be answered in the affirmative.

The inhomogeneous versions of SO(3,2) or rather the symplectic Sp(4,R) correspond to excentric and mildly disoriented systems.

As a special case of such systems we may take cylindrical lenses and systems made by them to realize Sp(2,R) and its extension.

When the considerations change from scalar wave optics to vector wave optics we need to augment the coordinates by the matrices g_1 and g_2 so that all quantities are expressed in terms of q_1 and q_2.

5. LIGHT RAYS IN WAVE OPTICS

The fundamental representation of Sp(2,R) is by 2x2 matrices:

$$\mathcal{L} = \frac{1}{2}(q_1^2 + q_2^2) \rightarrow \begin{pmatrix} 0 & 0 \\ i & 0 \end{pmatrix}$$

$$\mathcal{H} = \frac{1}{2}(p_1^2 + p_2^2) \rightarrow \begin{pmatrix} 0 & i \\ 0 & 0 \end{pmatrix}$$

$$\mathcal{M} = \frac{1}{2}(q_1 p_1 + p_2 q_2) \rightarrow \begin{pmatrix} \frac{-i}{2} & 0 \\ 0 & +\frac{i}{2} \end{pmatrix}$$

These imaginary matrices and the corresponding realization of the group elements by real 2x2 matrices have an elementary geometrical interpretation in terms of light rays: q and p correspond to their spatial coordinates and their directions. A thin lens may be thought of as introducing an abrupt change in direction of the rays proportional to the distance of the point of incidence from the center; while free propogation maintains the direction but displaces the coordinate proportional to the slope! But how come we have rays coming up in wave optics?

Note that the symplectic groups Sp(2,R) and Sp(4,R) are there whether we think in terms of rays or not: the 2x2 and 4x4 matrices provide faithful realizations of Sp(2,R) and Sp(4,R) respectively.

We can obtain pencils of rays in wave optics if you realize that the two-point functions

$$\Gamma(x,y) = \langle \phi^{\dagger}(x)\ \phi(y)\rangle$$

furnish the mutual coherence functions in optics which propogate from one value of z to another: in quantum optics they are the normal ordered expectation values of field operators. Given $\Gamma(x,y)$ we can define the Wolf function by the Wigner-Moyal transformation:

$$W(q,p) = \int e^{-ip\xi}\ \Gamma(q + \tfrac{1}{2}\xi,\ q - \tfrac{1}{2}\xi)\,d\xi\ .$$

From the properties of $\Gamma(x,y)$ we can deduce that the Wolf function is always real but it is not always nonnegative. Any time W(q,p) becomes negative typical wavelike effects (diffraction, interference) can be expected.

Given any pencil of light with a Wolf function W(q,p) which may or may not be nonnegative, the effect of encountering an optical element like a lens or a segment of free propogation the new pencil of light is represented by a new Wolf function W'(q,p). The new pencil is obtained from the old pencil by pointwise ray transformation:

$$W'(q',p') = W(q,p).$$

We have thus the simple but startling result that the pencils in wave optics are transferred by an optical system in a manner simply described by geometrical optical transformation of the rays.

When mirrors are introduced we may generalize these considerations by path ordering the propogation and viewing the axis as being reflected by the mirrors, with a corresponding redefinition of the light cone coordinates. With this understanding a paraxial beam continues to be paraxial.

6. SQUEEZED LIGHT AND ESOTERICA

In the considerations so far the system elements were passive and lossless. The mutual coherence functions behave as convex cones, in the sense that whenever $\Gamma^{I}(x,y)$ and $\Gamma^{II}(x,y)$ are mutual coherence functions

$$\lambda_1\Gamma^{I}(x,y) + \lambda_2\Gamma^{II}(x,y)\ ;\ \lambda_1 > 0,\ \lambda_2 > 0\ ,$$

is also an acceptable mutual coherence function. Identical convex cone property obtains for the Wolf functions. We may therefore deal with generating elements of these convex cones. The corresponding Wolf functions describe elementary pencils of light.

So far we have discussed the transformations of one such pencil into another by optical elements. These transform vector potentials into vector potentials and are therefore of the form of general linear canonical coordinate transformations. But from mechanics we know of more general canonical transformations. For a single mode we may work in terms of the creation and annihilation operators and ask for linear canonical transformations amongst them. We write

$$a \to \Lambda_{11}\, a + \Lambda_{12} a^{\dagger} \; ; \quad a^{\dagger} \to \Lambda_{21}\, a + \Lambda_{22}\, a^{\dagger} \, .$$

with

$$\Lambda_{11}^{*} = \Lambda_{22} \, , \; \Lambda_{12}^{*} = \Lambda_{21} \; ; \; \Lambda_{11}\, \Lambda_{22} - \Lambda_{21}\, \Lambda_{12} = 1 \, .$$

These are again Sp(2,R) transformations and can be generated by the matrices

$$\frac{1}{2}\begin{pmatrix} 1 & 0 \\ 0 & -1 \end{pmatrix} , \; \frac{1}{2}\begin{pmatrix} 0 & i \\ i & 0 \end{pmatrix} , \quad \frac{1}{2}\begin{pmatrix} 0 & -1 \\ 1 & 0 \end{pmatrix}$$

This Lie algebra is realized by the generators

$$\frac{1}{4}\,(a^{\dagger}a + aa^{\dagger}) \, , \; \frac{i}{4}\,(a^{\dagger 2} - a^{2}) \, , \; \frac{1}{4}\,(a^{\dagger 2} + a^{2})$$

On the field ϕ (vector potential) and the conjugate momentum π (electric field) the transformations generated by SL(2,R) are real and include the natural time evolution mixing ϕ and π, as the scale transformations changing ϕ and π by reciprocal factors and the hyperbolic transformations between ϕ and π. Among these the scale transformations are most interesting since the uncertainty $\Delta\phi$ in the field variable ϕ, for example, could be made arbitrarily small though $\Delta\phi \cdot \Delta\pi$ could not be reduced. This process is called "squeezing". The simplest case of squeezing arises when we pass from a medium of high dielectric constant to one of low dielectric constant, though this is not easily amenable to experiment. In nonlinear optics such squeezed states have been identified and studied.

When mode mixing occurs we have more general linear canonical transformations amongst them. For two degrees of freedom we would have Sp(4,R), and, more generally for n degrees of freedom we would get Sp(2n,R). In addition to considering quantum electrodynamics of dielectrics which, in relation to empty space, realize squeezing transformations the optics of moving dielectrics realize one parameter subgroups of these more general transformations.

7. REMARKS

These considerations in optics are only some of the context in which group theory arises most naturally. There are many more which we have not discussed. One important example is the modification of a generalized Gaussian Schell beam: such a beam is defined by a mutual coherence function which is the exponential of a general quadratic form. Propogation changes this into another such beam; and these transformations have their own invariant, and the exponent transforms like a symmetric second rank tensor of the O(3,2) group. Similarly the propogation in an optical fibre with a refractive index varying quadratically with distance from center.

In addition to their mathematical elegance, group theoretic methods would also be useful in synthesis and design of systems.

ACKNOWLEDGEMENT

This work was supported by the Department of Energy gran number DE-FG05-85ER40200.

SYMMETRY-ADAPTATION AND SELECTION RULES FOR EFFECTIVE CRYSTAL FIELD HAMILTONIANS

J. A. Tuszyński

Department of Physics
Memorial University of Newfoundland
St. John's, Newfoundland, A1B 3X7, Canada

INTRODUCTION

The theory of nuclear and atomic shells owes its elegance to the group theoretic method of Racah[1]. He implemented a branching scheme leading to rotational subgroups of the unitary group $U(4\ell+2)$ of transformations among the spin-orbitals $|n\ell m_s m_\ell\rangle$[2]. This subgroup reduction scheme induces analogous transformations among the many-particle states of the nucleus or the electronic shells. This results in a convenient labelling according to the irreducible representations of the rotational subgroups[3] as $|\ell^N SLM_S M_L WU\tau\rangle$. The corresponding Lie algebra of generators consists of double tensor operators[4] $w^{k_1k_2}$ whose $(4\ell+2)^2$ components $w^{k_1k_2}_{q_1q_2}$ ($k_1 = 0, 1; -k_1 \leq q_1 \leq k_1; k_2 = 0, 1, \ldots, 2\ell; -k_2 \leq q_2 \leq k_2$) span the full unitary group $U(4\ell+2)$. Their definition via commutation relations with angular momenta reflects the rotation properties of spherical harmonics and is supplemented by the normalization condition[4]: $\langle s\ell||w^{k_1k_2}||s\ell\rangle = [k_1,k_2]^{\frac{1}{2}}$. An extension of these operators into the many-body formalism requires defining them as sums of single-particle operators: $W^{k_1k_2} \equiv \sum_{i=1}^{N} w_i^{k_1k_2}$. Consequently, the many-particle atomic or nuclear problem can be reduced to the evaluation of matrix elements of tensor operators between the eigenstates of compound angular momentum. In practice such calculations are facilitated by the use of Wigner-Eckart theorem and the recoupling properties of angular momenta. Furthermore, coefficients of fractional parentage[5] which relate N-particle states to (N-1)-particle states render the method almost algorithmic.

Tensor operators are not restricted to the free atom whose inherent symmetry is nearly spherical but are also suitable in the presence of crystal fields since linear combinations of tensor operators are invariant under point group operations[6]. Since the radial aspect of the free atom problem is solved self-consistently to a high degree of accuracy, the crystal field can usually be considered a small perturbation which is subject to detailed symmetry requirements.

The intention of this paper is to systematically derive an effective Hamiltonian in the presence of crystal fields in such a way as to incorporate relativistic effects and higher order perturbation corrections including configuration mixing. This Hamiltonian will then be conveniently represented as a symmetry-adapted series of one- and two-body double tensor operators whose matrix elements will be analyzed for selection rules.

THE EFFECTIVE HAMILTONIAN

In the relativistic quantum mechanics a free atom is described by the Dirac-Breit Hamiltonian[7]

$$H_{DB} = \sum_i (c\vec{\alpha}_i \cdot \vec{\pi}_i + \beta_i mc^2 - \frac{Ze^2}{r_i})$$

$$+ \sum_{i<j} \left[\frac{e^2}{r_{ij}} - \tfrac{1}{2}e^2\left(\frac{\vec{\alpha}_i \cdot \vec{\alpha}_j}{r_{ij}} + \frac{(\vec{\alpha}_i \cdot \vec{r}_{ij})(\vec{\alpha}_j \cdot \vec{r}_{ij})}{r_{ij}^3}\right)\right] \quad (1)$$

which is rather inconvenient to work with since it contains odd operators mixing the large F(r) and small G(r) components of the wavefunction $|\psi\rangle$. Successive applications of the Foldy-Wouthuysen transformation[8] result in an equivalent Hamiltonian which contains only even operators[9], i.e.: (i) of one-body type: the kinetic energy with a relativistic correction, the electrostatic potential, the Zeeman term, the spin-orbit operator and the Darwin term; and (ii) of two-body type: spin-spin, spin-other-orbit, orbit-orbit, contact interaction, i.e. transformed Breit operator. Self-consistent Dirac-Hartree-Fock computer programs[10] which calculate relativistic eigenfunctions $|\psi_\alpha\rangle$ and eigenenergies ε_α for multi-electron atoms are now readily available. They also provide various radial integrals which are useful in crystal field calculations.

The molecular Hamiltonian can be written as: $H = \sum_I h_{DB}(I) + \sum_{I<J} v(I,J)$ where I and J denote atoms (ions). If the interaction term v(I,J) is relatively weak, we can replace it by a mean-field type of operator called the crystal (ligand) field which acts on the central paramagnetic ion of the complex. Its form is[6]

$$h_{CF} = \sum_{kq} A_q^k r^k C_q^k \quad (2)$$

where C_q^k is a tesseral harmonic and the coefficients A_q^k are chosen in such a way as to reflect the point symmetry of the environment. Consequently, the effective relativistic Hamiltonian can be schematically represented as

$$H_R = H_o + \sum_i h'(s_i,\ell_i,C_i) + \sum_{i<j} h''(s_i,s_j,\ell_i,\ell_j,C_i,C_j) \quad (3)$$

where H_o is the spherically symmetric part assumed as a zeroth approximation, h' and h" are respectively one- and two-body perturbations depending on spins s_i, orbits ℓ_i and crystal fields C_i of the electrons which are denoted by i and j. It is convenient to unify notation by replacing an arbitrary tensor procuct of these operators by a double tensor operator according to

$$\{T^{k_1}(s) \times [T^{k_{21}}(\ell) \times T^{k_{22}}(C)]^{k_2}\}^k_q = b_1(\{k\})\, W^{\{k\}}_q \tag{4}$$

where $\{k\} \equiv [k_1(k_{21}k_{22})k_2]k$ and the distinction between k_{21} and k_{22} will later be shown to have important repercussions on symmetry-adaptation and selection rules. We have calculated the coefficient $b_1(\{k\})$ for arbitrary ranks and orbital quantum numbers ℓ and ℓ' of the states between which $W^{\{k\}}$ is supposed to act:

$$\begin{aligned} b_1(\{k\}) &= (-1)^{\ell'+k_2}\,[s,\ell,\ell']^{\frac{1}{2}}\,[k_1]^{-\frac{1}{2}} \\ &\times \begin{pmatrix} \ell & k_{22} & \ell' \\ 0 & 0 & 0 \end{pmatrix} \begin{Bmatrix} k_{21} & k_{22} & k_2 \\ \ell' & \ell & \ell \end{Bmatrix} [\delta(k_1,0) + \delta(k_1,1)\sqrt{s(s+1)}] \\ &\times \left\{ \delta(k_{21},0) + [1-\delta(k_{21},0)] \left| \frac{k_{21}!\,k_{21}!\,(2\ell+k_{21}+1)!}{2^{k_{21}}(2k_{21})!\,(2\ell-k_{21})!} \right|^{\frac{1}{2}} \right\} \end{aligned} \tag{5}$$

where $(\vdots\vdots\vdots)$ is a 3j-symbol, $\{\vdots\vdots\vdots\}$ is a 6j-symbol.

Defining two-body double tensor operators as

$$X^{\{kk'\}}_Q \equiv \sum_{i<j} [w^{\{k\}}_i \times w^{\{k'\}}_j]^K_Q \tag{6}$$

it can be shown that the two-body part of the effective Hamiltonian (3) contains terms $b_2(\{kk'\})\, X^{\{kk'\}}$ where b_2 is analogous to b_1 of (5), and it can further be proved that: $b_2(\{kk'\}) = b_1(\{k\})b_1(\{k'\})$. Finally, then the effective relativistic Hamiltonian is a series of double tensor operators

$$H_R = H_o + \sum_{\{k\}q} f(r)b_1(\{k\})W^{\{k\}}_q + \sum_{\{kk'\}Q} g(r,r')b_2(\{kk'\})\, X^{\{kk'\}}_Q \tag{7}$$

where $f(r)$ and $g(r,r')$ are the corresponding one- and two-body radial functions which are characteristic interaction strengths for each term.

In order to be able to use non-relativistic wavefunctions $|\psi\rangle$ we seek an equivalent Hamiltonian H_{NR} so that[11]: $(\psi|H_R|\psi') = \langle\psi|H_{NR}|\psi'\rangle$ where $|\psi)$ and $|\psi\rangle$ are related via[7]

$$|s\ell jm) = \begin{cases} \dfrac{F(r)}{r}\,|s\ell jm\rangle \\ \\ i\,\dfrac{G(r)}{r}\,|s\ell\pm 1jm\rangle \end{cases} \tag{8}$$

Wybourne[12] used this procedure to calculate the relativistic crystal field and we have derived transformation coefficients to replace arbitrary relativistic tensor operators by their non-relativistic equivalents

$$\begin{aligned} c_1(\{\bar{k}\}) &= b_1(\{k\})\,[k_1,k_2,k',k'',\ell]^{\frac{1}{2}}\,[\ell']^{-\frac{1}{2}} \\ &\times \sum_{j,j'} [j,j'] \begin{bmatrix} s & s & k_1 \\ \ell & \ell' & k_2 \\ j & j' & k \end{bmatrix} \begin{bmatrix} s & s & k' \\ \ell & \ell' & k'' \\ j & j' & k \end{bmatrix} \int_o^\infty f(r)(F_jF^*_{j'} + G_jG^*_{j'})dr \end{aligned} \tag{9}$$

where $\{\bar{k}\} \equiv [k'(\bar{k}_{21}k_{22})k'']k$ and $[\vdots\vdots\vdots]$ is a 9j-symbol. It can also be demonstrated that if $g(r,r') = f(r)f(r')$, then a corresponding two-body coefficient $c_2(\{\bar{k}\bar{\bar{k}}\})$ can be factorized as: $c_2(\{\bar{k}\bar{\bar{k}}\}) = c_1(\{\bar{k}\})\ c_1(\{\bar{\bar{k}}\})$. It is worth pointing out that apart from affecting the radial integrals, rèlativitiy introduces new tensorial ranks into the effective Hamiltonian. In fact, to each tensor operator with ranks $\{k\}$ in H_R there corresponds an equivalent combination of 3 tensor operators with ranks $\{\bar{k}\}$ in H_{NR} where: (i) $k' = k_1$, $\bar{k}_{21} = k_{21}$, $k'' = k_2$; (ii) $k' = |k_1-1|$, $\bar{k}_{21} = k_{21}+1$, $k'' = k_2+1$; and (iii) $k' = |k_1-1|$, $\bar{k}_{21} = k_{21}-1$, $k'' = k_2-1$. Consequently, to each two-body term $\{kk'\}$ in H_R there corresponds a linear combination of up to 9 terms with ranks $\{\bar{k}\bar{\bar{k}}\}$ in H_{NR}. Hence, the non-relativistic Hamiltonian can be symbolically written as

$$H_{NR} = H_o + \sum_{\{\bar{k}\}q} c_1(\{\bar{k}\})\ W_q^{\{\bar{k}\}} + \sum_{\{\bar{k}\bar{\bar{k}}\}Q} c_2(\{\bar{k}\bar{\bar{k}}\})\ X_Q^{\{\bar{k}\bar{\bar{k}}\}} \tag{10}$$

Finally, it is often advantageous, especially in zero-field splitting calculations, to include higher order perturbations in the form of the effective Hamiltonian which is then restricted to act within the ground state manifold $|\Psi_o\rangle$. If $H = H_o + H_1$ where $H_o|\Psi_o\rangle = \varepsilon_o|\Psi_o\rangle$ and $\varepsilon = \langle\Psi_o|H_{eff}|\Psi_o\rangle/\langle\Psi_o|\Psi_o\rangle$, then H_{eff} is given by

$$H_{eff} = H_o + H_1 + \sum_\alpha \frac{H_1|\Psi_\alpha\rangle\langle\Psi_\alpha|H_1}{\varepsilon_o-\varepsilon_\alpha} + \ldots \tag{11}$$

where H_1 covers both the one-body and two-body terms of H_{NR} of (10) and α runs over all excited states within and outside the ground configuration. We find the form of H_{eff} as

$$H_{eff} = H_o + \sum_{\{K\}q} [c_1(\{K\}) + d_{11}(\{K\})]\ W_q^{\{K\}}$$

$$+ \sum_{\{KK'\}Q} [c_2(\{KK'\}) + d_{12}(\{KK'\}) + d_{22}(\{KK'\})]\ X_Q^{\{KK'\}} \tag{12}$$

where the additional coefficients d_{nm} are obtained using the closure property as

$$d_{nm} = [2-\delta(m,n)] \sum_{\{k\}} \sum_{\{k'\}} [2-\delta(\{k\},\{k'\})][k,k',K_1,K_2]^{\frac{1}{2}}$$

$$\times \begin{bmatrix} k_1 & k_2 & k \\ k_1' & k_2' & k' \\ K_1 & K_2 & K \end{bmatrix} \left[\frac{c_m(\{k\})c_n(\{k'\})}{\Delta\varepsilon} + \sum_{\alpha\neq 0} c_m^\alpha(\{k\})\ c_n^\alpha(\{k'\}) \right.$$

$$\left. \times \left(\frac{1}{\varepsilon_\alpha-\varepsilon_o} - \frac{1}{\Delta\varepsilon}\right) \right] \tag{13}$$

where α runs over all excited configurations, $\Delta\varepsilon$ is the average energy of the ground configuration, and c_m^α and c_n^α are modified with respect to c_m and c_n by using the transition radial integrals between the ground configuration and the α-th configuration.

If the effective Hamiltonian H_{eff} is to properly reflect the symmetry of the system under consideration, it must satisfy the invariance requirement

$$g_\ell \, H_{eff} \, g_\ell^{-1} = H_{eff} \tag{14}$$

where g_ℓ denotes all symmetry elements, i.e. point group elements (rotations C_n, reflections σ_h, σ_v, σ_d and inversion i), time reversal θ, hermitian conjugation h and, if the ground state belongs to a half-filled shell, quasi-spin conjugation Q.

We start our analysis by investigating the conjugation properties of various tensor operators O_q^k with respect to parity reversal P(≡i), θ, h, Q and charge conjugation C. Each conjugation property can be written in general as

$$g_\ell \, O_q^k \, g_\ell^{-1} = (-1)^{e(\ell,0)} \, O_{\pm q}^k \tag{15}$$

where e(ℓ,0) is an appropriate exponent whose value depends on the particular type of the operator O_q^k and on the symmetry operation g_ℓ; +q on the right hand side of (15) appears only for $g_\ell = P$. It has been recently pointed out[13] that all single tensor operators can be divided into two classes: polar and axial. Polar operators are represented by electric tensors denoted $T_q^k(E)$, e.g. C_q^k, and they change under parity reversal. They do not conserve the magnitude of angular momentum. Axial operators are represented by magnetic tensors denoted $T_q^k(B)$, e.g. $T_q^k(s)$ or $T_q^k(\ell)$, and they are invariant with respect to parity reversal. They conserve the magnitude of angular momentum. Thus, only the former are capable of configuration mixing effects. The inequivalence of the two classes of operators has been fully realized by Watanabe[6] in his replacement theorem. Interesting consequences of this classification have also been found for

Table 1. Conjugation Properties of Various Tensor Operators

Type of Operator	Symmetry Operations and Exponents e				
	P	θ	C	Q	h
$T_q^k(E)$	k	$-q$	$k+q$	$k+q+1$	q
$T_q^k(B)$	0	$k-q$	$k+q$	$k+q+1$	q
$W_q^{\{k\}}$	k_{22}	$k+k_{22}-q$	$k+q$	$k_1+k_2+k+1+q$	$k_1+k_{21}+k_{22}+k_2+q$
$X_Q^{\{kk'\}}$	$k_{22}+k'_{22}$	$k_{22}+k'_{22}+K-Q$	$K+0$	$K+Q$	$k_{11}+k_{21}+k_{22}+k'_{11}+k'_{21}+k'_{22}+K+Q$

product tensor operators, in particular one- and two-body double tensor operators[14]. The conjugation properties of all these operators are summarized in Table 1. It can be concluded on its basis that: (i) only polar ranks affect parity-reversal, (ii) time-reversal requires the axial ranks to be even, (iii) in all cases $PC\theta \equiv I$, (iv) hermiticity of operators is satisfied when sums of the individual ranks are even. Invariance properties of polar tensor operators with respect to point group operations were analyzed in the past[15] and are summarized in Table 2. Both classes of tensor operators behave identically under rotations. Since for an arbitrary reflection $\sigma = i \times C_2$ where C_2 is a two-fold rotation about the axis perpendicular to the reflection plane and passing through the origin, we can readily derive point group invariants for axial tensor operators. We have summarized our results in Table 3. Both Tables 2 and 3 utilize tensor combinations analogous to tesseral harmonics[15], i.e.

$$C^k_q \equiv \frac{1}{\sqrt{2}} [T^k_{-q} + (-1)^q\, T^k_q]; \quad S^k_q \quad \frac{i}{\sqrt{2}} [T^k_{-q} - (-1)^q\, T^k_q] \quad q > 0$$

and

$$C^k_o \equiv T^k_o; \quad S^k_o = 0 \tag{16}$$

Both these Tables list only operators which are invariant under time reversal.

Within cubic groups axial operators distinguish only 2 inequivalent classes: (T, T_h) and (O, O_h, T_d). It can be concluded that unlike polar operators, axial operators cannot be used to discriminate between the individual 32 point groups but only between 11 subclasses.

SELECTION RULES

Having found tensorial invariants with respect to the various symmetry operations we now proceed to derive selection rules for their matrix elements. To this end we use the following identity:

$$<\psi|g_\ell^{-1}\ (g_\ell\, H_{eff}\, g_\ell^{-1})g_\ell|\psi'> \;=\; <\psi_\ell|H_{eff}|\psi'_\ell> \tag{17}$$

and the following properties of the eigenstates of angular momentum[5,16]

$$P|s\ell jm> \;=\; (-1)^\ell\ |s\ell jm> \tag{18}$$

$$\theta|s\ell jm> \;=\; (-1)^{j-m}\ |s\ell j{-m}> \tag{19}$$

$$Q|\ell^N SLM_S M_L Q M_Q> \;=\; (-1)^{Q-M_Q}\ |\ell^{4\ell+2-N}\ SLM_S M_L Q{-M_Q}> \tag{20}$$

First, conservation of angular momentum follows from the Wigner-Eckart theorem and results in a requirement on projections and in triangular rules on ranks. Then, we analyze reduced matrix elements of one- and two-body double tensor operators. The general case for one-body operators is

$$<\ell^{N-1}(S_1L_1\alpha_1),\ell',SLJ||W^{\{k\}}||\ell^{N-1}(S'_1L'_1\alpha'_1),\ell'',\ S'L'J'> \tag{21}$$

which includes a few special cases. The general case of two-body operators is

Table 2. Invariant Polar Tensor Operators for the 27 Noncubic Point Groups[15]

		TRICLINIC		MONOCLINIC			ORTHO-RHOMBIC			TETRAGONAL							TRIGONAL					HEXAGONAL						
k	q	C_1	S_2	C_2	C_{1h}	C_{2h}	D_2	C_{2v}	D_{2h}	C_4	S_4	C_{4h}	D_4	C_{4v}	D_{2d}	D_{4h}	C_3	S_6	D_3	C_{3v}	D_{3d}	C_6	C_{3h}	C_{6h}	D_6	C_{6v}	D_{3h}	D_{6h}
1	0	C		C				C		C				C			C			C		C				C		
2	0	C	C	C	C	C	C	C	C	C	C	C	C	C	C	C	C	C	C	C	C	C	C	C	C	C	C	C
3	0	C		C				C		C				C			C			C		C				C		
4	0	C	C	C	C	C	C	C	C	C	C	C	C	C	C	C	C	C	C	C	C	C	C	C	C	C	C	C
5	0	C		C				C		C				C			C			C		C				C		
6	0	C	C	C	C	C	C	C	C	C	C	C	C	C	C	C	C	C	C	C	C	C	C	C	C	C	C	C
1	1	CS			CS																							
2	1	CS	CS																									
3	1	CS			CS																							
4	1	CS	CS																									
5	1	CS			CS																							
6	1	CS	CS																									
2	2	CS	CS	CS	CS	CS	C	C	C																			
3	2	CS		CS			S	C			CS				S													
4	2	CS	CS	CS	CS	CS	C	C	C																			
5	2	CS		CS			S	C			CS				S													
6	2	CS	CS	CS	CS	CS	C	C	C																			
3	3	CS			CS												CS		S	C			CS				S	
4	3	CS	CS														CS	CS	C	C	C							
5	3	CS			CS												CS		S	C			CS				S	
6	3	CS	CS														CS	CS	C	C	C							
4	4	CS	CS	CS	CS	CS	C	C	C	CS	CS	CS	C	C	C	C												
5	4	CS		CS			S	C		CS			S	C														
6	4	CS	CS	CS	CS	CS	C	C	C	CS	CS	CS	C	C	C	C												
5	5	CS			CS																							
6	5	CS	CS																									
6	6	CS	CS	CS	CS	CS	C	C	C								CS	CS	C	C	C	CS	CS	CS	C	C	C	C

Table 3. Invariant Axial Tensor Operators for the 27 Noncubic Point Groups

		TRICLINIC	MONOCLINIC	ORTHO-RHOMBIC	TETRAGONAL		TRIGONAL		HEXAGONAL	
k	q	C_1 S_2	C_2 C_{1h} C_{2h}	D_2, D_{2h} C_{2v}	C_4, C_{4h} S_4	D_4, D_{4h} C_{4v}, D_{2d}	C_3, S_6	D_3, D_{3d} C_{3v}	C_6, C_{6h} C_{3h}	D_6, D_{6h} C_{6v}, D_{3h}
2	0	C	C	C	C	C	C	C	C	C
4	0	C	C	C	C	C	C	C	C	C
6	0	C	C	C	C	C	C	C	C	C
2	1	CS								
4	1	CS								
6	1	CS								
2	2	CS	CS	C						
4	2	CS	CS	C						
6	2	CS	CS	C						
4	3	CS					CS	C		
6	3	CS					CS	C		
4	4	CS	CS	C	CS	C				
6	4	CS	CS	C	CS	C				
6	5	CS								
6	6	CS	CS	C			CS	C	CS	C

$$\langle \ell^{N-2}(\overline{SL\alpha}), \ell'\ell''(S_2L_2); SLJ || X^{\{kk'\}} || \ell^{N-2}(\overline{\overline{SL\alpha}}), \ell'''\ell'^{v}(S_2'L_2'); S'L'J' \rangle \quad (22)$$

and it includes several special cases. The results of our analysis are summarized in Table 4. For simplicity we use $\Delta(\ell, k_{22}, \ell')$ which is defined as follows: $\Delta = 1$ when $(\ell + k_{22} + \ell')$ is even and $\Delta = 0$ either when $(\ell + k_{22} + \ell')$ is odd or when $k_{22} = 0$ for $\ell \neq \ell'$. It should be emphasized that time-reversal does not introduce any selection rules. We have developed computer programs which calculate the various transformation coefficients b_n, c_n, d_{nm} and also matrix elements given in (21) and (22). Applications of this procedure to practical situations are currently underway.

Table 4. Selection Rules For Matrix Elements of One- and Two-Body Double Tensor Operators

Symmetry	Selection Rules
I. One-Body Operators W:	
1. Conservation of Angular Momentum	(a) $M = q+M'$ (b) triangular rules
2. Hermiticity	$k_1+k_{21}+k_{22}+k$ even
3. Parity-Reversal	$\Delta(\ell',k_{22},\ell'') \neq 0$
(a) if $\ell' = \ell'' \neq \ell$	$\Delta(\ell,k_{22},\ell) \neq 0$ or $\Delta(\ell',k_{22}\ell') \neq 0$
4. Quasi-Spin Invariance ($N = 2\ell+1$; $\ell' = \ell'' = \ell$)	$k_1 + k_2$ even
II. Two-Body Operators X:	
1. Conservation of Angular Momentum	(a) $M = Q+M'$ (b) triangular rules
2. Hermiticity	(a) $k_1 + k_2 + K$ even (b) $k_{11}+k_{21}+k_{22}+k'_{11}+k'_{21}+k'_{22}+K$ even
3. Parity-Reversal	$\Delta(\ell',k_{22},\ell''')\Delta(\ell'',k'_{22},\ell'^{V}) \neq 0$
(a) if $\ell'' = \ell'^{V} = \ell$; $\ell' = \ell''' \neq \ell$	$\Delta(\ell,k_{22},\ell)\Delta(\ell,k'_{22},\ell) \neq 0$ or $\Delta(\ell,k_{22},\ell')\Delta(\ell,k'_{22},\ell') \neq 0$
(b) if $\ell' = \ell''' \neq \ell$; $\ell'' = \ell'^{V} \neq \ell$	$\Delta(\ell,k_{22},\ell)\Delta(\ell,k'_{22},\ell) \neq 0$ or $\Delta(\ell',k_{22},\ell')\Delta(\ell'',k'_{22},\ell'') \neq 0$ or $\Delta(\ell,k_{22},\ell)\Delta(\ell'',k'_{22},\ell'') \neq 0$ or $\Delta(\ell,k_{22},\ell)\Delta(\ell',k'_{22},\ell') \neq 0$
4. Quasi-Spin Conjugation ($N = 2\ell+1$; $\ell' = \ell'' = \ell''' = \ell'^{V} = \ell$)	K even

REFERENCES

1. G. Racah, Phys. Rev. 62: 438 (1942).
2. B. R. Judd, Symmetry properties of atomic structure, in: "Atomic Physics," V. W. Hughes et al., eds., Plenum Press, New York (1968).
3. B. R. Judd, Adv. Atom. Molec. Phys. 7: 251 (1971).
4. B. R. Judd, "Operator Techniques in Atomic Spectroscopy", McGraw-Hill, New York (1967).
5. B. R. Judd, "Second Quantization and Atomic Spectroscopy", The Johns Hopkins University Press, Baltimore (1967).
6. H. Watanabe, "Operator Methods in Ligand Field Theory", Prentice Hall, Englewood Cliffs (1966).
7. L. Armstrong, Jr., "Theory of the Hyperfine Structure of Free Atoms", Wiley-Interscience, New York (1971).
8. L. Armstrong, Jr., J. Math. Phys. 7: 1891 (1966).
9. Lr. Armstrong, Jr. and S. Feneuille, Adv. Atom. Molec. Phys. 10: 1 (1974).
10. J. P. Desclaux, Comp. Phys. Commun. 9: 31 (1975).
11. P. G. H. Sandars and J. Beck, Proc. Roy. Soc. Lon. A289: 97 (1965).
12. B. G. Wybourne, J. Chem. Phys. 43: 4506 (1965).
13. R. Chatterjee, J. A. Tuszyński and H. A. Buckmaster, Can. J. Phys. 61: 1613 (1983).
14. J. A. Tuszyński, Physica A131: 289 (1985).
15. J. L. Prather, "Atomic Energy Levels in Crystals", U.S. National Bureau of Standards, Washington, D.C. (1961).
16. D. M. Brink and G. R. Satchler, "Angular Momentum", Oxford University Press, London (1968).

SYMMETRY-BREAKING IN BIOLOGICAL CELLS

J. A. Tuszyński
Department of Physics
Memorial University of Newfoundland
St. John's, Nfld., A1B 3X7, Canada

BROKEN SYMMETRIES

The concept of symmetry breaking has played a prominent role in the understanding of novel phenomena in physics, chemistry, and recently, biology. Since a spontaneously broken symmetry is associated with a degenerate ground state which is not invariant under the full transformation group of the Hamiltonian, it has been described as a growth of complexity out of simplicity[1]. As demonstrated by Landau[2] any change in the symmetry of the system (i.e. its ground state) must be sudden and is manifested by the emergence of a non-zero order parameter in the unsymmetric phase. This ordered phase is maintained by gapless energy bosons called Goldstone modes which are instrumental in propagating the order over long ranges[3]. Depending on the system these collective excitations take the form of magnons, phonons, excitons, polarons, etc. They can take the system from one vacuum state to another. The long-range order introduces new length scales which in systems with complex order parameters, e.g. lasers, superfluids, superconductors, may manifest itself in phase coherence. The collective, mean-field behavior results in stability sometimes referred to as generalized rigidity[1]. However, certain deviations from spatial uniformity are present, as exemplified by collective excitations themselves and by critical fluctuations which may accompany them. Moreover, Goldstone bosons may condense creating topological singularities (defect structures) possibly solitonic in character. These singularities can have various dimensionalities (points, lines, planes), curvatures (spirals, spheres, vortices, flux lines, disclination lines) and eventually, if sufficiently numerous, they may bring about a destruction of order. Specific critical phenomena can be divided into 5 groups[1] according to the types of broken symmetries, i.e. (i) translations (e.g. crystal formation); (ii) gauge invariance (e.g. superfluids); (iii) time-reversal (e.g. magnets), (iv) point-inversion (ferroelastics) and, (v) local rotations (e.g. nematic liquid crystals).

Besides equilibrium phase transitions two other distinct categories of symmetry breaking effects exist. Nonlinear instabilities (e.g. Bénard instability, limit cycles) are characteristic of dissipative structures and involve transitions from chaos to spatial and temporal pattern formation as a result of an infinitesimal change in the control parameters. The other category is nonequilibrium phase transitions (e.g. laser action) and they represent a locally broken phase symmetry which is driven by a sufficiently strong external incoherent field. A precondition of this effect is that the internal structure exhibit certain qualities as for example two or more metastable states.

PROPERTIES OF BIOLOGICAL SYSTEMS

It is a generally accepted precept that both animate and inanimate matter obey the same laws of physics. Moreover, typical sample sizes used in the laboratory are not dissimilar and involve 10^{20}-10^{23} atoms in each case. However, the living organisms distinguish themselves by a high level of complexity and internal structure. They are typically composed of 10^{10} cells which, in turn, contain 10^{10} atoms each[4]. Much of their physiological and biochemical activity takes place in the membranes. Membranes are lipid bilayers containing proteins and enveloping cells and organelles. They are impermeable to water but serve as channels of ions, molecules and energy. The building blocks of all living matter are biomolecules containing typically ~ 10^3 atoms. Since they are highly anisotropic and inhomogeneous, they are often referred to as aperiodic crystals[4]. The most important molecules of life are nucleic acids (DNA, RNA) and proteins. Nucleic acids are quasi-linear and are built from 4 types of nucleotides. They exhibit remarkable transport and storage properties of particles and energy and are instrumental in the transfer of genetic information. Proteins are built from 20 types of aminoacids and, in contrast to crystals, have not only vibrational but also translational degrees of freedom. Since their ground state is highly degenerate with respect to the configurational coordinates, it allows for numerous conformational changes at no or little energetical expense. Both nucleic acids and proteins have low excitation energies, low ionization potentials, high electron affinities and high polarizabilities. Hence, they keenly participate in electron transfer.

Notwithstanding the great achievements of biochemistry in understanding the structure and function of living matter it appears that this discipline focuses too sharply on the individual components rather than on their collective modes of behavior. It was Szent-Gyorgi who first applied the principles of solid state physics to biology analyzing the energy bands of proteins. Condensed matter physics is especially well suited for such applications since it makes a very practical distinction between slowly changing modes (order parameters) and fast changing modes (slaved variables) thus reducing computational complexity.

The physicist may characterize biological cells as thermodynamically open systems which exchange matter and energy with the environment[6]. They form dissipative structures which do not thermalize all the input energy but use some of it to create and maintain order thereby lowering their internal entropy. Some of their degrees of freedom appear to be in thermal equilibrium (stable) while others are far from equilibrium (steady states). The latter are very susceptible to external perturbations and under proper conditions may respond nonlinearly (and selectively) to external stimuli. Biological cells are self-regulating and self-reproducing objects which exhibit a high degree of organization and cooperation in many instances taking the form of long-range communication which is still poorly understood. This internal organization is functional rather than spatial in its character. Most of these remarkable properties cannot be properly explained by conventional approaches. Since chemical forces act over short ranges and Coulomb forces are effectively screened by the abundant water molecules one is tempted[7] to search for long-range forces which are associated with symmetry-breaking phenomena. Various forms of equilibrium ordered phases have been detected in biosystems, e.g. biomagnetism[8], ferroelectricity[9], organic superconductivity or liquid crystalline phases of membranes[10]. However, they seem to be locally confined and do not reflect the physiological state of the system. It should be emphasized that many constituents of biosystems possess electric dipoles[11], e.g. cell membranes are strongly polarized by transmembrane fields of $10^7 \frac{V}{m}$ and they also contain polar peptides and proteins, cell surfaces have ionic double layers, due to hydrogen bonds DNA and RNA have large dipole moments, finally the many delocalized electrons during chemical reactions may significantly contribute to polarity. It was claimed by Fröhlich[7] that these various types of dipoles can execute high frequency longitudinal oscillations (10^9-10^{14} Hz). Their interactions may be the key to understanding broken symmetries in biological cells. In this paper we intend to briefly review the early phenomenological models of symmetry breaking in biology and then concentrate on the more recent microscopic approaches. We attempt to establish if they are generally plausible rather than to give a detailed mathematical presentation.

MACROSCOPIC MODELS

Initial attempts at describing symmetry-breaking effects in biosystems applied the methodology of Prigogine's reaction-diffusion equations[12]

$$\eta_t = N_\alpha(\eta(\vec{r}), \vec{\nabla}\eta) + F(t) \tag{1}$$

where η is a (possibly multicomponent) order parameter, N_α is a nonlinear function describing the evolution of the system and parametrized by a control parameter α and F(t) represents a fluctuating force. These types of equations possess a very rich structure of solutions which depend on the form of N_α and F and embrace spatial pattern formation in steady states, various time-dependent solutions such as: pulses, fronts, spiral waves, periodic wave trains, limit cycles and chaotic solutions. The earliest example is that of the Lotka-Volterra system used to describe predator-prey population dynamics. Enormously successful applications of this methodology cover many disciplines of life sciences, e.g. embryology[12,13,14] (morphogenesis of cells), ecology[12], cardiology (circulation of blood in the heart), neurology[14] (nerve conduction in excitory-inhibitory neuronetworks), etc.

Haken[15] developed a theory called synergetics which describes various systems far from thermodynamic equilibrium exhibiting such features as cooperation and self-organization. Mathematical models of synergetics are often patterned after the laser optics and involve systems of coupled nonlinear differential equations with instabilities around which one distinguishes bifurcating variables (order parameters) and non-bifurcating variables (slaved modes). The latter ones are eliminated using a reduction procedure which often results in a laser-type equation

$$\frac{dE}{dt} = \alpha E - \beta|E|^2E + F(t). \tag{2}$$

The onset of coherence is achieved by α becoming positive due to the population inversion occurring when the pump exceeds a critical rate. The coefficient β is positive throughout to stabilize the process and F(t) includes random noise and a coherent injected signal which locks the phase of E(t). Such description is amenable to biological cells since, in contrast to equilibrium phase transitions, biological order arises by an energy input.

Finally, a number of phenomenological models were inspired by Fröhlich[7] who placed emphasis on the oscillations of the electric dipoles of the membrane and postulated thier role as regulators of mitotic activity. Growth control could then result from interactions between elastic deformation η and dielectric polarization P. A slightly oversimplified stationary model has been developed[16] by a Landau expansion of the free energy F in terms of η and P which are considered coupled order parameters

$$F = F_o + A_2P^2 + A_4P^4 - E \cdot P + B_2\eta^2 + B_4\eta^4 + P^2\eta\ (C_1 + C_2\eta) \qquad (3)$$

A first order field-induced phase transition takes place when $E \geqq E_c(T)$ and $T < T_o$ between a paraelectric phase and a ferroelectric, deformed phase, provided the coupling coefficients C_1 and C_2 are large enough. This, then, could provide a feasible description of division mechanism as geared towards minimizing the surface energy. An extension of this model to include dynamic effects requires an addition of kinetic energy, ohmic dissipation, time-dependent driving force and spatial inhomogeneities. The resultant equations of motion for P and η have been derived[16] as coupled nonlinear partial differential equations

$$m\,P_{tt} + \gamma P_t - D_1\nabla^2 P = (A_2 + C_1\eta + C_2\eta^2)P + 2A_4P^3 - E(t) \qquad (4a)$$

$$M\,\eta_{tt} + \Gamma\eta_t - D_2\nabla^2\eta = (B_2 + C_2P^2)\eta + 2B_4\eta^3 + C_1P^2 \qquad (4b)$$

They clearly conform to the reaction-diffusion formalism of Prigogine. Based on somewhat similar model equations[17] which were solved numerically, it is expected that the form of their solutions strongly depends on the initial conditions and may display transition from periodicity to chaos.

MICROSCOPIC MODELS

As a starting point for the development of microscopic models of long-range order in biomembranes consider the longitudinal dipole oscillations u_i of their segments described using the Hamiltonian[18]

$$H_o = \frac{1}{2}\sum_{i=1}^{N} m_i(\dot{u}_i^2 + \omega_i^2 u_i^2) - \frac{1}{2}\sum_{i\neq j=i}^{N} u_i V_{ij} u_j \qquad (5)$$

where the interaction term was given by Fröhlich[7] as

$$V_{ij} = \gamma^2 e^2 \frac{\sqrt{z_i/m_i}\,\sqrt{z_j/m_j}}{R_{ij}^3\ \varepsilon'(\omega)} \qquad (6)$$

where z_i is the number of bound charged particles, R_{ij} is the instantaneous spacing, $R_{ij} = r_{ij}^o + u_i - u_j$ and $\varepsilon'(\omega)$ is the real part of the frequency-dependent dielectric constant of the whole medium. Precisely because $\varepsilon' = \varepsilon'(\omega)$ Fröhlich[7] showed for two interacting dipoles, and it was also demonstrated[18] for a regular chain of dipoles, that the effective interaction energy is at resonance ($\omega_i = \omega_o$ for all $1 \leqq i \leqq N$) of long-range type ($1/R^3$) while in the other situations it is only of a Van der Waals type ($1/R^6$). Assuming equal masses of dipoles and expanding R_{ij}^{-3} around equilibrium r^o_{ij} according to $R^{-3}_{ij} \cong (r^o_{ij})^{-3}$ $[1 - 3(\Delta u_{ij}/r^o_{ij}) + 6(\Delta u_{ij}/r^o_{ij})^2 + \ldots]$, where $\Delta u_{ij} = u_i - u_j$, yields for the nearest neighbor approximation the following nonlinear differential-difference equation

$$m_i \ddot{u}_i = f_1(u_{i+1} + u_{i-1} - 2u_i) + f_2[(u_{i+1} - u_i)^2 - (u_i - u_{i-1})^2]$$
$$+ f_3[(u_{i+1} - u_i)^3 - (u_i - u_{i-1})^3] + \ldots \qquad (7)$$

When $f_3 \cong 0$ this represents the equation of the famous Fermi-Pasta-Ulam problem[11] which displays a remarkable lack of thermalization of input energy into various modes but, rather, a storage of it in the lowest mode. This is understandable in light of Zabusky's proof that this anharmonic lattice equation is transformed into the KdV equation when the continuum limit is taken. A quantum mechanical treatment of this problem has been carried out[19] and also resulted in a KdV equation derived from the evolution equation for the expectation value of the annihilation operator within a coherent state. Hence, the one-soliton solution is interpreted as a Poisson-distributed superposition of phonons. Another recent study[20] showed that depending on the values of coupling constants f_1, f_2, f_3 one obtains a full range of behavior between the normal modes and local modes which includes sinusoidal, periodic, quasi-periodic and chaotic solutions. Furthermore, Pnevmatikos[21] demonstrated that in the long-wavelength limit eq. (7) with $f_3 \neq 0$ reduces to the generalized Boussinesq equation for u_x. Again, depending on the coefficients one finds among its solutions topological solitons (dislocations) and non-topologocial solitons (rarefactions or compressions). Since broken symmetries are often manifested by solitons, these results are very encouraging in the context of our model. In our case, however, not only do we have to evaluate the parameters f_1, f_2, f_3 from ab initio calculations but also we should investigate the role of the dielectric constant since all of these parameters depend on it. This, therefore, suggests a self-consistent calculation where the solutions of (7) depend on $\varepsilon'(\omega)$ and, in turn, $\varepsilon'(\omega)$ depends on the regime of the solutions since the dipoles are a part of the whole system. A calculation of $\varepsilon'(\omega)$ for biological membranes was undertaken before[22] but it envisaged the dipoles as harmonic oscillators which are coupled to a heat bath and an energy pump and lacked direct coupling amongst themselves. A more realistic approach must include nonlinear coupling between dipoles using the Hamiltonian (5) - (6). Then, one can assume

$$\varepsilon(\omega) = v_m\, \varepsilon_m(\omega) + (1 - v_m)\, \varepsilon_w(\omega) \qquad (8)$$

where m and w refer to the membrane dipoles and water molecules, respectively, and v denotes a volume fraction of each, For ε'_w we can adopt Debye's expression

$$\varepsilon'_w(\omega) = \varepsilon_\infty + \frac{\varepsilon_s - \varepsilon_\infty}{1 + \omega^2\tau^2} \qquad (9)$$

where all the constants are well-known and ε'_m must be calculated ab initio using the Hamiltonian (5) with an interaction $-\vec{D} \cdot \vec{E}(t)$ where $\vec{D}$ is the total dipole moment of the system; $\vec{D} = e \sum_i z_i \vec{u}_i$. Then,

$$\varepsilon''(i\omega) = \frac{2\pi}{V\hbar} [1 - \exp(-\beta\hbar\omega)] \int_{-\infty}^{\infty} dt \, \langle[\vec{D}, \vec{D}(t)]\rangle \exp(-i\omega t) \tag{10}$$

where $\vec{D}(t) = e^{iHt} \vec{D} e^{-iHt}$ and $\varepsilon'(\omega)$ can be obtained from $\varepsilon''(\omega)$ using the Kramers-Kronig theorem

$$\varepsilon'(\omega) = 1 + \frac{1}{\pi} P \int_{-\infty}^{+\infty} \frac{\varepsilon''(\nu)}{\nu-\omega} d\nu \tag{11}$$

If fragmentary experimental data and previous model calculations can be any indication, we can expect various resonant "dips" of $\varepsilon'(\omega)$ which, when returned to the equation of motion for u_i, could reinforce certain modes at the expense of others providing the necessary positive feedback.

As mentioned earlier, membranes lack the perfect translational invariance of crystals due to the presence of various macromolecules. Many of these molecules are more massive than our oscillating dipole segments and their concentration is rather low. The situation resembles that encountered in impurity scattering of insulating ferromagnets[23] and with nearest-neighbor interactions included only, our Hamiltonian (5) could be transformed into an analogue of Anderson's Hamiltonian. The question that should then be answered is whether in real biomembranes the impurity scattering is strong and results in disorder and localized bands of ω_i or it is weak and allows for the use of extended states and perturbation approach to the impurity scattering. To the best of our knowledge this aspect of the problem has not been dealt with and it requires careful consideration. We tend to believe that in most situations the impurity dipoles can be considered annealed and an effective Hamiltonian can be constructed by projecting them out. On attaining the mobility edge and forming energy bands, the effective Hamiltonian can be written in normal modes k. Using boson creation and annihilation operators for them, a_k^+ and a_k, respectively, and accounting for two-particle scattering processes one can assume the form of H_o as

$$H_o = \sum_k \hbar\omega_k a_k^+ a_k + \sum_{kk'q} V_q \, a_{k'+q}^+ \, a_{k-q}^+ \, a_k a_{k'} \tag{12}$$

As discussed earlier, it is conceivable that as a result of an interplay between the structural and dynamical parameters of the membrane on the one hand and the dielectric properties of the medium on the other hand, one or more modes can be strongly favoured. In analogy to superfluidity an approximate Hamiltonian can be written with pairing effects and some extra terms as

$$H_o \cong \sum_k (NV_o + \hbar\omega_k) \, a_k^+ a_k + \tfrac{1}{2} \sum_k [\sum_{k' \neq \pm k} V_{k-k'}^* \, \langle a_{k'}^+, a_{k'}^+ \rangle \, a_k a_k]$$

$$+ \tfrac{1}{2} \sum_k [\sum_{k' \neq \pm k} V_{k-k'} \, \langle a_{k'}, a_{k'} \rangle \, a_k^+ a_k^+] + \sum_{k'q}' V_{k'}^* \, \langle a_q a_{q-k'}^+ a_{k+k'}^+ \rangle \, a_k$$

$$+ \sum_{k'q}' V_{k'} \, \langle a_q^+ \, a_{q-k'} \, a_{k+k'} \rangle \, a_k^+ \tag{13}$$

where Σ' denotes summation over $k' \neq 0$, $q \neq -k$ and $k' \neq q - k$. This Hamiltonian can subsequently be diagonalized using two canonical transformations. First, the Bogolyubov transformation employs $D_k \equiv \exp(\gamma_k a_k^+ a_{-k}^+ - \gamma_k^* a_k a_{-k})$ to give

$$\eta_k = D_k a_k D_k^+ = \frac{a_k - g_k a_{-k}^+}{(1 - |g_k|^2)^{\frac{1}{2}}} \tag{14}$$

where $g_k = \gamma_k \tanh |\gamma_k| / |\gamma_k|$ and it yields for H_o

$$H_o = E_o + \sum_k \hbar\bar{\omega}_k (\eta_k^+ - \alpha_k^*)(\eta_k - \alpha_k) \tag{15}$$

where α_k is a complex number. Then, the transformed Hamiltonian can be fully diagonalized using $G(\alpha_k) = \exp(\alpha_k \eta_k^+ - \alpha_k^* \eta_k)$ so that

$$A_k = G(\alpha_k) \eta_k G^+(\alpha_k) = \eta_k - \alpha_k \tag{16}$$

and

$$H_o = E_o + \sum_k \hbar\bar{\omega}_k A_k^+ A_k \tag{17}$$

The last step is called a field translation operation and it gives the ground state of H_o in the form of a coherent state

$$|\Phi_o\rangle = \exp[\sum_k (\alpha_k a_k^+ - \alpha_k^* a_k)] \exp[\sum_\ell (\gamma_\ell a_\ell^+ a_{-\ell}^+ - \gamma_\ell^* a_\ell a_{-\ell})] |0\rangle \tag{18}$$

where $a_\ell |0\rangle = 0$. The excited states can be described as

$$|\{n_k\}\rangle = \prod_k (n_k!)^{-\frac{1}{2}} A_k^+ |\Phi_o\rangle \tag{19}$$

A number of interesting properties follow. First, the ground state is a broken symmetry state. It is degenerate with respect to the gauge transformation $\alpha_k \to \alpha_k e^{i\phi}$ and $g_k \to g_k e^{2i\phi}$. It is a state where η-bosons (paired quasiparticles) are condensed and their number in the condensate is $\langle\Phi_o|A_k^+ A_k|\Phi_o\rangle = |\alpha_k|^2$ and it must be macroscopic to correspond to Bose-Einstein condensation. This is a phase transition which can be described using the expectation value of the field operator as an order parameter ψ

$$\psi \equiv \langle\Phi_o|\Psi|\Phi_o\rangle = \sqrt{N_o}\, e^{i\phi} \tag{20}$$

where $\Psi \equiv N^{-\frac{1}{2}} \sum_k e^{i\vec{k}\cdot\vec{r}} A_k$ and N_o is the number of condensed particles and ϕ is their phase. Below a critical temperature $N_o \neq 0$ and it is reduced to 0 as $T \rightarrow T_c$ due to the reduction in the magnitudes of α_k as a result of thermal fluctuations. Below T_c the system is also characterized by off-diagonal long range order, i.e. $<\Psi^+(r)\Psi(r')> \cong <\Psi^+(r)><\Psi(r')>$ and long-range phase correlations, i.e. $<\hat{\phi}(\vec{r})\hat{\phi}(\vec{r}')> - <\hat{\phi}(\vec{r})><\hat{\phi}(\vec{r}')> \sim |\vec{r} - \vec{r}'|^{-1}$. These properties bring the phenomena of superfluidity, superconductivity and laser action together and have an enormous appeal as possible prototypes for the models of metabolic activity in biological cells. This is even more enticing in light of the fact that coherent states, as minimum uncertainty states, form a representation which is closest to the classical formalism. Hence both the amplitude and the phase of the order parameter ψ are macroscopic quantities. To remove the phase degeneracy a small phase-locking signal can be injected into the system by applying an external coherent field that couples to ψ. In our case at 0°K the average dipole moment in the ground state is calculated as

$$<\vec{d}>_o = zeN^{-\frac{1}{2}}\sqrt{\frac{\hbar}{2m}}\sum_k \omega_k^{-\frac{1}{2}} e^{i\vec{k}\vec{r}} (\alpha_k + \alpha^*_{-k}) \frac{1 + g_k}{\sqrt{1 - |g_k|^2}} \tag{21}$$

and it crucially depends on the existence of nonzero field translation. Future studies should focus on the feasibility of this approximation in biomembranes.

So far we have merely formulated an equilibrium description of the membrane without including very important perturbation effects due to the heat bath (which maintains a constant temperature in the system by providing a reservoir of energy) and due to an energy pump. Wu and Austin[25] suggested the following form of the perturbation Hamiltonian

$$H_1 = \sum_m \hbar\Omega_m b_m^+ b_m + \sum_p \hbar W_p P_p^+ P_p + \sum_{k\,m} \hbar(\lambda b_m^+ a_k + \lambda^* b_m a_k^+)$$

$$+ \frac{1}{2}\sum_{kk'm} \hbar(\chi a_k^+ a_{k'} b_m^+ + \chi^* a_k a_{k'}^+ b_m) + \sum_{k\,p} \hbar(\xi P_p a_k^+ + \xi^* P_p^+ a_k) \tag{22}$$

where b_m^+, b_m are boson ladder operators of the heat bath and P_p^+, P_p are boson ladder operators of the energy pump. The terms in H_1 describe scattering processes between one or two dipole modes of the membrane and single quantum excitations of the heat bath and the pump. They are likely to cause the lifetime of a condensate's mode to become finite. The main problem, however, is whether or not the condensate will be entirely or only partially depleted. The state of the heat bath is in all likelihood incoherent and the energy pump may have a coherent component but its frequency is many orders of magnitude lower that the 10^{11} Hz assumed by Fröhlich as supported by the membrane. Therefore, we believe that the phenomenon of biological coherence cannot be achieved by considering just the effects of H_1 on harmonic dipoles with $H_o \cong \sum_k \hbar\omega_k a_k^+ a_k$, as was done in the past. This would be analogous to a classical damped driven harmonic oscillator whose response is resonant at $\omega = \omega_o$ which would mean 10^{11} Hz for the pump's frequency. Thus, the full Hamiltonian for the membrane described as an open, far-from-equilibrium system should be

$$H = H_o + H_1 \tag{23}$$

where H_o is given in (12) and under critical conditions may become (13) while H_1 is that of (22). This Hamiltonian is very complicated and at this stage we can offer only fragmentary insights into its properties. In a recent paper[26] a similar Hamiltonian has been considered where a Bose condensed system is coupled bilinearly to a reservoir and to a coherent pump. The result is that the dispersion relation for ω_k as $k \to 0$ is still gapless and linear in k as $\omega_k \cong \pm vk - i\gamma k^2$ where v is the sound velocity. Although the damping causes the condensed particles to acquire a finite lifetime, the population of the condensate is depleted to a fraction of N_o but may still constitute a macroscopic number due to the pumping effect. Since $\gamma \sim N_o^{-\frac{1}{2}}$, the influence of the damping is reduced with increased condensate density. Glauber and Manko[27] considered a Hamiltonian describing a system of harmonic oscillators bilinearly coupled to heat baths and classical forcing fields. They diagonalized the Hamiltonian using a unitary transformation and introduced a coherent state representation to calculate the field translation. Their result is a combination of a translation due to the coupling with the heat baths and a response function to the field. The field translated state is time-dependent and stable but it differs from the thermodynamic equilibrium state.

For the Wu-Austin Hamiltonian $H \cong \sum_k \hbar\omega_k a_k^+ a_k + H_1$ a series of two canonical transformations was applied[28] to find the influence of the heat bath and energy pump modes on the dipole oscillations of the membrane. The resultant effective Hamiltonian is to the lowest order

$$H_{eff} \cong \sum_{kk'} \hbar w_{kk'} a_k^+ a_{k'} + \sum_{kk'q} \hbar\Delta_{kk'q} a_{k'+q}^+ a_{k-q}^+ a_k a_{k'} \qquad (24)$$

which introduces momentum non-conserving terms whose magnitude is proportional to $|\xi|^2$ and $|\lambda|^2$ and two-particle scattering terms proportional to $|\chi|^2$. Although the latter term resembles the two-body term of the membrane Hamiltonian H_o of (12), their roles are quite different. The almost continuum spectrum of heat bath modes which act incoherently will tend to destroy the long-range correlations between the condensed quasi-particles. The pump may counter the loss of momentum but the end result will depend on the balance of all three factors combined: the correlations of vibrating dipolar modes, the dissipative interactions with the heat bath and the pumping of energy into the membrane. In fact, Paul[29] calculated the magnitude of the field translation for the Wu-Austin Hamiltonian as

$$\alpha_k = - \frac{\chi\lambda^*}{2\hbar^2\bar{\omega}_k} \sum_{k'm} \frac{n(\hbar\bar{\omega}_{k'}) - n(\hbar\bar{\Omega}_m)}{\bar{\omega}_{k'} - \bar{\Omega}_m} \qquad (25)$$

where $n(\varepsilon) = [\exp(\beta\varepsilon) - 1]^{-1}$ is the equilibrium occupation number for energy level ε and $\bar{\omega}_k = \omega_k + \Sigma_r \pm i\Sigma_i$ where Σ_r and Σ_i are the real and imaginary parts of the self-energy for the dipoles and $\bar{\Omega}_m$ is defined similarly for the heat bath. It is thus conceivable that this field translation of (25) may act in the opposite direction to the one of (16). Therefore, the effects of self-organization can be destroyed as a result of dissipation or salvaged by sufficiently strong energy pumps.

A further insight may be gained by considering the rate equation for the occupation number of k-th dipolar mode $n_k = a_k^+ a_k$. According to the Heisenberg equation

$$\dot{n}_k = (i\hbar)^{-1} [n_k, H] \tag{26}$$

Calculating the expectation value of n_k within the ground state $|\Psi_o(t)>$ of the entire Hamiltonian $H_o + H_1$ and using time-dependent perturbation theory with the interactions adiabatically switched on the following rate equation is obtained in the lowest order of coupling constants[30]

$$\begin{aligned} \dot{\bar{n}}_k &= S - \phi(T,\omega_k) [\bar{n}_k \exp(\beta\hbar\omega_k) - \bar{n}_k - 1] \\ &- \sum_{k'}' \Lambda(T,\omega_k,\omega_{k'}) [\bar{n}_k(1 + \bar{n}_{k'})\exp(\beta\hbar(\omega_k - \omega_{k'})) - \bar{n}_{k'}(1 + \bar{n}_k)] \\ &- \sum_{k'} \Gamma\, \bar{n}_k \bar{n}_{k'} \end{aligned} \tag{26}$$

where $\beta = (kT)^{-1}$ $\phi \sim |\lambda|^2$; $\Lambda \sim |\chi|^2$, $S \sim |\xi|^2$ and Γ describes the effects due to the correlation terms. This, in fact, is a system of coupled non-linear ordinary equations whose exact solution is unknown. However, in the stationary state ($\dot{\bar{n}}_k = 0$) one obtains the following expression for the average occupation number

$$\bar{n}_k = \left(1 + \frac{S}{\phi + \sum_{k'} \Lambda \bar{n}_{k'}}\right) \frac{1}{\exp[\beta(\hbar\omega_k - \mu) - 1]} \tag{27}$$

where the chemical potential μ is a function of Λ, ϕ and Γ. It can be demonstrated[30] that provided $|\chi|^2$ is less than a threshold value $|\chi_o|^2$ and $|\lambda|^2$ is greater than another threshold value $|\lambda_o|^2$ which depends on χ, Bose condensation may not take place in the system. These requirements become more and more severe as the temperature increases. To see that this corresponds to a dramatic enhancement of a single mode ko we can rewrite the rate equation for it schematically as

$$\dot{\bar{n}}_{ko} = A\, \bar{n}_{ko} + B\, \bar{n}_{ko}^2 + C \tag{28}$$

and for the other modes as

$$\dot{\bar{n}}_k = A'\, \bar{n}_k + C' \tag{29}$$

where we have neglected the various couplings between modes. These equations resemble those of population dynamics of various species in Haken's theory of synergetics[15] and are also reminescent of the laser action. The first equation has a bifurcation point at $A = 0$ and if the condition of population inversion prevails ($A > 0$), in spite of the presence of the stabilizing term with $B < 0$, the population dynamics of n_{ko} is described by an exponential growth followed by asymptotic saturation to its limit of n_1

$$n_{ko}(t) = \frac{n_1(n_0 + n_2) + n_2(n_0 - n_1)\exp[(n_1 - n_2)Bt]}{n_0 + n_2 + (n_1 - n_0)\exp[(n_1 - n_2)Bt]} \quad (30)$$

where $n_o = n_{ko}(0)$ and n_1, n_2 ($n_1 > n_2$) are the two roots of the right hand side of (28). At the same time the other modes' dynamics is given by an exponential relaxation into an equilibrium value which is non-zero as long as the pump sustains it ($C' \neq 0$)

$$n_k(t) = n_k^o \exp(A't) - C'/A' \quad (31)$$

where $n_k^o = n_k(0)$, $A' < 0$ and $C' > 0$. The form of the rate equation for ko is intimately related to the spontaneous symmetry breaking in the system and strongly depends on the types of dissipation mechanisms present and the external pumping rate.

Finally, it is also of interest to analyze the dynamics of the order parameter. We first derive the Heisenberg equation for a_k using the full Hamiltonian H of (23). We then Fourier-transform both sides of the equation and eliminate the reservoir operators b_m^+, b_m by introducing[31] a white noise function $f(t)$ and a damping coefficient γ. The pump is treated for simplicity as a classical field $S(t)$. The resultant equation of motion for the order parameter ψ is to the lowest order of approximation a damped-driven nonlinear Schrödinger equation

$$i\psi_t = (\varepsilon + i\gamma)\psi + \alpha\psi_{xx} + \delta|\psi|^2\psi + f(t) + S(t) \quad (32)$$

where γ increases as a result of dissipation and δ decreases when the correlations diminish due to a decreasing density of the condensate. Equations similar to this have recently been studied[32] and the results indicate that for small values of γ and external fields a phase-locked soliton is a fixed point attractor. Increasing the values of these parameters results in the fixed point bifurcating to a periodic cycle and the soliton amplitude undergoing oscillations. A further increase in $S(t)$ leads to period doubling, bifurcations which terminate at a critical value and lead to chaotic solutions. As in all other aspects of the problem there appears to be a delicate balance between order and disorder. Achieving order in the system may be a result of the interplay between internal forces of correlation, external random disturbances and coherent pumping. This may manifest itself in long-range phase correlations, Bose condensation, non-zero order parameter, the existence of solitons, etc. So far there have been several types of experiments performed to verify the Fröhlich conjecture. They can be put in three categories: (i) ESR resonance; (ii) Raman scattering and (iii) irradiation observations. There seem to be indications that in metabolically active cells one or more resonant frequencies exist in the microwave region but no conclusive evidence has been found yet. Perhaps the most convincing results have been obtained in erythrocytes which exhibit anomalous light scattering, have a resonant line at 36 GHz and rapidly form clusters called rouleaux, all of it taking place only under a set of conditions corresponding to Fröhlich's prerequisites.

Another microscopic theory of biological order was proposed by Davydov[33] to explain the remarkable electron and energy transfer and storage on a molecular level. He considered quasi-linear molecular chains as e.g. peptides $(H\text{-}N\text{-}C{=}O)_n$, DNA, etc., and postulated the following Hamiltonian

$$H = \frac{1}{2}\sum_n [M\,\dot{x}_n^2 + K(\Delta x_n)^2] - \sum_n [D\,B_n^+ B_n + J\,(B_{n+1}^+ B_n + B_{n+1} B_n^+)] \qquad (33)$$

where Δx_n is the displacement at site n, B_n^+ and B_n are ladder operators of an exciton at this site and constants D and J are assumed to depend linearly on Δx_n. Davydov assumed the ground state $\Phi_o(t)$ of H as a coherent state which allowed him to calculate its amplitude A(x,t) from Hamilton's equations for the Hamiltonian function $H = \langle\Phi_o(t)|H|\Phi_o(t)\rangle$. In the continuum limit he obtained a nonlinear Schrödinger equation for the amplitude coupled to a wave equation for the displacement field. The solution to this system is a soliton propagation of the exciton field (called the Davydov soliton) followed by a displacement field proportional to $sech^2(x)$. Both propagate with a sobsonic velocity and have permanent profiles. The window for soliton formation has been foujd numerically as a narrow interval for the coupling constant J divided by the latttice spacing. Recently, a calculation was done[34] which included thermal effects in the form of white noise and ohmic dissipation. The results indicate that at room temperatures Davydov's self-trapped states are destroyed and chaotic solutions emerge instead.

CONCLUSIONS

Almost too decades after Fröhlich presented his theory of biological order more questions seem to be asked than answered about its validity. We have shown that it is in general plausible to expect microscopically broken symmetries in biological systems but conditions leading to these effects are very restricted and their precise formulation is unknown. Special attention must be given to the structural parameters of biomembranes in order to validate the assumption about strong correlations between dipole modes. The role of dissipation and external pumping must be analyzed at room temperatures requiring the use of a finite-temperature field theory for open systems. A suitable formalism has now been constructed[35] and it is hoped that it will soon be applied to this problem. Finally, more experiments are needed to find more information about the physical conditions of biological cells in various physiological situations.

ACKNOWLEDGEMENT

The author is greatly indebted to Dr. J. P. Whitehead for his many valuable comments.

REFERENCES

1. P. W. Anderson, "Basic Notions of Condensed Matter Physics", The Benjamin/Cummings, Menlo Park (1984).
2. L. D. Landau and E. M. Lifshitz, "Statistical Physics", Pergamon Press, London (1959).
3. H. Umezawa, H. Matsumoto and M. Tachiki, "Thermo Field Dynamics and Condensed States", North-Holland, Amsterdam (1982).
4. H. Frauenfelder, Helv. Phys. Acta 57: 165 (1984).
5. A. Szent-Györgi, Nature 148: 157 (1941).
6. H. Fröhlich, Nature 228: 1093 (1970).
7. H. Fröhlich, IEEE Trans. MIT 26: 613 (1978).
8. S. J. Williamson and L. Kaufman, J. Magn. Magn. Mat. 22: 129 (1981).

9. B. T. Matthias, Organic ferroelectricity, in: "From Theoretical Physics to Biology", M. Marois, ed., S. Karger, Basel (1973).
10. T. Izuyama and Y. Akutus, J. Phys. Soc. Jap. 51: 50 (1982).
11. H. A. Pohl, Coll. Phenom. 221 (1981).
12. I. Prigogine, "From Being to Becoming", Freeman, San Francisco (1980).
13. P. Ortoleva, Symmetry-breaking in far-from equilibrium order, in: "Symmetries in Science", B. Gruber and R. S. Millman, eds., Plenum Press, New York (1980).
14. J. D. Cowan, Symmetry-breaking in embryology and in neurology, ibid.
15. H. Haken, "Synergetics, an Introduction", Springer, Berlin (1980).
16. J. A. Tuszyński, Phys. Lett. A108: 177 (1985).
17. Z. Szabo and F. Kaiser, Z. Naturforsch. C37: 733 (1982).
18. J. A. Tuszyński, Phys. Lett. A107: 225 (1985).
19. Y. H. Ichikawa, N. Yajima and K. Takano, Prog. Theor. Phys. 55: 1723 (1976).
20. A. C. Scott, P. S. Lomdahl and J. C. Eilbeck, Chem. Phys. Lett. 113: 29 (1985).
21. S. N. Pnevmatikos, Solitons in nonlinear atomic chains, in: "Singularities and Dynamical Systems", S. N. Pnevmatikos, ed., Elsevier Science, Amsterdam (1985).
22. R. Paul, J. A. Tuszyński and R. Chatterjee, Phys. Rev. A30: 2676 (1984).
23. D. J. Thouless, Percolation and localization, in: "Ill-Condensed Matter", North-Holland, New York (1979).
24. F. W. Cummings and J. R. Johnston, Phys. Rev. 151: 105 (1966).
25. T. M. Wu and S. Austin, J. Theor. Biol. 71: 209 (1978).
26. H. Haug and H. H. Kranz, Z. Phys. B53: 151 (1983).
27. R. J. Glauber and V. I. Manko, Academy of Sciences of USSR Preprint #96, Moscow (1984).
28. J. A. Tuszyński, R. Paul, R. Chatterjee and S. R. Sreenivasan, Phys. Rev. 30: 2666 (1984).
29. R. Paul, Phys. Lett. A96: 263 (1983).
30. T. M. Wu and S. Austin, Phys. Lett. A65: 74 (1978).
31. M. Sargent III, M. O. Scully and W. E. Lamb, Jr., "Laser Physics", Addison-Wesley, London (1974).
32. N. Bekki and K. Nozaki, unpublished report (1984).
33. A. S. Davydov, "Biology and Quantum Mechanics", Pergamon, New York (1982).
34. P. S. Lomdahl and W. C. Kerr, Phys. Rev. Lett. 55: 1235 (1985).
35. T. Arimitsu and H. Umezawa, Prog. Theor. Phys. 74: 429 (1985).

A NON-EQUILIBRIUM THEORY FOR PHASE TRANSITIONS AND THE EXPANDING UNIVERSE

Hiroomi Umezawa

Department of Physcs
University of Alberta
Edmonton, Alberta T6G 2J1 Canada

INTRODUCTION

This paper presents a very brief sketch of three topics concerning thermo field dynamics (TFD), a quantum field theory with thermal degrees of freedom. The basic formalism of non-equilibrium TFD will be presented and then its application to a study of the growth of an ordered state (a state of broken symmetry) through the course of non-equilibrium thermal transition from a non-ordered state (i.e. a symmetric state) will be discussed. Lastly the application to a study of non-equilibrium expansion of a system such as the universe will be touched upon. Although further analysis is needed to obtain many physical results within these applications, these studies clarify several aspects of non-equilibrium TFD, which is in the early stage of its development. Summarizing, the purpose of this paper is to present the basic structure of non-equilibrium TFD by presenting its construction and its applications.

THE BASIC FORMULATION OF NON-EQUILIBRIUM TFD

Real time quantum field theory with thermal degrees of freedom has attracted much attention from many physicists. Among the many approaches, the thermo field dynamics (TFD)[1-6] most faithfully adopts the structure of the usual quantum field theory. Thus, TFD preserves the operator formalism realized in a linear vector space, the Green function formalism and the path-integral formalism. The formalism of TFD for equilibrium situations has been well established and has been applied to many problems by many physicists.[1] However, the development of non-equilibrium TFD has started only recently.

The most remarkable feature of TFD lies in the fact that TFD is built, not on any stochastic concept such as the probability distribution or mixed states, but on the concept of a pure state as is the case in the usual quantum field theory. In TFD any thermal situation is represented by a thermal state which is represented by a vector in a linear vector space called the thermal space. Any temporal development of a system is described as a change of the thermal state . The change of thermal state should be generated by a time-translation generator, that is by the Hamiltonian. Therefore, knowledge of the Hamiltonian together with the initial thermal state determines entirely the behavior of a system. In this sense, the two concepts, that is the thermal state and the Hamiltonian, form the most basic set of concepts in TFD. The thermal space consists of all of the thermal states including the thermal vacuum states. In equilibrium situations we need only one thermal vacuum. In this case the other states are those states which contain some non-thermally excited quanta. However, in the case of non-equilibrium situations, the thermal vacuum state itself changes in time, giving rise to an infinite choice of thermal vacua. Each of the thermal vacua is then specified by a condition called the thermal state condition, which has a form (certain operator)$|0\rangle = 0$ or $\langle 0|$(certain operators) $= 0$, where $|0\rangle$ and $\langle 0|$ are the thermal vacua. In order to be able to introduce such a condition without causing a decrease in the degrees of freedom, we should prepare the theory with an increased number of degrees of freedom. Therefore, in TFD, the number of degrees of freedom is doubled; each field is a thermal doublet. In this paper, for simplicity, we consider the operators, $a(\vec{k})$ and $a^{\dagger}(\vec{k})$, which satisfy the commutation relations of the harmonic oscillator type:

$$[a(\vec{k}),a^{\dagger}(\vec{\ell})]_{\sigma} \equiv a(\vec{k})a^{\dagger}(\vec{\ell}) - \sigma a^{\dagger}(\vec{\ell})a(\vec{k}) = \delta_{\vec{k}\vec{\ell}}\ , \text{ etc.},\quad \sigma = \pm 1. \tag{1}$$

With these operators are associated the so-called tilde operators, $\tilde{a}(k)$ and $\tilde{a}^{\dagger}(k)$, satisfying the same commutation relations as above and σ-commuting with non-tilde operators. In general, to any operator A is associated a tilde operator $\tilde{A}$. The tilde operation satisfies the so-called tilde conjugation rules, which state

$$[AB]^{\sim} = \tilde{A}\tilde{B}, \tag{2a}$$

$$[c_1A + c_2B]^{\sim} = c_1^{*}\tilde{A} + c_2^{*}\tilde{B}. \tag{2b}$$

Thus the concept of tilde conjugation becomes the third basic concept in TFD. Although TFD has been formulated in more complicated cases such as spin systems, Anderson models, etc.[7,8,9], in this paper, we consider the operators of the harmonic oscillator type only, because the consideration in this paper can easily be extended to more general kinds of operators algebras. Summarizing, the basic concepts in TFD are the tilde conjugation, the Hamiltonian and the thermal state condition. The initial thermal state conditions specifies the choice of initial thermal state, and then, the knowledge of the Hamiltonian determines the temporal development of the thermal state.

A central problem in TFD is the construction of the Hamiltonian. We find that the renormalized Hamiltonian for non-equilibrium processes are mostly time-dependent[10,11]; the time-independent renormalized Hamiltonian is very exceptional. Note that the famous Schwinger-Keldysh method[12] for perturbative calculation of non-equilibrium phenomena cannot treat a time-dependent unperturbed Hamiltonian, while non-equilibrium TFD is able to deal with such a complex and physically important situation.[11] This is a particular merit of the TFD . The main cause of the time-dependence of the Hamiltonian is a higher order loop corrections which make almost all the parameters such as the mass, coupling constant, the dissipative coefficient, etc., dependent on time. We will see this explicitly in the next section.

In this paper, we examine non-equilibrium TFD using the interaction representation in which the renormalized unperturbative Hamiltonian is bilinear in field operators. The effects of an interaction Hamiltonian will be taken into account as perturbative interaction effects.

In general, the renormalized unperturbated Hamiltonian contains a dissipative term. To understand this, we recall that all of the observable operators consist only of non-tilde operators; in other words, the tilde fields act as a kind of shadow field. Furthermore, the tilde field have negative energies. The dissipation is a result of a communication between the non-tilde and tilde fields. In reality there are many ways of realizing dissipation. However, whatever the cause of the dissipation, it can always be described phenomenogically as a communication between the tilde and non-tilde fields. We are going to see this explicitly in two examples.

We should recall also that the thermal average of an observable, say A, is given, not by the trace average, but by the thermal vacuum expectation value of A.

We are now ready to summarize the basic structure of the non-equilibrium TFD. The time dependence of the field operators is controlled by the Hamiltonian as

$$
\begin{aligned}
a(t,\mathbf{k}) &= \hat{S}^{-1}(t)a(\mathbf{k})\hat{S}(t) \\
\tilde{a}(t,\mathbf{k}) &= \hat{S}^{-1}(t)\tilde{a}(\mathbf{k})\hat{S}(t) \\
a^{\dagger\dagger}(t,\mathbf{k}) &= \hat{S}^{-1}(t)a^{\dagger}(\mathbf{k})\hat{S}(t) \\
\tilde{a}^{\dagger\dagger}(t,\mathbf{k}) &= \hat{S}^{-1}(t)\tilde{a}^{\dagger}(\mathbf{k})\hat{S}(t),
\end{aligned}
\tag{3}
$$

where

$$
i\partial_t\hat{S}(t) = \hat{H}_t\hat{S}(t) \tag{4}
$$

with $\hat{H}_t$ being the Hamiltonian at time t. Here and in the following use is made of the following notations. An operator consisting only of the non-tilde fields is denoted by a symbol without either tilde or hat (e.g. A), an operator consisting only of tilde fields is denoted by a symbol with tilde e.g. $(\tilde{A})$ and an operator which contains both of non-tilde and tilde fields is denoted by a symbol with hat e.g. $(\hat{A})$.

Note that in order for the second line in (3) to follow from the first line through tilde conjugation the Hamiltonian, $\hat{H}_t$, must have the property: $[i\hat{H}_t]^{\sim} = i\hat{H}_t$ which means that the Hamiltonian is Tildian.[13] Since the Hamiltonian contains a dissipative term, $\hat{S}$ is not unitary. This is a reason why we used the symbol $\dagger\dagger$ in the third and fourth lines of (3) instead of the $\dagger$. Although the Hamiltonian is not necessarily Hermitian, it should be Tildian. The general form of the bilinear Tildian Hamiltonian is [13]

$$\hat{H}_t = \sum_{\vec{k}} \omega(t,\vec{k})[a^\dagger(\vec{k})a(\vec{k}) - \tilde{a}^\dagger(\vec{k})\tilde{a}(\vec{k})] + i\hat{\Pi}_t, \tag{5}$$

$$\hat{\Pi}_t = \sum_{\vec{k}} \{\bar{\kappa}(t,\vec{k})[a^\dagger(\vec{k})a(\vec{k}) + \tilde{a}^\dagger(\vec{k})\tilde{a}(\vec{k})] + c_1(t,\vec{k})\tilde{a}(\vec{k})a(\vec{k}) + c_2(t,\vec{k})\tilde{a}^\dagger(\vec{k})a^\dagger(\vec{k})\} + \text{c-number}. \tag{6}$$

Here $\bar{\kappa}$, c_1 and c_2 are certain real functions of t and $\vec{k}$.[11]

It has been shown that the thermal state condition at time t in the interaction representation can be written as follows:

$$[a(t,\vec{k}) - f^\alpha(t,\vec{k})\tilde{a}^{\dagger\dagger}(t,\vec{k})]|0\rangle = 0, \tag{7a}$$

$$\langle 0|[a^{\dagger\dagger}(t,\vec{k}) - f^{1-\alpha}(t,\vec{k})\tilde{a}(t,\vec{k})] = 0, \tag{7b}$$

where f(t,k) is a real function of t and $\vec{k}$ and α is between 1 and 0. Physical results, that is the thermal vacuum expectation values of non-tilde operators, are independent of the choice of the power α. The freedom of α originates from the identity $\mathrm{Tr}[\rho^{1-\alpha} A\rho^\alpha] = \mathrm{Tr}[\rho A]$. The fact that the function f is real is related to the Tildian property of the Hamiltonian. When $\alpha = 1$ the representation is called the asymmetric representation, while when $\alpha = 1/2$ the representation is the symmetric one. Representations with other values of α are called intermediate representations. The real function f is related to the average number $n(t,\vec{k}) = \langle 0|a^{\dagger\dagger}(t,\vec{k})a(t,\vec{k})|0\rangle$ through the relation $n = f/[1-\sigma f]$. It can be shown that the time-independent Hamiltonian is possible only in the asymmetric representation. However, even in the asymmetric representation, the renormalized unperturbed Hamiltonian quite commonly becomes dependent on time due to the higher order loop corrections. Since the Schwinger-Keldysh method[12] is applicable only to a time-independent unperturbed Hamiltonian, we realize that the applicability of the Schwinger-Keldysh method to a system of quantum field is very much narrowed.

Let us now study the relation between the Hamiltonian and the thermal state conditions. It follows from (7) that[11]

$$\frac{\partial}{\partial t}[a(t,\vec{k}) - f^\alpha(t,\vec{k})\tilde{a}^{\dagger\dagger}(t,\vec{k})] = M(t,\vec{k})[a(t,\vec{k})-f^\alpha(t,\vec{k})\tilde{a}^{\dagger\dagger}(t,\vec{k})]$$

$$\frac{\partial}{\partial t}[a^{\dagger\dagger}(t,\vec{k})-f^{1-\alpha}(t,\vec{k})\tilde{a}(t,\vec{k})] = N(t,\vec{k})[a^{\dagger\dagger}(t,\vec{k})-f^{1-\alpha}(t,\vec{k})\tilde{a}(t,\vec{k})]$$

where M and N are certain functions of t and $\vec{k}$. These give

$$[a(\vec{k}) - f^\alpha(t,\vec{k})\tilde{a}^\dagger(\vec{k}),\hat{H}_t] - i\alpha f^{\alpha-1}(t,\vec{k})\dot{f}(t,\vec{k})\tilde{a}^\dagger(\vec{k}) = M(t,\vec{k})[a(\vec{k}) - f^\alpha(t,\vec{k})\tilde{a}^\dagger(\vec{k})] \tag{8a}$$

$$[a^\dagger(\vec{k}) - f^{1-\alpha}(t,\vec{k})\tilde{a}(\vec{k}),\hat{H}_t] - i(1-\alpha)f^{-\alpha}(t,\vec{k})\dot{f}(t,\vec{k})\tilde{a}(\vec{k}) = N(t,\vec{k})[a^\dagger(\vec{k}) - f^{1-\alpha}(t,\vec{k})\tilde{a}(\vec{k})]. \tag{8b}$$

These relations relate the Hamiltonian in (6) to the function $f(t,\vec{k})$.

We now restrict our attention to the asymmetric representation although it is easy to extend our consideration to cover other representations. In the asymmetric representation, that is $\alpha=1$, the relations in (8) leads to[11]

$$c_1 = -\frac{\sigma}{1+2\sigma n}[\dot{n} + 2\sigma\bar{\kappa}(1+\sigma n)], \tag{9}$$

$$c_2 = \frac{\sigma}{1+2\sigma n}[\dot{n} - 2\bar{\kappa}n]. \tag{10}$$

Since these two equations show that $c_1 = -2\bar{\kappa} - c_2$, c_1 is determined by $\bar{\kappa}$ and c_2. Then eq. (10) becomes the equation which determines n and therefore, f, when the initial values of n or f is known:

$$\dot{n}(t,\vec{k}) = -\kappa(t,\vec{k})n(t,\vec{k}) + \sigma c_2(t,\vec{k}), \tag{11}$$

$$\kappa(t,\vec{k}) \equiv -2(\bar{\kappa}(t,\vec{k}) + c_2(t,\vec{k})) = 2(\bar{\kappa}(t,\vec{k}) + c_1(t,\vec{k})). \tag{12}$$

According to (11) the number takes the simple exponential form, i.e. $n = a\exp(-\kappa t)+b$, when and only when, $\bar{\kappa}$ and c_2 and therefore also, c_1, are independent of time. However, as was pointed out previously, higher order loop corrections tend to make the renormalized quantities dependent on time. Therefore, in case of a system of field, the simple exponential form is very unusual. In general, the temporal behavior of n takes a much more complex form.

We are now going to see how the above aspects of the formalism of TFD help carry out practical calculations in a perturbational formalism. To facilitate the calculative method, we introduce the following operators and their tilde conjugate:[13]

$$\gamma_t(\vec{k}) = Z^{1/2}(t,\vec{k})[a(\vec{k}) - f(t,\vec{k})\tilde{a}^\dagger(\vec{k})] \tag{13a}$$

$$\gamma_t^{♀}(\vec{k}) = Z^{1/2}(t,\vec{k})[a^\dagger(\vec{k}) - \tilde{a}(\vec{k})]. \tag{13b}$$

The normalization factor is determined by the usual canonical σ-commutator between γ_t and $\gamma_t^{♀}$, which gives $Z = 1/[1+\sigma f] = 1+\sigma n$. Since the thermal state conditions give

$$\gamma_t|0(t)\rangle = \tilde{\gamma}_t|0(t)\rangle = 0, \quad \langle 0(t)|\gamma_t^{♀} = \langle 0(t)\tilde{\gamma}_t^{♀} = 0, \tag{14}$$

the operators γ_t and $\tilde{\gamma}_t$ are the annihilation operators while $\gamma_t^{♀}$ and $\tilde{\gamma}_t^{♀}$ act as the creation operators in the unpertured Schrödinger representation. It is remarkable that when the Hamiltonian is expressed in terms of these creation and annihilation operators, it takes on the extremely simple form:[11,14]

$$\hat{H}_t = \Sigma\ \omega[\gamma_t^{♀}\gamma_t - \tilde{\gamma}_t^{♀}\tilde{\gamma}_t] + i\hat{\Pi}_t, \tag{15}$$

$$\hat{\Pi}_t = \Sigma\ \{-\tfrac{1}{2}\kappa[\gamma_t^{♀}\gamma_t + \tilde{\gamma}_t^{♀}\tilde{\gamma}_t] + \frac{\dot{f}}{1-\sigma f}\gamma_t^{♀}\tilde{\gamma}_t^{♀}\}. \tag{16}$$

This indicates that $\langle 0|\hat{H}_t = 0$ and also that, except for the last term in $\hat{\Pi}_t$ all terms in H_t annihilate the ket-vacuum $|0(t)\rangle$.

This together with (4) determines the structure of $|0(t)\rangle$;

$$|0(t)\rangle = \exp[\Sigma \int_0^t d\tau \frac{\dot{n}(\tau,\vec{k})}{1+\sigma n(\tau,\vec{k})} \gamma_t^{\text{♀}}\tilde{\gamma}_t^{\text{♀}}]|0\rangle. \tag{17}$$

The annihilation and creation operators in the renormalized interaction representation are

$$\gamma(t,\vec{k}) = \hat{S}^{-1}(t)\gamma_t(\vec{k})\hat{S}(t), \quad \gamma^{\text{♀}}(t,\vec{k}) = \hat{S}^{-1}(t)\gamma_t^{\text{♀}}(\vec{k})\hat{S}(t) \tag{18}$$

and their tilde conjugate, because

$$\gamma(t,\vec{k})|0\rangle = \tilde{\gamma}(t,\vec{k})|0\rangle = 0, \quad \langle 0|\gamma^{\text{♀}}(t,\vec{k}) = \langle 0|\tilde{\gamma}^{\text{♀}}(t,\vec{k}) = 0. \tag{19}$$

Using the simple expression of the Hamiltonian in (15) with (16), we can identify the annihilation and creation operators as functions of time and momentum given as;

$$\gamma(t,\vec{k}) = Z^{-1/2}(t,\vec{k})Z^{1/2}(0,\vec{k}) \times \exp\{-\int_0^t d\tau(i\omega(\tau,\vec{k}) + \tfrac{1}{2}\kappa(\tau,\vec{k}))\}\gamma_0(\vec{k}) \tag{20a}$$

$$\gamma^{\text{♀}}(t,\vec{k}) = Z^{1/2}(t,\vec{k})Z^{-1/2}(0,\vec{k}) \times \exp\{\int_0^t d\tau(i\omega(\tau,\vec{k}) + \tfrac{1}{2}\kappa(\tau,\vec{k}))\}\gamma_0^{\text{♀}}(\vec{k}) \tag{20b}$$

where γ_0 and $\gamma_0^{\text{♀}}$ are respectively γ_t and $\gamma_t^{\text{♀}}$ at t=0. Note that $\gamma^{\text{♀}}(t,\vec{k})\gamma(t,\vec{k}) = \gamma_0^{\text{♀}}(\vec{k})\gamma_0(\vec{k})$. When we make use of the thermal doublet notation:

$$\begin{cases} a^\alpha;\ a^1 = a, & a^2 = \tilde{a}^{\dagger\dagger} \\ \bar{a}^\alpha;\ \bar{a}^1 = a^{\dagger\dagger}, & \bar{a}^2 = \tilde{a}\ , \end{cases} \qquad \begin{cases} \gamma^\alpha;\ \gamma^1 = \gamma, & \gamma^2 = \tilde{\gamma}^{\text{♀}} \\ \bar{\gamma}^\alpha;\ \bar{\gamma} = \gamma^{\text{♀}}, & \bar{\gamma}^2 = \tilde{\gamma}, \end{cases} \tag{21}$$

the transformation in (13) takes the simple form

$$a^\alpha(t,\vec{k}) = Z^{1/2}(t,\vec{k})B^{\alpha\beta}(t,\vec{k})\gamma^\beta(t,\vec{k})\ , \tag{22a}$$

$$\bar{a}^\alpha(t,\vec{k}) = Z^{1/2}(t,\vec{k})\bar{\gamma}^\beta(t,\vec{k})[\tau_3 B(t,\vec{k})\tau_3]^{\beta\alpha}, \tag{22b}$$

$$B(t,\vec{k}) = \begin{pmatrix} 1 & -f \\ -\sigma & 1 \end{pmatrix}, \tag{23}$$

The thermal vacuum expectation value of a product of operators a^α and $\bar{a}^\beta$ can be calculated by expressing them in terms of creation and annihilation operators by means of (22). For example we find;

$$\langle 0|T[a^\alpha(t,\vec{k}),\bar{a}^\beta(s,\vec{\ell})]|0\rangle = \delta_{\vec{k}\vec{\ell}} B^{\alpha\rho}(t,\vec{k})C^{\rho\sigma}(t,s;\vec{k})[\tau_3 B(s,\vec{k})\tau_3]^{\sigma\beta}\ , \tag{24}$$

$$C^{\rho\sigma}(t,s;\vec{k}) = \qquad \exp\ [-\int_s^t d\tau\ i\omega(\tau,\vec{k})]$$

$$\times \begin{pmatrix} \theta(t-s)Z(s,\vec{k})\exp\{\int_s^t d\tau - \frac{1}{2}\kappa(\tau,\vec{k})\} & 0 \\ 0 & \sigma\theta(s-t)Z(t,\vec{k})\exp\{-\int_s^t d\tau - \frac{1}{2}\kappa(\tau,\vec{k})\} \end{pmatrix} . \qquad (25)$$

When there is an interaction, $H_I - \tilde{H}_I$, in the system, we can form the Feynman diagrams, in which H_I and $-\tilde{H}_I$ act as vertices and the internal lines are given by (24). We then calculate the loop corrections and require that the bilinear terms due to these loop corrections should compensate the renormalization counter terms. This requirement determines the time dependent renormalized parameters $\omega(t)$, $\bar{\kappa}(t)$, $c_1(t)$ and $c_2(t)$ and this completes the time-dependent perturbation renormalization calculation formalism. In the following two sections we are going to see very briefly how this formalism can be applied to an analysis of non-equilibrium transitions.

I close this section with a brief comment about the construction of the effective Hamiltonian $\hat{H}_t$ out of a most basic one. The entire consideration should begin with the most basic Hamiltonian for the entire system. The basic Hamiltonian has the form $H - \tilde{H}$. In the usual quantum field theory it is well known that we should first identify the physical particles by calculating the self-energies, and then, the basic Hamiltonian is realized in the linear vector space of the physical particles which are usually chosen to be the asymptotic fields. In TFD the same method should be applied to the basic Hamiltonian. A significant new feature in TFD is that most of physical particles appear to be unstable due to the thermal effects caused by the shadow field, i.e. the tilde field. After the calculation of the self-energy, we drop the part of the Hamiltonian which is concerned with an unobserved part. This is called coarse graining. For example, the elimination of the reservoir is made through the calculation of self-energy of physical particles in the observed part of the systems due to its interaction with the reservoir. We have seen[13] how this coarse graining determines $\bar{\kappa}$, c_1 and c_2 in the effective Hamiltonian of the observed system. The example in the next section treats such a system. The example in the last section (i.e. the expansion of universe) does not require such a coarse graining process; a simple calculation of self-energies creates the effective Hamiltonian. The self-energy is a 2×2 -matrix associated with the thermal doublet space, and therefore, may contain the $a\tilde{a}$- and $a^\dagger\tilde{a}^\dagger$-terms which causes a non-equilibrium process.

DISORDER-ORDER TRANSITION

In this section we briefly examine the growth of an ordered state. Using the simple model of a complex scalar field, we try[10,14] to study how the order parameter grows through the course of the non-equilibrium transition from a non-ordered state to an ordered state. Here we use a crude approximation. A more precise analysis is currently in progress. We consider the model of complex scalar field with the Lagrangian

$$\mathcal{L} = -\partial_\mu\psi^\dagger\partial_\mu\psi + m_0^2\psi^\dagger\psi - \lambda(\psi^\dagger\psi)^2 \qquad (26)$$

The order parameter is given by $v = \langle 0|\psi|0\rangle$. In case of the equilibrium states, v is independent of time. In such a case, the ordered state (i.e. $v \neq 0$) appears in the temperature domain below the critical temperature, say T_c. We prepare an initial state at $T_o > T_c$, and follow its thermal transition to a final state with $T < T_c$. Thus, we have $v(\beta_o) = 0$, while $v(\beta) \neq 0$. Temperature is not well-defined in the middle of the non-equilibrium transition. The time-dependent order parameter will be denoted by v_t. Thus $v_{t=o} = v(\beta_o) = 0$ while $v_{t=\infty} = v(\beta)$.

Define the real fields ϕ and χ through the relation $\psi \equiv (\varphi + i\chi)/\sqrt{2}$ with $\varphi = \phi + \sqrt{2}\, v$. Then, we have $\langle 0|\phi|0\rangle = 0$ and $\langle 0|\varphi|0\rangle = \sqrt{2}\, v$. We can write

$$
\begin{aligned}
\mathscr{L} = &- \tfrac{1}{2} [\partial_\mu \phi \partial_\mu \phi + (-m_o^2 + 6\lambda v^2)\phi^2] \\
&- \tfrac{1}{2} [\partial_\mu \chi \partial_\mu \chi + (-m_o^2 + 2\lambda v^2)\chi^2] \\
&- \sqrt{2}\, v(-m_o^2 + 2\lambda v^2)\phi \\
&- \sqrt{2}\, \lambda v \phi(\phi^2 + \chi^2) - \tfrac{\lambda}{4} (\phi^2 + \chi^2)^2.
\end{aligned}
\tag{27}
$$

The tilde-conjugate of ϕ, χ and ψ are $\tilde{\phi}$, $\tilde{\chi}$ and $\tilde{\psi}$, respectively.

We use a simplified one-loop calculation, in which we calculate only the tag-type one-loop contributions given by $C_t \equiv \langle 0|\phi^2 + \chi^2|0\rangle$. Thus, the linear ϕ-term in $\phi(\phi^2 + \chi^2)$ is not given by $\langle 0|3\phi^2 + \chi^2|0\rangle\phi$ but by $\langle 0|\phi^2 + \chi^2|0\rangle\phi$. Then, the linear ϕ-term in the interaction Hamiltonian which is obtained from (27) is $\sqrt{2}\, v_t(-m_o^2 + 2\lambda v_t^2 + \lambda C_t)\phi$. On the other hand, the ϕ^2-term and the χ^2-term (the renormalized mass terms of the ϕ- and χ-field) in the Hamiltonian are $(1/2)m_t^2\phi^2$ and $(1/2)\mu_t^2\chi^2$ with

$$
m_t^2 = -m_o^2 + 6\lambda v_t^2 + \lambda C_t \tag{28}
$$

$$
\mu_t^2 = -m_o^2 + 2\lambda v_t^2 + \lambda C_t . \tag{29}
$$

In the equilibrium state at $t=0$ and $t=\infty$ the linear ϕ-term should vanish. At $t=0$, this is achieved by $v_{t=0} = 0$, while, at $t=\infty$, this leads to $[-m_o^2 + 2\lambda v_t^2 + \lambda C_t]_{t=\infty} = 0$, which together with (28) and (29) gives $m_{t=\infty}^2 = 4\lambda\, v_{t=\infty}^2$ and $\mu_{t=\infty}^2 = 0$. The latter equation implies that χ at $t=\infty$ is the Goldstone boson. In the middle of the non-equilibrium process, χ becomes massive.

To calculate v_t, it is convenient to come back to $\varphi \equiv \phi + \sqrt{2}\, v_t$ so that $\langle 0|\varphi|0\rangle = \sqrt{2}\, v_t$. Then, the renormalized unperturbed Hamiltonian in the approximation under consideration is obtained by considering those terms which are linear or bilinear in φ and χ.

$$
\begin{aligned}
H_{o,t} = &\tfrac{1}{2} [\pi_\varphi^2 + \vec{\nabla}\varphi \cdot \vec{\nabla}\varphi + m_t^2 \varphi^2] \\
&+ \tfrac{1}{2} [\pi_\chi^2 + \vec{\nabla}\chi \cdot \vec{\nabla}\chi + \mu_t^2 \chi^2]
\end{aligned}
$$

$$+ \sqrt{Z}\, v_t(-m_o^2 - m_t^2 + 2\lambda v_t^2 + \lambda C_t)\varphi. \tag{30}$$

At t=0 the linear φ-term vanishes because $v_{t=0} = 0$, while, at $t=\infty$, the linear φ-term becomes $-\sqrt{Z}\, v_t m_t^2 \varphi$ which compensates the $\sqrt{Z}\, v_t m_t^2 \varphi$ in the mass term $(1/2)m_t^2\varphi^2$. Thus, $H_{o,t}$ is purely bilinear in ϕ at t=0 or $t=\infty$, as is expected from the fact that the states at t=0 and $t=\infty$ are equilibrium ones.

We write φ and χ as

$$\varphi(x) = \int \frac{d^3k}{(2\pi)^3} \frac{1}{\sqrt{2\omega(t,k)}} [\alpha_k(t)e^{i\vec{k}\vec{x}} + \alpha_k^{\dagger\dagger}(t)e^{-i\vec{k}\vec{x}}] \tag{31a}$$

$$\chi(x) = \int \frac{d^3k}{(2\pi)^3} \frac{1}{\sqrt{2\varepsilon(t,k)}} [\beta_k(t)e^{i\vec{k}\vec{x}} + \beta_k^{\dagger\dagger}(t)e^{-i\vec{k}\vec{x}}]. \tag{31b}$$

We then have

$$\pi_\varphi(x) = i\int \frac{d^3k}{(2\pi)^3} \frac{\omega(t,k)}{\sqrt{2\omega(t,k)}} [\alpha_k(t)e^{i\vec{k}\vec{x}} - \alpha_k^{\dagger\dagger}(t)e^{-i\vec{k}\vec{x}}], \text{ etc.} \tag{32}$$

Then

$$H_{o,t} = \int d^3k\, [\omega(t,k)\alpha_k^{\dagger\dagger}(t)\alpha_k(t) + \varepsilon(t,k)\beta_k^{\dagger\dagger}(t)\beta_k(t)]$$
$$+ \sqrt{Z}\, v_t(-m_o^2 - m_t^2 + 2\lambda v_t^2 + \lambda C_t)\varphi. \tag{33}$$

Recalling the consideration in section 2, the total unperturbed renormalized Hamiltonian is

$$\hat{H}_t = H_{o,t} - \tilde{H}_{o,t} + i\hat{\Pi}_t^\varphi + i\hat{\Pi}_t^\chi \tag{34}$$

where the structure of $\hat{\Pi}_t^\varphi(t)$ or $\hat{\Pi}_t^\chi(t)$ are given by (6) with a and $\tilde{a}$ being replaced by α and $\tilde{\alpha}$ or β and $\tilde{\beta}$. Since we are considering the system to be controlled by a reservoir with the temperature $T = 1/\beta$, the parameters $\bar{\kappa}$, c_1 and c_2 in (6) are independent of time. This is because the approximation under consideration (i.e. the tag-loop correction) does not modify these parameters. Therefore, the particle numbers defined by $n_k^\alpha(t) = \langle 0|\alpha_k^{\dagger\dagger}(t)\alpha_k(t)|0\rangle$ and $n_k^\beta(t) = \langle 0|\beta_k^{\dagger\dagger}(t)\beta_k(t)|0\rangle$ have the simple exponential form when the linear term in (33) is ignored:

$$n_k^\alpha(t) = n_k^\alpha(\infty) + [n_k^\alpha(0) - n_k^\alpha(\infty)]e^{-\kappa_k^\alpha t}, \text{ etc.} \tag{35}$$

Feeding (31) into $C_t = \langle 0|\phi^2 + \chi^2|0\rangle$, we obtain

$$C_t = \int \frac{d^3k}{(2\Pi)^3} \{\frac{1}{2\omega(t,k)} [1 + 2n_k^\alpha(t)]$$
$$+ \frac{1}{2\varepsilon(t,k)} [1 + 2n_k^\beta(t)]\}. \tag{36}$$

Then (28) and (29) determine the time-dependent masses. Since $\hat{H}_t$ generates the time translation of φ, we have

$$\partial_t\alpha_k(t) = -i[\alpha_k(t),\hat{H}_t] + \frac{m_t\dot{m}_t}{[\omega(t,k)]^2}\, \dot{\alpha}_{-k}^{\dagger\dagger}(t), \text{ etc.} \tag{37}$$

where the last term originates from the time-dependent factor, $1/\sqrt{2\omega(t,\vec{k})}$, in (31a). The thermal vacuum expectation value of both sides of (37) with $\vec{k}=0$ gives the differential equation for v_t. The equation is highly non-linear in v_t due to the φ-linear term in (33). A crude calculation gives[10]

$$v_t = v(\beta)[1 - \cos(\int_0^t d\tau\, m_\tau)e^{-\kappa_0^\alpha t}]. \tag{38}$$

A more rigorous treatment of this subject is currently in progress.

THE EXPANDING UNIVERSE

As an example of a self-consistent non-equilibrium process, we briefly sketch a treatment[15] of the expanding universe by considering a simple model of an interacting complex scalar ψ and real scalar ϕ. The central idea is to solve the model by using the Robertson-Walker line element; $ds^2 = dt^2 - a^2(t)d\vec{x}^2$ with the time dependent scaling factor a(t) which is the cause of the non-equilibrium nature of the system. In this sense our calculation may be considered as an extension of the work of Paul R. Anderson.[16] The extension comes from our full consideration of the non-equilibrium thermal effect. The model is defined by the action

$$\begin{aligned} S &= \int d^4x\, \sqrt{-g}\, \{\frac{1}{2}\, [g^{\mu\nu}\partial_\mu\phi\partial_\nu\phi - m_0^2\phi^2] \\ &\quad + [g^{\mu\nu}\partial_\mu\psi^\dagger\partial_\nu\psi - M_0^2\psi^\dagger\psi] - \lambda_0\psi^\dagger\psi\phi\} \\ &= \int d^4x\mathcal{L}(x) \end{aligned} \tag{39}$$

where $g_{00} = 1$, $g_{ij} = -a^2(t)\delta_{ij}$ and $g_{0i} = g_{i0} = 0$. Changing the definition of the fields as $a^{3/2}(t)\psi \to \psi$ and $a^{3/2}(t)\phi \to \phi$, we can write $\mathcal{L}(x)$ as $\mathcal{L}(x) = \mathcal{L}_0 + \mathcal{L}_I$, where

$$\begin{aligned} \mathcal{L}_0 &= Z_\psi[\dot{\psi}^\dagger\dot{\psi} - \psi^\dagger\varepsilon^2(-i\vec{\nabla},t)\psi] \\ &\quad + \frac{1}{2}\, Z_\phi[\dot{\phi}^2 - \phi\omega^2(-i\vec{\nabla},t)\phi] \end{aligned} \tag{40}$$

$$\begin{aligned} \mathcal{L}_I &= -\lambda(t)\psi^\dagger\psi\phi - \delta Z_\psi\dot{\psi}^\dagger\dot{\psi} + \psi^\dagger\delta\varepsilon^2(-i\vec{\nabla},t)\psi \\ &\quad - \frac{1}{2}\, \delta Z_\phi\dot{\phi}^2 + \phi\delta\omega^2(-i\vec{\nabla},t)\phi - \lambda(t)\delta Z_1\psi^\dagger\psi\phi \end{aligned} \tag{41}$$

where $\delta Z_\psi = Z_\psi - 1$, $\delta Z_\phi = Z_\phi - 1$ and

$$\varepsilon^2(\vec{k},t) = a^{-2}(t)\vec{k}^2 + M_0^2 - \frac{3}{2}\frac{d}{dt}(\frac{\dot{a}}{a}) - \frac{9}{4}(\frac{\dot{a}}{a})^2 + \delta\varepsilon^2(\vec{k},t) \tag{42a}$$

$$\omega^2(\vec{k},t) = a^{-2}(t)\vec{k}^2 + m_0^2 - \frac{3}{2}\frac{d}{dt}(\frac{\dot{a}}{a}) - \frac{9}{4}(\frac{\dot{a}}{a})^2 + \delta\omega^2(\vec{k},t) \tag{42b}$$

$$\lambda(t) = \lambda_0 a^{-3/2}(t) Z_\psi^{-1} Z_\phi^{-1/2} Z_1. \tag{43}$$

Obviously, $\delta\varepsilon^2$ and $\delta\omega^2$ are the self-energy counter terms, while

Z_ψ and Z_ϕ are the wave-function renormalization factors and Z_1 is the vertex renormalization factor. Now, construct the Hamiltonian H_o from $\mathcal{L}_o$. Then, the total Hamiltonian $\hat{H}_t$ is $H_o - \tilde{H}_o - + i\hat{\Pi}_t$ where $\hat{\Pi}_t = \hat{\Pi}_t^\phi + \hat{\Pi}_t^\psi$, in which $\hat{\Pi}_t^\phi$ and $\hat{\Pi}_t^\psi$ have the form (6) for ϕ- and ψ-fields respectively. Their coefficients are denoted by $\bar{\kappa}^\phi$, $\bar{\kappa}^\psi$, c_1^ϕ, c_1^ψ, c_2^ϕ and c_2^ψ. These time-dependent coefficients and $\delta\varepsilon^2$ and $\delta\omega^2$ should be determined by a calculation of loop corrections in which the vertices are given by $\hat{H}_I = -\mathcal{L}_I + \tilde{\mathcal{L}}_I$ and the Feynman-type internal line is given by the two point functions in (24). We use the approximation in which only the one-loop self-energy diagrams are considered. Thus $Z_1=1$. Since ψ and ϕ are the thermal doublet fields (ψ^α; $\alpha=1,2$, ϕ^α; $\alpha=1,2$), the self-energy diagrams, $\Sigma_\psi(t_1,t_2;\vec{x}_1-\vec{x}_2)$ and $\Sigma_\phi(t_1,t_2;\vec{x}_1-\vec{x}_2)$ are 2×2 matrices. In $\bar{\psi}^\alpha(x_1)\Sigma_\psi^{\alpha\beta}(t_1,t_2,\vec{x}_1-\vec{x}_2)\times\psi(x_2)$, the diagonal elements of $\Sigma^{\alpha\beta}$ determine $\delta\varepsilon^2$ and $\bar{\kappa}^\psi$, while the off-diagonal elements c_1^ψ and c_2^ψ. The same is true for the ϕ-field. An immediate question may be to ask how these coefficients which are functions of one time come out of the two-time function $\Sigma^{\alpha\beta}(t_1,t_2;\vec{x}_1-\vec{x}_2)$. The answer to this question may be found in the following consideration. The time ordered product which contains Σ in $\int d^4x\ \delta\mathcal{L}$ yields the following two terms,

$$\int_0^\infty dt_1\, d^3x_1 \int_0^{t_1} dt_2\, d^3x_2\ \bar{\psi}^\alpha(x_1)\Sigma_\psi^{\alpha\beta}(t_1,t_2;\vec{x}_1-\vec{x}_2)\psi^\beta(x_2)$$

$$+\int_0^\infty dt_2\, d^3x_2 \int_0^{t_2} dt_1\, d^3x_2\ \psi^\beta(x_2)\Sigma_\psi^{\alpha\beta}(t_1,t_2;\vec{x}_1-\vec{x}_2)\bar{\psi}^\alpha(x_1),$$

which implies that Σ_ψ contributes to the effective Hamiltonian through the term[15]

$$\delta\hat{H}_t = \int d^3x \int_0^t dt'dx'[\theta(t-t')\bar{\psi}^\alpha(\vec{x},t)\Sigma^{\alpha\beta}(t,t';\vec{x}-\vec{x}')\psi^\beta(x')$$
$$+\ \theta(t'-t)\psi^\beta(\vec{x},t)\Sigma^{\alpha\beta}(t',t;\vec{x}'-\vec{x})\bar{\psi}^\alpha(x')] \qquad (44)$$

which determines the time-dependent coefficients ε, ω, $\bar{\kappa}$, c_1 and c_2.

There is one more step needed in order to calculate the effective Hamiltonian. We should express ψ (and ϕ) in terms of the operators $a(\vec{k})$, etc., because $\hat{\Pi}_t$ in (6) is expressed in terms of these operators. This is done using the usual quantum field method. Since the unperturbed renormalized field ψ satisfies $[(\partial/\partial t)^2 + \varepsilon^2(-i\vec{\nabla},t)]\psi = 0$, we write

$$\psi(x) = (2\pi)^{-3/2}\int d^3k[e^{i\vec{k}\vec{x}}u(\vec{k},t)a(\vec{k}) + e^{-i\vec{k}\vec{x}}u^*(\vec{k},t)b^\dagger(\vec{k})] \quad (45)$$

with the wave function satisfying $[(\partial/\partial t)^2 + \varepsilon^2(\vec{k},t)]u(\vec{k},t) = 0$. When the two c-number functions, f and g, satisfy this equation, $(f\cdot g) \equiv \dot{f}^*g - f^*\dot{g}$ is independent of time. Considering this fact, we orthonormalizes the wave functions by $(u(\vec{k},t)\cdot u(\vec{\ell},t)) = \delta_{\vec{k}\vec{\ell}}$, etc. An <u>approximate</u> form of u is $\left(1/\sqrt{2\varepsilon(\vec{k},t)}\right)\exp\{-i\int_0^t dt'\varepsilon(\vec{k},t')\}$.

We have calculated $\hat{\Pi}_t$ using (44). A remarkable feature[15] of our result is that $\bar{\kappa}(\vec{k},t)$, $c_1(\vec{k},t)$ and $c_2(\vec{k},t)$ for ψ and ϕ thus calculated satisfy the general conditions (9) and (10), even though the approach used here is very different from the one which gave rise to these relations in the first place.

To obtain physical results, we must determine the Robertson-Walker coefficient a(t). To derive the self-consistent equation for a(t), we should calculate the thermal vacuum expectation value of the energy-stress tensor and feed the result into the Einstein equation. This work is currently in progress.

The author would like to thank H. Matsumoto, T. Arimitsu, J. Pradko, M. Guida and N. Yamamoto for valuable discussions and collaborations.

References

1. H. Umezawa, H. Matsumoto and M. Tachiki, "Thermo Field Dynamics and Condensed States", North-Holland Publisher, Amsterdam, New York, London, 1982.
2. T. Arimitsu and H. Umezawa, Prog. Theor. Phys. 74 429 (1985).
   T. Arimitsu and H. Umezawa, "Non-Equilibrium Thermo Field Dynamics", preprint, Univ. of Alberta, (1985).
3. H. Umezawa, and T. Arimitsu, Proceeding of the Bielefeld Encounters in Physics and Mathematics, ed. M.C. Gutzwiller et al. (World Scientific, Singapore, 1986).
4. T. Arimitsu and H. Umezawa, "Unperturbed Representation in Non-Equilibrium Thermo Field Dynamics", to be published in J. Phys. Soc. Japan 55 (1986) No. 5.
5. T. Arimitsu, J. Pradko and H. Umezawa, "Generating Functional Methods in Non-Equilibrium Thermo Field Dynamics", to be published in Physica A, (1986).
6. H. Matsumoto, "Thermo Field Dynamics - Thermal State Condition and Quantum Algebra", to be published in the Proceeding for the Symposium "Progress in Quantum Field Theory" held at Positano, Italy, July, 1985.
7. H. Matsumoto and H. Umezawa, Phys. Rev. B31, 4433 (1985).
8. M. Suzuki, "Progress in Quantum Field Theory", eds. H. Ezawa and S. Kamefuchi (North-Holland, 1985).
9. T. Tominaga, J. Pradko, M. Ban, T. Arimitsu and H. Umezawa, "Axiomatic Treatment of Spin System in TFD", preprint, Univ. of Alberta, (1986).
10. H. Umezawa and T. Arimitsu, "Time-Dependent Renormalization in Thermo Field Dynamics", to be published in Prog. Theor. Phys. Supplement (1986).
11. T. Arimitsu, M. Guida and H. Umezawa, "Time-Dependent Hamiltonian and Thermal State Condition in Non-Equilibrium Thermo Field Dynamics", preprint, Univ. of Alberta, (1986).
12. J. Schwinger, J. Math. Phys. 2, 407 (1961)
    L.V. Keldysh, Soviet Phys. JETP 20, 1018 (1965)
    R.A. Craig, J. Math. Phys. 9, 605 (1968)
    R. Mills, "Propagators for Many-Particle Systems (Gordon and Breach Science Publisher, New York, 1969).
13. T. Arimitsu and H. Umezawa, "General Structure of Non-Equilibrium Thermo Field Dynamics", preprint, Univ. of Alberta, (1985).

14. T. Arimitsu, Y. Sudo and H. Umezawa, "Dynamical Rearrangement of the Thermal Vacuum in Thermo Field Dynamics", preprint, Tsukuba University, (1986).
15. H. Matsumoto, H. Umezawa and N. Yamamoto, "Non-Equilibrium Thermo Field Dynamical Treatment of the Expanding Universe", preprint, Univ. of Alberta (1986).
16. P. Anderson, Phys. Rev. D28, 271 (1983); D29, 615 (1984); D32, 1302 (1985).

SYMMETRIES AND SPECIAL FUNCTIONS

N.Ya. Vilenkin* and A.U. Klimyk**

*Mathematical Department, The Correspondence Pedagogical Institute (MGZPI), Moscow, 109004 USSR

**Institute for Theoretical Physics, Kiev-130 252130 USSR

In the 18-th and 19-th centuries there appeared a great number of types of special functions to solve the equations of mathematical physics and to calculate the integrals. Many of them turned out to be special or limiting cases of the hypergeometric function $F(\alpha,\beta;\gamma;x)$, introduced in 1769 by L. Euler and scrutinized at the beginning of the 19-th century by Gauss. Gauss' work triggered a flow of investigations which established different recurrent relations, differential equations, integral representations, generating functions, addition and multiplication theorems, asymptotic expansions for the hypergeometric function and its associates (Legendre, Gegenbauer, Hermite, Laguerre, Chebyshev polynomials; Bessel, Neumann, Macdonald, Whittaker functions, etc.), sought for relations between these functions, and calculated puzzling integrals involving them, etc. Books containing hundreds of pages were devoted to studies of some classes of special functions.

The fullest account of the results on special functions obtained by the middle of the 20-th century is given by a five volume collection published by the "Bateman Project" group ("Higher Transcendental Functions" and "Integral Transformations").

On the face of it, the entire set of the results resembles a chaotic collection of formulas in which every proposition is proved by crafty analytic transformations, incomprehensible substitutions and other technique of the "analytical kitchen". It seems to be almost impossible to introduce any order in this chaos, to elucidate the meaning and the depth of the formulas obtained, to understand their relationship, their role in other fields of mathematics, the reasons for their origin. Of course, the mathematicians drive to unify the matters under study, to view them unambiguously has also influenced this field of science. Some sections in the theory of special functions were first subjected to unification. In the second half of the 19-th century P. Chebyshev constructed a general theory of orthogonal polynomials that enabled a unified treatment of the results concerned with Legendre, Gegenbauer, Laguerre and Hermite polynomials and made it possible to introduce new classes of orthogonal polynomials, specifically, orthogonal polynomials of a discrete variable associated with point mass distribution. These contributions established the relationship of the orthogonal polynomial theory with continued fractions, Jacobi matrices, mechanical quadratures and other fields of mathematics.

Another line to unify the theory of special functions rested on the general theory of analytic functions created in the middle of the 19-th century. The creation of this theory made it possible to construct an analytical theory of linear differential equations that include the hypergeometric differential equation

$$x(1-x)y'' + [\gamma - (\alpha+\beta+1)x]y' - \alpha\beta y = 0, \qquad (1)$$

one of its solutions being $F(\alpha,\beta;\gamma;x)$. The theory studies linear transformations of the solutions that arise from by-passing the singular points of the equation (for (1), these points are $0,1,\infty$). These linear transformations form a group of monodromy of a given equation. In the case of equation (1) we thus get the relations that connect linearly the values of the hypergeometric functions at the points x, $1/x$, $1-x$, $1/(1-x)$, $x/(1-x)$, $(x-1)/x$ (these six linear-fractional transformations permute the singular points $0,1,\infty$ of equation (1)). Different functions related to the hypergeometric one satisfy the equations that follow from (1) when the singular points are confluent.

We mention another direction to unify the theory of special functions that is based on employing integral transformations, in particular, the Laplace, Fourier Mellin transformations. Specifically, the generalized hypergeometric function

$${}_pF_q(\alpha_1,\ldots,\alpha_p;\gamma_1,\ldots,\gamma_q;x) = \sum_{n=0}^{\infty} \frac{(\alpha_1)_n \cdots (\alpha_p)_n}{(\gamma_1)_n \cdots (\gamma_q)_n} \frac{x^n}{n!}$$

where $(\alpha)_n = \alpha(\alpha+1)(\alpha+2)\ldots(\alpha+n-1)$, can be derived from the function ${}_0F_0(x) = \exp x$ by a successive application of the Laplace transformation and its inverse. Use of the general properties of this transformation, in particular the convolution theorem, enables us to derive different identities for special functions.

However, a really unified view on the theory of the basic classes of special functions (the exceptions are Lame and Matier functions) was established by employing the considerations that belong to a field of mathematics seemingly quite far from the subject under consideration, the theory of representations of Lie groups, i.e., in fact, the considerations concerned with the symmetry and homogeneity of some objects in multidimensional geometry. The objects include, in particular, spheres, hyperboloids and paraboloids in multidimensional spaces (real, complex, quaternion and even octave ones) as well as their generalizations, for example, Stiefel manifold, homogeneous cones, homogeneous complex regions, etc. Generally, these manifolds are homogeneous (that is, there are transformation groups G which act transitively upon them) and for any pair of points they allow a symmetry (permuting these points) with respect to G-invariant metric. The spaces characterized by these properties are called symmetric. E. Cartan who introduced this notion studied Riemannian symmetric spaces and showed that they are characterized by the compactness of the stationary subgroups of their points (spaces having noncompact stationary subgroups are pseudo-Riemannian, i.e. their metric is given by an indeterminate quadratic differential form).

The relationship between special functions and the geometry of homogeneous spaces is based on the following facts. The special functions most often arise when the equations of mathematical physics are solved by the method of separation of variables in a certain coordinate system. The most important equations are invariant under some transformation groups (for example, the Laplace equation is invariant under the group of motions of Euclidean space R^n, the wave equation under the group of linear transformations that preserve a quadratic form $x_0^2 - x_1^2 - \ldots - x_n^2$, the Maxwell equation under the

Poincare group, etc.). But the Laplace operator coincides, up to a constant factor, with the operator $\lim_{r\to 0} [S(x,r,f)-f(x)]/r^2$, where $S(x,r,f)$ is the average value of function f on the sphere with center x and radius r. It is therefore defined in a natural way on symmetric spaces, giving on them G-invariant differential operators. This allows us to construct on such spaces the analogues of the classical differential equations of mathematical physics. When the variables are separated the spaces are fibered into coordinate surfaces which, in turn, are symmetric spaces. The special functions arise precisely when the eigenfunctions of invariant differential operators (in particular, the Laplace operator and its generalizations) are sought for, and it is therefore clear that their properties should involve the invariance of the operators under transformations of group G. Just the eigenfunction of an invariant operator transforms under the action of $g \in G$ into an eigenfunction that corresponds to the same eigenvalue. The linear transformation $T(g)$ is thus defined in the space of such eigenfunctions, and here the equality $T(g_1)T(g_2) = T(g_1g_2)$ is valid. The correspondence $g \to T(g)$ is called <u>a representation</u> of the group G. So, we throw a bridge between the differential operators invariant under the action of some group G and the representation of the group.

In addition to differential equations, we can also consider integral equations on homogeneous spaces whose kernels are G-invariant, i.e. $K(x,y) = K(gx,gy)$ (in particular, equations with a kernel that depends only on an invariant distance $\rho(x,y)$ between x and y). If, besides, the measure μ is also G-invariant, the action of the group G gives its representation in the space of the eigenfunctions of the integral equation

$$\int K(x,y)f(y)d\mu(y) = \lambda f(x).$$

Let us remind some basic definitions of the theory of group representations. The representation T of group G is called <u>irreducible</u> if the space L where the operators $T(g)$ are acting has no nontrivial closed subspaces invariant under all these operators. It is called <u>unitary</u> if L is a Hilbert space and all the operators $T(g)$ are unitary. The representations are generally considered with an accuracy to <u>equivalency</u>, i.e. to a replacement of $T(g)$ by $AT(g)A^{-1}$ (A is called an <u>intertwining operator</u> for $T(g)$ and $AT(g)A^{-1}$). Using two representations T_1 and T_2 of group G it is possible to construct their <u>direct sum</u> $T_1 + T_2$ and <u>tensor product</u> $T_1 \otimes T_2$, which acts in the space $\mathrm{Lin}(L_2', L_1)$ and maps $B \in \mathrm{Lin}(L_2', L_1)$ into $T_1(g)BT_2'(g)$. The representations are most often given as shifts in the spaces of functions on homogeneous spaces. These representations are, generally, reducible, and to separate minimal invariant subspaces, in the space of functions additional conditions are imposed on the functions. For example, the irreducible representations of group $U(n)$ of unitary matrices of the n-th order are constructed in space $L(\ell_1,\dots,\ell_n)$, $\ell_1 \geq \dots \geq \ell_n$ of polynomials of the elements of the matrix of the n-th order $x = (x_{ij})$ that have power ℓ_1 in the elements of the first line, power ℓ_2 in the minors of the second order composed of the elements of two first lines, etc. Here $T(g)p(x) = p(xg)$. A set of integers $(\ell_1,\ell_2,\dots,\ell_n)$ is called the <u>highest weight</u> of this representation. In a similar way it is possible to describe irreducible representations of the other classical Lie groups ($SO(n)$, $Sp(n)$, etc.), moreover, in addition to the requirement of homogeneity in minors, the conditions that the functions become zero under the action of certain invariant differential operators (for example, the harmonicity for group $SO(n)$) are also imposed.

E. Cartan developed a general theory of zonal spherical functions on compact symmetric spaces X, i.e. the functions constant under the action of a stationary subgroup H of some point and such that their shifts generate a subspace in which an irreducible representation of this group is realized. If $G = SO(3)$, $H = SO(2)$, X is a sphere, and the zonal spherical functions coincide with the classical polynomials introduced by Legendre and Laplace. A system of zonal spherical functions is orthogonal with respect to an invariant measure on X. A similar theory is constructed on locally compact symmetric Riemannian spaces, but then the set of zonal spherical functions has the cardinality of continuum and their orthogonality is interpreted in the sense same as in the theory of Fourier integral. For example, in the case when X is a two-sheeted hyperboloid we obtain a set of Legendre functions. It is to be mentioned that the relationship between the special functions and the theory of invariants - one of the fore-runners of the theory of group representations was mentioned before Cartan. The methods used by Cartan were based on the ideas employed by H. Weyl and F. Peter to prove the general theorem that the matrix elements of irreducible unitary representations of a compact group G form a complete orthogonal set of functions on G - this theorem explains the orthogonality of many systems of special functions.

It is to be noted that the group-theoretical approach allows us to explain the separation of a class of elementary functions. All of them are generated by constants and functions x, e^x, $\ell n\, x$, $\cos x = \mathrm{Re}\, e^{ix}$, arc cos x using arithmetical operations and function composition. The functions separated are closely related to homomorphisms of the simplest groups R, R_+, U(1) (specifically, e^x gives the isomorphism of R onto R_+, $\ell n\, x$ gives its inverse, e^{ix} the homomorphism of R onto U(1), etc.). The elementary character of the above functions follows from the fact that these groups are commutative and their irreducible representations are one-dimensional; therefore, the relevant properties of the functions are formulated in a very simple manner.

The elementary functions e^{inx}, $n \in \mathbb{Z}$, associated with homomorphisms $U(1) \to U(1)$ underlie the harmonic analysis on a circle (Fourier series), and the functions $e^{i\lambda x}$, $\lambda \in R$, that give homomorphisms of R into U(1) underlie the harmonic analysis on R (Fourier integrals). Applying more complicated commutative groups, we get other classes of orthonormalized systems of functions. For example, the homomorphisms of a diadic group (direct product of a denumerable set of cyclic groups of second order) into U(1) lead to a system of Walsh functions widely used currently in information transmission studies. Its generalizations are systems of functions available in the theory of homomorphisms of arbitrary zero-dimensional compact commutative groups into U(1) (see [1]). This class of functions obeys the basic propositions of the theory of trigonometric series including such elegant propositions as the theorem on convergence almost everywhere of Fourier series for the functions of $L^2(U(1))$. The theory of these and other systems of functions (for example, trigonometric functions of an infinite set of variables), as well as the theory of almost periodic functions are based on the theory of characters (continuous homomorphisms into U(1)) of commutative locally compact groups.

Another aspect of the relationship between the special functions and the theory of group representations has become important for theoretical physics. To solve the differential equations available in quantum mechanics, it was necessary to use the symmetry of the physical systems under study, i.e. transformation groups that leave invariant some important characteristics of these systems (for example, potential in the Shrödinger equation). Since the solutions of these equations for some particular cases (for example, for the harmonic oscillator) could be expressed in terms of the special functions, it was necessary to establish a relationship between the theory of these functions and the transformation group that leave invariant the physical systems studied. Here we must mention Wigner's contributions. In those years the group theory was little known to physicists of classical school (astronomer J. Jince even thought that the physicists would never need it), and this period in the development of theoretical physics was named "the Gruppenpest". Spectroscopy studies began to make increasingly wider use of such concepts as Clebsch-Gordan coefficients, Racah coefficients and more general symbols related to a decomposition into irreducible representations of the tensor product of representations of groups. The requirements of relativistic physics advanced the task of studying the representations of noncompact noncommutative Lie groups, in particular, of the Lorentz group $SO_o(3,1)$ and its three-dimensional analogue $SO_o(2,1)$. These investigations resulted in a theory of infinite-dimensional representations of semi-simple Lie groups (V. Bargmann, I. Gel'fand and M. Naimark, Harish-Chandra), and then of nilpotent (A. Kirillov) and solvable (Auslander) Lie groups.

While studying the matrix elements of irreducible unitary representations of group $SO_o(2,1)$, V. Bargmann found that they are expressed through the hypergeometric functions, moreover, the matrix elements of representations of a discrete series are expressed through a particular case of this function (Jacobi polynomials). The same polynomials are used to express the matrix elements of irreducible representations of the group SO(3) which is a compact real form of the group SO(3,C) - a complexification of the group $SO_o(2,1)$. "Straightening" groups $SO_o(2,1)$ and SO(3), we obtain a group ISO(2) of motions of an Euclidean plane. The matrix elements of irreducible unitary representations of this group are expressed through the Bessel function.

So the theory of the classes of special functions most important for applications - the hypergeometric function and the Bessel function - turned out to be associated with the representations of the simplest noncommutative Lie groups SO(3), $SO_o(2,1)$, ISO(2). Cartan's theory of zonal spherical functions constructed earlier was also associated with the matrix elements of representations, namely, of the representations of class 1, i.e. such that their space has a single vector ξ_o invariant under the operators T(h), $h \in H$. If we take ξ_o to be one of the basis vectors, the corresponding matrix element $(T(g)\xi_o,\xi_o)$ will be constant on two-sided coset space with respect to the subgroup H, and so it gives a zonal spherical function on X = G/H. Similarly, matrix elements such as $(T(g)\xi_o,\xi)$ and $(T(g)\xi,\xi_o)$ are expressed through associated spherical functions.

We now use Jacobi polynomials to show how the theory of special functions is developed in group-theoretical terms. We first choose the space of representation of the group SU(2) locally isomorphic to SO(3). It is the set of polynomials of degree 2l of x. Then to every element $g \in SU(2)$ we assign an operator

$$T(g)p(x) = (\beta x+\bar{\alpha})p\left(\frac{\alpha x-\bar{\beta}}{\beta x+\bar{\alpha}}\right), \quad g = \begin{pmatrix} \alpha & \beta \\ -\bar{\beta} & \bar{\alpha} \end{pmatrix}$$

It is easy to see that if we set $(x^k,x^k) = (\ell-k)!(\ell+k)!$, this representation is unitary. Its matrix elements are expressed by the formula $t^{\ell}_{mn}(g) = (T(g)x^n,x^m)$. Replacing standard polynomials by trigonometric ones we express the scalar product in integral form, obtaining the integral representation of matrix elements. Because the analyst prefers to deal not with functions on a group, but with functions of numerical arguments, we parametrize the group SU(2) using, for example, Euler angles ϕ,θ,ψ. Then to the elements of a one-parametric subgroup $\{g(\phi,0,0)\}$ there correspond diagonal matrices with the functions exp $ik\phi$, $-\ell \le k \le \ell$, on the diagonal, and to the elements of a subgroup $\{g(0,\theta,0)\}$ there correspond matrices which consist of the functions $P^{\ell}_{mn}(\cos\theta) \equiv t^{\ell}_{mn}(g(0,\theta,0))$. Also, $P^{\ell}_{mn}(x)$ is insignificantly different from Jacobi polynomials, and therefore the properties of the Jacobi polynomials are trivially derived from the properties of the functions $P^{\ell}_{mn}(x)$.

The first of these properties is the addition theorem that follows directly from the equality $T_\ell(g_1g_2) = T_\ell(g_1)T_\ell(g_2)$ and the matrix multiplication rules: for example,

$$t^{\ell}_{mn}(g_1g_2) = \sum_k t^{\ell}_{mk}(g_1)t^{\ell}_{kn}(g).$$

Under a corresponding choice of elements g_1 and g_2 we get an addition theorem for Jacobi polynomials whose right-hand side has the form of a sum of functions exp $ik\phi$ multiplied by a product of two such polynomials. We thus derive a multiplication theorem that expresses the product of two Jacobi polynomials of cos θ_1 and cos θ_2 as an integral over some circle - the orbit of subgroup SO(2). As mentioned above, the Laplace operator is expressed by such an average, and so we can use the multiplication theorem to obtain a Laplace equation for matrix elements, and then use the equation to obtain a second-order differential equation for Jacobi polynomials.

The same differential equation may be obtained in a different way. By writing down the addition theorem for infinitely small g_2 (i.e. linearizing it), we get recurrent relations for matrix elements, and thus also for Jacobi polynomials. These relations can be combined so that their successive application produces an initial function with an additional numerical factor. Removing the brackets we get the desired equation. Other recurrent relations for Jacobi polynomials are associated with the tensor multiplication of representations. An orthonormalized basis in the space $L_1 \otimes L_2$ of the representation $T_{\ell_1} \otimes T_{\ell_2}$ can be chosen in two ways: first, using vectors $e_i \otimes f_j$, where $\{e_i\}$ (accordingly, $\{f_j\}$) is an orthonormalized basis in L_1 (accordingly, in L_2); secondly, using a basis $\{h^{\ell}_k\}$ derived by decomposing the representation $T_{\ell_1} \otimes T_{\ell_2}$ into irreducible components, and then choosing a corresponding orthonormalized basis in their spaces. Denote by $C\ (\ell_1,\ell_2,\ell;i,j,k)$ matrix elements of transition from one basis to another - these elements are called the Clebsch-Gordan coefficients. If one of the representations T_{ℓ_1}, T_{ℓ_2} of the group SU(2) is identical (to every rotation there corresponds its matrix), we can use the Clebsch-Gordon coefficients to obtain new recurrent relations for Jacobi polynomials. The general theorems of linear algebra and the above orthogonality of matrix elements with respect to an invariant measure on a group yield the relations that express the integral of a product of three matrix elements as a product of two Clebsch-Gordan coefficients. This gives the expressions for the integral of a product of three Jacobi polynomials.

The fact that the representations T_{ℓ} include all (up to equivalency) irreducible unitary representations of SU(2) implies the completeness of an orthogonal system of matrix elements and thus the completeness of systems of Jacobi polynomials. In a similar way we conclude that the systems of Legendre polynomials and the associated Legendre polynomials are also complete.

The matrix elements of irreducible unitary representations of the group SU(1,1), locally isomorphic to $SO_0(2,1)$, are expressed via hypergeometric functions. Therefore, the theory of hypergeometric functions is developed along similar lines. This theory, however, is more sophisticated. First of all, instead of one series there are several different series of irreducible unitary representations: continuous, two discrete and complementary ones, moreover, the decomposition of functions of $L^2(G)$ does not involve the representations of a complementary series, although they are involved in the formulation of an analogue of the Bochner theorem on positive definite functions on this group. The complementary series appears because with increasing radius of a sphere in SU(1,1) its measure increases exponentially, but not in a power-like manner, and therefore we have an appreciable difference in the properties of functions of $L^2(G)$ and $L^1(G)$. Next, the representations of a continuous series are involved in Plancherel formula continuously (similar to the Fourier integral but not the Fourier series). Finally, the theory of

Clebsch-Gordan coefficients becomes too complicated, because we have to deal with representations of various series; moreover, the decompositions of some tensor products can contain components with multiplicity 2.

It is to be noted that neither the theory of matrix elements of representations of SU(1,1) in a basis that diagonalizes a compact subgroup U(1), nor the theory of matrix elements of representations of SU(2) provide a comprehensive way to derive the properties of the hypergeometric function. To obtain the properties of arbitrary hypergeometric functions, it is necessary to choose continual basis composed of generalized functions of representation spaces (similar to the basis $e^{i\lambda x}$ in $L^2(R)$), for example, the bases that diagonalize a subgroup of hyperbolic rotations or a subgroup of translations (orispheric rotations). In these bases the representation operators are given by integral operators whose kernels are expressed by the hypergeometric function or cylindrical Hankel functions. Applying for these kernels the methods used above for the matrix elements we obtain relations for these functions including "continual addition theorems" in which the integration is over the parameters of the functions, but not over their arguments. Ordinary addition theorems are derived from them using residue theorems. It is of interest to consider "mixed bases", i.e. a decomposition of the result of action of representation operators on the elements of one basis into the elements of the other. As a result, we get the Whittaker functions, the Laguerre polynomials, the Pollaczek polynomials and different relations connecting these functions with the hypergeometric function. We note that the Whittaker functions and the Laguerre polynomials also appear when we study the matrix elements of irreducible representations of the group S_4 of triangular third-order matrices which is an extension of the Heisenberg group - the simplest in the class of nilpotent groups. A series of new relations for the special functions arises when we realize the representations using boson creation and annihilation operators.

The group representation theory gives an insight not only into the properties of classical special functions of a continuous argument, but also into the properties of special functions of a discrete variable introduced by Chebyshev. Since the representation matrix $T_\ell(g)$ is unitary for any g, the totality of its lines forms an orthonormalized system of vectors, i.e. functions of a discrete variable. In the case of group SU(2) these functions are expressed through Krawtchouk polynomials - polynomials of a discrete variable orthogonal under binomial distribution. Similarly, the matrix elements of irreducible representations of a discrete series of SU(1,1) lead to Meixner polynomials, and those of group S_4 to Charlier polynomials. The properties of the polynomials in reference follow from the properties of the matrix elements derived by group-theoretical methods. We note that in exactly the same way it is possible to obtain continual analogues of the above systems of polynomials of a discrete variable - for this purpose we must regard the kernels of the operators of corresponding representations as functions of the "continual number" of a column.

The Clebsch-Gordan coefficients are also the matrix elements of unitary operators (because the bases $\{e_i \otimes f_j\}$ and $\{h_k^\ell\}$ are orthonormalized) and therefore lead us to systems of orthogonal functions of a discrete variable. These coefficients are expressed through Hahn polynomials closely associated with the generalized hypergeometric function ${}_3F_2(\alpha,\beta,\gamma;\delta,\varepsilon;1)$ of a unit argument. Replacing in the relations the Clebsch-Gordan coefficients by Hahn polynomials, and the matrix elements by Krawtchouk polynomials, we get analogues of the addition theorems for these polynomials.

The Racah coefficients are still more complicated functions of a discrete argument. They appear when we compare the decompositions of the tensor product $T_{\ell_1} \times T_{\ell_2} \times T_{\ell_3}$ according to a scheme $(T_{\ell_1} \times T_{\ell_2}) \times T_{\ell_3}$, on the one hand, and according to a scheme $T_{\ell_1} \otimes (T_{\ell_2} \otimes T_{\ell_3})$, on the other hand, and are the coefficients of the matrix of the transition between the resulting bases. The Racah coefficients are associated with the polynomials of a discrete variable introduced by Askey and Wilson and expressed through a generalized hypergeometric function ${}_4F_3(\ldots;1)$ of a unit argument. Note that for the group S_4 both the Clebsch-Gordan coefficients and the Racah coefficients are expressed through ${}_2F_1(\alpha,\beta;\gamma;x)$, and this allows us to derive a number of relations between this function, the Laguerre polynomials and the Charlier polynomials.

If in the expression for ${}_3F_2(\alpha,\beta,\gamma;\delta,\varepsilon;1)$ we replace γ and δ by $r\gamma$ and $r\delta$ and then let r tend to infinity, we obtain ${}_2F_1(\alpha,\beta;\gamma;\gamma/\varepsilon)$ in the limit. We thus get the relations that express Jacobi polynomials as the limits of Clebsch-Gordon coefficients. A similar reasoning allows us to express the Clebsch-Gordon coefficients as the limits of Racah coefficients when some of the parameters tend to infinity. The possibility of such limiting transitions explains a far-reaching analogy between the theories of Jacobi polynomials and Clebsch-Gordon coefficients; as a matter of fact, every formula related to Jacobi polynomials (or, in the physicist's language, to Wigner d-functions) corresponds to a formula for Clebsch-Gordon coefficients, moreover, the differentiation operators are replaced by the difference operators, and the powers, by "combinatorial powers" $(\alpha)^{(n)} = \alpha(\alpha-1)(\alpha-2)\ldots(\alpha-n+1)$. It would be of interest to elucidate deeper reasons for this peculiar duality first observed by Gel'fand. The same possibility to get Jacobi polynomials from Racah coefficients results in a peculiar geometry of these coefficients that changes to Euclidean geometry in the limit. In some "Pickwick" sense of the word this geometry may be regarded as the result of quantization of the Euclidean geometry.

Studies of separate kinds of integral transformations are concerned with the theory of representations of $SO_0(2,1)$. They occur as intertwining operators between different realizations of these representations. Specifically, the irreducible representations of the group $SO_0(2,1)$ can, on the one hand, be realized in the space of homogeneous functions on a cone $x_0^2 - x_1^2 - x_2^2 = 0$, and, on the other hand, in the space of the functions on hyperbolids $x_0^2 - x_1^2 - x_2^2 = \pm 1$ which are restrictions onto them of the homogeneous solutions of a wave equation. These realizations are intertwined by the analogue of Poisson transformation called a Gel'fand-Graev transformation:

$$f(\xi) = \int_{[x,x]=\pm 1} f(x)[x,\xi]^{\sigma} dx$$

Writing down this transformation and its inverse in corresponding systems of coordinates on the cone and the hyperboloids, we get integral Fock-Mehler, Cantorovich-Lebedev and other transformations, as well as formulas to invert these transformations. This aspect of applications of the group representation theory to special functions joins an integral geometry in the sense of Gel'fand - a replacement of functional spaces on manifolds by other functional spaces derived by integrating the functions over some families of submanifolds.

The same type of transformations also includes a Radon transformation studied by S. Helgason. To every function on some space of constant curvature it assigns its integrals over completely geodesic manifolds of codimension 1 (in particular, transformation of the integration over hyperplanes in the Lobachevski space, and in an imaginary Lobachevski space). This transformation transforms the spherical functions of a given space into the spherical functions of another space of constant curvature (for example, the spherical functions of Lobachevski space into the spherical functions of an imaginary Lobachevski space and vice versa).

The above examples of applying the group representation theory to study special functions are models of a much more general theory. Namely, let G be a noncompact semisimple Lie group with a finite centre and let K be its maximal compact subgroup. As Cartan showed, homogeneous space $X = G/K$ can be imbedded into a group G so that there holds the decomposition $G = K\Theta$, where Θ is the image of X. For example, if $G = SL(n,C)$, then $K = SU(n)$, and Θ is the space of positively definite hermitian matrices. Denote by A a maximal commutative subgroup contained in Θ and satisfying some regularity condition (in our case this is a subgroup of diagonal matrices with positive elements), and by M a centralizer of A in K, i.e. a set of elements m of K such that $m\alpha = \alpha m$, $\alpha \in A$ (in our case this is a set of unitary diagonal matrices). Finally, we denote by M' a normalizer of A in K, i.e. a set $m' \in K$ such that $m'A = Am'$ (in our case M' is obtained by extending M with help of a permutation group), by W a group M'/M (a Weyl group), and by N a maximal nilpotent subgroup in G related in a definite way to A (in our case N is a subgroup of upper triangular matrices with unities on the main diagonal). We then have the following equalities: $G = KAK$ and $G = KAN$ (Cartan and Iwasawa decompositions).

We give a one-dimensional representation χ of group A and ρ finite-dimensional representation of group M that acts in the space L; we consider L-valued functions on G that satisfy the condition $f(mang) = \chi(\alpha)\rho(m)f(g)$ and the condition of square integrability of their restriction onto K. The equality $T_{\chi\rho}(g_o)f(g) = f(gg_o)$ gives a representation of the group G in the space of these functions. Generally speaking, these representations are irreducible (there are exceptions when χ satisfies some integral conditions), and they exhaust almost all the irreducible representations of G.

The functions on which the representation $T_{\chi\rho}$ is defined are unambiguously defined by their values on K. Therefore, the matrix elements $t_{rs}(k)$ of irreducible representations of K that satisfy the condition of covariance under representation ρ of subgroup M form a basis of the space of the representation $T_{\chi\rho}$. The matrix elements of the operators $T_{\chi\rho}(\alpha)$, $\alpha \in A$, with respect to this basis can naturally be called special functions that correspond to a given representation. They depend on a smaller number of arguments than the number of parameters in group G and, because the group A is commutative, are reduced to functions of one argument.

In addition to an Iwasawa decomposition, its generalizations are considered, resulting in new classes of special functions. Study of the behaviour of special functions at integral values of representation parameters is concerned with considering the discrete series of unitary representations.

Complexifying the group G and taking the compact real form of the complex group, we obtain a compact Lie group G_k (in our case $G_k = SU(n)\times SU(n)$); the homogeneous space G_k/K is called dual to $X = G/K$. The matrix elements of irreduciable representations of the group G_k can be obtained from the matrix elements of the representations $T_{\chi\rho}$ of G by a relevant analytical continuation and unitarization using intertwining operators [2]. The unitarization operator is diagonal for the groups $SO_o(n,1)$, $SU(n,1)$ [3] and for most degenerate representations [4-7]. The groups G and G_k are associated with another group which is a semi-direct product of subgroup K and a space tangent to Θ. The matrix elements of irreducible representations of this group are obtained by a limiting transition from the matrix elements of representations of the groups G and G_k, which is concerned with "straightening" the space Θ.

If we know the matrix elements of representations of the above groups, we can calculate the Clebsch-Gordon coefficients, the Racah coefficients, etc., obtaining corresponding functions of discrete variables. Unfortunately, although the above programme of constructing new classes of special functions is fairly transparent theoretically, its realization runs into considerable computational difficulties, because resulting expressions rapidly become cumbersome and hardly controllable. Dimensionality recursion enables us to write down the integral representations of matrix elements and to express recursively matrix elements of class s in terms of matrix elements of class s-1 for the groups SO(n) and $SO_0(n,1)$ (see [8,9]). We also mention papers by A. and L. Rosenbloom in which the matrix elements are expressed as solutions of linear differential equations with matrix coefficients. The resulting expressions may be written down as special functions with matrix parameters. This allows us to reduce the difficulties of calculating the matrix elements to the problem of writing the above parameters canonically [10].

In some cases it is possible to find fairly convenient expressions for the matrix elements. Sometimes this results in their being expressed through the already available special functions and thus facilitates establishing new properties of these functions. In other cases we express the matrix elements through functions though unfamiliar in mathematical analysis, but having a fairly clear form. The need to study new classes of functions follows from the fact that the presently available functions are insufficient to describe the matrix elements and the Clebsch-Gordan coefficients of the representations of such physically important groups as the Lorentz group, Poincare group ISO(3,1), the conformal group $SO_0(4,2)$ or the group SU(2,2) linked with the twistor theory.

The available functions may be used to express the "zero column" matrix elements of class 1 representations of the groups SO(n), $SO_0(n,1)$ and ISO(n) (i.e. representations $T_{\chi\rho}$, where ρ is a unit representation; they can be realized in the space of scalar functions on X = SO(n)/SO(n-1); the "zero column" involves matrix elements $(T_{\chi\rho}\xi_0,\xi_0)$, where ξ_0 is a constant function on X). For the group SO(n), these matrix elements are expressed through Gegenbauer polynomials, and here the change of the dimensionality leads only to shifted polynomial indices. A similar role for the group $SO_0(n,1)$ is played by the functions of a cone, and for the group ISO(n) by the Bessel functions. As a result, we get different relations for these functions. Interesting formulas are derived when n tends to infinity - the matrix elements under a corresponding normalization transform into the Hermite polynomials.

Choosing a basis in the space of class 1 representations is related to choosing a certain system of subgroups over which a successive reduction is performed, or, what basically amounts to the same thing, to choosing a certain system of coordinates on homogeneous spaces. Different choices can be described by graphs of special kind - the trees. In [11] the systems of coordinates on spheres and hyperboloids obtained by this method were described, in which the Laplace operator enables a separation of variables; the relevant eigenfunctions of this operator are expressed through hypergeometric and cylindric functions. Many papers dealt with the matrix elements that correspond to transitions from one tree to another ("branch transplantation") which are expressed through hypergeometric functions, Hahn polynomials and other functions and this results in new relations for these functions [12, 13].

Relatively simple expressions for the matrix elements of class 1 irreducible representations of the groups SO(n), $SO_0(n,1)$, ISO(n) are obtained in [14]. These matrix elements are expressed through functions unfamiliar in mathematical analysis. The further development of this theory leads us to a study of peculiar generalizations of hypergeometric functions.

In general, the study of the matrix elements of irreducible representations of semi-simple Lie groups and related inhomogeneous groups leads us to different generalizations of the concept of a hypergeometric functions. The most convenient basis to calculate these matrix elements was proposed by Gel'fand and Zetlin [15,16]. The basis for the group SL(n,C) is constructed as follows. We restrict a finite-dimensional representation T of weight $(\ell_1,\ell_2,\ldots,\ell_n)$ of this group to the subgroup SL(n-1,C) (this representation is constructed in the same way as the representation of the same weight of the group U(n)). A restricted representation is broken down into irreducible representations of the subgroup the weights $(\ell_1',\ldots,\ell_{n-1}')$ of which satisfy the following betweenness condition:

$$\ell_1 \geq \ell_1' \geq \ell_2 \geq \ell_2' \geq \cdots \geq \ell_{n-1} \geq \ell_n.$$

Continuing this process, we come to Gel'fand-Zetlin patterns where the betweenness condition is satisfied for every two neighbouring lines. These patterns number the basis vectors in the representation space. Gel'fand and Zetlin wrote down in these bases the matrix elements of relevant representations of Lie algebra, and later Gel'fand and Graev wrote down the matrix elements of representations of the group itselt [17]. For matrices such as $E_n + tE_{n,n-1}$ the matrix elements are expressed by the generalization of beta-function introduced by these authors. Another generalization of beta-function is given in the book [18]. The expression for matrix elements of representations of the group SU(n) is given in [19]. They contain an interesting generalization of hypergeometric function. The rows of Gel'fand-Zetlin patterns are parameters of this function. The generalization of these results to the case of the group U(n,1) is given in [3].

The zonal spherical functions for homogeneous spaces

$$X_1 = SU(p+q)/S(U(p)\times U(q)), \quad X_2 = SU(p,q)/S(U(p)\times U(q))$$

and for that triple to them were calculated in [20]. The answer obtained has the form of the ratio of two determinants, and here the enumerator involves a Wandermond determinant, and the denominator involves a determinant composed of special functions (Gegenbauer polynomials, hypergeometric functions or Bessel functions) of arguments that have an obvious geometric meaning. Specifically, X_1 can be realized as the space of p-dimensional subspaces of C^{p+q}. Then the arguments are cosines of the stationary angles between these subspaces and a fixed subspace spanned by basis vectors $e_1,\ldots,e_p$. The expressions obtained in [20] can be expanded into series of the ratios of two determinants resembling a Wandermond determinant. This suggests us the idea to consider the generalizations of hypergeometric functions in which the powers of the variables are replaced by determinants (see also [21]).

Another way to generalize hypergeometric functions is to consider the functions of a matrix argument that figure in the papers on many-dimensional statistics (see [22-24]). In [25], the theory of such special functions is constructed using the Laplace transformation of the functions of a matrix argument introduced by Bochner. The relationship between these functions and the theory of representations of block-triangular groups is investigated in [26]. The spherical functions of a matrix argument were studied by Maass [27-29]. The further development of this theory resulted in a construction of the theory of special functions on homogeneous cones. In [30], the generalizations of gamma- and beta-functions, as well as some types of hypergeometric functions were discussed and the relationship of this theory with the functions of many complex variables was indicated. The functions on matrix cones were treated in connection with the representations of orthogonal and pseudoorthogonal groups [31].

The theory of other analogues of a hypergeometric function is related to the representations of Chevallier groups over Galois fields, in particular a group of unimodular matrices with elements from such fields. The zonal spherical functions of these representations are expressed through basis hypergeometric functions whose coefficients involve, instead of the factorials, expressions such as $(q^n-1)(q^{n-1}-1)\ldots(q-1)$. The fractions that contain such expressions appear when we calculate the number of k-dimensional subspaces in n-dimensional linear space over a Galois field and are analogues of binomial coefficients. Some spherical functions are expressed through q-analogues of Hahn polynomials, this allowing us to derive an addition theorem for these polynomials [32-35]. As far as we know, the theory of Clebsch-Gordan and Racah coefficients for the representations of Chevallier groups as well as the general theory of matrix elements of these representations have not yet been developed.

It is of great interest that the same special functions of a discrete variable which appear in group representation theory also figure in the branch of discrete mathematics generated over the last few decades and called "algebraic combinatorial analysis". This branch of mathematics that can be called a "group theory without groups" embraces the algebraic theory of graphs, the algebraic theory of codes, etc., and the methods applied go, to a great extent, back to the theory of finite groups. For example, Krawtchouk polynomials appear in connection with the Hemming metric, Hahn polynomials are related to the Johnson metric, and Askey-Wilson polynomials exhaust the functions related to a very general class of P- and Q-polynomial associative schemes [36]. It is likely that there should be the same relationship between the orthogonal polynomials of a real argument and the continual analogues of associative schemes whose theory has not yet been constructed. We note that the same orthogonal polynomials also occur in some points covered by the theories of random processes [37].

We have so far been considering the role of group representations in the theory of special functions. There is, however, another aspect of the relationship between representations and special functions which is not often taken into account. Expressing the matrix elements of group representations through special functions enables us to get a deeper insight into the representations themselves, to understand the relationship between representations of various groups.

Let us first consider some examples of compact group representation theory. Let T^χ be an irreducible representation of SO(n), U(n) or Sp(n) that has class 1 with respect to subgroups SO(n-1), $U(n-1)\times U(1)$, $Sp(n-1)\times Sp(1)$, respectively, and A_k a one-parameter subgroup of real rotations $g(\theta)$ in a plane (n-1,n). In the case of each of these three groups the matrix elements of a "zero" column $t^\chi_{no}(g(\theta))$ of the operator $T^\chi(g(\theta))$ are simply expressed through Jacobi polynomials $P_n^{(\alpha,\beta)}(\cos\theta)$ with integer or half-integer indices α and β. These polynomials are used to express the matrix elements of representations of the group $SU(2) \sim SO(3)$. Consequently, the matrix elements $t^\chi_{no}(g(\theta))$ are expressed through Wigner d-functions, i.e. through the functions $P^\ell_{mn}(x)$ mentioned above. There is also a relationship between the matrix elements $t^\chi_{no}(g(\theta))$ of the representation operators of the groups SO(n), U(n), Sp(n). Specifically, the matrix elements $t^\chi_{no}(g(\theta))$, $\chi = (m,0,\ldots,0,-m)$, for the group U(n) are multiples of corresponding matrix elements $t^\chi_{n'o}(g(\theta))$, $\chi = (2m,0,\ldots,0)$, for the group SO(2n) in a coordinate system that corresponds to restriction to the subgroup $SO(2n-2)\times SO(2)$ and the matrix elements $t^\chi_{no}(g(\theta))$, $\chi = (m,m,0,\ldots,0)$, for the group Sp(n) are multiples of the matrix elements $t^\chi_{n'o}(g(\theta))$, $\chi = (2m,0,\ldots,0)$, for the group SO(4n) in a coordinate system that corresponds to a restriction to the subgroup $SO(4n-4)\times SO(4)$.

Next, the matrix elements $t^\chi_{kk'r}(g(\theta))$ of the operators of the representations T^χ, $\chi = (m,0,\ldots,0)$ of all groups U(n), n = 3,4,5,..., in the Gel'fand-Zetlin basis are the same and equal to corresponding Wigner d-functions:

$$t^{\chi}_{kk'r}(g(\theta)) = d^{(m-r)/2}_{(k-(m+r)/2,k'-(m+r)/2}(2\theta),$$

where k and k' correspond to the highest weights (k,0,...,0), (k',0,...,0) of the subgroup U(n-1), and r to the highest weight of the subgroup U(n-2). A more general statement is formulated as follows. In the Gel'fand-Zetlin basis the matrix elements $t^{\chi}_{mm'}(g(\theta))$ of the operators of the representations T^{χ}, $\chi = (m_1,\ldots,m_i,0,\ldots,0)$ of the groups U(n), n = i+1,i+2,..., are the same; the matrix elements $t^{\chi}_{mm'}(g(\theta))$ of the operators of the representations T^{χ}, $\chi = (m_1,\ldots,m_i,0,\ldots,0,m_i',\ldots,m_1')$, of the group U(n) coincide with corresponding matrix elements of the operators of the representations $T^{\chi'}$, $\chi' = (m_1,\ldots,m_i,0,\ldots,0,m_i'+1,\ldots,m_1'+1)$, of U(n-1). Similar relationships exist between the matrix elements of representations of the groups SO(n).

The propositions formulated on the relationship between representation operators of different groups may become crucial in solving one of the fundamental problems in elementary particle physics and quantum field theory: what is the nature of unitary symmetry?

We now turn to corresponding problems in the theory of representations of noncompact simple Lie groups. Pseudo-Riemannian symmetric spaces

$$SO_o(p,q)/SO_o(p,q-1), \tag{2'}$$

$$SU(p,q)/S(U(p,q-1)\times U(1)), \tag{2''}$$

$$Sp(p,q)/Sp(p,q-1)\times Sp(1) \tag{2'''}$$

have rank 1. Consequently, their points are associated with elements $kg(\alpha)$ of subgroup KA, where K is a maximal compact subgroup in $SO_o(p,q)$, U(p,q), Sp(p,q), and A is a one-parametric subgroup of real hyperbolic rotations. For the associated spherical functions $t^{\chi}_{MO}(kg(\alpha))$ (χ is given by one continuous parameter τ) on the spaces (2')-(2''') in a coordinate system that corresponds to the subgroup K we have separation of variables:

$$t^{\chi}_{MO}(kg(\alpha)) = t^{\sigma}_{M'O}(k)\, t^{\chi}_{mO}(g(\alpha)).$$

It turns out that the function $t^{\chi}_{mO}(g(\alpha))$ for the symmetric space (2''') is the same as a corresponding function $t^{\chi}_{m'O}(g(\alpha))$ for the space SU(2p,2q)/S(U(2p, 2q-1) × U(1)), and the function $t^{\chi}_{mO}(g(\alpha))$ for the space (2'') is the same as a corresponding function $t^{\chi}_{m'O}(g(\alpha))$ for the space $SO_o(2p,2q)/SO_o(2p,2q-1)$. This reduces harmonic analysis on the spaces (2'') and (2''') to that on the spaces (2') with even p and q. In particular, there is a correspondence between the representations of discrete square integrable series on the spaces

$$Sp(p,q)/Sp(p,q-1)\times Sp(1) \quad\text{and}\quad SO_o(4p,4q)/SO_o(4p,4q-1)$$

as well as between the representations of discrete series on the spaces

$$SU(p,q)/S(U(p,q-1)\times U(1)) \quad\text{and}\quad SO_o(2p,2q)/SO_o(2p,2q-1).$$

The functions $t^{\chi}_{mO}(g(\alpha))$ for the space $\chi = SO_o(p,q)/SO_o(p,q-1)$ are expressed through the matrix elements $\sigma_{mn}(\text{ch}\,\alpha)$ of representations of the group $SU(1,1) \sim SO_o(2,1)$:

$$t^{\chi}_{(kk')O}(g(\alpha)) = c(\text{th}\,\alpha)^{1-p/2}(\text{ch}\,\alpha)^{-(p+q-4)/2}\; \sigma_{rr'}(\text{ch}\,\alpha),$$

where k and k' correspond to the highest weights (k,0,...,0), (k',0,...,0) of representations of the subgroups SO(p) and SO(q), c is independent of α and related to Plancherel measure on $L^2(X)$;

$$\sigma = \tau + \frac{p+q-4}{2}, \quad r = \frac{k+k'}{2} + \frac{p+q-4}{2}, \quad r' = \frac{k-k'}{2} - \frac{p-q}{2}$$

(τ is the number that characterizes χ). The harmonic analysis on $SO_0(p,q)/SO_0(p,q-1)$ is therefore connected with the harmonic analysis on SU(1,1). In other words, the problems of harmonic analysis on $SO_0(p,q)/SO_0(p,q-1)$ can be solved using the harmonic analysis on the group SU(1,1). This fact seems to admit a generalization to pseudo-Riemannian symmetric spaces of higher rank: the harmonic analysis on G/H of rank r is related to harmonic analysis on some real noncompact simple group G' of real rank r.

In the present paper we have not covered all the relations between Lie groups and special functions. A very interesting aspect of the problem has been developed in a series of papers by Miller who, in particular, established the relationship between Lie groups and Lame functions. The relationship however makes no use of the restrictions of representations over a chain of subgroups, which is characteristic for the approach described in the present paper. The relationship between group representations and automorphic functions, as well as teta-functions have not been elucidated [38].

We conclude by noting that the problems related to a group-theoretical approach to the theory of special functions are now in intensive study and attract the attention of the specialists in different fields of mathematics. One of the most interesting problems that can find a wide range of applications is concerned with reasonable generalizations of hypergeometric functions.

REFERENCES

1. G. N. Ageev, N. Ja. Vilenkin, G. M. Javadov, A. I. Rubinshtein, "Multiplicative Systems of Functions and Harmonic Analysis on Nul-dimensional Groups," ELM, Baku (1981) (in Russian).
2. A. U. Klimyk, "Matrix Elements and Clebsch-Gordon Coefficients of Group Representations," Naukova Dumka, Kiev (1979) (in Russian).
3. A. U. Klimyk, A. M. Gavrilik, J. Math. Phys., 20:1624 (1979).
4. B. Gruber, A. U. Klimyk, J. Math. Phys., 22:2762 (1981).
5. A. U. Klimyk, B. Gruber, J. Math. Phys., 25:743 (1984).
6. A. U. Klimyk, B. Gruber, J. Math. Phys., 23:1399 (1982).
7. V. F. Molchanov, Mat. Sb., 99:139 (1976).
8. N. Ja. Vilenkin, DAN SSSR, 113:16, No. 1, (1957).
9. I. I. Kachurik, A. U. Klimyk, Reps. Math. Phys., 20:333 (1984).
10. A. V. Rosenbloom, L. V. Rosenbloom, Izv. AN BSSR, No. 4, 44 (1980).
11. N. Ja. Vilenkin, Mat. Sb., 68:432 (1965).
12. M. S. Kildjushov, Soviet Nucl. Phys., 15:197 (1972).
13. S. K. Suslov, Soviet Nucl. Phys., 38:1367 (1983).
14. N. Ya. Vilenkin, Trudy Mosk. Mat. Obsch., 12:185 (1963).
15. I. M. Gel'fand, M. L. Zetlin, DAN SSSR, 71:825 (1950).
16. I. M. Gel'fand, M. L. Zetlin, DAN SSSR, 71:1017 (1950).
17. I. M. Gel'fand, M. I. Graev, Isv. AN SSSR, 29:1329 (1965).
18. I. M. Gel'fand, M. I. Graev, N. Ja. Vilenkin, "Generalized Functions," Vol. 5, Academic Press, New York (1966).
19. N. Ja. Vilenkin, Sb. Nauchn. Trudov Mosk. Ped. Inst., 39:77 (1974).
20. F. A. Beresin, F. I. Karpelevich, DAN SSSR, 118:9 (1958).
21. J. D. Louck, L. C. Biedenharn, J. Math. Anal. Appl., 59:423 (1977).
22. A. T. James, Ann. Math. Statistics, 25:40 (1954).
23. A. T. James, Ann. Math. Statistics, 39:1711 (1968).
24. A. T. James, A. G. Constantine, Proc. London Math. Soc. (3), 29:174 (1974).

25. C. S. Herz, Ann. Math., 61:474 (1955).
26. N. Ja. Vilenkin, V. I. Paranuk, in "Some Problems of Mathematics and Physics," Krasnodar (1969), p. 52.
27. H. Maass, J. Indian Math. Soc., 20:117 (1956).
28. H. Maass, Math. Annalen, 135:391 (1958).
29. H. Maass, Math. Annalen, 137:142 (1959).
30. S. G. Gindikin, Uspechi Mat. Nauk, 19:3 No. 4, (1964).
31. N. J. Vilenkin, L. M. Klesova, A. P. Pavliyk, in "Group-theoretical Methods in Physics," Vol. 1, Nauka, Moscow (1980), p. 40.
32. D. Stanton, Amer. J. Math., 102:625 (1980).
33. D. Stanton, Geom. Dedicata, 10:403 (1981).
34. C. F. Duncl, Indiana Univ. Math. J., 25:335 (1976).
35. C. F. Duncl, Monats. Math., 85:5 (1977).
36. E. Bannai, T. Ito, Algebraic Combinatorics. I, (1984).
37. P. J. Feinsilver, Lect. Notes Math. 696 (1978).
38. G. Lions, M. Vergne, The Weil representations, Maslov index and theta series, Birkhauser, Basel, 1980.

OBSERVABLE MANIFESTATIONS OF INVARIANCE IN CONDENSED MATTER AND BIOLOGICAL SYSTEMS

Giuseppe Vitiello

Dipartimento di Fisica dell'Università, 84100 Salerno
Italia *
Istituto Nazionale di Fisica Nucleare, Sezione di Napoli
Italia

Abstract. Ordered patterns, low energy theorems, extended objects as bags, vortices, dislocations, etc. are described as observable manifestations of the dynamical rearrangement of the theory invariance. In spontaneously broken symmetry theories a preminent role is played by Inönü-Wigner group contraction which determines the symmetry group relevant to the observations in condensed matter physics as well as in biological systems.

One reason why symmetries play a fundamental role in Science is in the fact that when a symmetry is discovered many phenomena and observations which seem to have no common roots or appear as not to be described on the same common ground, finally can be classified in classes or families due to the discovered symmetry. Search for symmetries is in fact, in most cases, search for unification: Nature presents many different phenomena and events and, of the same phenomenon, many aspects and different faces, so that it usually seems too much difficult to understand such differences and give a reason for their appearance. So, in some sense, one tries to simplify his own life by collecting in classes and families and multiplets possibly related phenomena. We know that this sometime can be done with great success, and on the basis of this success one hopes in the possibility of finding soon or later a basic unifying dynamics dictated by a symmetry principle relating the most disparate aspects of Nature. In this search any time that a symmetry is esthablished, one moves a step towards a world whose dynamics is more and more determined by symmetries. Any success in discovering a new symmetry appears thus as a success of a great capability of synthesis. However, understanding of Nature also requires understanding of the mechanisms by which the basic symmetries turn out to be observed as "broken symmetries", so that elements of the same family or multiplet are seen to be distinct among themselves by different values of masses, charges, etc.. Moreover, the possibility of distinguishing

----
*Postal address

among elements of the same family allows the construction of ordered structures, which would be impossible in the presence of an exact symmetry. Thus, one has also to understand how observed ordered structures are generated.

A possible picture is that the basic symmetric dynamics can generate, under suitable boundary conditions, different observable realizations of the original symmetry. The variety of phenomena and ordered structures one observes could be thus seen as the realization of dynamical rearrangements of a basic symmetry.

In Quantum Field Theory (QFT) the Heisenberg field equations are the ones which are invariant under the basic symmetry group and the symmetry properties of these equations are referred to as the invariance of the theory. The dynamics is however not completely specified by the equations, since Heisenberg fields are operator fields and thus the vector space where they are to be realized must be assigned. This assignement is not a trivial one in QFT. In Quantum Mechanics, due to von Neumann theorem, there is unitary equivalence among the Hilbert spaces which are representations of the canonical commutation relations; therefore any choice of the Hilbert space is physically equivalent to another one, since unitary equivalence means the same norm and thus the same expectation values for observables. In QFT due to the existence of infinitely many degrees of freedom, the von Neumann theorem cannot be applied and one gets infinitely many unitarily inequivalent representations of the canonical commutation relations [1]. Consequently, a different assignement of Hilbert space means in general a different physics. Exactly in this non-triviality of the assignement of the state space consists the richness of QFT, which therefore can accomodate different physical realizations of the same original dynamics. Once the Heisenberg field equations are given, it is most convenient to choose as state space the Hilbert space for asymptotic fields, i.e. those fields corresponding to observable particles, charges and currents, in terms of which observable states are described. Of course, the task of the theory is to compute observable quantities, and therefore the complete structure of the physical state space is not known at beginning. Thus, one is led to self-consistent conditions and methods of LSZ formalism in QFT.

As the breakdown of the dynamical symmetry is an observable effect, it gives strong constraints on the choice of the Hilbert space. Once this choice is done, one has to build the mapping which allows to operate with Heisenberg fields on the chosen space for asymptotic fields[2]; in this way a mapping is esthablished between the two level language "built in" in QFT: The asymptotic fields language and the Heisenberg fields language. This mapping, known also as the Haag expansion [3], is determined by the dynamics and is crucially constrained by the fact that asymptotic states must be wave--packet states, since they are localized states. This dynamical mapping is moreover controlled by the invariance of the theory: invariance in fact requires that when the Heisenberg fields undergo the transformations of the basic symmetry group, say G, the asymptotic fields must correspondingly transform under some group of transformations, say G', which characterizes the observable states. Such a group G' can be in general different from the basic symmetry group G, as, for example, always happens in the case of spontaneous breakdown of symmetry. In this last case, furthermore, the observable group G' must describe also the ordered patterns created by the lack of symmetry due to the breakdown. The dynamical rearrangement of symmetry is therefore described by the insurgence at the level of the observations of such a group G' under which the equations for the asymptotic fields are symmetric. Since spontaneous breakdown implies the existence of gapless modes[4] (the Goldstone modes) among the asymptotic fields, the equations for these modes will be invariant under transformations which shift the corresponding fields. These transformations constitute an Abelian subgroup of the asymptotic fields group G'. The theory of the dynamical rearrangement of the symmetry has been much developed and we refer to ref. 2 and to works quoted therein. One of the main results is that[5,6] the group G', which appears in the observations, is the Inönü-Wigner group contraction[7] of the basic invariance group G in most of the physically interesting cases. Systems as ferromagnets, crystals, superconductors provide beautiful examples of symmetry rearrangement and group contraction[2,5]. The ordered patterns they exhibit can be accounted by the boson condensation induced through the shift transformations of the Goldstone fields.

It should be noticed that extreme examples of dynamical rearrangement of symmetry can be given[2,8] where the group G' is not the contraction of the invariance group G; in this cases, however, the set of the generators of G' can be enlarged and the group contraction of G is still obtained. It can be shown[2,8] that when G is a compact Lie group, G' is the group contraction of G when the unbroken part of G, under which the vacuum is invariant, is a maximal subgroup of G.

It has also been proved[5,9,10] that the origin of the rearrangement leading to group contraction is due to the infrared Goldstone bosons which are missing in local observations. When their contribution is taken into account by integrating over the whole system volume, the original symmetry group G is recovered. Observations are intrinsically local and the volume V of the system appears thus very large. Therefore, terms of the order of $1/V$ with $V \longrightarrow \infty$ are always missing in observations. When spontaneous breakdown of symmetry occurs, the Haag expansion of the generators of G in terms of asymptotic fields which are related to observations is equivalent to the limiting procedure $1/V \longrightarrow 0$ and this leads to group contraction. The generators, whose commutators go to zero as $1/V \longrightarrow 0$, are the ones which induce the shift of the Goldstone modes and thus control their condensation in the ground state. This relation between the whole system volume and the Goldstone modes reminds us that these modes are indeed collective excitations involving the whole system dynamics. Their condensation appears as a coherent state. Observable ordered structures and patterns are macroscopic manifestations of this coherence.

Another recent result[11,12] is the one which shows that contraction of group representations leads to non linear realizations[13] of the invariance group in theories with spontaneous breakdown of symmetry.

This result is an interesting one since non linear realizations provide a powerful tool of investigation in effective Lagrangian theories; it can be shown that the contraction parameter can be interpreted as the expansion parameter in phenomenological Lagrangian theories. On the other hand,we also know that low energy theorems as Adler theorem[14] in soft pion physics and Dyson theorem[15] in ferromagnets, which state that soft modes do not contribute to the scattering matrix, can be derived as observable consequences of the dynamical rearrangement of symmetry[2,10,16].

In theories with gauge fields the so called Anderson-Higgs-Kibble mechanism (AHK)[17] takes place, by which Goldstone bosons disappear from the physical spectrum and gauge vector fields become massive. Neverthless, the dynamical rearrangement of symmetry still occurs[18,19] and contraction of group representations leads again to non linear realizations of the invariance group[11,12].

The structure of the physical vacuum is controlled by the boson condensation of the gapless Goldstone. When this condensation is constant in space one has an homogeneous structure; however, one can also have a spacially localized condensation so that localized structures appear. Bags in elementary particle physics, votices, dislocations, surface singularities or other kind of soliton-like extended objects are examples of such localized structures. It can be shown[2] that extended objects with topological singularities can be created only by condensation of gapless modes, as Goldstone particles are. This explains why such a kind of objects are observed only in systems with ordered ground state.

Examples of rearrangement of SU(2)[9,10], SU(3)[20], chiral SU(2) x SU(2)[2], SU(n) x SU(n)[16], as well as general theorems for SO(n) and SU(n)[5] have been studied by means of techniques as path-integral formalism[5,9,10], group theoretical methods[5,22,23], projective geometry analysis[6], in many models of physical interest, in high energy physics as well as in many body physics.

In recent years,the formalism of QFT has been used also in the study of living matter[24,25,26,27]. Biological systems are higly organized structures, not only in their spacial arrangement, but also at a dynamical level as time ordered sequence of the biochemical reactions show. On the other hand, since the physical laws that rule the interactions among atoms and molecules in non-living matter also apply to atoms and molecules of living matter, we see that biological systems are good candidates to be studied in the framework of spontaneous breakdown of symmetry. Coherence and collective modes , so relevant in many body physics, must also play a crucial role in living matter.

Much attention has been paied to relevant problems as energy transfer along alfa- helix proteins[28], energy storage and dissipativity[29], self-ordering and water polarization[30].

In the QFT approach[24,25] the living system is schematically assumed to be a set of macromolecules surrounded by water molecules. The water molecules are seen as a set of electrical dipoles governed by a rotationally invariant Lagrangian. The experimental evidence[31] of a non-zero polarization of the ground state of the system is taken as the evidence of spontaneous breakdown of the rotational symmetry. The macromolecules are assumed to support a non linear dynamics by which non coherent energy supply from the outside (e.g. under the form of biochemical energy in ATP reaction)is transformed in a solitary wave travelling without dissipation on the macromolecule chains. The role of this wave is to trigger the polarization of the neighbouring water dipoles in a preferred direction,thus breaking the rotational symmetry. The water electret state is thus created and sustained by gapless polarization waves (the Goldstone modes) which possibly correspond to Fröhlich coherent modes[30].

Since the low energy theorem excludes observable effects coming from low momentum modes, the system is stable against external perturbations exciting soft polarization modes. Also, the coherent condensation of the dipole wave quanta, namely the water polarization, is induced through the boson shift transformation which is an invariant transformation and thus there is no energy expence in setting up the water electret state; dissipativity, which characterizes biological systems, appears thus to be a macroscopic manifestation of the microscopic invariance law. It should be noticed that the absence of dissipation in the solitary waves on the macromolecule chains implies that the external energy supply is completely released to the outside when the waves collapse at the end of the chains. Also, note that in the present approach the interaction between the living system and the environment is expressed through the choice of the vacuum (i.e. the physical Hilbert space). Thus, although the living system is an open system, one does not need to specify explicitely the coupling of the open system with the heat bath (environment).

When the finite size of the system is to be taken into account, the dipole wave quanta acquire a non-zero effective mass. By assuming they behaves as an ideal gas of non interacting particles, temperature can be introduced in a natural way. A relation is thus obtained[27], which connects the size of the system and its temperature. On the other hand, the insurgence of a non-zero effective mass introduces an energy threshold which enhances the stability of the system against external perturbations.

Since the water electret state has a finite life-time the above scheme goes through cycles each one characterized by creation of solitary waves on macromolecules and subsequent ordering of water in the electret state.

Let us include in the picture the electromagnetic (e.m.) interaction among dipoles[27]. This corresponds to consider an U(1) local phase transformation group for the dipole field. On the other hand, since the system is not a rigid one, spontaneous breakdown of the global dipole phase is to be expected. In such a case the Anderson-Higgs-Kibble mechanism takes place and the e.m. field acquires a mass. Then, the propagation of the e.m. field in the system can be shown[27] to be controlled by the same equation which gives rise to self-focusing propagation in non linear optics[32]. The e.m. field thus propagates as confined in filaments which go through the system: outside these filaments dipole wave coherence is preserved, inside of them the system is in the normal state. A network of filaments can thus appear whose shape and extension is dynamically sustained. On the interfaces of filaments with the coherent region a strong field gradient is available. Suppose that in the medium molecules able to oscillate on frequencies f_{0k} are present. One can show[33] that the field gradient force acting on the molecule located on the filament boundary is relevant when the field frequency f equals f_{0k}. Also, the force is attractive or repulsive depending on the sign of $(f_{0k} - f)$. A molecular coating of the filament is thus created with a specific molecule being attracted for a given frequency of the field. Electrically active sites of the biomolecules could be the originating sites of filaments and additional filaments would start from subsequent polymerization, finally producing a rich dynamical network.

In conclusion, the AHK mechanism produces a selective set of forces and the resulting coating of the filaments could provide a description of the observed cytoplasm structure. In dead cells cytoplasm appears as an homogeneous solution. However, in living cells a complex network of filaments is observed[34], which modifies its shape and size under external influences. Moreover, biochemical reactions among ions and molecules mainly occur in a time ordered pattern on these filaments. This activity seem to be well represented by the picture due to AHK mechanism described above, where the filament network is of dynamical origin. Formal details can be found in ref. 27. We finally observe that there is a growing experimental evidence which supports the theoretical relevance of long range correlation forces in living matter[35]. Among others, the observed rouleaux formation in erytrocyte suspensions[36] strongly supports long range correlation among metabolically active red blood cells[33].

In conclusion, it seems to us that the present status of QFT combined with group theoretical methods strongly points to the possibility that few fundamental mechanisms underly a very large number of phenomena. Here, our task was to stress that many different properties of many body systems and living matter appear as observable manifestations of an unifying mathematical scheme where group contraction has a preminent role: dynamical rearrangement of symmetry and boson condensation in spontaneously broken symmetry theories.

REFERENCES

1. L.van Hove, Physica 18, 145 (1952).
K.O.Friedrichs,"Mathematical Aspects of the Quantum Theory of Fields" , Interscience Publishers (1953).
2. H.Matsumoto, M.Tachiki and H.Umezawa, "Thermo-field Dynamics and

Condensed States," North-Holland, Amsterdam (1982).
3. N.N.Bogoliubov, A.A.Logunov, I.T.Todorov, "Introduction to Axiomatic Quantum Field Theory", Benjamin, Cumming (1975).
4. J.Goldstone, Nuovo Cimento 19, 154 (1961).
J.Goldstone, A.Salam and S.Weinberg, Phys. Rev. 127, 965 (1962).
5. C. De Concini and G.Vitiello, Nucl. Phys. B116, 141 (1976).
6. C. De Concini and G.Vitiello, Phys. Lett. 70B, 355 (1977).
7. E.Inönü and E.P.Wigner, Proc. Nat. Acad. Sci. US 39, 510 (1953).
I.E.Segal, Duke Math. J. 18, 221 (1953).
8. H.Matsumoto, N.J.Papastamatiou and H.Umezawa, Phys. Rev. D13, 1054 (1976).
M.Hongoh, H.Matsumoto and H.Umezawa, Prog. Theor. Phys. 65, 315 (1981).
9. H.Matsumoto, H.Umezawa, G.Vitiello and J.K.Wyly, Phys. Rev. D9, 2806 (1974).
10. M.N.Shah, H.Umezawa and G.Vitiello, Phys. Rev. B10, 4724 (1974).
11. E.Celeghini, P.Magnollay, M.Tarlini and G.Vitiello, Phys. Lett. 162B, 133 (1985).
12. E.Celeghini, M.Tarlini and G.Vitiello, Relation between Group Contraction and Nonlinear Realizations, in Proc. XXI Winter School in Theoretical Physics, Karpacz, Poland, 1985, eds.L.Michel and J.Mozrzymas, World Scient. Publisher, 1985.
13. S.Coleman, J.Wess and B.Zumino, Phys. Rev. 177, 2239 (1969).
C.G.Callan Jr., S.Coleman, J.Wess, and B.Zumino, Phys. Rev. 177, 2247 (1969).
14. S.L.Adler, Phys. Rev. 137B, 1022 (1965); 139B, 1638 (1965).
15. F.J.Dyson, Phys. Rev. 102, 1217 (1956).
16. Y.Fujimoto and N.J.Papastamatiou, Nuovo Cimento 40A, 468 (1977); 48A, 24 (1978).
17. P.W.Anderson, Phys. Rev. 110, 827 (1966).
P.W.Higgs, Phys. Rev. 145, 1156 (1966).
T.W.Kibble, Phys. Rev. 155, 1554 (1967).
18. H.Matsumoto, N.J.Papastamatiou, H.Umezawa and G.Vitiello, Nucl. Phys. B97, 61 (1975).
19. T.Kugo and I.Ojima, Prog. Theor. Phys. 61, 294 (1979).
20. G.Vitiello, Phys. Lett. 58A, 293 (1976).
21. J.Joos and E.Weimar, Nuovo Cimento 32A, 283 (1976).
22. E. Weimar, Acta Phys. Austriaca 48, 201 (1978).
23. E.Celeghini, M.Tarlini and G.Vitiello, Nuovo Cimento 84A, 19 (1984).
24. E.Del Giudice, S.Doglia, M.Milani and G.Vitiello, Phys. Lett. 95A, 508 (1983); Nucl. Phys. 251B [FS 13], 375 (1985).
25. E.Del Giudice, S.Doglia, M.Milani and G.Vitiello, in "Modern Bioelectrochemistry", eds. F.Guttmann and H.Keyzer, Plenum Press, N.Y. (1986); in "Nonlinear Electrodynamics in Biological Systems", eds.W.R.Adey and A.F.Lawrence, Plenum Press, N.Y. (1984).
26. J.A.Tuszyński, R.Paul, R.Chatterjee and S.R.Sreenivasan, Phys. Rev. 30A, 2666 (1984).
R.Paul, Phys Lett. 96A, 263 (1983).
T.M.Wu and S.Austin, Phys. Lett. 64A 15& (1977); 65A, 74 (1978); 73A, 266 (1979).
J.Chela-Flores, J.Theor.Biol. 117,107 (1985).
27. E.Del Giudice, S.Doglia, M.Milani and G.Vitiello, Electromagnetic Field and Spontaneous Symmetry Breaking in Biological Matter, preprint 1986.

28.A.S.Davydov,"Biology and Quantum Mechanics", Pergamon,Oxford (1982).
29.I.Prigogine and G.Nicolis,"Self-organization in non--equilibrium Systems; from dissipative structures to order through fluctuations", Wiley, N.Y. (1977).
30.H.Fröhlich, Rivista del Nuovo Cimento 7,399 (1977); Advances in Electron Physics, ed. L.Marton and C.Marton, vol.53,85 (1980).
31.S.Celaschi and S.Mascarenhas, Biophys. J. 20,273 (1977).
J.B.Hasted, H.M.Millany and D.Rosen, J.Chem. Soc. Faraday Trans. 77, 2289 (1981).
G.Albanese, A.Deriu and F.Ugozzoli, in Proc. Int. Conf. on the Application of Mössbauer Effect, Alma Ata, USSR, 1983.
32.Y.R.Shen,Prog. Quant. Electr. 4,1 (1975).
33.E.Del Giudice, S.Doglia and M.Milani, Phys. Lett. 90A, 104 (1982); in "Nonlinear Electrodynamics in Biological Systems", eds. W.R.Adey and A.F.Lawrence, Plenum Press, N.Y. (1984); in "Interactions between Electromagnetic Fields and Cells", eds. A.Chiabrera, C.Nicolini and H.Schwan, Plenum Press, N.Y. (1985).
34.I.I.Wolosewick and K.R.Porter, J.Cell Biol. 82, 114 (1979).
A.Hoglund, R.Karlsson, E.Arro, B.Fredrikssonn and U.Lindberg, J. Muscle Res. Cell Motility 1, 127 (1980).
J.S.Clegg, Colletive Phen. 3, 289 (1981).
R.L.Margolis and L.Wilson, Nature 293, 705 (1981).
35.H.Fröhlich and F.Kremer, eds.,"Coherent Excitations in Biological Systems", Springer, Berlin (1983).
36.S.Rowlands, L.S.Sewchand and E.G.Enns, Phys. Lett. 87A, 256 (1982); 93A, 363 (1983).

RELATIVISTIC SCATTERING OF COMPOSITE PARTICLES

Raj Wilson*

Department of Mathematics
University of Texas
San Antonio, Texas 78285

0. INTRODUCTION

In the New Testaments we read "Through faith we understand that the worlds were framed by the word of God, so that things which are seen were not made of things which do appear" - Hebrews 11:3. Perhaps this biblical verse may aptly describe the present day scientific endeavors in probing composite particles and in understanding the structure of matter. Whether this verse implies the belief on the so-called quark confinement, or on unknown symmetries, or on the notion of wholeness and the implicate order or on something else is left to our reader's discretion. However, the verse does seem to set up the pace of our continuing scientific discoveries as we observe that what we could not "see" in the past can be seen at the present and possibly what we cannot "see" at the present may be seen in the future and so on and on, till we understand the constituents of matter in terms of elementary particles which may be made up of things which may not appear.

From the onset of research on the structure of atom, scattering is considered to be an effective tool to probe a composite object although there may also exist the possibility to obtain certain information on constituents from spectroscopy. In any case, the need for a relativistic scattering theory is long wanted in order to enhance our understanding and to make in particular predictions. The scattering of the so-called elementary Dirac particles may be understood in terms of quantum electrodynamics within the framework of local quantum field theory. However, for the scattering of composite particles such as hadrons and atoms there exist several theoretical models with partial success of same or different-degree. More than two decades ago Barut[1] introduced an algebraic scattering matrix theory based on infinite dimensional irreducible representations of the dynamical group SO(4,2). One of the basic advantages of this theory over other theoretical models is that it sets up a unified description of hadrons and atoms. Furthermore, the Dirac theory for elementary particles may be consistently reformulated[2] using finite dimensional representations of the same dynamical group SO(4,2).

The theory based on the dynamical group SO(4,2) has been studied and

*In collaboration with A. O. Barut and A. Inomata.

applied extensively in particular by Barut[3], Fronsdal[4] and their callaborators and by Nambu[5]. Especially the theory has been successfully used to study the structure of proton, which is not a Dirac particle, in elastic, inelastic and deep-inelastic scattering regions (γp, ep, πp scatterings, etc.). Among several salient features it is worth pointing out at this point in time that this is the first theoretical model which has predicted the dipole behaviour of the magnetic form factor of proton and it is the only algebraic model which consistently explain the experimental data on the structure and the asymptotic properties of proton and pion in the inelastic as well as in the deep-inelastic scattering regions. Furthermore, the theory has also been very successfully used for the scattering of atoms such as γH, eH scatterings, etc.

In this paper[6] we very briefly describe the simple case of the scattering of pion with external scalar fields, interchanging infinite number of Reggeons. In Section 1 we briefly give the algebraic background while the Section 2 describes the computation of scattering amplitudes. In the last section we briefly summarize the results.

1: ALGEBRA AND REPRESENTATIONS OF SO(4,2)

The dynamical group SO(4,2) is generated by fifteen operators: $L_{\alpha\beta}$; α, β = 1, 2, 3, 4, 5 = 0, 6; $L_{\alpha\beta} = -L_{\beta\alpha}$ satisfying the Lie product:

$$\left[L_{\alpha\beta}, L_{\gamma\delta}\right] = i\left[g_{\alpha\delta}L_{\beta\gamma} + g_{\beta\gamma}L_{\alpha\delta} - g_{\alpha\gamma}L_{\beta\delta} - g_{\beta\delta}L_{\alpha\gamma}\right] ;$$

$$g_{\alpha\alpha} = (-1, -1, -1, -1, +1, +1): \ i = \sqrt{-1} \tag{1.1}$$

The Lorentz subgroup SO(3,1) is generated by $L_{\mu\nu}$; $\mu\nu$ = 1, 2, 3, 5 = 0. One of the groups SO[4,2] adjoint to SO(4,2) is generated by

$$L^{\pm} = \{F^{\pm}_{\sigma}\} \ ; \ F^{\pm}_{\sigma} = L_{5\sigma} \pm L_{3\sigma} \qquad \sigma = 1, 2, 4, 6$$

$$L^{o} = \{L_{\sigma\rho}\} \oplus Q; \ Q = - iL_{35} \tag{1.2}$$

such that the algebra L of SO[4,2]

$$L = L^{-} \oplus L^{o} \oplus L^{+}$$

form a Lie triple system satisfying,

$$[L^{o}, L^{o}] \subset L^{o} \ ; \qquad [L^{\pm}, L^{\pm}] = 0;$$

$$[L^{o}, L^{\pm}] \subset L^{\pm} \ ; \qquad [L^{+}, L^{-}] \subset L^{o}. \tag{1.3}$$

From (1.1) we find

$$[F^{\pm}_{\sigma}, F^{\pm}_{\rho}] = 0 \ ; \ [F^{\pm}_{\sigma}, L_{\rho\kappa}] = i[g_{\sigma\rho}F^{\pm}_{\kappa} - g_{\sigma\kappa}F^{\pm}_{\rho}],$$

$$[F^{\pm}_{\sigma}, L_{35}] = \pm i F^{\pm}_{\sigma}; [F^{+}_{\sigma}, F^{-}_{\rho}] = -2i[L_{\sigma\rho} + g_{\sigma\rho}L_{35}],$$

$$[L_{\sigma\rho}, L_{\kappa\lambda}] = i\,[g_{\sigma\lambda}L_{\rho\kappa} + g_{\rho\kappa}L_{\sigma\lambda} - g_{\sigma\kappa}L_{\rho\lambda} - g_{\rho\lambda}L_{\sigma\kappa}]. \tag{1.4}$$

The rank of SO(4,2) is three and its most general unitary representations are characterized in terms of nine real parameters - three from the Cartan subalgebra and six from the enveloping algebra. However, for the most degenerate unitary representations one may impose algebraically the Barut-Böhm-Umezawa condition:

$$\{L_{\alpha\beta}, L^{\alpha\gamma}\} = -2(1-\ell_0^2)g^\gamma_\beta \tag{1.5}$$

which correspondingly reduces the nine real parameters into four-three from the Cartan subalgebra and one from the enveloping albebra. We next consider four representation spaces spanned by the respective basis states - the so-called group states, according to the following four group reduction chains:

$$\text{Chain 1:} \Rightarrow \underset{\ell_o}{\overset{}{SO(4,2)}} \supset \underset{\ell_o}{SO(4,1)} \supset \underset{\ell_o, n}{SO(4)} \supset \underset{\ell}{SO(3)} \supset \underset{m}{SO(2)} \Rightarrow \phi^{\ell_o}_{n\ell m}$$

$$\ell_o = 0, \pm \tfrac{1}{2}, \pm 1, \ldots ; \; n = 1 + |\ell_o|, 2 + |\ell_o|, \ldots ;$$

$$\ell = |\ell_o| + 1 + |\ell_o|, \ldots, n-1; \; m = -\ell, \ldots, +\ell.$$

$$\text{Chain 2:} \Rightarrow \underset{\ell_o}{SO(4,2)} \supset \underset{\ell_o}{SO(3,2)} \supset \underset{\ell_o, \nu}{SO(3,1)} \supset \underset{\lambda}{SO(3)} \supset \underset{m}{SO(2)} \Rightarrow \phi^{\ell_o}_{\nu\lambda m}$$

$$\nu \in (-\infty, \infty); \; \lambda = |\ell_o|, 1 + |\ell_o|, \ldots ;$$

$$m = -\lambda, \ldots, +\lambda.$$

$$\text{Chain 3:} \Rightarrow \underset{\ell_o}{SO(4,2)} \supset \underset{\ell_o}{SO(3,2)} \supset \underset{\ell_o, \eta}{SO(3,1)} \supset \underset{j}{SO(3)} \supset \underset{m}{SO(2)} \Rightarrow \phi^{\ell_o}_{\eta j m}$$

$$\eta \in (-\infty, \infty), \; j = |\ell_o|, 1 + |\ell_o|, \ldots$$

$$m = -j, \ldots, +j.$$

$$\text{Chain 4:} \Rightarrow \underset{\ell_o}{SO[4,2]} \supset \underset{\ell_o}{E[3,1]} \supset \underset{\ell_o, \tau}{E[3]} \supset \underset{\lambda}{SO[3]} \supset \underset{m}{SO[2]} \Rightarrow \phi^{\ell}_{\tau\lambda m}$$

$$\tau \in (-\infty, \infty); \; \lambda = |\ell_o|, 1 + |\ell_o|, \ldots$$

$$m = -\lambda, \ldots, +\lambda \tag{1.6}$$

Here, the two SO(3,1) subgroups in Chains 3 and 4 are different and are related by a rotation in 3 - 4 space.

Furthermore, the computations of scattering amplitudes that will be described in the next section involve several representation functions especially of the subgroups SO(3),[7,8] SO(2,1),[7-10] E[2],[7] SO(4),[11] SO(3,1)[12] and E[3][7,13]. We define them as below:

1. Wigner function for SO(3):

$$[L_i, L_j] = i\,\varepsilon_{ijk} L_k ;$$

$$(\phi_{\ell m}, \exp(-i\theta L_2)\phi_{\ell n}) = d^{\ell}_{mn}(\theta)$$

$$= \alpha^{\ell}_{mn} \oint dx\, x^{m-\ell-1}\, [x\cos(\theta/2) + i\sin(\theta/2)]^{\ell-n} [\cos(\theta/2) + ix\sin(\theta/2)]^{\ell+n}$$

$$\alpha^{\ell}_{mn} = \frac{i^{m-n}}{2\pi i}\left[\frac{(\ell-m)!(\ell+m)!}{(\ell-n!(\ell+n)!}\right]^{\frac{1}{2}}, \qquad (1.7a)$$

2. Bargmann function for SO(2,1) - discrete basis:

$$[L_1, L_2] = -iL_3, \; [L_1, L_3] = -iL_2, \; [L_2, L_3] = iL_1;$$

$$(\phi_{km}, \exp(-i\theta L_2)\phi_{kn}) = V^{k}_{mn}(\theta) \qquad (1.7b)$$

$$= \alpha^{k}_{mn} \oint dx\, x^{m-k-1} [x\cosh(\theta/2) - \sinh(\theta/2)]^{k-n} [\cosh(\theta/2) - x\sinh(\theta/2)]^{k+n},$$

3. Barut-Phillips function for SO(2,1) - continuous basis:

$$(\phi_{k\mu}, \exp(-i\theta L_2)\phi_{k\nu}) = V^{k}_{\mu\nu}(\theta)$$

= linear combination of (1.7b) and its complex conjugate with $m \to i\mu$, $n \to i\nu$. (1.7c)

4. Inönü - Wigner function for E[2]:

$$[L_1, L_2] = 0, \; [L_1, L_3] = -iL_2, \; [L_2, L_3] = iL_1;$$

$$(\phi_{\tau m}, \exp(-izL_2)\phi_{\tau n}) = J_{mn}(\tau z)$$

$$= \frac{(-1)^{m-n}}{2\pi i} \oint dx\, x^{-m+n-1} \exp[\tfrac{1}{2}\tau z(x - \tfrac{1}{x})] \qquad (1.7d)$$

5. Biedenharn function for SO(4):

$$[L_i, L_j] = i\,\varepsilon_{ijk} L_k, \; [A_i, A_j] = i\,\varepsilon_{ijk} A_k, \; [L_i\, A_j] = i\,\varepsilon_{ijk} A_k,$$

$$\left(\phi^{\hat{\ell}\ell_o}_{\ell m}, \exp(-i\theta A_3)\phi^{\hat{\ell}\ell_o}_{\lambda m}\right) = D^{[\hat{\ell},\ell_o]}_{\ell m\lambda}(\theta); \; \hat{\ell} = n-1.$$

$$= \alpha^{n}_{\ell\lambda} \int_{-1}^{+1} dx(\cos\theta - ix\sin\theta)^{n-1}\, d^{\ell^*}_{\ell_o m}(x)\, d^{\lambda}_{\ell_o m}[x(\theta)]$$

$$x(\theta) = \frac{x\cos\theta - i\sin\theta}{\cos\theta - ix\sin\theta}$$

$$\alpha^{n}_{\ell\lambda} = \tfrac{1}{2}\left[(2\ell+1)(2\lambda+1)\,\frac{(n+\ell)!(n-\ell-1)!}{(n+\lambda)!(n-\lambda-1)!}\right]^{\frac{1}{2}}, \qquad (1.7e)$$

6. Dolginov - Toptygin function for SO(3,1):

$$[L_i, L_j] = i\varepsilon_{ijk}L_k,\ [A_i, A_j] = -i\varepsilon_{ijk}A_k,\ [L_i, A_j] = i\varepsilon_{ijk}A_k;$$

$$\left(\phi^{\hat{\ell}\,\ell_o}_{\ell m}, \exp(-i\theta A_3)\,\phi^{\hat{\ell}\,\ell_o}_{\lambda m}\right) = D^{[\hat{\ell},\ell_o]}_{\ell m\lambda}(\theta);\ \hat{\ell} = i\nu - 1.$$

$$= \alpha^{i\nu}_{\ell\lambda}\int_{-1}^{+1} dx(\cosh\theta - x\sinh\theta)^{i\nu-1}\, d^{\ell^*}_{\ell_o m}(x)\ d^{\lambda}_{\ell_o m}[x(-i\theta)] \tag{1.7f}$$

7. Vilenkin-Akim-Levin function for E[3]:

$$[L_i, L_j] = i\,\varepsilon_{ijk}L_k,\ [A_i, A_j] = 0,\ [L_i, A_j] = i\,\varepsilon_{ijk}A_k;$$

$$\left(\phi^{\ell_o}_{\ell m}(\tau), \exp(-i\,z\,A_3)\,\phi^{\ell_o}_{\lambda m}(\tau)\right) = J^{\ell_o}_{\ell m\lambda}(\tau z)$$

$$= \tfrac{1}{2}[(2\ell+1)(2\lambda+1)]^{\frac{1}{2}}\int_{-1}^{+1} dx\,\exp(-i\tau z x)\ d^{\ell^*}_{\ell_o m}(x)\ d^{\lambda}_{\ell_o m}(x) \tag{1.7g}$$

Finally, there exist several sum rules connecting these representation functions which are needed for exact computation of the scattering amplitudes. We give here only one, for example, which is obtained by taking the matrix elements of Lorentz boost between two arbitrary SO(4,2) canonical physical states. If $G = \exp(-i\theta_f L_{45})\exp(-i\xi L_{35})\exp(-i\theta_i L_{45})$, then

$$\left(\phi^{\ell_o}_{n_f\ell_f m},\ G\ \phi^{\ell_o}_{n_i\ell_i m}\right) = \int_{-\infty}^{+\infty} d\mu\ V^{\ell_f+1}_{n_f\mu}(\alpha_1)\ D^{[i\mu-1,\ \ell_o]}_{\ell_f m\ell_i}(\beta_1)\ V^{\ell_i+1}_{\mu n_i}(\gamma_1)$$

$$= \sum_L D^{[n_f-1,\ell_o]}_{\ell_f mL}(\alpha_2)\ V^{L+1}_{\eta_f n_i}(\beta_2)\ D^{[n_i-1,\ell_o]}_{Lm\ell_i}(\gamma_2) \tag{1.8}$$

where $L = |\ell_o|,\ 1 + |\ell_o|,\ \ldots\ .\ \mathrm{Min}\{n_f-1,\ n_i-1\}$, $\alpha_1 = \theta_f - i\frac{\pi}{2}$, $\beta_1 = \xi$, $\gamma_1 = \theta_i + i\frac{\pi}{2}$ and the relations among $\{\alpha_1, \beta_1, \gamma_1\}$ and $\{\alpha_2, \beta_2, \gamma_2\}$ can be obtained from Euler rotations.

2. SCATTERING AMPLITUDES

The scattering amplitude for the scattering of pion with external scalar fields is given by

$$A(s,t) = \oint \langle \Phi_f, p_f, k_f | \Phi, p\rangle\, G(p)\, \langle \Phi, p | \Phi_i, p_i, k_i\rangle \tag{2.1}$$

where the Mandelstam variables: $s = p_\mu p^\mu = (p_i - k_i)^2 = (p_f - k_f)^2$, $t = (p_i - p_f)^2 = (k_i - k_f)^2$ and $u = (p_i - k_f)^2 = (p_f - k_i)^2$; $|\Phi_i\rangle$ and $|\Phi_f\rangle$

are infinite multiplet one-particle physical states of the external particles and $|\Phi\rangle$ is that of the exchange particle and the Green's function

$$[G(p)]^{-1} = \langle \Phi, p | \Lambda(p) | \Phi, p \rangle \tag{2.2}$$

where the wave operator $\Lambda(p)$ is given by

$$\Lambda(p) = \Gamma_\mu p^\mu + \frac{1}{2m_2} [p_\mu p^\mu - m_- m_+] L_{46} - \frac{g}{2m_2} [p_\mu p^\mu - m_-^2] ;$$

$$m_\pm = m_1 \pm m_2 . \tag{2.3}$$

In (2.3), $\Gamma_\mu = \{L_{16}, L_{26}, L_{36}, L_{56}\}$ and L_{46} are the operators of SO(4,2), m_1 and m_2 are the masses of constituents of pion, and g is a constant parameter. The wave operator $\Lambda(p)$ is diagonalized by defining physical states for different momentum orbits and these physical states $|\phi\rangle$ are expressed in terms of group states defined under the four reduction chains given in (1.6). Thus we obtain,

time-like: discrete $\quad p_\mu p^\mu > 0 \Rightarrow$ Chain-1; $\quad |\Phi\rangle = \frac{1}{N_n} \exp(-i\theta_n L_{45}) |\phi^{\ell_o}_{n\ell m}\rangle ,$

time-like: continuous $\quad p_\mu p^\mu > 0 \Rightarrow$ Chain-2; $\quad |\Phi\rangle = \frac{1}{N_\nu} \exp(-i\theta_\nu L_{45}) |\phi^{\ell_o}_{\nu\lambda m}\rangle ,$

space-like: continuous $\quad p_\mu p^\mu < 0 \Rightarrow$ Chain-3; $\quad |\Phi\rangle = \frac{1}{N_\eta} \exp(-i\theta_\eta L_{34}) |\phi^{\ell_o}_{\eta jm}\rangle ,$

light-like: continuous $\quad p_\mu p^\mu = 0 \Rightarrow$ Chain-4; $\quad |\Phi\rangle = \frac{1}{N_\tau} \exp(-i\theta_\tau F_4^\pm) |\phi^{\ell_o}_{\tau\lambda m}\rangle , \quad (2.4)$

where N's are normalization constants, $n = Sp(L_{56})$ $\nu = Sp(L_{46})$, $\eta = Sp(L_{36})$ and $\tau = Sp(F_6^\pm = L_{56} \pm L_{36})$. The group parameter θ gives a measure for the composite nature of the particles and its energy dependency is given by

$$\exp(2\theta) = \frac{2m_2\sqrt{p_\mu p^\mu} + p_\mu p_\mu - m_- m_+}{2m_2\sqrt{p_\mu p^\mu} - p_\mu p^\mu + m_- m_+} . \tag{2.5}$$

For elementary particles $\theta \to 0$ and hence $p_\mu p^\mu = m_1^2 - m_2^2$ indicating that one of the internal masses vanishes or both masses are equal depending on the momentum orbit.

The diagonalization of the wave operator (2.3) gives the mass spectrum:

$$p_\mu p^\mu = m_1^2 + m_2^2 + 2m_1 m_2 [1 - 2g^2/[g^2 + (Sp\Gamma)^2] \tag{2.6}$$

where $\Gamma = \{L_{56}, L_{46}, L_{36}\}$. We now define the general Lorentz boost transformation $U(L_p)$ such that

$$|\Phi, p\rangle = U(L_p) |\Phi\rangle ;$$

$$U(L_p) = \exp(-\hat{\phi} L_{12}) \exp(-i\hat{\theta} L_{31}) \exp(i\hat{\phi} L_{12}) \exp(-\xi L_{35}) ;$$

$$p_\mu = \{p \cos\hat{\phi} \sin\hat{\theta}, p \sin\hat{\phi} \sin\hat{\theta}, p \cos\hat{\theta}, p_o\}; \; p_\mu p^\mu = s,$$

$$p = \theta(s)\sqrt{s} \quad \sinh\xi + \theta(s)\sqrt{s} \;\; \cosh\xi \; + \delta_{s,o}\kappa \exp(\xi),$$

$$p_o = \theta(s)\sqrt{s}\ \cosh\xi + \theta(-s)\sqrt{s}\ \sinh\xi + \delta_{s,o}\kappa\exp(\xi),$$
$$\mathrm{Re}\,\hat{\phi} \in [-\pi, \pi],\ \mathrm{Im}\,\hat{\phi} \in (-\infty, \infty),\ \kappa \text{ is a constant.}$$
$$\mathrm{Re}\,\hat{\theta} \in [\theta, \pi],\ \mathrm{Im}\,\hat{\theta} \in (-\infty, \infty),$$
$$\mathrm{Re}\,\xi \in [0, \infty),\ \mathrm{Im}\,\xi \in [-\pi, \pi], \tag{2.7}$$

We now substitute (2.2), (2.4) and (2.7) in (2.1) for pion with $n_i = n_f = 1$, $\ell_i = \ell_f = 0$, $m_i = m_f = 0$ and carry out the computation by using the properties and sum rules of the representation functions (1.7). We then express the final results as partial wave expansions. Thus we obtain,

$$A(s>0,\ t<0,\ \theta_n) = \sum_{n=0}^{\infty} \gamma_n(s)\ C_n^1(\cos\omega),$$
$$= \sum_{\ell=0}^{\infty} (2\ell+1)\ \beta_\ell(s)\ P_\ell(\cos\omega)$$
$$= \sum_{\ell=0}^{\infty} (2\ell+1)\ \alpha_\ell(s)\ P_\ell(\cos\theta) \tag{2.8}$$

$$A(s>0,\ t<0,\ \theta_\nu) = \frac{i}{2}\int_{\hat{\Gamma}} \frac{dn}{\sin(\pi n)}\gamma(n,s)\ C_n^1(-\cos\omega),$$
$$= \frac{i}{2}\int_{\Gamma} \frac{d\ell}{\sin(\pi\ell)}\ (2\ell+1)\beta(\ell,s)\ P_\ell(-\cos\omega),$$
$$= \frac{i}{2}\int_{\Gamma} \frac{d\ell}{\sin\pi\ell}\ (2\ell+1)\ \alpha(\ell,\ s)\ P_\ell(-\cos\theta)$$

$$\hat{\Gamma} = \{n:\ n = -1 + \varepsilon + iy;\quad \varepsilon \in (0, 1),\ y \in (-\infty, \infty)\}$$
$$\Gamma = \{\ell:\ \ell = -\tfrac{1}{2} + \varepsilon + iy;\quad \varepsilon \in (0, \tfrac{1}{2}),\ y \in (-\infty, \infty)\}\ . \tag{2.9}$$

$$A(s<0,\ t>0,\ \theta_\eta) = \frac{i}{2}\int_{\hat{\Gamma}} \frac{dn}{\sin(\pi n)}\ \gamma(n, s<0)C_n^1(-\cos\omega),$$
$$= \frac{i}{2}\int_{\Gamma} \frac{d\ell}{\sin(\pi\ell)}\ (2\ell+1)\beta(\ell,\ s<0)\ P_\ell(-\cos\omega),$$
$$= \frac{i}{2}\int_{\Gamma} \frac{d\ell}{\sin(\pi\ell)}\ (2\ell+1)\gamma(\ell,\ s<0)\ P\ (-\cos\theta)$$

$$A(s=0,\ t,\ \theta_\tau) = \int_{-\infty}^{\infty} d\tau\ f(\tau,\ \theta_\tau)\ I_{\frac{1}{2}}(rR)$$

In (2.8 - 2.10) γ's, β's and α's are functions of s and they are explicitly known; $\cos\omega$ and $\cos\theta$ are functions of s and t; C_n^λ $(\cos\theta)$ and $P_\ell(\cos\theta)$ are Gegenbauer and Legendre functions respectively. In (2.11), $f(\tau, \theta_\tau)$ R and r are functions of τ and t, and $I_\nu(z)$ is Bessel function of imaginary argument. From (2.8) - (2.10) one can establish analytic continuations of the three contributions as given by the following commutative diagram:

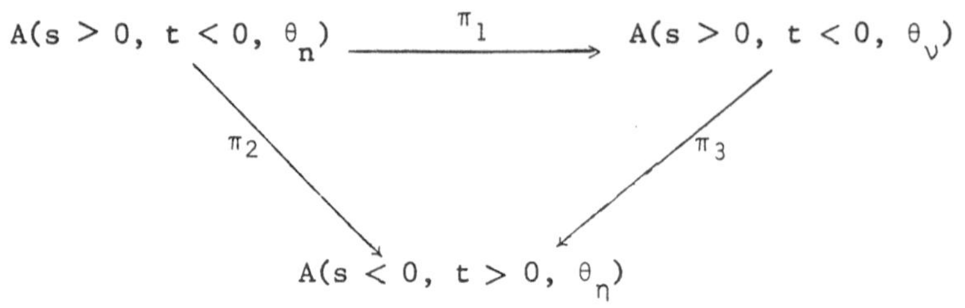

$$\pi_1: \quad n \to i\nu, \ \theta_n \to \theta_2 + i\frac{\pi}{2}$$

$$\pi_2: \quad n \to i\eta, \ \theta_n \to i\theta_\eta ,$$

$$\pi_3: \quad \nu \to \eta \ , \ \theta_\nu \to i(\theta_\eta - \frac{\pi}{2}) \qquad (2.12)$$

Furthermore, the light-like contribution may be obtained as the limiting case of time-like and space-like continuous contributions. This limit corresponds to contraction in representation space similar to Celeghini-Tarlini contraction.[14] That is,

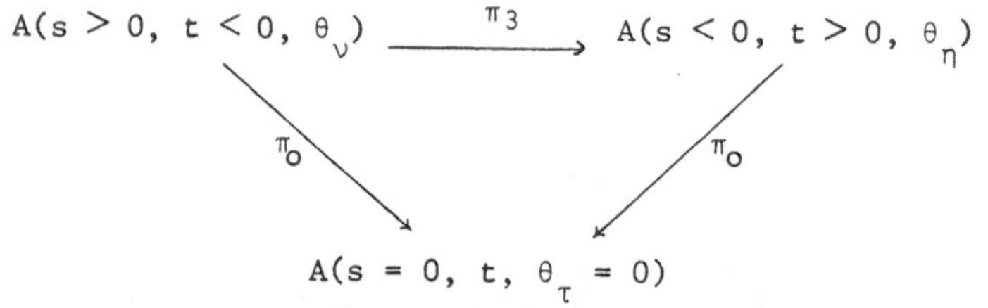

(2.13)

$$\pi_o: \quad s \to 0, \ (\nu,\eta) \to \infty, \ m_- m_+ \to 0, \ (\nu,\eta)(m_- m_+) \to i\tau, \theta_\nu \to 0, \ \theta_\eta \to \frac{\pi}{2} .$$

The three contributions to the scattering amplitude may be interpreted as follows: The time-like discrete contribution $A(s>0,\ t<0,\ \theta_n)$ for $s \in (m_-^2,\ m_+^2)$ is due to the exchange of infinite number of time-like Reggeons (poles) while the time-like continuum contribution $A(s > 0,\ t < 0,\ \theta_\nu)$ for $s \in [m_+^2, \infty)$ is due to the exchange of time-like two-particle states of mass m_1 and m_2 in the same direction. This latter contribution represents the right-hand cut. The space-like contribution $A(s < 0,\ t > 0,\ \theta_\eta)$ for $s \in (-\infty,\ m_-^2)$ is due to the exchange of space-like two-particle states of mass m_1 and m_2 in the opposite directions, and this contribution represents the left-hand cut in the s-plane. In the next section we summarize the important results derived from these contributions.

3. SUMMARY OF RESULTS

1. The scattering amplitudes as given in (2.8 - 2.10) possess SO(4) symmetry (or SO(3,1) as the case may be) and are similar to the generalized partial-wave expansions studied by Toller. In fact, any scattering amplitude which incorporates SO(4) symmetry is expressible in terms of Biedenharn function. Furthermore, the partial wave amplitude for t = 0 obtained from the absorptive part of the scattering amplitude satisfying the Bethe-Salpeter integral equation is in agreement with our partial wave amplitude

$\alpha_\ell(s)$ given by (2.8)

2. The three contributions to the total scattering amplitude give the usual Regge representation. We observe that the time-like discrete amplitude provides the Regge pole contributions while the time-like and space-like continuum contributions provide the expected background integral. Such a representation is usually obtained by performing the Watson-Sommerfeld transform on the complex ℓ-plane on the partial wave amplitude (2.8). In our case we obtain the Regge representation without the Watson-Sommerfeld transform. Furthermore, the existence of Lorentz-Toller poles, daughter trajectories can be transparently seen in our amplitudes.

3. The scattering amplitudes given by (2.8 - 2.10) show correct analyticity properties as given by Mandelstam double spectral representation. We summarize these properties below:

s-plane and θ(tilt-parameter)-plane:

$$\begin{aligned}
&\text{poles:} && s \in (m_-^2, m_+^2) && \Rightarrow \theta \in (-\infty, \infty). \\
&\text{normal branch points:} && s = m_\pm^2 && \Rightarrow \theta = \pm\infty \\
&\text{right-cut:} && s \in [m_+^2, \infty) && \Rightarrow \theta \in [\infty \to \pm i\tfrac{\pi}{2}) \\
&\text{left-cut:} && s \in (-\infty, m_-^2] && \Rightarrow \theta \in (\pm i\tfrac{\pi}{2} \leftarrow -\infty] \\
&\text{anomalous singular points:} && s = s_\pm && \Rightarrow \theta = \text{arc tanh}\, a
\end{aligned} \tag{3.1}$$

t-plane:

$$\begin{aligned}
&\text{normal singular points:} && t = 0 \Rightarrow \cos\omega = 1 \\
& && t = (m_+ \pm m_-)^2, \text{ absent} \\
&\text{singular point:} && t = -t_0 \Rightarrow \cos\omega = -1 \\
&\text{cut:} && t \in (-\infty, -t_0] \\
&\text{anomalous singular point:} && t = t_1
\end{aligned} \tag{3.2}$$

In (3.1) and (3.2), $s_\pm$, a, t_0, t_1 are known fixed points.

4. Strictly speaking, Mandelstam representation requires that the total amplitude to be analytic in the $\cos\omega = z$ - plane with a right-cut beginning at some point $z = z_0 > 1$ instead of at $z = 1$ and a left-cut beginning at $z = -\hat{z} < -1$ instead of at $z = -1$. In our case such analyticity behaviour is easily achieved by generalizing the method introduced by Khuri in obtaining the modified Regge representation.

5. The Regge pole term of our amplitude is expressed as a power series expansion in the variable t and then by using Rainville's lemma for series manipulation we can explain the contribution due to modified Regge-poles and satellite poles.

6. Our amplitude shows correct-Regge asymptotic behaviour: $A(s,t) \to g(s)t^{\alpha_1(s)}$ as $t \to \infty$ with $\alpha_1(s)$ giving the right-most pole. Furthermore, the total amplitude satisfies the positivity condition by Gribov-Pomeranchuk inequalities, namely

$$\frac{\partial^n}{\partial t^n} \operatorname{Im} A(s,t)\Big|_{t=0} \geqslant 0 \qquad n = 1, 2, 3, \ldots .$$

7. Finally, we observe that the space-like part of the amplitude is due to the tachyon[15] contribution which is essential for the full analyticity of the amplitude in the s-plane. Thus the space-like contribution turns out to be a virtue rather than a disease. In a more realistic scattering process which incorporates crossing-symmetry, the acausal effects due to the space-like contributions may most probably play a crucial role. Algebraically, crossing-symmetry requires the representations of the complexification $SO(4,2)_{\mathbb{C}}$ rather than that of its real form. In the words of Jung[16] "We shall naturally look around in vain in the macrophysical world for acausal events, for the simple reason that we can not imagine events that are connected non-causally and are capable of a non-causal explanation. But that does not mean that such events do not exist. Their existence - or at least their possibility - follows logically from the premise of statistical truth." The calculations given in this paper seem to indicate that perhaps we should believe in the acausal space-like contributions due to synchronicity in time as much as we believe in the causal time-like contributions due to synchronicity in space in order to formulate a complete relativistic scattering theory for composite particles.

ACKNOWLEDGEMENT

The author is very thankful to Proseddor Bruno Gruber for his kind invitation and for his encouragements.

REFERENCES

1. A. O. Barut, Symmetry Principles at High Energy, ed. B. Kursunoglu (W. H. Freeman, San Francisco, 1964); High-Energy Physics and Elementary Particles (International Atomic Energy Agency, Vienna, 1965); Non-Compact Groups in Particle Physics, ed. Y. Chow (W. A. Benjamin, New York, 1966); Scattering Theory, ed. A. O. Barut (Gordon and Breach, New York, 1969); Group Systems and Many body Physics, ed. P. Kramer (Vieweg, Braunschweig, 1980).

2. A. O. Barut, Phys. Rev. B135, 839 (1964); Phys. Rev. Letts. 20,893 (1968); R. Wilson, Nucl. Phys. B68, 157 (1974).

3. A. O. Barut, Dynamical Groups and Generalized Symmetries in Quantum Theory (University of Canterbury Press, Christ Church, 1972) and the references therein.
A. O. Barut and R. Raczka, Theory of Group Representations and Applications (Polish Scientific, Warszawa, 1980) and the references therein.
A. Inomata and D. Peak, Prog. Theor. Phys. 42, 134 (1969).
R. F. Peierls and R. Wilson, Ann. Phys. 81, 15 (1973).
A. O. Barut and R. Wilson, Phys. Rev. A13, 918 (1976).

4. C. Fronsdal, Phys. Rev. 156, 1653, 1665 (1967); 168, 1845 (1968); 171, 1811 (1968); 179, 1513 (1969); 182, 1564 (1969); 185, 1768 (1969); C. Fronsdal and L. E. Lundberg, Phys. Rev. D1 3247 (1970); C. Fronsdal and R. W. Huff, Phys. Rev. D3,933 (1971); C. Fronsdal, Phys. Rev. D4, 1689 (1971).

5. Y. Nambu, Prog. Theor. Phys. Suppl. 37-38, 368 (1966); Phys. Rev. 160, 1171 (1967).

6. A. O. Barut, A. Inomata and R. Wilson, Relativistic Quantum Theory of Two Interacting Composite Particles - I: Compton Scattering, to be published. This paper contains all the details and extensive references.

7. N. Ja. Wilenkin, Special Functions and the theory of Group Representations, (American Math. Soc., Providence, 1968).
W. Miller, Jr., Lie Theory and Special Functions (Academic, New York, 1968).

8. L. C. Biedenharn and J. D. Louck, Angular Momentum in Quantum Physics, (Addison-Wesley, Mass. 1981).

9. V. Bargmann, Ann. Maths. 48, 568(1947).

10. A. O. Barut and E. C. Phillips, Comm. Math. Phys. 8, 52 (1968).

11. L. C. Biedenharn, J. Math. Phys. 2 , 433 (1961).

12. A. Z. Dolginov and I. N. Toptygin, Sov. Phys. JETP 8, 550 (1959).
A. Sciarrino and M. Toller, J. Math. Phys. 8, 1252 (1967).

13. N. Ja. Vilenkin, E. L. Akim and A. A. Levin, Dokl. Akad. Nauk SSSR 112, 987 (1957), (in Russian),
W. Miller, Jr., Comm. Pure App. Math. 17 527, (1964),
S. Ström, Ark. Fysik 30, 267 (1965).

14. E. Celeghini and M. Tarlini, Nuovo Cimento B61, 265 (1982); B65, 172 (1982); B68, 133 (1982).

15. Tachyons, Monopoles and Related Topics, ed. E. Recami (North-Holland, Amsterdam, 1978).

16. C. G. Jung, Synchronicity (Princeton University, Princeton, 1960).

EMPIRICAL EVIDENCES FOR FERMION DYNAMICAL SYMMETRIES IN NUCLEI

Cheng-Li Wu

Department of Physics and Atmospheric Science
Drexel University,
Philadelphia, Pennsylvania,19104

INTRODUCTION

In the previous talk (Feng, in these proceedings)[1] a Fermion Dynamical Symmetry (FDS) model [2] has been presented. In this fermion model, all the well known collective modes (i. e. rotation, vabration and γ-soft), which have been found from the boson dynamical symmetries in the IBM (Arima and Iachello)[3], are rediscovered from the underlying fermion level.

There are basically two types of fermion dynamical symmetries in the FDS model: Sp(6) x *SU* (2) and SO(8) $\times$*SU* (2). The symmetry *SU* (2) is associated with the abnormal parity level in each major shell where only S pairs are introduced in the model while Sp(6) and SO(8) symmetries are associated with the normal parity levels in each major shell. For the low spin states, only the seniority zero states of *SU*(2) are considered (no broken pairs). Thus, the low lying structure of nuclei are mainly determined by Sp(6) or SO(8) symmetry. As has been shown in the previous talk, the SO(8) symmetry has three limiting symmetries: SO(5)$\times$SU(2), SO(6) and SO(7) while Sp(6) has SU(3) and SU(2)$\times$SO(3) dynamical symmetry chains (see eqs.(3)-(4) of Feng's talk)[1]. Of these 5 limits, SU(3) and SO(6) correspond exactly to the IBM SU(3) and SO(6) limits including both the energy formulas and γ - trasition rates, except in the FDS model, only the number of pairs in the normal parity levels (denoted as N_1) should be counted for the SU(3) and SO(6) instead of the total number of pairs (N) in the IBM case. The nucleon pairs in the abnormal level (denoted as N_0) are responsible for the *SU* (2) symmetry. Furthermore, the SU(3) irreducible representation with $\lambda+\mu > 2\Omega_1/3$ are forbidden due to the Pauli principle. The other three limits, although there are no exact counterparts in the IBM, do have the same energy formulas as the IBM U(5) limit.

Because of these similarities, the empirical manifestation of boson symmetries in nuclei could, in principle, be taken as indirect evidence for the corresponding fermion symmetries. Nevertheless, it would be more convincing if one could identify a fermion symmetry exhibiting properties different from any of the boson symmetries, and establish its empirical existence.

In this talk I am going to show you what are the properties, which exhibit the fermion dynamical symmetries predicted by the FDS model, and what are the experimental evidences to support these predictions.

Table-1 The Correspondence between the Fermion Dynamical Symmetries, Boson Dynamical Symmetries and the Nuclear Collective Modes

FDS		Collective mode		IBM
$Sp(6) \supset SU(3)$	$\longleftrightarrow$	Rotation	$\longleftrightarrow$	$U(6) \supset SU(3)$
$SO(8) \supset SO(6)$	$\longleftrightarrow$	γ- Soft	$\longleftrightarrow$	$U(6) \supset SO(6)$
$Sp(6) \supset SU(2) \times SO(3)$ $SO(8) \supset SO(5) \times SU(2)$ $SO(8) \supset SO(7)$	$\longleftrightarrow$	Vibration	$\longleftrightarrow$	$U(6) \supset U(5)$

THE EVIDENCE FOR THE SO(8) AND Sp(6) SYMMETRIES

In the last 30 years, prior to the IBM, whenever one deals with the so-called collective motion in nuclei, one always has to introduce the concept of deformation. In this language, a nucleus is treated as a deformed core plus a few valence nucleons. The connection between the collective modes of nuclear motion and the deformation was established: If the deformation is small, then the collective motion is vibrational-like; if it is large, then the nucleus is rotational-like. The IBM describes the nuclear collective motion in a completely different way. It is the first model to establish the connection between these various collective modes of nuclear motion and the multipole-chain dynamical symmetries of nuclei.

Although this connection is only achieved phenomenologically under the boson approximation, it nevertheless creates an entirely new direction to study nuclear collective motion. The succeses of the IBM seem to suggest that the nuclear collective motion can be understood from the symmetry point of view and can be described by the dynamical symmetries of the interacting many-body system. Thus the IBM goes one step further towards the microscopic description in the sense that all the well known collective modes: rotation, vibration, γ -soft ,etc come very naturally from the dynamical symmetries of the many- body interactions without having to introduce the classcal concept of deformation. Of course, the next step is to ask, what is the related dynamical symmetry at the underlying fermion level? The FDS model gives the answer as shown in table-1.

The FDS model not only reveal the underlying fermion symmetry but also establishes the connection between the symmetries and the shell structure. As one can see from the shell table in the previous talk given by Feng (table-2), for a given nucleus, whether it is k-active ($k=1$) or i-active ($i=3/2$) is entirely governed by which shell the valence nucleons occupy. If the open shell has $k=1$, it possesses the Sp(6) symmetry; if $i=3/2$, then it is the SO(8) symmetry. Thus it will lead to a conclusion that the nuclei in actinide region must possess Sp(6) symmetry since the valence neutrons and protons occupy shell 7 and 8 where both shells are k-active ($k=1$); nuclei with both neutrons and protons occupying shell 6 ($50 \leq (N,Z) \leq 82$) must possess SO(8) symmetry. For the rare-earth nuclei, the protons and neutrons possess different symmetries: SO(8) for protons and Sp(6) for neutrons.

For a given Sp(6) (SO(8)) symmetry, there are of course several dynamical symmetry chains: whether it is SU(3) (SO(6)) or SU(2) × SO(3) (SO(5)xSU(2)or SO(7)) limit will depends on the detail of the interactions and the number of valence nucleons. Within the same major shell, nuclei may vary from one symmetry limit to another as the number of nucleons changes. For a specific nucleus, it may not be at any one of these ideal limits. However, there is very inportant point here which is that **the highest symmetry (i.e. Sp(6) or SO(8)) does not depend on such details.** Once an open shell is given, the highest symmetry is fixed. Note that only Sp(6) symmetry has an SU(3) limit which has the characteristics of a good rotor. The SO(8) symmetry can never appear as a good rotor since it contains no SU(3) limit. Therefore the occurence of rotational nuclei can serve as a signal to distinguish whether a shell possesses Sp(6) or SO(8) symmetry. This distinguishing feature can be empirically checked.

Experimentally, we have known for a long time that the best rotational nuclei regions are the actinides and the rare-earths and no typical rotational nuclei have been found in the mass region with $50 \leq (N, Z) \leq 82$ and we do not know the precise reason for this. This mystery seems to be revealed now since the actinide and rare-earth are the only two regions which have the valence nucleons in shells with Sp(6) symmetry, while in the region of $50 \leq (N, Z) \leq 82$ the valence nucleons possess SO(8) symmetry. The latter implies that nuclei in this region can vary among the three dynamical symmetries SO(5)×SU(2) , SO(7) and SO(6) limits but never go to SU(3).

The situation in rare-earth region is more complicated then the actinide region, because the valence protons and neutrons occupy different physical shells with different symmetries. Thus in addition to the SU(3) limit there may be other possible limits which could occur and the neutron-proton interaction may mix them thereby making the rare-earth nuclei have more rich variety. However, since here we are only concerned with the possibility of the occurence of rotational nuclei, such varieties will not affect the conclusion of the existence of the rotational-like nuclei in this region.

Accoding to the FDS model, there are another two mass regions which could have Sp(6) symmetry: one is the s-d shell and the other is the f-p shell (shell 5, see table-2 in Feng's talk [1]) Empirically, these two regions are just the regions where some light nuclei show collective behaviors. Some rotational nuclei have also been found in this region (e. g. ^{24}Mg). However, the

E4/E2 ratio in these two mass regions is usually around 2.5--3 and never goes up to 3.3. It is not surprising that these rotational nuclei are not as typical as those in the heavier mass regions since these two shells have k=1 and i=3/2, and both k and i could be active. Besides the Sp(6), there are two other possible dynamical symmetries, (SO(8) and SU(3)xSO(6)). This obviously renders the situation more complicated. Nevertheless, the global picture of the occurence of rotational nuclei in various mass regions is in good argreement with the predictions of the FDS model. Thus, this can serve as one of the evidences of the Sp(6) and SO(8) fermion dynamical symmetries.

Besides the qualitative evidences we have presented in the previous paragraph, the Sp(6) and SO(8) symmetries can be studied more quantitatively. To this end, we have checked the B(E2) ratio

$$R_{22} = B(\ E2,\ 2^+_\gamma \rightarrow 0^+_g\ ) \,/\, B(\ E2, 2^+_\gamma \rightarrow 2^+_g\ ) \qquad (1)$$

systematically. In eq.(1), the subscripts γ and g stand for the quasi γ-band and the ground band respectively. For the rotational limit, quasi γ-band is just the γ-band. In other cases, 2^+_γ refers to the second 2^+ state while 2_g is the first 2^+ state. In the SU(3) limit, the transition from γ band to ground band is forbidden and the ratio can be estimated by the so-called "Alaga rule " which is 0.7 for R_{22}. In the SU(2) limit, 2^+_γ is the so-called 2-phonon state (seniority ν=4) and 2^+_g is the 1-phonon state (ν =2). Because of the seniority selection rule, the γ transition $2^+_\gamma \rightarrow 2^+_g$ ($\Delta\nu$ = 4) is forbidden and hence R_{22}=0. Since the two limits SU(3) and SU(2) are dynamical symmetries of theSp(6) symmetry, therefore, the values of R_{22} for this highest symmetry must range from 0 to 0.7. For the highest symmetry SO(8), we have quite a different situation. In this case, because of the fact that all the three limits contain SO(5) as a subgroup (see eq. (3) of Feng 's talk)[1], the state 2^+_γ is always the second 2^+ state with SO(5) quantum number $\tau = 2$ and 2^+_g is always the first 2^+ state with $\tau = 1$. Of course, the ground state 0^+_g must have $\tau = 0$. According to the τ selection rule ($\Delta\tau = \pm 1$), the transition $2^+_\gamma \rightarrow 0^+_g$ is forbidden. Hence R_{22} should be zero for nuclei which possess the SO(8) symmrtry. **Note that in this case R_{22} must always be zero independent of which subgroup chain they belong to.** Such a nice property of R_{22} that its values depend only on the highest symmetry Sp(6) or SO(8) and do not depend on the details of the subgroup chains makes it a good detector of the fermion dynamical symmetry of each shell.

In Fig.1 we have displayed all the available data of R_{22} for the actinides and the rare-earths. On the right hand side of this figure, the points belong to the actinides. We see that R_{22} is very small (<0.1) only for the near doubly closed shell nuclei (e.g. ^{212}Po and ^{214}Po). This is what we expected because these nuclei must be close to the SU(2) (vibration) symmetry. As the mass of the nuclei go far away from the closed shells, the values of R_{22} approach the Alaga rule of 0.7. This indicates that shells 7 and 8 are indeed in possession of the Sp(6) symmetry. On the left hand side of Fig.1, the data belong to rare-earth nuclei with neutron number N > 82. We see the same behavior as the actinides. This again demonstrates that shell 7 (where the valence neutrons occupy) possesses Sp(6) symmetry.

In the Fig.2 we show the R_{22} data for lighter rare-earths (on the right hand side) and the data for nuclei with proton number $Z \leq 50$ and neutron number $N \leq 82$ (on the left hand side of the figure). It is very interesting to note that some of the heavier isotopes of the lighter rare-earth nuclei have neutrons (N > 82) occupying shell 7 (e. g. ^{142}Ba, $^{144-148}$Nd, $^{146-154}$Sm and $^{148-158}$Gd). In Fig.2, these nuclei are denoted by black dots. It is seen that only these black dots can go from zero to about 0.54 (^{158}Gd). For the isotopes of these nuclei with no neutrons occupying shell 7 (denoted by open circles), namely both the neutrons and protons occupy shell 6 (e. g. $^{118-130}$Te, $^{120-136}$Xe, $^{130-136}$Ba, $^{130-140}$Ce, $^{138-142}$Nd and $^{142-144}$Sm), the R_{22} values never go above 0.1. Thus it clearly indicates that shell 6 possesses the SO(8) symmetry. This conclusion holds also for nuclei with $Z \leq 50$ (shown on the left hand side of Fig. 2). These nuclei have valence proton holes occupying shell 5 and valence neutrons occupying shell 6. Although in shell 5, as we have mentioned above, the highest symmetry can either be Sp(6), SO(8) or SO(6)×SU(3). Since the valence neutrons in shell 6 possess SO(8) symmetry, it is likely that due to the strong n-p interactions, SO(8) symmetry from the proton part will be "picked up" and form $SO^{\nu}(8) \times SO^{\pi}(8) \supset SO^{\nu+\pi}(8)$. If this happens, then R_{22} for these nuclei shoud all be very small (theoretically it should be zero). As one can see from Fig. 2, indeed, the data appear that way. All the open circles are "condensed" at the bottom of the figure. The R_{22} values are all less than 0.1 with almost no exception for the nuclei in this region.

There is only one exception from the available data which we can find and that is ^{138}Ba. Its second 2^+ state is at 2.218 MeV with a branching ratio $(2^+_\gamma \rightarrow 0^+_g) / (2^+_\gamma \rightarrow 2^+_g)$ equal to 98 /1.9. This branching ratio is somewhat unusual and causes the value of R_{22} (= 0.28) for ^{138}Ba to be an order of magnitude larger than its neighborhood values. This large value of R22 is in contradiction to the conclusion of shell 6 possessing SO(8) symmetry. The recent shell model calculations of ^{138}Ba indicates that there are several 2^+ states adjacent to the 2^+_γ state with almost degenerate energies (Ji and Wildenthal, 1986) [4]. Thus the 2.218 Mev γ-ray may, in fact, correspond to several transitions. Not clearly distinguishing them may cause an overestimation of the branching ratio of $2^+_\gamma \rightarrow 0^+_g$ transition. Of course, before coming to a final conclusion, further experimental and theoretical study is required.

There are some recent experimental studies on $^{128-132}$Nd and $^{126-128}$Ce nuclei (Varley,1985)[5] which seem to suggest that they may have large deformation ($\varepsilon \cong 0.3$). This result is due to the measurement of a large E4 / E2 (~3.0). However, there are only very few states measured and therefore the data is insufficient to distinguish as to whether it is indeed in contradiction with the conclusion of shell 6 possessing SO(8) symmetry or not.

As was shown by Ginocchio (Ginocchio,1982)[6], if the term $[\,D^{\dagger} \cdot \tilde{D}\,]^2$ is introduced and the quadrupole operator is redefined as Q,

$$Q = P^2 + \chi\,[\,D^{\dagger} \cdot \tilde{D}\,]^2 \qquad (2)$$

where χ is a parameter and

$$P^2 = \sqrt{2}\,(2k+1) \times [\,b^{\dagger}_{k3/2}\,\tilde{b}_{k3/2}\,]^2 \qquad (3)$$

$$D^{\dagger} = \sqrt{2}\,(2k+1) \times [\,b^{\dagger}_{k3/2}\,b^{\dagger}_{k3/2}\,]^2 \qquad (4)$$

then it is possible to have large deformation (i.e. large quadropole moment) and rotation-like spectrum by mixing different representations of the SO(6), and yet it is still within the SO(8) symmetry. Thus, by only looking at ,say, the E4 /E2 ratio (or deformation), it is quite dangerous to say that whether these nuclei possess SO(8) or Sp(6) symmetry. One really has to look at the B(E2) values as well. In our opinion, the most convincing indicator is R_{22}. In any case, this quantity should be zero if it is SO(8) symmetry, and go from 0 to 0.7 if it is Sp(6) symmetry. Unfortunately, the data for $^{128-132}$Nd and $^{126-128}$Ce are not available at this stage for us to calculate R_{22}. It would be very interesting to see the further measurements of these nuclei.

Of course, it should be remembered that all the above discussions about the symmetries are based strictly on a prerequisite that there are no broken pairs, i. e. SD seniority u = 0. In other words, all the inactive parts of the angular momenta are assumed to be frozen (coupled to zero). If under some circumstances the degrees of freedom of the inactive parts have to be taken into account, then the SD subspace is no longer a good subspace and neither Sp(6) nor SO(8) symmetry is preserved. Nevertheless, from the data we have shown here, it is unlikely to be the case. At least for the low lying states of most nuclei in the region we have studied, the SD subspace seem to be a good subspace. The data do show that shell 7 and shell 8 possess Sp(6) symmetry while shell 6 possess SO(8) symmetry.

THE EVIDENCE OF SO(7) SYMMETRY

Another remarkable difference between boson dynamical symmetries (IBM) and the fermion dynamical symmetries (FDS) is that there are three vibrational modes in the FDS: SU(2)×SO(3), SO(5)×SU(2) and SO(7) limits, while there is only one (U(5) limit) in the IBM. Establishing any one of these three limits and showing the difference from the IBM U(5) limit experimentally can serve as another experimental evidence of the fermion dynamical symmetries.

As I have pointed out at the very begining, as far as the spectrum is concerned, the three FDS vibration modes have the same formulas as that obtained from the IBM U(5) limit. It has fhe following form (Arima and Ichello, 1976 eq. (4.21))[3]:

$$\Delta E = \varepsilon N_d + \alpha/2\ N_d(N_d-1) + \beta(N_d-\tau)(N_d+\tau+3) + \gamma(L(L+1)-6N_d) \qquad (5)$$

As a phenomenological model, the IBM treats ε, α, β and γ as parameters to fit data. In principle, any values whatsoever of these parameters are permited. In the FDS model, however, all the parameters are related to the parameters of nuclear force. Different dynamical symmetries can have different expression as shown in table 2. The related FDS model Hamitonian is as follows:

$$H = e_0 n_0 + e_1 n_1 + G_0 S^{\dagger}\cdot S + G_0 \mathcal{S}^{\dagger}\cdot\mathcal{S}$$

$$+ g_0(S^{\dagger}\cdot\mathcal{S} + \mathcal{S}^{\dagger}\cdot S) + G_2 D^{\dagger}\cdot D + \sum_r B_r P_r^{\dagger}\cdot P_r \qquad (6)$$

Among these three limits, the SO(7) limit is the most interesting one. If one just look at the spectrum of each nucleus individually, one may not see the difference. They all behave like an anharmonic vibrator. Of course, they do have some tiny differences as one can see from table 2. For example, the SU(2)×SO(3) limit does not have τ dependence ($\beta = 0$) but stronger L dependence since the strength of the quadrupole interaction (B_2) is usually larger than B_1 and B_3 ; SO(5)×SU(2) limit is just the opposite; while the SO(7) limit has relatively weak τ and L dependences. However, these differences are too small to be differentiated experimentally. **The major difference** here is the N_1 dependence. It appears that when one looks at a set of nuclei (isotopes or isotones), for SU(2)×SO(3) and SO(5)×SU(2) limits, the spectrum will be unchanged,

Table 2. The Energy Formular of Vibration Limit

Parameter	SU(2) ×SO(3) limit	SO(5) × SU(2) limit	SO(7) limit
ε	$\Omega_1(G_2-G_0) - \Omega_0 g_0 + 9(B_1-B_2)/4$	$\Omega_1(G_2-G_0) - \Omega_0 g_0 + 6B_1/5 + 14B_3/5 - 4B_2$	$[G_2\Omega_1 - G_0(\Omega_1+4) + 6B_1/5 + 14B_3/5] - 2(G_2-G_0)(N_1-1)$
α	$2(G_0-G_2)$	$2(G_0-G_2+B_3-B_2)$	$2(B_3-B_2)$
β	0	B_2-B_3	G_2-B_3
γ	$3/8(B_1-B_2)$	$1/5(B_1-B_3)$	$1/5(B_1-B_3)$

1) For the SO(7) limit, Nd should be understood as number of D pairs denoted as $\bar{\kappa}$ hereafter.
2) For the SU(2) ×SO(3) and SO(5) × SU(2) limit, Nd should be understood as V /2 where V is seniority.
3) Ω_1 and Ω_0 are the pair degeneracy of the normal parity levels and abnormal parity level in one major shell respectively; N_1 is the number of pairs in normal parity levels.

while the SO(7) spectrum will be compressed linearly as N_1 increases, since ε (the energy difference between a D-pair and a S-pair) is decreasing linearly as N_1 increases (see table 2). This feature makes SO(7) quite distinquishable from the other fermion vibration modes and detectable experimentally.

The difference between the three fermion vibration modes and the IBM U(5) limit can be seen more clearly from the B(E2) γ - transition rates as shown in table 3. The B(E2)'s of FDS model differ from IBM ones by a Pauli factor. For the SO(5)×SU(2) limit, thePauli factor usually is less then 1. The situation of SU(2)×SO(3) , which we do not show in the table, is very similer to the SO(5)×SU(2) case. While for the SO(7) case it is different. The Pauli factor usually is larger than 1. When $\Omega_1 \rightarrow \infty$, the Pauli factors go to 1, all ofthem reduce to IBM U(5) limit. Experimentally examing these Pauli factors will provide a clear evidence of the fermion dynamical symmetries.

The SO(7) is particularly interesting since, unlike others, it does not correspond to a static structure but rather to one which inherently varies with mass. In Fig. 3 we show the ratio of the overlaps of SO(7) ground state wavefounction with these of SO(6) and SO(5). It indicates that, as number of nucleus increases, the SO(7) wavefunction becomes more and more close to the SO(6) limit.

This structural evolution is also clearly evident in Fig.4 which shows the effective intrinsic state deformation β_{int}. The β_{int} values of 0 and 1 describe the SO(5)×SU(2) (IBM U(5)) and SO(6) (IBM SO(6)) limits. For the SO(7) limit , the β_{int} is just the intermidiate between the SO(5)×SU(2) and SO(6) and tends to the latter as the number of nucleons increases.

Dynamically, the SO(7) symmetry can be easily understood by considering the fermion Hamitonian underlying it which contains residual interactions written in terms of pairing and multipole interaction terms (eq.(6)). For simplicity, let us neglect the quadrupole pairing terms (G_2=0) which generally is expected to be small. The Hamitonian can be reexpressed in term of Casimir operators as follows:

$$H = H_0 + (G_0 - g_0) C_{SU2} + g_0 C_{SU2} + (B_3 - B_2) C_{SO5} + B_2 C_{SO6} + (B_1 - B_3) L^2/5 \quad (7)$$

where H_0 only depends on number operators n_0 and n_1. The definition of Casimir operators are listed in table 4. When the fermion residual interaction is dominated by monopole pairing (B_2=0), it leads to SO(5)×SU(2) limit; when quadrupole interaction dominates (set $G_0 = g_0 = 0$), it leads to SO(6) symmetry, and when the pairing and quadrupole interaction in the normal parity leves are of equal strength ($G_0 = B_2$),' an SO(7) symmetry is generated. Actually, when $G_0 = B_2$, the Hamitonian is a strong mixture of SO(5)×SU(2) and SO(6) symmetry :

$$H = H_0 + g_0 (C_{SU2} - C_{SU2}) + G_0 (C_{SU2} + C_{SO6}) + (B_3 - G_0) C_{SO5} + (B1-B3) L^2 / 5 \quad (8)$$

Note that

$$C_{SU2} + C_{SO6} = C_{SO8} + C_{SO5} - C_{SO7} + S_0 (S_0 - 1) \quad (9)$$

if $g_0 = 0$, we obtain the SO(7) Hamitonian:

$$H = H_0' - G_0 C_{SO7} + B_3 C_{SO5} + (B_1 - B_3) L^2/5 \quad (10)$$

where

$$H_0' = H_0 + B_2 (C_{SO8} + S_0 (S_0 - 1)) \quad (11)$$

Thus, from the dynamical point of view, the SO(7) limit does describe the structure of nuclei during the transition from SO(5)xSU(2) towards to SO(6) limit.

The E4 /E2 ratio also indicates that . From table 2 , the excitation energy for the SO(7) limit is:

$$E = - G_0 \bar{\kappa} (\Omega 1 - 2 N_1 + \bar{\kappa} + 5) + B_3 \tau (\tau + 3) + 1/5 (B_1 - B_3) L (L + 1) \quad (12)$$

Neglecting the last two terms in eq. (11), we can estimate the lower limit of the E4 /E2 ratio:

$$E4/E2 = 2 \left(1 + 1/(\Omega_1 - 2 N_1 + 6) \right) \quad (13)$$

It ranges from 2 (when $N_1 << \Omega_1$ and $\Omega_1 >> 1$) to 2.33 (when $N_1 = \Omega_1 / 2$) as N_1 increases. Comparing with the SO(5)×SU(2) and SO(6) limits, which are around 2 and 2.5 respectively , it again shows the structure evolution from the SO(5)×SU(2) limit towards the SO(6) limit. All these properties of the SO(7) limit suggest that although its spectrum very similar to the U(5) limit, it does not correspond to an IBM symmetry but rather describes a transitional sequence which evolves between structures closely related to two of those symmetries, U(5) and SO(6). Relatively speaking, the SO(5) ×SU(2) and SU(2)×SO(3) symmetry are more close to the IBM U(5) limit, since they correspond to a static structure of an anharmonic vibrators, although there are still some differences due to the Pauli effects. Thus, we may regard them as the underlying fermion dynamical symmetry of the IBM U(5) limit in the shells possessing SP(6) and S(8) symmetry respectirely. However, SO(7) has no boson counterpart in the IBM. Experimentally identifying this new mode will be a direct evidence of the fermion dynamical symmetries.

Now the question is where to find this new mode? Note that only the SO(8) symmetry contains SO(7) limit. Therefore, SO(7) nuclei, if they exist, should be found at the mass regions either $50 \leq (N, Z) \leq 80$ or $50 \leq N \leq 80$ and $28 \leq Z \leq 50$. In the former case, both valence protons and neutrons fill shell 6 possessing SO(8) symmetry. In the latter case, neutrons fill shell 6 and protons fill shell 5 and hopefully $SO^{\pi}(8)\times SO^{\nu}(8)$ symmetry will be pickup due to the strong n-p interaction as we have discussed in last section. It turns out that, Ru and Pd isotopes in the latter mass region are the best evidence. In $50 \le (N, Z) \le 80$ region, the isotones of N=74 seem to show some indication of SO(7) behavior. However it goes too quickly to the well known A=130 SO(6) region and thus the SO(7) can not be fully shown . The reason why Ru and Pd is the better region to show SO(7) symmetry may somehow relate to the fact that the abnormal level in shell 5 is well separated from the normal parity levels so that one of the condition for SO(7) to appear, i.e. $g_0 = 0$, could be satisfied approximately. While in the shell 7, the abnormal level is deep into the normal levels, the condition may be hard to achieve.

A crucial aspect of the SO(7) specturm is that the energy will decrease linearly with increasing mass, reaching a minimum at mid shell. This feature resembles a transitional region from a vibrator towards γ-soft rotor . In the Pd and Ru region, the spectra do show this behavior. The recent numerical IBA calaulations have identified these nuclei as intermediate between the IBM U(5) and SO(6) limits (Stachel et al., 1984)[9]. The FDS model, however, can describe them as the SO(7) fermion dynamical symmetry with analytical formulas for both energies and γ-transitions. It is an appealing one ,of caurse, since transitional regions are especially difiicult to treat, and yet here the properties are obtained analytically. Yet it also presents a difficulty in attempting to identify the SO(7) symmetry empirically: it is essential to distinguish possible regions from phase transitions towards a deformed rotor. For example, the Ba-Gd nuclei near A=150 are merely a prelude to the deformed rare-earth region. However, in such a region the $E_{4^+_1} / E_{2^+_1}$ ratio will attain values ≥ 3.0 and branching ratio such as R_{22} which we have discussed in the previous section approach Alga value (0.7) while they vanish in the SO(7) case. In the $^{104-110}Pd$ and $^{98-104}Ru$ candidates for SO(7) , this ratio is <0.05. Similarly, the ratio $B(E2, 3^+_1 \rightarrow 2^+_1) / B(E2, 3^+_1 \rightarrow 4^+_1)$ approaches 2.5 in a transition to the rotor, but vanishes in SO(7). It is <0.1 (0.2) in these Pd (Ru) isotopes.

The application of the SO(7) symmetry to Pd and Ru is shown in Figs 4-6. The parameter values adopted for Ru (Pd) were: $-G_0 = 45\ (47.5)$ Kev, $b_3 = 5.3$ Kev $1/5\,(b1 - b_3) = 7.2$ Kev. (the latter two for both Ru and Pd). In the present approach the proton and neutron shells are combined (as in IBM -1), thus $\Omega_1 = \Omega_1^{\pi} + \Omega_1^{\nu} = 16$. The pair number in normal parity levels is parameterized as $N_1 = \alpha N$, where α represents the fraction of the 2N valence particles (neutrons plus proton holes) that occupy the normal parity levels. By fitting the spectrum, α is obtained: 0.96 for Ru and 0.91 for Pd. However, this parameter is not very sensitive . Choosing α=0.8 for both Ru and Pd, for example, the results will not change very much. Also note that, as expected, b3, $1/5\,(b_1 - b_3) << G_0$. The agreement with experiment in both figs. 4 and 5 is quite good. It is important to emphasize that the parameters, which are identical in Ru and Pd, were held constant for each set of isotopes. The most characteristic aspect of SO(7), the smooth decrease in energies with mass, is nicely reproduced. The slope changes predicted (and partially observed) for ^{106}Ru and ^{112}Pd arise because the normal parity orbits are effectively filled past midshell (i.e., $N1 > \Omega_1/2$). Of course, the midshell point is somewhat less ascertainable in the present approach since it depends on the number of nucleons assigned to the unique parity orbit and is parameterized by α. The empirical upturn tends to occur slightly later than predicted.

Figure 6 shows the $B(E2, 2^+_1 \rightarrow 0^+_1)$ values and the ratio $R_0 = B(E2, 0^+_2 \rightarrow 2^+_1) / B(E2, 2^+_1 \rightarrow 0^+_1)$. The SO(7) predictions here involve no new parameters (except for a normalization of the former quantity at N=6). The predicted $B(E2, 2^+_1 \rightarrow 0^+_1)$ values are in excellent agreement with the data for N<8. R_0 is a crucial indicator since the decay of the 0^+_2 level discriminates between vibrator and O(6)-like limits. The SO(7) predictions reproduce the data remarkably well and point to the transition from SO(5)-like towards SO(6)-like. Other observables such as $R_4 = B(E2, 4^+_1 \rightarrow 2^+_1) / B(E2, 2^+_1 \rightarrow 0^+_1)$ are also in good agreement with the SO(7) values .

Of course, there are deviations from this idealized picture. The triplet energy levels, especially 0^+_2, are predicted at slightly lower energies than observed, and the yrast levels in the heaviest isotopes shown drop well below the predictions. In Ru, the same evolution is indicated by the ratio R_{22} which reaches 0.1 in ^{108}Ru . Moreover, the $E4^+_1/ E2^+_1$ ratio in $^{106, 106}Ru$ is 2.65 and 2.75 whereas for $^{104-110}Pd$ and $^{98-104}Ru$ it lies in the narrow interval from 2.14—2.48. This deviation and the rise in 0^+_2 state signal a transition from SO(7) towards to SO(6) at the heavier end of the Pd and Ru isotopes.

SUMMARY

In this talk, the empirical evidences of fermion dynamical symmetries have been presented which show that shell 7 and shell 8 possesses the SP(6) symmetry while shell 6 possesses the SO(8) symmetry. For the nuclei with neutrons occupying shell 6 and protons occupying shell 5, the SO(8) symmetry is still preserved although shell 5 it self could have more complicated symmetries. In the latest region , $^{104-110}Pd$ and $^{98-104}Ru$ isotopes are found to be good empirical realizations of the SO(7) subgroup chain in the SO(8) symmetry. This is the first example of a fermion dynamical symmetry in heavy nuclei that is not simply analogous to an established boson symmetry. Moreover, it is the first observed example of a symmetry that incorporates a variable structure applicable to nuclear transition regions. Its empirical verification is encouraging for the IBA and points toward a more fundamental origin for the observed symmetries of the latter.

ACKNOWLEDGEMENTS

This work is in cooperation with Da Hsuan Feng, R.F.Casten, J.N. Ginocchio and Xiao-Ling Han. I would like to take this opportunity to express my thanks to my collaborators. I should also thank Professor Bruno Gruber for his warm hospitality during the conference. Discussions with F. Iachello, J. Draayer, D. D. Warner, J.-Q. Chen, M. Guidry, A. Aprahamian, and J.Stachel are gratefully acknowledged. Research has been supported by the USDOE, the Chinese Science Foundation and the NSF.

Table 3 The B (E2) Formulas

B(E2)	U_5	$SO_5 \times SU_2$	SO_7	SO_6
B_2/α_2^2	N	$N_1 \frac{\Omega_1 - N_1}{\Omega_1 - 1}$	$N_1 \frac{\Omega_1 + 6 - N_1}{\Omega_1 + 7 - 2N_1}$	$N_1 \frac{N_1 + 4}{5}$
B_4/α_2^2	2(N−1)	$2(N_1-1) \frac{\Omega_1 - 1 - N_1}{\Omega_1 - 3}$	$2(N_1-1) \frac{\Omega_1 + 7 - N_1}{\Omega_1 + 9 - 2N_1}$	$2(N_1-1) \frac{N_1+5}{7}$
$B_{2'}/\alpha_2^2$	2(N−1)	$2(N_1-1) \frac{\Omega_1 - 1 - N_1}{\Omega_1 - 3}$	$2(N_1-1) \frac{\Omega_1 + 7 - N_1}{\Omega_1 + 9 - 2N_1}$	$2(N_1-1) \frac{N_1+5}{7}$
$B_{0'}/\alpha_2^2$	2(N−1)	$2(N_1-1) \frac{\Omega_1 - 1 - N_1}{\Omega_1 - 3} \frac{\Omega_1 + 4}{\Omega_1 - 1}$	$2(N_1-1) \frac{\Omega_1 + 2 - 2N}{\Omega_1 + 7 - 2N} \frac{\Omega_1 + 7 - N_1}{\Omega_1 + 9 - 2N_1}$	0
R_0	$2(\frac{N-1}{N})$	$2(\frac{N_1-1}{N_1}) \frac{\Omega_1 - 1 - N_1}{\Omega_1 - 3} \frac{\Omega_1 + 4}{\Omega_1 - N}$	$2(\frac{N_1-1}{N_1}) \frac{\Omega_1 + 7 - N_1}{\Omega_1 + 9 - 2N_1} \frac{\Omega_1 + 2 - 2N_1}{\Omega_1 + 6 - N_1}$	0

1) $B_L \equiv B(E2, L_1 \to (L-2)_1)$; $B_{2'} \equiv B(E2, 2_2 \to 2_1)$; $B_{0'} \equiv B(E2, 0_2 \to 2_1)$; $R_0 \equiv B_{0'} / B_2$.

2) The quadrupole transition operator is $\alpha_2 P^2$ for the FDS model and $\alpha_2(d's + s'd)$ for the IBM

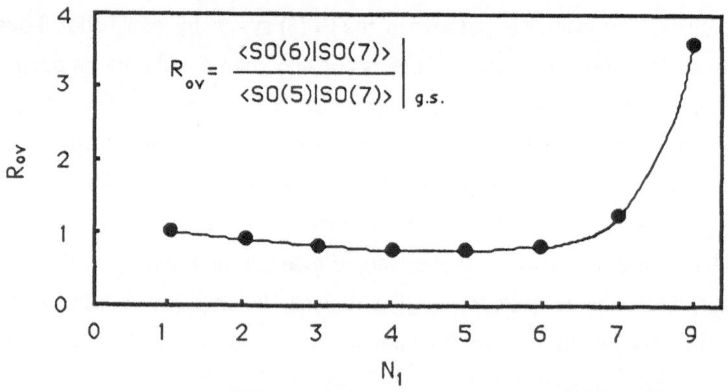

Fig. 3 The Overlap of the SO(7) Ground State Wavefuction.

Table 4 The Casimir Operators in The FDS Model

Groups	Casimir Operators	Eigenvalues	quantum numbers
SU_2	$C_{SU2} = S^{\dagger}\cdot S + S_0(S_0-1)$	$\frac{1}{4}(\Omega_1-v_1)(\Omega_1-v_1+2)$	v_1
SU_2	$C_{SU2} = S^{\dagger}\cdot S + S_0(S_0-1)$	$\frac{1}{4}(\Omega-v)(\Omega-v+2)$	v
SO_5	$C_{SO5} = \sum_{r=1,3} p^{r\dagger}\cdot p^{r}$	$\tau(\tau+3)$	τ
SO_6	$C_{SO6} = \sum_{r=1,2,3} p^{r\dagger}\cdot p^{r}$	$\sigma(\sigma+4)$	σ
SO_7	$C_{SO7} = D^{\dagger}\cdot D + \sum_{r=1,3} p^{r\dagger}\cdot p^{r} + S_0(S_0-6)$	$\frac{1}{4}\Omega_1(\Omega_1+12) - [N_1 + v_D(\Omega_1 - v_D + 5)]$	$N_1\ v_D$
SO_8	$C_{SO8} = S^{\dagger}\cdot S + D^{\dagger}\cdot D + \sum_{r=1,2,3} p^{r\dagger}\cdot p^{r} + S_0(S_0-6)$	$\frac{1}{4}(\Omega_1-u)(\Omega_1-u+12) + \phi(\rho_1,\rho_2,\rho_3)$	u ρ_1,ρ_2,ρ_3

1) The eigenvalue of SO(7) casimir operator is only given for $u=0$ case, where $v_D=$ $N1 - \bar{\kappa}$ and $\bar{\kappa}$ is the number of D pairs .

2) $\phi(\rho_1,\rho_2,\rho_3) = (\rho_1^2+\rho_2^2)/2 + (\rho_1+\rho_3)(\rho_1+\rho_3+4\rho_2+12)/4 + \rho_2(\rho_2+4)$

3) $s' = S' + S'$, S' and S' are the S pair creation operaters for normal parity levels and abnorma parity level respectively.

4) $s_0=(n-\Omega)/2$, $S_0=(n_1-\Omega_1)/2$, $S_0=(n_0-\Omega_0)/2$, $n=n_1+n_0$, $\Omega=\Omega_1+\Omega_0$.

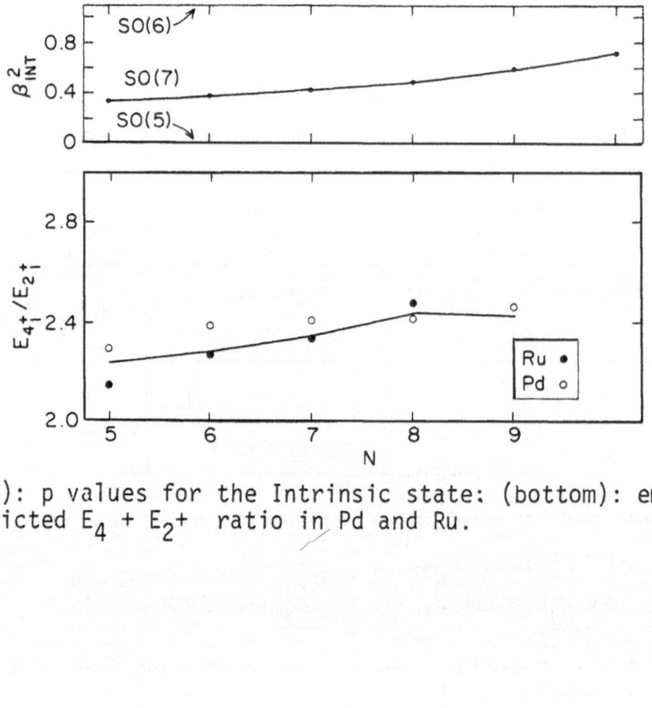

Fig. 4 (Top): p values for the Intrinsic state: (bottom): empicical and predicted $E_4 + E_2+$ ratio in Pd and Ru.

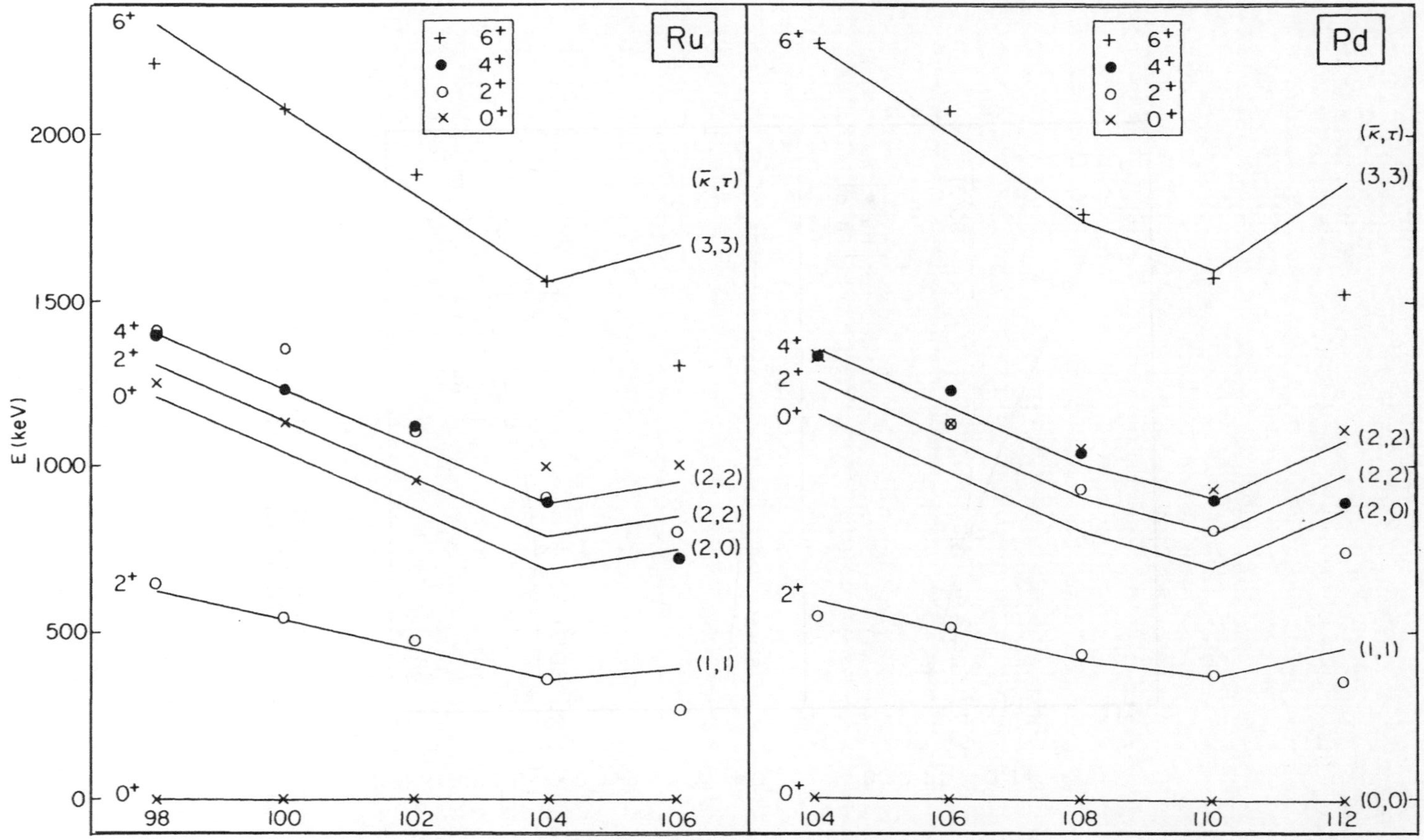

Fig. 5 Empirical [8-10] and predicted energy levels for Ru and Pd. The parameters, given in the text, are held constant for each set of isotopes.

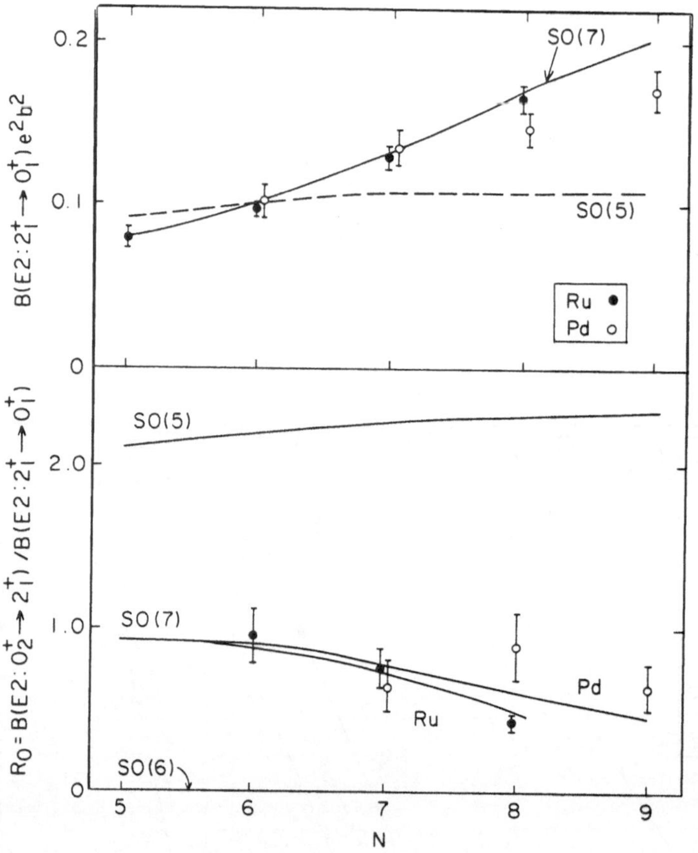

Fig. 6 Empicical 11-16 and predicted B(E2) values. Seperate curves are shown for Ru and Pd if they differ significantly.

REFERENCES

1. Da Hsuan Feng, " Fermion Daynamical Symmetry (FDS) Model, Nuclear Shell Model And Collective Nuclear Structure Physics " and the references therein, in this Proceedings (1986).
2. C. L. Wu, D.H. Feng, X.-G. Chen, J.-Q. Chen, and M. W. Guidry, Phys. Lett. **168B**, 313 (1986) ; Phys. Rev. C (submitted); M. W. Guidry, C. L. Wu, D.H. Feng, J. N. Ginocchio, X.-G. Chen, and J.-Q. Chen, Phys. Lett. (submitted).
3. A. Arima, and F. Iachello, Phys. Rev.Lett., **35** (1975) 1069; **99** (1976) 253; **111**, 201 (1978).
4. Xian-Dong Ji and B. H. Wildenthal, private communication (1986).
5. B. J.Varly, AIP conference Proceedings No.125, Capture Gamma-RaySpectroscopy and Related Topics-1984 , Edited by S. Raman, (American Institute of New York, 1985), P.709.
6. Ginocchio, J. N., Ann. Phys. (N.Y.),**126**(1980)234; Nucl. Phys. **A384** ,112 (1982).
7. Table of Isotpes, Eidted by C.Micheal Lederer and Virginia S. Shirley, seventh edition (Wiley-Interscience, New York, 1978).
8. M. Sakai, At. Data and Nucl. Data Tables **31**,399 (1984).
9. J. Stachel, P. Van Isacker, and K. Heyde, Phys. Rev. **C25**,650 (1982); J. Stachel et al.,Z. Phys. **316A**, 105 (1984) and private communication.
10. P. Bucurescu et al., preprint,Cent. Inst. of Phys. Bucharest, and M. Luontama et al., Jyvaskyla preprint JYFL 12/85.
11. P. M. Endt, At. Data and Nucl. Data Tables **26**, 47 (1981).
12. F. K. McGowan et al., Nucl. Phys. **A113,** 529 (1968).
13. S. Lansberger et al., Phys. Rev. **C21**, 588 (1980).
14. R. Robinson et al., Nucl. Phys. **A124**, 553 (1969).
15. L. Hasselgren et al., UUIP-957 (1977).
16. A.Christy et al., Nucl. Phys. **A142,** 591 (1970).

$SO(2\ell + 1) \supset\ ?\ \supset SO_L(3)$ IN GROUP CHAINS FOR L-S COUPLING

Z. Y. Wu,* C. P. Sun,* L. Zhang,* and B. F. Li**

*Northeast Normal University, China

**Jilin University, China

In his famous article of 1949,[1] G. Racah pointed out that there exists a proper subgroup of SO(7) which properly contains $SO_L(3)$ in group chains for L-S coupling. $SO(7) \supset G_2 \supset SO_L(3)$. An answer to the question "$SO(2\ell+1) \supset\ ?\ \supset SO_L(3)$ for an arbitrary ℓ" is given with a straightforward proof in this paper. "No, except the case for $\ell = 3$."

1. NOTATION AND REFORMULATION OF SOME WELL-ESTABLISHED FACTS

Following B. R. Judd,[2] we use the following notation for triple irreducible tensors of group $SU_Q(2) \times SU_S(2) \times SO_L(3)$

$$A^{q\ s\ \ell}_{m_q m_s m_\ell} = \begin{cases} A^{+\ s\ \ell}_{m_s m_\ell}, & \text{for } m_q = \frac{1}{2} \\ (-1)^{s-m_s+\ell-m_\ell} A^{s\ \ell}_{\overline{m}_s \overline{m}_\ell}, & \text{for } m_q = -\frac{1}{2} \end{cases} \tag{1}$$

$$X^{Q\ S\ L}_{M_Q M_S M_L}(\ell\ell) = \sum A^{q\ s\ \ell}_{m_q m_s m_\ell} A^{q\ s\ \ell}_{m'_q m'_s m'_\ell} \langle qm_q\ qm'_q \mid QM_Q\rangle \times \langle sm_s\ sm'_s \mid SM_S\rangle\langle \ell m_\ell\ \ell m'_\ell \mid LM_L\rangle \tag{2}$$

where $q = s = 1/2$. By making use of the anticommutation relation

$$\left\{ A^{q\ s\ \ell}_{m_q m_s m_\ell},\ A^{q\ s\ \ell}_{m'_q m'_s m'_\ell} \right\} = \delta_{\overline{m}_q m'_q}\, \delta_{\overline{m}_s m'_s}\, \delta_{\overline{m}_\ell m'_\ell} (-1)^{q+m_q+s-m_s+\ell-m_\ell}. \tag{3}$$

It can be shown that

$$X^{Q\ S\ L}_{M_Q M_S M_L}(\ell\ell) = -2\sqrt{2\ell+1}\, \delta_{Q0}\delta_{S0}\delta_{L0} + (-1)^{1+Q+S+L} X^{Q\ S\ L}_{M_Q M_S M_L}(\ell\ell). \tag{4}$$

In the following we will denote by $\mathcal{W}^{00L}(\ell\ell)$ the linear closure spanned by $\{X^{00L}_{00M_L} \mid M_L = -L,-L+1,\dots,L\}$. We are going to prove that

LEMMA. Any subalgebra of $so(2\ell+1)$ containing $so_L(3)$ must be a direct sum of $so_L(3)$ with some linear closures $\mathcal{W}^{00L}(\ell\ell)$ $(L = 3,5,\dots,2\ell-1)$.

Let Y be a subalgebra of $so(2\ell+1)$ properly containing $so_L(3)$. Then there must be in Y a linear combination

$$\sum_{L=1,3,\dots}^{2\ell-1} \sum_{M_L=-L}^{L} C^L_{M_L} X^{00L}_{00M_L}(\ell\ell) \in Y$$

with at least one non-vanishing coefficient $C^L_{M_L} \neq 0$ $(L>1)$. As $so_L(3) \subset Y$, and Y is closed under linear operation, it follows that

$$X = \sum_{L=3,5,\dots}^{2\ell-1} \sum_{M_L=-L}^{L} C^L_{M_L} X^{00L}_{00M_L}(\ell\ell) \in Y$$

Let $C^{L'}_{M'_L} \neq 0$ and $C^L_{M_L} = 0$ $\forall L > L'$. We rewrite the above expression as

$$X = \sum_{L=3,5}^{L'} \sum_{M_L=-L}^{L} C^L_{M_L} X^{00L}_{00M_L}(\ell\ell) \in Y \tag{10}$$

By applying $(\mathrm{ad}\, L_-)^{2L'} (\mathrm{ad}\, L_+)^{L'-M'_L}$ to X we get the following non-zero element of Y

$$(\mathrm{ad}\, L_-)^{2L'} (\mathrm{ad}\, L_+)^{L'-M'_L} X = \alpha C^{L'}_{M'_L} X^{00L'}_{00-L'}(\ell\ell) \in Y \tag{11}$$

where α is a non-zero number, for the only term in (10) which can survive $(\mathrm{ad}\, L_-)^{2L'} (\mathrm{ad}\, L_+)^{L'-M'_L}$ is $C^{L'}_{M'_L} X^{00L'}_{00M'_L}(\ell\ell)$ while all other terms are killed by it.

The fact that Y is closed under Lie-product has been made use of.

For the same reason, $X^{00L'}_{00-L'+1}(\ell\ell), \dots, X^{00L'}_{00L'}(\ell\ell)$ are elements of Y. Therefore the linear closure $\mathcal{W}^{00L'}(\ell\ell)$ is a part of Y.

Subtracting $\sum_{M_L=-L'}^{L'} C^{L'}_{M_L} X^{00L'}_{00M_L}(\ell\ell) \in Y$ from (10), we get

$$X' = \sum_{L=3,5}^{L'-2} \sum_{M_L=-L}^{L} C^L_{M_L} X^{00L}_{00M_L}(\ell\ell) \in Y \tag{12}$$

If $X' \neq 0$, let $C^{L''}_{M''_L} \neq 0$ and $C^L_{M_L} = 0$ $\forall L'' < L < L'$. It can be shown in the same way that the linear closure $\mathcal{W}^{00L''}(\ell\ell)$ is a part of Y.

Continuing in this way we reach the conclusion: if $Y \supset so_L(3)$, and Y is a subalgebra of $so(2\ell+1)$, and $\sum_L \sum_{M_L} A^L_{M_L} X^{00L}_{00M_L}(\ell\ell) \in Y$ then

In particular

$$X^{000}_{000}(\ell\,\ell) = -\sqrt{2\ell+1} \tag{5}$$

and

$$X^{Q\ S\ L}_{M_Q M_S M_L}(\ell\ell) = 0, \quad \text{if} \quad Q + S + L = \text{even} > 0. \tag{6}$$

It can also be shown[3] that

$$\left[X^{Q_1\ S_1\ L_1}_{M_{Q1} M_{S1} M_{L1}}(\ell\ell),\ X^{Q_2\ S_2\ L_2}_{M_{Q2} M_{S2} M_{L2}}(\ell\ell)\right] = \sum \prod_{i=1}^{3}$$

$$[(2Q_i+1)(2S_i+1)(2L_i+1)]^{1/2}(-1)^{Q_3-M_{Q3}+S_3-M_{S3}+L_3-M_{L3}} \times$$

$$\begin{Bmatrix} Q_1 & Q_2 & Q_3 \\ 1/2 & 1/2 & 1/2 \end{Bmatrix} \begin{Bmatrix} S_1 & S_2 & S_3 \\ 1/2 & 1/2 & 1/2 \end{Bmatrix} \begin{Bmatrix} L_1 & L_2 & L_3 \\ \ell & \ell & \ell \end{Bmatrix} \begin{pmatrix} Q_1 & Q_2 & Q_3 \\ M_{Q1} & M_{Q2} & \overline{M_{Q3}} \end{pmatrix} \times$$

$$\begin{pmatrix} S_1 & S_2 & S_3 \\ M_{S1} & M_{S2} & \overline{M_{S3}} \end{pmatrix} \begin{pmatrix} L_1 & L_2 & L_3 \\ M_{L1} & M_{L2} & \overline{M_{L3}} \end{pmatrix} (-1)^{Q_3+S_3+L_3}[1-(-1)^{Q_1+S_1+L_1}] \times$$

$$[1+(-1)^{Q_2+S_2+L_2+Q_3+S_3+L_3}]\, X^{Q_3\ S_3\ L_3}_{M_{Q3} M_{S3} M_{L3}}(\ell\ell) \tag{7}$$

where summation is taken over Q_3, M_{Q3}, S_3, M_{S3}, L_3, M_{L3}. In all possible group chains for L-S coupling, even for L-L coupling [4], there is a common part, that is ...$SO(2\ell+1) \supset \ldots \supset SO_L(3)$. We are going to show that there is no proper subgroup of $SO(2\ell+1)$ which properly contains $SO_L(3)$ except for $\ell = 3$.

2. EXISTENCE CRITERION FOR THE PROPER SUBGROUP OF SO(2ℓ+1) WHICH PROPERLY CONTAINS $SO_L(3)$

Since group $SU_Q(2) \times SU_S(2) \times SO_L(3)$ is simply reducible, the restriction of the total orbital angular momentum to ℓ-shell is a numerical multiple of $X^{001}_{00M_L}(\ell\ell)$,

$$L_0 = c\, X^{001}_{000}(\ell\ell) \in so_L(3)$$

$$\frac{\pm L\pm}{\sqrt{2}} = c\, X^{001}_{00\pm1}(\ell\ell) \in so_L(3) \tag{8}$$

$SO(2\ell+1)$ is generated by $\{X^{00L}_{00M_L}(\ell\ell) \mid L = \text{odd}\}$ while $SO_L(3)$ is generated by $\{X^{001}_{00M_L}(\ell\ell)\}$.

$$[L_0, X^{00L}_{00M_L}(\ell\ell)] = M_L\, X^{00L}_{00M_L}(\ell\ell),$$

$$[L_\pm, X^{00L}_{00M_L}(\ell\ell)] = \sqrt{(L\mp M_L)(L\pm M_L+1)}\, X^{00L}_{00M_L\pm1}(\ell\ell). \tag{9}$$

$$A^{L'}_{M'_L} \neq 0 \Rightarrow \mathcal{W}^{00L'}(\ell\ell) \subset Y.$$

It can be easily seen from this that Y must be a direct sum of $so_L(3)$ with some linear closures $\mathcal{W}^{00L}(\ell\ell)$ $(L = 3,5,\ldots,2\ell-1)$.

There are now only a finite number of such direct sums. We need only to check their closedness under Lie product to find out if they are Lie algebras or not. We know that

$$\begin{aligned} &\mathcal{W}^{001}(\ell\ell) = so_L(3), \\ &[\mathcal{W}^{001}(\ell\ell), \mathcal{W}^{001}(\ell\ell)] \subset \mathcal{W}^{001}(\ell\ell), \\ &[\mathcal{W}^{001}(\ell\ell), \mathcal{W}^{00L}(\ell\ell)] \subset \mathcal{W}^{00L}(\ell\ell). \end{aligned} \tag{13}$$

Therefore in order to check the closedness of $\mathcal{W}^{001}(\ell\ell) \oplus \mathcal{W}^{00L}(\ell\ell)$, we need only to check whether

$$[\mathcal{W}^{00L}(\ell\ell), \mathcal{W}^{00L}(\ell\ell)] \subset \mathcal{W}^{001}(\ell\ell) \dotplus \mathcal{W}^{00L}(\ell\ell) \tag{14}$$

is true or not.

Considering the commutation relation (7) which now assumes the following form

$$\begin{aligned} &[X^{00L_1}_{00M_{L_1}}(\ell\ell), X^{00L_2}_{00M_{L_2}}(\ell\ell)] = \sum_{L_3 M_{L_3}} 4(2L_1+1)^{1/2} \times \\ &(2L_2+1)^{1/2}(2L_3+1)^{1/2}(-1)^{M_{L_3}} \begin{Bmatrix} L_1 & L_2 & L_3 \\ \ell & \ell & \ell \end{Bmatrix} \begin{pmatrix} L_1 & L_2 & L_3 \\ M_{L_1} & M_{L_2} & M_{L_3} \end{pmatrix} \\ &X^{00L_3}_{00M_{L_3}}(\ell\ell) \end{aligned} \tag{15}$$

We reduce the above question to whether

$$\begin{Bmatrix} L & L & L_3 \\ \ell & \ell & \ell \end{Bmatrix} = 0 \quad \forall\, L_3 \neq 1, L \tag{16}$$

is true or not.

If (16) is true, then $\mathcal{W}^{001}(\ell\ell) \dotplus \mathcal{W}^{00L}(\ell\ell)$ is a subalgebra of $so(2\ell+1)$ properly containing $so_L(3)$.

If (16) is not true, say, $\begin{Bmatrix} L & L & L' \\ \ell & \ell & \ell \end{Bmatrix} \neq 0$, then we have to check $\begin{Bmatrix} L' & L' & L_3 \\ \ell & \ell & \ell \end{Bmatrix} \overset{?}{=} 0$, $\begin{Bmatrix} L & L' & L_3 \\ \ell & \ell & \ell \end{Bmatrix} \overset{?}{=} 0$, $\forall\, L_3 \neq 1, L, L'$ to determine whether $\mathcal{W}^{001}(\ell\ell) \dotplus \mathcal{W}^{00L}(\ell\ell) \dotplus \mathcal{W}^{00L'}(\ell\ell)$ is closed or not (under Lie product).

Thus the existence of proper subalgebras of so(2 +1) which properly contains $so_L(3)$ is equivalent to the existence of proper subsets S of $\{1,3,5,\ldots,2\ell-1\}$ such that

$$\begin{Bmatrix} L_1 & L_2 & L_3 \\ \ell & \ell & \ell \end{Bmatrix} = 0, \quad \forall\, L_1, L_2 \in S,\ L_3 \notin S \tag{17}$$

3. CONCLUSION

The values of $\begin{Bmatrix} L_1 & L_2 & L_3 \\ \ell & \ell & \ell \end{Bmatrix}$ show that there don't exist enough vanishing relevant 6j-symbols for a proper subalgebra of $so(2\ell+1)$ which properly contains $so_L(3)$ for all ℓ (> 1) but 3.

REFERENCES

1. G. Racah, Phys. Rev. 76:1352 (1949)
2. B. R. Judd, Topics in Atomic Theory (Caxton Press, Christchurch, 1970).
3. J. Z. Sun, B. F. Li, Irreducible Tensors in Quantum Chemistry (Science Press, Beijing, to be published).
4. B. Gruber, M. Samuel Thomas, Kinam Vol. 2, 133-159 (1980).

THE SUPERSYMMETRY OF THE DIRAC-YAND-MILLS OPERATOR AND SOME APPLICATIONS

L. O'Raifeartaigh

Dublin Institute for Advanced Studies
10 Burlington Road
Dublin 4, Ireland

1. Introduction

Although supersymmetry first became known within the context of string-theory(1) and field theory(2), and has thus come to be associated with Fermi-Bose symmetry, it is actually a much broader concept(3). In this paper the broader concept of supersymmetry is defined, and it is pointed out that in this broader sense, there exists in nature a fundamental operator, namely the square $\not{D}^2$ of the Dirac, or Yang-Mills operator $\not{D} \equiv \gamma^\mu \partial_\mu + \gamma^\mu A_\mu$, where A_μ is the gauge-potential, which is supersymmetric. It is then shown how the supersymmetry of $\not{D}^2$ can be put to good use by applying it to two different problems, namely

(i) and constructing the negative modes of unstable finite-energy monopoles(4), and

(ii) giving a simple derivation of the relationship between the chiral anomaly and the index theorem (which relatés the zero-modes of $\not{D}^2$ to the flux) and of the index theorem itself in the 2-dimensional sphere.

As a final illustration of the power of the method, a result is quoted in which the supersymmetry of $\not{D}^2$ has been used to generalize the results (ii) to the more difficult, but more realistic, case of Euclidean manifolds. In this case the $\not{D}^2$-operator has a continuous spectrum and the flux acquires a fractional part that can be identified with (Bohm-Aharonov-like) scattering phase-shifts.

2. The Supersymmetric Structure

The general definition of supersymmetry is not completely fixed, but for our purposes the following definition will suffice. Let Γ be an idempotent self-adjoint operator ($\Gamma^2 = 1$), Q_r, $r=1...n$ a set of self-adjoint operators which anti-commute with Γ, and H the sum of the squares of the Q_r,

$$H = \sum_{r=1}^{n} Q_r^2 \ . \tag{2.1}$$

Then H evidently commutes with Γ, and if it commutes with the Q_r i.e.

$$[H, Q_r] = 0, \qquad r=1...n \tag{2.2}$$

then H is said to be supersymmetric. If, in addition, the quantities

$$G_{rs} = \{Q_r, Q_s\}, = 1\ldots n, \tag{2.3}$$

are the generators of a Lie algebra, the Q_r are said to generate an associated supersymmetric algebra. Note that if the operators Q_r are decomposed into their left and right handed parts with respect to r by suitable projection operators i.e.

$$Q_r^{\pm} = P_{\pm} Q_r P_{\mp}, \qquad P_{\pm} = (1 \pm \Gamma)|2, \tag{2.4}$$

then one may also write

$$H = \sum_r \{Q_r^+, Q_r^-\}, \qquad G_{rs} = \{Q_r^+, Q_s^-\}, \qquad Q_r^{+2} = Q_r^{-2} = 0. \tag{2.5}$$

Examples of supersymmetric systems are:

(a) One-dimensional Supersymmetric Quantum Mechanics[3]

$$Q = (\frac{\partial}{\partial x} + U(x))\sigma^+, \quad Q^+ = (-\frac{\partial}{\partial x} + U(x))\sigma^-,$$

$$H = -\frac{\partial^2}{\partial x^2} + U^2(x) + \sigma_3 \frac{\partial U(x)}{\partial x}, \tag{2.6}$$

where $\sigma_3, \sigma^{\pm}$ are the Pauli matrices.

(b) Supersymmetric Chiral Scalar Field Theory[2]

$$Q_\alpha = \frac{\partial}{\alpha \partial_\alpha} + (\theta\gamma^\mu \partial_\mu)_\alpha, \quad (C\gamma^\mu P_\mu)_{\alpha\beta} = \{Q_\alpha, Q_\beta\}, \quad \alpha = 1\ldots.4, \tag{2.7}$$

where θ_α are anti-commuting Grassmann variables belonging to the Majorana (real Dirac) representation of the Lorentz group, C is the charge conjugtion matrix and P_μ the 4-momentum.

(c) Differential Geometry[3], where Q_r is the outer derivative ∂_r for n-forms, $n \leq d$, Q_r^+ is its dual ∂_r^*, and H is the Laplacian $\sum_r \partial_r \partial_r^* + \partial_r^* \partial_r$, $r = 1\ldots d$.

It is clear from the construction that the supersymmetric operator H is a direct sum of the form

$$H = H_+ \oplus H_- = (\sum_r Q_r^- Q_r^+) \oplus (\sum_r Q_r^+ Q_r^-), \quad \text{or}$$

$$H = \begin{bmatrix} \sum_r D_r D_r^+ & 0 \\ 0 & \sum_r D_r^+ D_r \end{bmatrix} \qquad \text{for } \Gamma = \begin{bmatrix} 1 & 0 \\ 0 & -1 \end{bmatrix}, \tag{2.8}$$

where D_r, D_r^+ denote the restrictions of the operators Q_r^-, Q_r^+ to the subspaces on which they are non-trivial, and that because of the form of $H_{\pm}$ the spectra of D_r, D_r^+ are closely related. For example, if there is only one Q and the spectrum of H is discrete, then since

$$QQ^+|\lambda\rangle = \lambda|\lambda\rangle \implies (Q^+Q)|\lambda\rangle' = \lambda|\lambda\rangle' \quad \text{where } |\lambda\rangle' = Q^+|\lambda\rangle \ , \tag{2.9}$$

the non-zero parts of the spectra of $H_\pm$ must be identical. These constraints on the spectra of $H_\pm$ show that supersymmetric operators are rather exceptional.

As in the case of ordinary symmetries, supersymmetry is said to be spontaneously broken if the ground states $|o\rangle$ (eigenstates of H with lowest eigenvalue) are not supersymmetric i.e. if

$$Q_r|o\rangle \neq 0 \quad \text{or } Q_r^+|o\rangle \neq 0 \quad \text{for some } r \ . \tag{2.10}$$

From the construction of H in (2.1), however, one sees that this will be the case if, and only if, the lowest eigenvalue of H is not zero

$$H|O\rangle = \lambda_o|o\rangle, \qquad \lambda_o \neq 0 \ . \tag{2.11}$$

Thus the criterion for supersymmetry breaking is extremely simple. Note that when the symmetry is unbroken the equations for the ground state reduce to

$$D^+|o\rangle_+ = 0 \quad \text{and} \quad D|o\rangle_- = 0 \quad \text{where } |o\rangle_\pm = P_\pm|o\rangle, \tag{2.12}$$

and these equations are linear in D, D^+.

3. Supersymmetry of the Dirac-Yang-Mills Operator

Apart from the phenomenological Fermi-Bose symmetry of nuclear physics, discussed in other talks at this conference, there is no evidence as yet for a Fermi-Bose symmetry in nature, in particular in particle physics or cosmology. But this does not mean that there is no-supersymmetry in nature, since, as we have seen, the concept of supersymmetry is more general, and, as we shall now see, the square of even-dimensional Dirac-Yang-Mills operator, $\not{D}^2$, is supersymmetric in the more general sense.

To see this let $\gamma_{2n+1} = (i)^n \varepsilon_{\mu\nu\lambda} \ldots \sigma \gamma^\mu \gamma^\nu \gamma^\lambda \ldots \gamma^\delta$ denote the (self-adjoint) product of all 2n γ-matrices in d=2n dimensions, and let $P_\pm$ denote the projection operators on the ±1 eigenspaces of γ_{2n+1}. (Note that $\gamma_{2n+1}^2 = 1$). Then choosing the idempotent operator Γ of section 2 to be γ_{2n+1} and the operators $Q^\pm$ of that section to be

$$D^\pm = P_\pm \not{D} P_\mp \ , \qquad \{\not{D}, \gamma_5\} = 0, \tag{3.1}$$

one sees at once that

$$\not{D}^2 = \{D^+, D^-\} \ , \tag{3.2}$$

and thus $\not{D}^2$ is supersymmetric, as required. In particular in 2 and 4 dimensions, respectively, one has

$$D^{\pm} = D_1 \pm iD_2, \qquad \not{D}^2 = \begin{bmatrix} D^2 + B & o \\ o & D^2 - B \end{bmatrix} , \tag{3.3}$$

$$\text{and} \qquad D^{\pm} = D_o \pm i\vec{\sigma}\cdot\vec{D}, \qquad \not{D}^2 = \begin{bmatrix} D^2 + \sigma.\ (B-E) & o \\ o & D^2-(B+E)\cdot\sigma \end{bmatrix} , \tag{3.4}$$

where E,B are the conventional electric and magnetic fields and $\vec{\sigma}$ are the Pauli-matrices.

As already mentioned, the ground state is of particular interest and may be written as

$$D^{+}\psi_{+} = 0 \qquad \text{and} \qquad D^{-}\psi_{-} = 0. \tag{3.5}$$

Because these equations are linear in $D_{\pm}$ they are relatively easy to solve. For example in 2 Euclidean dimensions with a Maxwell magnetic field they read

$$[(\partial_x + i\partial_y) + i\ (A_x + iA_y)]\psi_{+}=0 \quad \text{and} \quad [(\partial_x - i\partial_y) + i\ (A_x - iA_y)]\psi_{-}=0 \tag{3.6}$$

respectively, and the general solutions are easily seen to be

$$\psi_{+} = (x+iy)^m \exp(-\chi) \text{ and } \quad \psi_{-} = (x-iy)^n \exp(-\chi) , \tag{3.7}$$

where m,n are any positive integers, and χ is a pseudo-scalar potential for the vector-potential $\vec{A}$, i.e.

$$A_i = \epsilon_{ij}\partial_j\chi \qquad (\text{so } \Delta\dot{\chi} = B), \tag{3.8}$$

B being the (pseudo-scalar) magnetic field in two dimensions.

In view of the use of Dirac supersymmetry in Professor Ginochio's talk, it should perhaps be added that, since any γ-matrix can be substituted for γ_{2n+1} there will be an analogous supersymmetry for each γ and that used by Professor Ginochio is $\gamma = \gamma_o$. However, the supersymmetries for the different γ's, will have different physical interpretations and applications. Thus while the γ_5 - supersymmetry above for n=2 is Lorentz-invariant but parity-violating and requires a 4-vector $Q(=\gamma^{\mu}D_{\mu})$ with which to anti-commute, the γ_o-supersymmetry is rotation and parity conserving but Lorentz-violating, and requires a Q of the form $\vec{\gamma}.\vec{v} + \gamma_5 s$, where $\vec{v}$ is a 3-vector and s a pseudo-scalar, with which to anti-commute. The choice of $(\vec{v},s)$ found useful for nuclear physics is apparently $\vec{v} = \vec{\nabla} + \vec{E}$, s = 0

4. Application 1. Instability of Monopoles

Although the supersymmetry of $\not{D}^2$ is interesting, one may ask whether it is of any practical use, for example in deriving older results in a simpler and more elegant manner, or in deriving new results. We now wish to give some examples which show how useful it can be. In the present section we consider as example the problem of monopole instability[4].

The monopoles in question are the static, finite-energy solutions of the Yang-Mills-Higgs Hamiltonian system

$$H = \mathrm{tr}\int d^3x\{B^2 + (D\Phi)^2 + V(\Phi)\}, \qquad \vec{B} = \vec{\nabla}\times\vec{A} + [A,A], \quad D\Phi = \nabla\Phi + [A,\Phi] \quad , \tag{4.1}$$

and, as is well-known[4], these are non-trivial field configurations, if the topological (monopole) charge

$$Q = \int d^3x\ \vec{\nabla}\cdot(\vec{B},\Phi) = \int d\Omega(B,\ \Omega), \qquad \vec{B} = (\frac{\vec{x}}{r})B \quad , \tag{4.2}$$

(which must be an integer) is not identically zero. The question, however, is whether these configurations (for which $H<\infty$ because of finite-energy and $\delta H=0$ because of the Euler-Lagrange equations) satisfy the stability condition $\delta^2 H \geqslant 0$. In other words, the question is whether the extremal solutions for (4.1) correspond to minima of H or merely to saddle-points.

It is usual to consider variations of the gauge-field only ($\delta\Phi = 0$) and to choose the gauge so that $D\cdot\delta A = 0$. Since the variations in the B-field and covariant derivative for $\delta\Phi = 0$ are

$$\delta B = \nabla\times\delta A + [A,\delta A] = D\times\delta A, \quad \delta^2 B = [\delta A,\delta A], \quad \delta D\Phi = [\delta A,\Phi], \tag{4.3}$$

where x denotes the usual three-dimensional cross-product, one easily sees that the second variation $\delta^2 H$ of H may be written as

$$\delta^2 H = \mathrm{tr}\int d^3x\ \{(D\times\delta A)^2 + [\Phi,\delta A]^2 + B(\delta A\times\delta A)\}, \quad (D\cdot\delta A = 0). \tag{4.4}$$

Now since the first two terms in (4.4) are positive and the second and third terms decrease and increase respectively as $r\to\infty$, negative modes of $\delta^2 H$ are more likely for large r, and hence it is usual to restrict the variations δA of A to those with support only for $r \geqslant R$, where R is so large that the asymptotic expressions for the fields become valid. In that case (4.4) may be written as

$$\delta^2 H = \int d\Omega\{(D\times\delta A)^2 + \tfrac{1}{4}(\delta A)^2 + B(\delta A\delta A)\}\ , \tag{4.5}$$

where the integral is over the sphere at infinity (renormalized so as to become the unit sphere) and the 'mass'-term $(\delta A)^2$ comes from minimizing the radial variation[9].

Now it can be shown[10] that the mass-gap for the operator Dx in (4.5) is at least as large as $2|B|$, so for any state corresponding to a non-zero value of D, the first two terms in (4.5) dominate the third term, and there are no zero modes. On the other hand, for the zeros of D the the third term in (4.5) dominates the second, and so for any δA corresponding to negative B, there is a negative mode. In other words, one can show that

$$\text{Negative Modes of } \delta^2 H \iff \text{Zero Modes of } D\times \quad (\text{for } B<0) \quad . \tag{4.6}$$

Thus to find the negative modes of $\delta^2 H$ one has only to solve the equations

$$D\times\delta A = 0 \, , \qquad D\cdot\delta A = 0 \qquad , \tag{4.7}$$

(for $B<0$). For this purpose it is very convenient to use conformal co-ordinates on the sphere (i.e. coordinates such that $\sqrt{g}g^{\alpha\beta}=\delta_{\alpha\beta}$ where $g_{\alpha\beta}$ is the metric tensor e.g. stereograpic coordinates $x = \tan\theta/2\cos\phi$, $y=\tan\theta/2\sin\phi, ds^2 = (1+\rho^2)^{-2}(dx^2+dy^2)$, $\rho^2 = x^2+y^2$, where (θ,ϕ) are the usual polar angles. Then eqns. (4.7) reduce to the 'flat-space' equations

$$D_1a_2-D_2a_1=0,\ D_1a_1+D_2a_2=0 \quad \text{or} \quad D\psi=0 \text{ where } D=D_1+iD_2,\ \psi=a_1-ia_2 \tag{4.8}$$

and the norm for a reduces to the 'flat-space' norm

$$\int dxdy a_\alpha a_\alpha = \int dxdg\sqrt{g}g^{\alpha\beta}a_\alpha a_\beta . \tag{4.9}$$

One then sees that the negative-mode equations (4.7) are just the same equations (3.5) as we had in section (3) for the supersymmetric ground-state of the Dirac operator. Thus they have the general solution (3.7) i.e.

$$\psi = (x+iy)^m e^{-\chi} = \rho^m e^{im\phi}e^{-\chi}, \tag{4.10}$$

but with the special conditions that in the present (monopole) case they must be square-integrable with respect to the norm (4.9), and that the B-field, and hence the χ-field, are the known monopole fields. These are, respectively,

$$B=2q\sin\theta = 2q\rho(1+\rho^2) \text{ and } \chi = 2q\ln(1+\rho^2) \tag{4.11}$$

where q is the magnetic charge, and in deriving the expression for χ we have used (3.8). From (4.10) and (4.11) one sees at once that the negative-mode solutions are

$$\psi = \rho^m e^{im\phi}(1+\rho^2)^{-q} = e^{im\phi}(\sin\theta/2)^m(\cos\theta/2)^{q-m},\ 0\leqslant m\leqslant 2q-2 \tag{4.12}$$

where the range of m is determined by the square-integrability at $\rho=0$ (north pole) and $\rho=\infty$ (south pole). Thus one finds that there are actually $2q-1$ independent negative modes. This agrees with other methods of counting, but in contrast to the other methods the supersymmetric method yields the explicit form (4.12) of the negative modes. (Note that the south pole $\rho=\infty$ can be included in (4.12) by using two-coordinate patches $x^2+y^2=\rho^2$ and ρ^{-2} for $\rho\gtrless 1$ respectively, with transition functions exp $2iq\phi$ at $\rho=1$ (equator). Then the solution is (4.12) in the northern hemisphere and $e^{i(2q-m)\phi}(\cos\theta/2)^m(\sin\theta/2)^{q-m}$ in the southern hemisphere).

5. Application 2. The Global U(1)-Anomaly and the Index Theorem

It is easy to verify that if one considers the generator Δ of fermion loops(11) in the presence of an external gauge-potential A_μ i.e.

$$\Delta = \ln\det(\not{D}+M) \quad , \tag{5.1}$$

where $M=m+i\gamma_5 n$ is a chiral-covariant mass-term introduced to remove the infra-red divergence, then the chiral variation $\partial\Delta/\partial\alpha$ of Δ with respect to constant (global) chiral variations $\psi \to (\exp i\gamma_5\alpha)\psi$, is given by the formula

$$\frac{\partial\Delta}{\partial\alpha} = \mathrm{tr}(\not{D}+M)^{-1}\delta M = \mathrm{tr}\left(\frac{\rho^2}{\rho^2+\not{D}^2}\gamma_5\right) \text{ where } \delta M=(-n+i\gamma_5 m) \text{ and } \rho^2=m^2+n^2 . \tag{5.2}$$

The fact that even in the limit $\rho^2\to 0$ the right hand side of (5.2) is not zero is known as the (global, U(1)) chiral anomaly[6], and if one lets $\pm$ denote the $\gamma_5=\pm1$ projections of $\not{D}^2$ one sees that it may be written as

$$\frac{\partial\Delta}{\partial\alpha} = \mathrm{tr}\left[\frac{\rho^2}{\rho^2+\not{D}_+^2} - \frac{\rho^2}{\rho^2+\not{D}_-^2}\right] . \tag{5.3}$$

Now if the spectrum of $\not{D}^2$ is discrete (as happens typically for a compact manifold) then, on account of the supersymmetry of $\not{D}^2$ the non-zero eigenvalues of $\not{D}_+^2$ and $\not{D}_-^2$ cancel, as discussed in section 2. In that case the only contribution to (5.3) comes from the zero modes. Then (5.2) reduces to

$$\frac{\partial\Delta}{\partial\alpha} = (n_+ - n_-) , \tag{5.4}$$

where $n_\pm$ are the multiplicities of the respective zero modes. In this way the supersymmetry of $\not{D}^2$ can be used to show that the global chiral anomaly is actually due to the chiral asymmetry of the ground states of $\not{D}^2$.

But one can go further. The Atiyah-Singer index theorem[5] (for this case) states that the quantity (n_+-n_-) is equal to a flux-integral of the gauge-field i.e.

$$(n_+ - n_-) = \int\phi(x) = \int ds^\mu\phi_\mu(x) \quad \text{where}$$

$$\phi(x) = \partial_\mu\phi_\mu(x) = \varepsilon_{\mu\nu\lambda}\ldots{}_{\sigma\tau}F_{\mu\nu}\ldots F_{\sigma\tau} . \tag{5.5}$$

This implies, of course, that the anomaly is equal to the flux integral, which is in agreement with direct calculations made in the context of field theory[12]. What interests us here, however, is not the result (5.5) in itself, but rather the fact that it can also be obtained in a direct and simple manner by using the supersymmetry of $\not{D}^2$. We shall now illustrate this for the 2-dimensional sphere.

For this, one recalls from section 2 that the ground states of $\not{D}^2$ are just those which satisfy

$$D_\pm\psi = 0, \tag{5.6}$$

where the $D_\pm$ are the operators defined in that section, except that here the $D_\pm$ operators are defined on the sphere instead of flat-space. However, if one uses stereographic coordinates, then, as in section 4, the operators are formally the same as in flat-space and so equations (5.6) reduce to the 'flat-space' equations

$$(\partial_z + A_z)\psi = 0 \quad \text{and} \quad (\partial_{\bar{z}} - A_{\bar{z}})\psi = 0, \tag{5.7}$$

respectively. However, because ψ is a spinor, not a vector, its norm is

$$N_\psi = \int(d^2x\sqrt{g})g^{-1/4}|\psi|^2 = \int d^2x g^{1/4}|\psi|^2 = \int \frac{d^2x}{(1+\rho^2)}|\psi|^2, \tag{5.8}$$

and not $\int d^2x|\psi|^2$ as in section 4. (Recall that in 2-dimensions every real irreducible tensor, or real representation of SO(2), is 2-dimensional, so the distinction between spins appears only as the sth power of $g^{1/4}$ in the inner-product).

Now if eq. (5.7) is written in the polar form

$$[\partial_\rho \pm \frac{1}{\rho}(\frac{1}{i}\partial_\phi + A_\phi)]\psi = 0 \quad , \tag{5.9}$$

one obtains from it, without even solving it as a differential equation, an expansion for the radial log-derivative of ψ in terms of the flux and the angular momentum m, namely,

$$\frac{\partial \ell n \psi}{\partial \ell n \rho} = m \pm A\phi \quad , \tag{5.10}$$

and this is already sufficient to test for square-integrability. Indeed from (5.8) and (5.10) one sees that the square-integrability conditions are just

$$0 \lesssim m \lesssim \Phi - 1 \quad \text{where} \quad \Phi = (\frac{1}{2\pi} \; A_\phi d_\phi)_{\rho=\infty} \quad , \tag{5.11}$$

is the total flux. From these inequalities it is clear that for the supersymmetric sector with the same sign as $\Phi(\varepsilon\Phi>1)$ there are no ground state solutions and (recalling that m and Φ must be integers) for the other sector there are just $|\Phi|$ ground state solutions (corresponding to $m = 0,1..|\Phi|-1$). This shows that the index of $\not{D}$ is $|\Phi|$, which is just the AS result for this problem. However, as in the monopole example of the previous section, the supersymmetric method yields not only the index itself, but an explicit expression for the bound states. In particular, it shows that there is a contribution of just one unit to the index (i.e. there is just one bound state) for each angular momentum $m \lesssim |\Phi|-1$.

6. Application 3. The Non-Compact Version of the Previous Results

So far, the supersymmetry of $\not{D}^2$ has only been used to rederive some well-known results, though in a more direct, explicit and simple manner than before. Hence I should like to conclude by quoting a result which is relatively new, and is not easily derived without using supersymmetry, namely the generalization of the results of the previous section to the case when the underlying manifold is not compact, but Euclidean. Such a generalization is not trivial because in Euclidean space the spectrum of $\not{D}^2$ has, in general, not only the usual bound states at the origin but a continuum which is not bounded away from the origin. For this situation the generalization of the results of section 5 turns out to be

$$\frac{\partial \Delta}{\partial \alpha} = \oint \phi_\mu(x) ds^\mu = (n_+ - n_-) + \frac{1}{\pi}(\delta_+ - \delta_-) \quad , \tag{6.1}$$

where $n_\pm$ are the multiplicities of the left - and right - handed zero modes as before, and $\delta_\pm$ are the left - and right - handed scattering phase shifts at zero energy. Thus the anomaly and the flux-integral are fractional and the proper part of the fraction is given by the phase-shift. (In two dimensions $\delta_\pm$ are actually the Bohm-Aharonov phase shifts). I shall not give the proof of (6.1) here, as that would require another article of the same length, but merely quote the result as an illustration of the power of the supersymmetry of $\not{D}^2$, and refer the reader to the literature(7).

References

1. P. Ramond, Phys. Rev. D3, (1971) 2415,
J. Scherk, Rev. Mod. Phys. 47 (1975) 1,
F. Gliozzi, D. Olive, J. Scherk, Nucl. Phys. B122 (1977) 253.
2. J. Wess, B. Zumino, Phys. Lett. 49B 1974 52, Nucl. Phys. B78 (1974) 1,
P. Fayet, S. Ferrara, Phys. Reports 32 (1977) 249,
J. Bagger, J. Wess, Supersymmetry and Supergravity (Princeton Univ. Press 1983).
3. E. Witten, Nucl. Phys. B185 (1981) 513, J. Diff. Geom. 17 (1982) 661,
D. Lancaster, Nuovo Cim. 79A (1984) 28.
4. G. 't Hooft, Nucl. Phys. B79 (1974) 276,
A. Polyakov, JETP Lett. 20 (1974) 194,
P. Goddard, D. Olive, Rep. Prog. Phys. 41 (1978) 1357,
Monopoles in Quantum Field Theory (eds. N. Craigie, P. Goddard, W. Nahm, World Scientific, Singapore 1982).
5. M. Atiyah, I. Singer, Ann. Math 87 (1968) 485, 546,
K. Fujikawa, Phys. Rev. D21 (1980) 2848.
6. J. Bell, R. Jackiw, Nuovo Cim. 60A (1969) 47,
S. Adler, Phys. Rev. 177 (1969) 2426,
B. Zumino, Y. Wu, A. Zee, Nucl. Phys. B239 (1984) 477,
K. Huang, Quarks, Leptons and Gauge Fields (World Scientific, Singapore 1982).
7. R. Musto, L. O'Raifeartaigh, A. Wipf, Phys. Lett. (in print),
R. Blankenbecler, D. Boyanovsky, Phys. Rev. D31 (1985) 3234,
T. Jaroszewicz, Harvard Univ. Preprint 1986.
8. L. Alvarez-Gaumé, Comm. Math. Phys. 90 (1983) 161.
9. R. Brandt, F. Neri, Nucl. Phys. B161 (1979) 253,
S. Coleman, CERN Lecture Notes (1979),
P. Goddard, D. Olive, Nucl. Phys. B19 (1981) 528,
E. Weinberg, Nucl. Phys. B167 (1980) 500,
A. Balachandran et al. Phys. Rev. 29D (1984) 2919, 2936,
P. Horváthy, L. O'Raifeartaigh, DIAS Preprint 1986.
10. M. Atiyah, R. Bott, Phil. Trans. R. Soc. Lond. A308 (1982) 523.
11. P. Ramond, Field Theory, (Addison-Wesley, Reading, MA 1981).

INDEX